ANTIOXIDANTS IN THERAPY AND PREVENTIVE MEDICINE

ADVANCES IN EXPERIMENTAL MEDICINE AND BIOLOGY

Recent Volumes in this Series

A Continuation Order Plan is available for this series. A continuation order will bring delivery of each new volume immediately upon publication. Volumes are billed only upon actual shipment. For further information please contact the publisher.

ANTIOXIDANTS IN THERAPY AND PREVENTIVE MEDICINE

Edited by

Ingrid Emerit

Centre National de la Recherche Scientifique
and University of Paris
Paris, France

Lester Packer

University of California, Berkeley
Berkeley, California

and

Christian Auclair

Institut Gustave–Roussy
Villejuif, France

PLENUM PRESS • NEW YORK AND LONDON

Library of Congress Cataloging in Publication Data

Society for Free Radical Research Winter Meeting on Free Radicals in Medicine: Current Status of Antioxidant Therapy (1988: Paris, France)
   Antioxidants in therapy and preventive medicine / edited by Ingrid Emerit, Lester Packer, and Christian Auclair.
       p.   cm. — (Advances in experimental medicine and biology; v. 264)
   Proceedings of the Society for Free Radical Research Winter Meeting on Free Radicals in Medicine: Current Status of Antioxidant Therapy, held December 9–10, 1988, in Paris, France — T.p. verso.
   Includes bibliographical references.
   ISBN 0-306-43407-5
   1. Antioxidants — Therapeutic use — Testing — Congresses. 2. Active oxygen — Pathophysiology — Congresses. I. Emerit, Ingrid. II. Packer, Lester. III. Auclair, Christian. IV. Society for Free Radical Research. V. Title. VI. Series.
   [DNLM: 1. Antioxidants — therapeutic use — congresses. 2. Free Radicals — congresses. 3. Superoxide — adverse effects — congresses. 4. Superoxide Dismutase — therapeutic use — congresses. WD1 AD559 v. 264 / QV 800 S678a 1988]
RM666.A555S65   1988
615′.7 — dc20
DNLM/DLC   89-70966
for Library of Congress   CIP

Proceedings of the Society for Free Radical Research Winter Meeting
on Free Radicals in Medicine: Current Status of Antioxidant Therapy,
held December 9–10, 1988, in Paris, France

© 1990 Plenum Press, New York
A Division of Plenum Publishing Corporation
233 Spring Street, New York, N.Y. 10013

Printed in the United States of America

PREFACE

Twenty years ago, the enzyme superoxide dismutase which uses the superoxide radical anion as its specific substrate was reported. With this discovery was born a new scientific field, in which oxygen, necessary for aerobic life on this planet, had to be considered also in terms of its toxicity and stresses. This stimulated the search for knowledge of active oxygen species in biology and medicine. Superoxide and other reactive oxygen species are now implicated in many disease processes. Major advances have been achieved during these past years with respect to free radical generation and mechanisms of free radical action in causing tissue injury. In parallel, the possibility of influencing free radical related disease processes by antioxidant treatment was studied in various in vitro and in vivo systems. This was the unique theme of a conference organized in Paris by the Society for Free Radical Research (December 9-10, 1988) which brought together experts from basic sciences and clinicians in order to evaluate the current status of antioxidant therapy.

The conference emphasized fundamental processes in antioxidant action. Among the major topics were superoxide dismutase (SOD) and low molecular weight substances with such activity, called SOD mimics. Other antioxidant enzymes were also considered. Antioxidant vitamins, in particular vitamins E and C, other naturally occurring antioxidants and various synthetic antioxidants were included in the presentations as there is now a rapidly developing series of compounds with potentially interesting clinical applications.

Clinical trials and results obtained in animal models were considered together with the use of antioxidant therapy in various diseases. Examples are ischemia-reperfusion injury, rheumatic, pulmonary, neuromuscular and ocular diseases, pancreatitis, cystic fibrosis and diabetes. Another important topic was the possibility to influence various dermatological disorders, photooxidative stress and cutaneous aging. Radiation and chemotherapy of cancer were discussed in terms of minimizing side effects by antioxidant therapy.

Many reports of clinical trials were preliminary. There is a great need for controlled trials. Nevertheless, the overall impression at the conference and in these proceedings is optimistic for future developments.

We gratefully acknowledge support for the conference which was provided by the following sponsors:

Institut de Recherches Internationales Servier France
Fondation IPSEN p.la Recherche Therapeutique France
Henkel Corp. USA and Henkel KG FRG
Nestle SA Switzerland
Laboratoires Chauvin Blache France
Laboratoires Sandoz France
Nattermann GMBH FRG
Laboratories Labcatal France
Parfums Rochas Paris France

Lutsia Paris France
Groupement des Enterprises Francaises dans la Lutte
contre le Cancer Paris - Ile de France

The Editors

CONTENTS

SUPEROXIDE DISMUTASE

COPPER COMPLEXES AND OTHER SOD MIMICKS

ANTIOXIDANT VITAMINS

OTHER NATURAL AND SYNTHETIC ANTIOXIDANTS

RADIATION AND CHEMOTHERAPY

EXTRACELLULAR-SUPEROXIDE DISMUTASE, DISTRIBUTION IN THE BODY
AND THERAPEUTIC APPLICATIONS

Stefan L Marklund and Kurt Karlsson

Department of Clinical Chemistry
Umeå University Hospital
S-901 85 UMEÅ, SWEDEN

Extracellular-superoxide dismutase (EC 1.15.1.1, EC-SOD) is a secretory, tetrameric, Cu and Zn-containing glycoprotein with a subunit molecular weight of about 30 kDa.[1,2] EC-SOD is the major SOD isoenzyme in extracellular fluids.[3,4] Although it is the least predominant SOD isoenzyme in the tissues, 90 to 99 % of the EC-SOD in the body of mammals is located to the extravascular space of tissues.[5,6]

A prominent feature of EC-SOD is its affinity for heparin.[1] Upon chromatography on Heparin-Sepharose, plasma EC-SOD from man,[7] pig, cat, mouse, guinea pig and rabbit[8] can be divided into at least three fractions; A without, B with weak heparin-affinity and C which elutes relatively late in a NaCl gradient. In rat plasma, however, only A and B fractions can be demonstrated (Figure 1). EC-SOD from tissues is mainly composed of forms with high heparin-affinity. The binding to heparin is of electrostatic nature. Since EC-SOD carries a net negative charge at neutral pH (the isoelectric point is 4,50[9]), the binding to the strongly negatively charged heparin must be mediated by a cluster of positively charged amino acid residues in the enzyme. Such a cluster occurs in the very hydrophilic carboxy-terminal end of EC-SOD C, which contains three lysines, six arginine residues and one histidine residue among the last 20 amino acid residues.[10] The differences between EC-SOD A, B and C appear to reside in this region.

Intravenous injections of heparin in man and in most other mammals lead to a prompt release of EC-SOD C to plasma.[7,8] No post-heparin increase in plasma EC-SOD activity was seen in the rat, apparently because of the absence of the C fraction in this species (Figure 1). Human EC-SOD C injected intravenously into rabbits is within 5 to 10 min to 95 % sequestered from the blood.[11]

Antioxidants in Therapy and Preventive Medicine
Edited by I. Emerit *et al.*
Plenum Press, New York, 1990

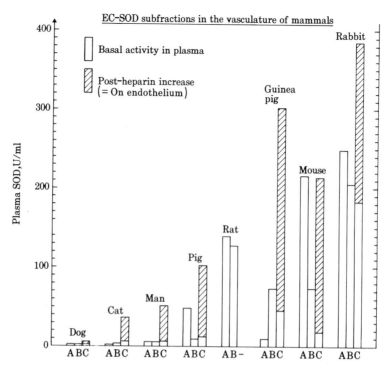

Figure 1. Pre- and postheparin plasma samples were separated
on Heparin-Sepharose into A, B and C frac-
tions.[7,8] The figure presents the means of the
results from two individuals of each species.
200 IU heparin per kg body weight was injected
intravenously to the human beings and 2000 IU
heparin per kg body weight to the other mammals. In the pig, maximal EC-SOD C release is
given by around 1000 IU/kg and 200 IU/kg gives
around 50 % release.[12] The data for EC-SOD C
release in man presented in the figure were
transformed to maximal release assuming that
the dose response relationship was the same as
in the pig. The open columns show the basal
plasma EC-SOD A,B and C activities and the
striped columns the post-heparin increase. The
activites were determined with the very sensitive direct KO_2 assay.[14] One unit in this assay
corresponds to 1/40 unit in the xanthine oxida-
se-CyC assay.[15]

Subsequent injection of heparin leads to an immediate release
of all sequestered EC-SOD C back to plasma. The rapid sequest-
ering of injected exogenous. EC-SOD C and the rapid heparin-induced release to plasma of both sequestered exogenous EC-SOD[11]
and endogenous EC-SOD,[7,8] indicates that EC-SOD C forms an
equilibrium between the plasma phase and heparin-analogues on
the vessel endothelium. EC-SOD A and B seem to primarily reside in the plasma. The wide diversity of EC-SOD in the vascular system of mammals with regard to total amount, division
into subfractions and distribution between plasma and endothe-

lium, indicates that the pathogenic potential of superoxide radicals in the extracellular space might vary much between species. Man appears to belong to the group of species with relatively little protection by the EC-SOD system in the vascular tree.

EC-SOD C added to the culture medium has been shown to bind reversibly to the cell surfaces of a large panel of investigated anchorage-dependent cell lines, including endothelial cells.[13] The binding capacities were very high, and at maximal binding the EC-SOD C activity associated with the exterior of the cells was several-fold higher than the endogenous intracellular SOD activity (CuZn SOD and Mn SOD). Heparan sulfate proteoglycan was the EC-SOD C ligand in the cell cultures. Suspension-growing cell lines were all weaker binders. Blood monomorphonuclear leukocytes and platelets bound little EC-SOD C, whereas no significant binding to neutrophil leukocytes, to erythrocytes and to E. coli could be demonstrated. Taken together, the findings further support the notion that EC-SOD C in the blood forms an equilibrium between plasma and heparan sulfate in the glycocalyx of the endothelium. Furthermore, tissue EC-SOD is probably distributed between heparan sulfate on the surface of most cell types in the organs and in the interstitial matrix. The binding pattern suggests that EC-SOD C has the potential to protect most normal cells in the body and the interstitial matrix, without protecting microorganisms lacking affinity, and without interfering with superoxide radicals produced at the surface of activated neutrophil leukocytes.

Therapeutic studies with recombinant human EC-SOD C have recently been initiated in various disease models. E.g., in a regional ischemia (30 min)- reperfusion model in Langendorff perfused rat hearts, r-EC-SOD C significantly reduced creatine kinase release. In a second study, isolated rat hearts were exposed to 15 min global ischemia followed by reperfusion. The hearts were perfused with the spin trap PBN, and trapped radicals were determined with EPR in the effluent. R-EC-SOD C plus catalase significantly reduced the amount of radicals trapped, possibly more so than the same activity of bovine CuZn SOD plus catalase. (Collaboration with PO Sjöquist, L Carlsson, Hässle AB, Mölndal, Sweden).

REFERENCES

1. S.L. Marklund, Human copper-containing superoxide dismutase of high molecular weight, Proc. Natl. Acad. Sci. USA. 79:7634-7638 (1982).
2. L. Tibell, K. Hjalmarsson, T. Edlund, G. Skogman, Å. Engström and S.L. Marklund, Expression of human extracellular-superoxide dismutase in chinese hamster ovary cells and characterization of the product, Proc. Natl. Acad. Sci. USA. 84:6634-6638 (1987).
3. S.L. Marklund, E. Holme and L. Hellner, Superoxide dismutase in extracellular fluids, Clin. Chim. Acta, 126: 41-51 (1982).
4. S.L. Marklund, A. Bjelle and L.-G. Elmqvist, Superoxide dismutase isoenzymes of the synovial fluid in rheumatoid arthritis and in reactive arthritides, Ann. Rheum. Dis. 45:847-851 (1986).

5. S.L. Marklund, Extracellular-superoxide dismutase in human tissues and human cell lines, J. Clin. Invest. 74: 1398-1403 (1984).

6. S.L. Marklund, Extracellular-superoxide dismutase and other superoxide dismutase isoenzymes in tissues from nine mammalian species, Biochem. J. 222:649-655 (1984).

7. K. Karlsson and S.L. Marklund, Heparin-induced release of extracellular-superoxide dismutase to human blood plasma, Biochem. J. 242:55-59 (1987).

8. K. Karlsson and S.L. Marklund, Extracellular-superoxide dismutase in the vascular system of mammals. Biochem. J. 255:223-228 (1988).

9. S.L. Marklund, EC-SOD, molecular composition and distribution in the body, in: "Oxidative damage and related enzymes", G. Rotilio and J.V. Bannister, eds. Life Chemistry Reports Supplement Series, Harwood Academic Publishers, pp. 411-416 (1984).

10. K. Hjalmarsson, S.L. Marklund, Å. Engström and T. Edlund, Isolation and sequence of complementary DNA encoding human extracellular-superoxide dismutase, Proc. Natl. Acad. Sci. USA, 84:6340-6344 (1987).

11. K. Karlsson and S.L. Marklund, Plasma clearance of human extracellular-superoxide dismutase C in rabbits, J. Clin. Invest. 82:762-766 (1988).

12. K. Karlsson and S.L. Marklund, Heparin-, dextran sulfate- and protamine-induced release of extracellular-superoxide dismutase to plasma in pigs, Biochim. Biophys. Acta, 967:110-114 (1988).

13. K. Karlsson and S.L. Marklund, Binding of human extracellular-superoxide dismutase C to cultured cell lines and to blood cells, Lab. Invest, 6:659-666 (1989).

14. S.L. Marklund, Direct assay with potassium superoxide, in: "Handbook of Methods for Oxygen Radical Research", R. Greenwald, ed. CRC press, pp. 249-255 (1985).

15. J.M. McCord and Fridovich, I, J. Biol. Chem. 244:6049-6055 (1969).

TARGETING SOD BY GENE AND PROTEIN ENGINEERING AND

INHIBITION OF OXIDATIVE STRESS IN VARIOUS DISEASES

MASAYASU INOUE and NOBUKAZU WATANABE

Department of Biochemistry
Kumamoto University Medical School
2-2-1 Honjo, Kumamoto 860, JAPAN

A large number of human-type enzymes and bioactive peptides has been produced by gene engineering technique. Based on molecular mechanisms for various diseases, some patients have been treated successfully with such recombinant enzymes and peptide hormones without immunological complications and nonspecific side effects. However, recent studies revealed that there are many cases in which such recombinant products failed to show therapeutic effect predominantly due to their unfavorable in vivo behavior. It should be noted that enzymes and bioactive peptides could have their metabolic impact only when sufficient amounts of these molecules were localized in appropriate compartments during the time required for metabolic modulation. Thus, targeting enzymes and bioactive peptides to their appropriate sites of action is prerequisite to normalize pathologic metabolism in a patient.

Reactive oxygen species play a critical role in cellular defence mechanisms, such as bactericidal action. However, they also result in oxidative tissue injury as observed in various diseases, such as inflammation. The hazardous effects of these reactive species on cultured cells can be inhibited by adding SOD and/or catalase to the medium, suggesting that reactive oxygens occurring in an extracellular space might play a critical role in oxidative cell injury. However, clinical application of these enzymes has considerable limitations due to their unfavorable behavior in vivo. For example, when injected intravenously, both recombinant and naturally occurring Cu^{++}/Zn^{++}-SOD are rapidly removed from the circulation (1). Furthermore, these hazardous oxygens rapidly inactivate a wide variety of biomolecules and impair cell functions, they should be scavenged at or near the site of generation. Thus, in addition to extend the plasma half life of SOD, the enzyme should be targeted close to the injured site of a tissue. The present work describes a new strategy for targeting SOD by gene and protein engineering methods.

I   pH-Sensitive SOD derivatives that circulate bound to albumin with prolonged in vivo half life

Since renal glomerular filtration is predominantly responsible for the rapid removal of SOD from the circulation, its plasma half life would be prolonged by increasing its molecular size. In fact, this can be achieved by covalently linking high molecular weight compounds, such

Antioxidants in Therapy and Preventive Medicine
Edited by I. Emerit *et al.*
Plenum Press, New York, 1990

5

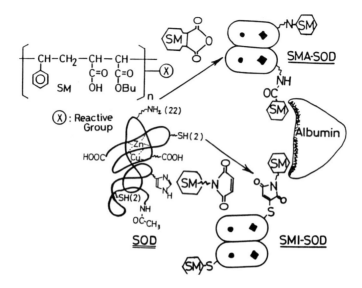

Fig. 1   Synthesis of SM-SOD that circulates bound to albumin
    Free SH group of Cys-111 or lysyl amino group of recombinant human SOD was covalently linked to maleimido- (SMI-SOD) or carboxyl-group of SM (SMA-SOD), respectively.

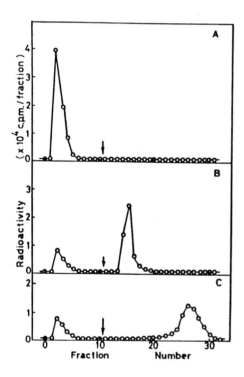

Fig. 2   Binding of SM-SOD to albumin   Radiolabeled enzyme samples (100,000 cpm) were subjected to affinity chromatography on an albumin Sepharose column (0.8 x 4 cm) which was equilibrated with PBS. When the column was washed with PBS, unmodified SOD was eluted in an unbound fraction (A). Under identical conditions, SM-SOD bound to the column and was eluted by 10 mM warfarin (B) or 0.5% SDS (C). Chromatographic profiles of both types of Sm-SOD samples were similar with each other.

as polyethylene glycol (1,2). However, the increase in their molecular size should be minimized not to decrease the diffusion rates of the derivatized enzymes through the interstitial space of an injured tissue. During the study of ligand-albumin interaction, we found that some hydrophobic anions, such as poly(styrene co-maleic acid butyl ester)(SM, MW = 1,600), reversibly bound to the warfarin site on albumin and circulated for long time with a half life of 8 h (3,4). To test whether binding to albumin is also effective for extending plasma half life of enzymes and bioactive peptides, we synthesized SOD derivatives (SM-SOD) by linking two mol SM to 1 mol SOD; SM was linked to the enzyme via its Cys-111 or the lysyl amino groups (Fig. 1) (4-6). When subjected to affinity chromatography on an albumin-Sepharose column, SM-SOD but not SOD bound to the column and was eluted by a solution containing 10 mM warfarin (Fig. 2). Although the bound SM-SOD was also eluted from the column by sodium dodecyl sulfate (SDS), a potent denaturant, the elution occurred more slowly than with warfarin. These results suggest that SM-SOD bound to the warfarin site on albumin with high affinity (6).

To test whether SM-SOD also forms a dissociable complex with albumin, radioactive enzyme samples were injected intravenously to the rat (Fig. 3). As expected, SM-SOD circulated bound to albumin with a half life of 6 h while SOD rapidly disappeared from the circulation with a half life of 4 min. To regulate the in vivo behavior of SM-SOD, the amphipathic nature of SM moiety was controled by changing the extent of esterification of its maleic acid residues. When the remaining carboxyl groups of SM moiety were protonated, the lipophilic nature of SM-SOD

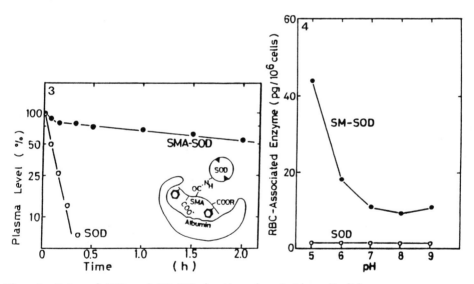

Fig. 3   Fate of SOD and SM-SOD in the circulation (left)
Under pentobarbital anesthesia, animals were injected with 1 mg of radioactive SOD or SM-SOD and time dependent changes in plasma radioactivity was determined. SM-SOD predominantly bound to albumin and circulated with a half life of 6 h while SOD was rapidly filtered by the kidney and disappeared from the circulation with a half life of 4 min.
Fig. 4   pH-Dependent binding of SM-SOD to cell membrane surface (right)
Human erythrocytes were incubated in PBS with radioactive SOD or SM-SOD. After changing pH of the medium, the incubated mixtures were centrifuged and the cell associated radioactivity was determined. When the medium pH was decreased lower than 7, SM-SOD but not SOD bound to erythrocyte membrane surface.

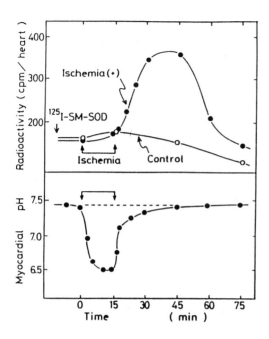

Fig. 5   Changes in myocardial pH and SM-SOD levels during ischemia and reflow

Under pentobarbital anesthesia and artificial respiration, radioactive SOD or SM-SOD was injected intravenously to the rat. Then, the left anterior descending artery was occluded for 15 min followed by reflow for 60 min. At the indicated times, myocardial pH (B) was determined by a needle-type electrode. Animals were exanguinated by bleeding and the heart was excised. After perfusion of the heart with ice-cold buffer, tissue-associated radioactivity (A) was determined. Open circles, control animals; closed circles, ischemic animals.

(60% of its maleic acid residues were esterified) markedly increased. Due to such pH-sensitive property of SM-SOD, it bound to membrane/lipid bilayer if the medium pH was decreased (Fig. 4) (6,7). Since local pH of an injured tissue decreases predominantly due to perturbation of mitochondrial energy metabolism and enhanced glycolysis, SM-SOD might also bind to plasma membrane surface of an acidified tissue. To test whether pH-dependent binding of SM-SOD to cell surface membranes also occur in an injured tissue, changes in myocardial pH and cardiac levels of SM-SOD were observed in the heart whose left anterior descending artery was transiently occluded (Fig. 5) (8). Immediately after occlusion, the local pH decreased rapidly from 7.4 to 6.5. However, the decreased pH again increased to 7.2 within 3 min after reflow; it returned to normal levels about 30 min after reflow. Although cardiac level of SM-SOD remained unchanged during ischemia, it increased rapidly after reflow. Accumulation of SM-SOD in the reflowed heart reached a

Table I   Inhibition of lethal arrhythmias of the rat by SM-SOD

Experimental groups	(n)	PVCs (/30 min)	Incidence(%) VT	Vf	Duration(sec) VT	Vf	Mortality (%)
group I	(6)	95.± 25	100	100	112 ± 20	465 ± 172	66.7
group II	(7)	158 ± 41	100	71	110 ± 28	88 ± 54[*]	14.2[*]
group III	(6)	147 ± 35	100	67	99 ± 15	148 ± 95[*]	16.7[*]

Under pentobarbital anesthesia and artificial respiration, the left anterior descending artery was occluded and ECG was monitored for 30 min. Animals were intravenously administered with saline (group I) or 5 mg/kg of SM-SOD (group II) 15 min before occlusion. Group III was administered with SM-SOD 5 min after occlusion. Mortality was compared at 6 h after occlusion. Data for PVCs and the duration of arrhythmias show mean ± SEM derived from 6-7 animals. PVCs, premature ventricular complexes; VT, ventricular tachycardia; Vf, ventricular fibrillation; n, number of animals. *Significantly different from group I (P<0.01).

maximum 30 min after reflow and returned to control levels thereafter. Thus, pH-dependent binding of SM-SOD to the surface membranes of the injured myocardium seems to occur in vivo. To test whether oxygen radicals play a critical role in ischemic tissue injury, the effect of SM-SOD on lethal arrhythmias was observed. When the left anterior descending artery was occluded transiently (15 min), a marked ventricular tachycardia and fibrillation was elicited immediately after reflow. Administration of SM-SOD, but not SOD, blocked the reperfusion arrhythmias almost completely (6,8). When the descending artery was occluded permanently, lethal arrhythmias were elicited and 67% of animals died within 30 min (Table I). Administration of SM-SOD, but not SOD, before or after occlusion significantly decreased the occurrence of lethal arrhythmias. To our surprise, the mortality of animals decreased from 67 to 14-18% in SM-SOD-treated groups. These results suggest that superoxide radical and/or its hazardous metabolites occurring in an extracellular space, which is readily accesible to the circulation, might play a critical role in ischemic heart disease and that SM-SOD would be an useful inhibitor for oxygen-induced tissue injury.

II Targeting SOD by gene engineering method   Although aerobic energy metabolism could easily be perturbed in some diseases with acute onset, such as myocardial infarction, no significant decrease in local pH of a tissue would occur in many other diseases, particularly in chronic diseases. Thus, a method for targeting SOD to the site of injury independent from changes in local pH should be developed. Since cells and tissues have specific marker proteins on their surface, these membranous proteins might be useful for targeting an agent to particular cell surface. Vascular endothelial cells play a critical role in protecting parenchymal cells in both acute and chronic diseases, such as brain edema and atherosclerosis. Thus, scavenging superoxide radicals close to endothelial cell membranes might be important for protecting

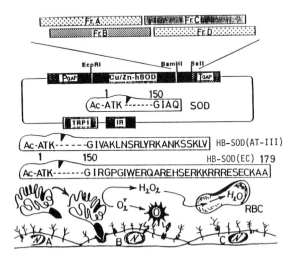

Fig. 6   Construction of plasmid and amino acid sequence for HB-SOD
    Fr. A-D, synthetic DNA fragments encoding heparin binding peptide; PGAP, glyceraldehyde-3-phosphate dehydrogenase promotor; HB-SOD (EC), extracellular-type; HB-SOD (AT-III), antithrombin-type.

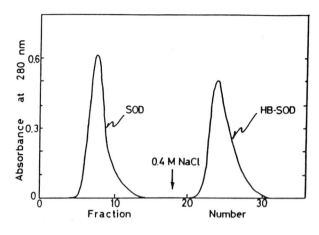

Fig. 7 Binding of HB-SOD to heparin-Sepharose
        The enzyme samples dissolved in 1 ml of 20 mM phosphate buffer, pH 7.4, containing 0.15 M NaCl (PBS) were applied to a heparin-Sepharose column (0.5 x 2 cm) equilibrated with PBS. After washing with PBS, the bound enzyme was eluted by 0.4 M NaCl and determined for absorbance at 280 nm.

tissues from oxidative stress. Since vascular endothelial cells have unique membrane proteins on their surface, such as heparan sulfate (9), enzymes and bioactive peptides could be targeted onto their surface by way of binding to their specific proteoglycans. To dismutate superoxide radicals efficiently on vascular endothelial cell surface, a fusion protein (HB-SOD) consisting from human Cu^{++}/Zn^{++}-SOD and a heparin-binding peptide was constructed by gene engineering method (Fig. 6) (10); the C-terminal portion of HB-SOD possesses a basic domain structure identical to the heparin-binding site of either human antithrombin III (11) or extracellular SOD (12). Both HB-SOD preparations expressed in yeasts bound to a heparin-Sepharose column and were eluted by a high salt concentraiton (Fig. 7). Immunocytochemical examination revealed that, when injected intravenously, a significant fraction of HB-SOD bound to vascular endothelial cell surface. Both carrageenin-induced paw edema (Fig. 8) and cold-induced brain edema were markedly inhibited by HB-SOD (10). These results suggest that superoxide radicals might play a critical role in the pathogenesis of vasogenic edema.
        To test whether dismutation of superoxide radicals on the outer surface of vascular endothelial cells also decreases oxygen toxicity in ischemic tissues, effect of HB-SOD on postischemic reflow arrhythmias was observed. When the left anterior descending artery of the rat was

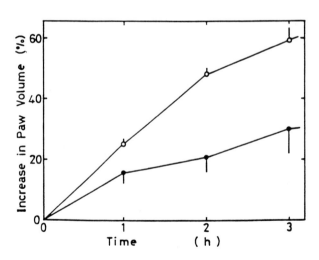

Fig. 8   Inhibition of carrageenin-induced paw edema by HB-SOD
        Under light ether anesthesia, rats were intravenously injected with 0.2 ml of saline (0) or 5 mg/Kg of HB-SOD (0). After 5 min, 2 mg of carrageenin in 0.1 ml saline was injected in the paw and time dependent change in paw volume was measured.

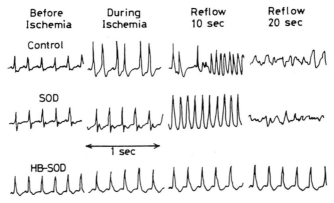

Fig. 9 Inhibition of postischemic reflow arrythmias by HB-SOD
        Animals were injected with either saline or 5 mg/kg of SOD or HB-
SOD. Then, the left anterior descending artery was occluded for 10 min.

occluded transiently, marked arrhythmias were elicited immediately after
reflow by an HB-SOD inhibitable mechanism (Fig. 9) (10).   Preliminary
experiments revealed that HB-SOD inhibited ischemic injury of liver,
stomach and small intestine. These observations suggest that superoxide
radical and/or its metabolites occurring on the outer surface of
endothelial cells might play a critical role in the pathogenesis of
vasogenic edema and ischemic tissue injury and that the site-directed SOD
derivatives might be useful for decreasing oxidative tissue injury in
various diseases (Fig. 10). This methodology for targeting SOD by gene
and protein engineering to subcellular sites for action permits studies
on the mechanism for oxygen toxicity in vivo.

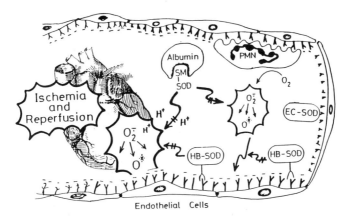

Fig. 10 Inhibition of oxidative stress by site-directed SOD derivatives

REFERENCES

1 P. Pyatak, A. Abuchowski, F. Davis, Preparation of polyethylene glycol: superoxide adduct, and an examination of its blood circulating life and anti-inflammatory activity, Res. Commun. Chem. Pathol. Pharmacol., 29: 113-127 (1980).

2 C. Beauchamp, S. Gonias, D. Menapace, S. Pizzo, A new procedure for the synthesis of polyethylene glycol-protein adducts; Effects on function, receptor recognition, and clearance of superoxide dismutase, lactoferrin, and $_2$-macroglobulin, Anal. Biochem., 131: 25-33 (1983).

3 M. Inoue, Metabolism and transport of amphipathic molecules in analbuminemic rats and human subjects, Hepatology, 5: 892-898 (1985).

4 M. Inoue, Albumin, a biovehicle for amphipathic molecules in the circulation, Seikagaku, 59: 441-447 (1987).

5 T. Ogino, M. Inoue, Y. Ando, M. Awai, H. Maeda, Y. Morino, Chemical modification of superoxide dismutase: Extension of plasma half life of the enzyme through its reversible binding to the circulating albumin, Int. J. Peptide Protein Res., 32: 153-159 (1988).

6 M. Inoue, N. Watanabe, Y. Morino, Synthesis of a superoxide dismutase derivative that circulates bound to albumin and accumulates in a tissue whose pH is decreased, Biochemistry, 28: 6619-6624 (1989).

7 M. Inoue, M. Hirota, Y. Ando, T. Ogino, I. Ebashi, M. Akagi, Y. Morino, Role of oxygen radicals in the pathogenesis of acute gastric mucosal lesion, in: "Free Radicals in Digestive Diseases", M. Tsuchiya, ed., pp83-98, Excerpta Medica, New York (1988)

8 N. Watanabe, M. Inoue, Y. Morino, Inhibition of postischemic reperfusion arrhythmias by an SOD derivative that circulates bound to albumin with prolonged in vivo half life, Biochem. pharmacol., in press (1989).

9 J. Marcum, D. Atha, L. Fritze, P. Nawroth, D. Stern, R. Rosenberg, Cloned bovine aortic endothelial cells synthesize anticoagulantly active heparan sulfate proteoglycan, J. Biol. Chem., 261: 7507-7517 (1986).

10 M. Inoue, N. Watanabe, Y. Tanaka, T. Amachi, J. Sasaki, Inhibition of oxidative tissue injury by targeting SOD to endothelial cell surface. Cell Struct. Funct. in press (1989).

11 J. Smith, D. Knauer, A heparin binding site in antithrombin III, J. Biol. Chem., 262: 11964-11972 (1987).

12 K. Hjalmarsson, S. Marklund, A. Engstrm, T. Edlund, Isolation and sequence of cDNA encoding human extracellular superoxide dismutase, Proc. Natl. Acad. Sci. USA, 84: 6340-6344 (1987).

THE ROLE OF SUPEROXIDE DISMUTASE (SOD) IN PREVENTING POSTISCHEMIC SPINAL CORD INJURY

K. Grabitz, E. Freye, R. Prior*, R. Kolvenbach and
W. Sandmann

Dept. of Vascular Surgery
*Dept. of Neuropathology
Heinrich Heine Universität Düsseldorf, F.R.G.

Introduction

The incidence of postischemic paraparesis or paraplegia following thoracoabdominal aneurysm repair ranges between 4 and 30% (Crawford et al., 1986). Inspite numerous surgical techniques prolonged hypoxia of the lower parts of the spinal cord cannot always be avoided. Thus, application of pharmacological agents during ischemia and/or the reperfusion period for the reduction of spinal cord injury is appealing and has been the goal of numerous investigations (Faden et al., 1984; Lim et al., 1986; Oldfield et al., 1982) The following study was designed to evaluate the protective effect of the antioxidant superoxide dismutase (SOD) in order to minimize the so called "reperfusion injury". In the animal study spinal cord function was measured by somatosensory evoked potentials (SEP) in order to determine the effectiveness of therapy. Additionally, the neurological outcome and the hiostological findings were compared to a non treated group.

Material and methods

23 beagle dogs, weighting 15-25 kg, were divided into two groups. In group 1 (control; n=11) the descending thoracic aorta was clamped distal to the origin of the left subclavian artery during enflurane N_2O/O_2-anesthesia while the animals were ventilated to normocarbia. The clamping period lasted for 60 minutes. After recovery from anesthesia the animals were observed for 7 days with regard to motor control of the hind limbs. The neurological outcome was graded using Tarlov's criteria of functional neurologic recovery (0= no voluntary movement of the hind limbs; 1= perceptible movement of joints; 2= good movement of joints but inability to stand; 3= ability to stand and walk; 4= complete recovery). After assessing the neurological results every day the means of the group were computed. The animals of group 2 (SOD; n=12) were treated in a similar manner except that 5 minutes prior to declamping a bolus of

Antioxidants in Therapy and Preventive Medicine
Edited by I. Emerit *et al.*
Plenum Press, New York, 1990

superoxide dismutase (Peroxinorm^R) 1 mg/kg was given into the aortic arch. This was followed by continuous intraaortic infusion of SOD 0,4 mg/kg/minute in order to scavenge the potential release of free oxygen radicals in the following 25 minutes. For intraoperative examination of the ascending nervous pathways, somatosensory evoked potentials were derived from spinally evoked potentials using a special peridural catheter (bipacing cath) (Kaschner et al., 1984) with a pulse of 0,2 ms duration, 0,1 mA above motor threshold. 256 sweeps were fed into a computer (Lifescan^R, Neurometrics; San Diego; USA) with a band width of 30-1500Hz, 3db cut-off, and a sampling rate of 3,2 kHz. Evoked potentials were derived from Cz-FpZ electrode positioning (10/20 system) and the major deflection peak around 50 ms post stimuli was computed by means of cursor position using peak to peak difference.

For histological examination the animals were sacrificed 7 days postoperatively and rostro-caudal extension and the total area of grey matter necrosis in the single most damaged section was determined.

Results

Cross-clamping of the thoracoabdominal aorta resulted in a loss of amplitude of somatosensory evoked potentials within 15-20 minutes. In comparison to the control-group reperfusion in the SOD-treated group was characterized by an earlier return of the previously depressed amplitude (table 1). This was also reflected by a decrease of the duration of total amplitude loss being significantly shorter in the SOD-group (p<0.05; Kruskal-Wallis-test).

Table 1   Somatosensory Evoked Potentials Changes During and After Aortic Cross-Clamping (Mean Minute ± SD; *=p <0.05)

	control-group	SOD-group
Loss of evoked potential	14 ± 7.3	15.5 ± 7.2
Recovery of potential	36 ± 24.7	15.9 ± 10.3*
Duration of potential loss	72 ± 32.5	58.6 ± 14.5*

In addition to the results of somatosensory evoked potential monitoring, neurological recovery of the animals in the SOD-group was better than in the native group (fig. 1). In the SOD-treated group, four out of twelve dogs (33%) showed complete neurologic recovery, three dogs sustained grade 2 motor function deficit (= good movements at joints but inability to stand) and the remaining five animals demonstrated complete spastic paraplegia. In contrast, only one dog of the native group (9%) showed good neurologic recovery (grade 3 = ability to stand and walk). Two dogs sustained grade 1 or grade 2 motor function deficit and eight animals became completely paraplegic. The results of histological evaluation confirmed these clinical findings. Four out of twelve dogs of the treated group presented without morphologically detectable spinal cord alterations. In other cases, slight cellular infiltrations, mild edema or circumscribed infarctions corresponded to an overall limited degree of ischemic spinal cord changes. In contrast, the majority of the non treated animals had partial and in some cases total necrosis of the lumbo-sacral grey matter indicating irreversible spinal cord injury.

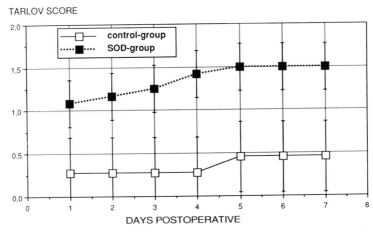

TARLOV SCORE

Fig. 1   Neurological outcome, graded during 7 days using Tarlov's score of functional neurologic recovery (mean ± SEM)

Discussion

Reperfusion injury mediated by free oxygen radicals is believed to be one of the major reasons causing spinal cord damage after thoracic aortic cross-clamping. This can be interpreted from studies of ischemia-reperfusion injury in cerebro-vascular damage (Yatsu, 1987). It can also be assumed that cerebral and spinal cord ischemia are comparable, since the pathophysiological mechanims occurring in cortex and spinal cord neurons are similar. Pretreatment with free oxygen radical scavengers such as SOD, catalase and mannitol may prevent the morphologic and functional changes associated with ischemia and reperfusion. Superoxide anions can effecticvely be eliminated by SOD, thus preventing its transformation to the highly reactive hydroxyl radicals (Hess and Manson, 1984; Southorn, 1988). Our results in the SOD-treated group compared to the non treated animals, indicate, that reperfusion injury is at least partly responsible for postoperative ischemic paraplegia. Although the clinical neurologic results show only a moderate difference between the protected and the control-group, measurements of neuromonitoring using evoked potentials and in addition the histological findings support the assumption of a preventive effect. Similar beneficial results had been already reported (Lim et al., 1986). Aside from the usual protective measurements during aortic cross-clamping, SOD also seems a potentionally interesting medicament for use in man. Further controlled studies should be undertaken for the prevention of paraplegia in surgery for thoracoabdominal aneurysms using SOD. In regard to neurological monitoring, evoked potentials which have been shown to be a good parameter for the evaluation of sensory function deficits (Laschinger et al., 1987), also proved to be a reliable tool in the present study. Thus, conclusions can be drawn from those data to determine the prognostic outcome after aortic cross-clamping.

References

Crawford, E.S., Crawford, J.L., Safi, H.J., Coselli, J.S., Hess, K.R., Brooks, B., Norton, H.J. and Glaeser, D.H., 1986, Thoracoabdominal aortic aneurysms: Preoperative and intraoperative factors determining immediate and long-term results of operations in 605 patients. J Vasc Surg, 3: 389

Faden, F.J., Jacobs, T.P., and Smith, M.T., 1984, Evaluation of the calcium channel antagonist nimopidine in experimental spinal cord ischemia. J Neurosurg, 60: 796

Hess, M.L. and Manson, N.H., 1984, Molecular oxygen: Friend and foe. The role of the oxygen free radical system in the calcium paradox, the oxygen paradox and ischemia/reperfusion injury. J Mol Cell Cardiol, 16: 969

Kaschner, A.G. Sandmann, W., and Larkamp, H., 1984, Percutaneous flexible bipolar epidural neuroelectrode for spinal cord stimulation. J Neurosurg, 60: 1317

Laschinger, J.C., Cunningham, J.N., Cooper, M.M., Baumann, F:G., and Spencer, F.C., 1987, Monitoring of somatosensory evoked potentials during procedures on the thoracoabdominal aorta. J Thoracc Cardiovasc Surg, 94: 260

Lim, K.H., Connoly, M., Rose, D., Siegmann, F., Jacobowitz, I., Acinapura, A., and Cunningham, J.N., 1986, Prevention of reperfusion of the ischemic spinal cord: Use of recombinant superoxide dismutase. Ann Thorac Surg, 42: 282

Oldfield, E.H., Plunkett, R.J., Nylander, W.A., and Meacham, W.F., 1982, Barbiturate protection in acute experimental spinal cord ischemia. J Neurosurg, 56: 511

Southorn, P., 1988, Free radicals in Medicine. Mayo Clinic Proceed, 63: 390

Yatsu, F.M., 1987, Clinical aspects of reversible cerebral ischemia, in: "Cerebral ischemia and hemorheology," Hartmann, Springer Verlag, Berlin, Heidelberg

SOD IN RAT MODELS OF SHOCK AND ORGAN FAILURE

Heinz Redl, Camille Lieners, Soheyl Bahrami, Günther Schlag, Ignaz P.T. van Bebber*, R. Jan A. Goris*

Ludwig Boltzmann Institute for Experimental and Clinical Traumatology, Vienna, Austria
* Department of General Surgery St. Radboud, University Hospital, Nijmegen, The Netherlands

1. INTRODUCTION

Posttraumatic or postoperative (multi) organ failure is the result of a multitude of reactions similar to (generalized) inflammatory situations. The fuel for this reaction cascade is either derived from continuous activation of humoral cascades (e.g. complement), continuous overshooting phagocytosis or toxic products of bacterial cells (e.g. endotoxin).

Endotoxemia occurs post trauma as well as after major surgery in the critically ill patient, and is frequently without a septic focus. Already 1959 Fine et al. mentioned that the intestine is the source of endotoxin (LPS). This is in accordance with the report by Meakins and Marshall (1986), who state that the "motor" of multi organ failure syndrome (MOFS) is the gastrointestinal tract. The reason for this breakdown of the intestinal mucosa and bacterial translocation is hypoxia (Younes et al., 1984) and reperfusion injury together with radical release by granulocytes (Grisham et al., 1986). Endotoxin itself also perpetuates this process by a positive feedback loop (Deitch et al., 1987). In our ongoing baboon model of hypovolemic-traumatic shock (Pretorius et al., 1987), we can directly demonstrate this "bacterial translocation" (Redl and Schlag, 1989a), we also find gastrointestinal bleeding and signs of lipid peroxidation (Redl and Schlag, 1989a). We have previously demonstrated such posttraumatic lipid peroxidation also in dogs (Lieners et al., 1989; Schlag and Redl, 1986).

Experimentally organ failure can be induced by LPS administration (i.v., i.p.) in several animal species, of which we have selected rats and sheep (Bahrami et al., 1989b).

On the other hand organ failure clinically (Goris et al., 1985) is seen without clear evidence of bacterial involvement or the necessity of endotoxin. Therefore a model to mimick this "sepsis-like" state was set up, too (Goris et al., 1986).

Similar to the clinical situation this model leads to MOF after about 12 days with a severe acute phase on the 2nd - 3rd days and a severe late phase (> 9 days). This effect is achieved by intraperitoneal application of zymosan in paraffine oil. Zymosan activates the

Antioxidants in Therapy and Preventive Medicine
Edited by I. Emerit *et al.*
Plenum Press, New York, 1990

alternative complement pathway and stimulates continuous phagocytosis by granulocytes and macrophages.

Accordingly organ failure is thought to be the final result of cell injury from the action of oxygen radicals, proteases, and other mediators. Phagocytes and ischemic peripheral tissue (xanthine oxidase action) are believed to be the major source for oxygen radicals, with the endothelial cell as the main target organ.

2. ENDOTOXIN EXPERIMENTS

2.1. Analysis of lipid peroxidation

In a pilot study with Sprague Dawley rats (n = 4), we catheterized the jugular vein and administered endotoxin (Difco E. coli 026:B6) 5 mg/kg over four hours and measured thiobarbituric reactive (TBAR) substances and fluorescence (for methods see page 4).

	TBAR subst. (nmol MDA/ml)	Fluorescence (FU/ml)
0	1.56 ± 0.1	89.2 ± 29
4	3.24 ± 0.3	139.0 ± 25
24	1.63 ± 0.4	117.0 ± 50

We found a minimal trend towards increased lipid peroxidation products 4^h after start of LPS administration at the time of most severe leukopenia and coagulation disorders (results not shown). At 24^h (or 48^h) no increased TBAR were found.

Similar to the results in rats $\geq 24^h$ we do not find an indication of lipid peroxidation in the ovine endotoxin model (unpublished results), which is in contrast to the reports of Demling et al. (1986).

Although we could not clearly prove the involvement of radicals in endotoxin shock from this pilot experiments, we started to use antioxidative therapy in long term (48^h) experiments based upon several positive reports in the literature (Kunimoto et al., 1987; Yoshikawa et al., 1983).

2.2. SOD effects

2.2.1. Methods

In male Sprague Dawley rats with a mean body weight of 250 g, endotoxin (Difco E. coli 026:B6) was administered intraperitoneally under halothane anaesthesia at a dose of 14 or 17.5 mg/kg BW (Bahrami et al., 1989b), which leads to a mortality of 70 - 90 %.

The following therapeutical measures were chosen in 3 sets of experiments (10 rats in each group).

1) a) SOD (Peroxinorm, Grünenthal) 1 mg/kg S.C. 1^h pre LPS
   b) SOD   10 mg/kg 5 mg i.p. 1^h pre LPS
                                        + 5 mg s.c. 4^h post LPS
   c) LPS only (14 mg/kg)

2) a) LPS +SOD-Albumin Conjugates (as described in the next section)
      + Catalase Albumin Conjugates
            13500 U/kg SOD i.p. 1^h pre LPS
         + 120000 U/kg Catalase i.p. 1^h pre LPS
   b) LPS only (17.5 mg/kg)

3) a) LPS + 15-Aminosteroid U74006F (UpJohn) acting
      as iron chelator   12 mg/kg 30'pre LPS
                          + 12 mg/kg 4^h post LPS
   b) LPS only (14 mg/kg)

2.2.2. Results

Mortality (Death/Total)

		24^h	48^h
1) a) SOD 1 mg		7/ 9	8/ 9
b) SOD 10 mg		10/10	10/10
c) Co		6/ 8	6/ 8
2) a) SOD/Catalase Conj.		6/10	7/10
b) Co		6/10	7/10
3) a) Aminosteroid		9/10	9/10
b) Co		8/10	8/10

Thus no protective effect could be observed and TBAR material measured in the SOD/CAT group (24^h - 1.65 ± 0.24 nmol/ml, 48^h - 1.51 ± 0.31, n = 4) was not different from the above cited pilot experiments.

3. STERILE ZYMOSAN INDUCED PERITONITIS

3.1. Lipid peroxidation analysis

In contrast to the endotoxin experiments we have previously demonstrated the involvement of oxygen radicals in the experimental peritonitis situation (see 3.2.1. Methods) (van Bebber et al., 1989). Briefly the following was found.

A triphasic illness with maximal clinical signs at day 2 and 14 occured. 25 % of the animals died within 2 days. At the same time the lowest complement CH$_{50}$ level coincided with the smallest number of circulating PMN and their highest activation state (unstimulated and PMA stimulated whole blood superoxide production). At the same time there was evidence of lipid peroxidation (elevated TBAR material in lung, liver and plasma). We concluded that the severe inflammatory response was probably the result of excessive toxic oxygen radical release by leukocytes.

3.2. SOD effects

Therefore we tried in the same model protective measures against oxygen radicals and the associated lipid peroxidation. Superoxide dismutase and catalase are most suitable for this purpose, as they are at the top level in the reaction cascade to prevent a radical chain reaction. However, their half-life in vivo is very short (minutes), which makes application problematic. We increased the half life of SOD and catalase to several hours by binding the antioxidant enzymes to a carrier, albumin.

3.2.1. Methods

Male Wistar rats with an average weight of 260 g were used and injected with 100 mg Zymosan / 100 g body weight in 4 ml paraffin to induce sterile peritonitis.

Bovine liver SOD (Peroxinorm, Grünenthal, FRG) and catalase (Sigma Ltd, USA) were conjugated to human albumin (Immuno, Austria) using a modification of a previously described method (Wong et al., 1980).

Conjugates were separated by gel filtration and fractions referring to a MW of 150 - 250 kDa were pooled.

SOD activity was determined using the inhibition of nitrite formation described by Elstner and Heupel (1976). Catalase activity was assessed by recording the decrease of H_2O_2 absorption at 240 nm as described by Aebi (1984).

Thiobarbituric Acid Reactive material (TBAR) was determined as previously described (Lieners et al., 1989). Conjugated dienes were quantitated by measuring the O.D. at 233 nm in hexane after extraction with chloroform-methanol. Quantitative fluorescence data were determined in the same extract by measuring the emission intensity at 435 nm after excitation at 355 nm. Data were correlated to the amount of total lipid present in the extract (Mangold and Bezzegh, 1974) and expressed as O.D. 233 nm/mg total lipid, respectively fluorescence units-FU/mg total lipid.

Groups		treatment	observation time
Group 1	zymosan	none	1 day
Group 2	zymosan	none	2 days
Group 3	zymosan	SOD/CAT	1 day
Group 4	zymosan	SOD/CAT	2 days
Group 5	control rats	none	

Three hours before administering zymosan, the rats were anaesthetized with ether and received an aseptic intraperitoneal injection of the conjugates. The dosis was 25,000 U/kg SOD and 75,000 U/kg catalase (conjugated to albumin). Identical concentrations of conjugates were additionally injected subcutaneously 8 and 24 hours after inoculation of zymosan.

The statistical evaluation (group comparisons) was performed after testing for homogeneous distribution (Bartlett test) with ANOVA and linear contrasts according to Scheffe. The H-test according to Wilcoxon, was performed with heterogeneous distributions. $p < 0.05$ was selected as the level of significance.

3.2.2. Results

Peak activites of conjugated SOD and catalase were seen after 3 - 4 hours for i.p. and after 6 - 8 hours for s.c. administration. The absolute values were higher with i.p. application. Even after 48 hours significant enzymatic activities were measured, while native enzymes were not detected after 8 hours even when a ten-fold higher concentration was supplied.

All animals with zymosan application showed the classical symptoms of severe illness, i.e. lethargy, anorexia, loss of hemorrhagic fluid from the nose and conjuctivae and liquid stools. Non-treated animals had

MORTALITY TABLE (%)

	Group 1	Group 2	Group 3	Group 4
Group:	Zymosan	Zymosan	Z.+ Conjug.	Z.+ Conjug.
Day				
1	43 % (6/14)	0 % (0/14)	71 % (10/14)	14 % (2/13)
2	-	14 % (2/14)	-	54 % (9/13)

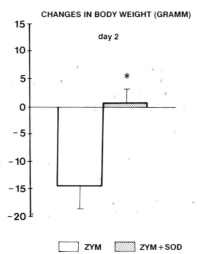

Fig. 1. Change of body weight. (*) p < 0.05 compared to zymosan group.

21

a significantly higher loss of body weight (p < 0.05) after two days (Fig. 1). Due to the great variability of experimental sets a comparison between groups for mortality is not appropriate (Mortality Table).

The leukocyte count was significantly decreased on days 1 and 2 (Fig.2). There was considerable group differences in the differential count, with the lowest percentage of PMN in the untreated zymosan group.

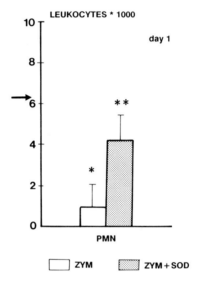

Fig. 2. Peripheral leukocyte count. (*) p < 0.05 compared to controls, (**) p < 0.05 compared to zymosan group.

Elevated levels of plasma lipid peroxidation products are found with all measured parameters in the zymosan-induced inflammatory state, which is more pronounced on the second day of the experiment. Significant inhibition of lipid peroxidation is achieved by applying SOD/CAT conjugates. TBAR-M increased in all groups, especially on day 2 in the untreated group (Fig.3).

The same tendency is seen for conjugated dienes and fluorescent substances (Fig. 3).

Three-dimensional fluorescent techniques revealed a shift in the fluorescent pattern in all animals treated with zymosan (not shown).

4. GENERAL DISCUSSION

Although the molecular weight is rather high, the conjugate uptake via the i.p. and s.c.routes is quite satisfactory and ensures a measurable enzyme activity level over a certain period of time.

Lipid peroxidation was attenuated by application of enzyme conjugates. Beside TBAR, the formation of fluorescent substances (EX 355 nm EM 435 nm) was a sensitive parameter for lipid peroxidation and probably a product of HNE in addition to proteins and phospholipids. Esterbauer et al. (1986) first described production of this fluorophore after exposing LDL to oxidative stress in vitro. We demonstrated its occurrence in whole plasma samples after hypovolemic-traumatic shock (Lieners et al., 1989).

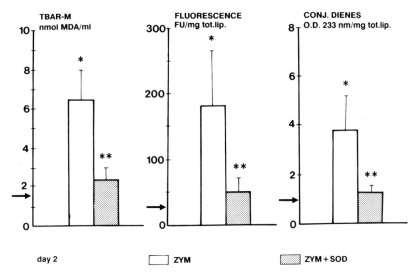

Fig. 3. Lipid peroxidation products TBAR-M, conjugated dienes and fluorescent substances in plasma. (*) p < 0.05 compared to zymosan group. The arrows represent levels in control rats.

The SOD conjugate considerably reduced a major drop of circulating leukocytes at 24ʰ. This may be attributed to the radical trapping effect of SOD, and also of catalase, which may have prevented massive formation of chemotactic compounds (McCord, 1980; Curzio et al., 1986) and migration of PMN into the lung.

Weight loss was found to be significantly improved by conjugate application, which is clinically the most important finding. The increased mortality with SOD should not be overemphasized, since there was no consistent mortality in the model.

Although the antioxidative therapy was partially successful, the question remains whether oxygen radical formation is the key event in this inflammatory chain reaction.

In our endotoxin experiments we are convinced, that radical formation is not a key issue. We believe that with LPS infusion in the rat coagulation disorders (DIC) are prevalent, antioxidative treatment is not effective, but anticoagulants (Bahrami et al., 1989a) or cortico-steroids (Bahrami et al., 1987) are a successful therapeutical approach.

Among several radical scavenging systems studied in similar models, McKechnie et al. (1986) also found no protective effect of SOD and catalase, while Yoshikawa et al. (1983), Kumimoto et al. (1987) and Schneider et al. (1989) found salutary effects. These controversial results are not fully explainable at the present time.

The iron chelator system 15-aminosteroid was also not effective, probably also because of the prevalent DIC effects, while the same substance was effective in ischemic situations (Natale et al., 1989).

In contrast to our findings in endotoxemia, with continuous activation of phagocytes either by massive complement activation from CVF (Till and Ward, 1987) or by zymosan in our peritonitis model, lipid peroxidation is found and antioxidative therapy is at least partially effective.

Our preliminary data in humans (Lieners, 1988; Redl and Schlag, 1989b) also demonstrate increased plasma levels of lipid peroxidation products in plasma after polytrauma and Nerlich et al. (1986) found increased TBAR levels in lungs of patients dying after trauma. These data together with findings of reduced anti-oxidative plasma capacity in certain patients conditions (Wayner et al., 1985), such as reduced tocopherol levels after burn injuries (Paubert-Braquet et al., 1988), and increased hydroperoxide levels in ARDS patients (Frei et al., 1988), also support the concept of clinical antioxidative therapy.

Finally, from pilot experiments in baboons (n = 3), we also learned that GI bleeding after polytrauma could be avoided by SOD administration (Dosis 24 mg/kg/h).

Thus SOD could have a positive effect against mediator activation in the sequence endotoxin release, cytokine formation, phagocyte activation, and radical action.

In addition SOD could also diminish gut derived endotoxin liberation as suggest by Granger and Parks (1983), thus interrupting a detrimental feed back loop, which otherwise could lead to organ failure.

REFERENCES

Aebi, H., 1984, Catalase in vitro. Methods in enzymology, in: "Oxygen Radicals in Biological Systems," L. Packer, ed., Academic Press, Orlando.

Bahrami, S., Paul, E., Redl, H., and Schlag, G., 1989a, Therapeutic modalities to ameliorate endotoxin induced DIC in the rats, in: "Progress in Clinical and Biological Research, Vol. 308: Second Vienna Shock Forum," G. Schlag, H. Redl, eds., Alan R. Liss Inc, New York, pp 977.

Bahrami, S., Redl, H., Thurnher, M., Vogl, C., Paul, E., Schießer, A., and Schlag G., 1989b, Effect of PAF-antagonists in endotoxin shock - ovine and rat exerpiments, in: "Progress in Clinical and Biological Research, Vol. 308: Second Vienna Shock Forum," G. Schlag, H. Redl, eds., Alan R. Liss Inc, New York, pp 931.

Bahrami, S., Schießer, A., Redl, H., and Schlag, G., 1987, Comparison of different corticosteroids in rat endotoxemia, in: "Progress in Clinical and Biological Research, First Vienne Shock Forum, Vol. 236-B: Monitoring and Treatment of Shock, G. Schlag, H. Redl, eds., Alan R. Liss, pp 237.

Curzio, M., Esterbauer, H., DiMauro, C., Cecchini, G., and Dianzani, M., 1986, Chemotactic activity of lipid peroxidation product 4-hydroxynonenal and homologous hydroxyalkenals, Biol. Chem. Hoppe Seyler 367:321.

Deitch, E.A., Berg, R., and Specian R, 1987, Endotoxin promotes the translocation of bacteria from the gut, Arch. Surg. 122:185.

Demling, R.H., Lalonde, C., Jin, L.J., Tyan, P., and Fox, R., 1986, Endotoxemia causes increased lung tissue lipid peroxidation in unanesthetized sheep, J. Appl. Physiol. 60:2094.

Elstern, E.F., and Heupel, A., 1976, Inhibition of nitrite formation from hydroxylammoniumchloride: A simple assay for superoxide dismutase, Anal. Biochem. 70:616.

Esterbauer, H., Koller, E., Slee, R.G., and Koster, J.F., 1986, Possible involvement of the lipid peroxidation product 4 hydroxy-nonenal in the formation of fluorescent chromolipids, Biochem. J. 239:405.

Fine, J., Rutenburg, S.H., and Schweinburg F.R., 1959, The role of the RES in hemorrhagic shock, J. Exp. Med. 110:547.

Frei, B., Stocker, R., and Ames, B.N., 1988, Antioxidant defenses and lipid peroxidation in human blood plasma. Proc. Natl. Acad. Sci. USA 85:9748.

Goris, R.J.A., Boekholtz W.K.F., van Bebber, I.P.T., Nuytinck, J.K.S., and Schillings, P.H.M., 1986, Multiple organ failure and sepsis without bacteria, Arch. Surg. 121:897.

Goris, R.J.A., te Boekhorst, T.P.A., Nuytinck, J.K.S., and Gimbrere, J.S.F. (1985), Multiple-organ failure. Generalized autodestructive inflammation? Arch. Surg. 120:1109.

Granger, D.N., and Parks, D.A., 1983, Role of oxygen radicals in the pathogenesis of intestinal ischemia, Physiologist 26:159.

Grisham, M.B., Hernandez , L.A., and Granger, D.N., 1986, Xanthine oxidase and neutrophils infiltration in intestinal ischemia, Am. J. Physiol. 251:G567.

Kunimoto, F., Morita, T., Ogawa, R., and Fujita, T., 1987, Inhibition of lipid peroxidation improves survival rate of endotoxemic rats, Circ. Shock 21:15.

Lieners, C.F.J., 1988, Lipidperoxidation - Analytik und Beeinflussung durch antioxidativ wirksame Enzyme, Dissertation, Technische Universität Wien.

Lieners, C.F.J., Redl, H., Molnar, H., Fürst, W., Hallström, S., and Schlag, G., 1989, Lipidperoxidation in a canine model of hypo-volemic-traumatic shock, in: "Progress in Clinical and Biological Research, Vol. 308: Second Vienna Shock Forum," G. Schlag, H. Redl, eds., Alan R. Liss Inc, New York, pp 345.

Mangold, H.K., Bezzegh, T., 1984, Absorptimetric determination of total lipids in serum, according to Zöllner, Kirsch, in: "Routine Methods of Lipids Analysis, Clinical Biochemistry, Principles and Methods, Vol. II, H.C. Curtius, M. Roth, eds., de Gruyter, Berlin.

McCord, J.M., 1980, Oxygen free radicals: release and activities. A superoxide activated chemotactic factor and its role in inflammatory processes. Agents Actions 10:522.

McKechnie, K., Furman, B.L., and Parratt, J.R., 1986, Modification by oxygen freee radical scavengers of the metabolic and cardiovascular effects of endotoxin infusion in conscious rats, Circ. Shock 19: 429.

Meakins, J.L., and Marshall, J.C., 1986, Multi-organ-failure syndrome. The gastrointestinal tract: the "motor" of MOF, Arch. Surg. 121: 196.

Natale, J.E., Schott, R.J., Hall, E.D., Braughler, J.M., and D'Alecy L.G., 1989, The 21-aminosteroid U74006F reduces systemic lipid peroxidation, improves neurologic function, and reduces mortality after cardiopulmonary arrest in dogs, in: "Progress in Clinical and Biological Research, Vol. 308: Second Vienna Shock Forum," G. Schlag, H. Redl, eds., Alan R. Liss Inc, New York, pp 891.

Nerlich, M.L., Seidel, J., Regel, G., Nerlich, A.G., and Sturm, J.A., 1986), Klinisch experimentelle Untersuchungen zum oxidativen Membranschaden nach schwerem Trauma, Langenbecks Arch. Chir. Suppl: 217.

Paubert-Braquet, M., Lavaud, P., Bellanger, L., Deby, C., and Guilbaud J., 1986, Depression of plasmatic tocopherol levels and platelet-activating factor (PAF)-induced platelet and polymorphonuclear leukocyte aggregation in burn-injured patients, in: "Abstractbook - Second Vienne Shock Forum, Vienna.

Pretorius, J.P., Schlag, G., Redl, H., Botha, W.S., Goosen, D.J., Bosman, H., and van Eeden, A.F., 1987, The 'lung in shock'as a result of hypovolemic-traumatic shock in baboons, J. Trauma 27:1344.

Redl, H., and Schlag, G., 1989a, Humoral and cellular activation in blood leading to organ failure, in: "First Wiggers-Bernard Conference on Shock, Sepsis, and Organ Failure," G. Schlag, H. Redl, eds., Alan R. Liss Inc., New York, in press.

Redl, H., and Schlag, G., 1989b, Möglichkeiten der biochemischen Analytik im posttraumatischen und postoperativen Organversagen. Intensivmed., in press.

Schlag, G., and Redl, H., 1986, Oxygen radicals in hypovolemic traumatic shock, in: "Oxygen Free Radicals in Shock," G.P. Novelli, F. Ursini, eds., S. Karger, Basel.

Schneider, J., Friderichs, E., and Giertz H., 1989, Protection by recombinant human superoxide dismutase in lethal rat endotoxemia, in: "Progress in Clinical and Biological Research, Vol. 308: Second Vienna Shock Forum," G. Schlag, H. Redl, eds., Alan R. Liss Inc, New York, pp 913.

Till, G.O., and Ward, P.A., 1987, Oxygen radicals and lipid peroxidation in experimental shock, in: "Progress in Clinical and Biological Research, First Vienna Shock Forum, Vol. 236-A: Pathophysiological Role of Mediators and Mediator Inhibitors in Shock, G. Schlag, H. Redl, eds., Alan R. Liss Inc, New York, pp 235.

Van Bebber, I.P.T., Boekholtz, W.K.F., Goris, R.J.A., Schillings, P.H.M., Dinges, H.P., Bahrami, S., Redl, H., and Schlag, G., 1989, Neutrophil function and lipid peroxidation in a rat model of multiple organ failure, J. Surg. Res., in press.

Wayner, D.D.M., Burton, G.W., Ingold, K.U., and Locke, S., 1985, Quantitative measurement of the total, peroxyl radical trapping antioxidant capability of human blood plasma by controlled peroxidation: the important contribution made by plasmaq proteins, Febs. Lett. 187:33.

Wong, K., Cleland, L.G., and Poznansky, M.J., 1980, Enhanced anti-inflammatory effect and reduced immunogenicity of bovine liver superoxide dismutase by conjugation with homologous albumin, Agents Actions 10:231.

Yoshikawa, T., Murakami, M., Yoshida, N., Seto, O., and Kondo, M., 1983, Effects of superoxide dismutase and catalase on disseminated intravascular coagultion in rats, Thromb. Haemost. 50:869.

Younes, M., Schoenberg, M.H., Jung, H., Fredholm, B.B., Haglund, U., and Schildberg, F.W., (1984), Oxidative tissue damage following regional intestinal ischemia and reperfusion in the cat, Res. Exp. Med. 184:259.

THE PROTECTIVE ROLE OF SUPEROXIDE DISMUTASE DURING ACTIVATION OF

RIBONUCLEOTIDE REDUCTASE

Marc FONTECAVE* and Peter REICHARD**

* Université René Descartes, CNRS U.A. 400, 45 rue des
Saints-Pères, 75270 Paris Cedex 06, France
**Department of Biochemistry, Karolinska Institute
S-10401 Stockholm, Sweden

INTRODUCTION

Ribonucleotide reductase (ribonucleoside diphosphate
reductase, EC 1.17.4.1) is a key enzyme in the process of DNA
synthesis, in all living cells[1]. It catalyzes and regulates the
production of the four common deoxyribonucleotides, the
precursors of DNA, from the corresponding ribonucleotides, the
hydroxyl group at the 2' position of the ribose moiety being
substituted by a hydrogen atom.

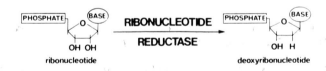

The relative amounts of the deoxyribonucleotides are controlled
mainly by the complicated allosteric regulation of the
reductase. In this chapter, we first briefly describe the
structural background for this essential protein and then, in
more detail, we discuss experiments that led to the discovery
and characterization of a new biological mechanism for its
regulation, in Escherichia Coli. In this context, we will focus
on the unexpected finding that superoxide dismutase (SOD) was an
integral part of the regulatory system.

STRUCTURE OF E. COLI RIBONUCLEOTIDE REDUCTASE

Ribonucleotide reductase from E. Coli, which is the prototype
of all known mammalian, bacterial and viral reductases, consists
of two non identical homodimer subunits, B1 and B2 (Figure 1)[2].

Antioxidants in Therapy and Preventive Medicine
Edited by I. Emerit *et al.*
Plenum Press, New York, 1990

29

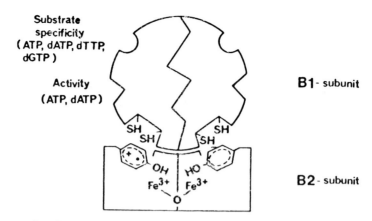

Figure 1. Structure of E. Coli ribonucleotide reductase

Protein B1, with its active dithiols, provides the reducing
equivalents for the reaction and has two binding sites for the
substrates and four binding sites for allosteric effectors.
Protein B2 contains a binuclear non-heme iron center, where the
two high-spin Fe(III) ions are antiferromagnetically coupled by
a μ-oxo bridge[2], as well as a stable free radical which has been
located to a specific tyrosine residue of the polypeptide chain
(Tyr-122)[3]. This radical gives a characteristic EPR signal. The
activity of B2 is absolutely dependent on the presence of the
radical, which has been suggested to be involved in the first
step of the reaction, the radicalar abstraction of the hydrogen
atom at the 3' position of the ribose[4]. MetB2 is an inactive form
of the protein which lacks the radical but still contains the
iron center.

ACTIVATION OF RIBONUCLEOTIDE REDUCTASE: ENZYMATIC OXIDATION OF A
SPECIFIC TYROSINE RESIDUE OF PROTEIN B2

Our contribution to the understanding of the regulation of
ribonucleotide reductase, in E. Coli, was concerned with the
following question: how does the cell introduce a tyrosyl
radical into protein B2, a process absolutely required for
providing the cell with a fully active enzyme? Recently, it was
found that wild type E. Coli extracts contain an enzyme system
capable of transforming the inactive form, metB2, with a normal
tyrosine residue, into the active radical-containing form of B2[5].
The reaction depended on the presence of dithiothreitol, Mg^{2+} and
also oxygen, which is not surprising since oxidizing equivalents
are required for the oxidation of the tyrosine residue (Figure
2). Moreover, addition of pyridine nucleotides and flavins
greatly stimulated the reaction. It was suggested that this
"radicalization" of a specific tyrosine residue of protein B2
might have a regulatory function.

SUPEROXIDE DISMUTASE PARTICIPATES IN THE ACTIVATION OF PROTEIN B2

When we began with the fractionation of the bacterial activating system it soon turned out that we were dealing with a very complex system in which several proteins participate in the overall reaction. It is not the proper place to describe in detail our efforts in the purification and characterization of this system. Thus, a summary of the purification scheme and of our present knowledge concerning the various components required for the reaction is given in Figure 2. It shows that three protein fractions, A,b and c, after DEAE and gel exclusion chromatography, were found to be required together for the reaction. Rechromatography of fraction A on DEAE separated three active peaks (A_0, A_1 and A_2). Fraction A_0 was obtained in pure form after two further purification steps involving chromatography on hydroxylapatite and Ultropac TSK 3000 SW. From its amino acid sequence, A_0 was identified as manganese superoxide dismutase (Mn-SOD)[6]. This explained why fraction A contained three interchangeable active proteins since they corresponded to the three forms of SOD present in E. Coli, the Mn-SOD, the Fe-SOD and the hybrid Mn-Fe-SOD. Moreover, commercial SOD from beef erythrocytes, for example, could substitute for fraction A.

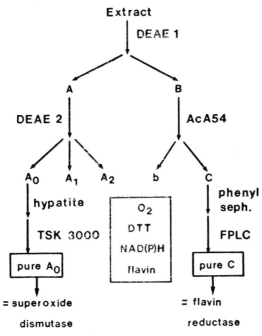

Figure 2. Purification scheme for the various protein fractions required for activation of ribonucleotide reductase in E. Coli.

The availability of E. Coli mutants completely devoid of SOD activity made possible a genetic approach to test for the requirement for SOD during metB2 activation. In table 1, it is shown that a crude extract from such a mutant strain has a greatly dimished capacity to activate metB2, when compared to an extract from wild type E. Coli. Moreover addition of exogenous SOD to the first extract increased its activity approximately 3-fold. Those results support the importance of SOD for the activation of ribonucleotide reductase under physiological conditions[7].

Table 1. MetB2 activation by extracts from wild type E. Coli C600 and from E. Coli GC4468 lacking SOD activity

| Bacterial strain | Specific activity[a] | |
	No SOD	+ SOD
GC4468 (-SOD)	0.25	0.70
C600 (wild type)	1.4	1.3

(a) Units were defined and determined as described in ref.7.E. Coli strain GC4468 was a gift from D. Touati (France).Specific activity is defined as units per milligram of protein.

INACTIVATION OF PROTEIN B2 BY OXYGEN RADICALS AND HYDROGEN PEROXIDE

When we obtained fraction c in pure form after chromatography on Phenyl-Sepharose and on Mono Q and identified it as an NAD(P)H:flavin oxidoreductase[8], an enzyme which catalyzes the reduction of the flavins, FMN, FAD and riboflavin, by NADPH and NADH, the function of SOD during activation of protein B2 became more evident . Actually, reduced flavins are known to be able to efficiently transfer reducing equivalents to oxygen and SOD is considered to prevent the toxic effects of oxygen radicals. In spin trapping experiments using DMPO as a spin trap, we have shown that large amounts of superoxide radicals, identified by the characteristic EPR signal of the corresponding DMPO adduct, are formed during aerobic incubation of the flavin reductase with NADPH and FMN. Addition of SOD suppressed the signal but a new EPR signal appeared, characteristic of the DMPO adduct of the hydroxyl radical. No signal could be detected if both SOD and catalase were present in the reaction mixture[7].

One likely target for those radicals during activation of metB2 is the protein B2 itself and in particular its tyrosyl free radical. Table 2 shows that hydrogen peroxide and hydroxyl radicals, generated from H2O2 by the Fenton reaction, are very harmful to B2[7]. Catalase but also the very crude fraction b provided a full protection. B2 was also found to be very sensitive to superoxide radicals derived from the flavin reductase activity. Addition of SOD did not protect B2, certainly because SOD activity results in hydrogen peroxide

Table 2. Inactivation of protein B2 by reduced
oxygen species.

Additions			Activity (%)
None			100
+ H2O2			45
	+ catalase		92
	+ fraction b		95
	+ Fe2+		8
+ flavin reductase			67
	+ SOD		66
	+ SOD + catalase		91
	+ SOD + fraction b		100

B2 activity was determined by CDP reduction, after
incubation in the presence of the various
indicated compounds, as described in ref.7. Where
indicated, flavin reductase was added in the
presence of NADPH and FMN.

Table 3. Loss of the tyrosyl radical during
inactivation of B2 with reduced oxygen
species.

Additions		EPR signal amplitude (%)
None		100
+ H2O2		31
+ flavin reductase		47
	+ SOD	31
	+ catalase	35
	+ SOD + catalase	90

B2 was incubated for 15 min with the various
indicated compounds, as described in ref.7. After
freezing the sample its EPR spectrum was recorded.

production. This is actually supported by the finding that B2 was efficiently protected when both SOD and catalase were present in the reaction mixture. Table 3 shows that inactivation of B2 by H2O2 and oxygen radicals was paralleled by a destruction of the tyrosyl radical[7].

GENERAL CONCLUSIONS

Activation of ribonucleotide reductase, consisting of the introduction of the tyrosyl radical into the inactive form, metB2, is achieved at least in E. Coli by a complex system where several proteins are participating in the reaction. The following figure shows our present understanding of the specific requirement for SOD. The flavin reductase occupies a central position, directly responsible for the formation of the tyrosyl radical. Very recently, it has been suggested that the role of the flavin reductase was to provide the system with reduced flavins for the reduction of the iron center[9]. Then it is during reoxidation of the ferrous center by oxygen that the tyrosyl radical is formed. However, the flavin reductase activity is also responsible for the production of byproducts, superoxide radicals, hydrogen peroxide and hydroxyl radicals, very harmful to B2. Thus protective systems are needed. This is the function of SOD and maybe of fraction b, at least partly since catalase could not substitute for fraction b during activation of metB2. Understanding the role of fraction b awaits its purification and characterization.

Moreover, since ribonucleotide reductase is sensitive to oxygen radicals, it has to be considered as a possible target during the various cellular events which lead to an oxidative stress. As an example, it has been recently shown that thiosemicarbazones or their iron complexes, which are the most powerful inhibitors of ribonucleotide reductase, might inactivate the enzyme and inhibit DNA synthesis because of their ability to redox-cycle and produce oxygen radicals[10].

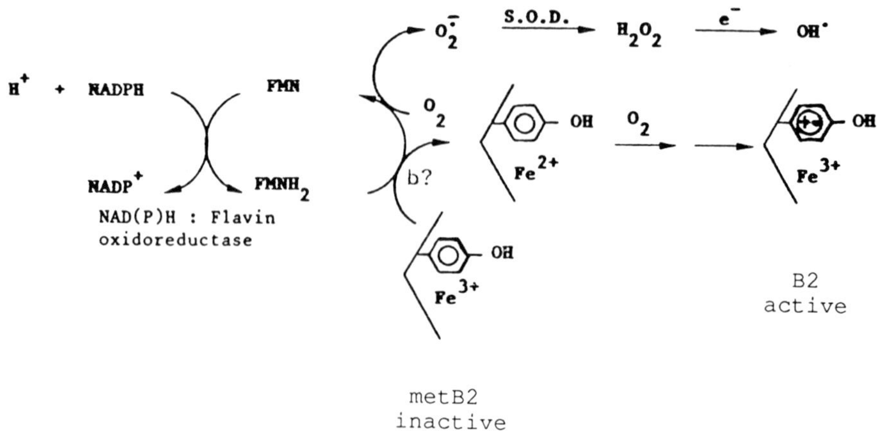

Figure 3. Mechanism of radical introduction in protein B2

REFERENCES

1. P. Reichard, Interactions between deoxyribonucleotide and DNA synthesis, _Ann. Rev. Biochem._ 57:349 (1988).
2. P. Reichard and A. Ehrenberg, Ribonucleotide reductase- a radical enzyme, _Science_ 221:514 (1983).
3. A. Larsson and B.M. Sjoberg, Identification of the stable free radical tyrosine residue in ribonucleotide reductase, _EMBO J._ 5:2037 (1986).
4. G.W. Ashley and J. Stubbe, Current ideas on the chemical mechanism of ribonucleotide reductases, _Pharm. Ther._ 30:301 (1985).
5. T. Barlow, R. Eliasson, A. Platz, P. Reichard and B.M. Sjoberg, Enzymic modification of a tyrosine residue to a stable free radical in ribonucleotide reductase, _P.N.A.S._ 80:1492 (1983).
6. R. Eliasson, H. Jornvall and P. Reichard, Superoxide dismutase participates in the enzymatic formation of the tyrosine radical of ribonucleotide reductase from E. Coli, _P.N.A.S._ 83:2373 (1986).
7. M. Fontecave, A. Graslund and P. Reichard, The function of superoxide dismutase during the enzymatic formation of the free radical of ribonucleotide reductase, _J. Biol. Chem._ 262:12332 (1987).
8. M. Fontecave, R. Eliasson and P. Reichard, NAD(P)H:Flavin oxidoreductase of E. Coli: a ferric iron reductase participating in the generation of the free radical of ribonucleotide reductase, _J. Biol. Chem._ 262:12325 (1987).
9. M. Fontecave, R. Eliasson and P. Reichard, _J. Biol. Chem._ in press.
10. L. Thelander and A. Graslund, Mechanism of inhibition of mammalian ribonucleotide reductase by the iron chelate of 1-formylisoquinoline thiosemicarbazone, _J. Biol. Chem._ 258:4063 (1983).

EFFECT OF SUPEROXIDE DISMUTASE ON THE AUTOXIDATION OF HYDROQUINONES FORMED

DURING DT-DIAPHORASE CATALYSIS AND GLUTATHIONE NUCLEOPHILIC ADDITION

Enrique Cadenas and Lars Ernster

Department of Pathology II, University of Linköping, 581 85 Linköping
and Department of Biochemistry, University of Stockholm, 106 91
Stockholm, Sweden

Electron-transfer reactions as well as the generation and reactivity of free radicals in biological systems are controlled by thermodynamic-, kinetic-, and environmental factors[1]. The redox chemistry of hydro- and semi-quinones is in large extent determined by the physico-chemical properties of the molecule, such as the reduction potential and the influence on it of the substitution pattern, by environmental factors, such as pH, solvent cage, solvation energy, and medium polarity, and by kinetic factors, which can allow a reaction -otherwise thermodynamically unlikely- to proceed by exerting a modification on its equilibrium.

The autoxidation of semiquinones -represented by the overall equation: $Q^{\cdot-} + O_2 \Leftrightarrow Q + O_2^{\cdot-}$- is accelerated by superoxide dismutase[2,3]; pulse radiolysis studies indicated that the equilibrium of the autoxidation reaction is driven towards the right upon removal of $O_2^{\cdot-}$ by superoxide dismutase[1]. This, along with the reduction of the aminopyrine cation radical by GSH[4], set examples of kinetically -rather than thermodynamically- controlled reactions.

The effect of superoxide dismutase on hydroquinone autoxidation, on the other hand, is controversial, because the enzyme has been shown -depending on the type of hydroquinone- to inhibit or stimulate O_2 consumption and H_2O_2 formation linked to the autoxidation reaction[3,5]. Superoxide dismutase has been reported to inhibit the autoxidation of trihydroxybenzene[6], leucoflavins[7], divicine and dialuric acid[8,9], 6-hydroxy-dopamine[10], dopamine hydroquinone[11], and pyrogallol[12].

Recent studies indicated that Cu-Zn superoxide dismutase could either inhibit or enhance hydroquinone autoxidation[3,5,13]. Analysis of these effects of superoxide dismutase requires step-wise consideration of the processes involved in the overall hydroquinone autoxidation: [a] the two-electron reduction of the quinone as accomplished during DT-diaphorase catalysis[3,5] or GSH nucleophilic addition[13]. [b] One-electron oxidation of the hydroquinone to yield a semiquinone intermediate; this process can involve different redox transi-

Antioxidants in Therapy and Preventive Medicine
Edited by I. Emerit *et al.*
Plenum Press, New York, 1990

37

tions and the relative contribution of each individual reaction to the overall process will be determined by thermodynamic-, kinetic-, and environmental factors[1]. [c] One-electron auto-xidation of the semiquinone with formation of $O_2^{\cdot-}$. These relationships are illustrated in the scheme below, in which the state of protonation of the intermediate species is intentionally ambiguous; likewise, a single process -autoxidation- was used to account for the $Q^{\cdot-} \Leftrightarrow Q$ transition, for it is relevant to the further evaluation of the role of superoxide dismutase on hydroquinone autoxidation.

TWO-ELECTRON REDUCTION	ONE-ELECTRON REDOX TRANSITIONS

DT-diaphorase catalysis	Auto-oxidation	Autoxidation
1,4-Reductive addition	Oxidation by superoxide	
	Disproportionation	
	Cross-oxidation	
	Oxidation by metals	

Two-electron transfer to quinones: DT-diaphorase catalysis and nucleophilic addition

Two-electron transfer processes to quinoid compounds are mainly encompassed by DT-diaphorase and 1,4-reductive addition, e.g., reactions with sulfur nucleophiles such as GSH. The former activity can be formally understood in terms of a hydride transfer from the flavoprotein to a two-electron acceptor (reaction 1)[14],

$$[1]$$

whereas the reaction between quinoid compounds and nucleophiles is a 1,4-reductive addition of the Michael type[15] (reaction 2). This reaction proceeds with formation of a transition state anion, which further leads to the generation of a primary thioether sulfide, costumarily termed -for the case of GSH as sulfur nucleophile- hydroquinone-glutathione conjugate.

$$[2]$$

One-electron oxidation of hydroquinones: formation of semiquinone intermediates

The hydroquinones or hydroquinone thioether derivatives formed by the processes outlined above can undergo one-electron transfer reactions with formation of semiquinone intermediates. The reactions listed below contemplate only a limited number of the possible redox transitions, but they are relevant to the overall process inasmuch they can be considered to contribute to the initiation of hydroquinone autoxidation.

Electron transfer to O_2 leads to the formation of a semiquinone intermediate and $O_2^{\cdot-}$ (reaction 3). This reaction proceeds at slow rates, because the reduction potential of the intermediate step $[E(Q^{\cdot-}/Q^{2-})]$ brings forward a thermodynamic restriction for electron transfer to O_2, for -almost invariably- $E(Q^{\cdot-}/Q^{2-}) > E(O_2/O_2^{\cdot-})$. Hydroquinones are generally believed to be stable compounds, their autoxidation restricted by the unfavourable reduction potential at the intermediate step $[E(Q^{\cdot-}/Q^{2-})]$ and by their high pK values. Because of the latter property, two-electron reduced quinones are, at pH 7, in a protonated state -as hydroquinones- and therefore they do not participate readily in electron-transfer reactions[16].

$$[3]$$

At variance with reaction 3, oxidation of a hydroquinone by $O_2^{\cdot-}$ (reaction 4) should proceed at a much faster rate, for $[E(Q^{\cdot-}/Q^{2-}) \ll E(O_2^{\cdot-},2H^+/H_2O_2)]$. An analogous reaction was studied with catechols, ascorbic acid, and dihydrophenazine leading to the postulation of a common mechanism via a sequential proton-hydrogen transfer[17].

$$[4]$$

One-electron oxidation of hydroquinones by other quinones is implied in disproportionation (reaction 5) and cross-oxidation (reaction 6) reactions. Because of the thermodynamic restrictions implied in reaction 3 as initiation of hydroquinone autoxidation, reactions 5 and 6 -along with reaction 4- could function as initiation reactions leading to the formation of semiquinones.

$$[5]$$

The variables controlling the redox transitions subsequent to GSH nucleophilic addition to quinones (reaction 2) are complex; the prevalence of cross-oxidation or autoxidation reactions in the $GS-Q^{2-} \leftrightarrow GS-Q^{-}$ transition is partly determined by the relative concentrations and the reduction potential of the thioether derivative and the parent quinone, as well as of the O_2/O_2^- couple. Evidence for one-electron transfer reactions was obtained by the identification with the ESR technique of a glutahionyl-semiquinone adduct of p-benzoquinone and 1,4-naphthoquinone[18], the occurrence of which was accounted for in terms of cross-oxidation reactions as exemplified in reaction 6.

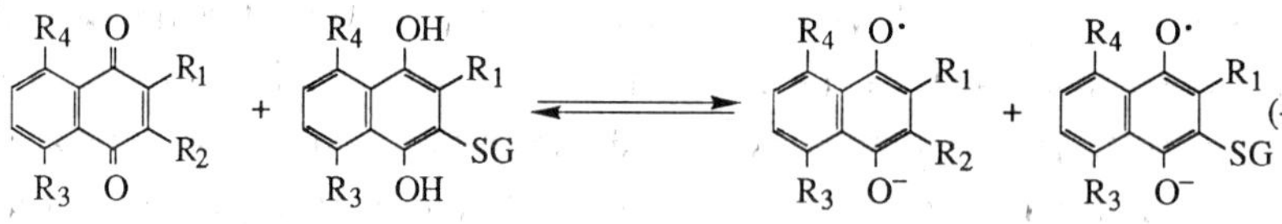

Autoxidation of semiquinone intermediates

Semiquinones -generated upon hydroquinone oxidation by the above reactions- autoxidize according to reaction 7. At variance with hydroquinone autoxidation (reaction 3), the

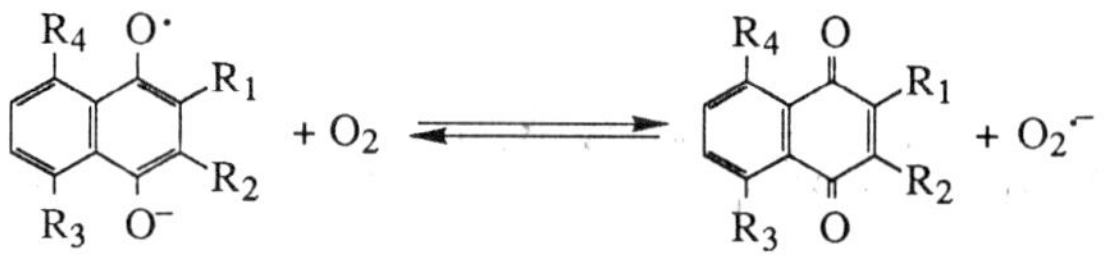

thermodynamic restrictions imposed upon semiquinone autoxidation are overcome by the more favourable reduction potentials of the couples involved $[E(Q/Q^{-}) \leq E(O_2/O_2^{-})]$. Thus -depending on the physico-chemical properties of the quinone- the autoxidation of the semiquinone form (reaction 7) should proceeds at rates several order of magnitude faster than that of the hydroquinone (reaction 3).

Role of superoxide dismutase on hydroquinone autoxidation

Hydroquinones of the p-benzo- and 1,4-naphthoquinone series, which are either unsubstituted or bearing $-CH_3$ substituents and which are formed during DT-diaphorase catalysis, are rather stable inasmuch as they autoxidize very slowly. However, the presence in the quinone molecule of $-OCH_3$, $-SG$, or $-OH$ substituents accelerates autoxidation of the hydroquinone formed by about 15-fold[19]. Similar considerations apply to the autoxidation of hydroquinone-thioether derivatives formed during the GSH nucleophilic addition to quinones: although the new $-SG$ substituent does not seem to produce significant changes in the reduction potential of the quinone, it enhances dramatically the rate of autoxidation. Glutathionyl-substituted p-benzohydroquinone autoxidize at a rate 8-fold higher than the parent compound, whereas the hydroxyl- and glutatuionyl-disubstituted compound does at a rate 350-fold higher[13].

Table I lists the effect of superoxide dismutase on the autoxidation of hydroquinones or hydroquinone-thioether derivatives as formed during DT-diaphorase catalysis and GSH nucleophilic addition, respectively.

Table I. *Effect of superoxide dismutase on the autoxidation of hydroquinones formed during DT-diaphorase catalysis and GSH nucleophilic addition*

Quinone	Effect of superoxide dismutase
I. DT-DIAPHORASE CATALYSIS	
1,4-Naphthoquinones	
Unsubstituted	
1,4-Naphthoquinone	−
Methyl-substituted	
2-Methyl-1,4-naphthoquinone	−
2,3-Dimethyl-1,4-naphthoquinone	−
Methoxyl-substituted	
2,3-Dimethoxyl-1,4-naphthoquinone	−
Glutathionyl-substituted	
3-Glutathionyl-1,4-naphthoquinone	−
2-Methyl-2-glutathionyl-1,4-naphthoquinone	−
Hydroxyl-substituted (quinoid ring)	
2-Hydroxy-1,4-naphthoquinone	−
Hydroxyl-substituted (benzenoid ring)	
5-Hydroxyl-1,4-naphthoquinone	+
2-Methyl-5-hydroxyl-1,4-naphthoquinone	+
5,8-Dihydroxyl-1,4-naphthoquinone	+
Hydroxyl- and glutathionyl-substituted	
3-Glutathionyl-5-hydroxyl-1,4-naphthoquinone	+
2-Methyl-3-glutathionyl-5-hydroxyl-1,4-naphthoquinone	+
3-Glutathionyl-5,8-dihydroxyl-1,4-naphthoquinone	+
1,2-Naphthoquinone	
Unsubstituted	+
II. 1,4-REDUCTIVE ADDITION OF GSH	
p-Benzoquinone	
Un- and methyl-substituted	
p-Benzoquinone	+
6-Methyl-p-benzoquinone	+
2,6-Dimethyl-p-benzoquinone	+
Epoxides	
2,3-Epoxy-p-benzoquinone	−
6-Methyl-2,3-epoxy-p-benzoquinone	−
2,6-Dimethyl-2,3-epoxy-p-benzoquinone	−

+ and −, stimulation and inhibition of hydroquinone autoxidation, respectively. Data from refs.[3,5,13].

The autoxidation of unsubstituted naphthohydroquinone, as well as those derivatives bearing $-CH_3$, $-OCH_3$, $-SG$, and $-OH$ (in the quinoid ring) substitutents, is inhibited by superoxide dismutase, whereas that of aromatic-ring, hydroxyl-substituted naphthohydroquinones is enhanced. Half-maximal inhibition of hydroquinone autoxidation during DT-diaphorase catalysis was obtained with concentrations of Cu-Zn superoxide dismutase within the catalytic range (2-6 nM)[3,5]. Similarly, superoxide dismutase suppressed autoxidation of the p-benzohydroquinone-thioether derivatives bearing a $-OH$ substituent (compounds with a very negative reduction potential and arising from the GSH nucleophilic addition to the corresponding p-benzoquinone epoxides), whereas it enhanced the autoxidation of unsubstituted or $-CH_3$-substituted p-benzohydroquinone-thioether derivatives[13].

The inhibition of hydroquinone autoxidation by superoxide dismutase can be analyzed in terms of: [1] a displacement of the equilibrium of the autoxidation reaction involving changes in the steady-state concentration of $O_2^{\cdot-}$ and in the reactions propagated by this species, and [2] a direct catalytic interaction of the enzyme involving semiquinone reduction at expense of $O_2^{\cdot-}$.

[1] The former possibility has been analyzed assuming that $O_2^{\cdot-}$ is the propagating species in a free radical chain reaction. The transition $Q^{\cdot-} \Leftrightarrow Q^{2-}$ ($2H^+$) -represented by the cycle encompassed in reactions $\underline{4}$ and $\underline{7}$, where $O_2^{\cdot-}$ is formed upon semiquinone autoxidation ($Q^{\cdot-} + O_2 \Leftrightarrow Q + O_2^{\cdot-}$) and consumed upon hydroquinone oxidation ($QH_2 + O_2^{\cdot-} \Leftrightarrow Q^{\cdot-} + H_2O_2$)- might set the basis for the rapid hydroquinone autoxidation observed in several model systems. As a result of this, $O_2^{\cdot-}$ is the propagating species within a free radical chain reaction and its steady-state concentration is low[6,8]. Thus, superoxide dismutase suppresses a chain reaction upon removal of $O_2^{\cdot-}$ by means of its classical $O_2^{\cdot-}$-dismutating activity[6,8]. Because the steady state concentration of $O_2^{\cdot-}$ -as stated above- is low, its disproportionation by superoxide dismutase does not result in H_2O_2 production but, mainly, on its suppression along with an accumulation of the hydroquinone form. This analysis has been applied to the autoxidation of dialuric acid, divicine, and isouramil[6] and 6-hydroxy-dopamine[8].

[2] The latter possibility refers to a direct catalytic interaction of superoxide dismutase with the products of the first autoxidation step (reaction $\underline{3}$; $Q^{2-} + O_2 \Leftrightarrow Q^{\cdot-} + O_2^{\cdot-}$). This activity is a mixed function of superoxide dismutase, which involves semiquinone reduction at expense of $O_2^{\cdot-}$ and which differs from the disproportionation of semiquinones[20] and peroxyl radicals[21]. The basic requirement would be the simultaneous breakdown of the organic compound into a semiquinone intermediate and $O_2^{\cdot-}$. Similar mixed function activities have been reported (see discussion ref.[22]), such as the oxidation of ferrocyanide by $O_2^{\cdot-}$.

A direct catalytic interaction could explain -in terms of specificity- the lack of inhibition of autoxidation of aromatic-ring, hydroxyl-substituted naphthohydroquinones (table I)[3,5], process that cannot be rationalized in terms of a removal of $O_2^{\cdot-}$ -as propagating species- from the $Q^{\cdot-} \Leftrightarrow Q^{2-}$ cycle. Mechanistically, a direct catalytic interaction of the intermediates of hydroquinone autoxidation with superoxide dismutase could be analyzed in terms of reactions in which alternate reduction and oxidation of Cu in the enzyme takes place (reactions 8 and 9)

$$SOD-Cu^{++} + O_2{}^- \Rightarrow SOD-Cu^+ + O_2 \qquad [8]$$

$$SOD-Cu^+ + Q^- + 2\,H^+ \Rightarrow SOD-Cu^{++} + QH_2 \qquad [9]$$

or formation of binary complex (reaction 10 and 11), in which Cu does not undergo valency changes.

$$SOD-Cu^{++} + O_2{}^- \Rightarrow SOD-Cu^{++}-O_2{}^- \qquad [10]$$

$$SOD-Cu^{++}-O_2{}^- + Q^- + 2\,H^+ \Rightarrow SOD-Cu^{++} + QH_2 \qquad [11]$$

Both set of equations, indicated by the overall reaction: $O_2{}^- + Q^- + 2H^+ \Rightarrow O_2 + QH_2$, are thermodynamically unfavourable, for $E(Cu^{++}/Cu^+)$, being +260 mV[23], is generally more positive than the one-electron reduction potential of the naphthoquinones tested $[E(Q^-/Q^{2-})]$. The formation of a ternary complex, $[{}^-Q-Cu^{++}-O_2{}^-]$, could, however, overcome these thermodynamic restrictions. In this context, the inhibition of adrenaline autoxidation by CuZn-superoxide dismutase has been attributed to the formation of a ternary complex between the activated substrate and the enzyme[24]. At variance with the native enzyme, the superoxide dismutase activity of copper-tyrosine complexes accelerated the formation of adrenochrome[24], thus indicating the requirement of both protein and coordinated copper to account for inhibition of adrenaline autoxidation.

The inhibition of hydroquinone autoxidation by superoxide dismutase reported here was observed with the Cu-Zn enzyme: experiments with the Mn-superoxide dismutase are required in order to evaluate more precisely this potential mixed function activity of superoxide dismutase.

Independent of the molecular mechanism(s) by which superoxide dismutase inhibits autoxidation of several hydroquinones, it is clear that this activity, connected to the two-electron transfer catalyzed by DT-diaphorase or -in some instances- to the GSH nucelophilic addition represents an efficient detoxication mechanism against quinone toxicity.

Acknowledgements – Supported by grant 2703-B89-01XA from the Swedish Cancer Foundation and grant 4481 from the Swedish Medical Research Council.

REFERENCES

1. Wardman, P. & Wilson, I., 1987, Control of the generation and reactions of free radicals in biological systems by kinetic and thermodynamic factors, *Free Radical Res. Commun.*, **2**: 217.
2. Winterbourn, C.C., 1981, Cytochrome *c* reduction by semiquinone radicals can be indirectly inhibited by superoxide dismutase, *Arch. Biochem. Biophys.*, **209**: 159.
3. Öllinger, K., Buffinton, G., Ernster, L. & Cadenas, E., 1989, Effect of superoxide dismutase on the autoxidation of substituted hydro- and semiquinones, *Biochem. J.*, submittted.
4. Wilson, I., Wardman, P., Cohen, G.M. & D'Arcy Doherty, M., 1986, *Biochem. Pharmacol.*, **35**: 21.
5. Cadenas, E., Mira, D., Brunmark, A., Segura-Aguilar, J., Lind, C. & Ernster, L., 1988, Effect of superoxide dismutase on the autoxidation of various hydroquinones. A possible role of superoxide dismutase as superoxide-semiquinone oxidoreductase. *Free Radical Biol. Med.* **5**: 79.

6. Greenlee, W.F., Sun, J.D. and Bus, J.S., 1981, A proposed mechanism of benzene toxicity: formation of reactive intermediates from polyphenol metabolites, *Toxicol. Appl. Pharmacol.*, **59**: 187.

7. Ballau, D., Palmer, G. & Massey, V., 1969, Direct demonstration of superoxide anion production during the oxidation of reduced flavin and its catalytic decomposition by erythrocuprein, *Biochem. Biophys. Res. Commun.*, **36**: 898.

8. Winterbourn, C.C., Cowden, W.B. & Sutton, H.C., 1989, Auto-oxidation of dialuric acid, divicine, and isouramil. Superoxide dependent and independent mechanism, *Biochem. Pharmacol.*, **38**: 611.

9. Munday, R., 1987, Dialuric acid autoxidation: Effect of transition metals on the reaction rate and on the generation of 'active oxygen' species, *Biochem. Pharmacol.*, **37**: 409.

10. Gee, P. and Davison, A.J., 1989, Intermediates in the aerobic oxidation of 6-hydroxy-dopamine: relative importance under different reaction conditions, *Free Radical Biol. Med.*, in press.

11. Segura-Aguilar, J. & Lind, C., 1989, On the mechanism of the Mn^{3+}-induced neurotoxicity of dopamine: prevention of quinone-derived oxygen toxicity by DT-diaphorase and superoxide dismutase, *Chem.-Biol. Interact.*, submitted.

12. Marklund, S. & Marklund, G., 1974, Involvement of superoxide anion radical in autoxidation of pyrogallol and a convenient assay for superoxide dismutase, *Eur. J. Biochem.* **47**: 469.

13. Brunmark, A. & Cadenas, E., 1988, Reductive addition of glutathione to *p*-benzoquinone, 2-hydroxy-*p*-benzoquinone, and *p*-benzoquinone epoxides. Effect of the hydroxy- and glutathionyl substituents on *p*-benzohydroquinone autoxidation, *Chem.-Biol. Interact.*, **68**: 273.

14. Iyanagi, T., 1987, On the mechanisms of one- and two-electron transfer by flavin enzymes, *Chem. Scr.*, **27A**: 31.

15. Finley, K.T., 1974, The addition and substitution chemistry of quinones, in *The Chemistry of Quinonoid Compounds*, S. Patai, ed., p. 877, John Wiley & Sons, London.

16. Stenken, S., 1979, Oxidation of phenolates and phenylenediamines by 2-alkanonyl radicals produced from 1,2-dihydroxy- and 1-hydroxy-2-alkoxylalkyl radicals, *J. Phys. Chem.*, **83**: 595.

17. Sawyer, D.T., Calderwood, T.S., Johlman, C.L. & Wilkins, C.L., 1985, Oxidation by superoxide ion of catechols, ascorbic acid, dihydrophenazine, and reduced flavins to their respective anion radicals. A common mechanism with a sequential proton-hydrogen atom transfer, *J. Org. Chem.*, **50**: 1409.

18. Gant, T.W., d'Arcy Doherty, M., Odowole, D., Sales, K.D. & Cohen, G.M., 1986, Semiquinone anion radicals formed by the reaction of quinones with glutathione or amino acids, *FEBS Letts.*, **201**: 296.

19. Buffinton, G.D., Öllinger, K., Brunmark, A. & Cadenas, E., 1989, DT-diaphorase-catalyzed reduction of 1,4-naphthoquinone derivatives and glutathionyl-quinone conjugates. Effect of substituents on autoxidation rates, *Biochem. J.*, **257**: 561.

20. Butler, J. & Hoey, B.M., 1986, The apparent inhibition of superoxide dismutase activity by quinones, *Free Radical Biol. Med.*, **2**: 77.

21. Willson, R.L., 1985, Organic peroxy radicals as ultimate agents in oxygen toxicity, in *Oxidative Stress*, H. Sies, ed., p. 41, Academic Press, London.

22. Fridovich, I., 1988, The biology of oxygen radicals: general concepts, in *Oxygen Radicals and Tissue Injury Symposium*, B. Halliwell, ed., p. 1, Fed. Am. Soc. Exp. Biol., Bethesda.

23. Barette, W.C., Sawyer, D.T., Fee, J.A. & Asada, K., 1983, Potentiometric titrations and oxidation-reduction potentials of several iron superoxide dismutase, *Biochemistry* **22**: 624.

24. Shubotz, L.M., Younes, M. & Weser, U., 1980, Transient reaction of 2Cu, 2Zn-superoxide dismutase with a superoxide-generating substrate, in *Chemical and Biochemical Aspects of Superoxide and Superoxide Dismutase*, J.V. Bannister & H.A.O. Hill, eds., p. 328, Elsevier/North Holland, Amsterdam.

SUPEROXIDE SCAVENGERS AND SOD OR SOD MIMICS

Gidon Czapski and Sara Goldstein

Department of Physical Chemistry
The Hebrew University of Jerusalem
Jerusalem 91904, Israel

The evidence for the participation of O_2^- in many biological processes has been grown rapidly since the discovery of superoxide dismutase (SOD) by McCord and Fridovich in 1969.[1] It has been demonstrated, in numerous systems, that O_2^- exhibits deleterious properties and is a very toxic entity.[2,3]

Due to the toxic role of O_2^- in various processes, e.g., heart events and anoxia,[4] there is an urge to attempt the prevention of its formation or at least lowering its concentration as fast as possible. Some of these methods are already applied clinically in cardiac events. The attempt to prevent O_2^- damage is being carried out along these lines:

a) Prevention of O_2^- formation
b) Lowering O_2^- concentration
c) Removal of O_2^-

The formation of O_2^- is believed to be initiated in reperfusion due to the oxidation of xanthine, catalysed by xanthine oxidase. This process is inhibited by Allopurinol.[5] The toxicity of O_2^- in some systems is believed to be mediated by iron compounds, through the Haber-Weiss cycle. Therefore the addition of desferal, which binds iron firmly, and the addition of catalase or peroxidase, which prevents the Haber-Weiss reaction can protect the systems from O_2^- toxicity.[6]

Another approach is to lower the concentration of O_2^- through the addition of O_2^- scavengers, which form non toxic products, or through the introduction of SOD or SOD mimics, which catalyse the destruction of O_2^- via its dismutation. Most of the SOD mimics and SOD's are either copper, iron or manganese compounds[7] although some metal free SOD mimics were discovered as well.[8]

The mechanism of the catalysis of SOD and SOD mimics is believed to proceed through the 'ping-pong' mechanism, where O_2^- first reduces a metal compound and then reoxidizes it. (The order of course can be reversed, oxidation first proceeds by reduction)

Antioxidants in Therapy and Preventive Medicine
Edited by I. Emerit *et al.*
Plenum Press, New York, 1990

$$Cu(II) + O_2^- \longrightarrow Cu(I) + O_2 \tag{1}$$

$$Cu(I) + O_2^- \xrightarrow{2H^+} Cu(II) + H_2O_2 \tag{2}$$

$$2O_2^- + 2H^+ \longrightarrow H_2O_2 + O_2 \tag{3}$$

The Cu(II)/Cu(I) couple can in this mechanism be replaced by the Fe(III)/Fe(II) or Mn(III)/Mn(II) couples, but the mechanism would be indistinguishable if the copper would oscillate between Cu(III) and Cu(II).

$$Cu(III) + O_2^- \longrightarrow Cu(II) + O_2 \tag{1a}$$

$$Cu(II) + O_2^- \xrightarrow{2H^+} Cu(III) + H_2O_2 \tag{2a}$$

These reaction sequences (reactions (1) or (2) and (1a) and (2a)) yield O_2^- dismutation catalysed by these catalysts with the rate equation

$$\frac{-d[O_2^-]}{dt} = k_{cat}[cat]_0[O_2^-] \tag{4}$$

$$\text{where } k_{cat} = \frac{2k_1k_2}{k_1+k_2} \tag{5}$$

Protection against O_2^- toxicity with scavengers, SOD or SOD mimics

In *in vivo* or *in vitro* systems, where O_2^- is formed and exhibits toxic effects, it is possible to protect these systems against O_2^- toxicity through the use of either O_2^- scavengers or SOD and SOD mimics. In all of these systems there is a competition between the deleterious effect of O_2^- reacting with the target (T) in reaction (6) and in reactions (7) and (8), where O_2^- reacts with a scavenger (S) or with the SOD or SOD mimics.

$$O_2^- + T \longrightarrow damage \tag{6}$$

$$O_2^- + S \longrightarrow P \tag{7}$$

$$2O_2^- + 2H^+ \xrightarrow{SOD \text{ or } SOD \text{ mimics}} H_2O_2 + O_2 \tag{8}$$

The prerequisites that S could serve as a protecting agent against O_2^- toxicity are:

1) That reaction (7) competes effectively with reation (6) or that $k_6[T] \ll k_7[S]$

2) That P, the product of reaction (7), is an entity which is not toxic by itself and is not a precursor of another toxic entity. There are some efficient scavengers for O_2^- which do not protect against O_2^- , e.g., some Fe(III) compounds, which after being reduced by O_2^- react subsequently with H_2O_2 and yield the very toxic OH.

The use of SOD or SOD mimics requires that $k_6[T] \ll k_{cat}[SOD]_0$, and that SOD is not depleted but regenerated through the cycle of reactions (1) and (2).

The difference between a scavenger and a catalyst

Both an O_2^- scavenger and a catalyst for O_2^- dismutation prevent the reaction of O_2^- with the target (T). What is the difference between an O_2^- scavenger and a catalyst for its destruction? The definition of a reactant's scavenger is clear and unambiguous – it is a substance reacting with it. The definitions and common accepted meaning of a catalyst are not as unique. In 1836, Berzelius[9] defined a catalyst as a substance influencing the reaction rate and remaining unchanged at the end of the reaction. Bell's definition was different. He defined a catalyst as a substance whose concentration appearing in the reaction rate equation in a higher power than it does in the stoichiometric equation.[10] This definition seems to discriminate between an O_2^- scavenger and a catalyst for its destruction.

Efficient SOD and SOD mimics operate through the 'ping-pong' mechanism described by equations (1)-(5). When $k_1 \sim k_2$, the catalyst is more efficient and $k_{cat} \sim k_1 \sim k_2$. Provided $k_1 \sim k_2 = 2\text{-}3\times10^9 M^{-1}s^{-1}$, k_{cat} reaches these values.[7]

The lower the value of k_{cat} is, one needs higher concentrations of the catalyst in order that $k_{cat}[cat] > k_6[T]$. For the typical experimetns where O_2^- is generated in a solution, from xanthine-xanthine oxidase, the flux of O_2^- production is ~1.0 μM/min and 10^{-8}M of SOD has a protective effect. In such systems, if one would replace SOD with a SOD mimic, for which $k_{cat} \sim 10^3\text{-}10^4 M^{-1}s^{-1}$ as in the case of the nonmetal SOD mimic OXANO found by Samuni et al.[8], one would require catalyst concentrations of the order of 6 mM. In this case, even after an hour less than 10^{-4}M of O_2^- would be formed, and if they would react with the SOD mimic, assuming $k_1 \sim k_2$, practically all the O_2^- will react in (1) and thus the catalyst would in reality serve as a scavenger.

Generally it is not necessary that the concentration of the catalyst should be much lower than that of the reactant, but it should be lower than the total amount of the reactant reacting over the time, otherwise the 'catalyst' practically operates only through reaction (1) and really serves as a scavenger. Thus, for the SOD function we do not require that $[SOD] < [O_2^-]_{ss}$, but it is required that $[SOD] < [O_2^-$ flux] x t. (t is the reaction time).

In addition, we need a regeneration reaction for the reduced SOD (reaction (2) or an alternative regeneration of the oxidized catalyst). However, if reaction (1) is reversible, then the back reaction

$$Cu(I) + O_2 \longrightarrow Cu(II) + O_2^- \qquad (-1)$$

competes with reaction (2). The faster reaction (-1) is, as compared to reaction (2), the less efficient is the catalyst. In this case, k_{cat} is given by[11]

$$k_{cat} = 2k_1k_2/(k_1+k_2+k_{-1}[O_2]/[O_2^-]) \qquad (9)$$

and k_{cat} depends on $[O_2]/[O_2^-]$.

As a result, the lower $[O_2^-]$ is, reaction (-1) becomes more important and the catalyst is less efficient as k_{cat} decreases.

SOD Mimics *in vivo* and *in vitro*

There are many SOD mimics, mainly copper and iron compounds, which are very efficient catalysts, compare to CuZnSOD while Cu^{2+} even exceeds the efficiency of CuZnSOD in aqueous solutions.[7] The ability of metal compounds to catalyze O_2^- dismutation in solutions, when measured with different assays, often yields results differing in orders of magnitude.[12] The differences were shown to be traced back to the competition between reactions (-1) and (2), which depends on $[O_2]/[O_2^-]$ as well as on the assay method[12] (due to different values of k_{-1}/k_2). These differences show up not only in assays, but also in *in vivo* experiments. The feature that a given compound in a solution has a very high SOD activity does not guarantee that this substance will be an efficient SOD mimic *in vivo*.[13]

An *in vivo* SOD mimic has to have the following properties:

1) The substance should not be toxic at concentrations needed for its SOD activity.
2) The metastable halflife should not be too short, as otherwise its activity would be limited to very short instances. This is one of the problems in the application of human CuZnSOD.
3) The SOD mimic should be able to reach the target region, thus it must have the ability to penetrate into the cells and its solubility in hydrophobic and hydrophobic regions may be critical.
4) The SOD mimic should retain *in vivo* the SOD activity observed in solutions.

The first three points are relatively easy to test, and are often obtainable. We will now concentrate on the last required point.

It was observed that several SOD mimics which were found to be very efficient in their SOD activity in solution, e.g., copper phenanthroline -($(op)_2Cu^{2+}$), failed to show this activity *in vivo*. On the contrary, $(op)_2Cu^{2+}$ *in vivo* enhances O_2^- toxicity rather than protect against it.[13]

The reason for this behaviour cannot be traced to the back reaction (-1), but to the fact that *in vivo* $(op)_2Cu^{2+}$ as well as $(op)_2Cu^+$ form ternary complexes with DNA.

$$(op)_2Cu^{2+} + DNA ===== DNA \equiv (op)_2Cu^{2+} \qquad (10)$$

$$(op)_2Cu^+ + DNA ===== DNA \equiv (op)_2Cu^+ \qquad (11)$$

These ternary complexes can react in reactions similar to reactions (1) and (2) but it has been shown that these reactions are very slow, if they occur at all.[15] Consequently, the ternary $DNA{\equiv}(op)_2Cu^{2+}$ is a poor SOD mimic.[15] Furthermore, it has been found that $DNA{\equiv}(op)_2Cu^+$ reacts with H_2O_2 to yield an oxidizing entity which can be either an $OH^{\cdot}$ radical or a Cu(III) compound. Either of these entities can subsequently react with DNA causing its degradation.

$$DNA{\equiv}(op)_2Cu^+ + H_2O_2 \longrightarrow (DNA{\equiv}(op)_2Cu^{2+} \cdots OH^{\cdot}) \text{ or } DNA{\equiv}(op)_2Cu^{3+}$$

DNA degradation (12)

Reaction (10) followed by the reduction of $DNA{\equiv}(op)_2Cu^{2+}$ with O_2^- and subsequently reaction (12) describes the site specific mechanism of O_2^- damage mediated by $(op)_2Cu^{2+}$.

To summarize, the properties required for an efficient SOD mimic in *in vivo* conditions are: The compound should be nontoxic, should have a relatively long metabolic halflife and should be able to penetrate into the cells. It should have a high SOD activity, and either not form ternary complexes with biological molecules, or if it forms such complexes, these complexes should retain the SOD activitry.

ACKNOWLEDGEMENT

This work was supported by grant 1409 of the Council of Tobacco Research and by the Israel Academy of Science.

REFERENCES

1. J.M. McCord, and I. Fridovich, Superoxide dismutase: an enzymatic function for erythrocuprein. J. Biol. Chem. 244:6049 (1969).
2. I. Fridovich, Superoxide radical and Superoxide dismutase. Acc. Chem. Res. 5:321 (1972).
3. G. Czapski, S. Goldstein, and D. Meyerstein, What is unique about superoxide toxicity as compared to other biological reductants? A hypothesis. Free Rad. Res. Comm. 4:231 (1988).
4. J.M. McCord, Free radicals and myocardial Ischemia: overview and outlook. Free Radical Biol. Med. 4:9 (1988).
5. T. Miura, D.M. Yellon, J. Kingma, and J.M. Downey, Protection afforded by Allopurinol in the first 24 hours of coronary occlusion is diminished after 48 hours. ibid. 4:25 (1988).
6. A.S. Manning, Reperfusion-induced Arrhythmias: Do free radicals play a critical role? ibid. 4:305 (1988).
7. S. Goldstein, and G. Czapski, The role and mechanism of metal ions and their complexes in enhancing damage in biological systems or in protecting these systems from the toxicity of O_2^- ibid. 2:3 (1986).
8. A. Samuni, C. Murali Krishna, P. Riesz, E. Finkelstein, and A. Russo, A novel metal-free low molecular weight superoxide dismutase mimic. J. Biol. Chem. 263:17921 (1988).

9. J.J. Berzelius, Jahresber. Chem. 15:237 (1836).

10. R.P. Bell, Acid Base Catalysis. Clarendon Press, Oxford (1941).

11. G. Czapski, and S. Goldstein, The uniqueness of superoxide dismutase (SOD). Why cannot most copper compounds substitute SOD *in vivo?* Free Rad. Res. Comm. 4:225 (1988).

12. S. Goldstein, C. Michel, W. Bors, M. Saran and G. Czapski, A critical revaluation of some assay methods for superoxide dismutase activity. Free Radical. Biol. and Med., 4:295 (1988).

13. J. Aronovitch, A. Samuni, D. Godinger, and G. Czapski, *In vivo* degradation of bacterial DNA by H_2O_2 and by o-phenanthroline in Super oxide and Superoxide Dismutase in Chemistry, Biology and Medicine; G. Rotilio, editor, p. 346, Elsevier Science Publishers, B.V., (1986).

14. S. Goldstein, and G. Czapski, Mechanism and reaction of some copper complexes in the presence of DNA with O_2^- H_2O_2 and molecular oxygen. J. Am. Chem. Soc. 108:2244 (1986).

REACTIVITY OF ACTIVE CENTRE ANALOGUES OF Cu_2Zn_2 SUPEROXIDE DISMUTASE

Ulrich Weser, Ralf Miesel and Margot Linss

Anorganische Biochemie
Physiologisch-Chemisches Institut der Universität Tübingen
Hoppe-Seyler-Str. 4
7400 Tübingen, FRG

ABSTRACT

Active centre analogues of Cu_2Zn_2 superoxide dismutase were devised and successfully employed. Emphasis was placed on the flexible nature of the superoxide mimicking compounds. Di-Schiff-bases proved most appropriate to fulfil these requirements. Both structural and functional aspects of the copper binding centre of the intact enzyme were met by these complexes. Nanomolar concentrations of copper coordinated in these complexes were sufficient to inhibit the K_3CrO_8 induced chemiluminescence identical to the reaction of Cu_2Zn_2 superoxide dismutase.

INTRODUCTION

There are many small M_r Cu-chelates which are known to exhibit superoxide dismutase like activities (1,2). In these complexes Cu(II) is coordinated in an acetate or biuret type structure. Both geometry and the first shell atoms are unsuitable to keep the transiently formed Cu(I) tenaciously bound to survive competitive biological chelators (3). By way of contrast, in the active centre of Cu_2Zn_2 superoxide dismutase Cu(II) is found in a square planar arrangement and bound to the unsaturated nitrogen atoms of his-44,-46,-61 and -118, respectively. Throughout the catalytic cycle the copper remains firmly bound. The geometry fluctuates from the distorted square planar arranged Cu(II) into tetrahedrally coordinated Cu(I). The flexibility of the protein backbone allows these changes without losing the metal. It was of interest to devise and prepare flexible mononuclear active centre analogues where copper will remain chelated in both the +II and +I form.

Antioxidants in Therapy and Preventive Medicine
Edited by I. Emerit *et al.*
Plenum Press, New York, 1990

Convenient chelators encouraged to be di-Schiff-bases prepared from dicarbonyls and a primary amine or by the reaction of an aliphatic di-amine with an aromatic N-heterocyclic aldehyde (4,5). The resulting tetradentate ligand provided 4 unsaturated nitrogen atoms which success-fully coordinated both Cu(II) and Cu(I). As in the case of Cu_2Zn_2 super-oxid dismutase the nitrogen atoms followed the transient conformational changes of the coordination sphere. The aliphatic C_4 unit originating from 1,4-diaminobutane was sufficient to allow these movements (Fig. 1).

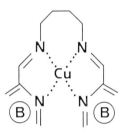

Figure 1.   Flexible active centre analogue of Cu_2Zn_2 superoxide dismutase
          B pyridine or imidazole

MAGNETIC BEHAVIOUR AND ELECTRONIC ABSORPTION

The EPR properties are strikingly similar to those of the native enzyme (Table I). g_{II}/A_{II} an empirical factor is a measure for the degree of tetrahedral distortion (8). Square planar arrangements usually have values between 105 cm and 134 cm. Above 134 cm and up to 250 cm a tetra-hedral distortion can be noticed. The observed 134 cm and 135 cm of the di-Schiff-base-Cu complexes are located at the borderline of such a tetrahedral arrangement and are quite similar to the values obtained with the intact copper proteins.

The less pronounced electronic absorption between 680 and 710 nm was redshifted and can be attributed to d-d transitions in a weak tetra-gonal field. The absorption coefficients ranged between 35 and 150 $M^{-1}cm^{-1}$ (Table II).

TABLE I. Comparison of EPR data of active site analogues and Cu_2Zn_2-superoxide dismutase.

Copper Chelate	$g_{\parallel}$	$g_{\perp}$	$g_{\parallel}/A_{\parallel}$ (cm)	$A_{\parallel}$ (G)	$x\ 10^{-4}(cm^{-1})$
Cu-Pu(Py)$_2$	2.226	2.040	134	160	166
Cu-Pu(Im)$_2$	2.234	2.047	135	160	166
Cu$_2$E$_2$Superoxide Dismutase (7)	2.268	2.067	146	146	155
Cu$_2$Zn$_2$Superoxide Dismutase (6)	2.268	2.087	160	134	142

OXIDATION-REDUCTION POTENTIAL

The oxidation-reduction potential of bovine Cu_2Zn_2superoxide dis-mutase is close to +0.27 V (9). In the present study cyclic voltammetry was used to determine the oxidation-reduction potential of the different Cu-complexes. Much to our surprise the redox behaviour of the employed CuPu(Py)$_2$(ClO$_4$)$_2$·0.5 H$_2$O was indistinguishable close at +0.21 V. As with the intact enzyme the redox cycle proved to be fully reversible. When pyridine was replaced by imidazole no such reversibility was seen. The thermodynamic stability of CuPu(Py)$_2$ was lgK=16.1. CuPu(Im)$_2$ was almost one order of magnitude more stable (lgK=17.1) compared to that of Cu(II) serum albumin (lgK=16.2).

TABLE II. Electronic absorption of di-Schiff-base Cu-chelates and Cu_2Zn_2superoxide dismutase.

Copper Complex	nm	$\varepsilon(M^{-1}cm^{-1})$
Cu$_2$Zn$_2$Superoxide Dismutase	680	150
Cu$_2$E$_2$Superoxide Dismutase	740	140
Cu-Pu(Py)$_2$	710	101
Cu-Pu(Im)$_2$	690	35

The two di-Schiff-base Cu-complexes displayed a marked superoxide dismutase mimicking activity. It was 40±20 times more pronounced compared to the data obtained using the earlier described biuret- or acetate-type Cu-complexes (10,11). Both the thermodynamic stability and the flexible nature of the di-Schiff-base copper complexes improved the enzymic activity. In the intact copper-protein where four unsaturated nitrogens are buried in the imidazolate moieties a convenient electron dislocation was possible. The diminished activity of the present active centre analogues may have its cause in the bridging of a saturated butylic moiety known to be rather reluctant for electron transport (Table III).

It was intriguing to realize the marked thermodynamic stability of the above Cu-di-Schiff-base complexes and their virtual identical oxidation-reduction properties compared to Cu_2Zn_2 superoxide dismutase. We do not fully understand why such a limited enzymic activity was noticed using the nitro tetrazolium blue reductase assay where $\cdot O_2^-$ is enzymically generated. It was of interest to compare the reactivities of these Cu-complexes and the intact enzyme using a completely different assay where all the beforementioned compounds were omitted. For this purpose the CrO_8^{3-} decay (12,13) developed in our laboratory proved most appropriate for characterization of these phenomena.

The reactivity of Cu_2Zn_2 superoxide dismutase, the former two di-Schiff-base Cu-complexes, Cu(salicylate)$_2$ and Cu(serine)$_2$ were compared in the course of the aqueous K_3CrO_8 decay. The CrO_8^{3-} decay led to transiently formed excited oxygen species which were monitored by chemiluminescence in the presence of luminol (Table IV).

TABLE III. Cu_2Zn_2 superoxide dismutase mimetic activity of some Cu-complexes.

Copper chelate	50% inhibition of formazane generation (µM Cu)
Cu_2Zn_2 Superoxide Dismutase	0.04
Cu-Pu(Py)$_2$	1.4
Cu-Pu(Im)$_2$	4.0
Cu-(Acethylsalicylate)$_2$ (10)	23.0
Cu-(Lysine)$_2$ (11)	86.0

TABLE IV. Reactivity of Cu-chelates during the CrO_8^{3-} decay.

Cu-Chelate	nM Cu(II) for 50 % inhibition of photonemission	Superoxide Dismutase units
Cu_2Zn_2 Superoxide Dismutase	3	1
Cu-Pu(Py)$_2$	3	1
Cu-Pu(Im)$_2$	32	10
Cu-(Salicylate)$_2$	300	97
Cu-(Serine)$_2$	950	306
CuSO$_4$	1000	323

The addition of formate to increase the $\cdot O_2^-$ concentration and the presence of mannitol as a possible $\cdot$OH scavenger did not affect the chemiluminescence. Thus, secondary reactions involving superoxide and $\cdot$OH leading to the observed chemiluminescence must be discarded. Singlet oxygen and/or transiently formed radical species including chromium-peroxide have to be assigned to the observed emission of photons. All ligands in concentrations up to 10 µM did not affect the chemiluminescence. It was intriguing to measure the identical reactivity of CuPu(Py)$_2$ compared to the intact Cu_2Zn_2 superoxide dismutase. Only 3 nM of CuPu(Py)$_2$ were required for 50 % inhibition of the initial chemiluminescence. When pyridine was replaced by imidazole 10-times higher concentrations of the Cu(II)-complex became necessary. Compared to Cu(Sal)$_2$, Cu(Ser)$_2$ and CuSO$_4$ the reactivity of either active centre analogue was remarkably pronounced. Sequential additions of 15 µM K$_3$CrO$_8$ three times each to the same incubation mixture containing CuPu(Py)$_2$, CuPu(Im)$_2$ or Cu_2Zn_2 superoxide dismutase did not change the rate of inhibition.

CONCLUSION

There are many different superoxide dismutase assays in which uncontrolled metal chelation of the different constituents involving enzymes, flavins, serum albumin or metal scavengers like EDTA is observed. Thus, the actual reactivity of superoxide dismutase mimicks is quite frequently obscured. Similar to pulse radiolytic measurements (1,2,14,15) the K$_3$CrO$_8$ decay is a suitable tool to compare the efficacy of the intact Cu-enzyme with that of the active centre analogues. Earlier reports from this laboratory (12) have shown that acetate- or biuret-type Cu(II) chelates reacted insignificantly during the CrO_8^{3-}-decay. By way of contrast, the two di-Schiff-base active centre analogues displayed a surprisingly identical activity in the chemiluminescence measurements. As they are also marked inhibitors of the hyaluronic acid depolymerisation their possible use as potent antiinflammatory compounds will be awaited with great interest.

ACKNOWLEDGEMENTS

Portions of the experimental section were aided by grants from the Deutsche Forschungsgemeinschaft. The expert technical help of Helga Heinemann is gratefully acknowlegded.

REFERENCES

1. U. Weser, K.-H. Sellinger, E. Lengfelder, W. Werner, and J. Strähle, Structure of $Cu_2(indomethacin)_4$ and the reaction with superoxide in aprotic systems, Biochim. Biophys. Acta, 631:232 (1980).

2. R. Brigelius, H.-J. Hartmann, W. Bors, M. Saran, E. Lengfelder, and U. Weser, Superoxide dismutase activity of $Cu(Tyr)_2$ and Cu,Co-erythrocuprein, Z. Physiol. Chem., 356:739 (1975).

3. U. Deuschle and U. Weser, Copper and inflammation, Prog. Clin. Biochem. Med., 2:99 (1985).

4. M. Linss and U. Weser, The di-Schiff-base of pryidine-2-aldehyde and 1,4-diaminobutane, a flexible Cu(I)/Cu(II) chelator of significant superoxide dismutase mimetic activity, Inorg. Chim. Acta, 125:117 (1986).

5. M. Linss and U. Weser, Redox behaviour and stability of active centre analogues of Cu_2Zn_2 superoxide dismutase, Inorg. Chim. Acta, 138:175 (1987).

6. G. Rotilio, A. Finazzi Agrò, L. Calabrese, F. Bossa, P. Guerrieri, and B. Mondovi, Studies of the metal sites of copper proteins. Ligands of copper in hemocuprein, Biochemistry, 10:616 (1971).

7. K.M. Beem, D.C. Richardson, and K.V. Rajagopalan, Metal sites of copper-zinc superoxide dismutase, Biochemistry, 16:1930 (1977).

8. U. Sakaguchi and A.W. Addison, Spectroscopic and redox studies of some copper(II) complexes with biomimetic donor atoms: implications for protein copper centers, J. Chem. Soc., Dalton Tran., 4:600 (1979).

9. G.D. Lawrence and D.T. Sawyer, Potentiometric titrations and oxidation-reduction potenials of manganese and copper-zinc superoxide dismutase, Biochemistry, 18:3045 (1979).

10. U. Weser, C. Richter, A. Wendel, and M. Younes, Reactivity of anti-inflammatory and superoxide dismutase active Cu(II)-salicylates, Bioinorg. Chem., 8:201 (1978).

11. M. Younes and U. Weser, Inhibition of nitroblue tetrazolium reduction by cuprein (superoxide dismutase), Cu(tyr)$_2$ and Cu(lys)$_2$, FEBS Lett., 61:209 (1976).

12. U. Weser, W. Paschen, and M. Younes, Singlet oxygen and superoxide dismutase (cuprein), Biochem. Biophys. Res. Commun., 66:769 (1975).

13. R. Miesel and U. Weser, Reactivity of active centre analogues of Cu$_2$Zn$_2$superoxide dismutase during the aqueous decay of K$_3$CrO$_8$, Inorg. Chim. Acta, in the press (1989).

14. R. Brigelius, R. Spöttl, W. Bors, E. Lengfelder, M. Saran, and U. Weser, Superoxide dismutase activity of low molecular weight Cu^{2+}-chelates studied by pulse radiolysis, FEBS Lett., 47:72 (1974).

15. M. Younes, E. Lengfelder, S. Zienau, and U. Weser, Pulse radiolytically generated superoxide and Cu(II)-salicylates, Biochem. Biophys. Res. Commun., 81:576 (1978).

SOD MIMICKING PROPERTIES OF COPPER (II) COMPLEXES: HEALTH SIDE EFFECTS

Gérard LAPLUYE

Université PARIS 7. Laboratoire de Chimie-Physique
2, place Jussieu 75251 PARIS CEDEX 05

Controlled regulation of the superoxide ion in living systems is an important challenge, due to the multiple pathological involvements of this radical ion.

Among the three naturally occurring superoxide dismutases (SODs) the copper zinc SOD has been extensively investigated, with as a consequence a good knowledge of its structure and behaviour.

I. Mechanism of O_2 dismutation by Cu Zn SOD

The dismutation proceeds in two steps at the level of the copper ion :

$$\text{Protein-Cu}^{2+} + O_2 \rightleftharpoons \text{Protein-Cu}^+ + O_2$$

In this first step the superoxide binds probably to the copper ion in axial position, the subsequent reaction being an inner sphere electron transfer.

$$\text{Protein-Cu}^+ + O_2 + 2 H^+ \rightarrow \text{Protein-Cu}^{2+} + H_2O_2$$

SOD is generally associated with a catalase, their partnership constituting a reciprocally protective set. Indeed SOD is inhibited by H_2O_2, catalase is inhibited by O_2.

Zn^{2+} plays at least a structural part in this system.

The SOD molecule is a dimer with a molecular weight of 31,500, it is composed of two sub units, each containing a Cu^{2+} and a Zn^{2+}ion. In each unit these ions are separated by an imidazole ring (fig.1), the distance between them being 0.6 nm.

Antioxidants in Therapy and Preventive Medicine
Edited by I. Emerit *et al.*
Plenum Press, New York, 1990

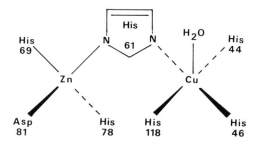

Fig. 1 - Ionic environment in Cu Zn SOD.

Coordination structures of the metal ions are slightly distorted for the tetrahedrally coordinated Zn, but more deeply altered for the square planar Cu. An exchangeable water molecule is set in axial position on Cu^{2+}. The copper ion inside the enzyme is strongly coordinated and directly accessible to O_2 and solvent. The eight histidines in each subunit are equally accessible to solvent molecules.

Research on Cu Zn SOD models (1,2,3,4) give some mechanistic information on the dynamics of substrate interactions with copper ions in the enzyme. The superoxide ion would be steered along a channel in the apoenzyme to the Cu^{2+} ion, in a non centrosymmetrical electric field due to the electric charge distribution in the polypeptidic structure. The positive charges on the lysine and histidine would play an important role in this driving effect, which would increase the diffusion controlled rate by about 40%.

There is yet some incertitude as to the origin of the proton in the second step of the dismutation.

II. SOD Mimics

Natural SOD is rather a large molecule. Its diffusibility through most of the living tissues is low and it is almost negligible through cellular membranes. In view of therapeutical uses, in which a higher ability of migration through tissues and a capacity of intracellular transference (e.g. in post-radiolysis treatments) are required, it would be advisable to get low molecular weight SOD mimics available. Such enzymatic systems should reach deeper and faster their targets. Moreover in some uses a reduced SOD activity or an increased stability could be required.

From a scientific stand point, specially designed SODs would constitute tools of value in the investigation of a field of entangled reactions. Among these the problem of a possible partnership between SOD and interleukines appears as an interesting one.

A lot of SOD mimics have been prepared, which are generally transition metal complexes.

III. Copper SOD mimics

Many of the SOD mimics reported are copper derivatives. But a thorough examination of their properties must be done, for some so called SOD mimics are the result of artefacts. Indeed the copper(II) ion is the most active SOD catalyst known. In the pH range excluding Cu^{2+} hydrolysis, its activity is four times higher than that of the Cu Zn SOD at identical copper concentrations. Its rate constant for O_2 dismutation is $k=8.10^9 \ M^{-1}s^{-1}$ (5) that of bovine Cu Zn SOD is $k=2.10^9 \ M^{-1}s^{-1}$ (6). When copper(II) ions are released from an unstable complex, their own dismuting activity reinforce or simply supply the deficiency of the complex. It must be remembered that metal complexes are characterized by some well defined domains of stability in which they constitute a minor, the major or the single species. The other species are either metal ions, different complexes or hydroxides. This situation is illustrated in fig. 2 which displays the species distribution as a function of pH for the Cu^{2+}-oxidized glutathione system (7). From this set of curves it is obvious that a SOD activity measurement on a sample of MA^{2-} complex at pH = 5 should be disturbed by the presence of the MHA^- complex and of a low content in Cu(II) ions.

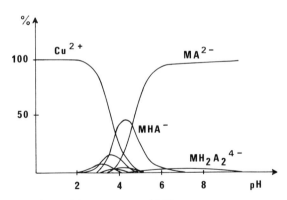

Fig. 2- Oxidized glutathione copper(II)
Species distribution as a function of pH at the 10^{-5} M concentrations, obtained by the extrapolated potentiometric measurements made at the 10^{-3} M concentrations. Percentages with respect to Cu^{2+}.

In the case of the glycyl-glycyl-histidine-Cu^{2+} system at pH = 5, the activity corresponding to the $MH_{-2}A$ complex would be more deeply screened by the presence of copper(II) ions (8) (fig. 3).

Moreover, hydrated Cu(II) ions are easily complexable and possess a short life-time in biological media, which contain many potential ligands, either soluble or immobilized ones. So unforeseen complexes are formed, which may possess or not the SOD activity.

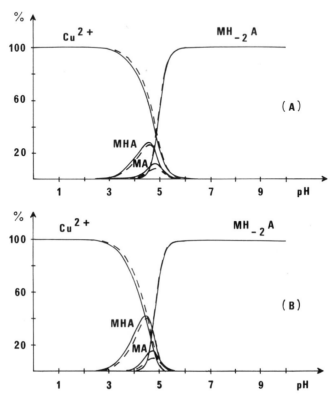

Fig. 3 - Glycyl-glycyl-L-Histidine copper(II)
Species distribution as a function of pH at $C_A = C_M = 2.10^{-3}$ M (Graph A) $C_A = 2 C_M = 4.10^{-3}$ M (Graph B). Percentages with respect to Cu^{2+}.

_____ SCOGS affinement method
--------- MINIQUAD affinement method

In other circumstances the observed dismutation of O_2 was assigned to a complex, when a careful examination did point out that it was due to decomposition products from the former substance. This is the case of the reported SOD activity of copper-D-Penicillamine (9). ROBERTSON & FRIDOVICH (10) demonstrated that this activity originated in simpler complexes obtained from decomposition of Cu-D-Penicillamine.

KLUG-ROTH & RABANI (11) have measured the dismuting reactivities of eight Cu(II)-aminoacid complexes (alanine, glutamic acid, glycine, hydroxyproline, methionine, proline, valine and glycylglycine) in the presence or not of formate and in the biological pH range. The observed activities are located between 10^6 and 10^7 $M^{-1}s^{-1}$. A higher value is encountered in the case of valine (k = 2.4 10^8) with a low ligand concentration. The presence of free copper ions could be at the origin of this value. No information is given concerning the structure of each complex, their stability diagram and consequently their location on the diagram. A more reliable study has been published by WEINSTEIN & BIELSKI (12). Of six Cu(II)-Histidine complexes examined, taking into account the distribution diagrams of the system, one only catalyzes the disproportionation of O_2 in the pH range 1 to 10. This active complex is $(Cu\ His_2H)^{3+}$, the proposed structure of which is given in fig. 4. This complex is the one among the six complexes in the copper-histidine system which may possess an exchangeable water molecule loosely bounded on the copper axial free position.

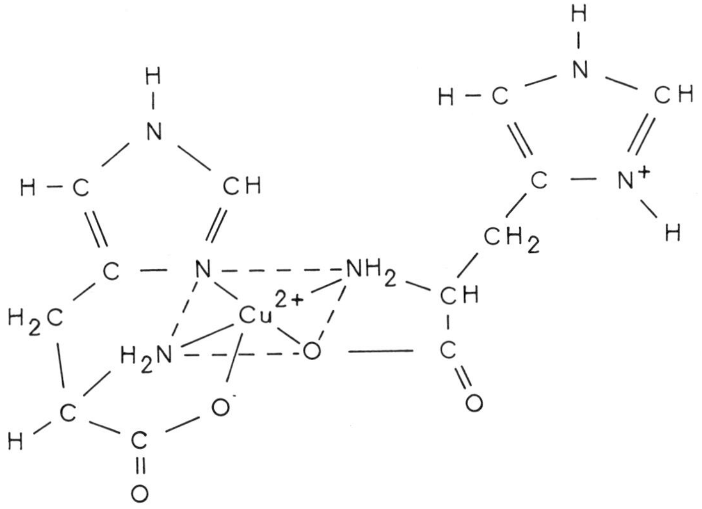

Fig. 4 - $(Cu\ His_2H)^{3+}$ proposed octahedral structure.

The SOD activity of ethylene bridged aminoacid-Cu(II) complexes has been investigated by KATZ & STENBERG (13). In these complexes, excepting the case of glycine, two aminoacid molecules would be bounded with ethylene by their amino nitrogen. These authors did not calculate the k values, but indicate comparative in-vitro activities. These are located in the following decreasing order : Leucine⌣ Alanine⌣ 2-methylalanine⟩ Isoleucine⟩ Valine⟩⟩ Glycine.

In an investigation on the copper-oxidized glutathione system
(7), we obtained a MA complex which is stable in the pH interval 6.5-9
(fig. 2) and presents the following SOD activities :

pH	7	8	9
k	$4.7.10^6$	6.10^6	$6.2.10^6$

These are moderate activities. In the proposed structure two exchangeable
water molecules are located on the copper ion (fig. 5).

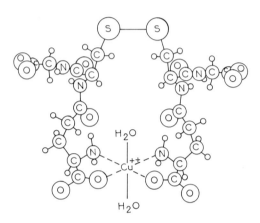

Fig. 5 - Oxidized glutathione Copper(II) chelate . MA complex.

The $MH_{-2}A$ chelate in the Cu(II)-glycylglycylhistidine system
(8) is stable in the pH range 6.5 to 10. It presents a higher activity
and appears to contain two exchangeable water molecules on Cu^{2+}. The
dismutation rate constant is k = 5.10^7 $M^{-1}s^{-1}$.
This chelate is represented in fig. 6.

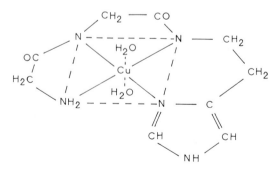

Fig. 6 - Cu(II)glycylglycylhistidine $MH_{-2}A$ chelate.

A chelate of Cu(II)-1.10 phenanthroline dismutates O_2 with a rate constant $k = 5.1.10^8$ $M^{-1}s^{-1}$ (14). Four copper(II)-arginine complexes react with O_2. Their domains of stability are located in different zones of the pH range 1.5 to 12.5 (15). The following rate constants are reported :

Cu Arg H^{2+} $k = 3.3.10^9$ M^{-1} s^{-1} Cu $(Arg\ H)_2^{2+}$ $k = 1.3.10^7$

Cu Arg_2H^+ $k = 1.0.10^6$ Cu Arg $^+$ $k = 1.0.10^8$

Though the influence of free Cu^{2+} has been taken into account, the k value for the first complex should be checked.

Copper(II)-salicylate chelates present SOD activity (16). It is the case also with an -picolinic acid copper complex (17) for which the structure $Cu(II)(pic)_2$ is proposed.

This is not an exhaustive enumeration of SOD active copper complexes.

IV. Requirements for SOD activity in complexes

From an examination of the structure-reactivity relations ruling O_2 dismutation and of the energetic features, it appears that seven conditions at least must be fullfilled in a metallic complex to give SOD activity :

1. The ionic or molecular complex comprises one or several multivalent transition metal ions ;

2. The rate of the redox process on the metal ion must be faster than the spontaneous rate of O_2 dismutation ($k = 8.10^4$ $M^{-1}s^{-1}$) ;

3. The complex must be water soluble ;

4. Its stability constant is sufficiently high to prevent Cu^{2+} release ;

5. The SOD activity must be maintained in the pH range 4.8-10 ;

6. The structure of the complex includes at least one coordination site accessible to O_2 and water molecules, with an easily exchangeable water molecule ;

7. The reduction potential must be included (18) in the interval -0.33 V corresponding to $E°(O_2/O_2)$ (using air pressure for the O_2 standard state) and 0.65 V corresponding to $E°$ $(^1O_2/O_2)$ in order to avoid singlet oxygen formation.

V. Health effects of copper complexes

1. Positive effects

These are discussed more extensively by other authors in this Conference. They concern the main properties and uses of natural SOD :

. Anti-inflammatory power,

. Treatment of ischemia,

- Prevention and treatment of reperfusion injury,
- Treatment of haemorrhagic shock,
- Treatment of hypoxia,
- Prevention of platelet induced micro-occlusions,
- Prevention and treatment of irradiation injuries,
- Attenuation of injuries by chemotherapic drugs in cancer treatments.

The available therapeutic techniques could include local or intravenous perfusions, transdermal delivery for low sized mimics, local applying to wounds, aerosols inhalation, etc...

Addition or grafting of tensioactive components to the copper complex could increase intra cellular and intratissular permeation. A prerequisite property must be the non toxic and non antigenic character.

2. Toxic side effects.

Unfortunately, some copper complexes could directly or indirectly, by their own chemical nature, induce damages in man or animals. Yet the major hazard should originate in the release of copper ions at the immediate neighbourhood of cellular organites. It could be at the origin of biochemical degradations leading to pathological consequences. This release may be achieved by dissociation of a low stability constant complex or by direct intake of copper ions. Such an effect should be taken into account and prevented by previous physical-chemical, biochemical and toxicological investigations on the proposed mimics.

The total amount of copper in the human body is about 50 mg, 10% of which are located in the liver. A casual intake of copper ions in excess of this amount may cause diarrhoea. Lung damages and dermatological effects are reported in the case of a copper intake during a long period. Genetic toxicity is assumed.

In vitro experiments on penicillinase (19) in the presence of O_2 and Cu^{2+} indicate the occurrence of an important damaging in this enzyme, whilst in the absence of Cu^{2+} the superoxide radical did not contribute toward enzyme inactivation. In the presence of H_2O_2 the damage is strongly amplified.

This phenomenon is explained by the formation of a copper penicillase complex which can react with O_2 in a redox process giving Cu^+. This reduced form would react with H_2O_2 according to a Fenton's reaction. The OH radicals damage the enzyme.

A similar redox process is supposed to occur when Cu^{2+} ions enter into contact with thiol groups on erythrocytes membranes (20), in which these functions are oxidized.

Cleavage of DNA may be achieved by a 1,10 phenanthroline Cu(II) chelate after reduction to the Cu(I) complex (21).

DNA degradation has been observed in the case of the simultaneous presence of a reducing agent and of one among the following complexes (1,10 Phenanthroline — Cu, 5-nitro 1,10 Phenanthroline or 2-2' bipyridine—Cu) (22). A ternary complex is formed with DNA (DNA—CuL_2) which catalyzes the Haber-Weiss reaction. The OH radicals are formed close to the copper ion in the complex and they attack DNA. There is no doubt that other copper complexes can operate according to the same mechanism.

On the other hand, it is known that under suitable conditions cupric ions stimulate lipid peroxidation. Lipid peroxides inhibit prostaglandin, with as a consequence a rise in the pro-aggregating action of thromboxane-A on platelets and an increased risk of platelet induced micro-occlusions.

This is not an exhaustive review of biochemical interactions involving copper ions. But it is evident from these examples that similar interactions could be at the origin of significant cytological and tissular injuries.

VI. Conclusion

The development of SOD mimics constitutes undoubtdly an exciting and promising field of research. But it must be kept in mind that the structure and energetic characteristics in the natural copper zinc superoxide dismutase are the result of a long adaptation to optimal conditions of activity, without damaging the neighbouring cells and organites.

The discovery of in-vitro SOD activity in a copper complex should be normally accompanied by an accurate determination of its structure and of its stability pH-domain. In-vivo experimentation should permit then to assess the efficiency and innocuity of this substance in view of therapeutical or research applications.

REFERENCES

(1) E. GETZOFF, J.A. TAINER, P.K. WEINER, P.A. KOLLMAN, J.S. RICHARDSON, D.C. RICHARDSON- Electrostatic recognition between superoxide and copper zinc superoxide dismutase. Nature 306,17, 287-90 (1983).
(2) S.A. ALLISON, J.A. Mc CAMMON- Dynamics of substrate binding to copper zinc SOD J. Phys.Chem. 89, 1072-74 (1985).
(3) S.A. ALLISON, G. GANTI, J.A. Mc CAMMON- Simulation of the diffusion-controlled reaction between superoxide and SOD. - I - Simple models. Biopolymers 24, 1323-36 (1985).
(4) A. DESIDERI, M. FALCONI, V. PARISI, S. MORANTE, G. ROTILIO- Is the activity linked electrostatic gradient of bovine Cu Zn superoxide

dismutases conserved in homologous enzymes irrespective of the number and distribution of charges ?- Free Radical in Biology and Medicine 5, 313-17 (1988).

(5) J. RABANI, D. KLUG-ROTH, J. LILIE- Pulse radiolytic investigations of the catalyzed disproportionation of peroxy radicals. Aqueous cupric ions- J. Phys-Chem. 77, 9, 1169-75 (1973).

(6) M. McADAM, E. FIELDEN, F. LAVELLE, L. CALABRESE, D. COCCO, G. ROTILIO- The involvement of the bridging imidazolate in the catalytic mechanism of action of bovine SOD. Biochem.J. 167, 271-4 (1977).

(7) J. HUET, M. JOUINI, L. ABELLO, G. LAPLUYE- Structural study of copper oligopeptide complexes - I - Oxidized glutathione-Cu(II) system. J.Chim.Phys. 81, 7/8, 505-11 (1984).

(8) M. JOUINI, G. LAPLUYE, J. HUET, R. JULIEN, C. FERRADINI- Catalytic activity of a copper (II) - Oxidized glutathione complex on aqueous superoxide ion dismutation- J. Inorg.Biochem. 26, 269-80 (1986).

(9) M. YOUNES, U. WESER- SOD activity of copper penicillamine : Possible involvement of Cu (I) stabilized sulphur radical. Biochem. Biophys. Research Comm. 78, 4, 1247-53 (1977).

(10) P. ROBERTSON Jr, I. FRIDOVICH- Does Copper-D-Penicillamine Catalyze the dismutation of O_2 ? Arch. Biochem.Biophys. 203, 2, 830-31 (1980).

(11) D. KLUG-ROTH, J. RABANI- Pulse radiolytic studies on reactions of aqueous superoxide radicals with copper(II) complexes. J.Phys.Chem. 80, 6, 588-91 (1976).

(12) J. WEINSTEIN, B.H.J. BIELSKI- Reaction of superoxide radicals with copper (II) histidine complexes. J.A.C.S. 102, 4916-19 (1980).

(13) B.M. KATZ, V.I. STENBERG- Catalytic activity of ethylene-bridged aminoacid- copper (II) complexes for the dismutation of superoxide. Polyhedron 4, 12, 2031-38 (1985).

(14) S. GOLDSTEIN, G. CZAPSKI- Mechanisms of the dismutation of superoxide catalyzed by the copper (II) phenanthroline complex and of the oxidation of the copper (I) phenanthroline complex by oxygen in aqueous solution. J.A.C.S. 105, 7276-80 (1983).

(15) D.E. CABELLI, B.H.J. BIELSKI, J. HOLCMAN- Interaction between copper (II)- arginine complexes and HO_2/O_2 radicals, a pulse radiolysis study. J.A.C.S. 109, 3665-69 (1987).

(16) U. WESER, C. RICHTER, A. WENDEL, M. YOUNES. Reactivity of antiinflammatory and superoxide dismutase active copper(II)-salicylates. Bioinorg. Chem. 8, 3, 201-13 (1978).

(17) W.H. BANNISTER, J.V. BANNISTER, A.J.F. SEARLE, P.J. THORNALLEY- The reaction of superoxide radicals with metal picolinate complexes. Inorganica Chim.Act. 78, 139-42 (1983).

(18) W.H. KOPPENOL, F. LEVINE, T.L. HATMAKER, J. EPP, J.D. RUSH- Catalysis of superoxide dismutation by manganese aminopolycarboxylate complexes. Arch. Biochem. Biophys. 251, 2, 594-99 (1986).

(19) A. SAMUNI, M. CHEVION, G. CZAPSKI- Unusual copper-induced sensitization of the biological damage due to superoxide radicals. J.Biol. Chem. 256, 24, 12632-35 (1981).

(20) K.S. KUMAR, C. ROWSE, P. HOCHSTEIN- Copper induced generation of superoxide in human red cell membrane- Biochem. Biophys. Research Comm. 83, 2, 587-92 (1978).

(21) K.A. REICH, L.E. MARSHALL, D.R. GRAHAM, D.S. SIGMAN- Cleavage of DNA by the 1,10-phenanthroline- copper ion complex. Superoxide mediates the reaction dependent on NADH and hydrogen peroxide. J.A.C.S. 103, 3582-84 (1981).

(22) S. GOLDSTEIN, G. CZAPSKI- Mechanisms of the reactions of some copper complexes in the presence of DNA with O_2, H_2O_2 and molecular oxygen. J.A.C.S. 108, 2244-50 (1986).

RADIATION RECOVERY AGENTS: Cu(II), Mn(II), Zn(II), OR Fe(III) 3,5-DIISOPROPYLSALICYLATE COMPLEXES FACILITATE RECOVERY FROM IONIZING RADIATION INDUCED RADICAL MEDIATED TISSUE DAMAGE

John J.R. Sorenson, Lee S.F. Soderberg,[1] Max L.Baker,[2] John B. Barnett,[1] Louis W. Chang,[3] Hamid Salari, and William M. Willingham [4]

Division of Medicinal Chemistry, Department of Biopharmaceutical Sciences, College of Pharmacy and Departments [1]Microbiology, and Immunology, [2]Radiology, and [3]Pathology, College of Medicine, University of Arkansas for Medical Sciences, Little Rock, Arkansas 72205, and [4]Chemistry Department, University of Arkansas at Pine Bluff, Pine Bluff, Arkansas 71601, U.S.A.

INTRODUCTION

Pathological consequences of human exposure to increasing doses of ionizing radiation: Hematopoietic, Gastrointestinal, and Central Nervous System syndromes, are rather well understood.[1] Bone marrow aplasia and loss of immunocompetency with low (2 to 10 Gray) doses of irradiation, gastrointestinal ulceration and systemic infection with higher radiation doses, and loss of brain function with still higher doses account for the diminishing duration of survival following exposure to increasing doses of ionizing radiation.

Ionizing radiation of tissues produces e^-aq, H, HO, HO_2, and O^-_2, as a result of homolytic bond breaking of water,[2-4] as well as other radicals resulting from homolytic cleavage of bonds in all other biochemical components of cells and extracellular matricies including essential metalloelement dependent enzymes. Since these radicals have millisecond half-lives it is unlikely that drugs can be developed to scavenge these radicals as they are formed. Surviving destructive effects of irradiation may be possible by facilitating recovery of immunocompetency as well as cellular and extracellular repair processes. Recognizing that many of these immune modulating and repair

Antioxidants in Therapy and Preventive Medicine
Edited by I. Emerit *et al.*
Plenum Press, New York, 1990

69

processes are Cu-, Fe-, Zn-, or Mn- dependent enzymes[5] we have studied complexes of these essential metalloelements as radiation recovery agents.

MATERIALS AND METHODS

Hydrates of Cu(II), Fe(III), Zn(II), and Mn(II) 3,5-diisopropylsalicylate complexes; $Cu(II)_2(3,5-DIPS)_4$, $Fe(III)(3,5-DIPS)_3$, $Zn(II)_2(3,5-DIPS)_4$, and $Mn(II)_2(3,5-DIPS)_4$, were prepared as described for the preparation of $Cu(II)_2(3,5-DIPS)_4(H_2O)_2$.[6,7] Hydrates of these complexes are much more easily suspended in the propylene glycol polyvinyl alcohol saline vehicle[7] and they were used exclusively. Eight to ten week old mice were treated and irradiated as described in the legend of each figure and survival at the end of a 30 day post-irradiation period was determined as a measure of radiation recovery. The Fischer Exact test was used to statistically compare results obtained for vehicle or complex-treated groups. Recovery of immunocompetency was determined using previously described methods.[7,8] Recovery of histopathological damage was determined at the light microscopic level following fixation with 10% buffered formalin, embedding in paraffin, and cutting thin sections which were mounted and stained with hematoxylin and eosin.

RESULTS

Following the original report that $Cu(II)_2(3,5-DIPS)_4$ given subcutaneously (sc) produced 58% survival in lethally irradiated mice[9] it was found that doses smaller than originally reported (250 mol/kg) actually accounted for the observed increase in survival.[10] As shown in Figure 1, doses ranging from 80 to 20 mol/kg were effective when given sc 3 hrs before whole-body irradiation and the 20 mol/kg dose was found to be the smallest effective dose when given 3 hrs before an $LD_{50/30}$ irradiation dose (8.0 Gy, 1.55 Gy/min). All irradiations were begun at 1200 hrs Central Standard Time.

As shown in Table 1, survival of $Cu(II)_2(3,5-DIPS)_4$ treated mice was accompanied by rapid recovery of immunocompetency. Seven days after exposure to 8 Gy irradiation, the number of viable bone marrow cells was reduced by 98% and spleen cells were reduced by 99%.[7] At 7 days, these mice had no detectable hemopoietic activity in the bone marrow or spleen. At this early time viable spleen cells had only minimal responses to T cell mitogens, concanavalin A (Con A) and phytohemagglutinin (PHA), or to the B cell mitogen, lipopolysaccharide (LPS). T-dependent antibody responses to sheep erythrocytes were also negligible at 7 days. This total lack of responsiveness suggests that $Cu(II)_2(3,5-DIPS)_4$ did not protect these cells from radiation-induced damage. At 14, 24, and 42 days after irradiation, almost all of these parameters were increased in $Cu(II)_2(3,5-DIPS)_4$-treated animals,

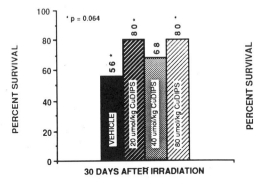

Figure1. Radiation recovery actvity of Cu(II)$_2$ (3,5-DIPS)$_4$ given sc to female C57BL/6N mice (25 per group) 3 hr before irradiation (8.0 Gy,1.55 Gy/min) at 1200 hrs.

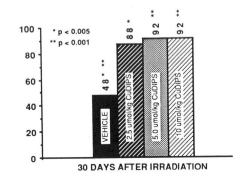

Figure 2. Radiation recovery activity of Cu(II)$_2$ (3,5-DIPS)$_4$ given sc to female C57BL/6N mice (25 per group) 3 hrs after irradiation (8.0 Gy, 1.55 Gy/min) at 1200 hrs.

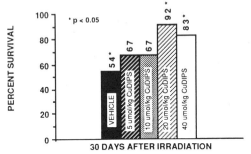

Figure 3. Radiation recovery activity of Cu(II)$_2$ (3,5-DIPS)$_4$ given sc to male C57BL/6N mice (24 per group) 3 hrs before irradiation(8.0 Gy, 1.55 Gy/min) at 1200 hours.

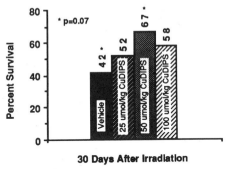

Figure 4. Radiation recovery activity of Cu(II)$_2$(3,5-DIPS)$_4$ given orally to male C57BL/6N mice(24 per group) 24 hrs before irradiation(8.0 Gy, 1.55 Gy/min) at 1200 hrs and 24 hrs after removing solid food. These mice were re-fed 21 hrs before irradiation.

including the number of competent myeloid progenitor cells in both bone marrow and spleen as well as specific and non-specific lymphocyte activation. These data suggest that $Cu(II)_2(3,5\text{-}DIPS)_4$ action in increasing survival of lethally irradiated mice was to accelerate recovery of hematopoietic and lymphoid cells in these animals, probably by stimulating progenitor or stem cell division.

Table 1. $Cu(II)_2(3,5\text{-}DIPS)_4$ Accelerates Recovery of Hemopoietic and Immune Activity Following Lethal Irradiation (% of Non-Irradiated Control $\pm$ SE)

		Days Following Irradiation			
	Dose[a]	7	14	24	42
Bone marrow					
CFU-C[b]	-	0.0(0.0)	0.2(0.0)	0.4(0.4)	16.7(2.4)
	+	0.0(0.0)	10.1(0.3)	14.9(7.5)	74.0(1.7)
Spleen					
CFU-C[b]	-	0.0(0.0)	0.0(0.0)	10.0(10.0)	-
	+	0.0(0.0)	25.0(15.0)	267.0(54.0)	-
Con A[c]	-	0.0(0.0)	0.1(0.0)	2.0(0.4)	26.4(0.2)
	+	0.0(0.0)	0.1(0.0)	26.7(2.1)	50.3(1.5)
PHA[c]	-	0.1(0.0)	1.8(0.1)	0.7(0.1)	48.7(3.7)
	+	0.2(0.0)	0.9(0.1)	13.5(2.8)	78.6(2.3)
LPS[c]	-	0.0(0.0)	3.5(0.5)	5.3(0.3)	25.9(0.5)
	+	0.0(0.0)	7.3(0.5)	42.9(1.8)	40.9(2.0)
Antibody[d]	-	0.2(0.1)	1.1(0.2)	5.0(1.0)	34.4(2.2)
	+	0.5(0.1)	14.5(2.0)	35.3(7.1)	103.0(10.2)

[a]Female mice were injected sc with vehicle (-) or 80 mol/kg $Cu(II)_2(3,5\text{-}DIPS)_4$ (+) 3 hr before exposure to 8 Gy (1.55 Gy/min) whole body irradiation. [b]1×10^5 bone marrow or 5×10^5 spleen cells were cultured in semi-solid agar cultures containing Dulbecco's minimal essential medium. Colonies of greater than 50 cells were counted on day 7. [c]Spleen cells (2×10^5/culture) were incubated with 2.5 g/ml concanavalin A (Con A), 25 g/ml phytohemagglutinin (PHA), or 10 g/ml lipopolysaccharide (LPS) for 72 hr. Proliferation was measured by the uptake of ^{3}H-thymidine over the final 4 hr of culture. [d]Mice were immunized intraperitoneally with 1×10^8 sheep red blood cells 5 days prior to assay. Plaque-forming cells were measured using a slide modification of the Jerne plaque assay.

Histological examination at the light microscopic level revealed that 80 μ mol/kg $Cu(II)_2(3,5\text{-}DIPS)_4$ treatment alone did not induce any histological changes in spleen, thymus, bone marrow, or liver at days 1, 7, 14, 24, or 42 following treatment when compared to tissues taken from non-irradiated vehicle-treated mice. Examination of tissues from vehicle-treated and irradiated mice (V-IR) revealed rapid destruction of lymphoid tissues in spleen, thymus, and bone marrow. Maximal cellular destruction in the spleen and thymus was observed by day 7 with some recovery by day 14. Bone marrow cell depletion was maximal by day 14. Significant recovery was observed by day 24 post-irridation. Although the 80 μmol/kg $Cu(II)_2(3,5-DIPS)_4$ treatment of irradiated mice (Cu-IR) did not prevent this cellular destruction there was a more rapid and complete recovery with $Cu(II)_2(3,5-DIPS)_4$ treatment as shown in Table 2. Treatment did not protect lymphoid tissues from radiation damage, however, it certainly facilitated cellular recovery.

Table 2. $Cu(II)_2(3,5-DIPS)_4$ mediated recovery (-) of radiation-induced spleen, thymus, and bone marrow histopathology (+).

Days	Spleen		Thymus		Bone Marrow	
	V-IR	Cu-IR	V-IR	Cu-IR	V-IR	Cu-IR
1	++	++	+++	+++	++	++
7	+++	+++	+++	++	+++	+++
14	++	++	++	+	++++	++/+
24-42	-	-a	±	±	+	-

[a]Increased fibrosis and generalized proliferation of lymphocytes obliterating lymphoid follicle formation were observed in some samples.

Copper$(II)_2(3,5-DIPS)_4$ was also found to increase survival when it was given sc after irradiation and the smallest effective dose was found to be 2.5 μmol/kg (Figure 2). Doses of 80 μ mol/kg and greater appeared to be toxic when given after irradiation since survivals were always less than vehicle-treated mice. The observed increases in survival when treatment was given after irradiation supports the hypothesis that this complex facilitates recovery from irradiation-induced tissue damage rather than scavenging radicals produced by irradiation.

Survival data presented in Figure 3 show that male mice were also protected when they were treated prior to irradiation and the smallest effective dose for males, 20 μ mol/kg, was the same as the dose providing maximum protection in female mice treated 3 hrs before irradiation (Figure 1).

Copper$(II)_2(3,5-DIPS)_4$ was also found to increase survival following oral treatment. A dose of 50 μmol/kg produced the greatest survival when given orally 24 hrs before irradiation (Figure 4) and

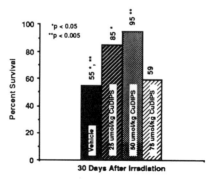

Figure 5. Radiation recovery activity of Cu(II)$_2$
(3,5-DIPS)$_4$ given orally to male C57BL/6N
mice(20 per group) 4 hrs before irradiation
(8.0 Gy/min, 1.55 Gy/min) at 1200 hrs and
24 hrs after removing solid food. These
mice were re-fed 1 hr before irradiation.

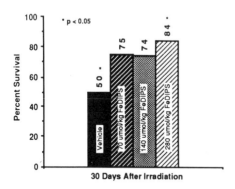

Figure 6. Radiation recovery activity of Fe(III)
(3,5-DIPS)$_3$ given sc to male C57BL/6N mice
(20 per group) 3 hr before irradiation (8.0 Gy,
1.55 Gy/min) at 1200 hours.

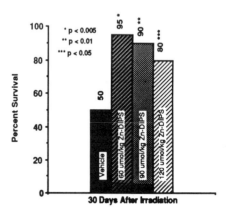

Figure 7. Radiation recovery activity of Zn(II)$_2$
(3,5-DIPS)$_4$given sc to male C57BL/6N mice
(20 per group) 3 hr before irradiation (8.0 Gy,
1.55 Gy/min) at 1200 hours.

this same dose produced the greatest survival when given orally 4 hrs before irradiation (Figure 5). Again, the largest doses 100 μ mol/kg given 24 hrs before irradiation and 75 μ mol/kg given 4 hrs before irradiation appeared to be toxic, decreasing survival.

Iron(III)(3,5-DIPS)$_3$ and Zn(II)$_2$(3,5-DIPS)$_4$ were also found to be effective in increasing survival in male mice as shown in Figures 6 and 7. The most effective dose of Fe(III)(3,5-DIPS)$_3$ was the highest dose tested to date, 280 μ mol/kg while smaller doses of Zn(II)$_2$(3,5-DIPS)$_4$ appear to be more effective than larger doses ranging up to 120μ mol/kg. The effectiveness of these complexes has not as yet been studied in female mice.

Manganese(II)$_2$(3,5-DIPS)$_4$, however, has been found to be effective in increasing survival in both male and female mice. From data presented in Figures 8 and 9 it appears that doses of 40 or 80 mol/kg were maximally effective in both sexes.

Table 3. Acute toxicity (LD$_{50/7}$ in mol/kg $\pm$ SD) of 3,5-diisopropylsalicylic acid and its Cu, Mn, Zn, and Fe complexes in male and female C57BL/6N mice as determined by Probit analysis.

	Female	Male
Cu(II)$_2$(3,5-DIPS)$_4$	261$\pm$36	91$\pm$13
Fe(III)(3,5-DIPS)$_3$	-	1,173$\pm$213
Zn(II)$_2$(3,5-DIPS)$_4$	-	381$\pm$75
Mn(II)$_2$(3,5-DIPS)$_4$	421$\pm$46	353$\pm$32
3,5-Diisopropylsalicylic Acid	1,176$\pm$158	1,144$\pm$149

Prior to using these compounds in survival studies we determined their acute toxicities (Table 3) in female and/or male mice to select reasonably safe doses for survival studies. Acute toxicity due to complex treatment and irradiation wherein death occurred in complex-treated and irradiated mice before death occurred in the irradiated and vehicle-treated mice was not observed. However it is clear from data in Table 3 that Cu(II)$_2$(3,5-DIPS)$_4$ is more toxic in male mice than female mice.

In spite of this difference the Therapeutic Indexes (TI=LD$_{50}$/ED$_{50}$) for female and male mice show that this complex is markedly safe for use as a radiation recovery agent. If the effective dose (ED$_{50}$) is conservatively defined as a dose that produces at least a 50% increase in survival compared to vehicle-treated mice, a 20

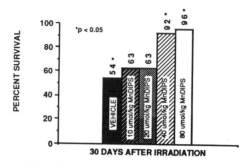

Figure 8. Radiation recovery activity of Mn(II)$_2$ (3,5-DIPS)$_4$ given sc to male C57BL/6N mice (24 per group) 3 hrs before irradiation (8.0 Gy, 1.55 Gy/min) at 1200 hours.

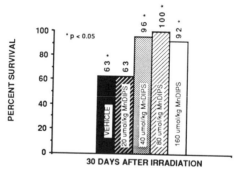

Figure 9. Radiation recovery activity of Mn(II)$_2$ (3,5-DIPS)$_4$ given sc to female C57BL/6N mice (24 per group) 3 hrs before irradiation (8.0 Gy, 1.55 Gy/min) at 1200 hours.

mol/kg dose of $Cu(II)_2(3,5-DIPS)_4$ given before irradiation has a TI of 4.6 and 13.1 respectively for male and female mice. If the effective dose is taken to be 2.5 mol/kg when given after irradiation then the TIs for male and female mice are 36.4 and 104.4 respectively. The TI for $Fe(III)(3,5-DIPS)_3$, based upon the present data, 1173/280, is 4.2 for male mice. The TI for $Zn(II)_2(3,5-DIPS)_4$, 381/60, is 6.4 for male mice. The TIs for $Mn(II)_2(3,5-DIPS)_4$, 421/40 or 353/40 respectively for female or male mice, are 10.5 or 8.8. Since an ideal TI is 10 wherein the acutely toxic dose is 10 times the effective dose, these conservative TI values demonstrate marked safety in use and further support studies of essential metalloelement complexes as radiation recovery agents.

ACKNOWLEDGEMENTS

J.R.J. S. is indebted to Labcatal Laboratorie (Montrough, France) for travel support enabling presentation of this paper. We are also indebted to the Max and Victoria Dreyfus Foundation, Elsa U. Pardee Foundation, the Denver Roller Corporation, and to the National Cancer Institute, Department of Health and Human Sciences, PHS grant number CA40380 for financial support.

REFERENCES

1. K.N.Prasad, Acute Radiation Syndromes, in: "Radiation Biology," D.J.Pizzarello, and Colombetti, L.G. ed., C.R.C. Press, Inc., Boca Raton (1982).
2. G. Czapski, Radiation Chemistry of Oxygenated Aqueous Solutions, Ann. Rev. Phys. Chem. 22:171 (1971).
3. L.W. Oberley, A.L. Lindgren, S.A. Baker, and R.H. Stevens, Super-oxide Ion as the Cause of the Oxygen Effect, Radiat. Res. 68:320 (1976).
4. H.P. Misra, and I. Fridovich, Superoxide Dismutase and the Oxygen Enhancement of Radiation Lethality, Arch. Biochem. Biophys. 176:577 (1976).
5. J.R.J. Sorenson, An Evaluation of Altered Copper, Iron, Magnesium, Manganese, and Zinc Concentrations in Rheumatoid Arthritis, Inorg. Perspec. Biol. Med. 2 (1978).
6. F.T.Greenaway, L.J.Norris, and J.R.J. Sorenson, Mononuclear and Binuclear Copper(II) complexes of 3-5-diisopropylsalicylic acid, Inorg. Chim. Acta, 145:279 (1988).
7. L.S.F. Soderberg, J.B. Barnett, M.L Baker, H. Salari, and J.R.J. Sorenson, Copper(II)3,5-Diisopropylsalicylate)$_2$ Accelerates Recovery of B and T Cell Reactivity Following Irradition, Scand. J. Immunol. 26:495 (1987).
8. L.S.F. Soderberg, J.B. Barnett, M.L. Baker, H. Salari, J.R.J. Sorenson, Copper(II)$_2$(3,5-diisopropylsalicylate)$_4$ Stimulates Hemopoiesis in Normal and Irradiated Mice, Expth. Hematol. 16:577 (1988).
9. J.R.J. Sorenson, Bis (3,5-diisopropylsalicylato) Copper(II), A Potent Radioprotectant With Superoxide Dismutase Mimetic Activity, J. Med. Chem. 27:1747 (1984).
10. J.R.J. Sorenson, H. Salari, M.L.Baker, W.M. Willingham, L.S.F. Soderberg, J.B. Barnett, and K.B. Bond, Radiation protection with $Cu(II)_2(3,5-DIPS)_4$, Recueil, 106:391 (1987).

EFFECTS OF GLYCYL-HISTIDYL-LYSYL CHELATED Cu(II)

ON FERRITIN DEPENDENT LIPID PEROXIDATION

D.M. Miller[1], D. DeSilva[1], L. Pickart[2] and S.D. Aust[1]

[1]Biotechnology Center
Utah State University
Logan, UT 84322-4430

[2]Procyte Corporation
2893 152nd Avenue NE
Redmond, WA 98052

ABSTRACT

The copper binding tripeptide, glycyl-L-histidyl-L-lysine [GHK:Cu(II)] has a plethora of biological effects related to the wound healing process. The presence of iron complexes in damaged tissues is detrimental to wound healing, due to local inflammation, as well as microbial infection mediated by iron. To test if the wound healing properties of GHK:Cu(II) are due to an affect on iron metabolism, we examined the effects of GHK:Cu(II) on iron catalyzed lipid peroxidation. GHK:Cu(II) inhibited lipid peroxidation only if the iron source was ferritin. Whereas GHK:Cu(II) inhibited ferritin iron release it did not exhibit significant superoxide dismutase-like or ceruloplasmin-like activity. We propose that GHK:Cu(II) binds to the channels of ferritin involved in iron release and physically prevents the release of Fe(II). Thus, a biological effect of GHK:Cu(II), possibly related to wound healing, may be the inhibition of ferritin iron release in damaged tissues, preventing inflammation and microbial infections.

INTRODUCTION

Glycyl-L-histidyl-L-lysine is a plasma tripeptide which has a high affinity for Cu(II) (1). Directly or indirectly, GHK:Cu(II) stimulates collagen synthesis (2), angiogenesis (3), and nerve outgrowth (4). GHK:Cu(II) posseses wound healing properties in animals (1) and in humans with skin ulcers (5). The compound is a potent chemoattractant for macrophages, mast cells, and monocytes; cells essential for the wound healing process (6). In addition, GHK:Cu(II) may be a SOD-like antioxidant, possibly related to its wound healing properties (5).

Antioxidants in Therapy and Preventive Medicine
Edited by I. Emerit *et al.*
Plenum Press, New York, 1990

79

After a tissue is wounded, neutrophils migrate into the damaged area and produce $O_2^{\cdot-}$, as part of the "respiratory burst". Low molecular weight chelatable iron accumulates within the wounded area, which microbial pathogens use for growth, leading to infection (7). The increase in chelatable iron, in conjunction with $O_2^{\cdot-}$, also may contribute to localized inflammation, mediated by iron catalyzed lipid peroxidation.

A significant source of iron for lipid peroxidation reactions in vivo is the ubiquitous iron storage protein ferritin (8). Ferritin stores iron as a ferric oxyhydroxide cluster, which will serve as a catalyst of lipid peroxidation only after the iron is released, by reduction and chelation (8). Although the physiological reductant(s) of ferritin iron is unknown, $O_2^{\cdot-}$ reduces ferritin iron, and facilitates its release (8). Superoxide also participates in subsequent iron catalyzed lipid peroxidation reactions (8).

Thus, since a connection between iron, inflammation and infection exists, we postulated that GHK:Cu(II) may exhibit its wound healing properties in vivo by affecting iron metabolism. To test this hypothesis, we examined the effects of GHK:Cu(II) on iron catalyzed lipid peroxidation. GHK:Cu(II) inhibits iron catalyzed lipid peroxidation only if the source of iron is ferritin, suggesting an interaction between GHK:Cu(II) and ferritin. Finally, since GHK:Cu(II) did not exhibit SOD- or ceruloplasmin (ferroxidase)-like activity, GHK:Cu(II) may represent a third class of copper containing antioxidants.

METHODS

All reagents were obtained from commercial sources except for glycyl-histidyl-lysine which was a generous gift from the Procyte Corp., and were of the highest purity possible. To minimize transition metal contamination, all reagents were prepared in purified water (Barnstead Nanopure II system; specific resistance of 18.0 mega ohms-cm). Metal chelates were prepared at the indicated molar ratio by titrating the appropriate metal with the chelator previously adjusted to pH 7 followed by readjustment of the pH to 7. All enzymes and proteins were chromatographed on Sephadex G-25 equilibrated with 50 mM NaCl to remove contaminating buffer salts, except for xanthine oxidase (XO) which was chromatographed on Sephadex G-100.

Rat liver microsomes were isolated as described by Thomas and Aust (9); microsomal phospholipids were isolated by an anaerobic Folch extraction (10). Microsomes were chromatographed on Sepharose CL-2B equilibrated with 20 mM Tris/0.15 M KCl (pH 7.4), to remove contaminating ferritin, SOD, catalase, and other contaminants (9). Malondialdehyde concentrations were determined by thiobarbituric acid reactivity as described by Buege and Aust (11). The release of Fe(II) from ferritin was measured by absorbance at 530 nm due to the tris bathophenanthroline disulfonate:Fe(II) complex ($\epsilon_{530\ nm}$ = 22,110 M^{-1} cm^{-1}), or by the absorbance at 564 nm due to the tris ferrozine:Fe(II) complex ($\epsilon_{564\ nm}$ = 27,900 M^{-1} cm^{-1}) (12). The specifics of each experiment are indicated in the table or figure legends.

RESULTS AND DISCUSSION

The effects of SOD, ceruloplasmin (CP), or GHK:Cu(II) on several model iron catalyzed lipid peroxidation systems are shown in Table I. Ceruloplasmin inhibited lipid peroxidation in every system tested. SOD only inhibited in systems which relied on O_2^- for iron reduction. GHK:Cu(II) chelate only inhibited ferritin dependent lipid peroxidation.

Table I. Effects of Cuproenzymes or GHK:Cu(II) on Iron Catalyzed Lipid Peroxidation.

| System | nmole MDA/min/ml | | | |
| | | | Additions | |
	none	GHK:Cu(II) 10 uM	SOD 50 units/ml	CP 0.5 uM
1) ADP:Fe^{2+}	0.4	0.4	0.2	0.1
2) ADP:Fe^{3+}, XO	1.7	1.3	0.2	0.1
3) Ferritin, XO*	0.4	0.2	0.1	0.1
4) Ferritin, PQ	0.4	0.1	0.1	0.1

All reaction mixtures were performed in 50 mM NaCl and included the following: 1) 850:50 μM ADP:Fe^{2+}, 1 umole lipid phosphate/ml; 2) 250:50 μM ADP:Fe^{3+}, 0.02 units/ml XO, 0.33 mM xanthine, 100 units/ml catalase, 1 umole lipid phosphate/ml; 3) 200 uM ferritin iron, 0.02 units/ml XO, 0.33 mM xanthine, 100 units/ml catalase, 1 umole lipid phosphate/ml; 4) 500 uM ferritin iron, 0.05 units/ml NADPH-cytochrome P450 reductase, 0.5 mM PQ, 0.2 mM NADPH, and 1 μmole lipid phosphate/ml. All incubations were conducted at 37°C in a Dubnoff metabolic shaker. *Ceruloplasmin concentration was 0.1 μM.

Both the xanthine\XO and the NADPH\PQ\NADPH-cytochrome P450 reductase systems rely, aerobically, on O_2^- to release iron from ferritin. Because GHK:Cu(II) reportedly has SOD-like activity (1), it might inhibit ferritin dependent lipid peroxidation by scavenging O_2^-. However, at equal XO activities, GHK:Cu(II) failed to inhibit ADP:Fe^{3+} and O_2^- dependent lipid peroxidation, whereas it did inhibit ferritin and O_2^- dependent lipid peroxidation. In addition, GHK:Cu^{2+} only slightly inhibited O_2^- dependent cytochrome c reduction (Figure 1), thereby discounting a SOD-like activity of GHK:Cu(II) in these studies. GHK:Cu(II) also did not catalyze the oxidation of Fe^{+2} or ADP:Fe(II) (data not shown). This is consistent with the inability of GHK:Cu(II) to inhibit non-ferritin dependent lipid peroxidation. Thus, the inhibition of ferritin dependent lipid peroxidation by GHK:Cu(II) is not due to O_2^- scavenging or Fe(II) oxidation.

Since GHK:Cu(II) inhibited ferritin dependent lipid peroxidation, we examined iron release from ferritin in the presence of GHK:Cu(II). As shown in Figure 2, GHK:Cu(II)

81

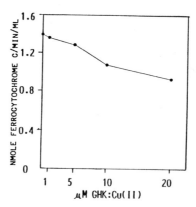

Figure 1. Inhibition of O_2^- dependent cytochrome c reduction by GHK:Cu(II). Reaction mixtures contained: 0.36 mM xanthine, 0.02 units/ml XO, 410 μM cytochrome c, 50 mM NaCl and the indicated concentrations of GHK:Cu(II). Cytochrome c reduction was measured at 37°C by the increase in absorbance at 550 nm ($\epsilon_{550\ nm}$ = 28,000 M^{-1} cm^{-1}).

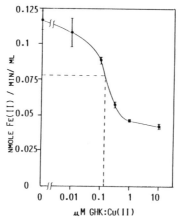

Figure 2. Inhibition of ferritin iron release by GHK:Cu(II). Reaction mixtures contained 200 uM ferritin iron, 0.36 mM xanthine, 0.01 units/ml XO, 100 units/ml catalase, 0.1 mM ferrozine, 50 mM NaCl and the indicated concentrations of GHK:Cu(II). Iron release was measured at 37°C by the increase in absorbance at 564 nm due to the ferrozine:Fe(II) complex.

inhibited O_2^- dependent ferritin iron release, catalyzed by xanthine\XO, with an approximate IC_{50} of 1.4 x 10^{-7} M. Iron release from ferritin by NADPH\PQ\NADPH-cytochrome P450 reductase was also inhibited by GHK:Cu(II) (data not shown). GHK:Cu(II) may inhibit iron release by transferring Cu^{+2} to ferritin, thus we compared the ability of $CuCl_2$ (Cu^{+2}) or GHK:Cu(II) to inhibit catalyzed xanthine/XO, ferritin iron release or NADPH\PQ\NADPH-cytochrome P450 reductase (Table II). Ferritin iron release was inhibited only slightly more by GHK:Cu(II) than by Cu^{+2}. However, as shown in Table III, a combination of Cu^{+2} and ferritin inhibits cytochrome c reduction, whereas GHK:Cu(II) and ferritin does not. Thus, at least part of the inhibition of ferritin iron release by

Table II.   Effects of Cu^{+2} or GHK:Cu(II) on Ferritin Iron Release.

	nmole Fe(II) released/min/ml Additions		
	none	1 μM GHK:Cu(II)	1 μM Cu^{+2}
Xanthine/XO	0.6	0.33	0.34

[1]Reaction mixtures contained: 0.36 mM xanthine, 0.05 units/ml XO, 500 units/ml catalase, 150 uM bathophenanthroline disulfonate, 200 μM ferritin iron and 50 mM NaCl. The assay was performed at 37°C.

Table III.   Effects of GHK:Cu(II) or Cu^{+2} on Superoxide Dependent Cytochrome c Reduction.

Additions	nmole ferrocytochrome c formed/min/ml	
None	1.18	(--)
1 μM GHK:Cu(II)	1.25	(+6%)
1 μM Cu^{+2}	1.00	(-15%)

Reaction mixtures contained: 0.36 mM xanthine, 1 milli unit /ml XO, 410 μM cytochrome c, 200 μM ferritin iron and 50 mM NaCl. The assay was conducted at 37°C.

Cu^{+2} may be due to O_2^{-} scavenging, an activity not exhibited by GHK:Cu(II) at the concentrations used.

It is unlikely that GHK:Cu(II) prevents O_2^{-}. from reducing ferritin iron, but instead inhibits Fe(II) release from ferritin. The ferritin protein molecule is composed of 24 subunits which form numerous channels through which iron is deposited and released (13). GHK:Cu(II) may bind to the channels responsible for iron release, thereby inhibiting that process. Because GHK:Cu(II) has a net positive charge, it is likely that it binds to the same cation channels through which iron is released, rather than the anion channels through which the superoxide anion must enter. In addition, it appears that Cu(II) remains chelated to GHK in the presence of ferritin, otherwise a mixture of GHK:Cu(II) and ferritin would behave like ferritin:Cu^{+2} complex, a phenonmenon not observed.

GHK:Cu(II) is reported to accelerate the wound healing process and prevent infections (1). Numerous studies have shown that a key factor in the onset of infection is the availability of host iron (7). Because a significant amount of iron in serum is ferritin (14), GHK:Cu(II) may interact with serum ferritin. Additionally, cellular ferritin may become available in wounded areas if significant cell lysis has occurred. Thus, the wound healing and anti-inflammatory properties of GHK:Cu(II) may be related, at least in part, to the inhibition of ferritin iron release, effectively inhibiting microbial growth and preventing deleterious oxidations, such as lipid peroxidation. If GHK:Cu(II) prevents the iron release from ferritin by blocking the iron channels, then GHK:Cu(II) can prevent microbial acquistion of iron from ferritin regardless of the iron reductant used.

In summary, GHK:Cu(II) inhibited ferrritin iron release and ferritin dependent lipid peroxidation by a mechanism not related to O_2^-. scavenging, or Fe(II) oxidation, but presumably by binding to and blocking the iron channels of ferritin.

REFERENCES

1. Pickart, L. and Lovejoy, S., Biological activity of human plasma copper-binding growth factor glycyl-L-histidyl-L-lysine, in "Methods of Enzymology" 147, 314-328 (1987).
2. Maquart, F.X., Pickart, L., Laurent, M., Gillery, P., Monboisse, J.C., and Borel, J.P., Stimulation of collagen synthesis in fibroblast cultures by the tripeptide-copper complex glycyl-L-histidyl-L-lysine-Cu(2+). FEBS Lett. 238:343 (1988).
3. Raju, K., Alessandri, G., and Gullino, P., Characterization of a chemoattractant for endothelium induced by angiogenesis effectors. Cancer Res. 44:1579 (1984).
4. Sensenbrenner, M., Jaros, G.G., Moonen, G., and Mandel, P., Effects of synthetic tripeptide on the differentiation of dissociated cerebral hemisphere nerve cells in culture. Neurobiology 5:207 (1975).
5. Maquart, F.X., Kalis, B., Pickart, L., Gillery, P., Monboisse, J.C., Salagnac, V., and Borel, J.P., Glycyl-histidyl-lysine: A copper binding peptide with wound healing properties. (Abstract) in A Hard Look at Collagen-Related Disorders, European Society for Dermatological Research, Liege, Feb. 16-18, 1989.
6. Zetter, B.R., Rasmussen, N., and Brown, L., An in vivo assay for chemoattactant activity. Lab. Invest. 53:362 (1985).
7. Weinberg, E.D. Iron withholding: A defense against infection and neoplasia. Physiol. Rev. 64:65 (1984).
9. Thomas, C.E. and Aust, S.D., Rat liver microsomal NADPH-dependent release of iron from ferritin and lipid peroxidation. J. Free Rad. Biol. Med. 1:293 (1985).
10. Folch, J., Lees, M., and Sloane-Stanley, G.H., A simple method for the isolation and purification of total lipids from animal tissues. J. Biol. Chem. 226:466 (1959).
11. Buege, J.A., and Aust, S.D., Microsomal lipid peroxidation, in Methods of Enzymology 52:302 (1978).
12. Stookey, L.L., Ferrozine - A new spectrophotometric reagent for iron. Anal. Chem. 42:779 (1970).
13. Theil, E.C. Ferritin: Structure, gene regulation, and cellular function in animals, plants, and microorganisms, in Ann. Rev. Biochem. 56:289 (1987).
14. Stevens, R.G., Jones, D.Y., Micozzi, M.S., and Taylor, P.R., Body iron stores and the risk of cancer. New England J. Med. 319:1047 (1988).

SOD-LIKE ACTIVITY OF 5-MEMBERED RING NITROXIDE SPIN LABELS

Amram Samuni*, Ahn Min, C. Murali Krishna, James B. Mitchell
and Angelo Russo

Radiation Oncology Branch, Clinical Oncology Program, Division of Cancer
Treatment, N.C.I., National Institutes of Health, Bethesda, MD. 20892.
*Molecular Biology, Hebrew University Medical School, Jerusalem, Israel

INTRODUCTION

Although superoxide anion mediated redox reactions have been extensively studied, only a few studies involving radical-radical reactions of superoxide with nitroxides have been so far reported[1-8]. Superoxide can act as a one-electron reductant with many species, including the biochemical and toxicological interesting reactions with ferricytochrome c [9], tetranitromethane[10], or nitroblue-tetrazolium[11]. Oxidation reactions involving superoxide are less common and usually involve complex multi-electron steps such as oxidation of sulfite or epinephrine[12] where $O_2^{\cdot}$ may serve as an initiator and propagator of free radical chain reactions. One of the few one-electron oxidations mediated by $O_2^{\cdot}$ is the conversion of secondary hydroxylamines into nitroxides[1]. The hydroxylamine, 2-ethyl-1-hydroxy-2,5,5-trimethyl-3-oxazolidine (OXANOH), is readily oxidized by $O_2^{\cdot}$ to a stable radical, 2-ethyl-2,5,5-trimethyl-3-oxazolidinoxyl (OXANO) and this oxidation has been employed to detect and assay superoxide[2]. Recently $O_2^{\cdot}$ has been found to also be capable of reducing OXANO to OXANOH[13]. Through the combined reduction and oxidation of $O_2^{\cdot}$, the OXANO / OXANOH couple exhibited a superoxide dismutase-like activity[13]. This low molecular weight, cell permeable, non-immunogenic, non-cytotoxic, metal-independent SOD-mimic was found, in the present study, to protect mammalian cells against oxidative damage. To determine if other nitroxides have similar properties we investigated a series of nitroxide derivatives for their reactions with $O_2^{\cdot}$ by using electron spin resonance spectrometry. The 5-membered ring nitroxides were reduced by $O_2^{\cdot}$ to the corresponding hydroxylamines, which in turn were re-oxidizable to the parent nitroxides. The results indicate that SOD-like activity is not limited to OXANO but is shared by nitroxide spin-labels in general.

Antioxidants in Therapy and Preventive Medicine
Edited by I. Emerit *et al.*
Plenum Press, New York, 1990

MATERIALS AND METHODS

Chemicals: Desferrioxamine (DFO), was a gift from Ciba Geigy. Hypoxanthine (HP) was purchased from Calbiochem - Boehringer Co.; 4-hydroxypyrazolo[3,4,-*d*]- pyrimidine (allopurinol); 3-carbamoyl-2,2,5,5-tetramethyl-3-pyrroline-1-yloxy (3-CTM3P), 3-carbamoyl-2,2,5,5-tetramethyl-pyrrolidine-1-yloxy (3-CTMP), 2,2,6,6-tetramethyl-piperidine -1-oxyl (TEMPO), 4-hydroxy-2,2,6,6-tetramethylpiperidine-1-oxyl (TEMPOL), 4-amino-2,2,6,6-tetramethyl-piperidine-1-oxyl (TEMPAMINE), p-toluene sulfonic acid, and 2-amino-2-methyl-1-propanol were purchased from Aldrich Chemical Co.; xanthine oxidase (EC 1.2.3.2 xanthine: oxygen oxidoreductase) (XO) grade III from Buttermilk, xanthine, superoxide dismutase (SOD), and ferricytochrome c were obtained from Sigma (St. Louis, MO). All chemicals were used without further purification. Distilled-deionized water was used throughout all experiments and unless otherwise stated the experiments were conducted at room temperature.

Synthesis of oxazolidine derivatives: 2-Ethyl-2,5,5-trimethyl-3-oxazolidinoxyl (OXANO) and 2-pentyl-2,5,5-trimethyl-3-oxazolidinoxyl (MP-DOX) were synthesized as previously described[14]. To produce the amine, a ketone was reacted with 2-amino-2-methyl-1-propanol in benzene in the presence of p-toluene sulfonic acid catalysis. The formation of the cyclic structure resulted in the elimination of water. The volume of water collected in a Dean Stark apparatus was monitored and used to gauge the reaction progress. The amines thus produced were purified through fractional distillation under reduced pressure, characterized through NMR, IR, UV, mass spectroscopy, and oxidized to the nitroxides using m-chloroperbenzoic acid. The nitroxides were purified by silica flash chromatography.

Electron Spin Resonance: Samples (0.05-0.1 ml) for ESR experiments were drawn by a syringe into a gas-permeable 0.81mm ID, 0.38mm wall thickness, and 15 cm long, Teflon capillary which was inserted into a 2.5mm ID quartz tube and placed in the ESR cavity. During the experiments, gases of desired compositions and temperature were blown around the sample without having to disturb the alignment of the sample. ESR spectra were recorded on a Varian E4 (or E9) X-band spectrometer, with field set at 3357G, 100KHz modulation frequency, 1G modulation amplitude and non-saturating microwave power.

Cell Survival Analysis: Survival of Chinese hamster V79 cells in tissue culture has been assessed clonogenically. Inoculated cells were incubated 12-16 hours prior to experimental procedures, exposed to varying nitroxide concentrations in complete medium at 37°C in the absence and the presence of the HX/XO O_2^- generating system. Following treatment, cells were trypsinized, rinsed, counted, and plated in triplicates for macroscopic colony formation. Following appropriate incubation periods, colonies were fixed, stained, and lastly counted with the aid of a dissecting microscope. Survival curves in the presence and absence of various concentrations of the SOD-mimic were compared.

RESULTS

Superoxide Reaction with Nitroxides: Superoxide has been previously shown to reduce OXANO to OXANOH[13]. To examine if the reaction was limited to that particular derivative or could be extended to other nitroxides, 10μM of the 5-membered-ring nitroxide 3-carbamoyl-2,2,5,5-tetramethyl-3-pyrroline-1-yloxy (3-CTM3P) was exposed, in phosphate buffer, pH7.4, at room temperature to the hypoxanthine/xanthine oxidase (HX/XO) O_2^- generating system[15], and the intensity of its ESR signal was monitored. The signal decayed but did not fully disappear. This spin-loss was shown to be metal-independent by the addition of 50μM DTPA to block any possible effect of adventitious transition metal ions. The decay of the 3-CTM3P signal required the simultaneous presence of molecular oxygen, XO enzyme, and HX (or xanthine), as neither two of them alone had any effect on the nitroxide signal. The addition of superoxide dismutase (SOD) to the reaction system prevented the spin loss, indicating that the spin-loss was mediated by O_2^-.

The Specificity of O_2^- Reaction with Nitroxide Spin-Labels: It was previously reported that superoxide reversibly adds to the N atom of nitroxides but does not reduce 6-membered ring nitroxide unless thiol compounds shift the equilibrium by reacting with the addition-intermediate. In fact it was concluded that thiol-induced reduction of nitroxides is mediated by superoxide radicals[3]. Since the reduction potential of OXANO reportedly does not differ from that of other stable spin probes such as 2,2,6,6-tetramethyl-pipiridine-1-oxyl (TEMPO)[14], we compared the effects of O_2^- on several other nitroxide spin-labels including 3-CTMP, 2-pentyl-2,5,5-trimethyl-3-oxazolidinoxyl (MP-DOX), as well as the 6-membered-ring derivatives TEMPO, TEMPOL, and TEMPAMINE (see Table 1). Upon exposure to HX/XO the ESR signal of the 5-membered ring nitroxides decayed though did not fully disappear. In contrast, none of the 6-membered ring nitroxides signals were affected by the HX/XO system. However, in the presence of 1mM glutathione the signals of 6-membered ring spin-labels quantitatively disappeared, as previously found[3].

The Recovery of the Nitroxide ESR Signal: The major, though not sole, mechanism by which nitroxides (RR'N-O) decay in biological systems, and within cells in particular, is through a one-electron reduction process yielding hydroxylamines[4,14,16] (RR'N-OH). To examine whether O_2^- reduces nitroxide to its respective hydroxylamine, the recovery of the ESR signal were studied. Immediately following the O_2^--induced decay (though not complete) of the nitroxide ESR signal, 10mM allopurinol (XO inhibitor) and 100μg/ml SOD were added to the reaction mixture to prevent any further effect of superoxide. When 0.5mM ferricyanide (or 0.1mM Cu(II) with 2mM H_2O_2) was added, the original ESR signal was restored, verifying that the spin-loss indeed resulted from RR'N-O reduction to RR'N-OH.

The Steady-State Concentrations Distribution of Nitroxide and their Hydroxylamines: The failure of superoxide to totally deplete the nitroxide signal, even

Table 1. Nitroxides structures and notations.

DMPO-OH

OXANO

MP-DOX

3-CTMP

3-CTM3P

TEMPO

TEMPAMINE

TEMPOL

upon prolonged exposures, was attributed to a concurrent re-oxidation of the generated hydroxylamine. This assumption agreed well with the reported oxidation of OXANOH by superoxide[1-3]. The O_2^- -induced depletion and regeneration of the nitroxide appeared to be coupled through reactions 1 and 2:

$$H^+ + O_2^- + RR'N\text{-}OH \xrightarrow{\hspace{1cm}} H_2O_2 + RR'N\dot{\text{-}}O \quad\quad \{1\}$$

$$H^+ + O_2^- + RR'N\dot{\text{-}}O \xleftrightarrow{\hspace{1cm}} O_2 + RR'N\text{-}OH \quad\quad \{2\}$$

Since O_2^- is continuously removed through reactions 1 and 2 while H_2O_2 and molecular oxygen are generated, the reaction mechanism does not represent a genuine chemical equilibrium but rather a steady state where three processes affect RR'N$\dot{\text{-}}$O concentration:

$$d[RR'N\dot{\text{-}}O]/dt = k_1[RR'N\text{-}OH][O_2^-] + k_{-2}[O_2][RR'N\text{-}OH] - k_2[RR'N\dot{\text{-}}O][O_2^-] \quad\quad \{3\}$$

Therefore, in the presence of both oxygen and O_2^-, a steady state is achievable (where $d[RR'N\dot{\text{-}}O]/dt=0$) resulting in a time-invariant ratio of [nitroxide] / [hydroxylamine]:

$$[RR'N\dot{\text{-}}O]/[RR'N\text{-}OH] = k_1/k_2 + k_{-2} \cdot [O_2] /k_2 \cdot [O_2^-] \quad\quad \{4\}$$

Therefore, particularly under conditions where $k_{-2}[O_2] \ll k_2[O_2^-]$, it was assumed that $[RR'N\dot{\text{-}}O]/[RR'N\text{-}OH]$ will approach k_1/k_2. To verify this assumption, $[RR'N\dot{\text{-}}O]$ was varied between 2 to 50µM, exposed to HX/XO for 100min, and the residual intensities of the nitroxide ESR signal were monitored. The fractional residual $[RR'N\dot{\text{-}}O]$ observed was practically the same for the various concentrations initially introduced, verifying that a steady-state nitroxide concentration, which depends on [RR'N·O]/[RR'N-OH] but not on [RR'N·O]$_{initial}$, was attained.

The pH-Dependence of RR'N·O / RR'N-OH Concentrations Ratio: Reactions 1 and 2 are anticipated to depend on [H+]. Therefore, to examine if k_1/k_2 might consequently vary with pH, RR'N·O was incubated with HX/XO at different pH values and the ESR signal intensity of the residual nitroxide was monitored. The results are summarized in Fig. 2 and show the pH-dependence of the concentration distribution [RR'N·O]$_{s.s.}$ / ([RR'N·O]$_{s.s.}$+ [RR'N-OH]$_{s.s.}$) for several nitroxides. Evidently the fractional residual [RR'N·O] increases with pH thus reflecting a respective increase in k_1/k_2 as previously found for the OXANO/OXANOH couple[13].

Nitroxide Protective Effect: To test for protective effects of nitroxide against O_2^--induced damage, monolayered Chinese hamster V79 cells were exposed for 1h in full growth medium, in the absence and in the presence of 1mM OXANO to various fluxes of O_2^-. The radicals were generated enzymatically using HX/XO and the cellular damage was assessed by clonogenically monitoring cell viability. In control experiments performed with OXANO (up to 5mM) in the absence of XO, no cytotoxic effect was observed. Experiments performed with several XO concentrations in the absence of OXANO resulted in various degrees of cell killing, not inhibitable by SOD as previously found[17]. In the presence of OXANO, however, no cell killing was observed (Fig. 2).

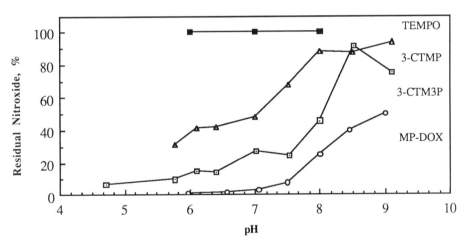

Fig. 1. The pH-dependence of the Steady-State Concentrations Distribution of [RR'N·O]/{[RR'N-OH]+[RR'N·O]} Under Superoxide Flux:
The steady-state residual ESR signal of RR'N·O observed following 50-100min incubation in 50mM phosphate buffer at different pH values of 1.2mM HX and 0.17U/ml XO enzyme, under air at room temperature, with 20μM of each of the nitroxides: TEMPO (■), 3-CTM3P (□), 3-CTMP (Δ), and MP-DOX (o).

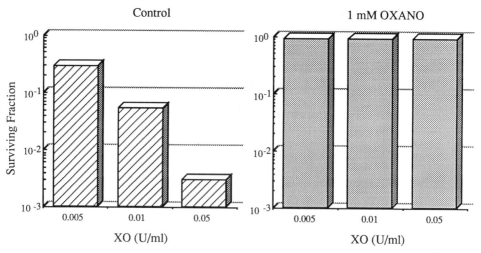

Fig. 2. The Effect of OXANO on Chinese Hamster V79 Cells Exposed to Hypoxanthine / Xanthine Oxidase (HX/XO). Monolayered cells were aerobically incubated for 1h at 37°C in full medium with 0.5mM HX in the presence of varying concentrations of XO, with and without 1mM OXANO, and their surviving fractions were determined.

DISCUSSION

The spin loss of the spin-label nitroxides, induced by HX/XO reaction system is inhibitable by SOD, indicating that it is mediated by O_2^-. Spin loss has been previously reported[8] also for spin adducts of 5,5-dimethyl-1-pyrroline-N-oxide (DMPO) including DMPO-OH, DMPO-CH$_3$ and DMPO-H. However, despite the similarity between DMPO spin-adduct nitroxides and stable 5-membered ring nitroxides, O_2^- affects these two groups of nitroxides differently. Unlike the irreversible spin-loss of DMPO spin-adducts, that of 5-membered ring stable nitroxides was never complete and always restorable, indicating that stable nitroxides are reduced by O_2^- to their corresponding hydroxylamines. The failure of O_2^- to totally reduce the stable nitroxides, agrees with previous reports of O_2^--induced oxidation of OXANOH (reaction 1), whereas the change of [RR'N-O]/([RR'N-O]+[RR'N-OH]) with pH (Fig. 1) reflects the pH-dependence of the rate constants of reactions 1 and 2, as was previously found for OXANO[13].

In spite of the similarity between the reduction potentials of OXANO and TEMPO[14], neither TEMPO nor other 6-member ring nitroxides were reduced by O_2^- unless thiols were present. This result is anticipated had these nitroxides not reacted directly with O_2^-. Alternatively the ESR signal would persist unchanged under O_2^- flux, if k_1 markedly exceeds k_2. i.e the hydroxylamine is instantaneously reoxidized. The present study indicates that $O_2^{\bar{\cdot}}$ reacts with nitroxides through several pathways: **a)** reducing 5-membered ring stable nitroxides to their respective hydroxylamines; **b)** reacting with DMPO spin-adducts converting them to ESR-silent products other than hydroxylamines[8]; and **c)** reducing nitroxides through a thiol-mediated reaction[3]. It is tempting to speculate that all or most of those processes are mediated through a common reaction intermediate as previously suggested[3]:

However, the present results are not sufficient to prove or exclude this speculation.

Most importantly, the present study shows that radical-radical reaction between stable nitroxide and $O_2^{\bar{\cdot}}$ is not limited to OXANO but is shared by other nitroxides which exhibit, therefore, SOD-like activity. Despite differences in charge, size, and lipophilicity the nitroxides studied readily react with $O_2^{\bar{\cdot}}$. This promising result should enable future synthetic modification of stable nitroxides in an attempt to achieve improved SOD-mimics. Naturally, not only the whole class of known spin-labels should now be screened for SOD-mimetic activity, but many new derivatives can be customized and synthesized. The biological implications and potential of these metal-free SOD-mimics might be far reaching. Nitroxide spin-labels are relatively stable within cells and can be made soluble in either/or/both polar and non-polar milieus. As these compounds are relatively non-toxic, they might substitute for and augment SOD protective functions both intra- and extracellularly as well as provide a means to selectively place SOD-like activity in desired organelles or cellular compartments.

Acknowledgment: This work was supported in part by a grant from the Israel-USA Binational Science Foundation.

REFERENCES

1. E.J. Rauckman, G.M. Rosen, and B.B. Kitchell, Superoxide radical as an intermediate in the oxidation of hydroxylamines by mixed function amine oxidase, Mol. Pharmacol.15:131 (1979).
2. G.M. Rosen, E. Finkelstein, and E.J. Rauckman, A method for the detection of superoxide in biological systems, Arch. Biochem. Biophys. 215:367 (1982).

3. E. Finkelstein, G.M. Rosen and, and E.J. Rauckman, Superoxide-dependent reduction of nitroxides by thiols, Biochim. Biophys. Acta 802:90 (1984).

4. S. Belkin, R.J. Mehlhorn, D. Hideg, O. Hankovszki, and L. Packer, Reduction and destruction rates of nitroxides spin probes, Arch. Biochem. Biophys. 256:232 (1987).

5. A. Rigo, E. Argese, R. Stevanato, E.F. Orsega, and P. Viglino, A new method of detecting O_2^- production, Inorg. Chim. Acta 24:L71 (1977).

6. R.J. Mehlhorn and L. Packer, Electron spin destruction methods for radical detection, Methods Enzymol. 105:215 (1984).

7. N.V. Blough, Electron paramagnetic resonance measurements of photochemical radical production in humic substance, Environ. Sci. Technol. 22:77 (1988).

8. A. Samuni, C.M. Krishna, P. Riesz, E. Finkelstein, and A. Russo, Superoxide reaction with nitroxide spin-adduct, Free Rad. Biol. Med.(in Press):(1989).

9. W.H. Koppenol, K.J. van Buuren, J. Butler, and R. Braams, The kinetics of the reduction of cytochrome c by the superoxide anion radical, Biochim. Biophys. Acta 449:157 (1976).

10. J. Rabani, W.A. Mulac, and M.S. Matheson, The pulse radiolysis of aqueous tetranitromethane. I. Rate constants and extinction coefficients of e_{aq}^- . II. Oxygenated solution, J. Phys. Chem. 69:53 (1965).

11. C. Beauchamp and I. Fridovich, Superoxide dismutase: improved assays and an assay applicable to acrylamide gels, Anal. Biochem. 44:276 (1971).

12. J.M. McCord and I. Fridovich, Superoxide dismutase. An enzymic function for erythrocuprein (hemocuprein), J. Biol.. Chem. 244:6049 (1969).

13. A. Samuni, C.M. Krishna, P. Riesz, E. Finkelstein, and A. Russo, A novel metal-free low molecular weight superoxide dismutase mimic, J. Biol. Chem. 263:17921 (1988).

14. G.M. Rosen, E.J. Rauckman, and K.W. Hanck, Selective bioreduction of nitroxides by rat liver microsomes, Toxicol. Lett. 1:71 (1977).

15. I. Fridovich, Quantitative aspects of the production of superoxide anion radical by milk xanthine oxidase, J. Biol. Chem. 245:4053 (1970).

16. H.M. Swartz, M. Sentjurc, and P.D. Morse II, Cellular metabolism of water-soluble nitroxides: effect on rate of reduction of cell/nitroxide ratio, oxygen concentrations and permeability of nitroxides, Biochim. Biophys. Acta 888:82 (1986).

17. H. Hiraishi, A. Terano, S-I. Ota, K.J. Ivey and T. Sugimoto, Oxygen metabolite-induced cytotoxicity to cultured rat gastic mucosal cells, Am. J. Physiol. 253:G40 (1987).

VITAMIN E IN BIOLOGICAL SYSTEMS

Lester Packer and Sharon Landvik

Department of Physiology-Anatomy, University of California
Berkeley, CA 94720; and The Vitamin E Research and Information
Service, 5325 South Ninth Avenue, La Grange, IL 60525, U.S.A.

Introduction

Since the discovery of vitamin E over 50 years ago, its role in human
health and its biological effectiveness have been under extensive research.
It has been established that vitamin E deficiency states exist in humans in
specific clinical situations. It is well accepted that vitamin E is the
major lipid-soluble antioxidant in biological systems, protecting structures
and functions of cell membranes from free radical damage. The impact of
free radicals on the aging process and development of damage or disease and
the protective role of vitamin E and other biological antioxidants in preventing or delaying these processes is an area of active research.

Antioxidant Defense Systems

A number of defense mechanisms in the body have evolved to limit the
levels of reactive oxygen species and the damage they produce. Vitamin E is
considered the first line of defense against lipid peroxidation, protecting
polyunsaturated fats in cell membranes from free radical attack through its
free radical scavenging activity in biomembranes at an early stage of lipid
peroxidation.[1,2] Selenium containing glutathione peroxidase destroys peroxides before they damage cell membranes.[3] Glutathione and vitamin C generate reduced vitamin E, and vitamin C can also function as an antioxidant by
scavenging free radicals in the aqueous phase of the cell.[4]

It has been demonstrated that vitamin E is the major, and probably the
only lipid-soluble, free radical chain-breaking antioxidant in adult human
blood plasma and red blood cells. Approximately a ten-fold increase in
alpha tocopherol intake is required to double the plasma alpha tocopherol
level, with a similar or smaller increase occurring in the tissues.[5] Body
tissues vary significantly in their concentration of alpha tocopherol. The
highest concentrations of vitamin E are in membrane-rich cell fractions,
such as the mitochondria and microsomes. (Table 1)[6]

Vitamin E Deficiency States

Any condition affecting lipid digestion, absorption or transport can
lead to vitamin E deficiency. Cholestatic liver disease and cystic fibrosis

Antioxidants in Therapy and Preventive Medicine
Edited by I. Emerit *et al.*
Plenum Press, New York, 1990

Table 1
d-Alpha Tocopherol Content of Normal
Human Tissues

Tissue	ug/g Fresh Weight
Adipose tissue	150
Adrenal	132
Testis	40
Pituitary	40
Platelets	30
Heart	20
Muscle	19
Liver	13
Ovary	11
Plasma	9.5
Uterus	9
Kidney	7
Erythrocytes	2.3

are the most common chronic malabsorption syndromes resulting in vitamin E deficiency and serum vitamin E levels are often undetectable in patients with abetalipoproteinemia due to severe fat malabsorption.[7,8] A progressive neurological syndrome has been identified in long-standing vitamin E deficiencies, documenting the importance of vitamin E in optimal development and maintenance of the integrity and function of the human nervous system and skeletal muscle.[9] Research has shown that even in patients with very severe vitamin E deficiency, the vitamin E requirement is not met by another exogenous or endogenous antioxidant.[10] Long-term vitamin E therapy has resulted in absence of clinical abnormalities in patients treated from early infancy and stabilization or improvement of neurological dysfunction in patients first treated with vitamin E in childhood.[11] Vitamin E deficiency is also observed in premature infants of low birth weight. However, research has documented that adequate serum vitamin E levels can be achieved with vitamin E administration.[12]

Vitamin E Requirements

The allowance for vitamin E is complicated by the large variations in susceptibility of dietary and tissue fatty acids to peroxidation and the fact that inadequacies of vitamin E are not easily demonstrated in healthy adult subjects. Interpretation of a long-term vitamin E study resulted in a proposed allowance equivalent to 30 I.U. vitamin E for the 1968 edition of the National Research Council's Recommended Dietary Allowances (RDA), which was decreased to 15 I.U. for subsequent RDA editions, partly because of the difficulty in obtaining 30 I.U. vitamin E from regular diets without supplementation. An individual's requirement for vitamin E may vary five-fold, depending upon the composition of the diet and/or tissue composition from previous dietary intake. Evaluation of vitamin E requirements to protect body tissues from peroxidation damage must consider dietary intake of polyunsaturated fat, as animal research has shown that the vitamin E requirement increased with a high dietary polyunsaturated fat intake, due to the increased peroxidative potential of body tissues. (Table 2)[1]

Vitamin E Role in Prevention of Free Radical Damage

Research studies using breath pentane excretion as an index of lipid peroxidation have shown that short-term vitamin E supplementation (1000 I.U. per day for 10 days) significantly reduced breath pentane output in healthy

Table 2
Relationship of Tocopherol Required to Prevent
Creatinuria to Calculated Relative Susceptibility
of Muscle Tissue Lipid to Peroxidation

Number of Double Bonds	Relative Oxidation Rate	Tocopherol Requirement*
1	0.025	0.3
2	1	2
3	2	3
4	4	4
5	6	5
6	8	6

Data obtained from rats fed diets with different fatty acid compositions.
*Except for monoene, the tocopherol requirement is proportional to the
number of double bonds in muscle lipids.

adults on a normal diet. It may thus be inferred that an undesirable
chronic level of lipid peroxidation occurs in cell membranes and other
tissue components which can be ameliorated by supplementation with vitamin E
as a free radical scavenger. These results may also be significant in light
of research evidence which demonstrates a role for free radical damage in
normal body processes and in disease states and a protective role for vitamin E in controlling peroxidation in body tissues.[2,13]

Aging

Research studies in animals and humans have shown an accumulation of
free radical damage in the process of aging or during development of degenerative diseases.[14,15] To study the role of free radical reactions in
cellular aging, researchers monitored effects of vitamin E deficiency in
rats on appearance of a protein in red blood cell membranes that eventually
displays a "senescent cell antigen" that signals the immune system to destroy the cell. Red blood cells of all ages from vitamin E-deficient rats
behaved like old red blood cells from control animals and the researchers
concluded that vitamin E deficiency caused premature aging of red blood
cells and that oxidation may accelerate cellular aging. (Table 3)[16]

Dietary supplementation with vitamin E and the other antioxidants can
significantly affect age-related changes in body tissues. Results of a
number of animal and human studies have shown protective effects of vitamin
E and other antioxidants on free radical activity and lipid peroxidation in
the aging process.[17-19] Evidence is increasing that lipid peroxidation may
have a significant influence in making aging less than the healthy and long
process it should be. Research data is accumulating that long-term damage
can be controlled with antioxidant defense systems, including vitamin E.

Table 3
Phagocytosis of Age-Separated Erythrocytes
from Normal and Vitamin E-Deficient Rats

Erythrocyte Fraction	% Phagocytosis	
	50 mg Vitamin E/kg Diet	0 mg Vitamin E/kg diet
Young	2 + 3	89 + 1
Middle-aged	1 + 1	88 + 1
Old	72 + 3	99 + 0
Unfractionated	13 + 3	87 + 1

Cancer

There is increasing research evidence that reactive oxygen species are involved in the process of cancer initiation and promotion in the body and in isolated cells. The increased incidence of cancer with advancing age may be due, at least in part, to the higher level of free radical reactions with age along with the diminishing ability of the immune system to eliminate altered cells.[20] Results of cell and animal research and limited epidemiological studies suggest that vitamin E and the other antioxidants alter cancer incidence and growth by acting as anti-carcinogens, scavenging free radicals or reacting with their products.

In cell culture studies, incubation of mouse melanoma cells with vitamin E succinate altered cell appearance to that of a more normal cell and also inhibited growth of cancer cells.[21] Vitamin E and selenium inhibited transformation of normal mouse embryo cells to cancerous cells after exposure to radiation and chemicals, with a combination of the two nutrients having a greater effect than either alone. (Table 4)[22]

In cell cultures and animal research, vitamins E and C, acting as scavengers of nitrite compounds, prevented the formation of cancer-producing nitrosamines in the stomach.[23] Studies of induced colon and oral tumors in experimental animals showed that tumor incidence was decreased in vitamin E-treated animals compared to controls.[24-26] In a study of induced mammary tumors in rats, vitamin E and selenium supplementation had a significant inhibitory response on tumor development.[27]

In a number of epidemiological studies, subjects with the highest serum levels of vitamin E and other antioxidants had a lower subsequent risk of certain cancers than subjects with lower serum antioxidant levels.[28-32] The evidence provided by epidemiological studies as well as more controlled animal studies suggests that vitamin E, alone or in combination with other antioxidants, reduces the incidence of certain cancers. Additional research will hopefully provide conclusive documentation on the specific role of vitamin E and the other antioxidants in cancer risk and prevention.

Arthritis

Animal and human studies have demonstrated increased free radical generation in arthritis. Free radical levels in the knee joint synovial membrane increased dramatically 2 weeks after production of adjuvant arthritis in a study in rats.[33] Significantly higher lipid peroxide levels in serum and synovial fluids have also been documented in patients with rheumatoid arthritis and osteoarthritis compared to controls.[34-35] Results of animal studies have demonstrated that vitamin E therapy has beneficial

Table 4
Effect of Vitamin E Succinate and Selenium
on Radiation-Induced Transformation of
Mouse Embryo Cells

Radiation Treatment (Rads)	Antioxidant Supplement	Survival Fraction	Transformed Foci/ Surviving Cells (No./No.)
Control	None	1.00	0/33,931
400	None	0.75	65/66,168
400	Selenium	0.79	16/86,152
400	Vitamin E	0.81	14/79,765
400	Selenium + Vitamin E	0.79	10/110,361

Table 5
Effects of Vitamin E Therapy in Osteoarthritis

Pain Relief	Vitamin E-Supplemented Patients (600 mg/day/ 10 days	Unsupplemented Patients
Significant improvement*	15(51.7%)	1 (4%)
Mild improvement	7(24.2%)	12 (48%)
No improvement	7(24.2%)	12 (48%)

*Significantly different

effects on symptoms of arthritis and that vitamin E deficiency increases severity of the disease.[33],[36] Limited human data suggests that vitamin E therapy is also beneficial in osteoarthritis. In studies of patients with osteoarthritis, vitamin E supplementation was significantly more effective than placebo in relieving pain. (Table 5)[37] Improvement of mobility was also greater in vitamin E-supplemented patients.[38]

Circulatory Conditions

Conditions associated with impaired blood circulation include excessive blood platelet aggregation, intermittent claudication and ischemia. Research has demonstrated that vitamin E may have a beneficial role in improvement of circulation and prevention of damage associated with these circulatory problems.

Excessive platelet aggregation is a significant factor in development of atherosclerosis and other vascular diseases and results of a number of cell culture and animal studies have shown that vitamin E inhibits blood platelet aggregation and production of prostaglandins, which further stimulate platelet clumping.[39-42] Conversely, vitamin E-deficient animals showed a significant increase in platelet aggregation.[43] Vitamin E is also believed to stimulate production of prostacyclin, a powerful antiplatelet aggregation substance.[44] In studies evaluating the effect of vitamin E administration on blood platelet aggregation in healthy adults, vitamin E produced a significant reduction in induced platelet aggregation.[40,45] In insulin-dependent diabetics, blood platelets were significantly lower in vitamin E and showed increased platelet aggregability compared to controls.[46-47] In studies of oral contraceptive users and patients with high blood lipid levels, vitamin E supplementation resulted in a reduction of the elevated platelet aggregation rates observed in these groups.[44,48-49] Since elevated platelet aggregation is linked to an increased risk for atherosclerosis, the role of vitamin E in controlling platelet adhesiveness may help reduce the tendency to develop cardiovascular disease, especially in high risk groups.

Intermittent claudication is usually associated with an underlying disorder characterized by the body's inability to supply adequate arterial blood to contracting muscles, thus depriving the muscles of adequate oxygen to meet increased demands during exercise.[50] Studies in patients with intermittent claudication have demonstrated a beneficial effect of vitamin E over an extended period of time.[51],[52] In a long term study of the results of vitamin E therapy, combined with an exercise regimen, there was significant improvement in walking distance performance and arterial blood flow and the researchers concluded that vitamin E is a proven adjunct in the treatment of intermittent claudication in occlusive arterial diseases of the leg. (Table 6)[53]

Table 6
Intermittent Claudication - Long Term Effects of
Vitamin E on Arterial Flow and Walking Distance

Arterial Flow	Controls (%)	Vitamin E-Treated (300 mg/day)(%)
Diminished	66	15
No change	15	12
Improved	19	73
Walking Distance		
Worse	16	5
No change	54	13
Improved	30	82

Ischemia due to circulatory problems, disease or surgical procedures can result in tissue death in the affected area. It has been suggested that the damage to cell structures and functions observed in ischemia is related to production of free radicals, especially during the reperfusion process.[54] Research on induced cerebral and liver ischemia in rats has shown that vitamin E is depleted during ischemia and that vitamin E therapy helps prevent peroxidation damage related to ischemia and reperfusion injury.[54-57] Free radical levels did not increase significantly during or after coronary bypass surgery in a group of coronary artery disease patients pretreated with 2000 I.U. vitamin E 12 hours before surgery; free radical levels progressively increased during surgery in unsupplemented patients. Blood vitamin E levels were significantly reduced following bypass surgery in controls but vitamin E pretreatment maintained blood levels of vitamin E while completely neutralizing free radicals generated during cardiopulmonary bypass.[58]

Cataract

It is commonly accepted that oxidative mechanisms have an important role in development of cataract. Animal research has demonstrated that vitamin E is able to arrest and reverse cataract to some degree, suggesting that lipid peroxidation is involved. In studies of isolated rat lenses, addition of vitamin E to the medium decreased the incidence of cataract development induced by various agents.[59-61] Vitamin E supplementation helped delay induced cataract formation in rats and rabbits, while extensive cataract development occurred in unsupplemented animals.[61,62] In a study of factors associated with development of senile cataract in adults 40-70 years old, data suggest that subjects with high plasma levels of at least 2 of the 3 antioxidant vitamins had a decreased cataract risk compared to subjects with low levels of one or more of these vitamins. (Table 7)[63] As research continues, evidence to date indicates that vitamin E delays or minimizes cataract development in isolated animal lenses and in animal studies and that high plasma antioxidant levels may reduce cataract risk in adults.

Strenuous Exercise

Oxygen consumption may rise several-fold with high levels of physical exercise and is associated with an increased rate of lipid peroxidation.[64] Research in animals has demonstrated a 2-3 fold increase in free radical concentrations in muscle and liver following exercise to exhaustion.[65] Studies in rats have shown that vitamin E is consumed by body tissues in periods of increased physical exercise and that vitamin E deficiency resulted in a significant decrease in endurance capacity.[64-68]

Table 7
Odds Ratios for Senile Cataract and Serum Antioxidant Status

	Serum Antioxidant Status		
Marker	Low	Middle	High
Vitamins E, C, and Beta Carotene	1	0.6	0.2*

Odds ratio controlled for age, sex, race and diabetes
*P <0.05

A protective effect of vitamin E supplementation against free radical accumulation associated with exercise was demonstrated in a study of human volunteers. While strenuous physical exercise resulted in increased production of pentane, a measure of lipid peroxidation, vitamin E supplementation significantly reduced pentane production at rest and during exercise.[69] In mountain climbers, prolonged exposure to high altitudes led to reduced physical performance and significantly increased breath pentane excretion, leading to the conclusion that high altitude mountain climbing results in considerable risk of free radical-induced cell damage. In contrast, physical performance capacity and breath pentane excretion did not change significantly in vitamin E-supplemented mountain climbers (400 IU/day) and the researchers concluded that vitamin E has a beneficial effect on physical performance and cell protection, at least at high altitude. (Table 8)[70]

Air Pollution

Two of the most damaging air pollutants, ozone and nitrogen dioxide, are present in very high amounts in heavily polluted air and can generate free radicals. Since the lung is a primary target for free radical damage from air pollutants, vitamin E may be an important component of the lung's defense against free radical-related injury, helping to protect against the ill effects of smog and smoke.[71] In a number of animal studies, vitamin E deficiency increased lung susceptibility to lipid peroxidation and associated toxicity from ozone and nitrogen dioxide, while vitamin E supplementation protected lung microsomes from lipid peroxidation.[71-73] Vitamin E-deficient rats exposed to acute levels of cigarette smoke had increased enzymatic activities in the lungs and a higher mortality rate than vitamin E-supplemented animals.[74]

Human studies have also demonstrated a protective role for vitamin E against harmful effects of pollution. In a group of smokers, lower respiratory tract fluid was relatively deficient in vitamin E. Vitamin E levels in lower respiratory tract fluid increased with daily intake of 2400 I.U. vitamin E for 3 weeks, but still remained much lower than baseline levels of nonsmokers, showing increased vitamin E utilization in their lung cells. The findings suggest that vitamin E may be an important lower respiratory tract antioxidant in the lung's defense against cigarette smoke.[75] In a study of healthy adult smokers, the level of lipid peroxidation was measured by

Table 8
Effect of Vitamin E Treatment on Percent Change in Pentane Exhalation of Mountain Climbers at High Altitude

	Vitamin E Treatment Group (400 mg/day)	Control Group
Median	-3.0%	+104.0%
Lower Quartile	-7.4%	+ 25.5%
Upper Quartile	+2.9%	+121.5%

Table 9
Breath Pentane Output in Smokers and Nonsmokers

Nonsmokers	7.6 $\pm$ 0.8
Smokers	16.3 $\pm$ 1.9
Vitamin E-supplemented smokers (800 I.U./day/ 2 weeks)	10.5 $\pm$ 1.0

breath pentane output. Baseline breath pentane output was significantly higher in smokers than nonsmokers and was suppressed by daily supplementation with 800 I.U. vitamin E. (Table 9)[76]

Until the causative agents of air pollution are eliminated completely, a major potential for lung damage and disease will continue. However, results of animal and human studies show that vitamin E can help assure adequate protection against common pollutants and minimize the risk of lung damage associated with our present lifestyle.

Summary

The evidence is increasing that free radical reactions are implicated in the development of cell damage and degenerative disease. Results of animal and human studies have demonstrated that vitamin E and the other antioxidants have a significant role in preventing or minimizing peroxidation damage in biological systems. As we await results of additional human studies, there is sufficient evidence to suggest that adequate antioxidant defense by vitamin E and the other antioxidants can provide protection from the increasingly high levels of free radicals present in the environment due to current lifestyles and the rising concentration of environmental pollutants.

References

1. M.K. Horwitt, Interpretations of Requirements for Thiamin, Riboflavin, Niacin-Tryptophan, and Vitamin E Plus Comments on Balance Studies and Vitamin B$_6$, Am J. Clin. Nutr. 44:973 (1986).
2. A. Van Gossum, et al., Decrease in Lipid Peroxidation Measured by Breath Pentane Output in Normals after Oral Supplementation with Vitamin E, Clin. Nutr. 7:53 (1988).
3. S. Krishnamurthy, The Intriguing Biological Role of Vitamin E, J. Chem. Ed. 60:465 (1983).
4. H. Wefers and H. Sies, The Protection by Ascorbate and Glutathione against Microsomal Lipid Peroxidation is Dependent on Vitamin E, Eur. J. Biochem. 174:353 (1988).
5. J.G. Bieri, L. Corash, and V.S. Hubbard, Medical uses of Vitamin E, New Eng. J. Med. 308:1063 (1983).
6. L.J. Machlin, Vitamin E, in: "Handbook of Vitamins," L.J. Machlin, ed., Marcel Dekker, Inc., New York (1984).
7. D. Carpenter, Vitamin E Deficiency, Seminars in Neurol. 5:283 (1985).
8. D.P.R. Muller, Vitamin E-Its Role in Neurological Function, Postgrad. Med. J. 62:107 (1986).
9. R.J. Sokol, Vitamin E Deficiency and Neurologic Disease, Ann. Rev. Nutr. 8:351 (1988).
10. K.U. Ingold, et al., Vitamin E Remains the Major Lipid-Soluble, Chain-Breaking Antioxidant in Human Plasma even in Individuals Suffering Severe Vitamin E Deficiency, Arch. Biochem. Biophys. 259:224 (1987).

11. A.T. Diplock, Vitamin E, Selenium and Free Radicals, <u>Med. Biol</u>. 62:78 (1984).
12. G.A. Little, et al., Vitamin E and the Prevention of Retinopathy of Prematurity, <u>Pediatrics</u> 76:315 (1985).
13. M. Lemoyne, et al., Breath Pentane Analysis as an Index of Lipid Peroxidation: A Functional Test of Vitamin E Status, <u>Am. J. Clin. Nutr</u>. 46:267 (1987).
14. R. Lubrano, et al., Relationship between Red Blood Cell Lipid Peroxidation, Plasma Hemoglobin, and Red Blood Cell Osmotic Resistance before and after Vitamin E Supplementation in Hemodialysis Patients, <u>Artif. Organs</u> 10:245 (1986).
15. K.A. Pritchard, et al., Triglyceride-Lowering Effect of Dietary Vitamin E in Streptozocin-Induced Diabetic Rats, <u>Diabetes</u> 35:278 (1986).
16. M.M.B. Kay, et al., Oxidation as a Possible Mechanism of Cellular Aging: Vitamin E Deficiency Causes Premature Aging and IgG Binding to Erythrocytes, <u>Proc. Natl. Acad. Sci</u>. 83:2463 (1986).
17. M. Meydani, C.P. Verdon, and J.B. Blumberg, Effect of Vitamin E, Selenium and Age on Lipid Peroxidation Events in Rat Cerebrum, <u>Nutr. Res</u>. 5:1227 (1985).
18. M. Meydani, J.B. Macauley and J.B. Blumberg, Influence of Dietary Vitamin E, Selenium and Age on Regional Distribution of Alpha Tocopherol in the Rat Brain, <u>Lipids</u> 21:786 (1986).
19. M. Wartanowicz, et al., The Effect of Alpha Tocopherol and Ascorbic Acid on the Serum Lipid Peroxide Level in Elderly People, <u>Ann. Nutr. Metab</u>. 28:186 (1984).
20. D. Harman, Free Radical Theory of Aging: The Free Radical Diseases, <u>Age</u> 7:111 (1984).
21. K.N. Prasad and J. Edwards-Prasad, Effects of Tocopherol (Vitamin E) Acid Succinate on Morphological Alterations and Growth Inhibition in Melanoma Cells in Culture, <u>Cancer Res</u>. 42:550 (1982).
22. C. Borek, et al., Selenium and Vitamin E Inhibit Radiogenic and Chemically Induced Transformation in Vitro via Different Mechanisms, <u>Proc. Natl. Acad. Sci</u>. 83: 1490 (1986).
23. S.R. Tannenbaum and W. Mergens, Reaction of Nitrite with Vitamins C and E, <u>Ann. N.Y. Acad. Sci</u>. 355:267 (1980).
24. M.G. Cook and P. McNamara, Effect of Dietary Vitamin E on Dimethyl-hydrazine-Induced Colonic Tumors in Mice, <u>Cancer Res</u>. 40:1329 (1980).
25. O. Odukoya, F. Hawach, and G. Shklar, Retardation of Experimental Oral Cancer by Topical Vitamin E, <u>Nutr. Cancer</u> 6:98 (1984).
26. G. Shklar, Oral Mucosal Carcinogenesis in Hamsters: Inhibition by Vitamin E, <u>JNCI</u> 68:791 (1982).
27. P.M. Horvath and C. Ip, Synergistic Effect of Vitamin E and Selenium in the Chemoprevention of Mammary Carcinogenesis in Rats, <u>Cancer Res</u>. 43:5335 (1983).
28. J.T. Salonen, et al., Risk of Cancer in Relation to Serum Concentrations of Selenium and Vitamins A and E: Matched Case-Control Analysis of Prospective Data, <u>Brit. Med. J</u>. 290:417 (1985).
29. P. Knekt, et al., Serum Vitamin E and Risk of Cancer among Finnish Men during a Ten-Year Follow-up, <u>Am. J. Epidemiol</u>. 127:28 (1988).
30. H.B. Stahelin, et al., Cancer, Vitamins and Plasma Lipids: Prospective Basel Study, <u>JNCI</u> 73:1463 (1984).
31. F.J. Kok, et al., Micronutrients and the Risk of Lung Cancer, <u>New Eng. J. Med</u>. 316:1416 (1987).
32. M.S. Menkes, et al., Serum Beta-Carotene, Vitamins A and E, Selenium and the Risk of Lung Cancer, <u>New Eng. J. Med</u>. 315:1250 (1986).
33. T. Yoshikawa, H. Tanaka, and M. Kondo, Effect of Vitamin E on Adjuvant Arthritis in Rats, <u>Biochem. Med</u>. 29:227 (1983).

34. M. Mezes and G. Bartosiewicz, Investigations on Vitamin E and Lipid Peroxide Status in Rheumatic Diseases, Clin. Rheumatol. 2:259 (1983).

35. D. Rowley, et al., Lipid Peroxidation in Rheumatoid Arthritis: Thiobarbituric Acid-Reactive Material and Catalytic Iron Salts in Synovial Fluid from Rheumatoid Patients, Clin. Sci. 66:691 (1984).

36. T. Yoshikawa, et al, Influence of Vitamin E on Rat Adjuvant Arthritis, Jap. J. Inflam. 2:146 (1982).

37. I. Machtey and L. Ouaknine, Tocopherol in Osteoarthritis: A Controlled Pilot Study, J. Am. Ger. Soc. 26:328 (1978).

38. G. Blankenhorn, Clinical Efficacy of Spondyvit (Vitamin E) in Activated Arthroses, Z. Orthop. 24:340 (1986).

39. M.A. Boogaerts, et al., Protective Effect of Vitamin E on Immune Triggered, Granulocyte Mediated Endothelial Injury, Thromb. Haemostas (Stuttgart) 51:89 (1984).

40. K.C. Srivastava, Vitamin E Exerts Antiaggregatory Effects without Inhibiting the Enzymes of the Arachidonic Acid Cascade in Platelets, Prostagland. Leukotrienes Med. 21:177 (1986).

41. C.W. Karpen, et al., Restoration of Prostacyclin/Thromboxane A_2 Balance in the Diabetic Rat-Influence of Dietary Vitamin E, Diabetes 31:947 (1982).

42. M. Ciavatti, et al., Vitamin E Prevents the Platelet Abnormalities Induced by Estrogen in Rats, Contraception 30:279 (1984).

43. G.H. McIntosh, et al., The Influence of Linoleate and Vitamin E from Sunflower Seed Oil on Platelet Function and Prostaglandin Production in the Common Marmoset Monkey, J. Nutr. Sci. Vitaminol. 33:299 (1987).

44. A. Szczeklik, et al., Dietary Supplementation with Vitamin E in Hyperlipoproteinemias: Effects on Plasma Lipid Peroxides, Antioxidant Activity, Prostacyclin Generation and Platelet Aggregability, Thromb. Haemostas (Stuttgart) 54:425 (1985).

45. M. Steiner, Effect of Alpha Tocopherol Administration on Platelet Function in Man, Thromb. Haemostas (Stuttgart) 49:73 (1983).

46. C.W. Karpen, et al., Production of 12-Hydroxyeicosatetraenoic Acid and Vitamin E Status in Platelets from Type I Human Diabetic Subjects, Diabetes 34:526 (1985).

47. C.W. Karpen, et al., Interrelation of Platelet Vitamin E and Thromboxane Synthesis in Type I Diabetes Mellitus, Diabetes 33:239 (1984).

48. M. Steiner, Effect of Vitamin E on Platelet Function and Thrombosis, Agents and Actions 22:357 (1987).

49. S. Renaud, et al., Influence of Vitamin E Administration on Platelet Functions in Hormonal Contraceptive Users, Contraception 36:347 (1987).

50. M.J. Pinsky, Treatment of Intermittent Claudication with Alpha Tocopherol, J. Am. Podiatry Assoc. 70:454 (1980).

51. K. Haeger, Walking Distance and Arterial Flow during Long Term Treatment of Intermittent Claudication with d-Alpha Tocopherol, J. Vasc. Dis. 2:280 (1973).

52. K. Haeger, Long Time Treatment of Intermittent Claudication with Vitamin E, Am. J. Clin. Nutr. 27:1179 (1974).

53. K. Haeger, Long Term Study of Alpha Tocopherol in Intermittent Claudication, Ann. N.Y. Acad. Sci. 393:369 (1982).

54. S. Marubayashi, et al., Role of Free Radicals in Ischemic Rat Liver Cell Injury: Prevention of Damage by Alpha Tocopherol Administration, Surgery 99:184 (1986).

55. M. Yamamoto, et al., A Possible Role of Lipid Peroxidation in Cellular Damages caused by Cerebral Ischemia and the Protective Effect of Alpha Tocopherol Administration, Stroke 14:977 (1983).

56. S. Yoshida, et al., Influence of Transient Ischemia on Lipid-Soluble Antioxidants, Free Fatty Acids and Energy Metabolites in Rat Brain, Brain Res. 245:307 (1982).

57. S. Fujimoto, et al., The Protective Effect of Vitamin E on Cerebral Ischemia, Surg. Neurol. 22:449 (1984).

58. N.C. Cavarocchi, et al., Superoxide Generation during Cardiopulmonary Bypass: Is there a Role for Vitamin E?, J. Surg. Res. 40:519 (1986).

59. W.M. Ross, et al., Radiation Cataract Formation Diminished by Vitamin E in Rat Lenses in Vitro, Exp. Eye Res. 36:645 (1983).

60. J.R. Trevithick, et al., Modelling Cortical Cataractogenesis: 2. In Vitro Effects on the Lens of Agents Preventing Glucose- and Sorbitol-Induced Cataracts, Can. J. Ophthalmol. 16:32 (1981).

61. M.O. Creighton, et al., Modelling Cortical Cataractogenesis. VII. Effects of Vitamin E Treatment on Galactose-Induced Cataracts, Exp. Eye Res. 40:213 (1985).

62. W.M. Ross, et al., Modelling Cortical Cataractogenesis: 3. In Vivo Effects of Vitamin E on Cataractogenesis in Diabetic Rats, Can. J. Ophthalmol. 17:61 (1982).

63. P.F. Jacques, et al., Antioxidant Status in Persons with and without Senile Cataract, Arch. Ophthalmol. 106:337 (1988).

64. L. Packer, Vitamin E, Physical Exercise and Tissue Damage in Animals, Med. Biol. 62:105 (1984).

65. K.J.A. Davies, A.T. Quintanilha, G.A. Brooks and L. Packer, Free Radicals and Tissue Damage Produced by Exercise, Biochem. Biophys. Res. Commun. 107:1198 (1982).

66. K.M. Aikawa, A.T. Quintanilha, B.O. de Lumen, G.A. Brooks and L. Packer, Exercise Endurance Training Alters Vitamin E Tissue Levels and Red-Blood-Cell Hemolysis in Rodents, Biosci. Rep. 4:253 (1984).

67. K. Gohil, L. Rothfuss, J. Lang and L. Packer, Effect of Exercise Training on Tissue Vitamin E and Ubiquinone Content, J. Appl. Physiol. 63:1638 (1987).

68. K. Gohil, L. Packer, B.O. de Lumen, G.A. Brooks and S.E. Terblanche, Vitamin E Deficiency and Vitamin C Supplements: Exercise and Mitochondrial Oxidation, J. Appl. Physiol. 60:1986 (1986).

69. C.J. Dillard, et al., Effects of Exercise, Vitamin E, and Ozone on Pulmonary Function and Lipid Peroxidation, J. Appl. Physiol.: Respirat. Environ. Exercise Physiol. 45:927 (1978).

70. I. Simon-Schnass and H. Pabst, Influence of Vitamin E on Physical Performance, Internat. J. Vit. Nutr. Res. 58:49 (1988).

71. A. Sevanian, A.D. Hacker, and N. Elsayed, Influence of Vitamin E and Nitrogen Dioxide on Lipid Peroxidation in Rat Lung and Liver Microsomes, Lipids 17:269 (1982).

72. D.P. Franco and S.G. Jenkinson, Rat Lung Microsomal Lipid Peroxidation: Effects of Vitamin E and Reduced Glutathione, J. Appl. Physiol. 61:785 (1986).

73. C.K. Chow, et al, Dietary Vitamin E and Pulmonary Biochemical and Morphological Alterations of Rats Exposed to 0.1 ppm Ozone, Environ. Res. 24:315 (1981).

74. C.K. Chow, et al, Dietary Vitamin E and Pulmonary Biochemical Responses of Rats to Cigarette Smoking. Environ. Res. 34:8 (1984).

75. E.R. Pacht, et el., Deficiency of Vitamin E in the Alveolar Fluid of Cigarette Smokers, J. Clin. Invest. 77:789 (1986).

76. R. Shariff, et al., Vitamin E Supplementation in Smokers, Am. J. Clin. Nutr. 47:758 (1988).

THE ROLE OF VITAMIN E IN THE HEPATOTOXICITY BY GLUTATHIONE DEPLETING

AGENTS

Alessandro F. Casini*, Emilia Maellaro, Barbara Del Bello
and Mario Comporti

Istituto di Patologia Generale dell'Università di Siena
Siena, and *Dipartimento di Medicina Sperimentale e Scienze
Biochimiche dell'Università di Perugia, Perugia, Italy

During the last decades it has been recognized (1-3) that peroxidation of cellular membranes is an important event in the pathogenetic mechanisms of the liver injury induced by chemicals, such as CCl_4 or $BrCCl_3$, which give, upon metabolism, reactive free radicals. The latter ones alkylate cellular macromolecules but do not induce glutathione (GSH) depletion. It was subsequently shown (4-9) that lipid peroxidation is also strictly associated with the liver necrosis induced by chemicals, such as bromobenzene and acetaminophen, which are converted to electrophilic intermediates giving extensive GSH conjugation and consequent GSH depletion. We have studied in particular the liver injury produced in vivo by three prototypical GSH depleting agents which undergo different fates in the liver cell: i) bromobenzene, that is metabolized by the microsomal monooxygenase system with consumption of NADPH (10-12); ii) allyl alcohol that is metabolized by the cytosolic enzyme alcohol dehydrogenase to acrolein and by aldehyde dehydrogenase to acrylic acid, with production of NADH (13,14); iii) and diethylmaleate which is mainly conjugated with GSH by GSH-transferases without previous metabolism (15).

In this study the relationships between two major antioxidant systems of the liver cell, namely vitamin E and GSH, were investigated (GSH, in fact, may be involved in the tocopherol regenerating system (16,17)). In particular, we investigated the effects of the loss of the antioxidant system represented by GSH under conditions of different availability of tocopherol in the liver cell; we also investigated the possibility that tocopherol may effectively replace GSH in the defence against the ultimate effects of oxidative stress. The variations of the other possible tocopherol regenerating system, ascorbic acid (18,19), were also investigated under the same experimental conditions.

To this end the intoxications with the three GSH depleting agents previously mentioned were performed in animals maintained on either a vitamin E deficient diet (Ditta Piccioni, Brescia, Italy) or the same diet supplemented with different amounts of vitamin E (30 and 65 mg/Kg). These different dietary regimens resulted in the following hepatic le-

Antioxidants in Therapy and Preventive Medicine
Edited by I. Emerit *et al.*
Plenum Press, New York, 1990

vels of vitamin E: 4-5, 85-120 and 230-300 pmoles/mg protein, obtained
with the vitamin E deficient diet (basal diet), with the basal diet
supplemented with 30 mg of vitamin E/Kg, and with the basal diet
supplemented with 65 mg of vitamin E/Kg, respectively.

Each group of animals (male NMRI albino mice, Charles River) was
fed on the proper diet for 40 days after weaning. The selenium content
of the basal diet was 0.11-0.12 ppm. The mice were starved overnight
before intoxication (starvation decreases the hepatic GSH stores).

The intoxication with bromobenzene (13 mmoles/Kg body wt., p.o.) of
mice maintained on the basal diet supplemented with 30 mg of vitamin
E/Kg gave results similar to those observed by our group (7,8) in
animals fed on a standard laboratory diet (a complete pellet diet,
Altromin-MT), and therefore are not reported in details. The GSH depletion was nearly maximal at 3 hr already. Lipid peroxidation (as evaluated by the malonic dialdehyde (MDA) content of the liver, after it
was checked that this parameter correlates with other parameters to
detect lipid peroxidation in vivo (20)) and liver necrosis (serum
transaminases) occurred in about all the animals at 15-17 hr. The hepa-

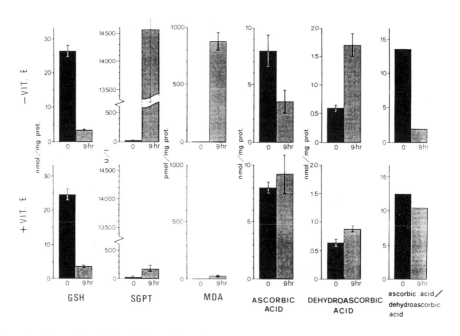

Fig. 1. Hepatic glutathione (GSH) depletion, liver necrosis (SGPT), lipid peroxidation (hepatic content of malonic dialdehyde (MDA)),
and variation of ascorbic to dehydroascorbic acid ratio after
bromobenzene intoxication in mice maintained on either a vitamin
E deficient diet (upper part) or the same diet supplemented with
65 mg of vitamin E/Kg (lower part). The hepatic vitamin E content in untreated controls (0 time) was 5.6 ± 0.9 and 228.4
± 23.6 pmoles/mg protein in vitamin E deficient and vitamin E
supplemented animals, respectively. Results are the means of 10
mice ± S.E.M.

tic content of vitamin E (which in the untreated control was almost identical to that of the control mice fed on the standard laboratory diet) significantly decreased even before the development of lipid peroxidation. Ascorbic acid was decreased at the latest time, dehydroascorbic acid was increased at all the times examined, so that the ascorbic/dehydroascorbic acid ratio showed the increase of the oxidated over the reduced form throughout the intoxication period.

In mice maintained on the vitamin E-deficient diet, bromobenzene intoxication (Fig. 1, upper part) caused the development of lipid peroxidation and liver necrosis much earlier (at 9 hr already) and in a much more severe way (SGPT, 14570 $\pm$ 2158 U/l versus 3718 $\pm$ 1270 U/l in mice fed the same diet supplemented with 30 mg of vitamin E/kg). Again ascorbic acid was decreased, dehydroascorbic acid was increased and the ascorbic/dehydroascorbic acid ratio was dramatically decreased.

Interestingly, only minor effects of bromobenzene were seen (Fig. 1, lower part) in animals fed the same diet supplemented with 65 mg of vitamin E/Kg, in spite of a comparable hepatic GSH-depletion. Even at 20 hr (data not reported) liver necrosis and lipid peroxidation were very low in this group of animals.

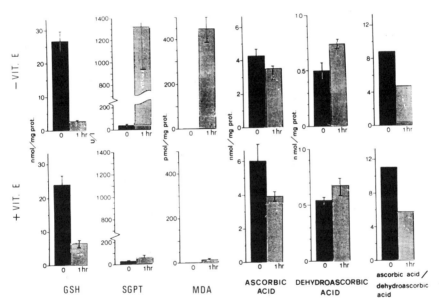

Fig. 2. Hepatic glutathione (GSH) depletion, liver necrosis (SGPT), lipid peroxidation (hepatic content of malonic dialdehyde (MDA)), and variation of ascorbic to dehydroascorbic acid ratio after allyl alcohol intoxication in mice maintained on either a vitamin E deficient diet (upper part) or the same diet supplemented with 65 mg of vitamin E/Kg (lower part). The hepatic vitamin E content in untreated controls (0 time) was 4.2 $\pm$ 0.7 and 306.8 $\pm$ 35.9 pmoles/mg protein in vitamin E deficient and vitamin E supplemented animals, respectively. Results are the means of 9 mice $\pm$ S.E.M.

Somewhat similar results were obtained with allyl alcohol. Since in bromobenzene intoxication the results obtained with the basal diet supplemented with 30 mg of vitamin E/kg were quite similar to those obtained with the standard laboratory diet, and since in preliminary experiments this was also true for allyl alcohol intoxication, we decided to carry out the other intoxications only in animals fed on both the vitamin E-deficient diet and the same diet supplemented with 65 mg of vitamin E/kg, and to compare the results with those obtained with the standard laboratory diet. With all the diets used the hepatic GSH depletion after allyl alcohol intoxication (1.2 mmoles/Kg body wt., i.p.) was maximal at 15 min already. In animals fed on the standard diet (not shown) lipid peroxidation and liver necrosis occurred at 2-4 hr. Vitamin E was unchanged at all the times examined, and again the ascorbic/dehydroascorbic acid ratio was dramatically decreased, due to a decrease in ascorbic and an increase in dehydroascorbic acid.

In animals fed on the vitamin E deficient diet and intoxicated with allyl alcohol (Fig. 2, upper part), lipid peroxidation and liver necrosis were well evident at 1 hr already, while, at the same time, both

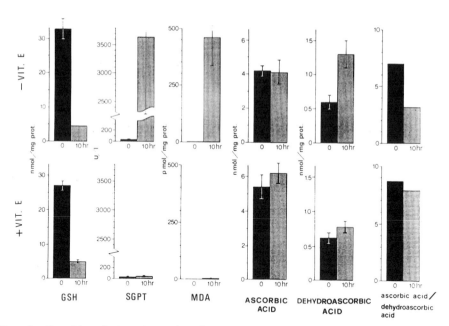

Fig. 3. Hepatic glutathione (GSH) depletion, liver necrosis (SGPT), lipid peroxidation (hepatic content of malonic dialdehyde (MDA)), and variation of ascorbic to dehydroascorbic acid ratio after diethylmaleate intoxication in mice maintained on either a vitamin E deficient diet (upper part) or the same diet supplemented with 65 mg of vitamin E/Kg (lower part). The hepatic vitamin E content in untreated controls (0 time) was 4.1 ± 0.9 and 270.4 ± 37.3 pmoles/mg protein in vitamin E deficient and vitamin E supplemented mice, respectively. Results are the means of 10 mice ± S.E.M.

phenomena were minimal if any in animals fed on the same diet supplemented with 65 mg of vitamin E (Fig. 2, lower part). Even at later times (2-3 hr) (not shown) lipid peroxidation and liver necrosis remained very low in this group of animals. The allyl alcohol-induced decrease in ascorbic/dehydroascorbic acid ratio, however, was not prevented by the diet with the highest level of vitamin E (65 mg/kg).

The most evident effects of the dietary status of the animals with respect to vitamin E were seen after diethylmaleate administration (12 mmoles/Kg body wt., p.o.). While in animals fed on the standard laboratory diet lipid peroxidation and liver necrosis occurred at 20 hr in 40% of the cases only (not shown), in vitamin E-deficient mice (Fig. 3, upper part) both phenomena occured at 10 hr already or even before and with a much higher frequency (nearly 70% of the cases). Dehydroascorbic acid was increased and again the ascorbic/dehydroascorbic acid ratio was markedly decreased. None of these effects, that is no lipid peroxidation, no liver necrosis and no change in the redox state of ascorbic acid were seen in animals fed on the vitamin E-rich (65 mg/kg) diet, in spite of a comparable hepatic GSH depletion.

It seems therefore that vitamin E is a key factor in the expression of the hepatotoxicity of GSH-depleting agents, since even under conditions of extreme GSH-depletion, an hepatic level of vitamin E not so much higher than that obtained with a complete standard diet, can prevent the pathological phenomena occurring in the liver cell as a direct or indirect consequence of GSH depletion.

ACKNOWLEDGEMENTS

This research was supported by a grant (no. 87.01242.44) from Italian CNR and by a grant from the Italian Ministry of Public Education for the Study of Liver Cirrhosis. Additional funds were derived from the Association for International Cancer Research (Great Britain).

REFERENCES

1. E. A. Glende, Jr., and R. O. Recknagel, Biochemical basis for the in vitro pro-oxidant action of carbon tetrachloride, Exp. Mol. Pathol. 11:172 (1969).

2. T. F. Slater, "Free Radical Mechanisms in Tissue Injury" Pion Limited, London (1972).

3. M. Comporti, Lipid peroxidation and cellular damage in toxic injury, Lab. Invest. 53:599 (1985).

4. I. Anundi, J. Högberg, and A. H. Stead, Glutathione depletion in isolated hepatocytes: its relation to lipid peroxidation and cell damage, Acta Pharmacol. Toxicol. 45:45 (1979).

5. A. Wendel, S. Feuerstein, and K. H. Konz, Acute paracetamol intoxication of starved mice leads to lipid peroxidation in vivo, Biochem. Pharmacol. 28:2051 (1979).

6. A. Casini, M. Giorli, R. J. Hyland, A. Serroni, D. Gilfor, and J. L. Farber, Mechanisms of cell injury in the killing of

cultured hepatocytes by bromobenzene, J. Biol. Chem. 257: 6721 (1982).

7. A. F. Casini, A. Pompella, and M. Comporti, Liver glutathione depletion induced by bromobenzene, iodobenzene, and diethylmaleate poisoning and its relation to lipid peroxidation and necrosis, Am. J. Pathol. 118:225 (1985).

8. A. F. Casini, E. Maellaro, A. Pompella, M. Ferrali, and M. Comporti, Lipid peroxidation, protein thiols and calcium homeostasis in bromobenzene-induced liver damage, Biochem. Pharmacol. 36:3689 (1987).

9. M. Comporti, Glutathione depleting agents and lipid peroxidation, Chem. Phys. Lipids 45:143 (1987).

10. D. J. Jollow, J. R. Mitchell, N. Zampaglione, and J. R. Gillette, Bromobenzene-induced liver necrosis. Protective role of glutathione and evidence for 3,4-bromobenzene oxide as the hepatotoxic metabolite, Pharmacology 11:151 (1974).

11. J. R. Mitchell, and D. J. Jollow, Metabolic activation of drugs to toxic substances, Gastroenterology 68:392 (1975).

12. H. Thor, P. Moldéus, R. Hermanson, J. Högberg, D. J. Reed, and S. Orrenius, Metabolic activation and hepatotoxicity, Arch. Biochem. Biophys. 188:122 (1978).

13. F. Serafini-Cessi, Conversion of allyl alcohol into acrolein by rat liver, Biochem. J. 128:1103 (1972).

14. J. M. Patel, W. P. Gordon, S. D. Nelson, and K. C. Leibman, Comparison of hepatic biotransformation and toxicity of allyl alcohol and $(1,1-^2H_2)$allyl alcohol in rats, Drug Metab. Dispos. 11:164 (1983).

15. E. Boyland, and L. F. Chasseaud, Enzyme-catalysed conjugation of glutathione with unsaturated compounds, Biochem. J. 104: 95 (1967).

16. C. C. Reddy, R. W. Scholz, C. E. Thomas, and E. J. Massaro, Evidence for a possible protein-dependent regeneration of vitamin E in rat liver microsomes, Ann. N. Y. Acad. Sci. 393:193 (1982).

17. P. B. McCay, E. K. Lai, S. R. Powell, and G. Breuggemann, Vitamin E functions as an electron shuttle for glutathione--depletion "free radical reductase" activity in biological membranes, Fed. Proc. 45:451 (1986).

18. J. E. Packer, T. F. Slater, and R. L. Willson, Direct observation of a free radical interaction between vitamin E and vitamin C, Nature (London) 278:737 (1979).

19. R. L. Willson, Free radical protection: why vitamin E, not vitamin C, ß-carotene or glutathione?, in: "Biology of Vitamin E. Ciba Foundation Symposium 101", p. 19, Pitman, London (1983).

20. A. Pompella, E. Maellaro, A. F. Casini, M. Ferrali, L. Ciccoli, and M. Comporti, Measurement of lipid peroxidation in vivo: a comparison of different procedures, Lipids 22: 206 (1987).

REGULATION OF LIPID PEROXIDATION BY GLUTATHIONE AND LIPOIC ACID:

INVOLVEMENT OF LIVER MICROSOMAL VITAMIN E FREE RADICAL REDUCTASE

Aalt Bast and Guido R.M.M. Haenen

Department of Pharmacochemistry, Faculty of Chemistry
Vrije Universiteit, De Boelelaan 1083, 1081 HV Amsterdam
The Netherlands

INTRODUCTION

Reduced glutathione (GSH) is known to play a pivotal role in the cellular oxidant defence system and is indispensable for preventing lipid peroxidation by free radicals and oxygen. In this respect several protective cytosolic and membrane bound enzymes have been described which utilize GSH as substrate or as cofactor. The respective contribution of each of these enzymes, like GSH S-transferases, selenium dependent GSH peroxidase and the selenium containing phospholipid hydroperoxide GSH peroxidase, to the oxidant defence has not been fully clarified yet[1].

The liver microsomal membrane forms an outstanding source of free radicals. Cytochrome P-450 and NADPH dependent cytochrome P-450 reductase are involved in lipid peroxidation [2]. Several physiological functions of cytochrome P-450 even require free radical generation of oxygen and xenobiotics[2]. The extensive microsomal free radical formation necessitates the presence of an effective radical scavenging system to prevent oxidative attack on microsomal membrane lipids and other microsomal components. Vitamin E is considered as a prominent antioxidant located within the membrane. Vitamin E traps lipid peroxyl radicals and shortens the length of the propagation phase of lipid peroxidation[3]. It may also protect against lipid peroxidation by scavenging radicals that induce peroxidation in the lipid membrane[4]. In addition, cytosolicly located GSH has been suggested to function as a cofactor for a putative membrane bound vitamin E free radical reductase which regenerates vitamin E from its α-chromanoxyl radical[5]. In this way vitamin E chain breaking antioxidant capacity is not overwhelmed easily by free radicals, since GSH obviates excessive consumption of vitamin E.

Such a rationalization in which vitamin E, vitamin E free radical reductase and cytochrome P-450 activities are linked and encompassed in the hepatic microsomal membrane entails, a key role for GSH. However, it is conceivable that too much GSH becomes utilized thus compromising cellular homeostasis which requires a certain intracellular level of GSH as well. We therefore investigated if an other endogenous occurring thiol, the disulfhydryl containing reduced lipoic acid can substitute for GSH in the prevention of microsomal lipid peroxidation.

MATERIALS AND METHODS

Male Wistar rats (Harlan Olec CPB, Zeist, the Netherlands), 200-20 g were killed by decapitation. Livers were removed and homogenized (1 : 2 w/v) in ice-cold phosphate buffer (50 mM, pH 7.4) containing 0.1 mM EDTA. Microsomes were prepared as described[5] and stored at -80°C. Before use the microsomes were thawed and diluted 5-fold with ice-cold Tris-HCl buffer (50 mM, pH 7.4) containing 150 mM KCl and washed twice by

Antioxidants in Therapy and Preventive Medicine
Edited by I. Emerit *et al.*
Plenum Press, New York, 1990

111

centrifugation at 115,000 x g (40 min.). Finally the pellet was resuspended in the Tris buffer and used. Incubations with microsomes (final concentration of 1/8 g liver/ml) were performed at 37°C, with shaking air being freely admitted in the Tris-HCl/KCl buffer. Compounds that were added were dissolved in the Tris buffer. Ascorbic acid, GSH and reduced lipoic acid were neutralized with KOH before addition. After a preincubation for 3 min. the reactions were started by adding a freshly prepared $FeSO_4$ solution. Experiments were conducted three times. Data presented in figures are representative examples.

Lipid peroxidation was assayed by measuring thiobarbituric (TBA) reactive material as reported previously[5]. In an aliquot of the incubation (0.3 ml) the reaction was stopped by mixing with ice-cold TBA-trichloroacetic-HCl-butylhydroxytoluene solution (2 ml)[5]. After heating (15 min., 80°C) and centrifugation (15 min.) the absorbance at 535 nm vs. 600 nm was determined.

GSH and oxidized glutathione (GSSG) were determined according to Hissin and Hilf[6].

Ascorbic acid concentrations were measured by absorbance at 265 nm (molar extinction coefficient $\varepsilon = 1.5 \times 10^4$ $M^{-1}cm^{-1}$. The reduction of dehydroascorbic acid was measured at room temperature.

RESULTS AND DISCUSSION

Fig. 1 shows that ascorbate (0.2 mM) added in combination with Fe^{3+} (10 μM) promotes lipid peroxidation. Ascorbate or Fe^{3+} added seperately in these concentrations donot stimulate lipid peroxidation (not shown). However Fe^{2+} produces substantial lipid peroxidation (fig. 2). The latter is probably explained by the gradual oxidation of Fe^{2+} to Fe^{3+} in time, which may lead to a Fe^{3+}/Fe^{2+} ratio which is capable of stimulating lipid peroxidation[7]. The observed pro-oxidant activity of ascorbate depends on the presence of iron ions and might be induced by the ascorbyl radical (dehydroascorbate radical anion) produced during the iron catalyzed autoxidation of ascorbate. However, it is more likely that the pro-oxidant activity of ascorbate has to be ascribed to its ability to reduce Fe^{3+}:

$$Fe^{3+} + \text{ascorbate} \rightarrow Fe^{2+} + \text{ascorbyl radical} + 2H^+$$

thus maintaining the critical Fe^{3+}/Fe^{2+} ratio required for lipid peroxidation to occur[8]. The ascorbyl radical is, because of resonance stabilisation, relatively non-reactive[9]. Its decay is mainly by disproportionation, thereby terminating the propagation of free radical reactions, resulting in the production of ascorbate and dehydroascorbate[9]

$$2 \text{ ascorbyl radicals} + 2H^+ \rightarrow \text{ascorbate} + \text{dehydroascorbate}$$

Alternatively, the ascorbyl radical may also reduce another Fe^{3+} ion[8]:

$$\text{ascorbyl radical} + Fe^{3+} \rightarrow \text{dehydroascorbate} + Fe^{2+}$$

GSH (1 mM) protects against lipid peroxidation (Fig. 1). It is known that this protection by GSH is concentration dependent[5]. However, GSH only delays the lipid peroxidation process. After some time the formation of TBA reactive material commences despite of the minimal consumption of GSH[5]. It has been suggested that GSH acts via a membrane bound vitamin E free radical reductase[5]. The transient character of the protection afforded by GSH can be explained by the vulnerability of the GSH dependent vitamin E free radical reductase for oxidative stress[5]. In particular sequestration of critical sulfhydryl moieties in the reductase by reactive aldehydes formed during lipid peroxidation, like 4-hydroxy-2,3-trans-nonenal might be involved[10].

Reduced lipoic acid, an eight carbon fatty acid with sulfhydryl groups on the sixth and eight carbon in the reduced form and a disulphide ring in the oxidized form, cannot replace GSH in its inhibition in microsomal lipid peroxidation (Fig.1). Rather, reduced lipoic acid slightly stimulates Fe^{2+}/ascorbate induced lipid peroxidation. This stimulation can be explained by taking into account that reduced lipoic acid can be regarded as a reductor comparable to ascorbate. Indeed, Fe^{2+} induced lipid peroxidation is stimulated by ascorbate and similarly by reduced lipoic acid (Fig. 2 and 3). Concentrations of ascorbate higher than 0.2 mM offer protection against lipid peroxidation. This might be due to an extensive reduction of Fe^{3+} by high concentrations of ascorbate thus leading to a Fe^{2+}/Fe^{3+} ratio

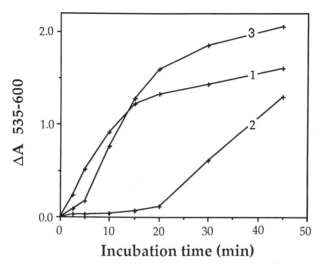

Fig. 1. Time course of lipid peroxidation. Hepatic microsomes were incubated with 0.2 mM ascorbate and (1) no addition (2) 1 mM GSH (3) 0.5 mM reduced lipoic acid. All reactions were started by addition of 10 μM Fe^{2+}.

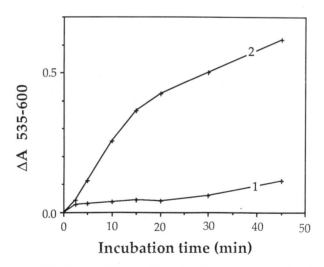

Fig. 2. Time course of lipid peroxidation. Hepatic microsomes were incubated without (1) or with (2) 0.5 mM reduced lipoic acid. The incubation reactions were started by the addition of 10 μM Fe^{2+}.

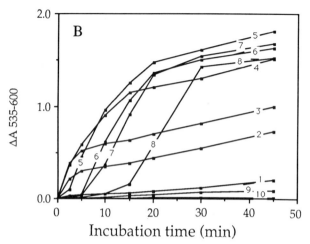

Fig. 3. Time course of lipid peroxidation in rat liver microsomes. The reaction was started by addition of $10\mu M$ Fe^{2+}. The concentrations of ascorbate were for curve 1-10 resp.: 0, 0.025, 0.1, 0.2, 0.35, 0.6, 1, 1.5, 2 and 3 mM.

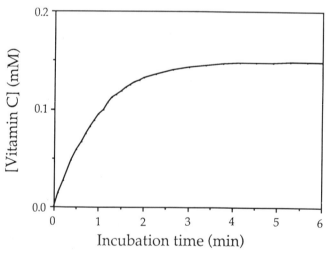

Fig. 4. Generation of ascorbate from 0.2 mM dehydroascorbate by 0.5 mM reduced lipoic acid.

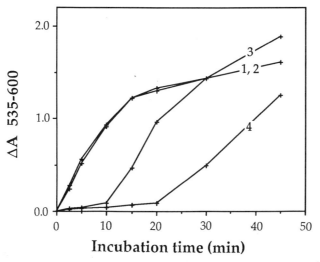

Fig. 5. Time course of lipid peroxidation. Hepatic microsomes were incubated with 0.2 mM ascorbate and (1) no addition (2) 0.5 mM GSSG (3) 0.5 mM reduced lipoic acid and 0.5 mM GSSG (4) 0.5 mM reduced lipoic acid and 1 mM GSH. The reactions were started by addition of 10 μM Fe^{2+}.

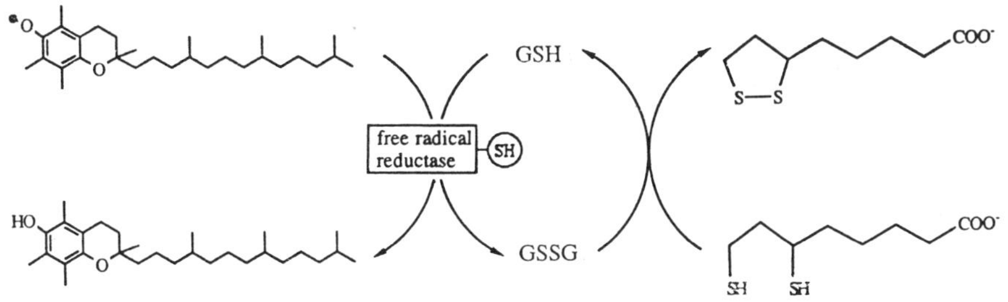

Fig. 6 Interplay between lipoic acid, glutathione, vitamin E and vitamin E free radical reductase in the protection against oxidative stress.

incapable to induce lipid peroxidation. The effect of simultaneous addition of ascorbic acid (0.2mM) and reduced lipoic acid (0.5mM) is comparable to the effect of ascorbic acid alone in a concentration higher than 0.2 mM (0.35 - 0.6 mM) (compare fig. 1 with fig. 3).

The pro-oxidant activity of reduced lipoic acid in the incubation system containing ascorbate, can also be attributed to the ability of reduced lipoic acid to regenerate ascorbate from dehydroascorbate. During lipid peroxidation dehydroascorbate is formed (vide supra) which is rapidly reduced by dihydrolipoic acid (Fig. 4).

Although GSSG (0.5 mM) alone did not protect against lipid peroxidation (Fig. 5), GSSG in combination with reduced lipoic acid (0.5 mM) gave a pronounced protection (Fig. 5) which was more than with reduced lipoic acid alone (Fig. 1). Heating the microsomes in boiling water for 90 s. abolished this protective effect by GSSG and reduced lipoic acid. The same characteristic is displayed for the protection by GSH alone[5]. We suggest that the protection by the combination of GSSG and reduced lipoic acid proceeds via the formation of GSH. Redox potentials of reduced lipoic acid / oxidized lipoic acid and GSH/GSSG allow the reduction of GSSG by reduced lipoic acid[11] as is indicated in table 1.

Table 1 Reduction of GSSG to GSH by reduced lipoic acid.

condition	GSH(mM)	GSSG(mM)
0.5mM GSSG	0.00±0.02	0.51±0.02
0.5mM reducedlipoic acid+ 0.5mM GSSG	0.42±0.02	0.27±0.02

Incubations were performed for 45 min at 37 ^{0}C in Tris-HCl/KCl buffer. Data represent the mean ±S.E. of three experiments.

The protection against lipid peroxidation in liver microsomes seems to occur predominantly via a GSH-dependent vitamin E free radical reductase which mediates a continuous regeneration of vitamin E (Fig. 6). The protection by reduced lipoic acid is most pronounced when added together with GSSG and appears to proceed via the formation of GSH. The slight pro-oxidant and anti-oxidant activity of dihydrolipoic acid in the Fe^{2+}/ascorbate induced lipid peroxidation might be ascribed to the capability of dihydrolipoic acid to reduce Fe^{3+} and dehydroascorbate.

Acknowledgement. The gift of dihydrolipoic acid by Asta-Werke (Frankfurt, FRG) is gratefully acknowledged.

LITERATURE

1. A. Bast and G.R.M.M. Haenen, Cytochrome P-450 and glutathione: What is the significance of their interrelationship in lipid peroxidation? Trends Biochem. Sci. 9:510 (1984).
2. A. Bast, Is formation of reactive oxygen by cytochrome P-450 perilous and predictable? Trends Pharmacol. Sci. 7:266 (1986).
3. L.A. Witting, Vitamin E and lipid antioxidants in free-radicals initiated reactions, in "Free Radicals in Biology Vol. IV", W.A. Pryor, ed., Academic Press, New York (1980).
4. K. Fukuzawa, S. Takase and H. Tsukatani, The effect of concentration on the antioxidant effectiveness of α-tocopherol in lipid peroxidation induced by superoxide free radicals. Arch. Biochem. Biophys. 240:117 (1985).
5. G.R.M.M. Haenen and A. Bast, Protection against lipid peroxidation by a microsomal glutathione-dependent labile factor. FEBS Lett. 159:24 (1983).
6. P.J. Hissin and R. Hilf, A fluorometric method for determination of oxidized and reduced glutathione in tissues. Anal. Biochem. 74:214 (1976).
7. G. Minotti and S.D. Aust, The requirement for iron (III) in the initiation of lipid peroxidation by iron (II) and hydrogen peroxide. J. Biol. Chem. 262:1098 (1987).
8. D.J. Kornburst and R.D. Mavis, Microsomal lipid peroxidation. Mol. Pharmacol. 17:400 (1980).
9. B.H.J. Bielski and H.W. Richter, Some properties of the ascorbate free radical. Ann. N.Y. Acad. Sci. 258:231 (1975).
10. G.R.M.M. Haenen, J.N.L. Tai Tin Tsoi, N.P.E. Vermeulen, H. Timmerman and A. Bast, 4-Hydroxy-2,3-trans-nonenal stimulates microsomal lipid peroxidation by reducing the glutathione-dependent protection. Arch. Biochem. Biophys. 259:449 (1987).
11. P.C. Jocelyn, "Biochemistry of the SH group", Academic Press, New York - London (1972).

QUENCHING OF SINGLET MOLECULAR OXYGEN BY TOCOPHEROLS

Stephan Kaiser, Paolo Di Mascio
Michael E. Murphy and Helmut Sies

Institut für Physiologische Chemie I
Universität Düsseldorf, Moorenstr.5
D-4000-Düsseldorf, FRG

INTRODUCTION

Tocopherols are known for their activity as biological antioxidants (1). Attention has focused on their function as free radical scavengers, but it has also been shown that tocopherols are efficient scavengers of singlet molecular oxygen ($^1O_2$) (2-4). The formation and possible pathological consequences of singlet oxygen in biological systems has been proposed to occur during photoexcitation of a variety of biological compounds and xenobiotics (5,6). In addition, $^1O_2$ can be produced by non-photochemical processes, such as enzymatic reactions catalyzed by lipoxygenase and chloroperoxidase (7,8) and has been implicated in the peroxidation of biological lipids (see 9). Furthermore, $^1O_2$ has been shown to be capable of inducing DNA damage (10).

Thus tocopherols are of biological interest for their $^1O_2$ quenching capability. This scavenging of $^1O_2$ by tocopherols includes physical quenching in which the excited state of oxygen is deactivated without light emission, as well as chemical quenching, leading to the production of various oxidation products. Physical quenching by energy transfer almost always predominates, the exact rate depending on solvent polarity. This has led to the suggestion that a charge-transfer intermediate might also be involved in the quenching process (2). In both free radical scavenging and $^1O_2$ quenching, α-tocopherol is more effective than the homologs in the sequence α-, β-, γ-, δ-tocopherol (11,12), in line with their relative in vivo biopotency (13).

Here we compared the relative physical and chemical quenching ability of the tocopherols toward $^1O_2$. This was achieved by using NDPO$_2$, the thermodissociable endoperoxide of 3,3'-(1,4-naphthylidene) dipropionate to generate the singlet oxygen (14), a methodology which offers advantages in comparison to other systems of $^1O_2$ generation. Generation and quenching of $^1O_2$ was directly monitored by measuring the monomol light emission signal detected in the infrared region by the use of a liquid nitrogen-cooled germanium photodiode detector.

$$O_2 \, (^1\Delta_2) \longrightarrow O_2 \, (^3\Sigma_g^-) + h\nu \, (1270 \text{ nm})$$

The decrease in concentration of the tocopherols due to chemical reaction with $^1O_2$ and the subsequent oxidation products were also examined by HPLC and electrochemical detection.

Antioxidants in Therapy and Preventive Medicine
Edited by I. Emerit *et al.*
Plenum Press, New York, 1990

117

MATERIALS AND METHODS

The tocopherol homologs were kindly provided by Dr. S. Wallat, Henkel KGaA (Düsseldorf, FRG). BHT and D_2O were from Sigma (Munich, FRG), all other chemicals were from Merck, Darmstadt, FRG.

Singlet oxygen was generated chemically using the thermodissociation of the endoperoxide of 3,3'-(1,4-naphthylidene) dipropionate ($NDPO_2$) (14). $NDPO_2$ dissociates yielding the 3,3'-(1,4-naphthylidene) dipropionate (NDP) and singlet molecular oxygen.

Infrared emission of $^1O_2$ was measured using a liquid-nitrogen cooled germanium photodiode detector sensitive in the spectral region of 800 nm to 1800 nm with a detector area of 0.25 cm^2 and a sapphire window. The Ge-diode signal was processed with a lock-in amplifier and an optical chopper at a frequency of 30 s^{-1}. Measurements were carried out in a 6 ml cuvette with mirrored walls. This allowed direct monitoring of both generation and quenching of $^1O_2$ by the use of an oscilloscope and a recorder (Fig. 1).

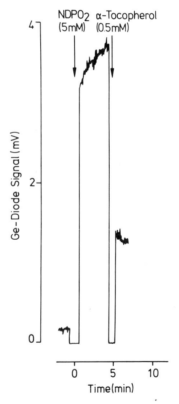

Fig. 1   Effect of tocopherol on monomol light emission of singlet oxygen generated by thermolysis of $NDPO_2$. At 37°C, ethanol/chloroform (1:1) was placed in a thermostated glass cuvette of 6 ml. 20 µl of a 0.4 M $NDPO_2$ solution, kept at 2°C, were added under constant stirring with a small magnetic bar. At the maximum of the signal, α-tocopherol dissolved in ethanol was added.

The quenching rate constant (k_q) was calculated according to Stern-Volmer plots, using the equation:

$$S/S_0 = 1 + (k_q + k_r) \, k_d^{-1} \, [T]$$

where S, S_0 is the chemiluminescence intensity in absence and prescence of the quencher, respectively, k_r the chemical reaction rate constant, k_d the singlet oxygen lifetime constant in the solvent and $[T]$ the tocopherol concentration; as $k_q \gg k_r$, k_r was negligible.

For determination of chemical reactivity tocopherols were incubated with $NDPO_2$ in $D_2O/ethanol$ (1:1) for various time intervals and subsequently extracted with n-hexane, dried under N_2 and then dissolved in HPLC buffer. Tocopherols and tocopherol oxidation products were determined by HPLC and electrochemical detection as described (15).

RESULTS

Physical Quenching of $^1O_2$ by Tocopherols

The physical quenching efficiency of the various tocopherols decreased in the order of α-, ß-, γ-, δ-tocopherol (Fig. 2). The corresponding half-quenching concentrations, i.e. the tocopherol concentration at which 50% of the generated $^1O_2$ was quenched, were 320, 340, 420 and 580 µM, respectively. The k_q values, calculated

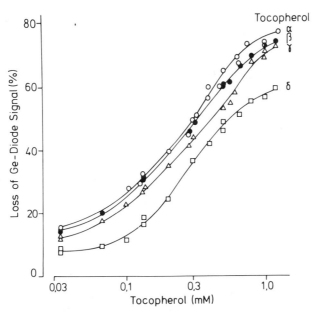

Fig. 2   Dependence of loss of germanium diode signal on the concentration of the different tocopherol homologs. Conditions as described in Fig. 1.

from these data, are 2.8, 2.7, 2.3 and $1.6 \cdot 10^8$ $M^{-1}s^{-1}$, respectively. This sequence is in agreement with the results obtained by others using different means of $^1O_2$ generation (12,16), and also in line with their relative effectiveness of preventing in vitro lipid peroxidation (11) and with their in vivo biological activity (13).

Replacing the $CHCl_3$ in the solvent with D_2O, thus creating a more polar solvent increased the quenching efficiency of α-tocopherol at concentrations of 0.3mM from 49.6 ± 0.8 % to 65.1 ± 0.3 % (n=4). This supports the suggestion that physical quenching of $^1O_2$ by tocopherols involves a charge-transfer mechanism (2). Furthermore, substitution of the hydroxyl-group in position 6 of the chromane ring for methyl-ether or an acetyl or succinyl ester groups abolished the $^1O_2$ quenching ability. The observed residual quenching with α-tocopherol methyl ether (Table 1) is explained by a 1% contamination with unesterified α-tocopherol, as detected electrochemically. Alkylation of the phenolic hydroxyl group should raise the oxidation potential and thus lower the quenching ability, if a charge-transfer mechanism is involved (2).

BHT did not show any significant quenching effect, thus excluding the interference of free radicals. On the other hand, the synthetic antioxidant diphenyl-p-phenylenediamine (DPPD) exhibited a quenching capacity almost identical to that of γ-tocopherol (Table 1).

Table 1 . Loss of 1270 nm light emission induced by different compounds.

	Loss of Ge-diode signal (%)
α-tocopherol	49.6 ± 0.8 (4)
ß-tocopherol	49.6 ± 1.2 (4)
γ-tocopherol	45.0 ± 1.3 (4)
δ-tocopherol	37.5 ± 0.9 (4)
α-tocopherol methyl ether	17.2 ± 0.3 (3)
α-tocopherol acetyl ester	no loss
α-tocopherol succinyl ester	no loss
DPPD	44.6 ± 1.1 (3)
BHT	6.9 ± 0.3 (3)

Conditions as described in materials and methods. Concentration of the compounds was 0.3 mM. The loss of Ge-diode signal is expressed in percent of the total light emitted by $^1O_2$ (Compare Fig. 2). Values are given as means $\pm$ S.E.M.(n exp.).

Chemical Reaction of $^1O_2$ with Tocopherols

As with physical quenching, chemical reaction differed considerably rates between the various tocopherols. However, the order of reactivity toward $^1O_2$ was α-, γ-, δ-, ß-tocopherol. At low concentrations of $NDPO_2$ α-tocopherol showed highest reactivity, while at $NDPO_2$ concentrations of more than 7 mM there was no significant difference between α-, γ- and δ-tocopherols. Interestingly, ß-tocopherol exhibited very low reactivity compared to the other tocopherols. However, at a reaction time of more than 2 h and at a $NDPO_2$ concentration of 20 mM all tocopherols had completely reacted, in contrast to the results reported for the microwave discharge system, measuring the loss of tocopherols in hexadecane as solvent (12). Examination of the reaction products indicated that they included quinones and quinone epoxides (Kaiser et. al., unpublished data).

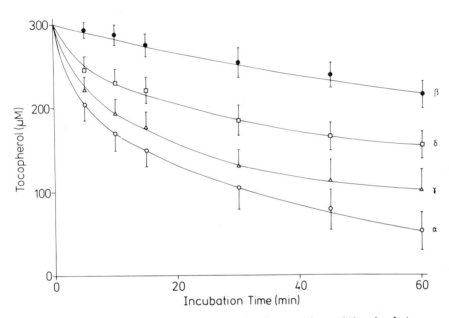

Fig. 3. Loss of tocopherol due to chemical reaction with singlet oxygen.
Values are expressed as the means from 4 experiments (± S.D.).
For incubation conditions see Materials and Methods.

DISCUSSION

The relative rates of physical quenching of singlet oxygen by the tocopherol homologs were found to be α-, β-, γ-, δ-tocopherol, in line with their biological effectiveness and consistent with results obtained elsewhere using other singlet oxygen generating systems (12,16). The physical quenching ability in the hydrophobic, less polar solvent of $C_2H_5OH/CHCl_3$ (1:1) was found to be slightly less than in the more polar conditions of C_2H_5OH/D_2O (1:1), supporting the possibility that the quenching of $^1O_2$ by tocopherols proceeds through a charge-transfer intermediate, as proposed in previous studies (2). Furthermore, esterification or ether formation at the 6-position in the chromane ring of α-tocopherol abolished the quenching ability, underscoring the requirement of a free hydroxyl group at position 6 in the quenching of $^1O_2$ by tocopherols.

The quenching ability of α-tocopherol in this study is about 50-60 times less than that of β-carotene determined in the same system, which compares well with previous data (3). β-carotene was able to quench 50 % of the generated singlet oxygen at a concentration of 5.2μM, and the calculated k_q value was $1.4 \cdot 10^{10}$ $M^{-1}s^{-1}$, (Kaiser et. al., unpublished data). However, taking into account that plasma β-carotene levels are 1/50 of those of α-tocopherol in humans (17,18), the overall quenching capacity of α-tocopherol in human plasma may be regarded to be similar or even equivalent to that of β-carotene. Regarding the increasing epidemiological evidence of an inverse relation between tocopherol plasma levels and certain types of cancer (19), and the observed DNA damaging effects of singlet oxygen, the quenching ability of tocopherols may become of importance in medicine. In this respect it is of interest to note that high intracellular levels of tocopherols are not only found in microsomes and mitochondria but also in the nucleus (20). However, the mutagenic and possible carcinogenic effects of singlet oxygen as well as a protective action of tocopherols against DNA damage in mammalian tissues have to be demonstrated in further research.

The various tocopherols also differed markedly in their chemical reaction rates with singlet oxygen. Following complete reaction with 4 mM NDPO$_2$ 18% α-tocopherol, 34% γ-tocopherol, 52% δ-tocopherol and 72% of β-tocopherol remained unreacted (compare Fig. 3).

The interesting difference of β-tocopherol regarding its physical quenching ability against its chemical reactivity toward singlet oxygen might be of physiological significance. β-Tocopherol has a quenching ability approaching that of α-tocopherol, but in contrast shows almost no chemical reactivity with singlet oxygen.

ACKNOWLEDGEMENTS

Excellent technical assistance was provided by U. Rabe. This work was supported by Deutsche Forschungsgemeinschaft, FRG, National Foundation for Cancer Research, USA, and by Fonds der Chemischen Industrie, FRG.

REFERENCES

1. Tappel, A.L., Vitamin E as the biological lipid antioxidant, Vitam. Horm. 20, 493-510 (1962)

2. Fahrenholtz, S.R., Doleiden, F.H., Trozzolo, A.M. and Lamola, A.A., On the quenching of singlet oxygen by α-tocopherol, Photochem. Photobiol. 20, 505-509 (1974)

3. Foote, C.S., Ching, T.Y. and Geller, G.G., Rates of reaction and quenching of α-tocopherol and singlet oxygen, Photochem. Photobiol. 20, 511-513 (1974)

4. Stevens, B., Small, R.D. and Perez, S.R., The photoperoxidation of unsaturated organic molecules - XIII. $O_2 {}^1\Delta_g$ quenching by α-tocopherol, Photochem. Photobiol. 20, 515-517 (1974)

5. Khan, A.U., Discovery of enzyme generation of singlet molecular oxygen: spectra of $(0,0) {}^1\Delta_g - {}^3\Sigma_g^-$ IR emission, J. Photochem. 25, 327-334 (1984)

6. Khan, A.U., Myeloperoxidase singlet oxygen generation detected by direct infrared electronic emission, Biochem. Biophys. Res. Commun. 122, 668-675 (1984)

7. Kanofsky, J.R. and Axelrod, B., Singlet oxygen production by soybean lipoxygenase isozymes, J. Biol. Chem. 261, 1099-1104 (1986)

8. Kanofsky, J.R., Singlet oxygen production by chloroperoxidase-hydrogen peroxide-halide systems, J. Biol. Chem., 5596-5599 (1984)

9. Sies, H., Biochemistry of oxidative stress, Angew. Chem. Int. Ed. Engl., 25, 1058-1071 (1986)

10. Di Mascio, P., Wefers, H., Do-Thi, H.P., Lafleur, M.V.M. and Sies, H. Singlet molecular oxygen causes loss of biological activity in plasmid and bacteriophage DNA and induces single strand breaks, Biochim. Biophys. Acta, 1007, 151-157 (1989)

11. Burton, G.W. and Ingold, K.U., Autooxidation of biological molecules. I. The antioxidant activity of vitamin E and related chain-breaking phenolic antioxidants in vitro, J. Am. Chem. Soc. 103, 6472-6477 (1981)

12. Neely, W.C., Martin, J.M. and Barker, S.A., Products and relative reaction rates of the oxidation of tocopherols with singlet molecular oxygen, Photochem. Photobiol. 48, 423-428 (1988)

13. Grams, G.W. and Eskins, K., Dye-sensitized photooxidation of tocopherols. Correlation between singlet oxygen reactivity and vitamin E activity, Biochemistry 11, 606-608 (1972)

14. Di Mascio, P. and Sies, H., Quantification of singlet oxygen generated by thermolysis of 3,3'-(1,4-naphthylidene) dipropionate, J. Am. Chem. Soc., 111, in press (1989)

15. Murphy, M.E. and Kehrer, J.P., Simultaneous measurement of tocopherols and tocopheryl quinones using high-performance liquid chromatography with redox-cycling electrochemical detection, J. Chromatogr. Biomed. Appl. 421, 71-82 (1987)

Yamauchi, R. and Matsushita, S., Quenching effect of tocopherols on the methyl linoleate photooxidation and their oxidation products, Agric. Biol. Chem. 41, 1425-1430 (1977)

16. Cavina, G., Gallinella, B., Porra, R. Pecora, P. and Saraci, C. Carotenoids, retinoids and α-tocopherol in human serum: identification and determination by reversed-phase HPLC, J. Pharm. Biomed. Anal. 6, 259-269 (1988)

18. Comstock, G.W., Menkes, M.S. Schober, S.E. Vuilleumier, J.P. and Helsing, K.J., Serum levels of retinol, ß-carotene and α-tocopherol in older adults, Am. J. Epidemiol. 127, 114-122 (1988)

19. Gey, K.F., Brubacher, G.B. and Stähelin, H.B., Plasma levels of antioxidant vitamins in relation to ischemic heart disease and cancer, Am. J. Clin. Nutr. 45, 1368-1377 (1987)

20. Machlin, L.W., Vitamin E, p.113, in Handbook of Vitamins, Machlin, L.W., ed., Marcel Dekker, New York (1984)

ANTIOXIDANT PROPERTIES OF VITAMIN E AND MEMBRANE PERMEABILITY IN HUMAN

FIBROBLAST CULTURES

Marc Conti, Martine Couturier, Frédérique Lemonnier and
Alain Lemonnier

Unité de Recherches d'Hépatologie Pédiatrique INSERM U 56
Hôpital de Bicêtre, 94270 Le Kremlin-Bicêtre, France

INTRODUCTION

α-tocopherol, the most active form of vitamin E (Vit. E) is considered as one of the chief cell components which maintains the structural and functional membrane integrity. It acts as a free radical scavenger preventing the peroxidation of membrane fatty acids. The role of Vit. E in cell permeability is widely studied using artificial membranes, but only a few investigations were devoted to this effect in cell cultures [1],[2].

We have used human fibroblast cultures as experimental model to specify the effects of Vit. E on 2 deoxy-D-glucose (2-DOG) uptake and on other parameters which might be implicated to maintain the membrane integrity : lipid peroxidation production, total superoxide dismutase activity (SOD) and intracellular cholesterol.

MATERIAL AND METHODS

Cell cultures were obtained from abdominal skin biopsies taken from infants requiring benign surgery. The culture medium was Eagle's minimal essential medium (MEM) supplemented with non essential amino acids and 10 % fetal calf serum.
- To measure 2-DOG uptake, we adapted the cluster tray method described by Gazzola et al. [3]. 8 cell strains were seeded in 24 multiplates, and supplemented or not with 10 µg Vit. E/ml during 3 days. At day 4, 2-deoxy-D-(1-^{3}H)-glucose uptake was determined after 1 min. incubation, using a scale of 16 concentrations ranging from 0.005 to 10 mmol/l.

For the three following parameters, 3 cell strains were seeded, and cells were split 1:2 every 3 or 4th day of culture. From the 5th to the 9th passage, the medium was supplemented or not with 4 concentrations of Vit. E : 0.2 ; 2.5 ; 10 and 50 µg/ml.
- Lipid peroxides [4] and Vit. E concentrations [5] were measured by HPLC separation with fluorescence detection. Lipid peroxides were expressed as malondialdehyde (MDA).
- Total superoxide dismutase activity was checked from the technique of Mc Cord and Fridowich [6].
- Total cholesterol was determined with an enzymatic colorimetric Boehringer Kit.

Antioxidants in Therapy and Preventive Medicine
Edited by I. Emerit *et al.*
Plenum Press, New York, 1990

RESULTS

2-DOG uptake. Addition of 10 µg/ml Vit. E to the culture medium during
72 h significantly reduced (p ≦ 0.01) 2-DOG uptake for all the concentrations used (Fig. 1). This decrease in 2-DOG uptake was inversely proportional to the rise in 2-DOG concentration (r = - 0.58 ; p ≦ 0.01).

Vitamin E. The intracellular Vit. E concentration was proportional to
the load of Vit. E up to the concentration of 10 µg/ml (Fig. 2d).

Lipid peroxides. The inhibition of lipid peroxide production was
related to the amount of Vit. E added in the medium (Fig. 2a). The formation
of MDA was reduced by an average of 36 % with 0.2 µg/ml and of 65 % with
2.5 µg/ml. However, this inhibition reached the same values, 74 and 75 %
with 10 and 50 µg vit. E/ml respectively.

Superoxide dismutase. Our results showed (Fig 2b) a slight increase in
SOD activity with 0.2 µg vit. E/ml and a decrease in SOD activity in rela-
tion with the amount of Vit. E added in the medium. Thus, with 50 µg vit.
E/ml the SOD activity was reduced by an average of 23 % (p ≦ 0.01).

Cholesterol. All the concentrations of Vit. E induced a slight decrease
of intracellular cholesterol (Fig. 2c). This variation was only significant
(p < 0.02) for the highest concentration, 50 µg vit. E/ml, with an average
decrease of 12 %.

Effects of vitamin E during the 5th to 9th passage period. Among the
different parameters studied, only the SOD activity decreased clearly in
relation with the passages (Fig. 3) principally with the highest concentration of Vit. E (50 µg/ml).

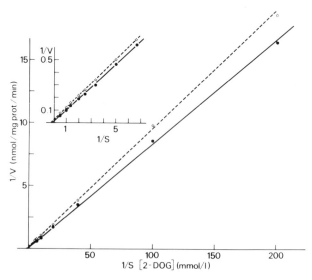

Fig. 1. Lineweaver plot of 2-deoxy-D-glucose uptake in human fibroblast cultures after 72 h of treatment with 10 µg/ml Vit. E
(✩) and in cultures not given this treatment (●). Each point
is the average of 11 to 16 values. The inset indicates the
continuation of 2-DOG uptake at high concentration.

DISCUSSION

When fibroblast cultures were treated with 10 µg/ml of α-tocopherol for 72 hours, 2-DOG uptake significantly decreased. These results are in opposite with those described previously [1], probably because experimental conditions and transport measurement are quite different. On the other hand, it has been recently shown that α-tocopherol reduced the permeability of artificial membranes and raised their resistance to protein-induced disruption [7]. From various findings, it can be recognized that the formation of a complex between α-tocopherol and unsaturated fatty acids decreases lipid peroxide production and maintains the membrane integrity [8]. Thus, in our study, the incorporation of Vit. E, reducing the lipid peroxidation, might modify the membrane structure and the concentration of some components of the membrane such as the cholesterol.

Little is known about the mechanism concerning the effects of lipid membrane changes on glucose transport [9]. It has been shown that glucose transport is related to the concentration and to the nature of the lipids added to or removed from the medium [10]. Whatever the mechanism of such interaction, any change in α-tocopherol and/or cholesterol concentration modifies membrane fluidity, which is now recognized to be one of the main regulator of uptake by cells [9].

The role of superoxide dismutase in the protection of the cell membrane is widely admitted. However, to our knowledge, there are no reports about the relation between the intracellular Vit. E and the antioxidant enzymes in

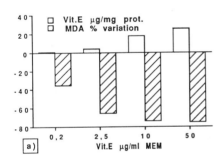

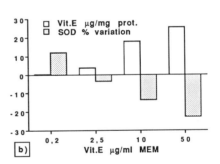

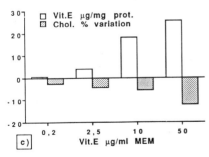

Vit. E µg/ml MEM	Vit. E µg/mg pr.	MDA nmol/mg pr.	SOD U/mg pr.	Chol. µmol/mg pr.
(MEM)	0,1±0,1	1,20±1,00	11,8±6,2	0,25±0,08
0,2	0,6±0,4	0,67±0,55	12,5±5,8	0,24±0,07
2,5	4,1±1,6	0,37±0,30	10,8±5,4	0,23±0,06
10	18,3±4,1	0,33±0,35	9,4±8,6	0,22±0,07
50	25,8±6,1	0,30±0,28	8,6±4,6	0,21±0,07

d) mg pr. = mg proteins in the total homogenate

Fig. 2. Effects of different concentrations of Vit. E on lipid peroxides a), SOD activity b) and cholesterol c). The results are expressed in percentage of variations with respect to the data obtained without Vit. E. The mean values ± SD of 15 determinations are shown in table d) included in the figure.

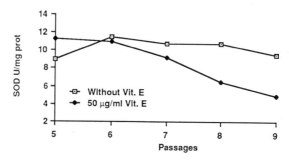

Fig. 3. Effects of Vit. E (50 μg/ml) on SOD activity during successive passages in fibroblast cultures.

cell cultures. In our experiment, high concentration of α-tocopherol induced a marked decrease of SOD activity. This last result is particularly interesting because it demonstrates the interdependence between the different antioxidant systems in human fibroblast cultures.

In conclusion, vitamin E added in the culture medium induced a decrease in lipid peroxidation, cholesterol concentration and SOD activity. These variations suggest changes in the membrane structure which might probably explain the decrease of 2-DOG uptake observed in our cell cultures.

REFERENCES

1. A.S.M. Giasuddin, A.T. Diplock, The influence of vitamin E and selenium on the growth and plasma membrane permeability of mouse fibroblasts in culture. Arch. Biochem. Biophys. 196:270 (1979).
2. B. Hennig, C. Enoch, and C.K. Chow, Protection by vitamin E against endothelial cell injury by linoleic acid hydroperoxides, Nutr. Res. 7:1253 (1987).
3. G.C. Gazzola, V. Dall'Asta, R. Franchi-Gazzola, M.F. White, The cluster-tray method for rapid measurement of solute fluxes in adherent cultured cells, Analyt. Biochem. 115:368 (1981).
4. J. Therasse, F. Lemonnier, Determination of plasma lipoperoxides by high-performance liquid chromatography, J. Chromat. (Biomed. Appl.) 413:237 (1987).
5. G.L. Catignani, J.G. Bieri, Simultaneous determination of retinol and α-tocopherol in serum or plasma by liquid chromatography, Clin. Chem. 29:708 (1983).
6. J.M. Mc Cord, and I. Fridovich, Superoxide dismutase, an enzymic function for erythrocuprein (hemocuprein), J. Biol. Chem. 244:6049 (1969).
7. M. Halks-Miller, L.S.S. Guo, R.L. Hamilton, Tocopherol-phospholipid liposomes : maximum content and stability to serum proteins, Lipids 20:195 (1985).
8. K. Fukuzawa, H. Chida, A. Tokumura, H. Tsukatani, Antioxidative effect of α-tocopherol incorporation into lecithin liposomes on ascorbic acid-Fe^{2+}-induced lipid peroxidation. Arch. Biochem. Biophys. 206:173 (1981).
9. I. Yuli, W. Wilbrandt, M. Shinitzky, Glucose transport through cell membranes of modified lipid fluidity, Biochemistry 20:4250 (1981).
10. T. Fujii, A. Tamura, H. Fujii, I. Miwa, J. Okuda, Effect of exogenous lipids incorporated into the membrane of human erythrocytes, on its glucose-transport activity. Biochem. Intern. 12:873 (1986).

VITAMIN E AND TUMOR GROWTH

Mariette GERBER[1], Sylvia RICHARDSON[2],
François FAVIER[3], André CRASTES DE PAULET[4]

1: INSERM, Centre Paul Lamarque, 34094,Montpellier
2: INSERM U 170, 94807, Villejuif; 3: INSERM, Service commun de cytométrie de flux.,
Avenue de la Cardonille, 34000, Montpellier; 4: Laboratoire de Biochimie, Faculté de
Médecine et Hôpital Lapeyronnie, 34000, Montpellier

INTRODUCTION

In a case-control study on breast cancer, we found (1) that plasma vitamin E levels were significantly higher in cases than in controls and that a statistically significant relative risk for breast cancer of 4.2 was associated to the highest quintile of plasma vitamin E. Moreover a lower plasma MDA level was observed in cases than in controls. This finding prompted us to undertake an experimental study in which mice with a transplanted mammary tumor are submitted to diets differing only by their content in vitamin E.

MATERIALS AND METHODS

A murine mammary tumor (EMT6) was transplanted in 3 groups of 20 female Balb/c at Day 0. The mice were sacrified on day 21 to 25.

The group A04 received a standard diet ad libitum, the group A a synthetic diet (%) Casein:24. Cellulose:16.5. Dextrose: 3.75. corn carbo-hydrate: 17.25. Vit.mix AIN 76 A: 2.2. enriched in vitamin E: 0.011. Mineral mix: AIN 76: 6.3. Zn CO3: 0.01. BHT: 0.010. Méthionine:0.3. The group B received an identical diet but without vitamin E. Linoleic acid (LA), 1% or 4%, was added in A and B diets on Day 0 as the only source of lipids.The caloric content of these diets was of 3,180 cal/kg with 1% linoleic acid and 3,420 cal/kg with 4%.

At the end of the experiment, blood, liver and tumors were collected for total lipids, total cholesterol,triglycerid and vitamin E measurements. In addition, fatty acid distribution was assessed in livers and tumors.

RESULTS

It has been verified that non tumor-transplanted mice were not affected by the A and B diets (normal intake, weight and hair, results not shown)

Two experiments with transplanted mice were performed the first one EMT6-I, with 1% LA, the second one EMT6-II with 4%.There exists a steatosis in EMT6-I demonstrated by the following results: In EMT6-I, A and B compared to A04 demonstrated a steatosis illustrated in Table 1 by a higher increase of weight, and in Table 2, by higher levels of total lipids (TL), total cholesterol (TC) and triglycerids (TG) in the liver.

Antioxidants in Therapy and Preventive Medicine
Edited by I. Emerit *et al.*
Plenum Press, New York, 1990

129

Table 1. Survey of the intake, body weight and tumor weight of EMT6 transplanted mice

| | EMT6-I | | | | EMT6-II | | | |
| | Weight g.. | | Intake g/D | Tumor Weight(g). | Weight g. | | Intake g/D | Tumor Weight(g) |
Per mouse:	J0	J21			J0	J21		
A04	13.3 ±0.7	17.2 ±0.4	3	1.9 ± 0.25	16.6 ± 0.5	18 ± 0.85	2.2	1.48 ± 0.75
A	13.4 ± 0.6	16.9 ± 0.7	2.7	1.5 ± 0.7	16.6 ± 0.6	15.5 ± 0.8	1.9	1 ± 0.5
B	13.3 ± 0.6	17.1 ± 0.7	3	1.5 ± 0.7	16.4 ± 0.6	15.2 ± 0.8	2	1 ± 0.5

Table 2. Lipid measurements in the livers of EMT6 transplanted mice

| | EMT6-I | | | EMT6-II | | |
	T.L[a].	T.C[b].	TG[c].	T.L[a].	T.C[b].	TG[c]
A04	40±1	2.7±0.5	21.5±4	45±2.3	3.2±0.2	20±5.4
A	53±6	3.5±1.2	33±7	41±3	3.8±0.4	19.4±1.6
B	47±8	2.9±0.05	31±11	39±0.8	2.9±0.2	19.1±0.1

[a]T.C. Total cholesterol, mg/g.
[b]T.L. Total lipids, mg/g.
[c]TG. Triglycerids, mg/g.

This could be explained by a deficit in LA, inducing an increase of the delta 9-desaturase, hence an increase in TG, (the same findings were observed in the normal mice receiving A or B diet supplemented with 1% of LA, results not shown). Consequently, the amount of LA was increased up to 4% in the second experiment. In this case, A and B mice lost some weight at the beginning of the experiment, (Table 1) and the levels of liver lipids came back to normal(Table 2).

In addition, it is shown on Table 1 that tumor weight is comparable between A and B group.

With regard to fatty acids, the amount of LA is always lower in the tumors of A and B groups than in the A04 group, whatever the percentage in the diet. (Table 3 and 4).

With regard to vitamin E , plasma levels are always higher in the A group (results not shown), and more so in the tumor (Table 5), where the ratio vit. E/TC is especially elevated in the A group, and in the EMT6-II experiment.

Table 3.Fatty acid percentage in the tumors of EMT6 transplanted mice. Experiment I (1% L A.)

	14:0	16:0	16:1	18:0	18:1	18:?	18:2	20:4	22:6
A04	1.78	29	8.35	8	32.4	-	8.5	2.4	1.2
	±0.01	±2.3	±0.1	±1.15	±3.4		±3.2	±2.1	±1.1
A	1.43	21.6	9	7.3	29	4.5	4.2	2.85	1.3
	±0.03	±0.3	±1.5	±2.8	±0.5	±0.65	±0.75	±1	±1.1
B	1.53	22.5	7.5	6.4	31.8	4.15	4.9	3.4	1.16
	±0.2	±1.3	±2.6	±0.5	±2.2	±0.25	±0.24	±0.7	±0.9

Table 4.Fatty acid percentage in the tumors of EMT6 transplanted mice. Experiment II (4% L A.)

	14:0	16:0	16:1	18:0	18:1	18 :?	18:2	20:4	22:6
A04	1.8	26	6.1	11.6	26	3.55	12.1	5.3	2.1
	±0.1	±0.4	±1.5	±3	±1	±0.5	±0.2	±0.1	±0.09
A	2.2	29.6	8.4	13	32	5.3	4.5	2	1.17
	±0.05	±2	±0.8	±1.5	±4	±0.2	±1.4	±1.5	±1.3
B	1.9	26.4	6.2	13.7	28.4	5.7	3.9	1.8	3.1
	±0.2	±1.8	±0.5	±1.1	±2.3	±0.3	±0.5	±0.8	±0.3

Table 5. Vitamin E and lipid measurements in the tumors of EMT6 transplanted mice

	EMT6-I				EMT6-II			
	[a]Vit E	[b]T.L.	[c]T.C.	[d]TG	[a]Vit E	[b]T.L.	[c]T.C.	[d]TG
A04	271	29	2.5	-	369	29	2.8	15
	±13	±9	±0.6		±1	±7	±0,4	± 5
A	500	36	3	19	691	27	2.7	12.5
	±0	±2	±0.4	±8	±1	±0.07	±0.3	-
B	212	32	2.6	31	195	24	2.6	10
	±36	±15	±0.9	±4	±0.4	±0.3	±0.07	-

[a]Vitamine E µg/g.
[b]T.C. Total cholesterol, mg/g.
[c]T.L. Total lipids, mg/g.
[d]TG. Triglycerids, mg/g.

When 4% linoleic acid was provided in the diet.(EMT6-II), the increase in total lipids, cholesterol and triglycerids disappeared (Table 1).but more important, survival was drastically different since survival was 50% of the B and A04 groups when such a difference was not observed in EMT6-I(results not shown).

In spite of this difference in survival between EMT6-I and EMT6-II, cytometry results were comparable and demonstrated a number of fluorescent cells in the G2 phase which was higher in A than in B, the difference between the two groups being higher in the EMT6-II.(Table 6)

Table 6. Assessment[a] of DNA content in the tumors of EMT6 transplanted mice

| | AO4 | | A | | B | | A | B |
	I+II	I+II	I+II	II			II	
G1	48 ± 6		39 ± 6		52 ± 11		36 ± 3,5	60 ± 9,6
S	25 ± 2,7		26 ± 2,5		32 ± 5,6		25 ± 3,8	30 ± 6
G2	27 ± 3,4		34 ± 6,3		14,5 ± 7,3		39,3 ± 0,7	9,8 ± 3,5

[a]evaluated by the percentage of fluorescent cells in each phase of the cell cycle.

DISCUSSION

These first experiments show that high vitamin E intake is reflected in the tumor content of this vitamin. Although there is no difference in tumor weight between groups receiving a diet enriched in vitamin E (A) or a diet without vitamin E (B), there is a difference in the DNA content evaluated by cytometry indicating a higher number of G2 phase tumor cells in group A than in group B.In addition, in the experiment conducted with 4% LA in the diet there is a difference in survival.

Other type of animal studies reported that high levels of vitamin E enhanced tumor growth (2) or tumor multiplicity (3). In addition, a recent report (4) demonstrates, using an in vitro system that PUFA, and especially gamma-linolenic acid, are toxic for a breast carcinoma cell line through peroxyl radical production and that vitamin E inhibited cell killing. This oxygen reactive species (ROS)-mediated tumor cytotoxicity might be induced by tumor necrosis factor alpha (5). Finally, it has been shown that fish oils inhibit the production of growth factor-like protein, and that its production is resumed after treatment by vitamin E (6). Thus, high levels of vitamin E could facilitate tumor growth by eliminating toxic radicals.

Our preliminary results support the hypothesis (7) of an increased tumor cell proliferation associated with vitamin E concentration in tumor, protecting tumor cell proliferation against the attack of lipid peroxides.

Acknowledgement: This work has been supported by "Ligue Française contre le Cancer"

References
1 M. Gerber., F. Cavallo, E.Marubini, S..Richardson, A. Barbieri, E. Capitelli, A. Costa, A. Crastes de Paulet, A.De Carli,U. Pastorino, H. Pujol. Liposoluble vitamins and lipid parameters in breast cancer. A joint study in northern Italy and southern France. Int. J. Cancer. 42: 49 (1988).
2 B. Toth, K. Patil Enhancing effect of vitamin E on murine intestinal tumorigenesis by 1,2-dimethylhydrazine dihydrochloride. JNCI ,70: 1107 (1983).
4 M. E Begin.,G.Ells,D., F.Hoorobin Polyunsaturated fatty acid-induced cytotoxicity against tumor cells and its relationship to lipidperoxidation., JNCI, 80,188 (1988).
5 N. Matthews, M.L.Neal, R.A. Fiera,S.K. Jackson and J.M. Stark, Are free radicals involved in tumor cytolysis by tumor necrosis factor. in "Tumor necrosis factor/cachectin and related cytokines".B. Bonavida, G.E. Gifford, H. Kirchner and L.J. Old, eds, Kaeger, Basel, (1988).
6 P.L. Fox and P.E. Di Corletto. Fish oil inhibit endothelial cell-production of platelet-derived growth factor - like protein. Science, 241:453 (1988).
7 T. Slater,C.Benedetto,G.W. Burton,K.H. Cheeseman,K.V.Ingold, J.T.Nodes, Lipid peroxidation in animal tumours: a disturbance in the control of cell division ? In:, "Icosanoïds and Cancer", H.Thaler-Dao, R. Paoletti, A. Crastes de Paulet, eds, Raven Press, New-York, (1984.

ABNORMAL SUSCEPTIBILITY TO LIPID PEROXIDATION OF PLASMA LDL AND ITS PREVENTION BY α-TOCOPHEROL DURING EXPERIMENTAL CHOLESTASIS IN THE RAT

L.G.ALCINDOR,H.ANTEBI,M.FADEL-KHADRA,M-C.PIOT
Y.GIUDICELLI AND R. NORDMANN

Faculté de Médecine Paris-Ouest
45 rue des Saints Pères 75006 PARIS
et CHI de POISSY (FRANCE)

INTRODUCTION

Thiobarbituric acid reactive substances (TBARS) represent an helpful index of lipid peroxidation in erythrocytes and lipoproteins(1).Peroxidation is depending on the status of the antioxidant protecting system including the vitamin E tissue level(2). In cholestatic states, biliary retention has been shown to induce:
- modifications of the lipoprotein structure and alterations of the plasma lipoprotein profile(3,4);
- lipid malabsorption and vitamin E deficiency(5).
These effects could account for an increase in blood lipid peroxidation and explain some pathological aspects of cholestasis(6) and dyslipoproteinemias(7.In this work we report increased susceptibility to lipid peroxidation of selectively precipitated low density lipoproteins(LDL) during cholestasis in the rat and its prevention by α-tocopherol administration .

MATERIAL AND METHODS

Eighteen laparotomized male Wistar rats were divided into three groups:
- Control , non cholestatic rats(A),
- Cholestatic (bile duct ligation) rats(B),
- Vitamin E treated cholestatic rats(C).These rats received once a week a subcutaneous injection of α-tocopheryl acetate(5 mg in micellar solution) (Ephynal-Roche).
Three weeks later:
1) Cholesterol and choline phospholipids were determined in plasma LDL and red blood cell membranes obtained by selective precipitation and solubilized in 0.015 N NaOH (8-11).
2) TBARS in LDL and erythrocyte membranes were assessed by the thiobarbituric acid reaction before and after lipid peroxidation induced by phenylhydrazine(0.6 mM)(12).Results are expressed as μmol TBARS/mmol cholesterol.
3) The susceptibility to lipid peroxidation was deduced from the

Antioxidants in Therapy and Preventive Medicine
Edited by I. Emerit *et al.*
Plenum Press, New York, 1990

difference between the TBARS content after and before exposure
to phenylhydrazine.

RESULTS

1) Cholesterol and choline phospholipids
    As shown in table 1,the cholesterol and phospholipid
content is increased by 71% and 100% respectively in LDL from
cholestatic rats.These changes result in a slight reduction
(-15%) of the cholesterol/choline phospholipids molar ratio as
compared to control animals.

Table 1.LDL-Cholesterol,LDL-phospholipids(mmol/l) and serum
        bilirubin (mmol/l) in control(NC),cholestatic(C) and
        vitamin E-treated cholestatic rats (EC).

	A	B	C
Bilirubin	≤ 10	175,00±40,00	150,00±25,00 n.s
LDL-Cholesterol	0,35±0,06	0,60± 0,06***	0,58± 0,05 n.s
LDL-Phospholipids	0,17±0,03	0,30± 0,03***	0,30± 0,05 n.s

Values are the mean ± S.D.of six determinations in each group
(B vs A : *** p ≤ 0.001)(C vs B : n.s = p ≥ 0.05).

    In erythrocyte membranes(table 2)the cholesterol content is
increased (+20%) in the cholestatic group.Like in LDL this
elevation is slighter than that of the choline phospholipids and
the mean value for the ratio cholesterol/choline phospholipids
is reduced.

Table 2.Erythrocyte membrane cholesterol and choline
        phospholipids in control (A),cholestatic (B) and
        vitamin E-treated cholestatic rats (C).

	A	B	C
Cholesterol	3.45±0.15	4.20±0.20***	3.95±0.30
Choline phospholipids	2.00±0.15	3.20±0.25***	2.90±0.15

Values are expressed in mmol / l of packed red blood cells and
represent the mean ± S.D. of six determinations in each group.
(B vs A : *** = p ≤ 0.001).

    As shown in tables 1 and 2 ,administration of tocopheryl
acetate to cholestatic rats failed to affect the LDL and red
blood cells membrane lipid content.

2) Thiobarbituric acid reactive substances
    - In LDL the TBARS content(table 3) is low in control rats.
Three weeks after the bile duct ligation a significant increase
(more than twofold) is observed.
    Induction of plasma lipid peroxidation by phenylhydrazine
results in a fivefold increase in LDL-TBARS in control rats as
shown in table 3.This increase is considerably higher (about
fiftyfold) in cholestatic rats.

Table 3.LDL-TBARS in control(A) and cholestatic rats (B)
21 days after bile duct ligation.

	A	B
Before	0.9±0.3	2.3±0.65 ***
After	4.3±1.6	115.0±20.5 ***

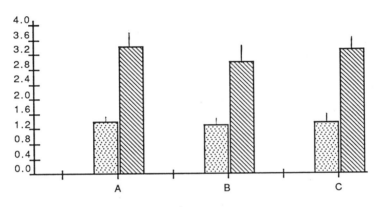

☐ BEFORE   ☒ AFTER

Fig 1.TBARS of red blood cell membranes   (μmol MDA/mmol
cholesterol).

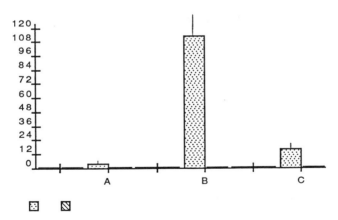

☐   ☒

Fig 2.LDL susceptibility to lipid peroxidation   (μmol MDA
/mmol LDL-cholesterol).

Values are expressed in μmol MDA / mmol LDL-Cholesterol before and after in vitro induction of lipid peroxidation.

-In erythrocyte membranes, TBARS values before and after induction of lipid peroxidation (fig 1) are not significantly different in normal and cholestatic rats.

The susceptibility in vitro induced to lipid peroxidation is considerably reduced in LDL after vitamin E administration to cholestatic rats (fig 2).

DISCUSSION AND CONCLUSION

Although cholestatis does not result in significant changes in the peroxide level in erythrocytes membranes, important abnormalities are observed in the LDL. The slight enrichment in choline phospholipids of the precipitated LDL observed in the cholestatic rats reflects the precedently reported (3-4) presence of LPX (4). These structurally abnormal lipoproteins are formed as a result of the bile influx into the plasma due to bile duct ligation. The lipid transport between lipoproteins and plasma membranes is therefore altered and modifications in the membrane lipid composition appear (13).

Since bile retention always elicits malabsorption of lipid and vitamin E(5), a deficiency of the antioxydant protection appears most likely responsible for the observed increase in LDL-TBARS of cholestatic rats.

In accordance with this interpretation is the beneficial effect of subcutaneous vitamin E administration which was found to reduce the LDL-TBARS and the LDL susceptibility to lipid peroxidationin the cholestatic rats.

In agreement with present concepts on the atherogenic role of oxidized LDL (14), it may be suggested from this study that the LDL-TBARS determination is of practical interest for the diagnosis of latent vitamin E deficiency (especially in normocholesterolemic atherogenic diseases and during LDL-cholesterol lowering therapy using bile acid binding resins).

REFERENCES

1) K.,Yagi,Assay for serum lipid peroxide level and its clinical significance in :"Lipid peroxides in biology and medicine",K.Yagi,ed.Academic Press.1980.
2) M.Miki,H.Tamai,M.Mino,Y.Yamamoto,and E.Niki,Free radicals chain oxidation of rat red blood cells by molecular oxygen and its inhibition by α-tocopherol,Arch.Biochem.BiophyS,258:373-389,1987.
3) D.Seidel,P.Alaupovic,R.H.Furman,A lipoprotein characterizing obstructive jaundice.I.Method for quantitative separation and identification of lipoproteins in jaundiced subjects,J.Clin.Invest.48,1211-1223,1969.
4) J.Picard,D.Veissiere,Abnormal serum lipoprotein in cholestasis identification and isolation,Clin.Chim.Acta,30,149-154,1970.
5) K.M.Brinkhous,E.D.Warner,Muscular dystrophy in biliary fistula dogs Possible relationship to vitamin E deficiency,Am. J.Pathol,98:81-86,1941.
6) V.G.Levy,"Cirrhose biliaire primitive de l'adulte",in"Précis des maladies du foie",J.Caroli,A,Ribet,A.,Paraf,ed Masson,Paris 1975.
7) A.M.Fogelman,I.Shechter,J.Saeger,M.Hokom,J.S.Child,P.A.Edward, Malondialdehyde alteration of low density lipoproteins leads to cholesteryl esters accumulation in human monocytes-macrophages, Proc,Natl,Acad,Sci USA 79:1712-1716,1980.

8) L.G.Alcindor, H.Aalam, J.Masliah, G.Thomas-Benhamou, M.Ouka, M-C.Piot, Le rapport de concentration molaire cholesterol sur phospholipides des LDL dans les hépatopathies, Méd.Chir.Dig. 10,673-674,1981.

9) M.Goudard, C.Aubry, L.G.Alcindor, Etude d'un réactif précipitant sélectivement les LDL.Vème Colloque international de Biologie Prospective .Ann.Biol .Clin., 40:466,1982.

10) L.G.Alcindor, M.Fadel-Khadra ,H.Antebi, M.C.Piot, M.J.Guerrito, Y.Giudicelli ,R.Nordmann, A rapid assay for cholesterol and phospholipids in erythrocyte membranes ,Ann .Biol.Clin.46:586,1988.

11) L.G.Alcindor, H.Antebi, L.De Paillette, L.Malagrida, D.Hillion , Y.Giudicelli, MDA des membranes erythrocytaires et LDL-MDA au cours de l'épuration extra-renale, Ann.Biol.Clin, 46,537,1988.

12) H.Ohkawa, N.Ohishi and K.Yagi, Assay for lipid peroxides in animal tissues by thiobarbituric acid reaction, Anal.Biochem 95:351-358,1979.

13) J.S.Owen, R.A.Hutton, M.J.Hope, D.S.Harry, K.R.Bruckdorfer, R.C.Day, N.McIntyre and J.A.Lucy, Lecithin:Cholesterol acyltransferase deficiency and cell membrane lipids and function in human liver disease Scand.J.Clin.Lab.Invest.38,suppl. 150:228-232,1978.

14) D.Steinberg, "Metabolism of lipoproteins and their role in the pathogenesis of atherosclerosis" in:Atherosclerosis Reviews,vol 18,J.Stokes III and M. Mancini ed,Raven Press limited New-York,1988.

THE ROLE OF OXIDATIVE MODIFICATION AND ANTIOXIDANTS

IN LDL METABOLISM AND ATHEROSCLEROSIS

Wendy Jessup*, Roger T. Dean**, Catherine V. de Whalley
Sara M. Rankin and David S. Leake

*Cell Biology Research Group
Brunel University
Uxbridge, U.K.

**Heart Research Institute
Sydney, NSW
Australia

Department of Pharmacology
King's College London
London, U.K.

INTRODUCTION

Recent studies have shown that low-density lipoprotein (LDL), when incubated with certain cell types in culture (including endothelial cells, smooth muscle cells and macrophages) is subject to a number of alterations in its physical and chemical properties[1]. Most interestingly, this 'modified' LDL is endocytosed by macrophages up to 20 times more rapidly than native LDL[1,2]. It is possible that the formation of foam cells from macrophages in the developing atherosclerotic plaque could be the result of the generation of similar 'modified' LDL particles by cells of the artery wall. Because the route for endocytosis of 'modified ' LDL largely bypasses the normal ApoB/E receptor, target cells such as the macrophage are unable to regulate their intake of this ligand and so accumulate large amounts of cholesteryl esters intracellularly.

The dependence of the cell-mediated modification of LDL on the presence of redox-active metals in the culture medium, and its inhibition by chain-breaking antioxidants, suggests that free radical attack and lipid peroxidation are essential features of the modification process[1].

LDL contains a number of endogenous antioxidants, including alpha-tocopherol, beta-carotene and retinyl stearate[3,4]. The concentration of such naturally occurring antioxidants should influence the susceptibility of individual LDL preparations to oxidative modification. We have examined this hypothesis by measuring lipid peroxidation, alpha-tocopherol consumption, and the formation of 'modified' high-uptake forms of LDL in samples which have been exposed to free radicals generated <u>in vitro</u>, or to a cell-mediated modification system.

Antioxidants in Therapy and Preventive Medicine
Edited by I. Emerit *et al.*
Plenum Press, New York, 1990

TECHNIQUES

Human LDL was prepared from serum of normal volunteers collected after overnight fasting. LDL was isolated by density centrifugation (1.019-1.050) using KBr solutions for density adjustments[5]. Labelling of LDL with ^{125}I was by the iodine monochloride method[6]. Solutions of LDL were dialysed against deoxygenated buffers containing 0.1mg/ml chloramphenicol and 1mg/ml EDTA or Chelex, as appropriate.

Oxygen-centred radicals were generated at RT^{o}C by steady-state radiolysis using the Brunel Biochemistry ^{60}Co source. LDL was used at 1.0 mg protein/ml in 10mM phosphate buffer (pH 7.2). Gassing with N_2O produced predominantly hydroxyl radicals; gassing with air in the presence of formate produced either superoxide (at pH 7.2) or hydroperoxyl (at pH 4.0) radicals[7].

Macrophage-mediated modification of LDL was performed as described by Rankin and Leake[2]. Productive modification was measured as the rate at which the modified LDL was degraded by a second set of macrophages[2].

Samples of LDL were extracted into heptane by the SDS method[8]. Alpha-tocopherol was determined by HPLC analysis and hydroperoxides by an automated version of the triiodide assay[9].

RESULTS AND DISCUSSION

1. The effects of defined oxygen-centred radicals on LDL oxidation and modification

Exposure of LDL to hydroxyl or hydroperoxyl radicals led to the dose-dependent production of lipid hydroperoxides. In both instances, the appearance of hydroperoxides was largely suppressed until all the endogenous alpha-tocopherol of the LDL samples was consumed. This observation is consistent with the hypothesis that such endogenous antioxidants contribute to the protection of LDL against oxidative damage.

Superoxide radical was unable to stimulate the peroxidation of LDL lipids, in agreement with previous studies which have shown that superoxide radicals are inactive in the peroxidation of lipids[10]. Nevertheless, superoxide exposure did lead to a gradual the depletion of alpha-tocopherol, and therefore might contribute indirectly to LDL oxidation in vivo by lowering its antioxidant content.

2. Antioxidant consumption and peroxidation during macrophage-mediated modification of LDL

LDL was incubated with resident mouse peritoneal macrophages. At intervals, samples of the medium were collected and assayed for alpha-tocopherol content, hydroperoxides, and the presence of high-uptake forms of LDL. The latter was assessed by measurement of the rate at which LDL was degraded by a second set of macrophages[2].

The alpha-tocopherol content of LDL was rapidly depleted by the modify-ing macrophages, so that after 4hr no detectable amounts remained. As the tocopherol level declined, the rate of LDL peroxidation accelerated. In corresponding cell-free control incubations, the decline in tocopherol levels and the formation of hydroperoxides was much slower, and no productive modification occurred in these samples of LDL.

The formation of high-uptake LDL did not occur until all of the alpha-tocopherol was consumed. This observation indicates that macrophage-mediated LDL modification is dependent on oxidation, and that the antioxidant content of LDL may be a significant factor in determining the susceptibility of LDL to modification in in vitro systems such as the one described here, and possibly also in vivo.

3. The effects of exogenous flavonoids on macrophage-mediated modification

Flavonoids have been reported to be free radical scavengers, inhibitors of lipid peroxidation, and of the activities of lipoxygenase, phospholipase A_2 and NADPH oxidase. We have found that low concentrations of flavonoids are potent inhibitors of macrophage-mediated oxidation of LDL[11,12]. For example, 1uM morin could completely prevent LDL peroxidation and the formation of high-uptake species by macrophages during a 24hr incubation. This was associated with maintenance of the alpha-tocopherol levels of these LDL samples. Since flavonoids also prevent cell-free oxidation of LDL by 10-100uM Cu^{2+}, their effects on cell-mediated modification may be related largely to their radical scavenging properties.

Thus exogenous antioxidants can protect LDL against oxidative damage. This may be of use the the development of protective agents for the suppression of atherogenesis in individuals identified as at high risk.

4. Modulation of the alpha-tocopherol content of LDL affects the progress of cell-mediated modification

LDL was prepared from plasma samples taken from a single donor immediately before and after a 3-day period of oral supplementation with d-alpha-tocopherol at 1.45g/day. The alpha-tocopherol content of the in vivo 'loaded' LDL was 2.5-fold higher than that of the corresponding control LDL. These LDL samples were then labelled with ^{125}I and incubated in a macrophage modification system. In the tocopherol-loaded LDL the disappearance of alpha-tocopherol, and the formation of high-uptake LDL were retarded by 2hr, compared with the corresponding unloaded control LDL. There was also a 2hr delay in the onset of lipid peroxidation.

Thus an increase in the alpha-tocopherol content of LDL apparently increases its resistance to oxidative modification. This is consistent with epidemiological data, which have shown and inverse correlation between the incidence of coronary death in several European populations and plasma antioxidant levels[13].

CONCLUSIONS

While endogenous alpha-tocopherol almost certainly has a role in the protection of LDL against oxidative stress, and perhaps also against atherogenesis, we believe that several other endogenous antioxidants are also important contributors to LDL defences in vivo. Several other lipid-soluble antioxidants have been measured in LDL[4], and it is quite possible that there are more species of which we are, as yet, unaware. In addition, the contribution of water soluble molecules, such as ascorbate, urate and bilirubin, to the protection of plasma lipids against oxidation have also been demonstrated recently[14]. Any approach to the prevention of atherosclerosis through suppression of LDL oxidation shoud include a consideration of the contributions of all of these natural systems to antioxidant protection in vivo, and an assessment of the most effective points for supplementation and/or intervention.

REFERENCES

1. Heinecke, J.W. (1987) Free Radical Biol. Med. $\underline{3}$, 65-73.
2. Rankin, S.M., Leake, D.S. (1988) Agents Actions Suppl. $\underline{26}$, 233-239.
3. Esterbauer, H., Jurgens, G., Quehenberger, O., Koller, E. (1987) J. Lipid Res. $\underline{28}$, 495-620.
4. Esterbauer, H., Striegl, G., Puhl, H., Rotheneder, M. (1989) Free Rad. Res. Comms., in press.
5. Havel, R.J., Eder, H.H., Bragdon, J.H. (1955) J. Clin. Invest. $\underline{34}$, 1345-53.
6. Bilheimer, D.W.S., Eisenberg, S., Levy, R.I. (1972) Biochim. Biophys. Acta $\underline{60}$, 212-221.
7. Willson, R.L. (1978) In: Biochemical mechanisms of liver injury. (Slater T.F.; Ed.) pp.123-224, Academic Press, N.Y.
8. Burton, G.W., Webb, A, Ingold, K.U. (1985) Lipids $\underline{20}$, 29-39.
9. Thomas, S.M., Jessup, W., Gebicki, J.M., Dean, R.T. (1989) Anal. Biochem. $\underline{176}$, 353-359.
10. Gebicki, J.M., Bielski, B.H.J. (1981) J. Am. Chem. Soc. $\underline{103}$, 7020-7022.
11. Rankin, S.M., Hoult, J.R.S., Leake, D.S. (1988) Brit. J. Pharmacol, in press.
12. Rankin, S.M., Hoult, J.R.S., Leake, D.S. (1988) Proc. 4th Int. Atherosclerosis Conf., Rome,; in press.
13. Gey, F. (1986) Biblthca Nutr. Dieta. $\underline{37}$, 53-91.
14. Frei, B., Stocker, R., Ames, B.N. (1988) Proc. Natl. Acad. Sci. USA $\underline{85}$, in press.

EFFECTS OF VITAMIN E TREATMENT IN CHOLESTATIC CHILDREN

Frédérique Lemonnier, Fernando Alvarez, François Babin,
Martine Couturier and Daniel Alagille

Unité de Recherche d'Hépatologie Pédiatrique INSERM U 56
Hôpital de Bicêtre, 94270 Le Kremlin-Bicêtre - France

INTRODUCTION

Several years ago, it has been shown [1],[2] that there was a relationship between vitamin E (Vit. E) deficiency and the neurological disorders observed in patients with chronic cholestasis and lipid malabsorption. We report here our clinical experience of eight years of Vit. E treatment (10 mg/kg of DL-α-tocopherol acetate, IM, every two weeks) in these patients.

Because Vit. E is considered as an important structural component of biological membranes, which acts as a free radical scavenger, we questioned if the Vit. E treatment could modify different biochemical parameters implicated in chronic cholestasis [3],[4]. In this way, we have determined in Vit. E treated children with syndromatic type-paucity of interlobular bile ducts (SPD) : Vit. E/total lipids ratio (Vit. E status), plasma lipid peroxide levels and plasma total and free fatty acid variations.

CLINICAL STUDIES

Patients

Twenty-three children with chronic cholestasis (18 SPD - Alagille's syndrome -, 1 non syndromatic PD, 3 Byler's disease, 1 biliary atresia) were followed during eight years. The protocol was started in 1980 and it was approved by the ethical committee of University Paris-Sud. Three groups of patients were compared. Group 1 included children with established neurologic signs at the beginning of Vit. E treatment, group 2 children without Vit. E therapy and group 3 children with prophylactic Vit. E treatment.

Results

Group 1. The results obtained for these children are reported in table 1. Patients 3, 6 and 7 improved muscle strength and they had less difficulty with writing and drawing. In patients 7 and 11 deep tendon reflexes reappeared in all limbs and in patient 10 only in upper limbs. Patients 3, 6, 7 and 9 showed partial recovery of oculomotor function. Patient 1 completely recovered from cerebellar dysfunction. In patients 3 and 7 ataxia was less evident. Thus, in patients with important neurological abnormalities, Vit. E therapy allowed a partial improvement.

Antioxidants in Therapy and Preventive Medicine
Edited by I. Emerit *et al.*
Plenum Press, New York, 1990

143

Table 1 - Children with established neurologic signs at the beginning of Vit. E treatment (group 1).

Patient	Age (yr)	At onset of treatment			At last examination	
		Peripheral neuropathy	Abnormal oculomotor function	Cerebellar syndrome	Age (yr)	Results
1	2 11/12	+	−	+	8 11/12	Stable
2	5 4/12	+	−	−	10 10/12	Improved
3	6 2/12	+	+	+	11 11/12	Improved
4	7 7/12	+	−	+	13	Stable
5	20	+	+	+	24	Stable *
6	13 11/12	+	+	−	20 3/12	Improved
7	16 3/12	+	+	+	21 7/12	Improved
8	11 3/12	+	+	+	18	Stable +
9	10 2/12	+	+	+	14 2/12	Improved +
10	6	+	−	−	11 1/12	Improved
11	3 2/12	+	−	−	8 2/12	Improved

* This patient had a liver transplantation at age 18 yrs
+ This child died of liver failure at age 14 yrs 6 m.

Table 2 - Children without Vit. E treatment (group 2).

Patient	First neurologic examination		Development of neurologic symptoms and beginning of vitamin E treatment	
	Age (yr)	Results	Age (yr)	Deep tendon reflexes
12	1 2/12	Normal	2 4/12	− (Lower limbs)
13	11/12	Normal	1 9/12	− (Upper, lower limbs)
14	11/12	Normal	1 11/12	− (Lower limbs)
15	1/12	Normal	2 4/12	− (Lower limbs)
16	1 2/12	Normal	2 6/12	− (Lower limbs)
17	1 1/12	Normal	1 6/12	− (Upper, lower limbs)

Group 2. The first signs of neurologic disorders in patients with chronic cholestasis were the development of abnormal deep tendon reflexes. Last neurological examination was normal in all the children between 1 yr 6 m to 3 yrs after the beginning of Vit. E (I.M.) treatment ; these patients had a normal H_2O_2 and NaN_3 hemolysis test and bilirubin was stable.

Group 3. Early Vit. E therapy (started in the first months of the life) prevents the development of neurological signs for all the treated children. Follow-up between 4 yrs 9 m and 7 yrs 5 m for six patients.

BIOCHEMICAL STUDIES

Patients and methods

Thirty-two patients with SPD were studied. This protocol was started in 1985. The mean age was 4 yrs (2 m – 17 yrs). Biliary acids and γGT were always found increased even in 11 patients with bilirubin (BIL) < 30 μmol/l. They received "Liprocil" (Sopharga Laboratories, France).

Table 3 - Effect of Vit. E treatment on lipid peroxide levels and fatty acid variations.

	Patient 1		Patient 2		Patient 3		Patient 4		Patient 5	
Vit E (months):	7	11	12	15	5	8	0	10	0	3
BIL µmol/l	16	20	12	5	46	37	149	120	325	333
LP nmol/ml	4.92	6.40	7.60	10	5.30	6.10	7.10	9.70	12.8	8.96
Vit E/LT mg/g	2.12	0.93	1.47	1.21	1.17	1.30	0.41	0.84	0.07	0.40
TFA										
C18:2 n-6 %	36.4	21.5	25.4	31.2	24.6	16.9	18.1	15.6	9.83	5.40
C20:4 n-6 %	9.43	9.32	8.58	8.47	8.52	8.29	5.15	5.51	5.38	3.78
FFA										
C18:2 n-6 %	33.9	19.2	20.31	24.0	19.2	11.05	10.2	13.4	7.07	5.79
C20:4 n-6 %	0.84	4.57	5.80	5.84	4.12	4.00	1.39	4.40	8.54	3.75

Twenty-seven patients were Vit. E treated for 2 months to 5 years. Five patients were tested in two occasions before or during the Vit. E treatment. Controls were 14 children without cholestasis.

Total plasma fatty acids (TFA) and free fatty acids (FFA) were analyzed by gas chromatography. Plasma lipid peroxides (LP) were determined using a spectrofluorimetric method [5] and were expressed in malondialdehyde (MDA). Plasma Vit. E was measured by HPLC [6].

Results

Fatty acid variations and lipid peroxides. We observed in our patients a significant increase in different lipid fractions [3] and in lipid peroxide concentrations : 8.80 ± 3.70 nmol/ml in the patients and 2.45 ± 0.65 nmol/ml in the controls. Fig. 1 shows the mean percentage composition of TFA in the controls and patients. In patients, we observed a significant rise in palmitic ac. C16:0 and palmitoleic ac. C16:1 n-7 balanced by a drop in linoleic ac. C18:2 n-6 (29.5 ± 6.1 % in the controls and 19.1 ± 8.03 % in the patients). In FFA we found principally a marked increase in arachidonic acid C20:4 n-6 (1.43 ± 0.85 % in the controls and 4.27 ± 2.24 % in the patients).

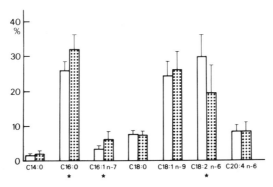

Fig. 1 - TFA composition of plasma total lipid extracts in controls (white column) and patients (striped column). * p ≤ 0.001 between patients and controls.

Vitamin E and lipid peroxides. Most of these patients (27) were Vit. E treated (mean plasma Vit. E = 10.55 $\pm$ 5.6 mg/1). However 1/4 of these children had a Vit. E (mg)/total lipids (TL.g) ratio < 0.80. We observed an inverse relationship between lipid peroxide levels and the Vit. E status (r = - 0.62 ; p $\leq$ 0.001).

Effects of several months of Vit. E treatment in five patients (Table 3). When one or several parameters were disturbed, we did not observe a clear improvement of these data with Vit. E therapy.

DISCUSSION AND CONCLUSION

Follow-up of eight years of Vit. E treatment started in the first weeks of life showed that this vitamin prevented the development of neurological disorders in these children. In patients with neurologic abnormalities, the evolution of neurologic symptoms under treatment showed, in most cases, an improvement essentially in relation to motor strength and oculomotor function. However, the effect of Vit. E was less evident on different biochemical parameters studied in children with Alagille's syndrome as previously suggested [7]. Thus, in despite this treatment, we observed significant fatty acid variations and high levels of lipid peroxides in comparison with the controls.

Several hypothesis [4] could be considered to explain these results : 1) a very important malabsorption of fatty acids principally the long chain fatty acids ; 2) other tissues (hepatocytes, bile ducts) and plasma lipoproteins could be subject to attack by free radicals ; 3) a defect in other antioxidant enzymatic systems which might emphasize lipid peroxidation.

In conclusion, further therapeutic assays, requiring for each patient different amounts of Vit. E in relation with plasma lipids and lipid peroxide levels could improve some biochemical parameters and consequently the cholestasis itself.

REFERENCES

1. M. A. Guggenheim, S. P. Ringel, A. Silverman, B. E. Grebert, Progressive neuromuscular disease in children with chronic cholestasis and vitamin E deficiency : diagnosis and treatment with alpha-tocopherol, J. Pediatr. 100:51 (1982).
2. F. Alvarez, P. Landrieu, P. Laget, F. Lemonnier, M. Odièvre and D. Alagille, Nervous and ocular disorders in children with cholestasis and vitamin A and E deficiencies, Hepatology 3:410 (1983).
3. F. Lemonnier, D. Cresteil, M. Fénéant, M. Couturier, O. Bernard and D. Alagille, Plasma lipid peroxides in cholestatic children, Acta Paediatr. Scand. 76:928 (1987).
4. F. Babin, F. Lemonnier, A. Coguelin, D. Alagille, A. Lemonnier, Plasma fatty acid composition and lipid peroxide levels in children with paucity of interlobular bile ducts, Ann. Nutr. Metab. 32:220 (1988).
5. K. Yagi, A simple fluorometric assay for lipoperoxide in blood plasma, Biochem. Med. 15:212 (1976).
6. G. L. Catignani, J. G. Bieri, Simultaneous determination of retinol and and α-tocopherol in serum or plasma by liquid chromatograpy, Clin. Chem. 29:708 (1983).
7. R. J. Sokol, J. E. Heubi, B. Mc Graw, W.F. Balistreri, Correction of vitamin E deficiency in children with chronic cholestasis. II - Effect on gastrointestinal and hepatic function, Hepatology 6:1263 (1986).

EFFICACY OF VITAMIN E AS A DRUG IN INFLAMMATORY JOINT

DISEASES

K.H. Schmidt   W. Bayer

Dept. of Surgery   Lab. of Spectral and
University of   Biological Analysis
Tuebingen, West Germany   Stuttgart, West Germany

Living organisms are composed of substances that are thermodynamically unstable towards oxygen. Proteins, lipids, polysaccharides and nucleid acids can react with molecular oxygen
and be deactivated. This interaction never occurs, however,
because of in-built inhibitory mechanisms; i.e. the reactions
proceed so slowly that in the course of normal observation
periods no noteworthy changes occur in the basic constituents
of the organism. As evolution has proceeded, this situation
has been promoted by the fact that the metabolism has
switched from anaerobic to aerobic processes for the purposes
of increasing performance. The reaction between oxygen and
hydrogen during oxidative phosphorylation has been restricted
in the mitochondria in such a way as to exploit the large
amounts of energy released by the oxohydrogen reaction and to
prevent the structure of the cell from being oxidatively damaged.

There is a certain paradox about the way in which the aerobic
metabolism developed.

The evolutionary processes that led to the aerobic metabolism
were accompanied by the development of antioxidative protection mechanisms that are activated when the delicate balance
of oxidative phosphorylation is upset and there is a possibility that very reactive oxidants might participate in side
reactions. Thus, the human metabolism is a fine balance between oxidation for the purpose of rapid energy release and
antioxidation for protecting the labile constituent materials
afforded by it. A major role is played by such antioxidative
nutrients as vitamins and trace elements.

It must be recognized that the permanent oxidative attack of
oxygen on cell material over the course of decades cannot be
compensated for and that aging is both inevitable and fatal
for every aerobic organism.

It is readily conceivable that the oxidative burden becomes
precarious when high performance is required, i.e., a high
turnover of energy occurs in the cells. Oxidation and anti-

Antioxidants in Therapy and Preventive Medicine   147
Edited by I. Emerit *et al.*
Plenum Press, New York, 1990

oxidation are therefore processes crucial to the performance of competitive sportsmen, labourers and those in an unfavourable climate.

A high turnover in energy increases the danger that the adaptive and regulatory mechanisms will be overburdened in the energy cascade of the respiratory chain, resulting in an accumulation of oxidants at various stages of the cascade that can react chemically with the immediate environment. Such a situation can arise under hypoxic conditions, for example, when oxygen, the ultimate electron acceptor, is missing at the end of the cascade. The electrons formed in the cascade then seek reaction partners in the membrane, for example, thereby giving rise to radicals.

Evolution has not only increased the efficiency of the metabolism by aerobic means and afforded it the protection of antioxidants but it has also developed a mechanism for warding off infections that operates by releasing the organism's strongest oxidants to oxidatively destroy alien information carriers. This occurs in the phagocytes, which are equipped with the appropriate tools. Antioxidative protection is crucial to these cells because, once they are activated, oxidative activity rises 100 fold. For this reason, phagocytes such as granulocytes and macrophages have special transport mechanisms in their membranes for pumping antioxidants such as L-ascorbic acid into the cells. Nevertheless, very often the phagocytes do not survive an encounter with infection and the granulocytes are often referred to as kamikaze cells.

Membrane structures are particularly susceptible to oxidative attack, which can result in a number of cellular changes.

The efficiency of transport processes, cell-cell interactions, mobility and motility, endocytosis, fusion and many other cellular functions presupposes an intact cell membrane.

The effects of oxidative processes on the chemotactic functions of the phagocytes is extremely well documented. Activation of the oxidative metabolism in phagocytes can render chemotaxis totally ineffective, although the reaction can be partially reversed by antioxidants. This is an important orientation point in the assessment of the antioxidative action of vitamins as immuno-modulators.

If we summarize all the substances that afford protection against oxidative attack by superoxide radical ions, hydrogen peroxide, hydroxyl radicals, singlet oxygen or its successor radicals, we can distiguish two large categories. One of them is enzymes and the other is low-molecular-weight antioxidants, of which the vitamins are an important sub-group. Certain trace elements such as copper, zinc and selenium are important components of the detoxifying enzymes and are indispensable to the activity of superoxide dismutase and glutathione peroxidase.

Essential vitamins are alpha-tocopherol and ascorbic acid and also the provitamin, beta-carotene. In addition, a certain role is played by several amino acids such as cystein and methionin and compounds such as uric acid and glutathione. The antioxidative vitamins and glutathione constitute a redox cascade that reduces the oxygen radicals, while beta-carotene is primarily responsible for detoxifying singlet oxygen.

The coupled redox system comprising tocopherol/ascorbate/glutathione is of major importance because oxidizing equivalents from the lipophilic compartment of the membrane are transferred to the hydrophilic cytoplasm where they are definitively reduced. An interesting fact here is that the ionic ascorbate can revert to the nonionic dehydroascorbate.

Every redox system, tocopherol/tocoquinone or ascorbate/dehydroascorbate or oxidized glutathione/reduced glutathione, contributes towards affording protection against the reactive oxygen species. Owing to the cascadal nature of radical chain reactions, a single antioxidative molecule is often enough to prevent several hundred subsequent oxidative reactions from occurring. This is what happens, for example, during the oxidation of fatty acid molecules in membranes: stored tocopherol there terminates the radical chain reaction.

It is difficult to extend this mechanistic knowledge of the action of antioxidative vitamins to draw definite conclusions about their importance in the prophylaxis and therapy of clinical syndromes because of the complexity of the causes and effects. When carefully conducted epidemiological studies and experimental research on biochemistry and pharmacology reveal a clear trend, considerably more credence is accorded to their conclusions.

The use of vitamin E in humans is indicated by deficiencies, the symptoms of which can often only be recognized with difficulty, and in times of increased demand, including pregnancy, lactation, special diets, exposure to radical formers, artificial respiration and dialysis.

Intake of vitamin E has also been beneficial in various genetic disorders, such as sickle cell anaemia, thalassaemia and certain enzyme defects in the oxidative metabolism. Blood vessel lesions involving inflammation, e.g. thrombophlebitis, and those involving degeneration, such as atherosclerosis, have proved responsive to vitamin E. Similar positive effects are achieved in the treatment of thrombocyte aggregation.

Vitamin E is also used in the therapy of inflamed joints and in preventing proliferative processes in connective tissue, such as mastopathies, contractures and keloids.

There have been reports in the literature of adjuvant effects of vitamin E in the treatment of hepatic disorders, in which the oxidative metabolism was improved and proliferation of the connective tissue was inhibited.

The beneficial effect of vitamin E on cell motility has also been successfully employed in improving the motility of spermatozoa.

Improvements in such pigment disorders as chloasma through the administration of vitamin E have been described in the literature.

More recent findings reveal a positive effect of vitamin E on premenstrual syndrome.

A placebo-controlled double blind study in patients with osteoarthritis showed a highly significant improvement of the pain associated with this disease. The highly dosed vitamin E therapy was characterized by antiphlogistic efficacy and at the same time excellent tolerance. Undesired side effects of vitamin E in combination with already existing standard treatments of rheumatic inflammatory joint diseases were not observed (1).

50 patients with osteoarthritis were randomely assigned to two groups and treated over a period of 6 weeks with vitamin E in a daily dose of 400 I.U. d-alpha-tocopherylacetate or the identical placebo. Vitamin E was superior to placebo with respect to the relief of pain at rest, pain during movement, and pressure-induced pain. Also the use of additional analgetic treatment was significantly reduced in the verum group. From the data of this study vitamin E has the potential to reduce the dose of standard antiphlogistic and analgetic treatment regimens.

Similar findings were obtained from another double blind study in which the clinical efficacy of vitamin E and Diclophenac were compared in patients with spondylitis ankylosans (Bechterew's disease). No significant difference could be found between high-dose oral antioxidant therapy by vitamin E and the traditional treatment regimen with the non steroidal antirheumatic remedy (2).

REFERENCES

(1) Blankenhorn, G.: Klinische Wirksamkeit von Vitamin E bei aktivierten Arthrosen. Z.Orthop. 124, 340-343 (1986).
(2) Klein, K.G., Blankenhorn, G.: Vergleich der klinischen Wirksamkeit von Vitamin E und Diclophenac-Natrium bei Spondylitis ankylosans (Morbus Bechterew). VitaMinSpur 2, 137-142 (1987).

VITAMIN E AND CORRELATED ANTIOXIDANTS:

A γ RADIOLYSIS STUDY

D. Jore, M.N. Kaouadji, and C. Ferradini

Laboratoire de Chimie Physique – Université René Descartes

45, rue des Saints Pères – F 75270 Paris Cedex 06 (France)

SUMMARY

γ irradiations of Vit.E-Vit.C aerated ethanolic solutions have been performed for several ratios (Vit.E)/(Vit.C) between 0.1 and 50. The obtained results show that Vit.C is able to regenerate Vit.E from its oxidized
radical, this regeneration being total for a ratio (Vit.E)/(Vit.C)$\geqslant$ 27 in
our conditions of irradiation. The ratio (Vit.E)/(Vit.C) seems to be the
main factor of this synergestic effect towards peroxyl radicals scavenging.

INTRODUCTION

α tocopherol (TH), major component of Vit.E (fig.1) is well known to represent the last possibility of preventing the membranes from peroxidation by
scavenging the LOO$^{\bullet}$ radicals involved in the peroxidation chains [1-2]. Vitamin E exhibits an OH phenolic group responsible of its antioxidant activity and a phytil side chan $C_{16}H_{33}$ which favors its location in the lipid
bilayer region. Numerous studies have been devoted to the monoelectronic
exchanges involves in this activity either in homogeneous solutions [3-4] or
in micellar systems [4-5]. In biological systems 1 Vit.E molecule seems able
to protect roughly 10^4 unsaturated fatty acids. Therefore the protective
effect of Vit.E could be enhanced by a synergestic effect of correlated
antioxidants. Tappel was the first to suggest a possible intervention of
Vit.C (AH_2), in the Vit.E radical mechanisms[6]. Such an interaction could
be presented by the following scheme :

LOO$^{\bullet}$ Vit.E   Vit.C   NADH

LOOH   Vit.E   Vit.C   NAD$^+$

Some papers were devoted to this effect in model systems[7-8]. We intended
to characterize quantitatively the Vit.E-Vit.C interaction using the
radiolysis method in aerated ethanolic medium, focusing on the influence
of the ratio Vit.E/Vit.C on the peroxyl radical scavenging efficiency.

Antioxidants in Therapy and Preventive Medicine
Edited by I. Emerit *et al.*
Plenum Press, New York, 1990

151

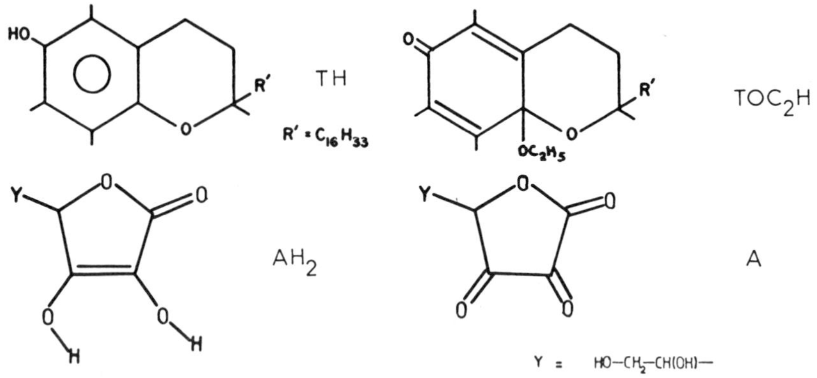

Fig. 1   Molecular structures : TH, TOC_2H_5, AH_2, A

MATERIAL AND METHODS

The ethanol used is Absolute Ethanol Normapur from Prolabo. α tocopherol, ascorbic acid and dehydroascorbic acid are Merck "for analysis" reagents. The γ irradiations have been performed with a ^{60}Co irradiator ($I = 1 \times 10^{18}$ eV.cm^{-3}.h^{-1}). The dosimetry has been determined by the Fricke's method. The doses have been used for the yields calculation (G values) without correction. The titration of the solutions has been made by spectrophotometry using a Beckman model 35 device.

$$\lambda \text{ max (TH)} = 292 \text{ nm} \qquad \varepsilon_{292} = 3.1 \times 10^3 \text{mol}^{-1}\text{l.cm}^{-1}$$
$$\lambda \text{ max (AH}_2) = 245 \text{ nm} \qquad \varepsilon_{245} = 8.5 \times 10^3 \text{mol}^{-1}\text{l.cm}^{-1}$$

RADIOLYSIS OF AERATED ETHANOLIC SOLUTIONS OF VIT.E AND VIT.C

Radiolysis of aerated ethanol provides homogeneous solutions of H_3C-CH(OH)OO$^\bullet$ (ROO$^\bullet$) model peroxyl radicals and of O_2^- superoxide anions with known yields [9] : $G(RO_2^\bullet) = 4.8$ molec/100eV $_-$ $G(O_2^-) = 1.7$ molec/100eV The radical oxidations of Vit.E and Vit.C have been studied in this way[10-11]. It could be shown that TH is oxidized by ROO$^\bullet$ radicals according to the following scheme :

(3) $\text{TH} + RO_2^\bullet \longrightarrow \text{T}^\bullet + \text{ROOH}$ $\qquad k_3 = 9.4 \times 10^4 \text{mol}^{-1}\text{l.s}^{-1}$

(4) $\text{T}^\bullet + RO_2^\bullet \xrightarrow{\text{H}^+} \text{T}^+ + \text{ROOH}$ $\qquad k_4 = 2.5 \times 10^6 \text{mol}^{-1}\text{l.s}^{-1}$

(5) $\text{T}^\bullet + \text{T}^\bullet \xrightarrow{\text{H}^+} \text{T}^+ + \text{TH}$ $\qquad k_5 = 1 \times 10^4 \text{mol}^{-1}\text{l.s}^{-1}$

(6) $\text{T}^+ + C_2H_5OH \longrightarrow TOC_2H_5 + \text{H}^+$

The oxidation yield G(TH) is equal to $G(RO_2^\bullet)/2 = 2.4$ molec/100eV for (TH) above 2×10^{-4} mol.l^{-1}. In the case of Vit.C[12] ROO$^\bullet$ and O_2^- oxidize AH$^-$ according to reactions 7 to 9

(7) $\text{AH}^- + RO_2^\bullet \longrightarrow \text{A}^{\overline{\bullet}} + \text{ROOH}$

(8) $\text{AH}^- + O_2^- \xrightarrow{\text{H}^+} \text{A}^{\overline{\bullet}} + H_2O_2$

(9) $\text{A}^{\overline{\bullet}} + \text{A}^{\overline{\bullet}} \xrightarrow{\text{H}^+} \text{AH}^- + \text{A}$

RADIOLYSIS OF VIT.E – VIT.C AERATED ETHANOLIC SOLUTIONS

γ irradiations of aerated TH-AH$^-$ ethanolic solutions have been performed

for several values of the ratio (TH)/(AH⁻) between 0.1 and 50. In all cases the evolution of the reaction is followed by observation at 292nm of
the optical density due to TH. This allows in each case the calculation
of the yield of oxidation of TH, G(TH). The obtained results are shown on
fig. 2. It can be observed that, according to the ratio (TH)/(AH⁻),G(TH)
increases, reaches a maximum for (TH)/(AH⁻)∿ 3 and decreases then to reach
zero for (TH)/(AH⁻) above 27.

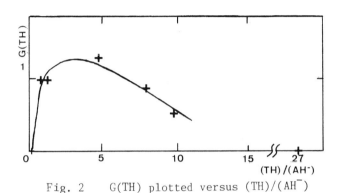

Fig. 2   G(TH) plotted versus (TH)/(AH⁻)

Ratio ⩾ 27 : In all those cases a dose lag is observed during which TH a-
mount is constant. This dose lag is proportionnal to (AH⁻) concentration.
After this lag, TH disappears with a yield G(TH) = 2.4 molec/100 eV equal
to $G(RO_2^{\bullet})/2$ (data not shown).
Ratio 1 : In this case, no dose lag could be observed. The value G(TH) is
then equal to 1 molec/100 eV for all experiments (see fig.2).
Ratio = 27 in the presence of A : In order to precise the possible influence of A in the medium due to reaction (9), we performed irradiations of
solutions with (TH)/(AH⁻)=27 and variable amounts of A between 7×10^{-8}
and 1×10^{-5} mol.l⁻¹. The results[13] show that the lag phase decreases according to (A) and disappears for (A)⩾1.2×10^{-6}mol.l⁻¹(data not shown).

DISCUSSION
The dose lag, proportionnal to (AH⁻), observed for TH/AH⁻ ⩾ 27 disappears
when the ratio decreases. This suggests that a competition between reactions (3) and (7) is not responsible of this lag phase. It can therefore
only be explained by reaction (10) which exhibits the regeneration of
Vit.E from its oxidized radical.When the ratio (TH)/(AH⁻) decreases,

(10) $T^{\bullet} + AH^{-} \longrightarrow TH + A^{\overline{\bullet}}$

another reaction such as (11) may compete with reaction (10)

(11) $T^{\bullet} + A \longrightarrow T^{+} + A^{\overline{\bullet}}$

The existence of reaction (11) is proved by the results obtained when A is added before irradiation, leading to the disappearance of the lag phase for $(TH)/(AH^-) = 27$. The calculation of $G(TH)$ for each value of the ratio $(TH)/(AH^-)$ was done by solving the system of differential equations corresponding to reactions (3) to (11). The following values :

$$k_7 = 1.2 \times 10^4 \, mol^{-1}.l.s^{-1} \qquad\qquad k_9 = 1 \times 10^2 \, mol^{-1}.l.s^{-1}$$
$$k_{10} = 5 \times 10^6 \, mol^{-1}.l.s^{-1} \qquad\qquad k_{11} = 2 \times 10^7 \, mol^{-1}.l.s^{-1}$$

gave the best agreement between experimental yield values and calculated curve as can be seen on fig.2.

CONCLUSION

This study proves that Vit.E can in fact be regenerated by Vit.C from its oxidized radical $(k_{AH^- + T^{\cdot}} = 5 \times 10^6 \, mol^{-1}.l.s^{-1})$. In our conditions of irradiations, this regeneration is total for a ratio $(TH)/(AH^-) \geqslant 27$. It seems that this ratio be the main factor involved in this synergestic effect.

REFERENCES

1. WITTIG, L.A., in Free radicals in Biology, Pryor, W.A., ed, Academic Press, New York, 4 (1980), 295.
2. BURTON, G.W., JOYCE, A. and INGOLD, K.U., Arch. Biochem. Biophys., 221 (1983), 281.
3. PACKER, J.E., SLATER, T.F., and WILLSON, R.L., Nature, 278 (1979),737.
4. SIMIC, M.G., in Oxygen and Oxy-Radicals in chemistry and Biology, Rodgers, M.A.J. and Powers E.L., eds, Academic Press,New York, (1981), 109.
5. PATTERSON, L.K., in Oxygen and Oxy-Radicals in chemistry and Biology, Rodgers,M.A.J. and Powers E.L.,eds,Academic Press,New York,(1981),89.
6. TAPPEL,A.L., Geriatrics, 23 (1968),97.
7. SCARPA, M., RIGO, A., MAIORINO, M., URSINI, F. and GREGOLIN, C., Biochim. Biophys. Acta, 801 (1984), 215.
8. LAMBELET, P. SAUCY, F. and LOLIGER, J., Experientia, 41 (1985), 1384.
9. FREEMAN, G.R., Radiation Chemistry of Ethanol, Nat. Stand. Ref. Data Serv., Nat. Bur. Stand. (U.S.) 48 (NSRDS-NBS 48), (1974).
10. JORE, D. and FERRADINI, C., Febs. Lett., 183 (1985), 299.
11. JORE, D., PATTERSON, L.K., and FERRADINI, C., J. Free Rad. Biol. Med., 2 (1986), 405.
12. KAOUADJI, M.N., JORE, D., PATTERSON L.K., and FERRADINI, C., Bioelectrochem. Bioenerg., 18 (1987), 59.
13. JORE, D., KAOUADJI, M.N., and FERRADINI, C., in Free radicals lipoproteins and Membrane lipids, NATO ASI Series, Crastes de Paulet, A., Douste-Blazy, L., and Paoletti, R., eds, (1989) (in press).

ASCORBATE: THE MOST EFFECTIVE ANTIOXIDANT IN HUMAN BLOOD PLASMA

Balz Frei, Roland Stocker,[*] Laura England, and Bruce N. Ames

Department of Biochemistry
University of California
Berkeley, CA 94720

INTRODUCTION

Living is like getting irradiated. This is because we are constantly exposed to oxidants such as superoxide radicals, hydrogen peroxide, hydroxyl radicals, and singlet oxygen. These reactive oxygen species are generated during normal oxidative metabolism, for example by spontaneous autoxidation of electron transport carriers in mitochondria, or as a result of the action of oxidases.[1] One of these oxidases, the NADPH oxidase of polymorphonuclear leukocytes (PMNs) (primarily neutrophils and eosinophils), is pivotal to the body's defense against pathogenic microorganisms. The immediate product of the stimulus-induced activation of the NADPH oxidase of PMNs is superoxide anion, whereas subsequent reactions form further oxidants including hydrogen peroxide, hypochlorite, and chloramines.[2,3] These oxidants not only kill the invading microorganisms, but also can cause considerable oxidative damage to the host himself. Other sources of oxidants to which we are constantly exposed include our diet, polluted air (particularly from smoking), natural radioactive gases, e.g. radon leaching from soils, and some drugs.[4,5]

One consequence of this life-long exposure to oxidants is peroxidative damage to lipids in cell membranes and lipoproteins. It is becoming increasingly evident that such peroxidative damage is relevant to many human diseases including atherosclerosis, cancer, rheumatoid arthritis, myocardial reoxygenation injury, and drug-associated toxicity,[6] as well as to the degenerative processes associated with aging. Preventing lipid peroxidation, therefore, could prove to be a very effective, yet simple, way of preventing these diseases and degenerative processes.

The body's defenses against oxidative stress include both, small antioxidant molecules and proteins. The proteins can act in

---

[*]Present address: Institute of Veterinary Virology, University of Berne, Länggass-Strasse 122, 3012 Berne, Switzerland.

Antioxidants in Therapy and Preventive Medicine
Edited by I. Emerit *et al.*
Plenum Press, New York, 1990

several ways: they either catalytically destroy oxidants (e.g. superoxide dismutase, catalase, glutathione peroxidase), or they scavenge oxidants in a sacrificial manner (e.g. albumin),[7] or they sequester transition metals in a way that prevents the metal ions from participating in free radical reactions (e.g. transferrin[8] and ceruloplasmin).[9] Human blood plasma, for example, contains a whole array of antioxidant defenses including ascorbate, urate, α-tocopherol, protein sulfhydryl groups, and bilirubin. We have chosen human blood plasma as a model to investigate the relative importance of these antioxidants in preventing lipid peroxidation under oxidative stress.[10]

EXPERIMENTAL DESIGN

Human blood plasma from healthy individuals was exposed to three types of oxidizing conditions: the water-soluble radical initiator 2,2'-azobis(2-amidinopropane) hydrochloride (AAPH), PMNs activated with phorbol 12-myristate 13-acetate, and the lipid-soluble radical initiator 2,2'-azobis(2,4-dimethylvaleronitrile) (AMVN). The radical-initiators, through thermal decomposition, generate peroxyl radicals at a constant rate in the aqueous phase and the lipids of plasma, respectively. Activated PMNs release a wide variety of oxidants (see above) following a single, so-called respiratory burst during which large amounts of superoxide radicals are generated by the activated NADPH oxidase.[2] The anti-oxidant defenses in human blood plasma under these three types of oxidizing conditions were investigated by measuring the temporal disappearance of endogenous ascorbate, urate, α-tocopherol, bilirubin, and sulfhydryl groups in relation to the appearance of various classes of hydroperoxides formed from endogenous lipids. Lipid peroxidation was measured with a sensitive and selective high-performance liquid chromatography/isoluminol chemiluminescence assay which detects the various classes of lipid hydroperoxides at plasma levels as low as 0.03 μM.[11,12]

ANTIOXIDANT DEFENSES IN HUMAN BLOOD PLASMA EXPOSED TO VARIOUS TYPES OF OXIDIZING CONDITIONS

Plasma Exposed to Aqueous Peroxyl Radicals

Exposure of plasma to aqueous peroxyl radicals generated at a constant rate of 3.0 μM/min leads to sequential depletion of endogenous ascorbate = sulfhydryl groups > bilirubin > urate > α-tocopherol within 300 min of incubation (Fig. 1).[10] Ascorbate is consumed completely within the first 50 min of incubation, a period during which no lipid hydroperoxides are formed in detectable amounts. (Normal human blood plasma does not contain detectable amounts of lipid hydroperoxides.)[12] After the complete consumption of ascorbate, micromolar concentrations of hydroperoxides of plasma phospholipids, cholesterol esters, and triglycerides appear simultaneously, even though sulfhydryl groups, bilirubin, urate, and α-tocopherol are still present at high concentrations (Fig. 1). Between 50 and 300 min of incubation, a period during which 750 μM aqueous peroxyl radicals are generated, a total of 273 μM lipid hydroperoxides are formed. Thus, during this period 36% of the peroxyl radicals generated react with lipids, whereas 64% are scavenged by the remaining antioxidants bilirubin, urate, sulfhydryl groups, and α-toco-

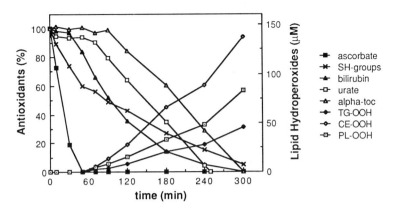

Fig. 1. *Antioxidant defenses and lipid peroxidation in human plasma exposed to the water-soluble radical initiator AAPH.* Plasma was incubated at 37°C in the presence of 50 mM AAPH. The levels of the antioxidants ascorbate (initial concentration 72 μM), sulfhydryl groups (SH-groups, 425 μM), bilirubin (18 μM), urate (225 μM), and α-tocopherol (alpha-toc, 32 μM) are given as % of the initial concentrations. The levels of the lipid hydroperoxides triglyceride hydroperoxide (TG-OOH), cholesterol ester hydroperoxide (CE-OOH), and phospholipid hydroperoxide (PL-OOH) are given in μM concentrations (right ordinate). From ref. 10.

pherol. Nonesterified fatty acids, the only lipid class in plasma not transported in lipoproteins but bound to albumin, are preserved from peroxidative damage even after the complete oxidation of ascorbate. We have shown previously that this is most likely due to site-specific antioxidant protection by albumin-bound bilirubin.[13]

The above results demonstrate that sulfhydryl groups, urate, and α-tocopherol lower considerably the rate of lipid peroxidation induced by aqueous peroxyl radicals, yet can not prevent formation of micromolar, *i.e.*, pathologically relevant,[6] concentrations of lipid hydroperoxides. In contrast, it appears that ascorbate can completely protect plasma lipids from detectable peroxidative damage. Indeed, in plasma deprived of endogenous ascorbate and urate (by gel filtration using a column centrifugation technique[14] that yields undiluted plasma) lipid peroxidation starts immediately upon generation of aqueous peroxyl radicals, despite the presence of protein sulfhydryl groups, albumin-bound bilirubin, and lipoprotein-associated α-tocopherol at levels unchanged by gel filtration. Adding back ascorbate, but not urate, restores the period of successful protection against detectable lipid peroxidation. During this period, the added ascorbate is consumed completely. Detectable lipid peroxidation is initiated immediately after the completion of ascorbate oxidation. Most importantly, addition of ascorbate to plasma after lipid peroxidation has already been initiated brings lipid peroxidation to a complete and transient standstill. In addition, depletion of the antioxidants bilirubin, urate, and α-tocopherol is also completely stopped by the addition of ascorbate. Only the sulfhydryl groups are not spared by ascorbate, most probably because they are oxidized by peroxyl radical-induced autoxidation. After complete consumption of the added ascorbate, all oxidation processes resume.

In summary, ascorbate is the only endogenous antioxidant in plasma that can completely protect the lipoproteins from detectable peroxidative damage induced by aqueous peroxyl radicals. Under this type of oxidative stress ascorbate is a much more effective antioxidant than bilirubin, urate, sulfhydryl groups, and α-tocopherol. Ascorbate appears to trap the peroxyl radicals in the aqueous phase with a rate constant large enough to intercept virtually all these radicals before they can diffuse into the plasma lipids. Once ascorbate has been consumed completely, the remaining water-soluble antioxidants urate, bilirubin, and the sulfhydryl groups can trap only part of the aqueous peroxyl radicals. The peroxyl radicals that escape the antioxidants in the aqueous phase diffuse into the plasma lipids, where they initiate lipid peroxidation. In contrast to the initiation step, propagation of peroxidation in the plasma lipids appears to be strongly inhibited, most likely by the lipid-soluble, chain-breaking antioxidant α-tocopherol. This is indicated by two observations: (i) α-tocopherol is consumed during lipid peroxidation, and (ii) lipid peroxidation comes to a complete standstill when ascorbate is added to plasma after lipid peroxidation has already been initiated (see above). It is important to note, however, that α-tocopherol, like urate and the sulfhydryl groups, can only lower the rate of detectable lipid peroxidation, but not completely prevent its initiation. Only ascorbate can do so. Finally, albumin-bound bilirubin appears to act more site-specifically, protecting albumin-bound nonesterified fatty acids from detectable peroxidative damage.

Plasma Exposed to Activated PMNs

Challenging plasma with oxidants released by activated PMNs leads to sequential depletion of ascorbate = sulfhydryl groups > urate, without depletion of bilirubin and α-tocopherol (Fig. 2).[10] Sulfhydryl groups (not shown) and urate (Fig. 2) are depleted only partially during the first 30 min of incubation. The cessation of their oxidation most probably reflects the cessation of oxidant production by the PMNs. Ascorbate is consumed immediately and very rapidly upon PMN activation. There is no detectable lipid peroxidation during ascorbate depletion. Once ascorbate has been consumed completely, detectable amounts of hydroperoxides of plasma phospholipids, cholesterol esters, and triglycerides are formed simultaneously, despite the presence of high concentrations of sulfhydryl groups, urate, bilirubin, and α-tocopherol (Fig. 2)

Thus, under this type of oxidative stress, too, ascorbate appears to be outstandingly effective and the only antioxidant in plasma capable of completely preventing detectable lipid peroxidation. This observation is relevant to and suggests the therapeutic use of ascorbate in inflammatory diseases and clinical conditions characterized by generalized PMN activation such as sepsis, shock, and trauma. It is particularly noteworthy that α-tocopherol is not consumed following activation of PMNs, despite

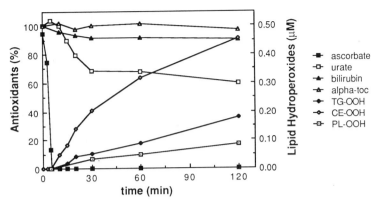

Fig. 2. *Antioxidant defenses and lipid peroxidation in human
plasma exposed to activated PMNs.* Plasma was incubated at 37°C with freshly prepared PMNs at ~1.3 x
10^7 cells/ml of plasma in the presence of 10 μg/ml of
cytochalasin B. The PMNs were stimulated by the addition of 6.8 μg/~10^7 cells of phorbol 12-myristate
13-acetate at time zero. The levels of the antioxidants ascorbate (initial concentration 58 μM),
bilirubin (12 μM), urate (152 μM), and α-tocopherol
(alpha-toc, 23 μM) are given as % of the initial concentrations. The ascorbate concentration prior to the
addition of PMNs was 43 μM. The levels of the lipid
hydroperoxides triglyceride hydroperoxide (TG-OOH),
cholesterol ester hydroperoxide (CE-OOH), and phospholipid hydroperoxide (PL-OOH) are given in μM concentrations (right ordinate). Note the difference in
the scales for lipid hydroperoxides between this
figure and Fig. 1. From ref. 10.

complete oxidation of ascorbate and initiation of lipid peroxidation. This observation not only excludes the possibility that
α-tocopherol is spared by ascorbate, but also indicates that
α-tocopherol is ineffective in scavenging the oxidants released by
PMNs.

Although very similar oxidation processes in plasma are
triggered by the aqueous peroxyl radicals and the activated PMNs,
some significant differences are observed. First, in the presence
of activated PMNs ascorbate depletion is much more rapid than with
the aqueous peroxyl radicals, despite a drastically lower rate of
subsequent lipid peroxidation in the presence of activated PMNs.
Second, as mentioned above, α-tocopherol does not become oxidized
following PMN activation. Third, bilirubin is not oxidized by the
oxidants released from PMNs despite partial depletion of urate,
whereas bilirubin is oxidized before urate by the aqueous peroxyl
radicals. These observations suggest that the effects of activated
PMNs on plasma antioxidants and lipids are not caused by aqueous
peroxyl radicals. A much more likely candidate is hypochlorite,
which is produced by activated PMNs[2] and which is known to be

scavenged by ascorbate[15] and urate,[16] but not albumin-bound bilirubin.[17]

Plasma Exposed to Lipid-Soluble Peroxyl Radicals

Challenging plasma with lipid-soluble peroxyl radicals generated at a constant rate leads to sequential depletion of ascorbate = α-tocopherol > bilirubin. No significant oxidation of urate and the sulfhydryl groups is observed (Fig. 3, lower panel).

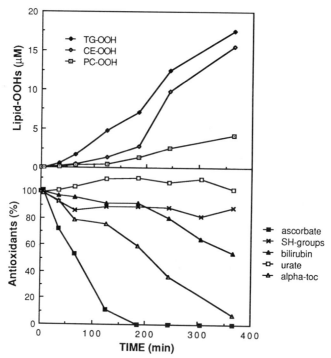

Fig. 3. *Antioxidant defenses and lipid peroxidation in human plasma exposed to the lipid-soluble radical initiator AMVN.* Plasma was incubated at $37^{\circ}C$ in the presence of 20 mM AMVN. The levels of the lipid hydroperoxides (lipid-OOHs) triglyceride hydroperoxide (TG-OOH), cholesterol ester hydroperoxide (CE-OOH), and phospholipid hydroperoxide (PC-OOH) are given in μM concentrations (upper panel).The levels of the antioxidants ascorbate (initial concentration 94 μM), sulfhydryl groups (SH-groups, 483 μM), bilirubin (6.0 μM), urate (335 μM), and α-tocopherol (alpha-toc, 32 μM) are given as % of the initial concentrations (lower panel).

Ascorbate is consumed more rapidly than α-tocopherol, possibly reflecting sparing of α-tocopherol in the lipids by interaction with ascorbate in the aqueous phase. Such interaction obviously does not take place between α-tocopherol and urate or the protein sulfhydryl groups, nor are these two water-soluble antioxidants able to otherwise intercept the peroxyl radicals formed in the lipids. Apparently, this does not hold true for bilirubin, which becomes oxidized after the complete consumption of ascorbate (Fig. 3).

In marked contrast to the above two types of oxidative stress (aqueous peroxyl radicals and activated PMNs), generation of peroxyl radicals in the lipids of plasma immediately and unsparingly induces lipid peroxidation (Fig. 3, upper panel). There is no lag phase preceding detectable lipid peroxidation due to antioxidant protection by ascorbate (c.f. Figs. 1 and 2). The lipid hydroperoxides formed by the lipid-soluble peroxyl radicals are primarily neutral lipid hydroperoxides (triglyceride hydroperoxides and cholesterol ester hydroperoxides), and only relatively few phospholipids become peroxidatively damaged (Fig. 3). In contrast, the relative amounts of the lipid hydroperoxide classes formed in the presence of aqueous peroxyl radicals (Fig. 1) approximately reflect the distribution of polyunsaturated fatty acyl side chains in the corresponding parent lipid classes.[18]

Again, it is interesting to note that α-tocopherol, considered the major lipid-soluble antioxidant in plasma,[19] is not able to prevent formation of micromolar, i.e., pathologically relevant,[6] concentrations of lipid hydroperoxides. For example, after consumption of 50% (or 16 μM) of endogenous α-tocopherol, about 14 μM lipid hydroperoxides have already been formed (Fig. 3). This observation, together with those reported above, suggests that α-tocopherol is relatively ineffective in preventing peroxidative damage to lipids in human blood plasma. It is conceivable that the main target of lipid peroxidation in plasma are not the lipids incorporated into the lipoproteins but those transported by transfer proteins between lipoproteins. The lipids bound to these transfer proteins might not be associated with α-tocopherol and thus would be expected to be more vulnerable to oxidative stress.

SUMMARY

Ascorbate is the only endogenous antioxidant in plasma that can completely protect the lipoproteins from detectable peroxidative damage induced by aqueous peroxyl radicals and the oxidants released from activated PMNs. In contrast to aqueous oxidants, lipid-soluble peroxyl radicals unsparingly induce detectable peroxidative damage to plasma lipids. However, under these conditions, too, ascorbate appears to belong to the first line of antioxidant defense.

Our findings strongly suggest that pathologically relevant lipid hydroperoxide formation consequent to acute or chronic leukocyte activation can be prevented by ascorbate supplementation, provided no free metal catalysts are present. Ascorbate should prove very helpful in the treatment and prevention of diseases and degenerative processes caused by oxidative stress.

ACKNOWLEDGMENTS

This work was supported by National Cancer Institute
Outstanding Investigator Grant CA39910 to B.N.A. and by National
Institute of Environmental Health Services Center Grant ES01896.
B.F. was supported in part by the Swiss National Foundation.

REFERENCES

1. B. Chance, H. Sies, and A. Boveris, Hydroperoxide metabolism
   in mammalian organs, *Physiol. Rev.* 59:527 (1979).
2. S. J. Klebanoff, Phagocytic cells: products of oxygen
   metabolism, *in*: "Inflammation: Basic Principles and
   Clinical Correlates," pp. 391, J. I. Gallin, I. M. Goldstein, and R. Snyderman, eds., Raven Press, New York,
   (1988).
3. S. J. Weiss, M. B. Lampert, and S. T. Test, Long-lived
   oxidants generated by human neutrophils: characterization
   and bioactivity, *Science* 222:625 (1983).
4. B. N. Ames, Dietary carcinogens and anticarcinogens, *Science*
   221:1256 (1983).
5. B. N. Ames, R. Magaw, and L. S. Gold, Ranking possible
   carcinogenic hazards, *Science* 236:271 (1987).
6. H. Esterbauer and K. H. Cheeseman, eds., "Lipid Peroxidation:
   Part II. Pathological Implications," *Chem. Phys. Lipids*
   45, Nos. 2-4 (1987).
7. B. Halliwell, Albumin-an important extracellular antioxidant?
   Biochem. Pharmacol. 37:569 (1988).
8. O. I. Aruoma and B. Halliwell, Superoxide-dependent and
   ascorbate-dependent formation of hydroxyl radicals from
   hydrogen peroxide in the presence of iron. Are lactoferrin
   and transferrin promoters of hydroxyl-radical generation?
   Biochem. J. 241:273 (1987).
9. J. M. C. Gutteridge, Antioxidant properties of caeruloplasmin
   towards iron- and copper-dependent oxygen radical
   formation, *FEBS Lett.* 157:37 (1983).
10. B. Frei, R. Stocker, and B. N. Ames, Antioxidant defenses and
    lipid peroxidation in human blood plasma, *Proc. Natl.
    Acad. Sci. USA* 85:9748 (1988).
11. Y. Yamamoto, M. H. Brodsky, J. C. Baker, and B. N. Ames,
    Detection and characterization of lipid hydroperoxides at
    picomole levels by high-performance liquid chromatography,
    Anal. Biochem. 160:7 (1987).
12. B. Frei, Y. Yamamoto, D. Niclas, and B. N. Ames, Evaluation of
    an isoluminol chemiluminescence assay for the detection of
    hydroperoxides in human blood plasma, *Anal. Biochem.*
    175:120 (1988).
13. R. Stocker, A. N. Glazer, and B. N. Ames, Antioxidant activity
    of albumin-bound bilirubin, *Proc. Natl. Acad. Sci. USA*
    84:5918 (1987).
14. E. Helmerhorst and G. B. Stokes, Microcentrifuge desalting: a
    rapid, quantitative method for desalting small amounts of
    protein, *Anal. Biochem.* 104:130 (1980).
15. B. Halliwell, M. Wasil, and M. Grootveld, Biologically
    significant scavenging of the myeloperoxidase-derived
    oxidant hypochlorous acid by ascorbic acid. Implications
    for antioxidant protection in the inflamed rheumatoid
    joint, *FEBS Lett.* 213:15 (1987).

16. M. Grootveld, B. Halliwell, and C. P. Moorhouse, Action of uric acid, allopurinol and oxypurinol on the myeloperoxidase-derived oxidant hypochlorous acid, *Free Rad. Res. Comms.* 4:69 (1987).
17. R. Stocker, A. Lai, E. Peterhans, and B. N. Ames, Antioxidant properties of bilirubin and biliverdin, *in*: "Medical, Biochemical and Chemical Aspects of Free Radicals," E. Niki and T. Yoshikawa, eds., Elsevier, Amsterdam, in press.
18. C. Lentner, "Geigy Scientific Tables," p. 122, Ciba-Geigy Limited, Basle (1984)
19. G. W. Burton, A. Joyce, and K. U. Ingold, First proof that vitamin E is the major lipid-soluble, chain-breaking antioxidant in human blood plasma, *Lancet* 2:327 (1982).

RADICAL CHEMISTRY OF FLAVONOID ANTIOXIDANTS

Wolf Bors, Werner Heller[§], Christa Michel and Manfred Saran

GSF Forschungszentrum
Institut für Strahlenbiologie and
Institut für Biochemische Pflanzenpathologie[§]
D-8042 Neuherberg, FRG

INTRODUCTION

Flavonoid aglycones, members of an ubiquitous class of plant phenols, have often been proposed to act as antioxidants.[1,2] More recently this activity has been specifically attributed to their radical-scavenging capabilities.[3-8] Compounds of various structural features have already been tested,[4,7,9,10] but only qualitative conclusions could be drawn.

Selective generation of individual radical species by pulse radiolysis and photolysis recently allowed the determination of rate constants of both radical attack and decay of the flavonoid-derived aroxyl radicals.[11-13] The spectral parameters of these aroxyl radicals have also been reported.[11] From these data we were able, for the first time, to derive a structure-activity relationship, which describes the structural requirements for optimal antioxidative and/or radical-scavenging efficiency.[14]

The present study, which includes two chromane model compounds, is an extension of these investigations, and it corroborates our original hypothesis on the kinetic behavior of flavonoid aroxyl radical derivatives.

MATERIALS AND METHODS

The compounds used in these experiments were 5,7-dihydroxy-4-chroman-on ($\underline{1}$), 5,7-dihydroxychromone ($\underline{2}$), 5,7,4'-trihydroxyflavanone (naringenin, $\underline{3}$) and 3,5,7,4'-tetrahydroxyflavanone (dihydrokaempferol, $\underline{4}$). The scheme also contains the structures of two other flavonoids, 5,7-dihydroxy-4'-methoxyflavone (acacetin, $\underline{5}$) and 3,5,7,3',4'-pentahydroxyflavone (quercetin, $\underline{6}$), which have been studied earlier[11] but are pertinent for the discussion.

Substances $\underline{1}$ and $\underline{2}$ were synthesized according to Heller et al.,[15] $\underline{3}$, $\underline{5}$ and $\underline{6}$ were from commercial sources and $\underline{4}$ was isolated from flowers of stock (<u>Alcea</u> <u>rosea</u>).[16]

Determination of $\underline{t}$-BuO· attack was assayed in the crocin system.[17] From the reaction with pulse-radiolytically generated azide (·N$_3$) radi-

Antioxidants in Therapy and Preventive Medicine
Edited by I. Emerit *et al.*
Plenum Press, New York, 1990

Scheme I - Structures of investigated compounds

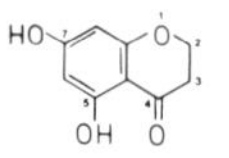

1

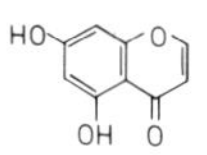

2

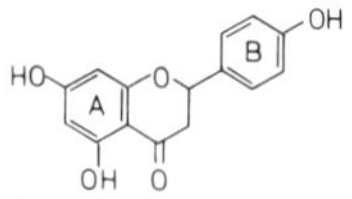

3

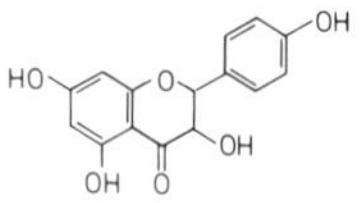

4

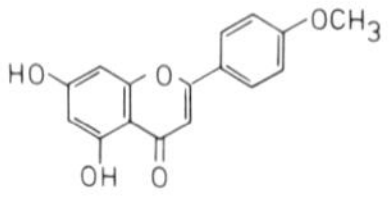

5

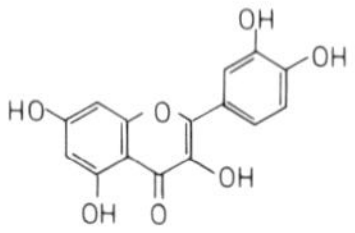

6

cals in alkaline aqueous solution we obtained the transient spectra and the kinetic data on formation and decay of the aroxyl radicals.[11]

RESULTS

Table I lists the rate constants for aroxyl radical formation of flavonoids and model compounds by the attack of the two electrophilic radicals t-BuO· and ·N3. Also listed are the spectral parameters of the aroxyl radicals used to determine the decay rate constants which, except for compound 4, reflected second-order processes.

Figure 1 shows the transient spectra of the aroxyl radicals formed at pH 11.3-11.5 after the attack of ·N3 radicals. The spectra of compounds 1/4 and 2/5 are grouped in pairs to demonstrate the close similarities. No such similarity exists for the spectra of naringenin (3) and quercetin (6).

DISCUSSION

Optimal antioxidative capacity of a given substance depends on two principles: (i) high absolute reactivity with radicals of different origin and (ii) relatively high stability of the intermediately formed 'antioxidant radical' to prevent participation in chain processes (a second-order decay in effect points to a dismutation reaction of two aroxyl radicals with each other). In a previous study we proposed that for flavonoids the following structural features should be responsible for optimal antioxidative function:[14]

- the o-dihydroxy or catechol group in the B-ring confers a high stability to the aroxyl radicals, possibly through hydrogen bonding;

Table I - Kinetic and spectral parameters of flavonoid aroxyl radicals

Substance	Rate Constant (10^{-8} $M^{-1}s^{-1}$)			Spectral Data[a]	
	t-BuO·	·N_3	decay (2k)	ε ($M^{-1}cm^{-1}$)	λ (nm)
1	0.21	36	6.96	5.300	285
2	0.96	39	2.26	4.000	285
3	2.67	52	7.72[b]	5.350	280
4	0.93	55	c	4.200	270
5[d]	1.30	28	5.00	6.600	325
6[d]	25.0	66	0.034	15.600	530

The data for t-BuO· were obtained after photolytic generation of these radicals and evaluation of the competitive inhibition of the bleaching of crocin.[17] All remaining kinetic and spectral data were obtained from pulse-radiolytic experiments.[11]

a values of peak absorption
b slow apparent second-order bleaching at 430 nm with k/ε = 5.5x10^4 cm/s (see discussion)
c first-order decay
d data from ref. 11

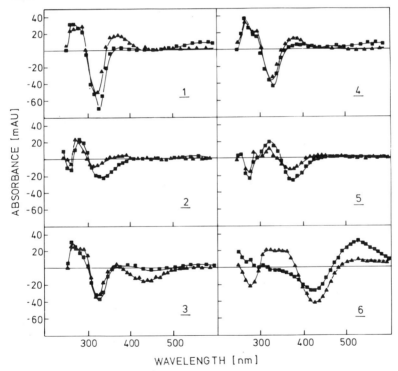

Figure 1 - Transient spectra of flavonoid aroxyl radicals

Dose-normalized transient spectra in alkaline aqueous solution, pH 11.3-11.5 (1 at pH 10.3), saturated with N_2O and containing 10 mM NaN_3. Concentration of substrates: 2, 4 - 30 /uM, 3, 5 - 38 /uM, 1, 6 - 45 /uM. (■) initial observation at 5 /us (5, 6) or 17.5 /us (1 - 4) after the pulse, (▲) final observation at 26 ms (5), 95.6 ms (1 - 4) or 360 ms (6). abscissa: wavelength in nm, ordinate: mAU per 10 Gy pulse dose.

- conjugation of the B-ring to the 4-oxo structure via a 2,3-double bond ensures extensive electron delocalization;
- hydrogen bonds between the 4-oxo and the 3- and 5-hydroxy groups allow electron delocalization from the oxo function to both substituents;
- highest electron delocalization is achieved by combination of all of these structural elements.

According to this hypothesis, quercetin (6) should be the most efficient flavonoid antioxidant.

In the present study we used four flavonoid model compounds, either devoid of the B-ring (1, 2) or lacking an o-dihydroxy or catechol structure in the B-ring (3, 4). Except for 2, all substances are saturated at carbons 2 and 3. Looking at the spectral and kinetic data, the results confirm our previous hypothesis for a structure-activity correlation.[14] As predicted, lack of a catechol structure in the B-ring confers high instability to all aroxyl radicals (dihydrokaempferol, 4, rapidly decays in a first-order process). A similar instability is observed with the chromane system which is devoid of a B-ring. Formation of these radicals is generally quite rapid with $\cdot N_3$ radicals, whereas t-BuO$\cdot$ radicals exhibit a broader range of rate constants, extending over two orders of magnitude. Radical formation with the flavanone naringenin (3) is almost three times faster than with the respective 3-hydroxy derivative dihydro-kaempferol (4). This result clearly shows that an **aliphatic** hydroxy group in 3 position does not contribute substantially to the redox behavior of flavanones - an important extension of our original hypothesis.

Comparing the transient spectra of the aroxyl radicals studied, the closest similarities exists between compound 1 and 4 as well as 2 and 5. According to its structure, naringenin (3) was expected to give spectra similar to 1 and 4. However, it exhibits a unique behavior which will be discussed in more detail.

It is obvious that saturation of carbons 2 and 3 blocks electron delocalization between the B-ring and the 4-oxo group. The individual chromophores in rings A and B absorb at similar wavelengths, however, the absorption of the 4-chromanone structure is much stronger than that of the B-ring. Lack of a dissociable 4'-hydroxy group also masks the spectral influence of the B-ring even in the presence of the conjugated 2,3-double bond. Consequently the transient spectrum of 5 is very similar to that of the chromone 2. In the case of quercetin (6), where maximal electron delocalization is possible, the major absorption band of the aroxyl radical is shifted to considerably higher wavelengths.

The transient spectrum of the naringenin (3) aroxyl radical shows a slow bleaching at 430 nm of apparent second order which is not observed with dihydrokaempferol (4), the corresponding 3-hydroxy derivative, or any of the other flavonoids studied.[11] This peculiar behavior can only be explained by taking into consideration that some flavanones undergo a base-catalyzed opening of the heterocyclic ring and exist in a pH-dependent equilibrium with the respective chalcones:[18]

We are currently investigating chalcone/flavanone model systems in which the equilibrium is shifted either towards the flavanone or the chalcone structure to verify this interpretation.

In conclusion, studying two model chromane structures, which are devoid of the flavonoid B-ring, and two flavanones lacking a B-ring catechol structure, we were able to confirm our original hypothesis on the structural requirements for efficient antioxidative behavior. Our experimental procedures were to react the phenolic compounds with the quite discriminating t-BuO· radical and the highly reactive and therefore less discriminating ·N$_3$ radicals. This approach, using selective photolytic (t-BuO·) or pulse-radiolytic (·N$_3$) radical generation is straight forward and superior to unspecific radical sources which allow only qualitative studies.

REFERENCES

1. J. Kuehnau, The flavonoids. A class of semi-essential food components: their role in human nutrition, Wld. Rev. Nutr. Diet. 24:117 (1976).
2. L.R. Dugan, Natural antioxidants, in: "Autoxidation in Food and Biological Systems," M.G. Simic, M. Karel, eds., Plenum Press, New York, p.261 (1980).
3. J. Baumann, G. Wurm, and F. von Bruchhausen, Hemmung der Prostaglandinsynthetase durch Flavonoide und Phenolderivate im Vergleich mit deren Superoxid-Radikalfängereigenschaften, Arch. Pharm. 313:330 (1980).
4. M. Damon, F. Michel, C. Le Doucen, and A. Crastes de Paulet, Action des flavonoides sur la liberation des espèces oxygénéés hautement réactives par les polymorphonucléaires. Etude par chimioluminescence, Bull. Liaison - Grp. Polyphenols 13:569 (1986).
5. J. Torel, J. Cillard, and P. Cillard, Antioxidative activity of flavonoids and reactivity with peroxy radicals, Phytochemistry 25:383 (1986).
6. S.R. Husain, J. Cillard, and P. Cillard, Hydroxyl radical scavenging activity of flavonoids, Phytochemistry 26:2489 (1987).
7. U. Takahama, Oxidation products of kaempferol by superoxide anion radical, Plant Cell Physiol. 28:953 (1987).
8. J. Robak, and R.J. Gryglewski, Flavonoids are scavengers of superoxide anions, Biochem. Pharmacol. 37:837 (1988).
9. B.J.F. Hudson, and J.I. Lewis, Polyhydroxy flavonoid antioxidants for edible oils. Structural criteria for acvtivity, Food Chem. 10:47 (1983).
10. A.K. Ratty, and N.P. Das, Effects of flavonoids on non-enzymatic lipid peroxidation: structure-activity relationship, Biochem. Med. Metab. Biol. 39:69 (1988).
11. W. Bors, and M. Saran, Radical scavenging by flavonoid antioxidants, Free Rad. Res. Comm. 2:289 (1987).
12. M. Erben-Russ, C. Michel, W. Bors, and M. Saran, Absolute rate constants for alkoxyl radical reactions in aqueous solution, J. Phys. Chem. 91:2362 (1987).
13. M. Erben-Russ, W. Bors, and M. Saran, Reactions of linoleic acid peroxyl radicals with phenolic antioxidants: a pulse radiolysis study, Int. J. Radiat. Biol. 52:393 (1987).
14. W. Bors, W. Heller, C. Michel, and M. Saran, in: "Oxygen Radicals in Biological Systems," Meth. Enzymol., L. Packer, A.N. Glazer, eds., Academic Press, Orlando, FL, in press (1989).
15. W. Heller, P. Andermatt, W.A. Schaad, and C. Tamm, Homoisoflavanone. IV. Neue Inhaltsstoffe der Eucomin Reihe von Eucomis bicolor, Helv. Chim. Acta 59:2048 (1976).

16. W. Heller, L. Britsch, G. Forkmann, and H. Grisebach, Leucoanthocyani-
    dins as intermediates in anthocyanidin biosynthesis in flowers of
    Matthiola incana R. Br., Planta 163:191 (1985).
17. W. Bors, C. Michel, and M. Saran, Inhibition of the bleaching of the
    carotenoid crocin. A rapid test for quantifying antioxidant acti-
    vity, Biochim. Biophys. Acta 796:312 (1984).
18. T.R. Sehadri, Interconversion of flavonoid compounds, in: "Chemistry
    of Flavonoid Compounds," T.A. Geissmann, ed., McMillan, New York,
    p.156 (1962).

SCAVENGING EFFECTS OF ASPALATHUS LINEALIS (ROOIBOS TEA) ON ACTIVE OXYGEN

SPECIES

T.YOSHIKAWA, Y.NAITO, H.OYAMADA, S.UEDA, T.TANIGAWA, T.TAKEMURA
S.SUGINO, and M.KONDO

First Department of Medicine, Kyoto Prefectural University of
Medicine, Kamigyo-ku, Kyoto 602, Japan

INTRODUCTION

Rooibos tea ia a totally unique South African product of the plant
Aspalathus linealis which is only produced in the Cadarberg mountains around
Clanwilliam. In South Africa it is mainly used as a substitute for the
Oriental black tea by people who enjoy it either hot or cold, or by those who
regard it as a healthy drink. Clinically, Rooibos tea is often prescribed
for nervous tension, allergies, stomach and digestive problems. For
evaluation of its antioxidant action, reactivity of Aspalathus linealis,
ascorbic acid, and quercetin which is one of non-glycosidically linked
flavonoids included in Asparathus linealis, to various reactive oxygen
species were assessed by electron spin resonance (ESR) spectrometry using
5,5-dimethyl-1-pyrroline-N-oxide (DMPO) as a spin trapper[1,2].

MATERIALS AND METHODS

Chemicals
    DMPO, xanthine oxidase, phorbol myristate acetate (PMA), and quercetin
were purchased from Sigma Chemical Company (St.Louis, MO).
Diethylenetriaminepentaacetic acid (DETAPAC), ascorbic acid, ferrous sulfate,
hydrogen peroxide, and hypoxanthine were obtained from Wako Chemical Indust.
Ltd. (Osaka). Ficoll-Paque was purchased from Pharmacia LKB (Uppsala).
Asparathus linealis was a gift from HF international Company (Osaka).
Freeze-dried water extraction of Asparathus linealis was produced by
following method. Mix it and water in the ratio of 1 : 5 and heat the
substances up to 95 °C. After leaving it for a while, take its supernatant,
which is to be filtrated and concentration.

Preparation of polymorphonuclear leukocytes (PMNs) suspension
    Human PMNs were isolated from heparinized venous blood of healthy
volunteers by dextran sedimentation followed by Ficoll-Paque separation and
hypotonic lysis of contaminating erythrocytes. PMNs preparations, which
contained 95-99% PMNs, were suspended in Hanks' balanced salt solution (HBSS)
at pH 7.4.

Superoxide and hydroxyl radicals generating system
    Superoxide anions were generated in the hypoxanthine-xanthine oxidase
(HX-XO) system and the PMA-stimulated PMN (PMA-PMN) system. The former

Antioxidants in Therapy and Preventive Medicine
Edited by I. Emerit *et al.*
Plenum Press, New York, 1990

comprised a solution of 0.5 mM of hypoxanthine, 0.1 mM of DETAPAC, 0.1 M of DMPO, and 0.05 U/ml of xanthine oxidase in 50 mM of phosphate buffer at pH 7.4. This mixture was incubated at 37°C for 2 min and then the record of ESR spectra started. The generation of superoxide from PMA-PMN system was measured in samples which contained 1.0×10^6/ml of PMN in HBSS, 0.1 mM of DETAPAC, 0.1 M of DMPO, and 400 ng/ml of PMA. PMNs suspension was preincubated with agent for 3 min, and then ESR spectra were obtained 2.5 min after stimulation.

Hydroxyl radicals were generated in the hydrogen peroxide-ferrous sulfate system (Fenton reaction). Spin trapping of hydroxyl radicals was undertaken by mixing 0.05 mM of ferrous sulfate, 0.01 or 0.001 M of DMPO, 0.1 mM of DETAPAC, 0.001 mM of hydrogen peroxide and 50 mM of phosphate buffer to reach a final volume of 0.2 ml.

ESR spectrometry

Reaction mixture were transferred to an ESR quartz cell and placed into the cavity of ESR spectrometer (JEOL-JES-FR80, JEOL Ltd., Tokyo), and the relative intensity of the signal of DMPO-OOH or DMPO-OH spin adduct was measured as the ratio to the intensity of Mn^{2+} signal. ESR spectra were recorded at 37°C with a field set 335.0+5.0 mT, modulation frequency 100 kHz, modulation amplitude 0.1 mT, sweep time 5 mT/min, response time 0.03 sec, and microwave power 8 mW when superoxide radicals were trapped, and 6 mW when trapped hydroxyl radicals.

RESULTS

Reactivity to superoxide anion radical

The relative intensity of DMPO-OOH spin adduct, generated either from HX-XO system or from PMA-PMN system, can be inhibited by Asparathus linealis : from HX-XO system, with 20.6 µg/ml concentration of IC_{50}, while from PMA-PMN system, with 20.1 µg/ml concentration of IC_{50}. The similar results could be expected in the case of inhibition of relative intensity of DMPO-OOH spin adduct by ascorbic acid : from HX-XO system, 20.1 µM concentration of IC_{50}, while from PMA-PMN system, with 20.4 µM concentration of IC_{50}. However, in the case of inhibition of the intensity of DMPO-OOH by quercetin, there was an evident difference in the concentration of IC_{50} : DMPO-OOH spin adduct from HX-XO system, with 16 µM concentration of IC_{50}, while DMPO-OOH adduct from PMA-PMN system, with 46 µM concentration of IC_{50}.

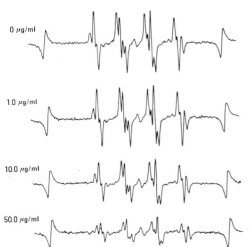

Fig.1
Effect of Aspalathus linealis on the ESR signal obtained from the hypoxanthine-xanthine oxidase system in the presence of 0.1 M of DMPO.

172

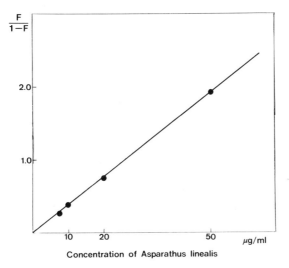

Concentration of Asparathus linealis

Fig.2   Relationship between the concentration of Aspalathus linealis and F/1-F. F indicates inhibition rate of the signal intensity of DMPO-OOH spin adduct in the presence of extracts of Aspalathus linealis.

Reactivity to hydroxyl radicals

Asparathus linealis inhibited the relative intensity of DMPO-OH spin adduct generated from ferrous sulfate-hydrogen peroxide system with 24.0 μg/ml concentration of IC_{50}. Quercetin showed no effect on inhibition of the relative intensity of DMPO-OH spin adduct with 0.001 M concentration of DMPO. The relative intensity of DMPO-OH spin adduct with 0.01 M concentration of DMPO was inhibited by the presence of 0.1 mM ascorbic acid, however, with 0.001 M DMPO, it was not inhibited by 0.001 mM ascorbic acid.

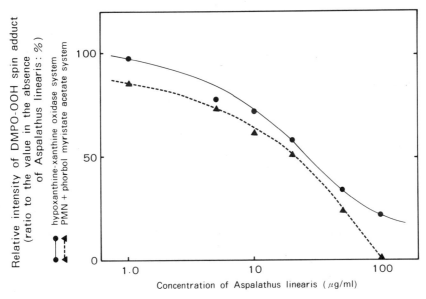

Fig. 3   Effect of Asparathus linealis on the ESR signal intensity of DMPO-OOH spin adduct generated from the hypoxanthine-xanthine oxidase system or PMA-stimulated PMNs system.

173

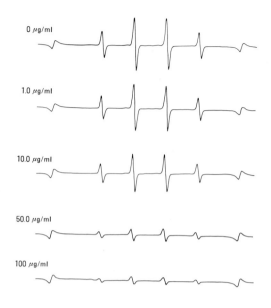

0 μg/ml

1.0 μg/ml

10.0 μg/ml

50.0 μg/ml

100 μg/ml

Fig.4
Effect of Aspalathus linealis on the ESR signal of DMPO-OH spin adduct generated from the hydrogen peroxide-ferrous sulfate reaction system in the presence of 0.01 M DMPO.

DISCUSSION

These studies demonstrated that extracts from Aspalathus linealis and ascorbic acid inhibited the DMPO-OOH signal intensity generated from HX-XO system as same degree as that from PMA-PMN system. This result suggests that the extracts and ascorbic acid scavenges superoxide radicals, and do not affect the respiratory burst activity of human polymorphonuclear leukocytes. It has been reportedly that quercetin, a major active component of Aspalathus linealis, is an inhibitor of xanthine oxidase[3], therefore, the superoxide scavenging activity of quercetin is not assessed exactly by our ESR assay system.

The relative intensity of DMPO-OH signal generated from Fenton reaction was inhibited by the presence of extracts of Aspalathus linealis, and not by quercetin. Hydroxyl radical scavenging activity of ascorbic acid was not estimated by our method, because the inhibition of signal intensity of DMPO-OH was caused not by competitive inhibition of ascorbic acid with DMPO, but by inhibition of the ferrous-hydrogen peroxides reaction.

REFERENCES

1)H.Miyagawa, T.Yoshikawa, T.Tanigawa, N.Yoshida, S.Sugino, M.Kondo, H.Nishikawa, and M.Kohno : Measurement of serum superoxide dismutase activity by electron spin resonance. J.Clin.Biochem.Nutr., 5: (1988).
2)T.Tanigawa, T.Yoshikawa, H.Miyagawa, S.Ueda, T.Takemura, K.Tainaka, Y.Morita, K.Itani, N.Yoshida, S.Sugino, and M.Kondo : Determination of superoxide generated by human polymorphonuclear leukocytes by ESR and its clinical application. Jap.J.Inflammation., 8: 443 (1988).
3)J.Robak and R.J.Gryglewski : Flavonoids are scavengers of superoxide anions. Biochem.Pharmacol., 37:837 (1988).

SELENIUM AND OXIDANT INJURY IN PATIENTS WITH CYSTIC FIBROSIS

M.J. RICHARD, B. AGUILANIU, J. ARNAUD,
J.P. GOUT, A. FAVIER

Laboratoire de Biochimie C, Pavillon D,
Service de Pédiatrie
C.H.R.U.G., BP 217X, 38043 Grenoble Cedex
France

INTRODUCTION

Malnutrition in cystic fibrosis (CF) has been associated with poor dietary intake, maldigestion and malabsorption of food, and increased nutrient requirements secondary to chronic infection[1]. Recent studies have suggested that nutritional status influenced the progression of disease and nutritional supplementation was frequently included in the comprehensive management of this desease. But, children with exocrine pancreatic insufficiency, as observed in CF, might have secondary deficiency of the essential micronutrients specially Selenium (Se) deficiency[2,3].

The present study was designed to investigate the relationship between nutritional status of trace elements in CF and other anti-oxidant micronutrients, anti-oxidant metalloenzymes and with lipid peroxidation products. On the basis of these data a Se supplementation was performed in four CF malnoutrished patients. The purpose of this second investigation was to find the amount of Se which will be effective in restoring normal serum Se level, and normalis activity of erythrocyte glutathione peroxidase (GPX).

PATIENTS and METHODS

First Study. Fourteen CF patients, aged 2 to 19 years, differing in the severity of disease, were studied and compared to age matched controls. Blood was collected in heparinized, trace element free tubes. Plasma trace elements : selenium, zinc (Zn), copper (Cu), and manganèse (Mn) were measured by direct electrothermal atomic absorption spectrometry. Plasma and erythrocyte Se GPX were determined according to Gunzler et al[4]. Vitamins E and A were assayed ,after specific extraction, by a fluorometric method. Plasma caroten was measured spectrophotometrically after cyclohexan extraction. Malondialdehyde (MDA) was measured by a fluorometric micromethod using thiobarbituric acid assay[5]. Lipid hydroperoxides (HPO) were determined by an enzymatic method[6]. For statistical analysis the Student's t-test was used.

Antioxidants in Therapy and Preventive Medicine
Edited by I. Emerit et al.
Plenum Press, New York, 1990

175

Second Study. Four CF patients, aged 9 – 19 years, who had been studied before, were treated. Selenium was given orally as sodium selenite in a daily dose of 200 µg during three months. The red blood cell GPX , the plasma GPX and MDA levels were measured before and during the treatment. Each patient was his own control.

RESULTS

First study

Plasma Se levels were significantly decreased in CF patients compared with the controls (0.44_0.22 vs 0.97_0.17 µmol.l⁻¹; p<0.001).(tab. 1). This deficiency was confirmed by both plasma GPX and erythrocyte GPX activities that were significantly decreased (p<0.001).We found a correlation between plasma Se and erythrocyte GPX (r=0.459,p<0.02). Despite oral supplementation , all patients had low serum retinol and tocopherol concentrations (p<0.001;p<0.02). Our study confirmed a real deficiency in caroten (p<0.001). The levels of all antioxidant vitamins were correlated with red blood cells GPX activity (Vit E: r=0.572,p<0.001; Vit A: r=0.490,p<0.01; caroten: r=0.638,p<0.001). Moreover MDA and HPO levels were significantly higher than those in the controls (p<0.001).

Table 1   The level of lipoperoxide and the antioxidant system in cystic fibrosis patients. Results are expressed as mean ± 1 standard deviation.

				CF Patients (n = 14)	Controls (n = 30)	
T	E	Selenium	µmol.l⁻¹	0.44 ± 0.22	0.97 ± 0.17	*
R	L	Zinc	µmol.l⁻¹	14.7 ± 1.8	16.7 ± 2.0	*
A	E	Copper	µmol.l⁻¹	22.3 ± 5.2	21.2 ± 3.9	
C	M	Manganese	nmol.l⁻¹	24.9 ± 10.0	20.5 ± 4.0	
E	N T S	Iron	µmol.l⁻¹	12.0 ± 5.7	22.0 ± 10.0	*
V I T A M I N S		Vitamin A	µmol.l⁻¹	2.2 ± 0.5	3.2 ± 0.9	*
		Beta Carotene	µmol.l⁻¹	0.27 (0-0.9)	3.3 ± 1.2	*
		Vitamin E	µmol.l⁻¹	22.1 ± 9.8	30.0 ± 9.0	**
		Vitamin B 9	nmol.l⁻¹	18.3 ± 8.2	22.5 ± 7.0	
		Vitamin B 12	pmol.l⁻¹	469 ± 240	369 ± 110	
Glutathione peroxidase						
Plasma			UI/l	239 ± 60	335 ± 19	*
Erythrocytes			UI/g Hb	10.9 ± 2.3	20.0 ± 1.7	*
Lipoperoxidation products						
MDA			µmol.l⁻¹	3.1 ± 0.6	2.5 ± 0.2	*
HPO			µmol.l⁻¹	181 ± 34	120 ± 19	*

T-test Student :   * (p<0.001)   ** (p<0.02)

Second study

The specific supplementation with sodium selenite resulted in an increase of plasma selenium and plasma GPX activity, as seen in Figure 1. Erythrocyte GPX increased more slowly. No conclusion concerning the rate of progression of the disease could be drawn from the clinical investigations. However, during treatment the plasma MDA levels seemed to decrease.

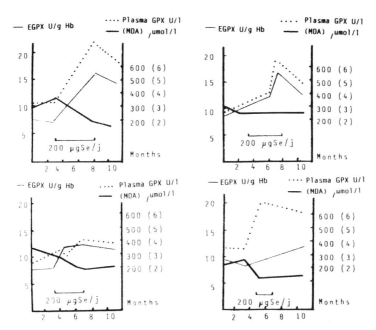

Plasma malonaldehyde (MDA) (——————), Plasma glutathione peroxidase (----------) and erythrocyte glutathione peroxidase (——————) changes before, during and after selenium treatment with a daily dose of 200 μg Se in four cystic fibrosis children.

DISCUSSION

Several authors[7,8] have shown that CF children have significantly lower plasma Se. It is not clear whether pancreatic insufficiency has a direct role in Se levels[9]. In the USA, Se status in children and adolescents with CF has been found to be essentially normal[10] or somewhat reduced[11]. On the other hand, our findings show that Se GPX activity of these CF infants is significantly lower than normal (50% of normal values). Such low levels are not in agreement with Foucauld[12] but for measuring Se GPX activity, as substrate, we used ter butyl hydroperoxide instead cumen. Vitamins A, E, and caroten deficiencies could potentiate the effects of Se deficiency. In the present study , the correlations found between GPX and antioxidant vitamins raised the hypothesis of a positive cooperation from all micronutrients implicated in free radical detoxication.

During Se supplementation, the Se GPX activities rose gradually . They decreased when the treatment stopped. A daily dose of 200 ug Se normalised the GPX activity in only two patients. In the first study we did not find a correlation between SeGPX activities and plasma lipoperoxides . However, when patients were treated with sodium selenite, MDA levels decreased. The small number of cases studied until now does not allow definite conclusions.

Altough Se is a potentially toxic nutrient, a supplementation in order to normalize plasma content and red blood cell GPX may be justified under the condition that plasma levels are monitored periodically. By its preventive effect, Se can limit peroxide induced damage in lipid structures, specially the deleterious effects on cellular membranes (leading to changes in fluidity and permeability of pulmonary parenchym), or on connective tissue elastase.

REFERENCES

1. Park R.W., Granr R.J.: Gastrointestinal manifestations of cystic fibrosis: A review. Gastroenterology 81:1149–1161.(1981)
2. Dworkin B., Newman L.J., Berezin S., Rosenthal W.S., Schwarz S.H., Weiss L. Low blood selenium levels in patients with cystic fibrosis compared to controls and healthy adults. J. P. E. N. 11:38-41 (1987).
3. Vancaillie Bertrand M., De Bieville F., Neisens H., Kerrebijn K., Fernandes J. and Degenhart H. Trace metals in cystic fibrosis. Acta Pediatr. Scand. 71:203-207 (1982).
4. W.A.Gunzler, H.Kremers, and L. Flohe. An improved coupled test procedure for glutathione peroxidase in blood. Z. Klin. Chem. Klin. Biochem. 12:444– 448 (1974).
5. S.B. Dousset, M. Trouilh, and M.J. Foglietti, Plasma malonaldehyde levels during myocardial infarction. Clin. Chim. Acta. 129:319-322 (1983).
6. R.L.Heath, A.L.Tappel A new sensitive assay for the measurement of hydroperoxydes. Anal. Biochem. 76: 184-191 (1976).
7.R.J. Stead, A.N. Redington, L.J. Hinks, B.E. Clayton, M.E. Hodson, J.C. Batten.Selenium deficiency and possible increased risk of carcinoma in adults with cystic fibrosis. Lancet. 19:862-863 (1985).
8. J.Neve, R. van Geffel, M.Hanocq and L.Molle. Plasma and erythrocyte Zinc, Copper and Selenium in cystic fibrosis. Acta Pediatr.Scand.72:437-440 (1983).
9. H.C.Heinrich, E.E.Gabbe, K.H.Bartels et al. Bioavaibility of food iron([57]Fe), vitamin B_{12}(^{60}Co), and protein bound selenomethionine (^{75}Se) in pancreatic exocrine insufficiency due to cystic fibrosis. Klin. Wochenschr. 55:595-601 (1977).
10. R.Castillo, C.Landon, K.Eckhardt, V.Morris, O.Levander and N.Lewiston. Selenium and vitamin E status in cystic fibrosis. J.Pediatr. 99:583-585 (1981).
11. J.D.Lloyd-Still, H.E.Ganther.Selenium and glutathione peroxidase levels in cystic fibrosis. Pediatrics 65:1010-12 (1980).
12. P.Foucaud, P.Therond, M.Marchand, F.Brion, J.F.Demelier, J.Navarro. Selenium et vitamine E au cours de la mucoviscidose. Arch.Fr.Pediatr. 45:383-386.(1988).

SELENIUM AND GLUTATHIONE PEROXIDASE ACTIVITY IN DIPLOID HUMAN
FIBROBLASTS: THEIR EFFECTS AGAINST H_2O_2 OR UV_B INDUCED
TOXICITY

M.J. RICHARD, P. FRAPPAT, J. ARNAUD, A. FAVIER

Laboratoire de Biochimie C, C.H.R.U.G., BP 217X
38043 Grenoble Cédex, France

INTRODUCTION

One of the proposed mechanisms of aging is generation
of free radicals that have been shown to react with lipids,
proteins, and DNA[1,2]. Trace elements by the fact of metalloenzymes (Zinc Copper Superoxide Dismutase (ZnCuSOD), Manganese Superoxide Dismutase (MnSOD), Selenium Glutathione
Peroxidase (SeGPX)) prevent such potentially toxic reactions.
It is well known[3], that selenium (Se) is necessary for
synthesis and activity of Glutathione peroxidase
(E.C.1.11.1.9). This primary cellular antioxidant enzyme catalyses the reduction of hydrogen peroxide and organic hydroperoxides at the expense of reduced glutathione (GSH) which
is oxidized to the disulfide (GSSG)[4].
We have been interested to study on cultured diploid
human skin fibroblasts, how selenium modulation in culture
medium induces a protection towards H_2O_2 or UV_B cytotoxicity.

MATERIAL AND METHODS

Cells and Selenium supply

Three normal cell lines of human fibroblasts were maintained for 4 subcultures in RPMI 1640 medium containing 10%
fetal bovine serum (FBS) or in the same Selenium supplemented
medium (25-500 μg.l^{-1} as sodium selenite). The cells were
then trypsined, washed twice with TRIS buffer pH 7.15 and
disrupted in bidistilled water.
In this lysate we determined :
    * Selenium by direct electrothermal atomic absorption
spectrometry (PERKIN ELMER 560 with HGA 500).
    * Se GPX by the method of Gunzler,[5] using ter butyl
hydroperoxide as substrat.
    * Malonaldehyde (MDA) by a modified fluorometric method
of DOUSSET[6].
    * Protein levels by the procedure developed by Lowry
    Results were expressed by gramme of proteins.

Antioxidants in Therapy and Preventive Medicine
Edited by I. Emerit *et al.*
Plenum Press, New York, 1990

Selenium and free radical cytotoxicity

After four subcultures in supplemented medium, cells were trypsined, collected, and counted. We divided them into four equal groups and plated them into Falcon T 25 flasks. Cells were incubated at 37°C in air:CO_2 (92.5:7.5) with appropriated Se supplemented medium. In all experiments, untreated skin fibroblasts served as controls.

Selenium and H_2O_2 toxicity

Cells were exposed to H_2O_2 10^{-4}M in tyrode buffer for 30 ₘin in the dark. Then, they were supplied with fresh normal ᵣ supplemented medium. After 2 days of incubation, we ᵥashed plates vigorously twice to remove non adherent cells. ᵥe determined survival responses by measuring the total proteins content of cells[7]. For each point 2 duplicate plates and 2 blanks (buffer only) were made. Results were expressed as proteins of assay / proteins of blank

Selenium and UV_B lethal effect

Human fibroblasts cultivated in Falcon T25 flasks as previously described ,were exposed to 0.12 J / cm^2 for 3 days. The irradiation time was 2 minutes per day. For exposure, culture medium was replaced by tyrode buffer. After each irradiation, cells were supplied with fresh medium (Se supplemented or not) and kept in the incubator until the next exposure. We investigated the survival 2 days after the last exposition. Results were expressed as protein of UV_B irradiated cells / protein of non irradiated cells.

RESULTS and DISCUSSION

Previous experiments[8] had shown that Selenium has a biphasic effect on cell growth. At the levels that we used cell growth is not inhibited (Fig 1).
The effect of Se supplementation is assessed by measuring the rise of intracellular Se and GPX (Fig 2).

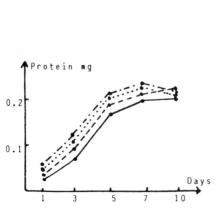

Fig. 1 Growths curve of in vitro human diploid fibroblasts after three weeks treatment with 50 ₘug/1^{-1} (•—·—•), 100 /ug.1^{-1} (•——•), 200 ₘug.1^{-1} (•······•). sodium selenite and without (•———•).

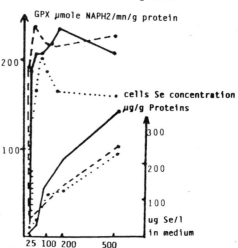

Fig. 2 Effect of differents levels of selenium in the growth medium on intracellular selenium and GSH-Px in human diploid fibroblast.

The concentration of total cellular Se increased with increasing levels of Se in the medium. GPX activity was also a function of Se supply in the cell culture medium until 100 µg.l⁻¹. With higher doses, GPX activity did not increase. Our results indicate that differential sensitivity to H_2O_2 is well correlated with fibroblast GPX activity (Fig 3).

Moreover the present study shows that the lethal effect of UV_B decreases when fibroblasts were grown in a Se supplemented culture medium (Fig 4).

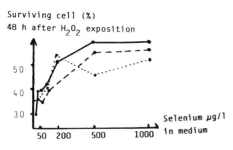

Surviving cell (%)
48 h after H_2O_2 exposition

Fig. 3   Effect of selenium concentration on H_2O_2 mediated fibroblast lysis. One curve represents means of two replicates values for each cell line of human diploid normal fibroblasts.

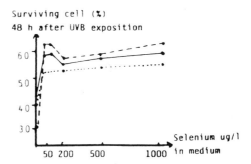

Surviving cell (%)
48 h after UVB exposition

Fig. 4   Preliminary study on selenium effect in cell culture on UVB irradiation. One curve represents means of two replicates values for each human diploid fibroblasts line.

Living cells are rich in polyunsaturated lipids and therefore intrinsically vulnerable. There is no doubt that under certain conditions they can autoxidize. It is also certain that the change is associated with a breakdown of biological function and structure. Lipid peroxides and other peroxidized products are powerfull enzyme inhibitors and membranes cease to be competent intercompartmental barriers[9]. The relation between aging and deseases involving free radical reactions seems to be a direct one[10]. Because the antioxidant enzymes are inducible[11], antioxidant adaptability and tissue tolerance against oxidative stress is possible. But this adaptability is only possible if antioxidant nutrients intake (Se, Vitamine E...) is adequate.

In this study we test the hypothesis that selenium deficient fibroblasts are more susceptible to oxidative cell damages. Hydrogen peroxide is present in the cells at concentration as low as 10^{-8} M but its production can be accelerated under certain conditions such as exposure to visible light, hyperbaric oxygen, and certain carcinogens. Moreover, this compound is produced by phagocytic cells and may cause fatal injury to other cells at sites of inflammation. Hoffman[12] investigated both the lethal and DNA damaging effects of H_2O_2 on human fibroblasts in culture, demonstrated a good correlation between cell killing and DNA damage.

The results of the present investigation indicate, that H_2O_2 induces cell lysis in fibroblast cultures, that can be prevented by Se supplementation. In addition, Se exerted a protective effect on UV- light induced changes in the fibroblasts. These results are in agreement with those of Thorling[13], who observed a better resistance of Se-supplemented mice against acute and subacute effects of UV-light than untreated animals.

181

The exact mechanisms by which Se exerts its protective role against carcinogenesis and aging is not known. It is conceivable that Se influences DNA polymerase or nucleotid kinase activities, as suggested by Medina[14]. Our in vitro model may be useful for further investigations in this field.

REFERENCES

1. S.K. Jain, Evidence for membrane lipid peroxidation during the in vivo aging of human erythrocytes, Biochim. biophys. Acta. 937:205-210 (1988).

2. T.F. Slater, Free radical mechanisms in tissue injury. Biochem. J., 222:1-15 (1984).

3. K. Takahashi, P.E. Newburger, and H.J. Cohen, Glutathione peroxidase protein: Absence in selenium deficiency states and correlation with enzymatic activity. J. Clin. Invest. 77:1402-1404 (1986)

4. G.F. Combs and S.B. Combs. in : "The role of Selenium in nutrition," Combs & Combs, Ed., Academic Press Inc. London (1986) pp 205-264.

5. W.A.Gunzler, H.Kremers, and L. Flohe. An improved coupled test procedure for glutathione peroxidase in blood. Z. Klin. Chem. Klin. Biochem. 12:444- 448 (1974).

6. S.B. Dousset, M. Trouilh, and M.J. Foglietti, Plasma malonaldehyde levels during myocardial infarction. Clin. Chim. Acta. 129:319-322 (1983).

7. Ch. Shopsis, G.J. Mackay, Semi automated assay for cell culture. Anal. Biochem. 140:104-107 (1984).

8. Ao Pung, Mei Zhao, and Yu Shu-yu, In vitro differential effects of sodium selenite on the growth of human hepatoma cells and human embryonic liver cells. Bio.Trace Elem. Res. 14:1-17 (1987).

9. J.F. Mead, Free radical mechanisms of lipid damage and consequences for cellular membranes. in: "Free Radicals in Biology," W.A. Pryor, ed., Academic Press.N.Y. (1976) 1: pp 51-66.

10. D. Herman, Free radicals in aging, Mol. Cell. Biochem. 84:155-161 (1988).

11. J. Chaudière, D. Gerard, M. Clement, J.M. Bourre. Induction of Selenium glutathione peroxidase by stimulation of metabolic hydrogen peroxide production in vivo. Biochem. Bioenergetics 18:247-256 (1987)

12. E. Hoffman, A.C. Mello Filho, and R. Meneghini, Correlation between cytotoxic effect of hydrogen peroxide and the yield of DNA strand breaks in cells of different species, Biochim. biophys. Acta. 781:234-238 (1984).

13. E.B. Thorling, K. Overvad, and P. Bjerring, Oral selenium inhibits skin reactions to UV light in hairless mice. Acta Path. Microbiol. Immunol. Scand. Sect. A., 91:81-83 (1983).

14. D. Medina, H.W. Lane, and C.M. Tracey, Selenium and mouse mammary tumorigenesis : an investigation of possible mechanisms. Cancer Res. suppl.43: 2460-2464 (1983).

SELENIUM THERAPY IN DOWN SYNDROME (DS):

A THEORY AND A CLINICAL TRIAL

Erkki Antila[1], Ulla-Riitta Nordberg[2], Eeva-Liisa
Syväoja[3], and Tuomas Westermarck[2]

[1]Dept. of Anatomy, Univ. of Helsinki,[2]Helsinki
Central Inst. for the Mentally Retarded, Kirkko-
nummi, and [3]Valio Finnish Co-operative Dairies'
Assoc., Research and Development Dept., Helsinki
Finland

INTRODUCTION.

The development of a DS child is genetically predestined to
express a multitude of clinical signs including mental retardation, premature aging and immunological disorders with hypo-
thyroid states. Thus attempts have been made to improve the
understanding of the etiopathogenesis of the syndrome and to
influence its progress.

Increased primary gene products which may contribute to the
pathology of DS include cytoplasmic Cu/Zn-superoxide dismutase
(SOD-1, Sinet 1982). A microduplication of chromosome 21 fragment containing the SOD-1 gene has been found in a karyotypical-
ly normal 18-month-old boy manifesting many typical DS features
(Huret et al.1987). Further evidence of the decisive role of
the SOD-1 gene in the pathology of DS has been derived from
studies performed with transgenic cell lines and mice carrying
the human SOD-1 gene (Avraham et al.1988).

Consistent with the gene dosage effect, enhanced SOD-1 activity,
leading to noxious concentrations of H_2O_2, is found in the
cerebral cortex of DS fetuses as well as in erythrocytes, blood
platelets, leukocytes and fibroblasts of DS patients . Elevated
formation of H_2O_2, which rapidly crosses cell membranes, results
in oxygen free radical stress through decomposition to a highly
reactive hydroxyl radical. H_2O_2 reaches extra and intracellular
compartments more easily than ordinary radicals. Therefore the
damage caused by the decomposition of H_2O_2 to hydroxyl radical
in iron (FeII) or copper(CuI) catalyzed reactions expands. Cytoplasmic glutathione(GSH) in concert with Se containing GSH-
peroxidase(Px), with NADPH generating pentosemonophosphate shunt
and with GSH-reductase form a powerful system reducing H_2O_2. The
overall redox state in many tissues, except that in the brain,
is corrected by an adaptive increase of GSHPx activity. This
means that the brain is especially susceptible to oxygen free
radical stress.

Our primary clinical trial on specific antioxidant therapy with
selenium rests on this theory reviewed recently (Balaz and

Antioxidants in Therapy and Preventive Medicine
Edited by I. Emerit *et al.*
Plenum Press, New York, 1990

Brooksbank,1985; Kedziora and Bartosz,1988; Antila and Wester-marck,1989). Because of difficulties in obtaining brain biopsies, variables found in plasma and erythrocyte samples, such as SOD-1 and GSHPx activity and vitamin E were used as indicators of antioxidant balance.

MATERIAL AND METHODS

24 DS patients of both sexes (m/f 0.4) aged 1 to 54 years (28.1±14.0 years) were chosen for the study. Only two (n2) youngest patients, age <5 years, were noninstitutionalized. Patients were given either 0.015-0.025 Se mg/kg/d as Na-sele-nite, placebo, or no preparation (control group).
Venous samples were taken before and after the supplementation period of 0.3 to 1.5 years. The plasma was removed after centri-fugation and the red cells (E) were washed twice with isotonic saline. E-GSHPx activity (U/g of Hb) was determined by the method of Beutler et al. (1977) and Paglia and Valentine (1967). The activity of E-SOD (U/g of Hb) was measured by a modification of Beuchamp and Fridovich's method (1971) using nitro blue tetrazolium in the xanthine oxidase system. The mean activity of the control samples was 3786+417 U/g of Hb.
Venous samples for vitamin E determinations were taken about 1 month after the beginning of the follow-up. Tocopherol levels were measured fluorometrically after HPLC separation (Syväoja et al.,1985). The results are expressed as α-tocopherol equi-valents, α-, β-, and γ-tocopherol (mg/100ml).

RESULTS AND DISCUSSION

The Se-supplementation increased E-GSHPx activity (from 25.7 to 32.1 U/gHb, n9) by 25% (61 % above normal). In the combined placebo and control group the mean activity decreased from 25.5 to 25.0 U/gHb (n.s.,n10). This effect, expressed as SOD-1/GSHPx-index, supports the notion that the functional balance between

SOD-1 / GSHPx INDEX

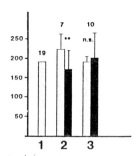

Fig.1. SOD-1/GSHPx index before and after follow-up period 1)normal reference population, 2) selenium (29.6±13.3 y) and 3)con-trol (30.1±10.7 y) groups excluding age group <5 y.

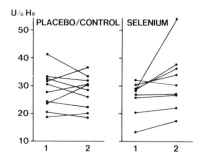

Fig.2. E-GSHPx activity in DS subjects before (1) and after (2)follow-up period (3mo-1.5y).

these enzymes depends greatly on the Se available. The mean SOD-1/GSHPx decreased by 23.9% ($p < 0.01$, n7) in the selenium group, whereas in the controls it increased (ns., n10). These results are shown in the Figs. 1 and 2. Plasma level of vitamin E, which may act synergistically with GSHPx, was similar in both groups (α-tocopherol equivalents 1.016 ± 0.352, n7/ 1.189 ± 0.393, n13 mg/100ml. Levels of α-, β- and γ-tocopherol were $1.009 \pm 0.350/1.173 \pm 0.392$, $0.012 \pm 0.011/0.025 \pm 0.018$, and $0.033 \pm 0.020/0.051 + 0.036$ (mg/100ml) respectively. The differences of tocopherol levels between the supplementation and control groups were nonsignificant. In healthy Finnish men, screened for selenium supplementation, plasma α-tocopherol averaged 0.97 ± 0.18 mg/100ml in subjects with low plasma selenium, which was significantly lower than that of the high plasma selenium group (1.16 ± 0.21 mg/100 ml) (Piironen et al. 1983). Furthermore, observations by Metcalfe et al. (1984) showed that no differences existed between α-tocopherol concentrations found in the fetal brains of controls and DS. The concentrations they measured in the brain, approximately 3 nmol/g of wet wt (1.3 ug/g), were about 10 times less than those we found in the plasma of DS patients. Although no definitive correlations between DS, GSHPx and vitamin E have been observed, there is some evidence of distortions of trace elements. Titanium found in DS erythrocytes may as Ti(IV) form stable compounds with hydrogen peroxides and probably with superoxide anions. This finding suggests insufficient protection by GSHPx. During selenium supplementation the primarily high blood mononuclear cell levels of copper decreased, whereas the concentration of iron and zinc was not affected (Johansson et al. 1989). Sensitivity to an adequate selenium supplementation was further corroborated by the significant decrease of SOD-1/GSHPx index in our experimental group during the follow-up period. Sinet et al. (1979) reported a strong positive correlation between E-GSHPx activity and IQ in DS. Anneren (1984) suggested that the higher plasma and erythrocyte selenium and E-GSHPx levels in girls with DS as compared with boys is causally connected to the significantly higher IQ scores of female than male DS subjects. A recent analysis on our patients indicated that E-GSHPx activity had a significant positive correlation to IQ only in a larger cohort of IQ>18, age>5years, (n18), but the increase of IQ after supplementation period of 0.8 to 1.5 years was insignificant (Timberg 1988). Therefore supplementation studies should be conducted on homogenous populations of DS subjects, in relation to age and environment. Since the distortions in DS affect embryonic and postnatal neurogenesis and differentiation , the therapy should be started as early as possible.

In conclusion we believe that, in spite of the heterogenity of the population, our DS patients have benefitted from the selenite supplementation through optimization of their antioxidant protection by GSHPx.

ACKNOWLEDGEMENTS

We wish to thank Mrs. Mervi Kulpakko for skilful technical assistance. This work was supported by the Wilhelm and Else Stockmann's Foundation, Finland.

REFERENCES

Anneren, G., 1984, Down's syndrome. A metabolic and endocrino-
logical study,Acta Univ.Upps.,Abstracts of Uppsala Disser-
tations from the Faculty of Medicine,483. 38 pp.,Uppsala.

Antila, E., and Westermarck, T., 1989, On the etiopathogenesis
of Down syndrome, Int.J.Devel.Biol. (in press).

Avraham, K.B., Schickler, M., Sapoznikov, D., Yarom, R., and
Groner, Y., 1988, Down's syndrome: abnormal neuromuscular
junction in tongue of transgenic mice with elevated levels
of human Cu/Zn-superoxide dismutase, Cell, 54:823.

Balazs,R.,and Brooksbank,B.W.L.,1985, Neurochemical approaches
to the pathogenesis of Down's syndrome, J.Ment.Def.Res.,29:1.

Beuchamp, C. and Fridovich, I., 1971, Superoxide dismutase:
improved assays and an assay applicable to acrylamide gels,
Anal. Biochem., 44:276

Beutler, E., Blume, K.D., Kaplan, J.C., Löhr, G.W., Ramot, B.
and Valentine, W., 1977, ICSH: Recommended methods for red
cell enzyme analysis, Br.J.Haematol. 35:331.

Huret, J.L., Delabar, J.M., Marlhens,F., Aurias,A., Nicole,A.,
Berthier, M., Tanzer, J., and Sinet, P.M., 1987, Down
syndrome with duplication of a region of chromosome 21
containing the CuZn superoxide dismutase gene without
karyotypic abnormality, Hum.Genet., 75:251.

Johansson, E., Lindh, U., Westermarck, T., Antila, E., and
Nordberg, U.-R., Nikkinen, P., Hyvönen-Dabek, M., 1989,
Deviation of elements and enzymes in blood cells in Down's
syndrome (trisomy 21), Biol.Trace Elem.Res.,(submitted).

Kedziora, J., and Bartosz, G., 1988, Down's syndrome: A pathology involving the lack of balance of reactive oxygen
species, Free Rad.Biol.Med.,4:317.

Metcalfe, T., Muller, D.P.R., and Brooksbank, B.W.L., 1984,
Vitamin E concentrations in brains from foetuses with
Down's syndrome. IRCS Med. Sci., 12: 121.

Paglia, D.E. and Valentine, W.N., 1967, Studies on the quantitative and qualitative characterization of erythrocyte glutathione peroxidase, J.Lab.Clin.Med., 70:158.

Piironen, V., Varo, P., Syväoja, E.-L., Salminen, K,
Koivistoinen, P. and Arvilommi, H., 1983, High-performance
liquid chromatographic determination of tocopherols and
tocotrienols and its application to diets and plasma of
Finnish men. II Applications, Internat. J. Vit. Nutr.
Res.,53:41.

Sinet, P.M., Lejeune, J., and Jerome, H., 1979, Trisomy 21
(Down's syndrome), glutathione peroxidase, hexose monophosphate shunt and IQ, Life Sci.,24:29.

Sinet, P.M., 1982, Metabolism of oxygen derivatives in Down's
syndrome, Ann.N.Y.Acad.Sci.,396:83.

Syväoja, E.-L., Salminen,K., Piironen,V., Varo,P., Kerojoki,O.,
and Koivistoinen,P., 1985, Tocopherol and tocotrienols in
finnish foods: fish and fish products, J.A.O.C.S.,62:1245.

Timberg, H., 1988, Antioksidatiivisen hoidon vaikutukset
kehitysvammaisten psyykkiseen suorituskykyyn ja kuntoi-
suustasoon. (A pro gradu study, Department of Psychology,
University of Helsinki).

PROTECTION BY SELENO-ORGANIC COMPOUND, EBSELEN, AGAINST ACUTE GASTRIC

MUCOSAL INJURY INDUCED BY ISCHEMIA-REPERFUSION IN RATS

S. Ueda, T. Yoshikawa, S. Takahashi, Y. Naito, H. Oyamada,
T. Takemura, Y. Morita, T. Tanigawa, S. Sugino, and M. Kondo

First Department of Medicine, Kyoto Prefectural University
of Medicine, Kamigyo-ku, Kyoto 602, Japan

INTRODUCTION

Several pathogenetic mechanisms have been suggested to account for acute gastric mucosal injury. Recently, Oxigen-derived free radicals and lipid peroxidation have been suggested to play a role in the pathogenesis of gastric mucosal injury[1,2,3]. Ischemia and reperfusion are of the greatest

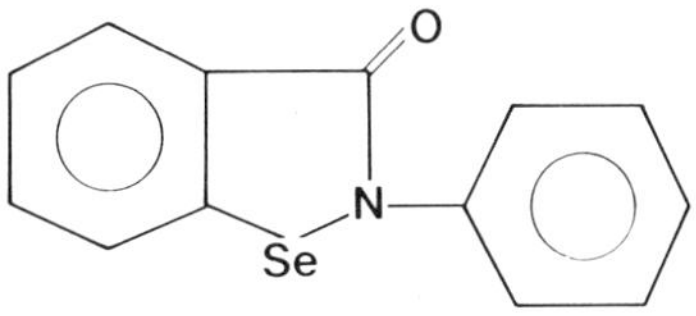

Ebselen

2-phenyl-1, 2-benzoisoselenazol-3 (2H)-one

$C_{13}H_9NOSe$: Mol wt : 274.18

C 56.95% : H 3.31%: N 5.11%

O 5.83% : Se 28.80%

Fig.1. Structure of seleno-organic compound, Ebselen.

importance in the pathology of many diseases. There has been great interest in the possible role of oxygen radical species in ischemia-reperfusion in the gastric mucosa[4]. A synthetic seleno-organic compound, 2-phenyl-1, 2-benzoisoselenazol-3(2H)-one(Ebselen), shows a glutathione peroxidase(GSH-Px)-like activity[5,6]. This study was designed to examine the protective effects of the agent, Ebselen against the gastric mucosal injury induced by ischemia-reperfusion.

MATERIALS AND METHODS

Male Splague-Dawley rats weighing about 200 g were used. The animals were not fed for 18h but allowed free access to water. Ischemia was created under intraperitoneal pentobarbital anesthesia(25 mg/kg) by applying small

Antioxidants in Therapy and Preventive Medicine
Edited by I. Emerit *et al.*
Plenum Press, New York, 1990

187

Table 1. Effects of Ebselen on the total area of erosions in the gastric mucosa induced by ischemia-reperfusion.

		Total area of erosions (mm^2)
Control		$21.4 \pm 4.1(16)$[1]
Ebselen	10mg/kg	$12.5 \pm 1.9(6)$
	50mg/kg	$9.4 \pm 2.0(7)$*
	100mg/kg	$7.5 \pm 2.7(6)$**
	200mg/kg	$4.3 \pm 0.7(15)$***

Each value represents the mean+SEM. *:$P<0.05$, **:$P<0.01$, ***:$P<0.001$ when compared to disease control group. 1)Number of rats used.

Table 2. Effects of Ebselen on increase of TBA reactants in gastric mucosa induced by ischemia-reperfusion.

		TBA reactants (n mol/g.wet weight)
Normal	control	$33.6 \pm 2.6(6)$*[1]
Disease	control	$48.4 \pm 3.4(10)$
Ebselen	10 mg/kg	$53.0 \pm 3.5(6)$
	50 mg/kg	$45.5 \pm 3.5(6)$
	100 mg/kg	$42.4 \pm 5.2(6)$
	200 mg/kg	$38.1 \pm 2.0(9)$*

Each value represents the mean+SEM. *:$p<0.05$ when compared to disease control group. 1)Number of rats used.

clamp to the celiac artery for 30 min. Reoxigenation was produced by removal of the clamps for 60 min. Ebselen(Daiichi Seiyaku Co. LTD.,Tokyo) dissolved in 0.5% carboxymethyl cellurose sodium solution(CMC, Wako Pure Chemical Industries Co.LTD., Osaka) was treated by gastric intubation 1h before the experiment. The control group was treated intragastric with 0.5% carboxymetyl cellurose sodium solution. Human SOD(Nippon Kayaku Co. LTD.,Tokyo) at 50,000 U/kg and catalase(Sigma Co. LTD., St.Louis,Mo.) at 90,000 U/kg were injected s.c. 1h before the ischemia, and 10,000 U/kg of SOD was injected i.v. just before the reoxygenation. The gastric mucosal lesions were expressed by the total area of the erosions as an ulcer index. Thiobarbituric acid(TBA) reactants, an index of lipid peroxidation, were measured in the gastric mucosa by the method of Ohkawa et al[7]. Gastric mucosal blood flow was measured by the laser-doppler blood flowmeter(Advance Co,Ltd.,Tokyo). The gastric acid secretion was measured 3h after ligation of duodenum and injecting Ebselen or 0.5 % CMC in the duodenum.

RESULTS

Gastric mucosal blood flow decreased to about 10 % of the initial value by clamping and pronptly recovered after the removal of clamping.

The gastric mucosal lesions and TBA reactants in the gastric mucosa were significantly increased after the reperfusion.

SOD and/or catalase could protect against the gastric mucosal injury and inhibited the increase of TBA reactants in the gastric mucosa.

Gastric mucosal injury was significantly inhibited by the treatment with Ebselen at 50, 100 and 200 mg/kg.

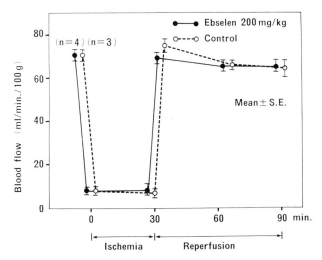

Fig.2. Gastric blood flow was measured by LASER-doppler blood flowmeter. Animals were treated with CMC(control) or 200 mg/kg
of Ebselen.

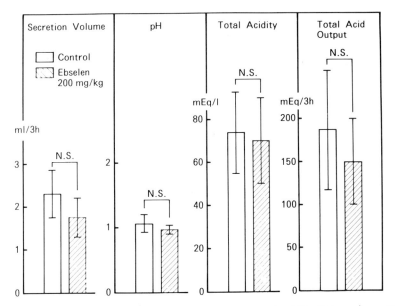

Fig.3. Gastric acid secretion was measured with 200 mg/kg of
Ebselen.

The increase of TBA reactants in the gastric mucosa were significantly inhibited in the case of the treatment with 200 mg/kg of Ebselen.

Ebselen at dose 200 mg/kg had no effect on the gastric blood flow and the gastric acid secretion.

DISCUSSION

SOD and catalase which are scavengers of the effects of superoxide(O_2^-) and hydrogen-peroxide(H_2O_2) could protect against the gastric mucosal injury and inhibited the increase of TBA reactants in gastric mucosa induced by ischemia reperfusion. This results suggest that the gastric mucosal injury induced by ischemia-reperfusion may be aggravated by oxygen derived free radicals. With ischemia and reperfusion, xanthine oxidase catalyzes the reaction between hypoxanthine and oxygen to form O_2^- and H_2O_2. The reactive and cytotoxic hydroxyl radical(HO^-) were formed from O_2^- and H_2O_2 in the prescence of trantion metals[8]. Oxygen-derived free radicals are generated in the cell during the reduction of molecular oxygen. Lipid peroxidation which mediated by active oxygen species is believed to be one of the important causes of cell membrane destruction and cell damage, and also may play an important role in the pathogenesis of gastric mucosal injury[1,2,3]. GSH-Px catalyzes the reaction between glutathione and hydrogenperoxide or a organic peroxide. This reaction may ptotect the cell from oxidative damage. Ebselen shows a glutathione peroxide(GSH-Px)-like activity. This study demonstrated that Ebselen can protect the gastric mucosal injury and inhibit the increase of TBA reactants in the gastric mucosa induced by ischemia-reperfusion without gultathione modifing the blood flow or gastric acid secretion. These results suggest that gastric mucosal injury induced by ischemia-reperfusion may be aggravated by oxygen radicals and lipid peroxidation. And that GSH-Px might play an important role in the protection against gastric mucosal injury induced by ischemia-reperfusion.

REFERENCES

1)T. Yoshikawa, N. Yoshida, H. Miyagawa, T. Takemura, T. Tanigawa, S. Sugino, M. Kondo, Role of lipid peroxidation in gastric mucosal lesions induced by burn shock in rats. J.Clin.Biochem.Nutr.2:163-170(1987).
2)T. Yoshikawa, H. Miyagawa, N. Yoshida, T. Tanigawa, Y. Kakimi, S. Sugino, M. Kondo, Role of free radicals in stress-induced gastric mucosal lesions in rats. Excepta Medica.98-103(1986).
3)D. A. Parks, G. B. Bulkley, D. N. Granger, Role of oxygen-derived free radicals in digestive tract diseases. Surgery.94:415-422(1983).
4)M. Itoh, P. H. Guth, Rule of oxygen-derived free radicals in heamorrhagicschock-induced gastric lesions in the rat. Gastroenterology.88:1162-1167(1985).
5)A. Muller, E. Cadenas, P. Graf, H. Sies, Glutathione peroxidase-like activity in vitro and antioxidant capacity of pz51(Ebselen). Biochemical Pharmacology.33:3235-3239(1984).
6)M. J. Parnham, S. Kindi, Effect of pz51(Ebselen) on glutathione peroxidaseand secretory activities of mouse macrophages. Biochemical Pharmacology. 33:3247-3250(1984).
7)H. Ohkawa, N. Ohishi, K. Yagi, Assay for lipid peroxides for animal tissue by thiobarbituric acid reaction. Anal Biochem.95:351-358(1979).
8)D. N. Granger, G. Rutili, J. M. McCord, Superoxide radicals in felin intestinal ischemia. Gastroenterology.81:22-29(1981).

The increase of TBA reactants in the gastric mucosa were significantly inhibited in the case of the treatment with 200 mg/kg of Ebselen.

Ebselen at dose 200 mg/kg had no effect on the gastric blood flow and the gastric acid secretion.

DISCUSSION

SOD and catalase which are scavengers of the effects of superoxide(O_2^-) and hydrogen-peroxide(H_2O_2) could protect against the gastric mucosal injury and inhibited the increase of TBA reactants in gastric mucosa induced by ischemia reperfusion. This results suggest that the gastric mucosal injury induced by ischemia-reperfusion may be aggravated by oxygen derived free radicals. With ischemia and reperfusion, xanthine oxidase catalyzes the reaction between hypoxanthine and oxygen to form O_2^- and H_2O_2. The reactive and cytotoxic hydroxyl radical(HO^-) were formed from O_2^- and H_2O_2 in the prescence of trantion metals[8]. Oxygen-derived free radicals are generated in the cell during the reduction of molecular oxygen. Lipid peroxidation which mediated by active oxygen species is believed to be one of the important causes of cell membrane destruction and cell damage, and also may play an important role in the pathogenesis of gastric mucosal injury[1,2,3]. GSH-Px catalyzes the reaction between glutathione and hydrogenperoxide or a organic peroxide. This reaction may ptotect the cell from oxidative damage. Ebselen shows a glutathione peroxide(GSH-Px)-like activity. This study demonstrated that Ebselen can protect the gastric mucosal injury and inhibit the increase of TBA reactants in the gastric mucosa induced by ischemia-reperfusion without gultathione modifing the blood flow or gastric acid secretion. These results suggest that gastric mucosal injury induced by ischemia-reperfusion may be aggravated by oxygen radicals and lipid peroxidation. And that GSH-Px might play an important role in the protection against gastric mucosal injury induced by ischemia-reperfusion.

REFERENCES

1)T. Yoshikawa, N. Yoshida, H. Miyagawa, T. Takemura, T. Tanigawa, S. Sugino, M. Kondo, Role of lipid peroxidation in gastric mucosal lesions induced by burn shock in rats. J.Clin.Biochem.Nutr.2:163-170(1987).
2)T. Yoshikawa, H. Miyagawa, N. Yoshida, T. Tanigawa, Y. Kakimi, S. Sugino, M. Kondo, Role of free radicals in stress-induced gastric mucosal lesions in rats. Excepta Medica.98-103(1986).
3)D. A. Parks, G. B. Bulkley, D. N. Granger, Role of oxygen-derived free radicals in digestive tract diseases. Surgery.94:415-422(1983).
4)M. Itoh, P. H. Guth, Rule of oxygen-derived free radicals in heamorrhagicschock-induced gastric lesions in the rat. Gastroenterology.88:1162-1167(1985).
5)A. Muller, E. Cadenas, P. Graf, H. Sies, Glutathione peroxidase-like activity in vitro and antioxidant capacity of pz51(Ebselen). Biochemical Pharmacology.33:3235-3239(1984).
6)M. J. Parnham, S. Kindi, Effect of pz51(Ebselen) on glutathione peroxidaseand secretory activities of mouse macrophages. Biochemical Pharmacology. 33:3247-3250(1984).
7)H. Ohkawa, N. Ohishi, K. Yagi, Assay for lipid peroxides for animal tissue by thiobarbituric acid reaction. Anal Biochem.95:351-358(1979).
8)D. N. Granger, G. Rutili, J. M. McCord, Superoxide radicals in felin intestinal ischemia. Gastroenterology.81:22-29(1981).

BIOLOGICAL ACTIVITIES AND CLINICAL POTENTIAL OF EBSELEN

Michael J. Parnham

Rhône-Poulenc/Nattermann, Cologne Research Centre

P.O. Box 350120, D-5000 Cologne 30, FRG

INTRODUCTION

Ebselen (2-phenyl-1,2-benzisoselenazol-3(2H)-one; Fig. 1) is a seleno-organic compound which exhibits catalytic GSH-dependent, hydroperoxide reducing properties, similar to the action of glutathione peroxidase (GSH-Px)[1-3].

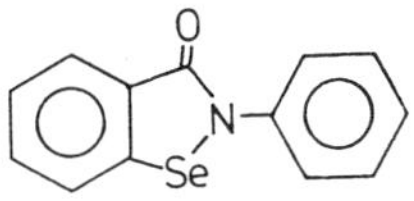

Fig. 1 Structure of ebselen

Recently, ebselen has been shown to exhibit a preferential reactivity with phospholipid hydroperoxides in vitro[4]. In addition, the compound inhibits 5-lipoxygenase and cyclo-oxygenase in vitro[5-6] and isomerises the inflammatory mediator leukotriene B_4 (LTB_4) to the biologically inactive trans-isomer[7]. The inactivation of hydroperoxides and/or the inhibition of arachidonic acid metabolic pathways probably account for other in vitro actions of ebselen, including inhibition of hepatic lipid peroxidation[8,9], inhibition of doxorubicin cytotoxicity[10] and inhibition of lymphocyte proliferation[11]. The more recent observation that ebselen inhibits the NADPH oxidase of granulocytes[12] probably results from the reactivity of ebselen with free thiol groups.

This spectrum of in vitro activities suggested that ebselen may be effective for the therapy of a variety of clinical conditions.

ANTI-INFLAMMATORY ACTIONS

Ebselen was initially detected as an inhibitor of acute paw oedema, induced in the rat by the injection of cobra venom factor (CVF) paw oedema; inhibitory activity of the compound in adjuvant arthritis in the rat was weak[13]. In comparison to classical non-steroidal anti-inflammatory drugs (NSAIDs), ebselen is a much more effective inhibitor of CVF

Antioxidants in Therapy and Preventive Medicine
Edited by I. Emerit *et al.*
Plenum Press, New York, 1990

193

paw oedema, but less potent as an inhibitor of the classical carrageenan-induced paw oedema, indicating the unusual profile of ebselen[14]. In
addition, ebselen, in contrast to NSAIDs, was found to be an effective
inhibitor of acute glucose-oxidase-induced monoarthritis in mice, an
H_2O_2-dependent inflammatory model[15], and on topical application, inhibited gingivitis in monkeys[16]. Consequently, the _in vivo_ anti-inflammatory activity of ebselen may well reflect its in vitro activity as an
hydroperoxide inactivator and distinguishes it from classical NSAIDs.
Ebselen, given intraperitoneally, also inhibited oedema and cellular infiltration in experimental alveolitis induced in rats with intratracheal
sephadex[17]. The authors proposed that inactivation of hydroperoxides
and/or lipoxygenase inhibition may have been the mechanism of action.

Table 1. Biological Activities of Ebselen _in vivo_

Species	Test	Inhibition ED_{50} (mg/kg)	Route	Ref
Rat	CVF paw oedema	56	oral	14
Mouse	Glucose oxidase monoarthritis	> 50	oral	15
Monkey	Dietary gingivitis	active at 5 %	topical	16
Rat	Gastric mucosal injury	50 - 100	oral	20
Rat	Exp. allergic neuritis	active at 10 - 100	oral	19
Rat	Experimental alveolitis	active at 10	i.p.	17
Mouse	Galactosamine/endotoxin liver damage	< 6	oral	22
Mouse	P.acnes/LPS liver injury (low dose)	< 10	oral	24
Rat	Galactosamine liver injury	< 30	oral	25
Rat	Ethanol liver injury	> 100	oral	25
Rat	Ischaemia reperfusion liver injury	< 30	oral	25
Rat	CCl_4 liver injury	ca. 100	oral	25
Rat	Ischaemic brain oedema	> 100	oral	27

Reactive oxygen species have been proposed to play a role in various
aspects of the inflammatory process, including cartilage breakdown,
immunoglobulin alteration to antigenic forms and disinhibition of proteases[18]. Blake et al.[18] have proposed that inhibition of reactive oxygen species, including hydroperoxides, may be an effective therapeutic
approach to rheumatoid arthritis. Autoimmune processes also play a role
in rheumatoid arthritis and ebselen has been found to be an inhibitor of
experimental allergic neuritis in rats, a T cell-mediated autoimmune
inflammatory response of the peripheral nerves[19]. This compound, therefore, offers promise for the therapy of rheumatoid arthritis. In this
respect, it is worth noting that, unlike NSAIDs, ebselen is an inhibitor
of gastric mucosal damage induced by various stimuli in the rat[20]. The
major mechanism of the anti-inflammatory activity of ebselen has yet to
be determined, but in initial phase I clinical trials, ebselen has been
reported to reduce plasma LTB_4 concentrations[21].

INHIBITION OF LIVER DAMAGE

Subsequent to the reports that ebselen inhibits 5-lipoxygenase in vitro, Wendel and Tiegs[22] studied the effect of ebselen on galactosamine/endotoxin-induced liver damage in mice. In contrast to NSAIDs, ebselen was a potent inhibitor, as measured by the increase in serum transaminases. Further studies indicated that ebselen was probably acting by inhibiting the synthesis of peptido-leukotrienes[23]. Akasaki et al.[24] also found ebselen to be an effective inhibitor of liver damage induced in mice by Propionobacterium acnes and lipopolysaccheride. Since LTD$_4$ could replace the requirement for lipopolysaccharide the authors also proposed that ebselen was acting as an inhibitor of peptido-leukotriene synthesis in this model.

While Wendel and Tiegs[22] found that ebselen was inactive as an inhibitor of drug-induced liver damage in mice, Hashizume et al.[25] observed in rats, that ebselen inhibits liver injury induced by galactosamine, ethanol, CCl$_4$ and ischaemia reperfusion. Consequently, in view of the proposal that leukotrienes are important mediators of liver injury[26], the study of ebselen in clinical liver disease appears warranted. At present ebselen is undergoing phase II clinical trials for this indication.

The finding that ebselen is an effective inhibitor of hepatic ischaemia reperfusion damage has received support from other studies in which ebselen was found to be a moderate inhibitor of ischaemic brain oedema in the rat[27]. It is thus possible that ebselen may also find therapeutic application in ischaemic conditions.

METABOLISM

The problem of the toxicity of selenium is one which continually confronts nutritionists in their studies on dietary selenium[28]. From the earliest animal experiments on ebselen it became clear that the compound had low toxicity. This was clarified by the studies of Wendel and colleagues[2] who showed that the selenium in ebselen is not available for incorporation into endogenous glutathione peroxidase in vivo. More recently, metabolic studies have shown that in the rat and man the major metabolites are the 2-glucuronyl-selenobenzanilide and various 2-methyl-seleno-benzanilides, formed following opening of the central ring structure of ebselen between the nitrogen and selenium[29,30]. No evidence for the formation of smaller, more toxic selenium-containing metabolites has been obtained. Classical selenium toxicity has, therefore, neither been observed nor is to be expected with ebselen.

CONCLUSIONS

Ebselen has been shown to be a catalytic inactivator of hydroperoxides and an inhibitor of leukotriene synthesis in vitro. Studies on inflammatory models have confirmed its unusual spectrum of activity, while experiments on liver injury models have indicated that the compound may act by inhibiting leukotriene synthesis and/or the effects of reperfusion of ischaemic areas. Ebselen is at present undergoing clinical trials with rheumatic arthritis and liver injury as major indications.

REFERENCES

1. A. Müller, E. Cadenas, P. Graf, and H. Sies, A novel biologically active seleno-organic compound I. Glutathione peroxidase-like activity *in vitro* and antioxidant capacity of PZ 51 (Ebselen), Biochem. Pharmacol. 33:3235 (1984).
2. A. Wendel, M. Fausel, H. Safayhi, G. Tiegs, and R. Otter, A novel biologically active seleno-organic compound II. Activity of PZ 51 in relation to glutathione peroxidase, Biochem. Pharmacol. 33:3241 (1984).
3. H. Fischer and N. Dereu, Mechanism of the catalytic reduction of hydroperoxides by ebselen: A selenium-77 NMR study, Bull. Soc. Chim. Belg. 96:757 (1987).
4. M. Maiorino, A. Roveri, M. Coassin, and F. Ursini, Kinetic mechanism and substrate specificity of glutathione peroxidase activity of ebselen (PZ 51), Biochem. Pharmacol. 37:2267 (1988).
5. M.J. Parnham and S. Kindt, A novel biologically active seleno-organic compound III. Effects of PZ 51 (Ebselen) on glutathione peroxidase and secretory activities of mouse macrophages, Biochem. Pharmacol. 33:3247 (1984).
6. H. Safayhi, G. Tiegs, and A. Wendel, A novel biologically active seleno-organic compound V. Inhibition by ebselen (PZ 51) of rat peritoneal neutrophil lipoxygenase, Biochem. Pharmacol. 34:2691 (1985)
7. P. Kuhl, H.O. Borbe, H. Fischer, A. Römer, and H. Safayhi, Ebselen reduces the formation of LTB_4 in human and porcine leukocytes by isomerisation to its 5S,12R-6-trans-isomer, Prostaglandins 31:1029 (1986).
8. A. Müller, H. Gabriel, and H. Sies, A novel biologically active seleno-organic compound IV. Protective glutathione-dependent effect of PZ 51 (ebselen) against ADP-Fe induced lipid peroxidation in isolated hepatocytes, Biochem. Pharmacol. 34:1185 (1985).
9. M. Hayashi and T.F. Slater, Inhibitory effects of ebselen on lipid peroxidation in rat liver microsomes, Free Rad. Res. Commun. 2:179 (1986).
10. J.H. Doroshow, Prevention of doxorubicin-induced killing of MCF-7 human breast cancer cells by oxygen radical scavengers and iron chelating agents, Biochem. Biophys. Res. Commun. 135:330 (1986).
11. W. Englberger, and M.J. Parnham, Inhibition by ebselen of macrophage eicosanoid generation and lymphocyte proliferation in vitro, Br. J. Pharmacol. 87:15P (1986).
12. S. Ichikawa, K. Omura, T. Katayama, N. Okamura, T. Ohtsuka, S. Ishibashi, and H. Masayasu, Inhibition of superoxide anion production in guinea pig polymorphomuclear leukocytes by a seleno-organic compound, ebselen, J. Pharmacobio-Dyn. 10:595 (1987).
13. M.J. Parnham, S. Leyck, N. Dereu, J. Winkelmann, and E. Graf, Ebselen (PZ 51): A GSH-peroxidase-like organoselenium compound with anti-inflammatory activity, Adv. Inflam. Res. 10:397 (1985).
14. M.J. Parnham, S. Leyck, P. Kuhl, J. Schalkwijk, and W.B. van den Berg, Ebselen: a new approach to the inhibition of peroxide-dependent inflammation, Int. J. Tissue React. 9:45 (1987).
15. J. Schalkwijk, W.B. van den Berg, L.B.A. van de Putte, and L.A.B. Joosten, An experimental model for hydrogen peroxide-induced tissue damage. Effects of a single inflammatory mediator on (peri)articular tissues, Arthr. Rheum. 29:532 (1986).

16. T.E. Van Dyke, L. Braswell, and S. Offenbacher, Inhibition of gingivitis by topical application of ebselen and rosmarinic acid, <u>Agents and Actions</u> 19: 376 (1986).

17. I.A. Cotgreave, U. Johansson, G. Westergren, P.W. Moldéus, and R. Brattsand, The anti-inflammatory activity of ebselen but not thiols in experimental alveolitis and bronchiolitis, <u>Agents and Actions</u> 24: 313 (1988).

18. D.R. Blake, R.E. Allen, and J. Lunec, Free radicals in biological systems - a review orientated to inflammatory processes, <u>Br. Med. Bull.</u> 43:371 (1987).

19. H.P. Hartung, B. Schäfer, K. Heininger, and K.V. Toyka, Interference with arachidonic acid metabolism suppresses experimental allergic neuritis, <u>Ann. Neurol.</u> 20:168 (1986).

20. S. Ueda, T. Yoshikawa, H. Oyamada, T. Takemura, Y. Morita, N. Yoshida, S. Sugino, and M. Kondo, Effects of ebselen on acute gastric mucosal injury in rats, <u>in</u>: "Abstracts, International Conference on Medical, Biochemical and Chemical Aspects of Free Radicals", p 228, Kyoto (1988).

21. K. Sakuma, H. Masayasu, H. Shibata, and S. Ashida, Ebselen, a seleno-organic compound, inhibits leukocyte aggregation and leukotriene B_4 production in human and rats, <u>in</u>: "Abstracts 4th International Symposium in Selenium in Biology and Medicine", p88, Univ. of Tübingen (1988).

22. A. Wendel, and G. Tiegs, A novel biologically active seleno-organic compound VI. Protection by ebselen (PZ 51) against galactosamine/endotoxin-induced hepatitis in mice, <u>Biochem. Pharmac.</u> 35:2115 (1986).

23. A. Wendel, G. Tiegs, C. Werner, and N. Dereu, Interaction of ebselen with hepatic arachidonate metabolism <u>in vivo</u>, <u>Phosphorus and Sulphur</u> 38:59 (1988).

24. M. Akasaki, T. Ikeda, F. Numata, Y. Kurebayashi, and W. Tsukada, Effect of ebselen (PZ 51) in liver failure induced by <u>Propionobacterium acnes</u> (<u>P.acnes</u>), <u>in</u>: Abstracts, 4th International Symposium on Selenium in Biology and Medicine., p 3, Univ. of Tübingen (1988).

25. T. Hashizume, S. Ide, and Y. Shimoto, Effect of ebselen (PZ 51) on experimental liver injury models, <u>in</u>: Abstracts, 4th International Symposium on Selenium in Biology and Medicine., p 38, Univ. of Tübingen (1988).

26. W. Hagmann, and D. Keppler, Leukotrienes and other eicosanoids in liver pathophysiology, <u>in</u>: "The Liver: Biology and Pathobiology" 2nd Ed. I.M. Arias, W.B. Jakoby, H. Popper, D. Schachter, and D.A. Schafritz, ed., Raven Press, New York, p 793 (1988).

27. J. Tanaka, and F. Yamada, Ebselen (PZ 51) inhibits the formation of ischemic brain edema, <u>in</u>: Abstracts, 4th International Symposium on Selenium in Biology and Medicine., p 38, Univ. of Tübingen (1988).

28. L.D. Koller, and J.H. Exon, The two faces of selenium - deficiency and toxicity - are similar in animals and man, <u>Can. J. Vet. Res.</u> 50:297 (1986).

29. A. Müller, H. Gabriel, H. Sies, R. Terlinden, H. Fischer, and A. Römer, A novel biologically active selenoorganic compound VII. Biotransformation of ebselen in perfused rat liver, <u>Biochem. Pharmacol.</u> 37:1103 (1988).

30. R. Terlinden, M. Feige, and A. Römer, Determination of the two major metabolites of ebselen in human plasma by high-performance liquid chromatography, <u>J. Chromatography</u> 430:438 (1988).

GLUTATHIONE LEVEL IN MICE BRAIN AFTER TESTOSTERONE ADMINISTRATION

Faik Atroshi[1], Lars Paulin[1], Tiina Paalanen[1], and
Tuomas Westermarck[2]

[1]Department of Pharmacology and Toxicology, College of
Veterinary Medicine. PO Box 6, SF 00550, and [2]Helsinki
Central Institution for the Mentally Retarded Kirkkonummi,
Finland

INTRODUCTION

Glutathione (GSH), an SH containing tripeptide, is a powerful
antioxidant protecting against free radical damage (Meister and Anderson,
1983). Glutathione has a role in the inactivation of a number of drugs
and in the metabolic processing of certain endogenous compounds, such as
estrogen, prostaglandins and leukotrienes. Such inhibitors and other
compounds that increase GSH synthesis, make it possible to effectively
manipulate the metabolism of this compound. GSH is also a coenzyme for
several enzymes (Meister, 1984). Since GSH serves effectively in the
detoxication of many drugs, the GSH status of an animal is of importance
in protection against toxicity. Munthe et.al. (1981) reported that the
erythrocyte GSH level increased in RA (Rheumatoid Arthritis) patients
receiving "2nd line drugs". This increase is suggested to precede
clinical improvement, and is thus potentially a unique biochemical
parameter for monitoring therapy. Recently, Igarashi et.al. (1984)
reported that hepatic glutathione S-transferase activity is enhanced
significantly in rats after testosterone administration. Since little
information is available concerning the behaviour of GSH in the mouse
brain after testosterone administration, we examined this effect.

MATERIALS AND METHODS

Male Balb/c mice (about 20 g body weight) were used. Animals were
allowed standard laboratory chow and water ad libitum and maintained on a
constant 12 h-ligth/12 h-dark lighting schedule. The mice were treated
with androgen by a subcutaneous injection of testosterone propionate
solution (4 mg/ml sesame oil) at a dose of 100 mg/kg three days before
death. The control animals received the vehicle alone. The whole brain of
the ether-anesthetized mice were excised quickly and chilled in saline at
0°C. The samples, usually 10 - 30 mg, were quickly weighed and
homogenized in 2.0 ml of 0.3 M phosphate buffer, pH 7.4 in a Potter
Elvehjelm type homogenizer. The homogenates were centrifuged for 5 min at
10,000 g. The resulting supernatant was then centrifuged at 20,000 g for
20 min at 4°C. The GSH contents were measured by the method described by
Atroshi and Sandholm (1982) using Ellman´ s (1959) reagent.

Antioxidants in Therapy and Preventive Medicine
Edited by I. Emerit *et al.*
Plenum Press, New York, 1990

The effect of the testosterone treatment on the glutathione content of the mice brain is shown in Fig. 1. A significant increase ($p < 0.05$) in the GSH content was noticed in the treated animals (2.24 ± 0.31 µmol/mg fresh wt) as compared with the controls (1.50 ± 0.17 µmol|mg fresh wt). A lot of investigation has been focused on the concentrations of glutathione in the brain and other tissues, mainly because this tripeptide is considered to have important functions in protecting cells against oxidative damage (Flohe´ et.al. 1984). In this capacity, reduced glutathione can either act as a free radical scavenger (Aebi and Suter, 1974), or reduce disulfides by thiol-disulfide exchange mechanisms (Eldjarn and Pihl, 1960). Several drugs are reported to elevate the activity of glutathione (Rosalki et.al., 1971). Powell (1984) reported that the treatment of rats with N-acetylcysteine (1 mg/kg, i.p.) increased whole brain reduced non-protein sulfhydryl content by 27%. The status of endogenous brain GSH plays an important role in determining the sensitivity of the brain to the development of O_2 toxicity. On the other hand, Hothersall (1985) reported that 6-aminonicotinamide treatment resulted in a decreased glutathione level in the rat nervous system. More recently, Mercurio and Combs (1985) demonstrated that anruthioglucose and D(-)- penicillamine hydrochloride inhibit selenium-dependent glutathione peroxidase from brain homogenate in vitro.

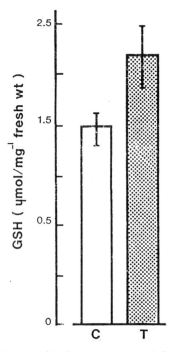

Fig. 1. Brain glutathione content was measured 3 days after treatment. The contents are expressed as means and standard deviations of 15 animals per group. C: control and T: treated.

The present study indicates that the activity of brain GSH in mice can be induced after a single dose of testosterone. However, the GSH values measured were somewhat lower, though not significantly, than those reported earlier (Mokrasch and Teschke, 1984). Changes in brain GSH concentration may either be reciprocal or involve alterations in the glutathione status. However, there seems to be several possible explanations for this discrepancy.

The discrepancy could be explained by the enhancement of glutathione reductase activity. However, oxidized glutathione (GSSG) levels were not included in the present analysis; thus it is not known whether the activity increased or decreased. Another explanation is provided by the possible involvement of γ-glutamyltransferase (γ-GTP). Glutathione synthesis involves two sequential enzymatic steps. The first synthetizing enzyme is γ-glutamylcysteine synthetase, which is probably rate-limiting (Meister, 1974), and the second one is glutathione synthetase. It has been shown that γ-GTP is primarily extracellular (Hahn et.al., 1978) and associated with the cell surface (Cheng et.al., 1978). Furthermore, it has been proposed that GSSG, and not GSH, is the preferred substrate for γ-GTP and that at least the initial reaction of GSSG metabolism is catalyzed by γ-GTP (Jones et.al., 1979). The physiological role of γ-GTP is suggested to be the reclamation of extracellular GSSG (Hahn et.al., 1978).

In conclusion the mechanisms that cause GSH to rise are not readily understood. Tissue destruction might be one source, but that would not increase the amount of GSH. However, it is known that GSH is virtually absent from extracellular tissue fluid, therefore an increase in total glutathione can be only partially explained by an unbalance of the GSH|GSSG ratio v formar. The higher GSH levels can be due either to an increase of GSH synthesis brought about by glutathione-synthetase stimulation, or a reduction of the GSH used to detoxify the free radicals produced by cell membrane lipid peroxidation. Moreover, both mechanisms counteract the GSH content feedback on the synthetizing enzymes (for ref. see Atroshi and Westermarck, 1987). Previously we have demonstrated (Atroshi and Westermarck, 1987) that GSH levels may be increased in erythrocytes of patients with neurological disorders such as neuronal ceroid lipofuscinosis and subacute sclerosing panencephalitis.

REFERENCES

Aebi,H. and Suter,H., 1974, In: Glutathione. Proceedings of the 16th Conference of the German Society of Biological Chemistry. Flohe, L.,Benöhr, H.Ch., Sies, H., Waller, H.D. and Wendel, A. (Eds.). Tübingen Thieme, Stuttgart, pp. 192.

Atroshi, F. and Sandholm, M., 1982, Red blood cells glutathione as a marker of milk production in Finn sheep,Res. Vet. Sci. 33, 256-259

Atroshi, F. and Westermarck, T., 1987, In: Free radicals, oxidant stress and drug action. Rice-Evans, C. (Ed.). Richelieu Press, London, pp. 419-424.

Cheng, S., Nassar, K. and Levy, D., 1978, gamma-glutamyl transpeptidase activity in normal, regenerating and malignant hepatocytes, FEBS Lett. 85, 310-312.

Eldjarn, L. and Pihl, A., 1960, In: Mechanisms in Radiobiology. Emrera, M. and Forsberg, H. (Eds.). Academic Press. New York, pp. 231.

Ellman, G.L., 1959, Tissue sulphhydryl groups, Arch. Biochem. 82, 70-77.

Flohe, L., Benöhr, H., Sies, H., Waller, H.D. and Wendel, A. (Eds.), 1974, In: Glutathione. Proceedings of the 16th Conference of the German Society of Biological Chemistry. Tübingen Thieme, Stuttgart, pp. 316.

Hahn, R., Wendel, A. and Flohe, L., 1978, The fade of extracellular glutathione in rat, Biochim. Biophys. Acta, 539, 324-337.

Hothersall, J., 1985, , Regional changes in glutathione and ascorbate levels in the rat nervous system resulting from 6-aminonicontinamide treatment,J. Neurochem. Suppl. 44, SI88.

Igarashi, T., Satoh, T., Ono, S., Iwashita, K., Hosokawa, M., Ueno, K. and Kitagawa, H., 1984,Effect of steroidal sex hormones on the sexrelated differences in the hepatic activities of gamma-glutamyl transpeptidase, glutathione S-transferase and glutathione peroxidase in rats, Res. Commun. Chem. path. Pharmacol. 45, 225-232.

Jones, D.P., Moldens, P., Stead, A.H., Ormstad, K., Jörnvall, H. and Orrenius, S., 1979, Metabolism of glutathione and a glutathione conjugate by isolated kidneys cells, J. Biol. Chem. 254, 2787-2792.

Meister, A., 1974, In: Glutathione: Metabolism and Function. Arias, J.M. and Jakoby, W.B. (Eds.). Raven Press, New York, pp. 35.

Meister, A., 1984, New aspects of glutathione biochemistry and transport. Selective alteration of glutathione metabolism, Nutrition Rev., 42, 397-410.

Meister,A. and Anderson, M.E., 1983, Glutathione Annu. Rev. Biochem. 52, 711-760.

Mercurio, S.D. and Combs, Jr.G.F., 1985, Drug-induced changes in selenium dependent glutathione peroxidase activity in the chick. J. Nutrition 115, 1459-1465.

Mokrasch, L.C. and Teschke, E.J., 1984, Glutathione content of cultured cells and rodent brain regions: Specific fluorometric assay. Anal. Biochem. 140, 506-509.

Munthe,E., Kass, E. and Jellum, E., 1981, D-penicillamine-induced increase in intracellular glutathione correlating to clinical response in rheumatoid arthritis. J. Reumatology Suppl. 7, 14.

Powell, S.R., 1984, The role of brain glutathione in the rate of developement of central nervous system oxygen toxicity. Dissertion Abstr. Inter. 44, 2396-B.

Rosalki, S.B., Tarlow, D. and Rau, D., 1971, Plasma gamma-glutamyl transpeptidase elevation in patient receiving enzyme inducing drugs Lancet 2, 376-377.

PROSTAGLANDINS, GLUTATHIONE METABOLISM, AND LIPID PEROXIDATION IN

RELATION TO INFLAMMATION IN BOVINE MASTITIS

Faik Atroshi[1], Satu Sankari[2], Aldo Rizzo, Tuomas Westermarck[3], and Jouko Parantainen

[1]Dept. of Pharmacology & Toxic., [2]Biochem., College of Veterinary Medicine., P.O. Box 6, Helsinki SF 00550, and [3]Helsinki Central Inst. Ment. Ret., Kirkkonummi Finland

INTRODUCTION

Bovine mastitis is an infectious disease causing inflammatory changes in the udder. Infection, inflammation and tissue injury are usually associated with lipid peroxidation and the formation of free radicals. We have studied the possible roles of some local tissue factors in mastitis. Among these, prostaglandins (PGs) are typical mediators of inflammation. The metabolism of glutathione (GSH) is in many ways involved in tissue protection particularly in limitation of excessive lipid peroxidation. GSH and GSH-enzymes like GSH-peroxidase (GSH-Px) are also closely involved in the metabolism of arachidonic acid to prostaglandins and other biologically active lipids.

CHANGES IN PROSTAGLANDIN LEVELS DURING MASTITIS

Prostaglandins (PGs) constitute a whole family of peroxidized lipids formed in most animal cells. Almost any kind of stimuli be it mechanical, chemical, physiological or traumatic, may initiate the formation of different kinds of PGs. Thus the particular importance of the local PG-impact is usually very difficult to evaluate. Some PGs, like PGEs and PGI_2 (prostacyclin) are potent vasodilators and they might participate in the generation of inflammatory symptoms. The main roles for PGEs and PGI_2 in inflammation may in fact be in generation of hyperalgesia, sensitization of the tissue to the irritant and pain producing activity of the amine and peptide type of mediators of inflammation[1]. On the other hand these PGs have marked tissue protective functions, e.g. in preventing vasoconstriction and platelet aggregation. Other prostanoids (PG-like substances) like PGD_2, $PGF_{2\alpha}$ and thromboxane A_2 (TXA_2) are mostly vasoconstrictors. Their formation may be associated with allergic and other reactions of hypersensitivity, and TXA_2 is a very potent aggregator of platelets.

The presence of PGs in the normal milk of different animal species is well established. Giri and coworkers[2], have demonstrated that milk levels of $PGF_{2\alpha}$ and TXB_2 (the metabolite of TXA_2) may increase markedly the in experimental bovine mastitis triggered by an infusion of bacterial endotoxin. Our studies in spontaneous bovine mastitis revealed that both the milk and blood levels of PGs were elevated, as in any typical

Antioxidants in Therapy and Preventive Medicine
Edited by I. Emerit *et al.*
Plenum Press, New York, 1990

203

inflammatory reaction[3,4]. The concentration of PGE_2, $PGF_{2\alpha}$ and TXB_2 in healthy milk samples were 44, 118, and 244 picograms/ml, and in mastitic samples 61, 135, and 357 pc/ml, correspondingly.

The sources of prostaglandins are not known very well and are a matter of speculation. Several possiblities are to be considered. First of all, bacterial toxins might have contributed to the PG-release, This was clearly demonstrated by Giri and coworkers[2], in experimental bovine mastitis, and similar mechanisms might operate in the spontaneous disease as well. Secondly, the production of PGs is greatly increased by polymorphonuclear leukocytes. Neutrophil invasion is a typical feature in mastitis, and in our results[3] the PGs correlated fairly well with the somatic cell counts (r values were between 0.63-0.68, P < 0.01). This was the best parameter predicting the PG-level. Thirdly, changes in tissue protein and electrolyte contents are factors that have marked effects on PG-production. Albumin is a typical factor increasing the formation of PGs, particularly $PGF_{2\alpha}$. In our material the blood samples from mastitic animals plasma PGS_2 was in a positive correlation with serum albumin (r = 0.43, P < 0.05), and finally, other factors that may have contributed are inflammatory mediators (e.g. monoamines and peptide hormones) and even mechanical stimulation of the sensitized udder during milking is a possible factor favouring PG-production.

Role of GSH-enzymes in mastitis

The tissue content of glutathione (GSH) is normally very high, in some tissues up to 5 millimolar level. The functions of GSH are often tissue protective, and there are numerous enzymes in which GSH plays a central role as a cofactor. Typical GSH-enzymes include GSH-peroxidase (GSH-Px), located in the circulation almost exclusively in the red cells, various GSH-transferases which possess peroxidase-like activity and bind chemicals and γ-glutamyl transferase which reflects the function of the liver and is involved in the transport of amino acids accross the cell membrane. GSH is also consumed by some cytochromes, most notably cytochrome P-450. Recently we have demonstrated the presence of cytochrome P-450 in severly inflamed udder tissues[5]. Together with GSH-Px cytochrome P-450 might participate in the proper handling of oxygen-free radicals in host defence (Fig 1).

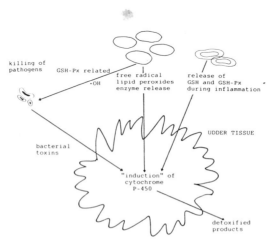

Fig.1 Possible roles of GSH and GSH enzymes, derived from erythrocytes or white cells, free radical related processes associated with microbial killing, and bacterial toxins in "induction" of cytochrome P-450 in inflamed udder tissue[5].

Furthermore, we have demonstrated[6] that tissue GSH was more than doubling in samples obtained from animals with mastitis when compared to the healthy ones. Like GSH also GSH-Px the seleno-enzyme, may have importance in tissue protection and microbial killing. The bacteriocidal activity of neutrophils depends on an adequate intake of selenium, the essential cofactor of GSH-Px[7]. Although the tissue content of GSH-Px was increased in the inflamed udder[6], in erythrocytes obtained from mastitic cows the enzyme levels were lowered[4]. It is possible that besides pyogenic bacteria, depletion of the enzyme from the erythrocytes could have increased the mucous components in the udder tissue. It is considered, that erythrocyte breakdown due to free radical formation and lipid-peroxidation related to inflammation could be responsible for the increase in the GSH-metabolism, and leak into the tissue along with the other plasma components and might well support prostaglandins formation (Fig. 2).

Free fatty acids involvement in mastitis

The whole concept free fatty acids (FFAs) implies the existence of a large group of lipid, without any specific attention paid to the individual components. Considering these lipids only as a mobile reservoir of fuel to be combusted in mitochondria may be rather fruitless, however, and much of the information may be missed. The degree of unsaturation is one aspect of particular interest. The same polyunsaturated fatty acids may have structural importance in the biological membranes, and when released with the other FFAs, they may be metabolized to biologically active compounds, including prostaglandins and leukotrienes (LTs), typical mediators of inflammation. Milk FFA content of unsaturated linoleic and linolenic acid are increased in mastitis[9], and these lipids may have importance in the killing of bacteria[10]. As a group these autacoids and their precursors are known as eicosanoids, i.e. metabolites of arachidonic acid and other polyunsaturated fatty acids. There is a growing body of evidence linking particularly these lipids with inflammation, host derence, and tissue destruction[11]. Recently, we have reported mastitis-related increase of FFA in milk[10]. By increasing the formation of PGs and leukotrienes, the FFA release may contribute to vasodilation, irritation of the tissue, sensitivity to pain (hyperalgesia), chemokinesis of polymorphonuclear neutrophils (PMN), and the peroxidative processes involved in the formation of oxygen free radicals.

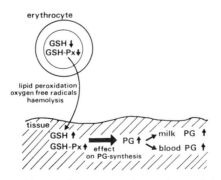

Fig.2 Possible interaction between erythrocyte and inflamed udder tissue during mastitis. Oxygen free radicals produced by hypoxia, bacterial toxins and phagocytosis increase lipid peroxidation and the peroxidative stress in the erythrocyte. GSH and GSH-Px are lowered which may impair resistance of the erythrocyte against haemolysis. The thiols released from the red cells may accumulate in the tissue favouring the metabolism of arachidonic acid to specific directions[8].

Briefly, mastitis is an inflammatory reaction caused by bacteria, but local tissue factors may also be important determining the resistance of the host. Glutathione and GSH-enzymes have regulatory and antioxidant roles in lipid peroxidation. GSH-peroxidase (GSH-Px) protects the tissues against free-radical damage. Free radicals may be important in aging and in various disorders including inflammation. Selenium administration is known to enhance the host immune response, which is clearly a factor in the defense against some disease states in animals. In neutrophils Se may be needed for the production of hydroxyl radical. GSH-S-transferase participate in several steps of the leukotriene (LT) synthesis. GSH-S-transferase and GSH also catalyze the formation of prostaglandins (PG) of the E type. We have studied GSH-metabolism and lipid peroxidation in relation to caprine and bovine mastitis. Prostaglandins and free fatty acids, including arachidonic acid, were elevated in blood and milk, while selenium and erythrocyte GSH-Px were decreased. Also cytochrome P-450, another GSH-enzyme, was detected. It is suggested that the elevation of GSH-enzymes may be an important factor in the regulation of lipid peroxidation, formation of prostaglandins, and the effectiveness of host defence.

REFERENCES

1. S. H. Ferreira, Site of analgesis action of aspirin-like drugs and opioids, in: "Mechanism of pain and analgetic compounds", R.E.Basset, eds., Raven Press, New York (1979).
2. S. N. Giri, Z. Chen, E. J. Carrol, R. Mueller, M. J. Schiedt, and L. Panico, Role of prostaglandins in pathologenesis of bovine mastitis indnced by Escherichia coli endotoxin, Am J Vet Res. 45:586 (1984).
3. F. Atroshi, J. Parantainen, R. Kangasniemi, and T. Österman, Milk prostaglandins and electrical conductivity in bovine mastitis, Vet Res Commun. 11:15 (1987).
4. F. Atroshi, J. Parantainen, S. Sankari, and T. Österman, Prostaglandins and glutathione peroxidase in bovine mastistis, Res Vet Sci. 40:361 (1986).
5. F. Atroshi, P. Kaipainen, and J. Parantainen, Evidence for the presence of cytochrome P-450 in mastitic bovine mammary gland, Pharm Res Commun. 19:673 (1987).
6. F. Atroshi, J. Parantainen, R. Kangasniemi, and S. Sankari, Sialic acid, glutathione metabolism, and electrical conductivity in bovine mastitic udder tissue, J Anim physiol a Anim Nutr. 58:200 (1987).
7. R. Boyne, and J. R. Arthur, Alterations of neutrophils function in Se-deficient cattle, J Compar Path. 89:151 (1979).
8. F. Atroshi, S. Sankari, T. Österman, R. Kangasniemi, and J. Parantainen, Possible interaction between prostaglandins and glutathione metabolism in bovine mastitis, in: "Proceedings of symposium on mastis control and hygienic production of milk", M.Sandholm, ed. Orion Corpuration Ltd, Espoo, Finland (1986).
9. F. Atroshi, A. Rizzo, T. Österman, and J. Parantainen, Free fatty acids and lipid peroxidation in normal and mastitic milk, J Vet. Med. (in Press).

10. J. S. Hogan, J. W. Pankey, and A. H. Duthie, Growth inhibition of mastitis pathogens by long-chain fatty acids, J Dairy Sci. 70:927 (1987).
11. F. Kuehl, and R. Egan, Prostaglandins, arachidonic acid, and inflammation, Science 210:978 (1980).

SULFUR CONTAINING COMPOUNDS AS ANTIOXIDANTS

Patrick M. DANSETTE, Amor SASSI,
Colette DESCHAMPS and Daniel MANSUY

Laboratoire de Chimie et de Biochimie Pharmacologiques et
Toxicologiques, Unité associée au CNRS en développement
concerté avec l'INSERM, UA 400 CNRS, Université René
Descartes, 45 rue des Saints-Pères 75270 PARIS CEDEX 06
FRANCE

ABSTRACT

Several sulfur containing antioxidants have been used for therapeutic
use as hepatoprotectors and radioprotectors. In particular the
dithiolthione SULFARLEM, used as anticholestatic drug for more than 40
years, is shown to be antioxidant; preliminary pharmacokinetic data in
man are presented, and hepatoprotective and radioprotective effect in
mice are demonstrated.

INTRODUCTION

The organism is constantly exposed to deleterious effect of oxygen
derived radical species, but many defensive mechanisms have evolved to
protect it against these radicals. The aim of our paper is to review a
certain number of sulfur containing drugs which have been proposed to
fight against these toxic radical processes.

Several enzymes are important for these defense mechanisms:
- Superoxide dismutase which transforms superoxide ion into
hydrogen peroxide
- Catalase which produces oxygen and water from hydrogen peroxide
- Glutathione peroxidase able to reduce hydrogen peroxide into
water, lipid hydroperoxides into fatty alcools and which oxidizes
glutathione (GSH) into GS-SG.
Therefore glutathione plays a pivotal role in the cell to detoxify
radicals and other electrophilic species, in particular with the help of
glutahione-S-transferases, and its level has to stay high in the cell
(the normal concentration of GSH in hepatocytes is about 5-6 mM).
The total amount of glutathione in the cell will depend of its
consumption and its production by:
- reduction of GS-SG catalysed by glutathione reductase at the
expense of NADH.
- its neosynthesis from endogenous or exogenous substrates which
should be able to penetrate into the cells.
- inhibition by antioxidants of its destruction.

Antioxidants in Therapy and Preventive Medicine
Edited by I. Emerit *et al.*
Plenum Press, New York, 1990

209

THIOL DERIVATIVES AS GLUTATHIONE PRECURSORS

In the last 20 years many compounds have been tested as precursors of GSH, or as inducers of the enzymes participating in its synthesis, since GSH itself does not penetrate well into the cells. As shown in Table 1, several thiol containing compounds are effective glutathione precursors or regenerators, but many cannot be used because although effective in vitro, they do not penetrate well in the cells in vivo or are not well absorbed, or they have minor or major adverse reactions. For instance methionine given at sufficiently high dose to protect against the toxicity of paracetamol overdose causes nausea and vomiting (Table 1).

Table 1. Glutathione precursors or regenerators

Substance	Properties and adverse effects	Use	Reference
S-adenosylmethionine	Does not penetrate well	A	
Dithiocarbamate	Used mostly as heavy metal trap	A,H	(5) 8444
Propylthiouracil			7770
Dimethylsulfoxide	only external use	A,H	3255
Cysteine	does not penetrate well	A	(5,6)
Methionine	idem, causes nausea, inefficient	A,H	5849
Cysteamine	penetrates, efficient as radioprotector and in paracetamol poisoning GI and CNS toxicity	A,H	(5) 2773
N-acetylcysteine	penetrates well, prodrug of cystein efficient within 10h, minor allergy	A,H	(2,3) 82
oxo-thiazolidine-carboxylate	penetrates well, substrate of oxoprolinase, prodrug of cysteine	A	(1)
Timonacic acid	penetrates well, used in man as hepatoprotector, overdose causes convulsions	A,H	9285

a) Used in animals only (A), or human (H)
b) Number of reference list or of Merck index 10th edition (1983).

Recently 2-oxothiazolidine-4-carboxylate ((Fig. 1) has been described by the group of Meister (1) as an excellent prodrug of cysteine since it is well absorbed and penetrates easily into the cell where oxoprolinase hydrolyses it to cysteine and CO_2. However this compound has not yet been tested extensively in human and therefore is far away of becoming a drug in human. On the contrary Timonacic acid (Fig. 1), which is also probably a prodrug of cysteine, is on the market for many years as hepatoprotective agent, and is used in France in hepatitis treatment. However overdosage can cause convulsions.

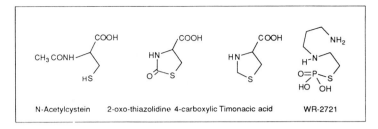

Fig. 1. Structures of glutathione precursors or regenerators.

The only drug from the list of table 1 which has been used extensively in human is N—acetylcysteine. This drug was shown by Mitchell (2) to clearly protect against the toxicity of CCl4 *in vivo* in mice and was used orally in human as lung fluidizing drug. Use of this drug as a perfusate has allowed to save patients after paracetamol poisoning. It was shown by Prescott (3) that, in man, perfusion within 10 hours after paracetamol intake was efficient in protecting against fulgurant hepatitis, the earliest being best. Only minor adverse reactions of this treatment (always performed in intensive care sections) are reported.

Cysteine and cysteamine have been used as radioprotectors in animals and human, but they have in human major inconvenience. More recently, a derivative of cysteamine has been proposed by the Walter Reed Institute, WR—2721 (Fig. 1), as a radioprotective agent; this compound is very efficient in animals but very little is published of its human use (classified informations?) (4).

Diethyldithiocarbamate has been marketed as a drug against heavy metal poisoning, as it is a very good chelator of lead or mercury. It has been shown that it also protects against the hepatotoxicity of CCl_4 or galactosamine (5) and that it is a radioprotector (4).

The dimer of diethyldithiocarbamate is disulfiram (Antabuse), a very well known inhibitor of alcohol dehydrogenase used in human, which has been shown also to have some protective effect in radical induced hepatotoxocity in animal (5) but at a too high dosage to be usefull in human.

ANTIOXIDANTS WHICH INCREASE GSH LEVEL

In contrast of all the preceding compounds which act as potential precursors of GSH, some compounds have been shown to induce the enzymes participating in glutathione synthesis (fig.2). In particular, compounds related to the lipoic acid amide (lipoamide), derivated from dithiocyclopentenes, has been shown to increase both glutathione and gluthathione transferase level: 1,3-dithiol like Malotilate and 1,2-dithiol-3-thiones like Sulfarlem (ADT) and Oltipraz (fig.2).

MALOTILATE has been marketed in Japan (KANTEC™) in 1985 with the claim that it protects the liver against several acute drug toxicities and as hepatoprotector" for the improvement of liver function in compensated liver cirrhosis", and that it "accelerates protein metabolism in liver, activates liver functions and suppress liver fibrosis". The posology used is high (0.75 - 2g per day) and many

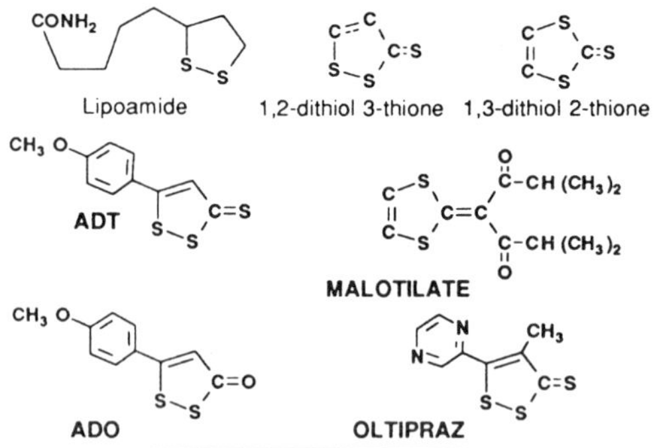

Fig. 2. Structures of analogs of lipoamide as hepatoprotectors.

adverse reactions have appeared, with some elevation of transaminases
during the first week, some increase of cholesterol level,
gastrointestinal problems, nausea and vomiting, and some allergic
reactions. The mechanism of action of malotilate is still not well
understood. It protects hepatocytes against cell damage in vitro and is
able to increase GHS transferase and GHS content in vivo (6). It also
prevents liver collagen accumulation (6,7) and fibroplast migration in
vitro.

1,2-DITHIOL-3-THIONES

Recently the group of Bueding has demonstrated that
1,2-dithiol-3-thione found in cabbage and other food and the two
derivatives, Oltipraz an antipaludic drug and Sulfarlem (ADT) an
anticholeretic agent marketed since 1947 and used as hepatoprotector and
hypersialic agent, were able to increase markedly the level of
glutathione and glutathione-S-transferase (8) and to protect rats
against the toxicity of CCl_4. Moreover this group has shown that
Oltipraz is a good antipromotor in the two steps skin carcinogenesis
model of Berenblum (9,10), and Talalay recently demonstrated that
dithiolthiones are inducers of quinone reductases (two electron
reduction without radical formation) and may inhibit a signal necessary
for one of the carcinogenic steps (11).

We have demonstrated that Oltipraz and Sulfarlem are good
antioxidants in vitro and in vivo in rat and mice (12,13) and that
Sulfarlem is able to protect mice against the hepatotoxicity and
mortality induced by overdoses of paracetamol (12).

Since the pharmacokinetics of this drug had not been investigated,
it was of interest to study its metabolism and its disposition. The
principal metabolites in vivo and in vitro were identified as shown on
Fig.3, and the potency of each as antioxidant in vitro and as
hepatoprotector in vivo was tested. The metabolite
5-anisyl-1,2-dithiol-3-one ADO and the derivatives having lost the
thione group were inactive as antioxidant but the conjugated
5-parahydroxyphenyl-1,2-dithiol-3-thione (sulfate and glucuronide) were
good antioxidants.

Thus preliminary pharmacokinetic data in mice are shown in Fig.4.
The concentration in plasma of dithiolthione derivatives was practically
constant between 2 and 3 h after treatment (100 mg per kg i.p. in corn

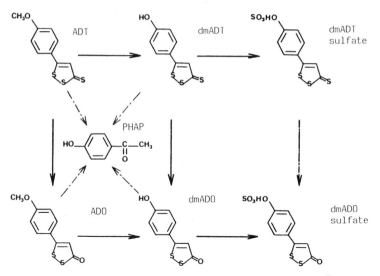

Fig. 3. Structures of the metabolites of ADT in rats and mice.

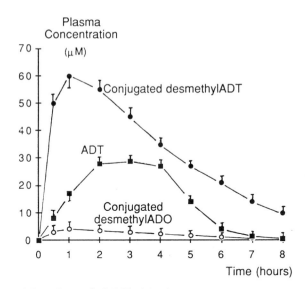

Fig. 4. Pharmacokinetics of SULFARLEM in mice.
Swiss NMRI female mice (25 g) were treated i.p. by 100 mg ADT/kg in corn oil, and determination of plasma concentration of metabolites was made by HPLC; each value is the mean $\pm$ s.e.m. of 3 different determinations in 3 different mice.

oil). Since ADT is a good hepatoprotector in mice against hepatotoxic agents known to cause oxidative stress, we thought that it could also be a trapping agent for radicals induced by radiations. The results of a radioprotection experiment where NMRI Swiss female mice were irradiated by gamma rays (cf. Sorenson, this volume) 2 h after treatment by corn oil or ADT in oil are shown in Fig.5 . ADT at the dose of 100 mg per kg was an excellent radioprotector, since, at the dose of 8.25 gray, 53% of the control mice died and only 11% of the treated mice. A similar study was repeated but treatment was done after irradiation. Preliminary data show that the drug is nearly as effective in these conditions.

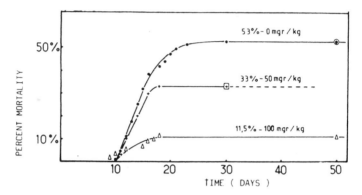

Fig. 5. Radioprotective effect of SULFARLEM (ADT).
NMRI Swiss female mice were treated i.p. by corn oil (0,25 ml sulfarlem in corn oil (100 mg/kg) and irradiated 2 to 2,5 h latter by gamma rays from a cobalt source (2,5 gray/min, 8,25 Gy). Then they were randomly distributed in cages and twice a day mortality was monitored.

CONCLUSION

In this paper we have first reviewed a certain number of thiol derivatives which are usefull hepatoprotectors or glutathione precursors, the only one used extensively being N-Acetylcysteine. Some heterocyclic sulfur derivatives have recently been presented as potential hepatoprotectors. Malotilate, a 1,3-dithiol, is still under clinical trials in Europe and the benefit/risk is not fully evaluated ; Sulfarlem, a 1,2-dithiol-3-thione, marketed for more than 40 years without adverse reactions, is effective as antioxidant and probably as radical trap in hepatotoxic reactions involving an oxidative stress, but it is also a good radioprotective agent in mice. It could perhaps be used as hepatoprotector after intoxication in human, and also it could be used as a preventive agent for people at risk of radiation exposure.

Acknowledgments

We are grateful to Aliette Deysine and Professor C. Ferradini for their decisive collaboration in the radioprotection experiments, and we thank Dr. M.O. Christen and Laboratoires LATEMA for their financial and scientific support.

REFERENCES

1) J. M. Williamson and A. Meister, Stimulation of hepatic glutathione formation by administration of L-2-oxothiazolidine-4-carboxylate, a 5-oxo-L-prolinase substrate, Proc.Nat.Acad.Sci. USA. 78:936-9 (1981).

2) B.H. Lauterburg, G. B. Corcoran and J.R. Mitchell, Mechanism of action of N-acetylcysteine in the protection against the hepatotoxicity of acetaminophen in the rats in vivo, J. Clin. Invest. 71:980-91 (1983).

3) L.F. Prescott, R.N. Illingworth, J.A.J.H. Critchley, M.J. Stewart, R.D. Adam and A.T. Proudfoot, Intravenous N-acetylcysteine: the treatment of choice for paracetamol poisoning, Brit. Med. J. 2:1087-1100 (1979).

4) M.R. Landauer, H.D. Davis, J.A. Dominitz and J.F. Weiss, Comparative behavioral toxicity of four sulhydryl radioprotective compounds in mice: WR-2721, cysteamine, diethyldithiocarbamate, and N-acetylcysteine, Pharmacol. Ther., 39:97-100 (1988).

5) M. Younes, C. Sause, C.P. Siegers, R. Lemoine, Effect of deferrioxamine and diethyldithiocarbamate on paracetamol-induced hepato-and nephrotoxicity.The role of lipid peroxidation, J. Appl. Toxicol. 8:261-5 (1988).

6) P.R. Ryle and J.M. Dumont, Malotilate: The new hope for a clinically effective agent for the treatment of liver disease, Alcohol-Alcohol, 22:121-141 (1987).

7) A. Poeschl, D. Rehn, J.M. Dumont, P.K. Mueller and G. Hennings, Malotilate produces collagen synthesis and cell migration activity of fibroplast in vitro, Biochem.Pharmacol. 36:3957-63 (1987).

8) S.S. Ansher, P. Dolan and E. Bueding, Chemoprotective effects of two dithiolthiones and of butylhydroxyanisole against carbon tetrachloride and acetaminophen toxicity, Hepatology 3:932-5 (1983).

9) T.W. Kensler, P.A. Egner, M.A. Trush, E. Bueding and J.D. Groopman, Modification of aflatoxine B1 binding to DNA in vivo in rats fed phenolic antioxidants, ethoxyquin and a dithiolthione, Carcinogenesis 6:759-63 (1985).

10) T.W. Kensler, P.A. Egner, P.M. Dolan, J.D. Groopman and B.D. Roebuck, Mechanism of protection against aflatoxin tumorigenicity in rats fed by 5-(2-pyrazinyl)-4-methyl-1,2-dithiol-3-thione (oltipraz) and related 1,2-dithiol-3-thiones and 1,2-dithiol-3-ones, Cancer Res. 47:4272-7 (1987).

11) P. Talalay, M.J. Delong and H.J. Prochaska, Indentification of a common chemical signal regulating the induction of enzymes that protect against chemical carcinogenesis, Proc. Nat. Acad. Sci. USA. 85:8261-5 (1988).

12) D. Mansuy, A. Sassi, P.M. Dansette and M. Plat, A new potent inhibitor of lipid peroxidation in vitro and in vivo, the hepatoprotective drug anisyldithiolthione, Biochem. Biophys. Res. Commun. 135:1015-21 (1986).

13) P.M. Dansette, A new potent inhibitor of lipid peroxidation in vitro and in vivo, the hepatoprotective drug anisyldithiolthione, in " Local immunity: Tissue fibrosis; immune cells and mediators", J.P. Revillard and N. Wierzbicki, ed., Fondation franco-allemande, Suresnes, France, 3:203-208 (1987).

14) P. Dansette, A. Sassi, M. Plat, M.O. Christen and D. Mansuy, Strong inhibition of lipid peroxidation in vitro and in vivo by dithiolthione derivatives, in "Drug metabolism from molecules to man", B.J. Benford, J.W. Bridges and G.G. Gibson, ed., Taylor and Francis, London, 755-9 (1987).

ZINC - A REDOX-INACTIVE METAL PROVIDES A NOVEL APPROACH FOR PROTECTION AGAINST METAL-MEDIATED FREE RADICAL INDUCED INJURY: STUDY OF PARAQUAT TOXICITY IN *E. COLI.*

Mordechai CHEVION[1,4] , Pnina KORBASHI[1], Joshua KATZHANDLER[2] and Paul SALTMAN[3]

The [1]Departments of Cellular Biochemistry and [2] Pharmaceutical Chemistry, Hebrew University of Jerusalem, ISRAEL 91010, the [3] Department of Biology, University of California at San Diego, La Jolla, CA 92093, and [4] The Molecular Toxicology Research Group, Oklahoma Medical Research Foundation, Oklahoma City, OK 73104

ABSTRACT

The essential mediatory role of copper and iron in a variety of free radical-induced injuries, including paraquat-induced biological damage has been recently demonstrated. It was postulated that these transition metals undergo cyclic redox reactions, and serve as centers for repeated production of hydroxyl radical, which are the ultimate deleterious agents. Additionally, we had presented evidence indicating efficient protection against paraquat toxicity by agents commonly employed (chelators, chemical scavengers and protecting enzymes).

In this study we have used the E. coli model in order to develop a new approach for protection against paraquat-induced metal-mediated cellular injury. It entails the administration of excess zinc (up to 50 fold over copper), which results in an inhibition of the toxic effect of paraquat. Lineweaver- Burk analysis demonstrates the competitive mode of this inhibition. The suggested mechanism involves the displacement of the redox-active copper (or iron) from its binding site and by this diverting the site of repeated production of free radicals. Thus, use of redox-inactive metals, which possess high similarity of their ligand chemistry, to that of iron and copper but are of relative low toxicity by themselves, should be considered for intervention in paraquat toxicity and in other metal-mediated free radical-induced injurious processes.

Introduction

Paraquat (1,1'-dimethyl-4-4' bipyridinium dichloride) (PQ^{+2}), also known as methyl viologen, is a widely used herbicide (1,2). The broad interest in PQ^{+2} arises from its agricultural, biological and chemical properties. Since the initiation of its commercial use in the early sixties it has caused many thousands of human deaths (3-5). PQ^{+2}

Antioxidants in Therapy and Preventive Medicine
Edited by I. Emerit *et al.*
Plenum Press, New York, 1990

217

toxicity has prompted extensive research using animal and bacterial models, both *in vivo* and *ex vivo* (6-10). These studies indicated that *in vivo* PQ^{+2} is reduced by a single electron to form the paraquat monocation radical, $PQ^{+}\cdot$, which, in turn reacts instantaneously with molecular oxygen to yield the superoxide radical (7). Recent investigations in chemical and biological systems, have demonstrated that the transition metals, copper or iron, could markedly enhance the damage exerted by superoxide radicals (11-14), or that due to paraquat (15-17) as well as in other cases where superoxide is involved (18-21). The proposed mechanism for the damage induced by paraquat is the "site-specific mechanism" (22).

These lines of evidence indicating the essential mediatory role of iron and copper in PQ^{+2}-toxicity (15-17, 23-26), prompted a recent approach for clinical intervention in treating PQ^{+2}-intoxicated individuals and for prevention of biological damage in other cases where the same "site-specific mechanism" is indicated (22). This approach includes the administration of a specific chelator/s in order to lower the levels of labile and redox available iron and/or copper in the tissue. This approach also provides an apparent advantage over the classical use of chemical radical-scavengers, whose efficiency depends on their local concentration at the site of radical production.

A new approach is being currently developed in our lab. This is based on the idea that displacement of redox-active metal from its binding site, will also shift the site of hydroxyl radical production. As hydroxyl radicals react with biological targets at the site of their formation, this displacement could provide protection (or enhancement) against the "site-specific" Fenton mechanism. Based on the similarity between the coordination chemistry of iron or copper, on the one hand, and zinc, on the other, and considering that zinc ions are redox-inactive (27-30), it is expected that zinc could displace copper(II) and iron(II) from their biological binding sites. Once it replaces copper, it will divert the site of formation and reaction of free radicals, and could lower the deleterious effects induced by paraquat. This study shows the protective effect of zinc(II) in paraquat-induced damage to *E. coli* B cells and demonstrates the competitive nature of this protection in copper-mediated cell inactivation.

Materials and Methods

Escherichia coli B [SR-9] were used throughout the experiments in a mode analogous to Korbashi et al. (16). Inactivation of bacterial cells was determined according to Korbashi et al. (16).

The complexes of copper and zinc were prepared by dissolving $CuSO_4 \cdot 5H_2O$ or $ZnSO_4 \cdot 7H_2O$ and the ligand in the desired ratio in triple distilled water; the solutions were then titrated to pH 7.0 by adding solid $NaHCO_3$ [B.D.H.].

RESULTS

Role of Copper Complexes

We had already shown (15-17) that exposure of *E. coli* B cells to 0.25 mM PQ^{+2} caused a 40% inactivation of the bacteria within 30 min. The addition of varying concentrations of $CuSO_4$ in combination with the PQ^{+2} led to a rapid, exponential inactivation of the cells, exceeding 2-3 decimal logarithmic units, within 20 min. Addition of Cu(II)-NTA caused a marked inhibition in the effect of this metal. Following 10 min of exposure to 0.25 mM PQ^{+2} and 0.5 µM of either Cu(II)-NTA or $CuSO_4$ the lethality was 51% or 97%, respectively. The effect of increasing concentrations of Cu(II)-NTA between 0.1 and 2.0 µM, on the rate of bacterial killing was studied. By plotting the initial slopes of the survival curves as a function of Cu(II)-NTA concentration, a hyperbolic curve is obtained (Fig. 1). These data were further treated by a Lineweaver-Burk analysis and the results are shown in Fig. 2. The value of Km, that concentration of Cu(II)-NTA for half maximal rate of inactivation, is 0.46 µM.

Inhibition of the Cu(II) Effect by Zn(II)

Zinc is a transition metal whose coordination chemistry has a high degree of similarity with copper (28). The effects of $ZnSO_4$, when added in the presence of $CuSO_4$, are shown in Fig. 3. The enhancement effect by copper on PQ^{+2} toxicity, in bacteria, could be reduced or prevented upon addition of 10 and 50 fold excess of zinc

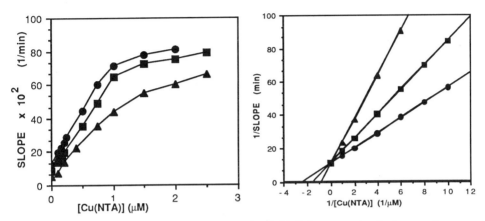

Fig 1 (left):Dependence of the bacterial killing rate upon varying Cu(II)-NTA concentrations in absence and presence of Zn(II)-NTA. ● no Zn(II)-NTA added; ■ 5 μM; Zn(II)-NTA; ▲ 25 μM Zn(II)-NTA.

Fig 2 (right): Double reciprocal plots [(initial killing rate)$^{-1}$ vs. (Cu(II)-NTA)$^{-1}$], showing the relationship of Zn(II)-NTA and Cu(II)-NTA to PQ-induced bacterial inactivation. ● no Zn(II)-NTA added; ■ 5 μM Zn(II)-NTA; ▲ 25 μM Zn(II)-NTA.

over copper. Addition of $ZnSO_4$ to PQ^{+2} alone caused a small decrease in PQ^{+2} toxicity. Incubation of cells with $ZnSO_4$ alone, for 30 min showed only marginal inactivation. For example, 100 μM zinc produced 7% inactivation during 30 min.

When similar experiments were carried out using the NTA complexes of Cu(II) and Zn(II), to compete for the more stable metal binding sites in or on the cell, the results of Fig. 4 were observed. There is a complete reversal of the bacterial killing effect of Cu(II)-NTA in the presence of a 50 fold excess of Zn(II)-NTA. In order to confirm the competitive relations between the zinc and the copper, Zn(II)-NTA at 5 μM and 25 μM were added with varying concentrations of Cu(II)-NTA over the range from 0 to 2.5 μM. The dependence of the initial slope of the bacterial killing curves as a function of Cu(II)-NTA concentration in the presence and absence of Zn(II)-NTA is presented in Fig. 1. These data were analyzed by Lineweaver-Burk graphics (Fig. 2). It appears that the zinc is a competitive inhibitor of copper in the PQ^{+2} -induced toxicity with *E. coli* B cells.

DISCUSSION

We have previously shown that in an *E. coli* system, PQ^{+2} toxicity is enhanced by redox transition metals, Cu(II) and Fe(II) (15-17). In an extension of the earlier studies, in this investigation we added Cu(II) as a complex with several chelators. The effects of Cu(II) on the survival curves of *E. coli* exposed to paraquat were similar for $CuSO_4$ ("aquo" complex) in which the metal is rather loosely bound, and for the other copper chelates with glycine, histidine and citrate. However, in the presence of Cu(NTA), which is a tighter copper complex (31), the bacterial killing is markedly reduced. We propose that copper ions which come off the NTA complex, bind to high affinity cellular sites. It seems that the number of such sites which could participate in paraquat-induced metal-mediated cellular injury is finite. This is indicated by the fact that the rate of bacterial killing reaches saturation at increasing Cu(II)-NTA concentrations.

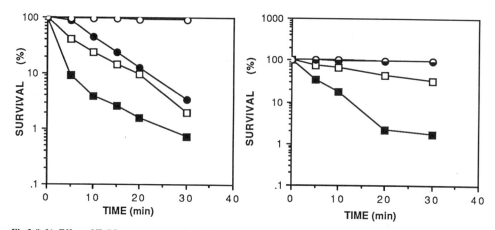

Fig 3 (left): Effect of ZnSO$_4$ on paraquat induced bacterial inactivation in the presence of added CuSO$_4$ [0.5 μM]. All the incubation systems contained 1 x 10^7 E. coli B cells/ml, PQ [0.25 mM] and glucose [0.5% w/v] in Hepes buffer [10 mM, pH 7.4]. ■ 0.5 μM CuSO$_4$; □ 0.5 μM CuSO$_4$ and 5 μM ZnSO$_4$; ● 0.5 μM CuSO$_4$ and 25 μM ZnSO$_4$; ○ 25 μM ZnSO$_4$; no paraquat added.

Fig 4 (right): Effect of Zn(II)-NTA addition on paraquat induced bacterial inactivation in the presence of added Cu(II)-NTA [2.0 μM]. All the incubation systems contained 1 x 10^7 E. coli B cells/ml, PQ [0.25 mM] and glucose [0.5% w/v] in Hepes buffer [10 mM, pH 7.4]. ■ 2.0 μM Cu(II)-NTA; □ 2.0 μM Cu(II)-NTA and 100 μM Zn(II)-NTA; ● 2.0 μM Cu(II)-NTA; no paraquat added; ○ 100 μM Zn(II)-NTA; no paraquat added.

The competitive mode of protection by Zn(II), against copper-mediated paraquat-induced bacterial killing, suggests that zinc competes with copper for sensitive sites, and when in excess, could displace copper from its binding sites. As zinc is a redox-inactive metal, it could *not* serve a center for a site-specific production of free radicals, like copper and iron, as seen in Scheme I, where R is a redox-active cellular binding site.

SCHEME I

$$R + Cu(II) \underset{}{\overset{K_{Cu}}{\rightleftharpoons}} \quad R\text{-}Cu(II) \xrightarrow{\text{Fenton Reaction}} \text{Free Radicals} \longrightarrow \text{Lethality}$$

$$+Zn(II) \Bigg\updownarrow K_{Zn}$$

$$R\text{-}Zn(II)$$

By analogy to an enzymatic reaction, the killing rate S (slope of the survival curves in the presence of Cu-(NTA) and Zn-(NTA) complexes) can be expressed by the following equation:

$$S = \frac{S_{max}\,[Cu(II)]}{K_{Cu}\left[1 + \dfrac{[Zn(II)]}{K_{Zn}}\right] + [Cu(II)]}$$

From the double reciprocal plot presented in Fig. 2, the corresponding values of K_{Cu}, S_{max}, and K_{Zn} were calculated to be 0.46 μM, 9.4 x 10^{-2} min $^{-1}$ and 12.85 μM, yielding K_{Zn}/K_{Cu} is 28. It is noteworthy that log K, the log of the dissociation constant of Cu(II)-NTA is -12.96, while that of Zn(II)-NTA is -10.66 (33). Therefore, for equimolar amounts of Zn(II)-NTA and Cu(II)-NTA, the ratio of ionic Zn(II) to Cu(II) is 200. Thus, under these conditions, the ratio R between the number of Zn-bound cellular sites to copper-bound sites is expected to be T = 200/26 = 7.6. In the presence of excess of Zn-(NTA), over Cu-(NTA), T is expected to further increase, and to represent a displacement of copper from specific cellular sites by zinc, and by this protect these sites. An alternative explanation, that is indistinguishable on the ground of these kinetic data, is that a ternary complex of copper *and* zinc with the cellular binding sites is formed, and by this the zinc interferes with the copper effect (32). A property shared by zinc, copper and iron is the ability of each to form complexes of various geometries (28). The possible protective effect of Zn, in iron-mediated (34,35) and in copper-mediated processes (27,29,30,36) has already been suggested. Likewise, the administration of zinc has been shown to inhibit iron-mediated lipid peroxidation (36,37); and to lower the incidence and severity of ischemia-induced arrhythmias in isolated rat heart; both processes are considered to involve free radicals.

In conclusion, the use of Zn ("push-mechanism") in a combination with an effective chelator for copper and iron ("push-pull" mechanism), in free radical-induced damage, should be considered in an effort to intervene in deleterious processes where the causative role of superoxide radical is indicated by the combined "push-pull mechanism".

ACKNOWLEDGEMENT

This study was supported by grants from the National Institutes of Health (ES-04296) and from the Gesselschaft fur Strahlen und Umweltforschung, Neuherberg, FRG D-8042. This study was conducted as a partial requirement for the Ph.D. program of Ms. P. Korbashi.

REFERENCES

1. Calderbank, A. (1968) Advn. Pest. Control Res. 8, 127- 131.
2. Dodge, A.D. (1971) Endeavour 111, 130-135.
3. Dasta, J.F. (1978) Am. J. Hosp. Pharm. 35, 1368-1372.
4. Autor, A.P. (1974) Life Sci. 14, 1309-1319.
5. Naito, H., and Yamashita, M. (1987) Human Toxicol. 6, 87-89.
6. Hassan, H.M., and Fridovich, I. (1978) J. Biol. Chem. 253, 8143-8148.
7. Hassan, H.M., and Fridovich, I. (1979) J. Biol. Chem. 254, 10846-10852.
8. Hassan, H.M., and Fridovich, I. (1979) Arch. Biochem. Biophys. 196, 385-395.
9. Delval, P.M., and Gillespie, D.J. (1985) Crit. Care Med. 13, 1056-1060.
10. Smith, L.L. (1986) Annu. Rev. Physiol. 48, 681-692.
11. McCord, J.M., and Day, E.D. (1978) FEBS Lett. 86, 139-142.
12. Halliwell, B. (1978) FEBS Lett. 92, 321-326.
13. Aust, S.D., Morehouse, L.A., and Thomas, C.E. (1985) J. Free Rad. Biol. Med. 1, 3-25.
14. Halliwell, B., and Gutteridge, J.M.C. (1986) Arch. Biochem. Biophys. 246, 501-514.11.
15. Kohen, R., and Chevion, M. (1985) Free Rad. Res. Commun. 1, 79-88.
16. Korbashi, P., Kohen, R., Katzhendler, J., and Chevion, M. (1985) J. Biol. Chem. 261, 12472-12476.
17. Kohen, R., and Chevion, M. (1988) Biochemistry 27, 2597-2603.
18. Samuni, A., Chevion, M., and Czapski, G. (1981) J. Biol. Chem. 265, 12632-12635.

19. Shinar, E., Navok, T., and Chevion, M. (1983) J. Biol. Chem. 258, 14778-14783.
20. Navok, T., and Chevion, M. (1984) Biochem. Biophys. Res. Commun. 122, 297-303.
21. Levine, R.L., Oliver, C.N., Fulks, M.R., and Stadtman, E.R. (1980) Proc. Natl. Acad. Sci. USA 78, 2120-2124.
22. Chevion, M. (1988) Free Rads Biol. Med. 5, 27-37.
23. Kohen, R., Korbashi, P., and Chevion, M. (1983) in: Paraquat Toxicity is Mediated by Transition Metal Ions, Abstract of the First Meeting of Israel Societies of Life Science, Jerusalem, October, 1983.
24. Sutton, H.C., and Winterbourn, C.C. (1984) Arch. Biochem. Biophys. 235, 106-115.
25. Winterbourn, C.C., and Sutton, H.C. (1984) Arch. Biochem. Biophys. 235, 116-126.
26. Kohen, R., and Chevion, M. (1985) Biochem. Pharmacol. 34, 1841-1843.
27. Hegetschweiler, K., Saltman, P., Dalvit, C., and Wright, P. (1987) Biochem. Biophys. Acta 912 (3), 384-397.
28. Cotton, F.A., and Wilkison, G. (1972) in: Advanced Inorganic Chemistry, New York: Interscience Publishing, 1972.
29. Eguchi, L.A., and Saltman, P. (1987) Inorg. Chem. 26, 3665-3669.
30. Eguchi, L.A., and Saltman, P. (1987) Inorg. Chem. 26, 3669-3672.
31. Stability Constants of Metal-Ion Complexes, Part B: Organic Ligands: IUPAC Chemical Data Series,No. 22, (Perrin, D.D., ed.) Pergamon Press, Oxford (1979).
32. Segel, I.H. (1975) in: Enzyme Kinetics, pp. 161-166, A Wiley-Interscience Publications, John Wiley and Sons, New York.
33. Anderegg, G. (1982) Pure Appl. Chem. 54, 2693-2758.
34. Willson, R.L. (1977) In: Iron Metabolism, Ciba Foundation Symposia, 51, 331-354.5
35. Willson, R.L. (1977) New Scientist, 1 December, 558-560.
36. Chvapil, M. (1973) Life Sci. 13, 1041-1049.
37. Girotti, A.W., Thomas, J.P., and Jordan, J.E. (1986) Arch. Biochem. Biophys. 251, 639-653.

ANTIOXIDATIVE ACTION OF ZINC-CARNOSINE COMPOUND Z-103

T.Tanigawa, T.Yoshikawa, Y.Naito, T.Yoneta, S.Ueda, H.Oyamada
T.Takemura, Y.Morita, K.Tainaka, N.Yoshida, S.Sugino, and
M.Kondo

First Department of Medicine
Kyoto Prefectural University of Medicine
Kyoto,602 JAPAN

INTRODUCTION

The objective of the present study was to examine the anti-free radical action of Z-103 in vitro. Z-103 is a novel synthetic compound of zinc and carnosine (β-alanyl-L-histidine) which has strong anti-ulcer action in many types of animal models[1] Free radicals have come to considered to be playing a role in the pathogenesis of gastric mucosal injuries[2,3]. The anti-free radical action of Z-103 were expected because zinc and carnosine have antioxidative properties[4,5]. To know its anti-free radical action in vitro will be significant to understand its mechanism of action as anti-ulcer drug.

MATERIALS AND METHODS

SOD-like Activity

SOD-like activity was measured by spin trapping and electron spin resonance(ESR)[6]. Dimethylpyrroline-N-oxide(DMPO)(Sigma) was used as a spin trap of superoxide, and hypoxanthine(100 μM) and xanthine oxidase (50 mU/ml)

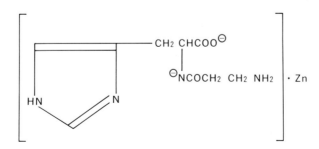

Fig. 1   The structure of Z-103.

Antioxidants in Therapy and Preventive Medicine
Edited by I. Emerit *et al.*
Plenum Press, New York, 1990

in PBS (pH 7.4) was used as a superoxide generating system. The signal was recorded by JES-FE-2XG ESR Spectrometer (JEOL, Tokyo). The ESR signal of the adduct (DMPO-OOH) was inhibited by SOD in a dose dependent manner, and a quantitation line could be obtained. From the inhibition rate, SOD-like activity of Z-103, $ZnSO_4$ and carnosine was measured.

Hydroxyl Radical Scavenging Activity or Inhibition of Fenton Reaction

The inhibition of the ESR signal of DMPO-OH obtained by diethylenetri-amine pentaacetic acid chelated ferrous iron (DETAPAC-Fe^{2+})(50 µM) and H_2O_2 (1 mM) in phosphate buffer (100 mM, pH 7.4) was observed. When the inhibition of the signal is due to the scavenging of hydroxyl radical by a sample, the inhibition rate should be dependent on the relative concentration of the sample v.s. DMPO, because DMPO and the sample compete in the reaction with hydroxyl radical. On the otherhand, when the inhibition is due to the inhibition of Fenton reaction, the inhibition rate should be dependent on the absolute concentration of a sample and independent on the concentration of DMPO. The cause of the inhibition of the signal was distinguished using two concentrations of DMPO (10^{-2} M, 10^{-3} M).

Inhibition of Superoxide Generation by Polymorphonuclear Leukocytes (PMN)

Superoxide generation by stimulated PMN was measured by ESR-spin trapping using DMPO (10^{-1} M).[7] Opsonized zymosan (OZ)(3 mg/ml) or phorbolmyristate acetate (PMA)(400 ng/ml) were used as stimulants. The effects of the presence of the agents and incubation (30 min, 37°C) with the agents followed by washing of PMN were examined.

Inhibition of Lipid Peroxidation

Autoxidation of rat brain homogenate. Rat brain homogenate in phosphate buffer (35 mM, pH7.4) was placed at 37°C in the presence or in the abscence of the agents. Lipid peroxidation was monitored by thiobarbituric acic (TBA) reacting substances.[8]

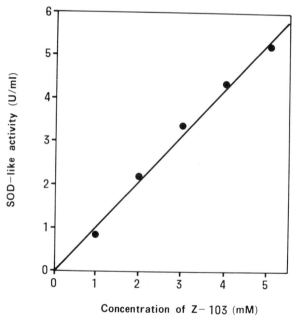

Fig. 2 SOD-like activity of Z-103. The activity was measured by the inhibition of DMPO-OOH ESR signal.

Lipid peroxidation of rat liver microsome. The lipid peroxidation of rat liver microsome induced by Fe^{2+} (5 μM) and NADPH (100 μM) was monitored by TBA reacting substances in the presence and the abscence of the agents.

Ferroxidase-like Activity

Decrease of Fe^{2+}. The concentration of ferrous iron in acetate buffer (136 mM, pH6.0) was monitored by 1,10-o-phenanthroline chelating assay. Ferrous iron decreases by oxidation to ferric iron, and the activity obtained by the rate of oxidation was expressed using ceruloplasmin as a standard.

O_2 consumption. YSI biological oxygen monitor type 5300 (YSI,Yellow Springs) was used for the measurement of O_2 pressure in Tris-HCl buffer (100 mM, pH 6.25). After the addition of $FeSO_4$, O_2 decreased by autoxidation of ferrous iron. The effects of the agents on O_2 consumption were examined.

RESULTS

Z-103 had weak SOD-like activity, but $ZnSO_4$ and carnosine did not (Fig. 2). Z-103 and $ZnSO_4$ inhibited the ESR signal of DMPO-OH in the Fenton system, and the inhibition was dependent on the concentration of the substances but independent on the concentration of DMPO, indicating that the inhibition was due to the inhibition of the Fenton reaction. Carnosine also inhibited the signal of DMPO-OH, but the inhibition was dependent on the relative concentration of carnosine v.s. DMPO, indicating that the inhibition was due to the competition of carnosine v.s. DMPO (Fig. 3)

Z-103 and $ZnSO_4$ inhibited superoxide generation by PMN. The inhibition could be observed not only in the presence of the substances but also by incubation with the substances followed by cell washing. Carnosine showed no effects on the generation of superoxide by PMN (Fig. 4).

Z-103 and $ZnSO_4$ but carnosine inhibited the lipid peroxidation in the two system we employed (Fig. 5).

Z-103 and $ZnSO_4$ had little effect on the decrease of ferrous iron or on oxygen consumption in the solution of ferrous sulfate. Carnosine accelerated the decrease of ferrous iron (0.209 U/ml of ferroxidase activity) and oxygen consumption in the solution of ferroussulfate, indicating the acceleration

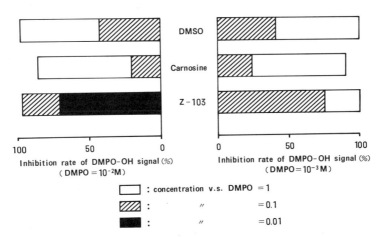

Fig. 3 The inhibition of the ESR signal of DMPO-OH in the Fenton system by Z-103 and carnosine. The concentration of DMPO was 10^{-2} M (left half), and 10^{-3} M (right half). The inhibition by dimethylsulfoxide (DMSO), a typical scavenger of hydroxyl radical is also showen for comparison.

Table   The summary of the results.

	Z-103	carnosine	ZnSO$_4$
SOD-like activity	+	-	-
·OH scavenging action	+?	+	+?
Inhibition of Fenton reaction	+	-	+
Inhibition of superoxide generation by PMN	+	-	+
Inhibition of lipid peroxidation	+	-	+
Ferroxidase-like activity	-	+	-

of autoxidation of ferrous iron by carnosine (Fig. 6).
The results are summarized in the Table.

DISCUSSION

Recent studies revealed that oxygen free radicals are implicated in the pathogenesis of the gastric mucosal lesions[2],[3]. A new therapeutic approach might be possible by using agents which has anti-free radical actions. Z-103 which has already reported to have strong anti-ulcer effect in many types of animal models[1] is a compound of zinc and carnosine, and both zinc and carnosine have antioxidative actions[4],[5]. So anti-free radical action of Z-103 had been expected. The present study has revealed that actually the agent can inhibit superoxide generation by PMN, the Fenton reaction, and lipid peroxidation, and scavenge superoxide. The reactivity of Z-103 with hydroxyl radical could not be examined because it inhibited the Fenton

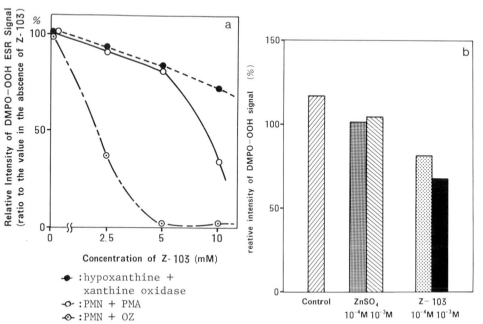

Fig. 4 (a) The effect of the presence of Z-103 on ESR signal of DMPO-OOH by stimulated PMN in comparison with that by hypoxanthine and xanthine oxidase system. (b) The effects of preincubation of PMN with Z-103 or ZnSO$_4$ followed by cell waxhing on superoxide generation by PMN.

226

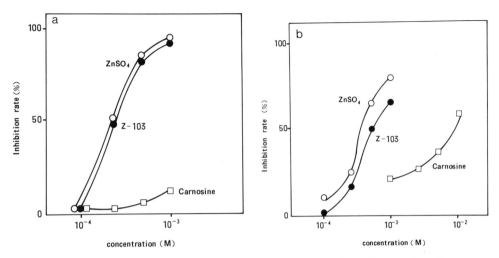

Fig. 5 Effects of Z-103, ZnSO$_4$ and carnosine on lipid peroxidation.
(a) Lipid peroxidation of rat brain homogenate. (b) Lipid peroxidation
of rat liver microsome.

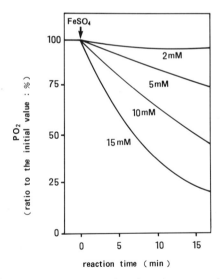

Fig. 6 Oxygen consumption by the autoxidation
of ferrous iron, and the effects of carnosine
on it at verious concentrations.

reaction, but similar reactivity as that of carnosine could be expected. $ZnSO_4$ inhibited the Fenton reaction, lipid peroxidation and superoxide generation by PMN. Carnosine accelerated the oxidation of ferrous iron and it was a good scavenger of hydroxyl radical. The strong anti-ulcer action of Z-103 in vivo might be explained by the combination of these anti-free radical actions.

CONCLUSIONS

Z-103 showed several anti-free radical actions in vitro. The mechanisms of its anti-ulcer action might be explained in part by these properties.

REFFERENCES

1. Naito,Y., Yoshikawa,T., Tanigawa,T.,et al. Protection by zinc N-(3-aminopropionyl)-L-histidine against rats gastric mucosal lesions and its anti-oxidative effects, in: International Conference on Medical, Biochemical and Chemical Aspects of Free Radicals (Abs.), Kyoto, 1988, pp. 207.
2. Yoshikawa,T., Miyagawa,H., et al. Increase of lipid peroxidation in rat gastric mucosal lesions induced by water-immersion restraint stress. J.Clin.Biochem.Nutr. 1:271-277, 1986.
3. Yoshikawa,T., Yoshida,N., et al. Role of lipid peroxidation in gastric mucosal lesions induced by burn shock in rats. J.Clin.Biochem.Nutr. 2: 163-170, 1987.
4. Willson,R.L. Vitamin, selenium, zinc and copper interactions in free radical protection against ill placed iron. Proceedings of the Nutrition Society. 46:27-34, 1987.
5. Kohen,R., Yamamoto,Y., Cundy,C., and Ames,B.N. Antioxidant activity of carnosine, homocarnosine, and anserine present in muscle and brain. Proc.Natl. Acad.Sci.USA, 85:3175-3179, 1988.
6. Miyagawa,H., Yoshikawa,T., Tanigawa,T., et al. Measurement of serum superoxide dismutase activity by electron spin resonance. J.Clin.Biochem. Nutr. 5:1-7, 1988.
7. Tanigawa,T., Yoshikawa,T., et al. Determination of superoxide generated by human polymorphonuclear leukocytes by electron spin resonance and chemiluminescence, in: Free Radicals in Digestive Diseases, Tsuchiya,M., Kawai,K., Kondo,M., and Yoshikawa,T. ed., Elsevier Science Publishers B.V., Amsterdam, 1988, pp. 37-42.
8. Stocks,J., Gutteridge,J.M.C., Sharp,R.J., and Dormandy,T.L. Assay using brain homogenate for measuring the antioxidant activity of biological fluids. Clin.Sci.Mol.Med. 47:215-222, 1974.

High Molecular Weight Forms of Deferoxamine:

Novel Therapeutic Agents for Treatment of Iron-Mediated Tissue Injury

Bo E. Hedlund*, Philip E. Hallaway* and John R. Mahoney[#]

*Biomedical Frontiers, Inc. and [#]University of Minnesota

Minneapolis, Minnesota, USA

INTRODUCTION

The iron chelator, deferoxamine (Desferal®), is presently used clinically for the treatment of acute and chronic iron toxicity. The chelator is an effective inhibitor of iron-catalyzed reactions leading to the formation of both oxygen and lipid-derived radicals. Deferoxamine (DFO) has been incorporated in a number of studies involving oxygen and lipid radical mediated reactions leading to tissue injury and has proven efficacious in ameliorating reperfusion injury.

The decompartmentalization of iron is probably a critical component of the molecular mechanism of reperfusion injury. Using a canine model of hemorrhagic shock, Mazur[1] demonstrated that, during global ischemia, sufficient quantities of iron are released to completely saturate transferrin binding capacity. The release of iron is likely to occur in any type of cell exposed to transient ischemia followed by reperfusion. Mechanistically it appears that ferritin-bound ferric iron is reduced to ferrous, "free" iron by the superoxide anion.[2,3] Recently, Paller and Hedlund[4] have demonstrated that significant quantities of free iron can be detected in the urine following renal ischemia and reperfusion. Ten-fold increases in urinary iron were observed following 60 minutes of renal ischemia. These investigators also demonstrated that DFO protects the kidney against reperfusion injury. Similarly, both in *ex vivo*[5,6] and *in vivo* models[7-9] involving ischemia/reperfusion injury to the heart, DFO clearly attenuates myocardial injury based on both functional and biochemical determinations. Efficacy has also been observed in models involving global ischemia. For example, in a rodent model of cardiovascular resuscitation,[10] treatment with DFO significantly improved survival. Recently, Sanan and collaborators[11], demonstrated that DFO improves liver function and enhances survival following hemorrhagic shock in dogs. The protective effect of a specific iron chelator in a variety of models strongly implicates "free" iron as an important component in the biochemical events leading to tissue injury following reperfusion.

Despite the encouraging results obtained using DFO as a therapeutic agent in animal models involving reperfusion injury of a variety of types, the clinical use of the drug is likely to be hampered by two undesirable physiological properties. First, the drug is excreted very rapidly, making it difficult to achieve vascular concentrations sufficiently high to effectively chelate toxic iron. Second, the drug is associated with considerable toxicity, particularly when given intravenously. The toxicity is manifested by severe hypotension and hemoconcentration. Such adverse effects have been demonstrated in several animal species as well as in man.[12,13]

Using hydroxyethyl starch or soluble dextran as carriers, we have synthesized a number of high molecular weight chelators by covalent attachment of deferoxamine to these polymers. These biocompatible, high molecular weight chelators have the following properties:

Antioxidants in Therapy and Preventive Medicine
Edited by I. Emerit *et al.*
Plenum Press, New York, 1990

1. Ability to chelate iron with affinity and specificity identical to those of the free chelator. Studies of reductive displacement of iron with gallium indicated that both forms of the conjugated-DFO have the same affinity for ferric iron as free DFO.

2. Conjugated-DFO inhibits iron-mediated lipid peroxidation as efficiently as the free drug. The degree to which the two forms of the chelator inhibited iron and hemoglobin mediated lipid peroxidation in murine brain homogenates was identical.

3. The vascular retention halftimes of conjugated-DFO are 10-30 times longer than that of free DFO.

4. Conjugated-DFO has a greatly reduced toxicity compared to that of the free drug. The LD_{50} of DFO administered IV to mice was 250 mg/kg. In contrast, the LD_{50} of dextran-DFO was approximately 4,000 mg/kg, and in this case the toxicity was likely due to the large volume needed to administer this dose.

The low toxicity and extended vascular retention time of the polysaccharide-deferoxamine conjugates make these compounds attractive potential therapeutic agents for use in clinical situations where "free" iron is involved in reactions leading to tissue injury.

Acute iron toxicity remains a significant problem in emergency medicine. Patients are usually infants who have accidently ingested supplemental iron tablets. Deferoxamine is the current drug of choice for treatment of acute iron intoxication. However, the amount of the drug that can be safely administered is limited by the adverse hemodynamic effects associated with the intravenous bolus administration of the drug. In this communication we will describe utilization of the high molecular weight chelators in a murine model of acute iron toxicity. Cohen and coworkers[14,15] have recently provided evidence that hydroxyl radicals are directly involved in the pathophysiology occurring secondary to ingestion of ferrous iron in rodents. Mechanistically, the tissue injury occurring during acute iron toxicity may result from similar toxic species formed during ischemia and ensuing reperfusion.

MATERIALS AND METHODS

The model of acute iron intoxication was established in male Swiss-Webster mice (25-30 g) obtained from Biolabs, St. Paul, Minnesota. Deferoxamine (DFO) was obtained as the mesylate salt, Desferal®, from Ciba-Geigy, Inc., Summit, New Jersey; dextran as Rheomacrodex®, a 10% W/V solution from Pharmacia Laboratories, Piscataway, New Jersey; hydroxyethyl starch as a 10% W/V solution in normal saline, from Dupont Critical Care, Waukeegan, Illinois. The complete description of the preparation of the high molecular weight chelators will be reported elsewhere[17]. Briefly, dextran or hydroxyethylstarch (in 10% solution w/v) was oxidized with 0.1 M sodium metaperiodate to yield reactive dialdehydes. Following removal of low molecular weight reaction products using dialysis or diafiltration, the activated polysaccharides were reacted with 0.1 M DFO. The Schiff's bases formed between the terminal amino group of the chelator and aldehyde groups on the polysaccharide were reduced with sodium cyanoborohydride. Remaining aldehyde groups were reduced with sodium borohydride. The high molecular weight conjugate was separated from reaction products by a second dialysis or diafiltration step. The presence of free DFO in the final preparation was determined by high pressure liquid chromatography (FPLC System, Pharmacia Biotechnology) using a Superose® 6 gel permeation column. The concentration of free DFO in the final preparations of the polysaccharide-DFO conjugate was less than 2%. The conjugates used in this report were dextran containing approximately 25% by weight of bound DFO and an hydroxyethyl starch derivative containing 15% by weight of bound DFO.

Model of acute iron intoxication in mice

The oral dose of iron was chosen after assessing the toxicity of doses from 5 to 15 μmol/g body weight. Ferrous sulfate was dissolved in deionized water and approximately 1 ml was administered by gavage. The dose of 10 μmol/g was chosen as the appropriate dose for this model. This dose is in agreement with a previously reported LD$_{50}$.[16] This dose was administered orally and followed by an intravenous injection (via the tail vein) of approximately 1 ml of either polymer (10% solution of dextran or hydroxyethyl starch in normal saline), free DFO and polymer or the conjugated-DFO in normal sterile saline. The dose of free DFO, which was limited by the toxicity of the drug, was 0.15 μmol/g. Higher doses of DFO given as a bolus injection resulted in mortality. The conjugated-DFO was given at a dose of 1.4 μmol/g (deferoxamine equivalents) in the case of dextran-DFO and 0.9 μmol/g (deferoxamine equivalents) in the case of hydroxyethyl starch-DFO. Thus, six- to ten-fold higher doses of chelator equivalents can easily be administered without adverse effects.

RESULTS

In order to establish the doses of iron causing mortality in Swiss Webster mice, we administered ferrous sulfate orally to groups of mice (Figure 1). The dose of iron for the acute oral model was chosen after administering 5, 7.5, 10 and 15 μmol Fe/g body weight. The dose of 10 μmol Fe/g body weight which caused approximately 75% mortality was chosen for the oral dose of iron. The results of these experiments are in good agreement with previously reported LD$_{50}$ values in another strain of mouse.[16]

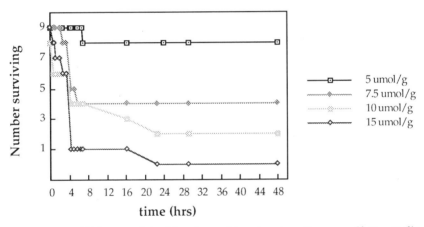

Figure 1. Establishment of the LD$_{50}$ for oral iron in mice. Ferrous sulfate was dissolved in deionized water for oral administration. All doses were given in a volume of approximately 1 ml.

Mortality following oral iron administration and IV drug treatment is demonstrated in Figure 2. The control animals were injected with either dextran or hydroxyethyl starch. In addition, free DFO dissolved in 10% polymer vehicle (dextran or hydroxyethyl starch) was administered at a dose of 0.15 μmol/g. Dextran-DFO or hydroxyethyl starch-DFO conjugates were injected IV at a dose of 1.4 and 0.9 μmol/g, respectively (deferoxamine equivalents). Both forms of conjugated-DFO were completely effective at preventing the mortality associated with oral administration of toxic amounts of iron.

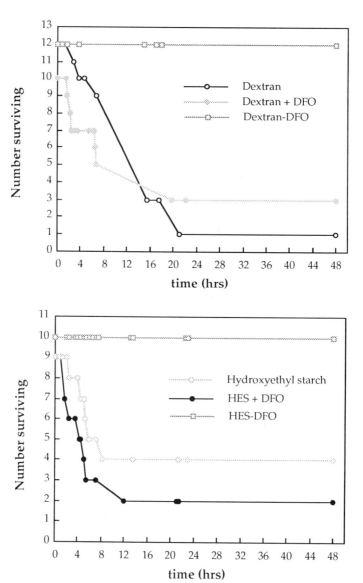

Figure 2. Mortality following oral iron administration and IV drug treatment. Ferrous sulfate was dissolved in deionized water and administered by gavage at a dose of 10 μmol/g. The control animals were injected with dextran or hydroxyethyl starch (HES) in sterile normal saline. Free DFO in dextran or hydroxyethyl starch in sterile normal saline was administered at a dose of 0.15 μmol DFO/g body weight. Dextran-DFO or hydroxyethyl starch-DFO conjugates were injected IV at a dose of 1.4 and 0.9 μmol/g, respectively (deferoxamine equivalents). All IV injections were approximately 1 ml.

DISCUSSION

Acute iron poisoning represents a significant number of the life-threatening accidental poisonings in children[18]. Currently, Desferal® is the drug of choice for treatment of acute iron poisoning. However, Desferal® has adverse effects on hemodynamics when administered as a bolus IV. Its ability to cause hypotension severely limits the maximum dose for treatment of acute iron toxicity. These experiments report the efficacy of novel high molecular weight forms of conjugated-deferoxamine in the treatment of experimental acute iron poisoning. It has recently been demonstrated that conjugated-DFO has several advantages over the free drug. The DFO-conjugates do not induce hemodynamic changes even when administered as an IV bolus [17]. The circulating half life of the DFO-conjugates is significantly longer than the free drug, which is rapidly metabolized and cleared from the circulation.[17] Most importantly, the DFO-conjugates are more effective in the treatment of acute iron toxicity than the free drug. DFO-conjugates were most effective when administered immediately following iron; however, they were also effective when given following a delay (data not shown). The proximate cause of death in iron poisoning is not known. A likely target is the liver, which generates activated oxygen species and accumulates much of the excess iron.[19]

Deferoxamine has recently been used to prevent oxidant injury during reperfusion of ischemic tissue [4-10]. This application requires administration of a large bolus of the drug in order to chelate available iron and render it catalytically unreactive.[20] Conjugated forms of deferoxamine would allow the infusion of higher amounts of drug without causing hypotension and may therefore be more effective in limiting iron-driven reperfusion injury. Another possible application for the conjugated forms of deferoxamine is in the treatment of chronic iron overload.

ACKNOWLEDGMENTS

The authors wish to thank Diane Konzen for her help with the preparation of this manuscript. This work was supported by grants R43- and R44-DK37207 from the National Institutes of Health.

Reprint requests should be addressed to Dr. Bo E. Hedlund, Biomedical Frontiers, Inc., 1095 10th Avenue S. E., Minneapolis, MN 55414 U.S.A.

REFERENCES

1. A. Mazur, S. Baez, and E. Shorr, The mechanism of iron release from ferritin as related to its biological properties, J. Biol. Chem. 213:147 (1955).

2. C. E. Thomas, L. A. Morehouse, and S. D. Aust, Ferritin and superoxide-dependent lipid peroxidation, J. Biol. Chem. 260:3275 (1985).

3. P. Biemond, A. J. G. Swaak, C. M. Biendorff, and J. F. Koster, Superoxide-dependent and independent mechanisms of iron mobilization from ferritin by xanthine oxidase, Biochem. J. 239:169 (1986).

4. M. S. Paller, and B. E. Hedlund, Role of iron in postischemic renal injury in the rat, Kidney Int. 34:474 (1980).

5. G. Ambrosio, J. L. Zweier, W. E. Jacobus, M. L. Weisfeldt, and J. T. Flaherty, Improvement of postischemic myocardial function and metabolism induced by administration of deferoxamine at the time of reflow: The role of iron in the pathogenesis of reperfusion injury, Circulation 76:906 (1987).

6. S. F. Badylak, A. Simmons, J. Turek, and C. F. Babbs, Protection from reperfusion injury in the isolated rat heart by postischemic deferoxamine and allopurinol administration, Cardiovasc. Res. 21:500 (1987).

7. R. Bolli, B. S. Patel, W. X. Zhu, P. G. O'Neill, M. L. Charlat, and R. Roberts, The iron chelator deferoxamine attenuates postischemic ventricular dysfunction, Am. J. Physiol. 253: H1372-H1380 (1987).

8. N. E. Farber, G. M. Vercellotti, H. S. Jacob, G. M. Pieper, and G. J. Gross, Evidence for a role of iron in functional and metabolic stunning in the canine heart, Circ. Res. 63:351 (1988).

9. B. R. Reddy, R. A. Kloner, and K. Przyklenk, Early treatment with deferoxamine limits myocardial ischemia/reperfusion injury, Free Rad. Biol. Med., in press (1989).

10. C. F. Babbs, The role of iron ions in the genesis of reperfusion injury following successful cardipulmonary resuscitation: Preliminary data and a biochemical hypothesis, Ann. Emerg. Med. 14:777 (1985).

11. S. Sanan, G. Sharma, R. Malhotra, D. P. Sanan, P. Jain and P. Vadhera: Protection by desferrioxamine against histopathological changes of the liver in the post-oligaemic of clinical haemomorrhagic shock in dogs: Correlation with improved survival rate and recovery. Free Rad. Res. Commun. 6:29 (1989).

12. C. F. Whitten, G. W. Gibson, M. H. Good, J. F. Goodwin, and A. J. Brough, Studies in acute iron poisoning. I. Desferrioxamine in the treatment of acute iron poisoning: Clinical observations, experimental studies, and theoretical considerations, Pediatrics 36:322 (1965).

13. C. F. Whitten, Y-C. Chen, and G. W. Gibson, Studies in acute iron poisoning: II. Further observations on desferrioxamine in the treatment of acute experimental iron poisoning, Pediatrics 38:102 (1966).

14. A. Slivka, J. Kang, and G. Cohen, Hydroxyl radicals and the toxicity of oral iron, Biochem. Pharmacol. 35:553 (1986).

15. J. O. Kang, A. Slivka, and G. Cohen, In vivo formation of hydroxyl radicals following intragastric administration of ferrous salts in rats, J. Inorg. Biochem. 35:55 (1989).

16. J. O. Hoppe, G. M. A. Marcelli, and M. L. Tainter, An experimental study of the toxicity of ferrous gluconate, Am. J. Med. Sci. 230:491 (1955).

17. P. E. Hallaway, J. W. Eaton, S. S. Panter, and B. E. Hedlund, Modulation of deferoxamine toxicity and clearance by covalent attachment to biocompatible polymers, Submitted for publication (1989).

18. T.L. Litovitz, B.F. Schmitz, N. Matyunas and T.G. Martin. 1987 Annual Report of the American Association of Poison Control Centers National Data Collection System. Am. J. Emer. Med. 6:479 (1988).

19. C. E. Ganote, and G. Nahara, Acute ferrous sulfate hepatotoxicity in rats. An electron microscopic and biochemical study, Lab. Invest. 28:426 (1973).

20. E. Graf, J. R. Mahoney, Jr., J. R. Bryant, and J. W. Eaton, Iron-catalyzed hydroxyl radical formation: Stringent requirement for iron-coordinated water, J. Biol. Chem. 295:3620 (1984).

OXYGEN TOXICITY: ROLE OF HYDROGEN PEROXIDE AND IRON

B.S. van Asbeck

Department of Medicine
University of Utrecht
3511 GV Utrecht
The Netherlands

INTRODUCTION

An important factor in tissue damage by toxic oxygen species is
the ability to increase the level of hydrogen peroxide. This intermediate of oxygen reduction is not only a precursor of species with a
higher reactivity, such as the hydroxyl radical, but it also controls
the process of inflammatation by its effect on the synthesis of vasoactive and chemotactic compounds. However, tissue injury by hydrogen
peroxide often, if not always, depends on the availability of catalytic
iron.

To be biologically active iron is bound to chelators. These are
molecules which form coordination compounds with metals. Since iron ions
in acidic aqueous solution are dissolved with six coordinating water
molecules, chelation results in replacement of H_2O by the chelator donor
atoms. This complex formation prevents hydrolysis at neutral pH, and the
formation of, catalytically inactive, aggregating polynuclear iron
hydroxide complexes. Iron chelators determine the standard redox
potential and, in particular, by the nature of the liganding atoms, the
oxidation state of the iron complex and therefore its ability to
catalyze the Haber-Weiss reaction. To be catalytically active iron
requires that at least one coordination site is occupied by a readily
dissociable ligand. Thus, depending on the nature and the concentration
of the chelating ligand, iron catalyzed oxygen toxicity can either be
prevented or promoted.

Antioxidants in Therapy and Preventive Medicine
Edited by I. Emerit *et al.*
Plenum Press, New York, 1990

This paper focuses on the role of hydrogen peroxide and iron in oxygen toxicity. Some aspects of treatment with sulfhydryl containing compounds and iron chelators will be discussed.

OXYGEN TOXICITY: ROLE OF H_2O_2

The primary function of oxygen is the acceptance of electrons, the standard redox potential (E_O' [O_2/H_2O]) being +0.8 Volt. Toxic oxygen intermediates can be produced during the reduction of O_2 to H_2O. These products, the superoxide anion (O_2^-), hydrogen peroxide (H_2O_2), and the hydroxyl radical ($\cdot OH$), can damage membranes and intracellular organelles and have been implicated in the pathogenesis of many diseases (1,2).

Superoxide is generated during normal cellular metabolism by the leakage of electrons onto oxygen. This may occur in the mitochrondrial electron transport chain, the cytochrome P-450 system, the arachidonic acid metabolism, and during the activity of enzymes such as monoamine oxidase and xanthine oxidase. Furthermore, autoxidation reactions of iron, copper, reduced flavins, and the oxidation of catecholamines are sources of O_2^- generation. In addition, oxygen radicals can be produced by phagocytic cells (3) when they become activated, for example in response to activated complement (4) or endotoxin (5). The superoxide anion can act as a reductant when it reacts with transition metals such as iron and copper, or as an oxidant in the dismutation reaction when H_2O_2 is produced. This latter reaction is more effective at lower pH, and the amount of H_2O_2 produced increases as the surrounding oxygen concentration is raised.

Of the three intermediates of oxygen reduction $\cdot OH$ is particularly noxious, being able to oxidize most organic compounds (E_O' [$\cdot OH/H_2O$] = +2.33 Volt) (6). Indeed, recent data indicate that the toxic effects of O_2^- and H_2O_2 are probably secondary to the generation of $\cdot OH$ (7). This species has a half life of 10^{-9} s and a diffusion radius of 2.3 nm (8). Damage depends on the site of $\cdot OH$ production; if close to DNA or membranes, the effect can be devastating to cells. However, the generation of $\cdot OH$ is dependent on the availability of H_2O_2 . Cells have potent defence systems which are specifically directed against H_2O_2 such as catalase and the glutathione redox-cycling system. On the other hand cells also contain superoxide dismutase (SOD), an enzyme that promotes the reduction of O_2^- to H_2O_2. These systems efficiently control the "peroxide tone" of cells and may prevent peroxide-driven reactions that can lead to cell injury and inflammation. In addition hydroxyl radical is not the only toxin that depends on H_2O_2, but also the generation of myeloperoxidase-catalyzed hypohalous acids (e.g. OCl^-) (9), the activation and inactivation of the arachidonic acid metabolism (10), bradykinin release (11), the synthesis of platelet-activating factor (12), inhibition of lymphocyte proliferation (13), and an increase in intracellular free calcium (14) are processes that can be influenced by H_2O_2. Indeed, sulfhydryl containing compounds such as glutathione (GSH), by which peroxides are inactivated, may decrease the antitumor effect of phagocytic cells (15), and the cytotoxic effects of various antineoplastic drugs (16), radiation (17), endotoxin (18), and hyperoxia (19,20).

Table 1. Inflammatory events which are influenced by H_2O_2

---

1   Generation of MPO[a]-catalyzed products (e.g. OCl^-)

2   $\cdot OH$ formation by reaction with iron

3   Arachidonic acid metabolism activation and inactivation

4   PAF[b] synthesis

5   Bradykinin release

6   Ca^{++} influx

7   Inhibition of lymphocyte proliferation

---

[a]Myeloperoxidase

[b]Platelet-activating factor

Because of these observations it seems an reasonable approach to focus on drugs that are able to control the level of H_2O_2. Since hydrogen peroxide can react with the sulfhydryl group of thiols resulting in its decomposition to water ($2R-SH + H_2O_2 \rightarrow RSSR + 2 H_2O$), compounds such as N-acetylcysteine may be of clinical value. In addition N-acetylcysteine contains L-cysteine which is intracellularly effectively utilized for the synthesis of the most abundant intracellular non-protein thiol glutathione (L-γ-glutamyl-L-cysteinyl-glycine) (21). It has been shown indeed that intraperitoneal injection of N-acetyl-L-cysteine increases liver glutathione in mice (21). N-acetylcysteine was also shown to effectively protect human fibroblasts against the toxic effects of tobacco smoke condensates (22), which contain high levels of oxygen radicals (23). In the same study N-acetylcysteine protected the isolated perfused lung against the GSH-depleting effect of tobacco smoke, suggesting at least an effective competetive antioxidant effect. Furthermore, intravenous administration of N-acetylcysteine in sheep has been shown to attenuate the endotoxin-mediated pathophysiologic changes in the lung and to substantially reduce the postendotoxin rise in lymph concentrations of thromboxane B_2 and 6-keto-prostaglandin $F_{1\alpha}$ (18). In addition, intratracheal insufflation of N-acetylcysteine in rats increased their lung GSH and protected the animals from hyperoxic (>95% O_2) lung injury (19).

The most important contribution of GSH, in its peroxide detoxifying effects lies in the enzymic redox-cycling. This provides a reversible antioxidant system of which the capacity not only depends on enzyme activities but also on the absolute level of the total glutathione content of cells (24). Since the latter is determined by the synthesis of glutathione, a process which can be enhanced by N-acetylcysteine (21), this drug may besides its direct oxygen radical scavenging effects, also indirectly potentiate the antioxidant capacity of cells.

It has to be emphasized, however, that tissue injury by H_2O_2 often, if not always, depends on the presence of a metal calalyst. Since iron is the most favorable, this metal plays a crucial role in tissue injury by oxygen radical-mediated reactions.

OXYGEN TOXICITY: ROLE OF IRON

Various metals which undergo univalent redox reactions can participate in the enzymatic and non-enzymatic oxidation and peroxidation of biological molecules. Of these metals, manganese, iron, cobalt, and copper are of biological importance. However, iron is the most effective catalyst of these oxidative processes, due to the fact than the concentration of iron in animal tissues is much higher than that of the other transition metals. Iron has the ability to exist in two stable oxidation states, Fe^{2+} and Fe^{3+}, and to form complexes for which the standard oxidation reduction potential of the Fe^{2+}/Fe^{3+} couple ranges from 0.35 Volt to -0.50 Volt. Although much less than the $^{\cdot}OH$, this means that it lies in the range of the strongest biological oxidants or reductants known (25). These physico-chemical properties of iron make the metal a very suitable catalyst in the Haber-Weiss reaction (26), which is expressed as follows:

$$
\begin{array}{llll}
Fe^{3+} + O_2^- & \rightarrow & Fe^{2+} + O_2 & (1) \\
Fe^{2+} + H_2O_2 & \rightarrow & Fe^{3+} + OH- +^{\cdot}OH & (2) \\
\hline
O_2^- + H_2O_2 & \rightarrow & O_2 + OH- +^{\cdot}OH & (3)
\end{array}
$$

Superoxide reacts with oxidized iron, the oxidation state in which the metal is stored in the tissues (equation 1). Equation 2 is also known as the Fenton reaction (27). Equation 3 is the classical Haber-Weiss reaction, which has no biological significance without a metal catalyst.

In addition, iron is involved in the function of many biological molecules (such as hemoglobin, various enzymes, and cytochromes) which may degrade to complexes that can catalyze the generation of $\cdot OH$. If in these molecules the iron ion is in the ferrous oxidation state (e.g in heme complex) formation with oxygen may occur via one of the free coordination sites. In this redox couple Fe^{2+} and O_2 alternately oxidize and reduce, respectively. The most simple expression of this reaction is as follows:

$$RFe^{2+} + O_2 \rightleftarrows RFe^{2+}...O_2 \rightleftarrows RFe^{3+}... O_2^- \rightleftarrows RFe^{3+} + O_2^-$$

The so formed O_2^- can be reduced to the more toxic oxygen metabolites H_2O_2 and $\cdot OH$ using another heme molecule for the Fenton reaction (28).

An example of iron-mediated, $\cdot OH$-induced damage is the injury of DNA by bleomycin (29):

$$DNA + BLM + Fe^{3+} - complex \xrightarrow{reductant} DNA-BLM-Fe^{2+}$$

$$DNA - BLM - Fe^{2+} \xrightarrow{O_2 \text{ species}} Damaged\ DNA-BLM+Fe^{3+} - complex$$

The bleomycin-bound Fe^{2+} is formed by the interaction of Fe^{3+}- complex (the oxidation state of the largest part of intracellular iron) with intracellular reducing equivalents such as ascorbic acid, GSH, and NAD(P)H. The injury to DNA is caused by $\cdot OH$ of which the formation is catalyzed by the BLM-Fe^{2+}-complex bound to DNA. This complex first reduces O_2 to O_2^- which in turn spontaneously, catalyzed by Fe^{2+} or by SOD, reduces to H_2O_2. The $\cdot OH$ is then formed in a Fenton type reaction. During the action of bleomycin not only DNA is damaged but there may also be injury to other cellular consituents, including the cytoplasmic membrane. Lungs are especially susceptible to bleomycin-mediated injury and acute lung injury is a potential risk of treatment with this drug.

For paraquat (PQ^{2+}),which is taken up by alveolar type 1 and type 2 epithelial cells via a polyamine uptake system (30) and then reduced to the paraquat cation radical ($PQ^{\cdot}$) by e.g. NAD(P)H (30), a similar iron-catalyzed oxygen radical-mediated toxic mechanism is proposed:

$$PQ^{+}\; +\; O2\; \rightarrow\; PQ^{2+}\; +\; O_2^{-}$$

$$O_2^{-}\; +\; Fe^{3+}\; \rightarrow\; O_2\; +\; Fe^{2+}$$

$$PQ^{+}\; +\; Fe^{3+}\; \rightarrow\; PQ^{2+}\; +\; Fe^{2+}$$

$$PQ^{+}\; +\; O_2^{-}\; +\; 2H^{+}\; \rightarrow\; PQ^{2+}\; +\; H_2O_2$$

$$O_2^{-}\; +\; O_2^{-}\; +\; 2H^{+}\; \rightarrow\; O_2\; +\; H_2O_2$$

The PQ^{+} reacts with oxygen to form O_2^{-} (31). Redox-cycling of iron can be sustained by either O_2^{-} (32) or PQ^{+} (33). Superoxide is reduced spontaneously or catalyzed by SOD, in both cases to form H_2O_2. In addition, O_2^{-} can also be reduced by the PQ^{+} to H_2O_2 (34). Hydrogen peroxide may then be reduced to $\cdot OH$ in an iron-catalyzed Fenton-type reaction.

In view of these mechanisms, one approach in the prevention of injury by toxic oxygen species, is the direct inhibition of the $\cdot OH$ formation using iron chelators. These are molecules which can form coordination compounds with iron (and possibly other metals). An electron acceptor (e.g. hydrogen ion, metal ions) forms a coordinate bond with an electron donor (a complexing or chelating ligand) that supplies the electron pair involved in bond formation (35). Since iron ions in aqueous solution are dissolved with six coordinating water molecules, chelation results in the replacement of H_2O by the iron-liganding donor atoms. To be catalytically active iron requires at least one coordination site that is occupied by a readily dissociable ligand (36-38). Thus, depending on the nature and the concentration of the chelating ligand, iron catalyzed oxygen toxicity can be prevented.

Table 2. Biological relevance of iron chelators

1	Maintain iron in a water soluble form
2	Determine the oxidation state of iron
3	Determine the standard redox potential of iron (E_o' [Fe^{3+}/Fe^{2+}])

As previously mentioned, inhibition of the generation of $\cdot OH$ by using iron chelators may be an approach to prevent oxygen radical-mediated tissue injury. However, chelates in which the iron is not completely encompased, thus leaving room for the binding of one or more water molecules so that the iron ion can exert its catalytic effect on oxygen toxicity) may increase $\cdot OH$ generation. Examples of such chelators are nitrilotriacetic acid NTA) (39) and citrate (40,41). In a 1:1 complex with Fe^{3+}, which leaves room for two and three coordinated water molecules, respectively, these iron chelators enchance oxygen toxicity (42). Another example is ethylenediaminetetraacetic acid (EDTA) which, although a hexadentate molecule, cannot occupy all of the six aquo positions of the iron ion because the six binding groups cannot stretch far enough (43). This permits the metal to be catalytically active via a seventh coordination possibility for water, induced by distortion of the usual coordination symmetry (43). In contrast, deferoxamine (DF), which is also a hexadentate ligand and has three hydroxamic acid moieties, completely "locks in" the iron ion in a 1:1 complex, binding to all of the six coordination sites (44). A similar "locking in" phenomenon, which offers a model for safe storage and transport of iron in biological systems, is shown by iron-binding proteins such as the transferrins and ferritin in vertebrates (45). These proteins have strong ferroxidase activity (25,46), the redox potential of iron in transferrin being approximately -400 mV, this prevents iron being an electron donor in the Fenton reaction.

Deferoxamine has indeed been shown to be protective in various models of oxygen-radical induced lipid peroxidation or cellular injury (47-63). The mechanism of this beneficial effect of DF is based on its blockade of the Haber-Weiss reaction:

$$O_2^- + H_2O_2 \quad - \quad \boxed{\begin{array}{c} \text{Fe-Salt} \\ \text{catalyst} \end{array}} \quad \rightarrow O_2 + OH^- + \cdot OH$$

$$O_2^- + H_2O_2 \quad - \quad \boxed{\text{DF}} \quad \rightarrow \text{no biological significance}$$

For example, the lethal effect of streptonigrin on cultures of *Escherichia coli*, a result of O_2^- production by streptonigrin, was enhanced by the addition of citrate and inhibited by deferoxamine (48). A similar effect was observed in paraquat-intoxicated mice. These animals animals survived if treated with deferoxamine but died earlier after the intoxication with paraquat if iron was also administered (49). Continuous intravenous infusion of deferoxamine also decreased mortality by paraquat in vitamin E-deficient rats (50). In this study, the vitamin E-deficient state of the animals may have been crucial for the effect of deferoxamine since it has been reported that high levels of vitamin E decrease DF-induced excretion of iron in rats (64). Preliminary studies in our laboratory using rats receiving the standard, vitamin E rich laboratory animal food support this possibility. In such animals deferoxamine does not seem to protect against paraquat. However, since the vitamin E level in lung tissue of normal fed rats is about three times higher than the concentration of vitamin E in lung tissue of humans (65), this observation does not exclude a beneficial effect of deferoxamine in human paraquat poisoning.

Several in vitro heart ischemia reperfusion studies also report the beneficial effect of deferoxamine. For example, the iron chelator improved the recovery of contractile function (51), and attenuated both myocardial dysfunction and metabolic abnormalities as well as a decrease in tissue edema (52). The drug also prevented ventricler fibrillation and normalized contractility in reperfused iron-loaded hearts (53). The mechanism of this observation is perhaps the same as that for the improved left ventricular function by deferoxamine in ß thalassemia (54).

In a recent clinical study deferoxamine reduced neutrophil-mediated free radical production during cardiopulmonary bypass (55). This contradicts the in vitro observation that neutrophil function, including the respiratory burst, is preserved in a medium containing deferoxamine (56). However, it is possible that deferoxamine, by protecting the vascular endothelium against injury, prevents the release of inflammatory substances such as platelet-activating factor, that "prime" the neutrophil for increased responsiveness to trigger their superoxide generating system. A similar mechanism could underlie the observation that iron chelation by deferoxamine showed a reduction in soft tissue swelling and bone erosion in the inflamed joints of rats (57).

Protection against cell lysis by deferoxamine has been shown in several in vitro studies. Erythrocyte (58) and endothelial cell (59) lysis by activated neutrophils can be prevented by the addition of deferoxamine to the incubation medium. Moreover, preincubation of endothelial cells in deferoxamine also protects against neutrophil cytotoxicity (60), suggesting that deferoxamine chelates the intracellular iron that would otherwise have been used for ˙OH generation.

Table 3. Antioxidant mechanism of deferoxamine

1. Inactivates catalytic activity of iron by:
   - occupations of the six aquo positions
     (no free aquated iron)
   - maintains iron in the ferric state
2. Direct scavenging of hydroxyl radical
   (diffusion limited)

Another antioxidant feature of deferoxamine is its capacity to react with superoxide (66,67), to form a relatively stable nitroxide (67).

The rection of deferoxamine with O_2^-, however, is very slow and is unlikely to influence the interpretation of experiments in which the
chelator is used (66). Deferoxamine also reacts with hydroxyl radical
and it is suggested that the interaction could be a source of error in
the explanation of results using the chelator (66). However, to form
$^\cdot$OH, which then might react with deferoxamine, iron in the ferrous
oxidation state is needed. Although deferoxamine binds the Fe^{2+}-ion
weakly (68), if at all (68,69), the chelator strongly promotes the
oxidation of the ferrous ion (68), thus maintaining iron in the ferric
state and inhibiting its reaction with H_2O_2. That is, iron sequestered
by deferoxamine will be exceptionally inert by catalyzing the Haber-
Weiss reaction.

By whatever mechanism, a deferoxamine-related decrease in oxygen
radicals may inhibit cell injury. In addition, this may also lead to an
inhibition of the activation and release of inflammatory mediators,
preserving tissues and organ functions.

REFERENCES

1. Freeman, B.A., Crapo, J.D.: Biology of disease. Free radicals and
    tissue injury. Lab Invest 47: 412-426, 1982.
2. Cross, C.E. e.a.: Oxygen radicals and human disease. Ann of Intern
    Med 107: 524-545, 1987.
3. Babior, B.M.: Oxygen-dependent microbial killing by phagocytes
    (two parts). N Engl J Med. 298: 659-668; 721-725, 1978.
4. Beauchamp, C., Fridovich, I.: A mechanism for the production of
    sthylene from methional. The generation of the hydroxyl radical
    by xanthine oxidase. J Biol Chem 245: 4641-4646, 1970.
5. Henricks, P.A.J., Van der Tol, M.E., Thyssen R.M.W., Van Asbeck,
    B.S., Verhoef J.: Escherichia coli lipopolisaccharides diminish
    and enhance cell function of human polymorphonuclear leukocytes.
    Infect Immun 41: 294-301, 1983.
6. Wrigglesworth, J.M., Baum, H.: The biochemical functions of iron.
    In: Iron in Biochemistry and Medicine. II. Jacobs A, Worwood M
    eds, Academic Press, London and New York, pp. 29-86, 1980.
7. Starke, P.E., Farber, J.L.: Ferric iron and superoxide are
    required for the killing of cultured hepatocytes by hydrogen
    peroxide. Evidence for the participation of hydroxyl radicals
    formed by an iron-catalyzed Haber-Weiss reaction. J Biol Chem 260:
    10099-10104, 1985.
8. Roots, R., Okada, S: Estimation of life times and diffusion
    distances of radicals involved in X-ray-induced DNA strand breaks
    or killing of mammalian cell. Radiat Res 64: 306-320, 1975.
9. Weiss, S.J.: Mechanisms of disease; tissue destruction by neutro-
    phils. N Engl J Med 320: 365-376, 1989.
10. Egan, R.W., Gale, P.H. Kuehl, F.A. Jr.: Reduction of hydroper-
    oxides in the prostaglandin biosynthetic pathway by a microsomal
    peroxidase. J Biol Chem 254: 3295-3302, 1979.
11. Rosenblum, W.I.: Hydroxyl radical mediates the endothelium-
    dependent relaxation produced by bradykinin in mouse cerebral
    arterioles. Circ Res. 61: 601-603, 1987.

12. Lewis, S.L., Whatley, R.E., Cain, P., McIntyre, T.M., Prescott, S.M., Zimmerman, G.A.: Hydrogen peroxide stimulates the synthesis of platelet-activating factor by endothelium and induces endothelial cell-dependent neutrophil adhesion. J Clin Invest. 82: 2045-2055, 1988.

13. Rush, D.N., McKenna, R.M., Walker, S.M., Bakkestad-Legare, P., Jeffrey, J.R.: Catalase increases lymphocyte proliferation in mixed lymphocyte culture. Transpl Proceed 20: 1271-1273, 1988.

14. Larsson, R., Cerutti, P.: Oxidants induce phosphorylation of ribosomal protein S6. J Biol chem 263, 17452-17458, 1988.

15. Arrick, B.A., Nathan, C.F., Griffith, O.W., Cohn, Z.A.: Glutathione depletion sensitizes tumor cells to oxidative cytolysis. J Biol Chem 257: 1231-1237, 1982.

16. Arrick, B.A., Nathan, C.F., Cohn, Z.A.: Inhibition of glutathione synthesis augments lysis of murine tumor cells by sulfhydryl-reactive antineoplastics. J Clin Invest 71: 258-267, 1983.

17. Mitchell, J.B., Russo, A.: Role of glutathione in radiation and drug induced cytotoxicity. Proceedings of the 13th L.H. Gray Conference, Brunel University, West London, 14-18 July, 1986. pp. 96.

18. Bernard, G.R., Lucht, W.D., Niedermeyer, M.E., Snapper, J.R., Ogletree, M.L., Brigham, K.L.: Effect of N-acetylcysteine on the pulmonary response to endotoxin in the awake sheep and upon in vitro granulocyte function. J Clin Invest 73:1772-1784, 1984.

19. Van Asbeck, B.S., Van der Wal, W.A.A., Heesbeen, E.C., Brandt, C.J.W.M., Vosmeer, J.W.G., Van Oirschot, J.F.L.M.: Crucial role for lung glutathione in protection against hyperoxia. Amer Rev Resp Dis 135: All 1987

20. Wagner, P.D., Mathieu-Costello, O., Bebout, D.E., Gray, A.T., Natterson, P.D., Glennow, G.: Protection against pulmonary O_2 toxicity by N-acetylcysteine. Eur Respir J 2: 116-126, 1989.

21. Williamson, J.M., Boettcher, B., Meister, A.: Intracellular cysteine delivery system that protects against toxicity by promoting glutathione synthesis. Proc Natl Acad Sci 79: 6246-6249, 1982.

22. Moldéus, P., Cotgreave, I.S., Berggren, M.: Lung protection by a thiol-containing antioxidant: N-acetylcysteine. Respiration 50, supp 1:31-42, 1986.

23. Nakayama, T., Kaneko, M., Kodama, M., Nagat, C.: Cigarette smoke induces DNA single-strand breaks in human cells. Nature 314: 462-464, 1985.

24. Meister, A.: Selective modification of glutathione metabolism. Science 220: 473-477, 1983.

25. Aisen P: Some physiochemical aspects of iron metabolism. In: Iron metabolism. Ciba Foundation Symposium. Elsevier: Exerpta Medica/North-Holland Inc, Amsterdam pp. 1-17, 1977.

26. Haber, F., Weiss, J: The catalytic decompensation of hydrogen peroxide by iron salts. Proc Roy Soc Lond (A) 147: 332-351, 1934.

27. Fenton HJH: Oxidation of tartaric acid in presence of iron. J Chem Soc 65: 899-910, 1894.

28. Sadrzadeh, S., Graf, E., Panter, S.S., Hallaway, P.E., Eaton, J.W.: Hemoglobin. A biologic fenton reagent. J Biol Chem 259:14354-14356, 1984.

29. Sausville, E.A., Peisach, J., Horwitz, S.B.: Effect of chelating agents and metal ions on the degradation of DNA by bleomycin. Chemistry 17: 2740-2746, 1978.

30. Smith, L.L., Rose, M.S., Wyatt, I.: The pathology and biochemistry of paraquat. London: Symposium on Oxygen Free Radicals and Tissue Damage: 321-431, 1976.

31. Bus, J.S., Gibson, J.E., Paraquat: model for oxidant-initiated toxicity. Environ Health Perspect 55: 37-46, 1984.

32. McCord, J.M., Day, E.D. Jr.: Superoxide-dependent production of

hydroxyl radical catalyzed by iron-EDTA complex. FEBS Lett 86: 139-142, 1978.

33. Land, E.J., Swallow, A.J.: Electron transfer from pyridinyl radicals to cytochrome c. Berl Bunsenges Phys Chem 79: 436-437, 1975.

34. Patterson, L.K., Small, R.D. Jr, Scaiano, L.C.: Reaction of paraquat radical cations with oxygen: a pulse radiolysis and laser photolysis study. Radiat Res 72: 218-225, 1977.

35. Martell, A.E.: The design and synthesis of chelating agents. In: development of iron chelators for clinical use. Martell AE, Anderson WF, Badman DG, eds. Elsevier/North-Holland, New York, Amsterdam, Oxford, pp. 67-131, 1981.

36. Dwyer, F.P.: Enzym-metal ion activation and catalytic phenomena with metal complexes. Chelating Agents and Metal Chelates. Dwyer F.P., Melloor D.P., eds. Academic Press, New York, London, pp. 335-382, 1964.

37. Graf, E., Mahoney, J.R., Bryant, R.G., Eaton, J.W.: Iron-catalyzed hydroxyl radical formation. J Biol Chem 259: 3620-3624, 1984.

38. Martell, A.E., Gustafson, R., Chaverek, S.: Metal chelate compounds in homogenous aqueous catalysis. Advances in Catalysis IX. Farkas A ed. Academic Press, New York, pp. 319-322, 1957.

39. Schwarzenbach, G., Heller, J.: Die Eisenkomplexe der Nitrolotriessigsäure. Helv Clin Acta 34: 1889-1901, 1951.

40. Spiro, T.G., Pape, L., Saltman, P.: The hydrolytic polymerization of ferric citrate. I. Chemistry of the polymer. J Am Chem Soc 89: 5555-5558, 1967.

41. Spiro, T.G., Bates, G., Saltman, P.: The hydrolytic polymerization of ferric citrate. II. The influence of excess citrate. J Am Chem Soc 89: 5559-5562, 1967.

42. Van Asbeck, B.S., Marx, J.J.M., Struyvenberg, A., Van Kats, J.H., Verhoef, J.: Effect of iron (III) in the presence of various ligands on the phagocytic and metabolic activity of human polymorphonuclear leukocytes. J of Immunol 132: 851-856, 1984.

43. Lind, M.D., Hamor, M.J., Hoard, J.L.: Sterochemistry of ethylene-diamine-tetraacetato complexes. Inorg Chem 3: 34-43, 1984.

44. Keberle, H.: The biochemistry of desferrioxamine and its relation to iron metabolism. Ann NY Acad Sci 119: 758-768, 1964.

45. Aisen, P.: Iron transport and storage proteins. Ann Rev Biochem 49: 357-393, 1980.

46. Bates, G.W., Workman, E.F. Jr., Schlabach, M.R.: Does transferrin exhibit ferroxidase activity. Biochem Bioph Res Com 50: 84-90, 1973.

47. Gutteridge, J.M.C., Richmond, R., Halliwell, B.: Inhibition of the iron-catalyzed fromation of hydroxyl radicals from superoxide and of lipid peroxidation by desferrioxamine. Biochem J 184: 469-472, 1979.

48. White, J.R., Yeowell, H.N.: Iron enhances the bacterial action of streptonigrin. Biochem Biophys Res Commun 106: 407-411, 1982.

49. Kohen, R., Chevion, M.: Paraquat toxicity is enhanced by iron and reduced by desferrioxamine in laboratory mice. Biochem Pharmacol 34: 1841-1843, 1985.

50. Van Asbeck, B.S., Hillen, F.C., Boonen, H.C.M., De Jong, Y., Dormans, J.A.M.A., Van der Wal, N.A.A., Marx, J.J.M., Sangster, B.: Continuous Intravenous Infusion of deferoxamine reduces mortality by paraquat in vitamin E-deficient rats. Am Rev Respir Dis 139: 769-773, 1989.

51. Bolli, R., Patel, B.S., Zhu, W., O'Neill, P.G., Hartley, C.J., Charlat, M.L., Roberts, R.: The iron chelator desferrioxamine attenuates postischemic ventricular dysfunction. Am Physiolog Soc: 1372-1380, 1987.

52. Farber, N.E., Vercellotti, G.M., Jacob, H.S., Pieper, G.M., Gross,

G.J.: Evidence for a role of iron-catalyzed oxidants in functional and metabolic stunning in the canine heart. Circ Res 63: 351-360, 1988.

53. Van der Kraaij AMM, Mostert LJ, Van Eijk HG, Koster JF: Ironload increases the susceptibility of rat hearts to oxygen reperfusion damage. Circulation 78: 442-449, 1988.

54. Grisaru, D., Goldfarb, A.W., Gotsman, M.S., Rachmilewitz, E.A., Hasin, Y.: Deferoxamine improves left ventricular function in - thalassemia. Arch Intern Med 146: 2344-2349, 1986.

55. Menasché, P., Pasquier, C., Bellucci, S., Lorente, P., Jaillon, P., Piwnica, A.: Deferoxamine reduces neutrophil-mediated free radical production during cardiopulmonary bypass in mann. J Thorac Cardiovasc Surg 96: 582-587, 1988.

56. Van Asbeck, B.S., Marx, J.J.M., Struyvenberg, A, Van Kats, J.H., Verhoef, J.: Deferoxamine enhances phagocytic function of human polymorphonuclear leukocytes. Blood 63: 714-720, 1984.

57. Andrews, F.J., Morris, C.J., Kondratowicz, G., Blake, D.R.: Effect of iron chelation on inflammatory joint disease. Ann of Rheum Dis 46: 327-333, 1987.

58. Vercellotti, G.M., Van Asbeck, B.S., Jacob, H.S.: Oxygen radical induced erythrocyte hemolysis by neutrophils: critical role of iron and lactoferrin. J Clin Invest 76: 956-962, 1985.

59. Ward, P.A., Till, G.O., Kunkel, R., Beauchamp, C.: Evidence for role of hydroxyl radical in complement and neutrophil-dependent tissue injury. J. Clin. Invest. 72: 789-801, 1983.

60. Gannon, D.E., J. Varani, Phan, S.H., Ward, J.H., Kaplan, J., Till, G.O., Simon, R.H., Ryan, U.S., Ward, P.A.: Source of iron in neutrophil-mediated killing of endothelial cells. Lab Invest 57: 37-44, 1987.

61. Fuller, B.J., Lunec, J., Healing, G., Simpkin, S., Green, C.J.: Reduction of susceptibility to lipid peroxidation by desferrioxamine in rabbit kidneys subjected to 24-hour cold ischemia and reperfusion. Transplantation 43: 604-606, 1986.

62. Menasché, P., Grousset, M.D.C., Gauduel, Y., Mouas, C., Pitwnica, A.: Prevention of hydroxyl radical formation: a critical concept for improving cardioplegia.Circulation 76 suppl. V: 180-185, 1987.

63. Cerchiari, E.L., Hoel, T.M., Safar, P., Sclabassi, S.J.: Protective effects of combined superoxide dismutase and deferoxamine on recovery of cerebral blood flow and function after cardiac arrest in dogs. Stroke 18: 869-878, 1987.

64. Hershko, C., Rachmilewitz, E.A.: The inhibitory effect of vitamin E on desferrioxamine-induced iron excretion in rats. Proc Soc Exp Biol Med 152: 249-252, 1976.

65. Slade, R., Stead, A.G., Graham, J.A., Hatch, G.E.: Comparison of lung antioxidant levels in humans and laboratory animals. Am Rev Respir Dis 131: 742-746, 1985.

66. Halliwell, B.: Evidence for a direct reaction between desferal and the superoxide radical. Biochem Pharm 34: 229-233, 1985.

67. Davies, M.J., Donkor, R., Dunster, C.A., Gee, C.A., Jonas, S., Willson, R.L.: Desferriozamine (desferal) and superoxide free radicals. Biochem J 246: 725-729, 1987.

68. Harris D.C., Aisen, P.: Facilitation of Fe(II) autoxidation by Fe(III) complexing agents. Biochim Biophys Acta 329: 156, 1973.

69. Goodwin, J.F., Whitten, C.F.: Chelation of ferrous sulphate solutions by desferrioxamine B. Nature 205: 281, 1965.

ANTIOXIDANT PROPERTIES OF AN ANTIISCHAEMIC AGENT : TRIMETAZIDINE

Pascale Clauser and Catherine Harpey

IRIS Neuilly sur Seine (France)

Trimetazidine (TMZ) has now been widely used throughout the world for more than ten years for the treatment of cellular manifestation of ischaemic disorders including coronary heart diseases, retinopathies and cochleo-vestibular diseases. Its efficacy has been assessed in double-blind studies versus placebo and versus reference drugs. In angina pectoris TMZ improves the patient's condition in diminishing the number of daily attacks, assessed by the objective improvement of the exercise test : longer total duration (+ 18 %), greater total work (+ 33 %) and delay in the 1 mm ST depression on the ECG which reflects a better tolerance to effort (Sellier, 1986). TMZ has also shown a very good tolerance and associated with other drugs, it allows to increase the therapeutical benefit (Brochier et al., 1986 ; Michaelides et al., 1987).

At the therapeutical dosage (1 mg/kg) it has no haemodynamic effects and consequently does not act as other drugs used in this pathology which either diminish the heart work and oxygen demand or increase its perfusion.

TMZ is described as a cellular antiischaemic agent as it has shown beneficial effects on the three major consequences of ischaemia : loss of energetic compounds, intracellular acidosis, production of free radicals.

Despite hypoxia or induced ischaemia, TMZ maintains homeostasis and cellular functions, thus inhibiting cytolysis as evidenced on different in vivo models of ischaemia : heart infarction after temporary coronary ligation (Fitoussi et al., 1985 ; Camilleri and Joseph, 1988) renal ischemia by renal pedicle clamping in the uninephrectomised rat (Catroux et al., 1986). In vitro, TMZ limited the extent of edema-induced histological lesions due to a period of severe hypoxia in the isolated rat hearts (Didier et al., 1984), and produced a significant improvement in action potential characteristics from isolated and stimulated guinea-pig left ventricle submitted to mild ischaemia and reperfusion (Honoré et al., 1986).

Using ^{31}P nuclear magnetic resonance (NMR) spectrometry Lavanchy et al. (1987) showed that despite severe hypoxia, TMZ antagonizes hypoxic-induced reduction in myocardial ATP levels, and significantly

Antioxidants in Therapy and Preventive Medicine
Edited by I. Emerit *et al.*
Plenum Press, New York, 1990

247

increases the levels of phosphocreatine. From low concentrations
(6.10^{-7} M) TMZ accelerated the reconstitution of energy pools during
reperfusion.

The authors also clearly demonstrated that during an episode of
ischaemia, the first effect induced by TMZ on the isolated rat heart
was a reduction of myocardial intracellular acidosis. More recently
Renaud (1988) studied the effect of the drug on the structure and
function of ionic carrier systems on plasma membranes of chick
embryonic heart and new-born rat heart as well as on cardiac cells of
the same origin, cultured in monolayers. He demonstrated that in
normal conditions TMZ does not act either directly or indirectly on
voltage-dependant Na^+ and Ca^{2+} channels or on (Na^+, K^+)-ATPase.
However under acid loaded conditions, TMZ reduced the accumulation of
Na^+ and Ca^{2+} in cardiac cells by antagonizing intracellular acidosis.

Several studies have assessed TMZ antioxydant activity :

Maridonneau-Parini and Harpey (1985) first demonstrated on human
red blood cells that TMZ, administered at a therapeutic dose for 7
days, exerted a potent antioxidant effect. The loss of intracellular
K^+ induced by oxygen free radicals (generated by phenazine
methosulphate added to the medium) in red cells as well as the
membrane content of peroxidated lipids were significantly reduced by
the drug.

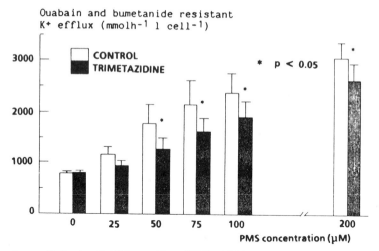

Fig. 1.   Effect of TMZ on the (PMS, DDC) - dependent increase
in passive K^+ permeability : ouabain and bumetanide
resistant K^+ efflux stimulated by various PMS
concentration was measured before ☐ and after ▓ 1
week treatment with TMZ in red cells from seven
healthy volunteers. * $p < 0.05$

Other experiments which assessed the antioxydant properties of
TMZ are presented in Harpey et al. (1987).In vitro on rat cultured
glomerular mesangial cells, Baud measured changes in hydrogen peroxide
(H_2O_2) production after activation with different stimuli (opsonised
zymosan or ionophore A23187). TMZ, at concentrations from 50 μM,
significantly reduced H_2O_2 production.

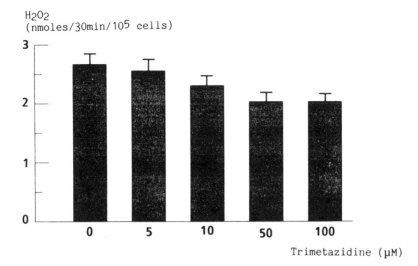

Fig. 2. Changes in H_2O_2 production on rat cultured mesangial cells 24 h after activation with zymozan. $p < 0.001$

Housset used pig aortic endothelial cells and evaluated free-radical-induced cytolysis by measuring LDH release elicited by glucose oxidase (an inducer of H_2O_2 generation). TMZ at 10^{-6} M and higher concentrations significantly reduced the release of LDH. Finally Piccinini induced cardiotoxicity in the guinea-pig after i.v. administration of doxorubicin (5 mg/kg), a generator of free radicals. TMZ (2.5 mg/kg, i.p., for 4 days) protected the animals from doxorubicin-induced cardiac toxicity.

Doly (1988) used isolated and perfused retinas from albino rat. This tissue is very rich in polyunsaturated fatty acids and is, therefore, very sensitive to the aggressive action of oxygen-derived free radicals, which the author generated by using a $FeSO_4$/Na ascorbate system. An electroretinogram recorded after a standard luminous stimulation was used to monitor the status of the retinal metabolism. Neoformed free radicals disturbed the genesis of the electroretinogram and decreased the positive "b wave" amplitude ; this effect was prevented by TMZ (2.5 mg/kg, i.p., for 5 days).

Guarnieri and Muscari (1988) produced a condition of heart hypertrophy by injection of monocrotaline in rats ; this drug decreases the mitochondrial function and increases generation of mitochondrial O_2^- radicals and lipoperoxydative damage. Concomittent treatment of the animals by TMZ (2.5 mg.kg^{-1}) improves the cardiac mitochondrial activity (table 1). Their results indicate that the cardiac submitochondrial particles in the rats treated with TMZ and monocrotaline produce less O_2^- than those in treated with MCT alone. Cambar and Catroux (see in this issue) also demonstrated that TMZ exerts antioxydant properties on their model.

After indirect evidence of TMZ antioxydant properties, Maupoil and Rochette, and Boucher and de Leiris tried a more direct approach to free radicals, using Electron Paramagnetic Resonance Spectroscopy and Spin Trapping Experiment (see in this issue).

In conclusion, TMZ when studied in vitro in free radicals generating systems (xanthine/xanthine oxydase, H_2O_2/FeEDTA, Cysteine/FeSO4) has only shown a slight direct antioxyradical effect but these conditions do not reproduce the enzymatic reactions which occur in biological systems. However TMZ has shown a protective effect against free radical deleterious effects in various in vitro and in vivo models. Moreover, more sophisticated methods which actually allow to evidence the free radical species in vitro, confirm that TMZ modifies the free radical production.

REFERENCES

Brochier M., Demange J., Ducloux G., Monpère C. and Warin J.F., 1986, Intérêt de l'association de la trimétazidine à un inhibiteur calcique dans le traitement de l'insuffisance coronarienne chronique, Annales de Cardio. et d'Angéiol., 35 : 49-56.

Camilleri J.P. and Joseph D., 1988, Effets de la trimétazidine (Vastarel 20 mg) sur l'infarctus expérimetal du rat perfusé, Arch. des Mal. du Coeur et des vaisseaux, 81 :371.

Catroux P., Dorian C., Harpey C. and Cambar J., 1986, Mise en évidence de l'effet protecteur de la trimétazidine vis-à-vis de l'enzymurie induite par clampage du pédicule rénal chez le rat, Néphrologie, 7 : 1240.

Didier J.P., Roux J., Violot D. and Justrabo E., 1984, Les effets de la trimétazidine sur le coeur isolé perfusé de rat en hypoxie : étude hémodynamique et histologique, Gaz. Med. France, 91 : 28-34.

Doly M., 1988, Rétine et radicaux libres, Ophtalm. Franç. n° spec. : 98-105.

Fitoussi M., Rochette L., Bralet J. and Harpey C., 1985, Incidences fonctionnelles et métaboliques d'un pré-traitement par la trimétazidine au niveau du coeur de rat, Symposium Biologie et Pathologie du Coeur et des Vaisseaux - Toulouse 28-29/03/85.

Guarnieri C. and Muscari C., 1988, Beneficial efects of trimetazidine on mitochondrial function and superoxide production in the cardiac muscle of monocrotaline-treated rats, Biochem. Pharmacol., 37 : 4685-4688

Harpey C., Labrid C., Baud L., Housset B., Maridonneau-Parini I., Piccinini F. and Goupit P., 1987, Evidence for antioxidant properties of trimetazidine, Xth Inter. Congress Pharmacol. Sydney 23-28/08/87.

Honoré E., Adamantidis M.M., Challice C.E. and Dupuis B.A., 1986, Cardioprotection by calcium antagonists, piridoxilate and trimetazidine, IRCS Med. Sci., 14 : 938-939.

Lavanchy N., Martin J. and Rossi A., 1987, Antiischemic effects of trimetazidine : ^{31}P-NMR spectroscopy in the isolated rat heart, Arch. Intern. Pharmacodyn. Ther., 286 : 97-110.

Maridonneau-Parini I. and Harpey C., 1985, Effects of trimetazidine on membrane damage induced by oxygen free radicals in human red cells, Br. J. Clin. Pharmac., 20 : 148-51.

Michaelides A.P., Vysoulis G.P., Bonoris P.E., Psaros TH. K., Papadopoulos P.D. and Toutouzas P.K., 1987, Beneficial effects of trimetazidine in patients with stable angina under B-blockers, Cardiovasc. Drugs Ther., 1 : 268.

Renaud J.F., 1988, Internal pH, Na+ and Ca2+ regulation by trimetazidine during cardiac cell acidosis, Cardiovasc. Drugs Ther., 1 : 677-686.

Sellier P., 1986, Effects de la trimétazidine sur les paramètres ergométriques dans l'angor d'effort, Arch. des Mal. du Coeur et des Vaisseaux, 9 : 1331-1336.

OXIDATIVE DAMAGE IN CHRONIC HEART FAILURE: PROTECTION BY

CAPTOPRIL THROUGH FREE RADICAL SCAVENGING?

M. Chopra, J. McMurray*, J. McLay*, A. Bridges**
N. Scott, W.E. Smith***, and J.J.F. Belch

*Dept. of Medicine, Clinical Pharmacology
**Dept. of Cardiology
***Ninewells Hospital, Dundee and Dept.of Chemistry
    Strathclyde University, Glasgow, Scotland

SUMMARY

The pathogenesis of heart failure is not yet fully understood. In animal models there is some evidence to suggest a role for free radicals (FRs). We have investigated malondialdehyde - LM in the plasma of patients with heart failure and found it to be raised when compared to controls. We present data to show that Captopril, a drug with an ACE inhibitory effect is a FR scavenger both in vitro and ex-vivo in patients with heart failure.

INTRODUCTION

Heart failure is a clinical state in which the heart is unable to provide sufficient blood for tissue metabolic needs. It has been suggested that the generation of free radicals (FRs) might be implicated in the pathogenesis of chronic heart failure (CHF). For example, in animal models heart failure secondary to adriamycin toxicity possibly results from myocardial injury induced by oxidative damage[1,2]. It has been shown that the FR producing activity of polymorphonuclear (PMN) leukocytes from dogs with heart failure is greater than that from control dogs[3]. There is evidence that in human CHF there is an increase in the production of prostaglandins, and their formation is associated with production of oxygen derived FRs through the arachidonic acid pathway[4]. Furthermore circulating catecholamine production increases in the failing heart and their auto-oxidation results in the generation of cytotoxic FRs[5]. FRs have also been implicated in the depression of cardiac contractility through inhibition of calcium transport and by decreasing the activity of Ca^{2+} ATPase of sarcoplasmic reticulum[6]. There is therefore some evidence to suggest an involvement of FRs in the genesis of myocardial damage in CHF.

Captopril, a derivative of the amino acid proline, has been shown to have beneficial cardiovascular effects in CHF[7,8]. These beneficial effects in CHF are believed to be related to inhibition of the angiotensin converting enzyme (ACE) responsible for converting angiotensin I to angiotensin II. However, other cardio protective effects of the drug have also been reported that may not be shared by other ACE Inhibitors[9,10,11]. Captopril has a sulphydryl group in its structure and since one of the effects of sulphydryl containing compounds is to scavenge FRs[12,13,14] we considered that captopril may be a FR scavenger. In view of the suggestion that CHF and FRs may be

Antioxidants in Therapy and Preventive Medicine
Edited by I. Emerit *et al.*
Plenum Press, New York, 1990

251

inter-related we have first of all investigated the possibility of FR pathology in human CHF. We have then studied the potential FR scavenging effect of captopril both in vitro and in patients with CHF treated with the drug.

PATIENTS AND METHODS

Study 1 - Patients with CHF

30 patients with CHF diagnosed by a team of cardiologists were included in the study. Patients with myocardial infarction within the previous 3 months, with obstructive valvular disease, chronic obstructive airways disease, myocarditis and unstable angina were excluded from the study. All patients gave informed consent and ethical permission was obtained. 10 mls of heparinised venous blood was sampled for plasma malondialdehyde-LM (MDA-LM) assay. An spectrophotometric assay described by Aust[15] was used for measuring MDA levels. Absorbance was converted to concentration using an extinction coefficient of 1.50×10^5 M^{-1} cm^{-1}.

Study II - Measurement of FR Scavenging effect of captopril in vitro

This was investigated using the technique described by Misra and Fridovich[16]. In this assay free radicals are generated by photo-oxidation of dianisidine sensitised by riboflavin. The photo-oxidation of dianisidine involves a complex series of FR chain reactions involving superoxide ion ($O_2 \cdot^-$) as the propagating species. A general FR scavenging compound has an inhibitory effect on this reaction leading to a decrease in the oxidised dianisidine measurable by UV/visible spectrophotometer. In contrast any compound which specifically scavenges $O_2 \cdot^-$ will have an augmentary effect on the assay. The effect of captopril on the assay was investigated at different final concentration of the drug in the reaction mixture viz 0.5, 1.0, 2.5 and 5.0 µg/ml. The details of this assay have been published in other studies[14,17].

Study III - Effect of captopril in CHF patients

30 patients with CHF were treated with captopril (6.25mg - 25 mg) and a further sample of blood was taken at one hour and between 2 to 3 hours after oral administration of the drug.

RESULTS

Table 1 shows the comparability of the two groups: patients with CHF and normal controls. As can be seen there is no significant difference between the two groups (Mann Whitney).

Table 1 Comparability of the groups

	Controls	CHF patients
n	15	30
Age (yrs) Median	60	62
IQR	55-63	60-65
Significance of difference from controls (MW)		N.S.
Sex M:F	8:7	15:10

The results of Figure 1 confirm the presence of increased FR activity in CHF, since there was statistically significant (Mann Whitney) increase in MDA-LM in CHF.

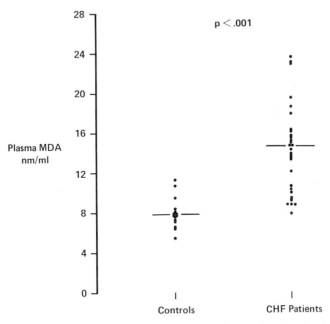

Figure 1   Plasma MDA levels in controls and chronic heart failure patients.   Significance of difference is given (Mann Whitney).

Table 2 shows that addition of captopril to the Misra and Fridovich assay produces an inhibitory effect on the assay, suggesting that it is a general FR scavenger.   This effect appeared to be dose dependant.

Table 2   Percentage Inhibition on the assay by Captopril

Concentration of drug in final solution µg/ml	% age Inhibition on the assay Mean ± S.D (n)
0.5	13.3 ± 2.0(7)
1.0	18.6 ± 1.3(7)
2.5	34.3 ± 1.6(7)
5.0	50.5 ± 1.8(7)

The results of Figure 2 show that at both one hour and between two to three hours there is a statistically significant (Wilcoxon paired rank sum test) fall in plasma MDA-LM levels after treatment with captopril.

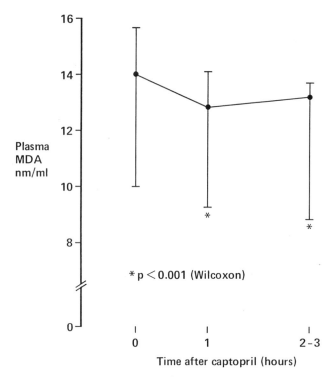

Figure 2   Plasma MDA (median, I.Q.R.) prior to and after oral administration of captopril.

DISCUSSION

This study documents the existence of oxidative damage in patients with heart failure as plasma MDA-LM levels were significantly raised in these patients when compared to controls. This suggests, for the first time, that FRs may be involved in the pathophysiology of human CHF. Furthermore, we have shown that captopril is a general FR scavenger, and this might be relevant to its therapeutic efficacy in CHF.

The peak blood level of captopril following oral administration has been reported at one hour and the average elimination $t^{1}/2$ after oral dose is 1.7 hours[18]. Our clinical work documents a fall in the plasma MDA-LM one hour after an oral dose of captopril. Although there is still a significant fall in MDA-LM 2-3 hours post-dose it tends to revert to baseline.

In conclusion therefore we suggest that free radical pathology exists in CHF. Captopril acts as a FR scavenger both in vitro & ex-vivo. The beneficial effect of captopril as a ACE inhibitor may therefore be further augmented by its scavenging activity making it superior to a drug with ACE inhibitory effect alone.

ACKNOWLEDGEMENTS

M.C. was supported by BHF grant 86/72(2D37). Captopril was kindly donated by Squibb and Sons Ltd.

REFERENCES

1. J. Goodman and P.L. Hochstein, Generation of free radicals and lipid peroxidation by redox cycling of adriamycin and duanomycin, Biochem. Biophys. Res. Comm. **77**:797 (1977).
2. P.C. Gervasi et al, Superoxide anion production by adriamycin from cardiac sarcosomes and by mitochondrial NADH DHase, Anticancer Res. **6**(5):1231 (1986).
3. K. Prasad, J. Kalra and B. Bharadwaj, Phagocytic activity in blood of dogs with chronic congestive heart failure, Clin. Invest. Med. **10**:1354 (1987).
4. V.J. Dzau et al, Prostaglandins in severe congestive heart failure. Relation to activation of the Renin-Angiotensin system and hyponatremia, N. Engl. J. Med. **310**:345 (1984).
5. D.G. Graham et al, Auto-oxidation versus covalent binding of quinones as the mechanism of toxicity of dopamine, 6-hydroxydopamine and related compounds towards C1300 neuroblastoma cells in vitro, Mol. Pharm. **14**:644 (1978).
6. M.L. Hess, E. Okabe and H.A. Kontos, Proton and free oxygen radical interaction with the calcium transport system of cardiac sarcoplasmic reticulum, J. Mol. Cell. Cardiol. **13**:767 (1981).
7. Captopril Multicentre Research Group, A placebo controlled trial of captopril in refractory chronic congestive heart failure, J. Am. Coll. Cardiol. **2**:755 (1983).
8. J.G.F Cleland et al, Captopril in heart failure. A double blind controlled trial, Br. Heart J. **52**:530 (1984).
9. G. Ertl, R.A. Kloner, W. Alexander and E. Braunwald, Limitation of experimental infarct size of an angiotensin - converting enzyme inhibitor, Circul. **65**:40 (1982).
10. W.H. Van Gilst, P.A. DeGraeff, H. Wesselingand C.D.J Dahangen, Reduction of reperfusion arrhythmias in the ischaemic isolated rat heart by angiotensin converting enzyme inhibitors: a comparison of captopril, enalpril and HOE 498, J. Cardiovasc. Pharm. **8**:722 (1986).
11. M.R.F Martin et al, Captopril: a new treatment for rheumatoid arthritis? Lancet I:1325 (1984).
12. W.H. Betts, L.G. Cleland, D.J. Gee and M.W. Whitehouse, Effects of D-pencillamine on a model of oxygen derived free radical mediated tissue damage, Actions **14**:283 (1984).
13. R.A. Cuperus, A.D. Muijsersand R. Weaver, Antiarthritic drugs containing thiol groups scavenge hypochloride and inhibit its formation by myeloperoxidase from human leucocytes: a therapeutic mechanism of these drugs in rheumatoid arthritis, Arth. Rheum. **28**:1228 (1985).
14. C.J. McNeil, et al, A relationship between thiols and the superoxide ion. FEBS Letts. **133**(1):175 (1981).
15. S.D. Aust, Lipid peroxidation, in: "Handbook of Methods for Oxygen Radical Research," R.A. Greenwald, ed., CRC Press Inc. Boca Raton, Florida, (1987).
16. H.P. Misra and I. Fredovich, Superoxide dismutase: A photochemical augmentation assay, Arch. Biochem. Biophys. **181**:308 (1977).
17. M. Chopra, J.J.F Belch, W.E. Smith, A comparison of the free radical scavenging activity of leucotrienes and prostaglandins, Free Rad. Res. Comm. **5**(2):95 (1988).
18. K.L. Duchin et al, Captopril kinetics, Clin. Pharm. Ther. **31**(4):452 (1982).

FREE RADICAL SCAVENGING CAPACITY AND ANTI-INFLAMMATORY ACTIVITY OF CBS-113 A

Claude Bonne*, Gérard Tissié, Hoa Ngo Trong, Elisabeth Latour and Claude Coquelet

Centre de Recherches, Laboratoire Chauvin, B.P. 1174, 34009 Montpellier cedex, France
* Laboratoire de Physiologie Cellulaire, Université Montpellier I, 15 avenue Charles Flahault, 34060 Montpellier cedex, France

CBS-113 A, 2-(2-Hydroxy-4-methylphenyl)aminothiazole hydrochloride, is a dual inhibitor of cyclooxygenase and lipoxygenase with topical anti-inflammatory activity (1). This compound is also a powerful free radical (FR) scavenger. The aim of this study was to design a pharmacological model of ocular inflammation for determining the respective contribution of the FR scavenging capacity and the inhibition of icosanoid production to the anti-inflammatory activity of this compound.

MATERIALS and METHODS

Free Radical Scavenging Activity

Reactivity of the drug with singlet oxygen ($^1O_2$) was determined with an adaptation of the method described by Ames et al.(2). Briefly, white light irradiation of a rose bengal solution produced singlet oxygen capable of oxidizing uric acid. The compound's capacity to inhibit the disappearance of uric acid was monitored spectrophotometrically at 292 nm.

Superoxide anion ($O_2^{\cdot-}$) was produced by a xanthine-xanthine oxidase system as described by Crapo et al.(3). The inhibition of cytochrome c reduction by the drug was followed spectrophotometrically at 550 nm.

Hydroxyl radical ($OH^{\cdot}$) produced by a Haber-Weiss reaction degraded the carbohydrate deoxyribose forming thiobarbituric acid reactants as reported by Gutteridge (4). The compound was tested for its efficiency in counteracting this degradation.

Peroxidation of phosphatidylcholine liposomes was induced by a Fenton reaction according to Liebler et al.(5). Drug protection was evaluated by measuring malonaldehyde production.

Antioxidants in Therapy and Preventive Medicine
Edited by I. Emerit *et al.*
Plenum Press, New York, 1990

257

Male New Zealand albino rabbits were used. Uveitis was induced by injection of FeSO$_4$ (0.3 mM) into the anterior chamber of the eye. Animals were pretreated with dimethylsulfoxide (DMSO = 0.3 ml Kg^{-1}) by intraveinous route 15 min before Fe^{2+} injection. CBS-113 A (0.01 %), indomethacin (0.1 %) and dexamethasone (0.1 %) were applied as eye drops prior to and after induction of inflammation at time −1 hr to + 1 hr every 15 min then at 3, 5, 7, 9 hrs. Polymorphonuclear leukocyte (PMN) count was determined in the aqueous humor 24 hrs after the FeSO4 injection.

RESULTS and DISCUSSION

In Vitro Studies

Figures 1 (a, b, c) report the capacity of CBS-113 A and reference compounds (tocopherol, ascorbate and DMSO) for scavenging singlet oxygen, superoxide anion and hydroxyl radical. The drug showed a significant and higher activity than salicylate and aspirin in the different tests in vitro. Data reported in Fig.1 (d) also show that CBS-113 A inhibited Fenton-induced peroxidation of phosphatidylcholine liposomes. In this test, the drug and tocopherol exhibited an equal activity, by contrast to non steroidal anti-inflammatory drugs (aspirin and indomethacin).

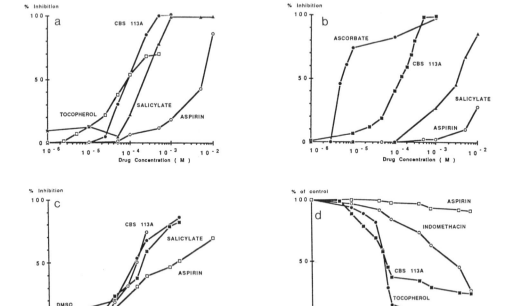

Fig. 1. Inhibition of oxygen-derived species reactivity with CBS-113 A and various compounds.
a) Singlet oxygen, b) Superoxide anion, c) Hydroxyl radical, d) Malonaldehyde formation in liposomes

Table 1. Fe^{2+}-induced Uveitis

Treatment	PMN (per ml x 10^6)	Inhibition (%)	n
Control	3.9 ± 0.6		4
DMSO (0.3ml Kg^{-1} – I.V.)	0.8 ± 0.1**	79	8
Vehicule	2.4 ± 0.5		8
CBS-113 A (0.01 %)	0.5 ± 0.1**	79	8
Vehicule	4.3 ± 1.6		4
Indomethacin (0.1 %)	2.0 ± 0.8*	53	6
Vehicule	1.1 ± 0.3		4
Dexamethasone (0.1%)	0.6 ± 0.3**	45	4

Statistical significance : * $p < 0.05$, ** $p < 0.01$ (Mann-Whitney test)

Fe^{2+}-induced Uveitis

Fe^{2+} injection into the aqueous humor which contains a high concentration of H_2O_2, induced an inflammatory reaction triggered by the generation of hydroxyl radical (OH·) via a Fenton reaction. DMSO, a potent OH·-scavenger, inhibited PMN migration when administered by general route. CBS-113 A at low concentrations (0.01%) in eye drop also very significantly decreased PMN accumulation in the aqueous humor.

By contrast, the non-radical scavenging anti-inflammatory drugs (indomethacin and dexamethasone) could inhibit the reaction only at a higher concentration (0.1 %) (Table 1).

In addition, CBS-113 A significantly lowered the protein level. By contrast, dexamethasone, at a dosage which eradicated icosanoid generation, did not suppress protein extravasation, whereas indomethacin increased it (data not shown). The results suggest that the free radical scavenging capacity of the drug could contribute to its original anti-inflammatory profile.

REFERENCES

1. C. Coquelet and C. Bonne, CBS-113 A, Drugs of the Future 12:525 (1987).

2. B.N. Ames, R. Cathcart, E. Schwiers, and P.Hochstein, Uric acid provides an antioxidant defense in humans against oxidants and radical-caused aging and cancer: a hypothesis, Proc. Natn. Acad. Sci. USA 78:6858 (1981).

3. J.D. Crapo, J.M. McCord, and I. Fridovich, Preparation and assay of superoxide dismutases, Meth. Enzym. 53:382 (1978).

4. J.M.C. Gutteridge, Reactivity of hydroxyl and hydroxyl-like radicals discriminated by release of thiobarbituric acid reactive material from deoxysugars, nucleosides and benzoate, Biochem. J. 224:761 (1984).

5. D.C. Liebler, D.S. Kling, and D.S. Rees, Antioxidant protection of phospholipid bilayers by α-tocopherol, J. Biol. Chem. 261:12114 (1986).

MECHANISM WHEREBY DIFFERENT ANTI-OXIDANTS AND LIPOXIGENASE PATHWAY INHIBITORS
SUPPRESS IL2 SYNTHESIS

Jacques DORNAND* and Mariette GERBER**

*CNRS, Laboratoire de Biochimie des Membranes, Ecole de Chimie, Rue de l'Ecole Normale MONTPELLIER, FRANCE
**INSERM, Centre Paul Lamarque MONTPELLIER, FRANCE

INTRODUCTION

It has been reported by several authors (1) and ourselves (2,3,4) that oxidative processes have been implicated in T cell responses. It was found that several types of anti-oxidants suppressed T cell response, but that LO inhibitors display specific characteristics when compared to the other anti-oxidants.

MATERIALS AND METHODS

The techniques used have been described in our previous papers (2,3,4) except for measurement of cytoplasmic pH which was measured fluorimetrically, exactly as described by Grinstein et al (5).

RESULTS

Anti-oxidant effect on IL2 production

At non-toxic concentrations, all the assayed anti-oxidants suppressed mouse T lymphocyte proliferation (data not shown) and this effect could be correlated with an inhibition IL2 secretion whatever the type of stimulation used (lectin or A23187). The concentration of each compound leading to the half inhibition of the synthesis is given in Table 1.

Anti-oxidant effect on [Ca++]i increase

We measured the effect of anti-oxidants on [Ca++]i rise triggered by the binding of the antigen (lectin) with its receptor. This interaction induces a message required for IL2 synthesis which can be bypassED with A23187. NDGA or BHA blocked the lectin induced [Ca++]i rise but not the [Ca++]i rise evoked by A23187. Other anti-oxidants did not display any inhibitory action on [Ca++]i rise resulting from either modes of stimulation.(Table 2).

Antioxidants in Therapy and Preventive Medicine
Edited by I. Emerit *et al.*
Plenum Press, New York, 1990

261

Table 1 Antioxidant concentrations promoting 50% inhibition of
IL2 synthesis in Con A or (PMA+A23187) murine
splenocytes

Antioxidants	Stimulus	
	Con A 2.5μg/ml	PMA+A23187 10ng/m+60ng/ml
Tiron (mM)	0.55 ± 0.05	0.75 ± 0.1
DABCO (mM)	0.87 ± 0.04	1.3 ± 0.2
DMSO (mM)	102 ± 15	130 ± 20
Thioure (mM)	30 ± 5	42 ± 3
Desferal (μM)	2.7 ± 0.5	2.9 ± 0.2
NDGA (μM)	8 ± 1	15 ± 2
BHA (μM)	15 ± 2	22 ± 4

Murine splenocytes were stimulated for 24 hours, in the
presence or in the absence of different concentrations of
antioxidants. Results are the mean of three different
determinations with standard deviation.

Table 2 Effect of different antioxidants on Con A or A23187 triggered [Ca++]i rise (nM) in murine
thymocytes

Preincubation	Preactivation	Post Con A	Post A 23187
Medium	155	299	385
Tiron (1.5 mM)	160	275	385
DABCO (3.2 mM)	165	262	398
Desferal (10μM)	160	278	400
DMSO (150 mM)	162	255	360
NDGA (20μM)	168	178	372
BHA (40μM)	158	150	368

Fluorescence of Quin2-AM loaded thymocytes was monitored after a 10mn preincubation with or
without the different antioxidants at the indicated concentration giving the baseline (preactivation).
Then 10ng/ml Con A or 60ng/ml A23187 were added to the cell suspension .The fluorescence was
determined 10mn later when a plateau was reached. [Ca++]i was calculated as previously described
(3). Results represent one typical experiment among many others.

Since NDGA and BHA inhibited T cell responses after PMA+ionophore stimulation although
they have no effect on ionophore induced [Ca++]i increase, they might also affect a PKC dependent
event.

Effect of anti-oxidants on PMA-induced pH change

When cells were preincubated for 10mn with 15μM NDGA before PMA addition, the
intracellular pH was not affected by the LO inhibitor and upon subsequent PMA addition, the expected
pH change was not observed (Fig. 1), indicating that the Na+H+ antiport cannot be activated.
Conversely to NDGA, the other anti-oxidants do not affect the PMA-induced intra-cellular pH change at
concentrations preventing IL2 synthesis Some of them (DMSO, DABCO) exert a partial suppressive
effect on the PMA-induced intracellular pH change, but at higher concentrations. These results show
that from all the anti-oxidants assayed, only NDGA is capable of specifically blocking the PMA-
induced activation of Na+H+ antiport.

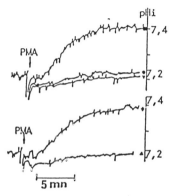

Figure 1. Effect of 15μM NDGA or 150mM on murine thymocyte intracellular pH change triggered by PMA. BCECF loaded thymocytes were suspended in buffer pH 7.2 at 37°C (■). In some experiments 100μM dimethylamiloride (▲) or 15μM NDGA (●) or 150mM DMSO (◆)or 100μM dimethylamiloride+ 15μM NDGA (▼) were present. When indicated 100ng PMA was added to the cell suspension 10 mn later. The fluorescence of the cell suspension was monitored at 528nM (emission wavelength 492 nM) and the thymocyte intracellular pH determined

<u>Effect of anti-oxidants on cell aggregation</u>

Interaction between accessory cell and T cell via adhesion molecules, occurs after the PKC-dependent phosphorylation of a T cell adhesion molecule, (6). It is shown on Fig. 2 that NDGA exerted a dose-dependent effect on cellular aggregation induced by PMA + A 23187, observed four hours after the addition of the activating agents. Conversely the other antioxidants have no effect on aggregation formation. This is exemplified with DMSO on Fig. 2-2.

DISCUSSION

All the assayed anti-oxidants, suppress murine thymocyte and splenocyte proliferation in a dose dependent manner (not shown). This effect results from a blockade of IL2 synthesis.

Conversely to LO inhibitors other anti-oxidants appear to have no effect on [Ca++]i rise and/or PKC dependent events. Thus, it is likely that they affect IL2 synthesis by acting either on a different PKC dependent event or after trans-membrane signals, maybe at the level of mRNA or protein synthesis.

On the contrary, it was found that LO inhibitors affected significantly Con A-induced [Ca++]i rise (and not ionophore-induced [Ca++]i rise), indicating that i) ligand-receptor binding was required to demonstrate the effect of anti-oxidants on [Ca++]i rise; ii) the inhibition of IL2 synthesis in PMA + ionophore conditions of activation might be consequent to an effect on another metabolic step. We show here that, indeed, LO inhibitors could also inhibit two PKC dependent events, Na^+H^+ antiport activation and cell adhesion.

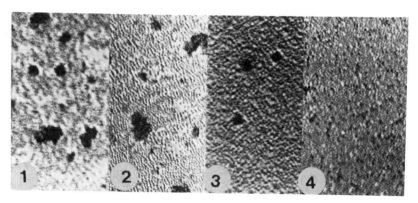

Figure 2. Inhibition of cluster formation of PMA + A 23187 activated lymphocytes by NDGA. Mouse thymocytes (25×10^6/ml) were stimulated by 10ng/ml PMA + 60ng/ml A23187 (1) in presence of 150mM DMSO (2) or 5µM NDGA (3) or 15µM NDGA (4). Pictures were taken 4 hours after the onset of activation. No aggregation was observed in control non stimulated cultures.

We propose that ligand receptor binding induces intracellular production of reactive oxygen species (ROS). In addition, O_2^- present in the oxidative burst of accessory cells could extracellularly increase this oxidative potential directed at the T cell membrane phospho-lipids. It is possible that these ROS triggered arachidonic acid (AA) cascade from which so far unknown metabolite(s) may affect directly the two second messengers. Extra-cellular metabolites, released by the AA cascade in accessory cells, might reinforce this role. Further work is needed to pinpoint the AA metabolite(s) eventually involved in T cell activation.

Acknowledgements: This work has been financially supported by "Association de Recherche sur les Cancer" and contract INSERM 89-2008.

REFERENCES

1 G. Chaudri, I.A.Clark, N.H.Hunt, W.B.Cowden, R.Ceredig Effect of antioxidants on primary alloantigen-induced T cell activation and proliferation. J.Immunol. 137:2646 (1986).

2.M.Gerber, D.Ball, F.Michel, A.Crastes de Paulet. Mechanism of enhancing effect of irradiation on production of IL 2. Immunol.Lett. 9: 279.(1985)

3.J.Dornand, C.Sekkat, J.C.Mani , M.Gerber Lipoxigenase inhibitors suppress IL2 synthesis: Relationship with [Ca++] rise and the events Immunol.Lett. 16: 101(1987)

4.C.Sekkat, J.Dornand, M.Gerber Oxidative phenomena are implicated inhuman T cell stimulation .Immunology, 63: 431.(1988)

5.S.Grinstein, S.Cohen, J.D.Goetz, A.Rothstein and E.W.Gelfand Characterisation of the activation of Na+/H+ exchange in lymphocytes by phorbolesters. Change in cytoplasmic pH dependence of the antiport. Proc. Natl. Acad.Sci. U.S.A. 82: 1429.(1985)

6.M.Patarroyo, P.G.Beatty, J.W. Fabre and GahmbergC.G. Identification of a cell surface protein complex mediating phorbol ester-induced adhesion (binding) among human mononuclear leukocytes. Scand. J. Immunol. 22, 171.(1985)

REACTIVE OXYGEN SPECIES RELEASED BY HUMAN BLOOD NEUTROPHILS

PIROXICAM-TREATED HEALTHY SUBJECTS

Marcelle Damon, Claude Chavis, Christian Le Doucen,
Francis Blotman *, and A. Crastes de Paulet

INSERM U 58, 60 rue de Navacelles and
* Hôpital Lapeyronie
34090 Montpellier, France

INTRODUCTION

The polymorphonuclear neutrophils (PMNs) are known to be the major infiltrating phagocytes in the synovia of inflammatory joints in patients with rheumatoid arthritis. A continued extravasation of peripheral blood PMNs occurs to maintain the high local concentration of these cells (1). Therefore, it seems to be important to study the ability of these blood leucocytes to respond to various stimuli for a better understanding of the effects of anti-inflammatory drugs.

When polymorphonuclear leukocytes interact with soluble or particulate matter, these cells develop a burst in oxydative metabolism generating reactive oxygen species (2). This production can be measured as luminol-amplified light emission or chemiluminescence (3).

In the present work, we investigated the intensity of the oxygen species release of stimulated peripheral blood neutrophils from healthy subjects. Knowing that non steroidal anti-inflammatory drugs (NSAIDs) were reported to affect the responses of neutrophils from subjects given these drugs in therapeutic doses (4), we studied the reactivity of neutrophils from healthy subjects treated with piroxicam. The oxygen species releasability was compared to the generation of two arachidonic metabolites which are known to participate to inflammatory reaction : LTB_4 and 5 HETE.

MATERIALS AND METHODS

Subjects

Blood was collected from 12 healthy male volunteers, aged 40-50 years, who had received no aspirin or NSAIDs treatment. Piroxicam (20 mg/day) was given during 5 days. The measurements were done before (D0) and after the last dose of piroxicam (D5).

The same measurements (D0 and D5) were realized on cells from 5 non-treated subjects.

Antioxidants in Therapy and Preventive Medicine
Edited by I. Emerit *et al.*
Plenum Press, New York, 1990

265

Preparation of neutrophil suspensions

Heparinized (10 units/ml) blood was obtained from subjects by venipuncture. Leukocytes were isolated by sedimentation on dextran : 15 ml of human blood were incubated at 37°C for 40 min with 1ml of dextran T 500 in NaCl 0.9 %. After centrifugation (5 min. at 400 g) purified suspensions of neutrophils were isolated by means of discontinuous isotonic Percoll gradients (5) and hypotonic lysis of remaining erytrocytes. Cells were then suspended in phosphate buffered saline (PBS 50 mM, pH 7.2 with $CaCl_2$ 2 x 10^{-3}M and $MgCl_2$ 5 x 10^{-4}M). Cell viability as shown by exclusion of the trypan blue was always > 96 % and the purity of neutrophils was > 95 %.

Study of oxygen species release

Oxygen species release was evaluated by luminol-enhanced chemi-luminescence (CL) (6). CL was measured with LKB Wallac luminometer model 1251, and recorded at 37°C every 120 sec for periods of up to 40 min ; integration time was 10 sec. After preincubation of cell suspension in the presence of luminol (10^{-3}M in 1 % BSA solution) for 10 min in luminometer, the background values were measured and stimuli were added to the reaction mixture. The cells were stimulated by opsonized zymosan (50 µg/ml), a formyl-peptide chemoattractant (FMLP, 10^{-6}M) and phorbol myristate acetate (PMA, 10^{-7}M). CL peak (maximal value - background value) obtained for each stimuli are expressed as mV/3 x 10^5 cells. Each value is the mean ± SD of triplicate assays.

HPLC analysis of arachidonic acid metabolites

Neutrophils were incubated in PBS for 5 min at 37°C with calcium ionophore A 23187 (2.5 µM). The reaction was stopped by cooling in ice, cell pellet was recovered by centrifugation and resuspended into 1 ml of O.9 % saline. An equal volume of MeOH was added to the incubation medium and to the cell suspension.

Arachidonic acid metabolites were studied by reverse phase HPLC (Waters apparatus) on C18 radial pak cartridge (7). Leukotrienes and HETEs were detected at 280 and 232 nm respectively and eluted with 2 different isocratic solvent systems at a flow rate of 1.2 ml/min : $MeOH/H_2O/AcOH$ 65/35/0.1 (v/v/v) for LTB_4, and $MeOH/H_2O/AcOH$ 75/25/0.01 (v/v/v) for 5 HETE. Reaction products were identified by comparison of their retention times with those of authentic standards. Leukotriene quantification was proceeded with PGB_2 as internal standard and 5 HETE concnentrations were calculated by comparing the area under the peak with that of known amounts of HETE. The results are expressed as pmol/10^7 cells. Each value is the mean ± SD of triplicate assays.

Statistical analysis

Statistical analysis was performed with a non-parametric test : Mann and Whitney U-Test).

RESULTS

CL analysis of oxygen species release

The CL peak values obtained for each stimulus are reported in figure 1a. No significant difference was observed between the subjects (D0). On the opposite, CL results distinguish 2 kinds of subjects according to their neutrophil responses to stimuli after piroxicam treatment (D5). In population 1 (n = 5), CL values were increased ; the values decreased after treatment in population 2 (n = 7). When the piroxicam effect was expressed as the percentages of D0 values (Fig.1b), a significant difference was observed (p < 0.01). No significant variation was observed in control subjects.

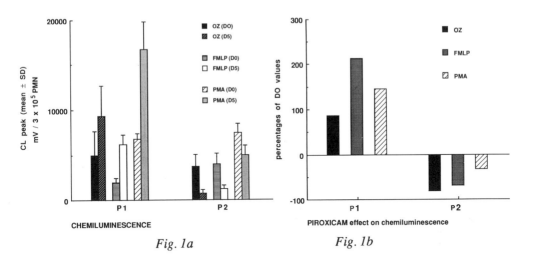

Fig. 1a

Fig. 1b

LTB$_4$ and 5 HETE concentrations

The metabolite levels are reported in figure 2a. A significant difference (p<0.03) was observed in the case of LTB$_4$ values obtained before treatment between population 1 and population 2. No difference was observed in the case of 5 HETE. On the opposite, the arachidonic acid metabolite generation was increased after piroxicam treatment in population 1, whereas, in population 2, piroxicam treatment promoted a decrease in LTB$_4$ and 5 HETE levels.A significant difference was observed between the 2 populations when the effect of piroxicam was expressed as percentages of D0 values (Fig.2b). These results correlate with those obtained by CL analysis. No significant variation was observed in control subjects.

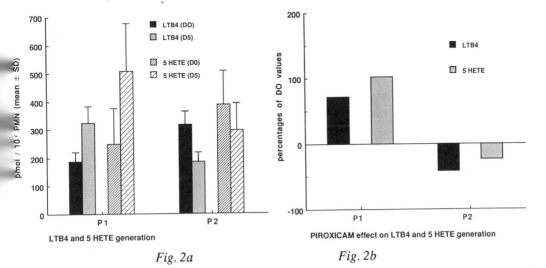

Fig. 2a

Fig. 2b

DISCUSSION

The present work focuses on the producion of oxygen species by neutrophils from healthy subjects and on the effect of a NSAID on oxygen activation. The release of oxygen was induced by phagocytable stimulus such as opsonized zymosan and surface active agents, such as FMLP and PMA. The activation state of the cells was evaluated by an analysis of their ability to generate 2 arachidonic acid metabolites from 5 lipoxygenase pathway after stimulation by the calcium ionophore A 23187.

Neutrophils from 12 healthy subjects showed a production of oxygen species which depended on the nature of the stimuli. No significant variation was observed between the subjects. Nevertheless, after piroxicam intake, the neutrophils from 5 subjects developed an increasing generation of oxygen species. On the opposite, in the case of 7 subjects a decrease was observed.

On the other hand, the analysis of LTB_4 and 5 HETE generation showed that there is no significant variation between the subjects before piroxicam treatment. On the opposite, the production of these 2 metabolites was increased in the case of 5 subjects, and decreased in the case of 7 subjects after piroxicam intakes. There is a correlation between the effect of piroxicam on the oxygen species release and on the generation of 5 lipoxygenase products.

In control subjects who were not treated by piroxicam, no significant difference was observed between the results obtained at 5 day intervals, in the case of oxygen species release as well as in the case of arachidonic acid metabolite production.

These results allow to distinguish 2 kinds of subjects through their neutrophil responses obtained after piroxicam treatment : in population 1 (n = 5) piroxicam promoted an increase of their ability to release oxygen species and to generate 5 lipoxygenase metabolites ; in population 2 (n =7), the treatment promoted a decrease of these neutrophil capacities.

This study indicates that the effect of piroxicam does not depend on the activation state of the neutrophils as it was determined before drug intake. On the other hand, piroxicam which is known to inhibit the release of oxygen species (8,9,10), could enhance this release. In the same way, this NSAID which is an inhibitor of cyclooxygenase pathway (11,12), promotes an enhancement of 5 lipoxygenase product generation. These findings are in agreement with those of Docherty et al. (13) who showed that the enhancement of lipoxygenase activity by cyclooxygenase inhibitors has been attributed to the shunting of arachidonic acid from cyclooxygenase to the lipoxygenase pathway or to the removal of inhibition on the lipoxygenase pathway by prostaglandins. Nevertheless, in some cases, piroxicam is able to decrease the generation of 5 lipoxygenase derivatives.

Piroxicam participates to the modulation of neutrophil function. This preliminary study showed that the modulation depends on the subject reactivity.

REFERENCES

1. K. A.Brown , J. D.Perry , C. Black , D. C. Dumonde Identification by cell electrophoresis of a subpopulation of polymorphonuclear cells which is increased in patients with rheumatoid arthritis and certain other rheumatological disorders. Ann. Rheum. Dis. 47:353-358 (1988).
2. C. Dahlgren. Polymorphonuclear-leukocyte chemiluminescence induced by formyl methionyl-leucyl-phenylalanine and phorbol myristate acetate : effects of catalase and superoxide dismutase. Agents and Actions, 21:104-112 (1987).
3. R.C. Allen. Phagocytic leukocyte oxygenation activities and chemiluminescence : a kinetic approach to analysis. In: "Methods in Enzymology", 133:449-507 (1986).

4. H. B.Kaplan , H. S.Edelson , H. M.Korchak , W. P.Given , S. Abramson , G. Weissmann Effects of non-steroidal anti-inflammatory agents on human neutrophil functions in vitro and in vivo. Biochem. Pharmacol. 33:371-378 (1984).

5. J. Verhagen, P. L. B. Bruynzel, J. A. Koedam, G. A. Wassink, M. de Boer, G. K. Terpstra, J. Kreukniet, G. A. Veldink, F. G. Vliegenthart. Specific leukotriene formation by purified human eosinophils and neutrophils. FEBS, 168:23-28 (1984).

6. M. P. Wymann, V. Von Tscharner, D. A. Derauleau, M. Baggiolim. Chemiluminescence detection of H_2O_2 produced by human neutrophils during the respiratory burst. Anal. Biochem. 165:371-378 (1987).

7. P.Borgeat , B.Samuelsson Arachidonic acid metabolism in polymorphonuclear leukocytes : unstable intermediate in formation of dihydroxy acids. Proc. Natl. Acad. Sci. USA , 76:3213-3217 (1979).

8. H. B. Kaplan, B. M. Babior, H. M. Korchak, D. P. Given, S. Abrhamson, G. Weissmann. Effects of non-steroidal anti-inflammatory agents on human neutrophil functions in vitro and in vivo. Biochem. Pharmacol. 33:371-378 (1984).

9. D. E. Van Epps, S. Greiwe, J. Potter, J.Goodwins. Alterations in neutrophil superoxide production following piroxide production following piroxicam therapy in patients with rheumatoid arthritis. Inflammation 11: 59-72 (1987).

10. S. Coli, D. Carruso, E. Tremoli, E. Stragliotto, G. Morazzoni, G. Galli. Effect of single oral administration of non steroidal anti-inflammatory drugs to healthy volunteers on arachidonic acid metabolism in peripheral polymorphonuclear and mononuclear leukocytes. Prostaglandins, Leukotriens and Essential Fatty Acids. 34:167-174 (1988).

11. S. Abramson, H. Edelson, H. Kaplan, R. Ludwig, G. Weissmann. Inhibition of neutrophil activation by non-steroidal anti-inflammatory drugs. Am. J. of Med., 15:3-6 (1984).

12. J. R. Walker, J. Harvey. Action of anti-inflammatory drugs on leukotriene and prostaglandin metabolism : relationships to asthma and other hypersensitivity reactions. In: "Advances in Inflammation Research", K. D. Rainsfort, G. P. Velo, eds., Raven Press, New-York, pp 227-238 (1984).

13. J. C. Docherty, T. W. Wilson. Indomethacin increases the formations of lipoxygenase products in calcium ionophore stimulated human neutrophils. Biochem. Biophys. Res. Comm. 148:534-538 (1987).

ANTIOXIDATIVE PROPERTIES OF MET- AND LEU-ENKEPHALIN

Tanja Marotti, Višnja Šverko, Ivo Hršak

Department of Experimental Biology and Medicine
Ruder Bošković Institute, Bijenička 54
41000 Zagreb, Yugoslavia

INTRODUCTION

Leukocytes have got receptors for opioid peptides and as such may be targets for immunomodulating activity by central neuroendocrine mechanisms (1). Opioid peptides alter adherence and chemotaxis of leukocytes (2). Phagocytosis by rat peritoneal macrophages was suppressed by Met-enkephalin (3). So far, Met- and Leu-enkephalin have been the most intensively studied opioid peptides. During phagocytosis, leukocytes experience a respiratory burst characterized by production of high levels of oxygen metabolites, such as superoxide anion (O_2^-) which results in highly toxic species that cause cell damage. We wanted, therefore, to examine whether Met- and Leu-enkephalin affected superoxide anion production of human polymorphonuclear neutrophils in vitro.

RESULTS AND DISCUSSION

Superoxide anion release (O_2^-) from polymorphonuclear (PMN) leukocytes isolated from peripheral blood of healthy donors was measured as superoxide dismutase (SOD) inhibitable reduction of ferrycytochrome C by the method of Johnson et al. (4). Purified PMNs were suspended at 1.8×10^6 cells/tube in phenol-free Hanks balanced salt solution (HBSS) to which Met- or Leu--enkephalin was added to achieve final opioid concentration of 10^{-2}, 10^{-4}, 10^{-6}, 10^{-8}, 10^{-10} and 10^{-12} mg/ml. Release of O_2^- in tubes containing Met- or Leu-enkephalin was compared with that from untreated cells of the same donor (Fig. 1).

Both opioid peptides suppressed superoxide anion release from PMNs of individuals with high baseline levels (80-110 nmol O_2^-/mg protein). Two peaks of maximal suppression (5-20 nmol O_2^-/mg protein) were observed: with Met-enkephalin at 10^{-2} and 10^{-8} mg/ml, and with Leu-enkephalin at 10^{-2} and 10^{-10} mg/ml. However, if O_2^- release was measured from PMNs of donors with medium baseline release (25-37 nmol O_2^-/mg protein), Met-enkephalin induced either suppression or stimulation, while Leu-enkephalin induced mainly suppression. Met-enkephalin induced suppression of O_2^- release at high (10^{-2} or 10^{-4} mg/ml) and at low concentration (10^{-8} or 10^{-10} mg/ml). Leu-

Antioxidants in Therapy and Preventive Medicine
Edited by I. Emerit *et al.*
Plenum Press, New York, 1990

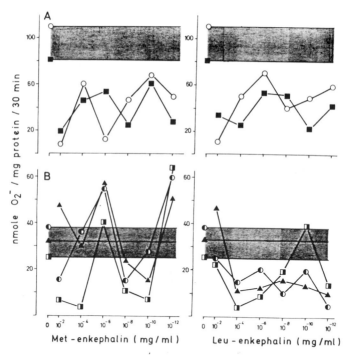

Fig. 1. Effect of Met- or Leu-enkephalin on O_2^- release by human
polymorphonuclear leukocytes

The data represent the mean value of three experiments and are
expressed as nmol of O_2^- per miligram of protein. Different
symbols represent PMNs from various donors (each symbol represents one donor). Shaded areas represent the range of control
(baseline) values of O_2^- release.
A - donors with high baseline O_2^- release
B - donors with medium baseline O_2^- release

-enkephalin induced suppression across a wide range of concentrations (from
10^{-4} to 10^{-12} mg/ml). In the presence of either Met- or Leu-enkephalin,
maximal release of O_2^- was only 3-15 nmol/mg protein.

Time dependance of O_2^- release was examined in the presence of 10^{-12}
mg/ml of Met- or Leu-enkephalin. This concentration corresponds to
physiological concentration of Met- and Leu-enkephalin in human plasma (5).
PMNs of healthy donors were incubated with the peptides for 10, 20, 30, 40,
50, or 60 minutes. Modulating effect appeared soon after addition of the
peptid but persisted at most up to 10 minutes (Fig. 2).

Thus, in vitro exposure to Met- or Leu-enkephalin alters superoxide
anion release by human polymorphonuclear leukocytes. Both peptides induced
the same range of suppression in PMNs with high baseline level of superoxide
anion release. This indicates that opioid peptides might serve as antioxidant agents in inflamatory processes, which as known, involve an increase
of "respiratory burst" products. Leu-enkephalin seems to be more potent
antioxidant than Met-enkephalin, since it suppressed O_2^- release from PMNs
with medium as well as high baseline level, whereas Met-enkephalin modulated only the activity of high baseline PMNs. However, the effect is very
short. Since enkephalins are aubject to fairly rapid enzymatic degradation
(6), efforts to stabilise the molecule are in progress in several laboratories, including ours.

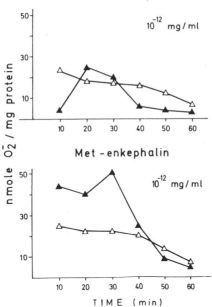

Fig. 2. Time dependence of O_2^- release by PMNs in the presence of Met- or Leu-enkephalin

The data represent the mean value of three experiments. Open triangles show the spontaneous O_2^- release, full triangles O_2^- release in the presence of Met- or Leu-enkephalin.

Our preliminary results indicate that opioid peptides deserve further attention as potential anti-inflammatory agents.

REFERENCES

1. J. M. Merishi and I. H. Mills, Opiate receptors on lymphocytes and platelets in man, Clin. Immunol. Immunopathol., 27:240 (1983).
2. S. L. Brown, S. Tokuda, L. C., Saland and D. E. Van Epps, Opioid peptide effects on leukocyte migration, in: "Enkephalins and Endorphins: Stress and the immune system", N. P. Plotnikoff, R. E. Faith, A. J. Murgo and R. A. Good, ed., Plenum press, New York and London (1986).
3. G. Foris, G. A. Madgyesi and M. Hauck, Bidirectional effect of Met-enkephalin on macrophage effector function, Mol. Cell Biochem., 69:127 (1986).
4. R. B. Johnston, C. A. Godzik and Z. A. Cohn, Increased superoxide anion production by immunologically activated and chemically elicitated macrophages, J. Exp. Med., 148:115 (1978).
5. M. F. Shanks, V. Clement-Jones, C.J. Linsell, P. E. Mullen, L. H. Rees and M. G. Besser, A study of 24-hour profiles of plasma Met-enkephalin in man, Brain Research, 212:403 (1981).
6. G. Roscetti, R. Possenti, E. Bossano and G. Roda, Mechanism of Leu-enkephalin hydrolysis in human plasma, Neurochem. Research, 10:1393 (1985).

PLATELET-ACTIVATING FACTOR AMPLIFIES TUMOUR NECROSIS FACTOR-INDUCED

SUPEROXIDE GENERATION BY HUMAN NEUTROPHILS

Monique Paubert-Braquet, Philipe Koltz, Jean Guilbaud
*David Hosford and *Pierre Braquet

Burn Centre, H.I.A. Percy, F 92141, Clamart, and *IHB,
17 avenue Descartes, 92350 Le Plessis-Robinson, France

INTRODUCTION

Inflammation is usually a tightly controlled process which confines tissue damage, prevents infection and assists in cellular regeneration. However, if the inflammatory response becomes unregulated, this normally beneficial local event may escalate into a wider malignant activity involving endothelial injury, excessive cell infiltration and vascular leakage. Such phenomena may underlie the microcirculatory damage observed in shock, sepsis, asthma, ischemia and graft rejection. Neutrophils (PMN) appear to play a particularly important role in inflammation and inflammatory disorders (1). In the inflammatory microenvironment, neutrophils become activated by various agonists, adhere to the endothelial surface and release lysosomal proteases. Activated PMN also undergo a "respiratory burst", which results in the reduction of molecular oxygen to superoxide (1). These toxic products released by PMN are capable of eliciting severe endothelial damage.

Among the agonistic mediators released in the inflammatory zone which modulate PMN function are platelet-activating factor (PAF) and tumour necrosis factor (TNF). PAF, structurally characterised as 1-0-alkyl-2(R)-acetyl-glycero-3-phosphocholine, is a potent autacoid mediator implicated in a diverse range of human pathologies (2). This alkyl phospholipid is now known to be produced by, and act on, a variety of cell types including eosinophils, monocytes, macrophages, platelets, endothelial cells and neutrophils themselves. Indeed, PAF is a potent chemotactic agent for neutrophils, inducing superoxide release, aggregation and degranulation in this cell type. In vivo the mediator evokes pulmonary sequestration of neutrophils, hypotension, bronchoconstriction, increased vascular permeability and plasma extravasation (2).

TNF is a polypeptide cytokine produced primarily by monocytes and macrophages in response to endotoxin and other immune and inflammatory stimuli (3). Injection of large doses of TNF mimick the effects of endotoxic shock in a variety of species and can lead to death via circulatory collapse, hemorrhage and organ failure (4). Thus, TNF also appears to exert many of its effects through vascular disruption. Similarly to PAF, this cytokine enhances both superoxide production and adherence in PMN and induces retraction of cultured endothelial cells (5).

Antioxidants in Therapy and Preventive Medicine
Edited by I. Emerit *et al.*
Plenum Press, New York, 1990

275

Although it is well established that endothelial cells produce PAF when stimulated with agonists such as thrombin, very recently, Camussi et al. (6) have shown that _in vitro_ TNF can also induce endothelial cells to synthesise PAF, the majority of which remains associated with the cells. Furthermore, Bonavida et al. (7) have reported that PAF can activate peripheral blood derived monocytes to produce and secrete TNF. In addition to PAF and TNF being able to elicit the production of each other from cell types intimately involved in inflammatory reactions, at very low concentrations both substances can also "prime" PMN to respond in an ehanced manner to subsequent agonistic stimuli that would otherwise be ineffectual. Following PAF (8) or TNF (9) priming, amplified PMN responses including aggregation, adhesiveness, superoxide production and elastase release have been observed using FMLP as the inducing agonist. The above data opens the possibility that in inflammatory responses, a positive self-generating feedback process may become operative between TNF and PAF, and that PAF associated to the endothelium may prime or amplify PMN responses elicited by TNF and other agonists.

Thus, we were interested to ascertain whether TNF-induced superoxide production by human PMN could be amplified by PAF and if so, whether structually unrelated specific PAF antagonists could modulate superoxide release elicited by these two mediators.

MATERIALS AND METHODS

Chemicals

Platelet-activating factor was purchased from Novabiochem Laboratories and Ficoll Paque from Pharmacia. N-Formyl-L-methionine-L-leucyl-L-phenylalanine (FMLP), cytochrome C, phosphate buffer solution and sodium chloride were obtained from Sigma Chemical Co. Tumour necrosis factor was supplied by Bio-Trans Co. (California). The PAF antagonists, BN 52021 (ginkgolide B), a 20-carbon cage terpene and BN 52111 (2-heptadecyl-2-methyl-4-[5(pyridinium-pentylcarbonyl-oxy-methyl)] 1,3-dioxolane bromide) were gifts from IHB Research Labs, Le Plessis-Robinson, France.

Preparation of PMN

Blood was centrifuged at 1000 x g for 25 min. The platelet-rich plasma was discarded, and the PMN fraction was obtained following 6 % dextran-500 sedimentation, centrifugation over Ficoll Paque and ammonium chloride (160 mM) + Tris (170 mM) treatment. PMN were then resuspended in a standard phosphate-buffered medium (PBS) containing (mM) : Na^+, 146 ; K^+, 4 ; Cl^-, 142 ; HPO_4^{2-}, 2.5 ; $H_2PO_4^-$, 0.5 ; Mg^{2+}, 1 ; glucose, 10 ; (pH 7.4 ; determined osmolarity $\sim$ 284 mOsM).

PMN were counted electronically on a Coulter Counter. The final preparation contained less than 5 % lymphocytes and less than 15 % thrombocytes. Trypan blue staining revealed 95 to 99 % viable cells. Electronic microscopy was carried out on this cell material.

PAF amplification of TNF-induced superoxide generation

TNF was diluted with 1 % bovine serum albumin (BSA) and the solutions used extemporaly. The reaction was performed in phosphate buffer saline (PBS). Cytochrome C (75 μM) was added to 4.10^6 cells at 37° C. TNF was then added to the medium at a concentration of 10 ng/ml and the cells incubated for 1 hour, after which time superoxide production was determined as described below. To examine the potential amplifying effect of PAF, the mediator (10^{-16} to 10^{-8} M) was added 10 minutes before the end of the reaction. The 1 hour incubation was terminated by centrifugation for 7 minutes at 4° C. The supernatant was removed for measurement of cytochrome C reduction. When PAF antagonists were used, they were added at a concentration of 10^{-6} M simultanously with the TNF at the start of the incubation period.

Results are expressed in nanomoles of superoxide anion :

Conc. of $O_2^{\cdot-}$ = $\dfrac{\text{Absorbance}}{21.1}$ x 1000 nM $O_2^{\cdot-}$ = $O_2^{\cdot-}/4.10^6$ cells/hr

where 21.1 μM^{-1} cm^{-1} = the absorption molecular coefficient (Σ)

Statistical Analysis

Results are expressed as mean $\pm$ standard error of the mean (SEM). Statistical analysis was carried out using the Student's test.

RESULTS

TNF-induced superoxide production

Dose-response effect of various concentrations of TNF

PMN (4.10^6 cells/0.5 ml) were incubated for 1 hour with concentrations of TNF ranging between 0.001 ng/ml to 1000 ng/ml. As shown in Figure 1, a dose-response relationship was observed between TNF concentration and superoxide generation. The maximium response was obtained with 10 ng/ml TNF where the $O_2^{\cdot}$ production was 4.45 $\pm$ 0.67 nmoles $O_2^{\cdot-}/4.10^6$ cells/hour. Thus, this concentration of TNF was selected for further studies on PMN with PAF.

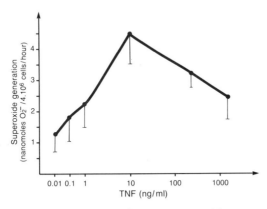

Figure 1   Dose response effect of PMN-superoxide generation induced by TNF (0.01 ng/ml - 1000 ng/ml).

Dose-response effect with various quantities of cells

Superoxide production was also dependent on cell number. With 4.10^6 cells/0.5 ml, the TNF-induced superoxide generation was 4.1 nmole/hour, which was considered adequate to detect any modulatory activity of PAF on superoxide production. Doubling the cell number only increased the superoxide release by 35%.

Time-course of TNF-induced superoxide production

To determine the time-course of TNF-induced superoxide generation we examined $O_2^{\cdot-}$ production by both control PMN (cells not stimulated with TNF) and by PMN incubated with 10 ng/ml TNF for 3 hours at 37 °C. As shown in Figure 2, superoxide generation by TNF-stimulated cells can be described by a parabolic curve ; $O_2^{\cdot-}$ production increased after 15 minutes, reached a maximum value at 90 minutes and then gradually declined. For the studies involving PAF, an incubation period of 1 hour was selected as this marked the onset of a relatively stable and maximal period of superoxide production.

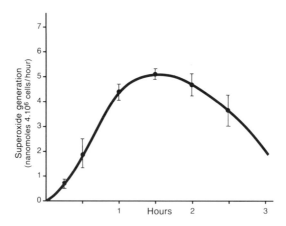

Figure 2   Time-course of TNF-induced superoxide production (10 ng/ml)

PAF-induced superoxide production

PMN were incubated with various concentrations of PAF (10^{-16} M to 10^{-8} M) for periods of between 5 minutes to 3 hours. No significant superoxide production was elicited by PAF at any of the concentrations examined, except 10^{-8} M which elicited the release of a marginally detectable ammount (less than 1 nmole/hr) of superoxide (Fig. 3, bottom curve). This lack of effect of PAF alone on superoxide production occurred regardless of the incubation time or cell number used.

PAF amplification of TNF-induced superoxide production

As shown in Figure 3, when PAF (10^{-16} to 10^{-8} M) was added for 10 minutes to neutrophils which had been previously incubated with 10 ng/ml TNF for 50 minutes, the $O_2^{\cdot-}$ production was significantly amplified (upper curve). Maximum amplification of TNF-induced $O_2^{\cdot-}$ production was elicited by 10^{-12} M PAF, which increased superoxide release by 25 - 30 % (upper curve) relative to that induced by TNF alone after 1 hr (dotted line). In this system, if PAF was added before the TNF at the start of the PMN incubation, no amplification was observed. Furthermore, the amplification effect required an intact PAF molecule since its biologic-ally inactive precursor and metabolite, lyso-PAF, had no effect on TNF-induced $O_2^{\cdot-}$ production. In addition the amplification by PAF was not reversed when the cells were washed twice to remove the excess of TNF before addition of PAF.

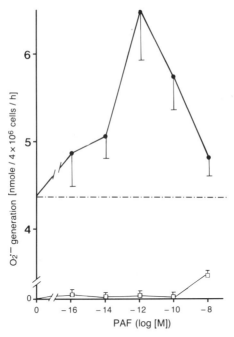

Figure 3   PAF amplification of TNF-induced superoxide generation in human neutrophils. Results show mean $\pm$ SEM (n = 10) ; PAF was introduced for 10 minutes before the end of the 1 hour incubation with TNF.
□ PAF alone ; ● TNF (10 ng/ml) + PAF.

Inhibition of PAF amplification by PAF antagonists

In order to assess the effects of PAF antagonists on the PAF-amplified TNF-induced superoxide production, two structurally unrelated specific antagonists were examined, the ginkgolide, BN 52021 (10), and BN 52111, which is a new constrained PAF-framework related antagonist containing a 1,3 dioxolane ring (11). These two potent antagonists inhibit [3H]-PAF binding to its rabbit platelet membrane receptor with

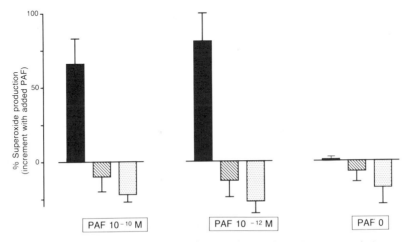

Figure 4   Effect of two chemically unrelated PAF antagonists on PAF-amplified TNF (10 ng/ml)-induced superoxide production in human neutrophils. Results show mean $\pm$ SEM, n = 5 ; both PAF antagonists were used at 10^{-6} M.   ■ Control ;   ▨ BN 52021 ;   ▤ BN 52111

IC$_{50}$ values of 0.25 uM and 35 nM, respectively. As shown in Figure 4, when added at a concentration of 1 uM concomitantly with TNF to the PMN at the start of the incubation, both PAF antagonists completely abolished the amplification of superoxide production induced by subsequent addition of 10^{-10} M PAF (left-hand graph) and 10^{-12} M PAF (centre graph). Indeed, the inhibition of superoxide release was greater than the amplification effect of PAF, indicating that the antagonists partially inhibited the superoxide production induced by TNF alone. This was confirmed by examining the effect of the two antagonists on TNF-induced superoxide production in PMN which received no exogenously added PAF. Figure 4 (right-hand graph) clearly shows that BN 52021 and BN 52111 inhibit 8 - 20 % of the superoxide release elicited by TNF alone, BN 52111 being the most potent antagonist in this respect.

DISCUSSION

Stimulation of PMN with various agonists leads to a respiratory burst with superoxide release, degranulation and cell adhesion. Both TNF and PAF are modulators of the activity of this cell type and play a crucial role in inflammation and host defenses. In our system, TNF induced a time- and dose-dependent production of superoxide from normal human PMN. PAF alone (10^{-16} M to 10^{-8} M) caused no significant release of superoxide from the cells, regardless of cell number or incubation time. However, if PAF was added 10 minutes prior to the end of the 1 hour incubation period with TNF, the superoxide production of the cells was significantly enhanced in comparison with that from cells incubated with TNF alone. Maximum amplification of superoxide production was observed with a concentration of 10^{-12} M PAF.

Several other workers have reported further PAF/TNF interactions in vitro. Valone and Epstein (12) have recently demonstrated that PMN stimulated with TNF are able to produce PAF in a biphasic fashion. The autacoid in the early peak (1-2 hours) is largely retained intra-cellularly, whereas the majority of the PAF in the late peak (6-8 hours) is released into the medium. Studies by Rola-Pleszczynski (13) have also shown that priming of human PMN for 18 hours with various concentrations of PAF (0.1 fM - 1 uM) markedly enhanced their subsequent TNF production in response to interleukin 1 (IL-1). Maximum priming activity was observed at 1 pM - 1 nM PAF, which resulted in a 2 to 3 fold increase in TNF production.

In the present study, PAF had no effect on $O_2^{\cdot-}$ release if it was added to the cells before the TNF at the start of the incubation period, and only a relatively small amplifying effect if it was added before 40 minutes or after 2 hours. The fact that PAF could not influence super-oxide production alone and that its maximum amplifying effect occurred within the period of maximum TNF-induced $O_2^{\cdot-}$ release indicates that PAF must augment some stage of the TNF-dependent $O_2^{\cdot-}$ generation process, which when producing maximum superoxide output can still be amplified to a further degree by a PAF dependent process. Although this mechanism remains undefined, it suggests that $O_2^{\cdot-}$ generation elicited solely by TNF may involve the production and participation of endogenous PAF.

The results obtained with the PAF antagonists support this hypo-thesis. The two structurally unrelated antagonists, BN 52021 and BN 52111, not only completely abolished the amplifying effect of PAF, but also partially inhibited the superoxide production evoked by TNF alone. The degree of this inhibition was related to the potency of the compound in antagonising [^{3}H]-PAF binding. The extremely potent PAF-related dioxolane-derived antagonist, BN 52111 (IC$_{50}$ = 35 nM), reduced TNF-induced $O_2^{\cdot-}$ release by 20 %, while the less potent ginkgolide PAF antagonist, BN 52021 (IC_{50} = 0.25 uM), reduced the superoxide production by only 8 %. This strongly suggests that TNF elicits PAF synthesis in the PMN and that this PAF contributes to the TNF-induced $O_2^{\cdot}$ production via a receptor mediated process. Interestingly, a recent study by Sun and Hsueh (14) has also shown that a PAF antagonist can inhibit TNF-induced effects. These latter authors found that <u>in vivo</u>, the PAF antagonist, SRI 63-119, could prevent TNF-induced bowel necrossis in the rat.

The observation that PAF has no amplifying effect if added before the TNF and only a small effect before (or after) the period of maximum TNF-induced $O_2^{\cdot}$ release may due to the short half-life of the mediator in the cell preparation. Neutrophils rapidly metabolise PAF to lyso-PAF by means of an acetyl-hydrolase (15), although the precise half-life of PAF in our system remains to be determined. Similar findings on the lack of an amplifying effect of PAF if administered before TNF have been obtained <u>in vivo</u>. Heuer and Letts (16) have shown that pretreatment of mice with either <u>S.Typhosa</u> endotoxin or TNF significantly enhanced the mortality induced by PAF. These effects occurred at doses at which PAF, endotoxin or TNF given alone did not significantly affect these parameters, however, the enhancing effect was not recorded when PAF was given prior to TNF.

In conlusion, it appears that an interaction between PAF and TNF, may play an important role in the modulation of neutrophil responses and therefore in inflammatory disorders. Although the biochemical mechanisms involved in the priming and amplification processes remain to be defined, the present study shows that PAF plays an integral role in TNF-induced superoxide production and thus suggests that PAF antagonists may offer some therapeutic value in conditions where this response is excessive and detrimental.

REFERENCES

1. H.L. Malech, and J.I. Gallin, 1987, Neutrophils in human diseases. <u>New Eng J Med</u>., 11:687.
2. P. Braquet, L. Touqui, T.S. Shen, and B.B. Vargaftig, 1987, Perspectives in platelet-activating factor research. <u>Pharmacol Reviews</u>., 39:97.
3. A. Cerami, and B. Beutler, 1988, The role of cachectin/TNF in endotoxic shock and cachexia. <u>Immunology Today</u>., 9:28.
4. K.J. Tracey, B. Beutler, S.F. Lowry, J. Merryweather, S. Wolpe, I.W. Milsark, R.J. Hariri, T.J. Fahey, A. Zentella, J.D. Albert, T. Shires, and A. Cerami, 1986, Shock and tissue injury induced by recombinant human cachectin. <u>Science</u>., 234:4704.
5. Y. Ozaki, T. Ohashi, Y. Niwa, and S. Kume, S., 1988, Effect of recombinant DNA-produced tumour necrosis factor on various parameters of neutrophil function. <u>Inflammation</u>., 12:297.
6. G. Camussi, F. Bussolino, G. Salvidio, and C. Baglioni, 1987, Tumour necrosis factor/cachectin stimulates peritoneal macrophages, polymorphonuclear neutrophils, and vascular endothelial cells to synthesize and release platelet-activating factor. <u>J Exp Med</u>., 166:1390.
7. B. Bonavida, J.M. Mencia-Huerta, and P. Braquet, 1989, Effect of platelet-activating factor (PAF) on monocyte activation and production of tumour necrosis factor (TNF). <u>Int Arch Allergy Appl Immunol</u>., 88:157.

8. G.M. Vercellotti, H.Q. Yin, K.S. Gustafson, P.O. Nelson, H.S. Jacob, 1988, Platelet activating factor primes neutrophil responses to agonists : role in promoting neutrophil-mediated endothelial damage. <u>Blood</u>., 71:1100.

9. R.L. Berkow, D. Wang, J.W. Larrich, R.W. Dodson, and T.H. Howard, 1987, Enhancement of human neutrophil superoxide production by preincubation with recombinant tumour necrosis factor. <u>J. Immunol</u>., 139:3783.

10. P. Braquet, 1987, The Ginkgolides : potent platelet-activating factor antagonists isolated from Ginkgo Biloba L. <u>Drugs of the Future</u>, 12:643.

11. C. Broquet, and P. Braquet, 1988, Compositions thérapeutiques à base de nouveaux aminoacylates d'acétal du glycerol. <u>Fr. Patent</u>. N° 2.616.326.

12. F.H. Valone, and L.B. Epstein, 1988, Biphasic platelet-activating factor synthesis by human monocytes stimulated with interleukin 1 beta (IL 1), tumor necrosis factor (TNF) or gamma interferon (IFN). <u>J. Immunol</u>., 11:3945.

13. M. Rola-Pleszczynski, 1988, Priming of human monocytes with PAF augments their production of tumour necrosis factor in response to interleukin 1. <u>J. Lipid Mediators</u>., in press.

14. X. Sun, and W. Hsueh, 1988, Bowel necrosis induced by tumor necrosis factor in rats is mediated by platelet-activating factor. <u>J Clin Invest</u>., 81:1328.

15. M.L. Blank, T.C. Lee, V. Fitzgerald, and F. Snyder, 1981, Anti-hypertensive activity of an alkyl ether analog of phosphatidyl-choline. <u>Biochem. Biophys. Res. Commun</u>., 29: 2472.

16. H. Heuer, and G. Letts, 1989, Priming of effects of PAF in vivo by tumor necrosis factor and endotoxin. <u>J. Lipid Mediators</u>., in press.

COMPARISON OF ANTIOXIDANT AND PROOXIDANT ACTIVITY OF VARIOUS

SYNTHETIC ANTIOXIDANTS

Regine Kahl, Sabine Weinke and Hermann Kappus*

Department of Clinical Pharmacology, University of
Göttingen, Robert-Koch-Str.40, D-3400 Göttingen, FRG
and *WE 15, FB 3, Free University of Berlin,
Augustenburger Platz 1, D-1000 Berlin 65, FRG

INTRODUCTION

In a classical sense, the term "antioxidant" denotes a
low molecular weight compound capable of terminating the
chain reaction of lipid peroxidation by scavenging lipid peroxy radicals. In biology and medicine, the term is now used
in a broader sense, including e.g. enzymes which detoxify
ROS* such as SOD, catalase or glutathione peroxidase, and
agents which inhibit the formation of specific ROS such as
iron chelators. Still, the classical antioxidants of which α-
tocopherol is a prototype have a role in experimental research on oxidative stress and probably in therapeutic regimens directed towards ROS-mediated diseases. Besides its
chain-breaking action in lipid autoxidation, these low molecular weight antioxidants may also react with the inorganic
oxygen radicals derived from partial reduction of molecular
oxygen. The most extensively studied compound in this respect
is α-tocopherol. Thus, it has been shown that chromanoxyl radical formation from α-tocopherol can be achieved by O_2^- and
by $OH\cdot$ [1-3]. Less numerous data exist on the reaction of synthetic antioxidants with O_2^- and $OH\cdot$ in model systems[4,5]

We are presently engaged in a comparative study[6-8] on
the potency in biological in vitro systems of a number of
synthetic food antioxidants known for their protective action
against a wide variety of mutagenic, carcinogenic and toxic
agents (for a review, see[9]). While major differences in potency towards microsomal lipid peroxidation were not observed
between the gallic acid ester antioxidants and the monophenolic compounds tested[7] they differed in their ability to exert

---

* Abbreviations: BHA, butylated hydroxyanisole; BHT, butylated hydroxytoluene; OG, octyl gallate; PG, propyl gallate; ROS, reactive oxygen species; SOD, superoxide dismutase; TBHQ, tert-butyl hydroquinone; TBQ, tert-butyl quinone;
XO, xanthine oxidase.

Antioxidants in Therapy and Preventive Medicine
Edited by I. Emerit *et al.*
Plenum Press, New York, 1990

"paradoxical" prooxidant effects. In the present study, we have examined the gallic acid ester antioxidants PG and OG and the monophenols BHA and BHT for their ability to modulate the concentration of $O_2^{\cdot-}$ and $OH\cdot$ produced in a set of in vitro assays including aqueous media, microsomes and isolated hepatocytes. Such studies may be helpful in selecting antioxidants for clinical use which lack adverse prooxidant activity.

METHODS

Conventionally prepared liver microsomes were obtained from phenobarbital-pretreated male Wistar rats. For the preparation of hepatocytes, untreated male Wistar rats (250-350 g body weight) were anesthetized with approximately 70 mg pentobarbital/kg i.p. and received 1000 I.U. heparin/kg. The livers were inserted into a recirculating perfusion system operated with Ca^{2+}-fortified Hepes buffer pH 7.6 containing 500 mg/l collagenase IV at 37°C for 12-15 min. The cells were then transferred to a washing solution and washed three times. Viability was tested by Trypan blue exclusion and was ≥ 85%. $O_2^{\cdot-}$ was determined by SOD-inhibitable reduction of native (xanthine oxidase system) or succinoylated (microsomes and hepatocytes) cytochrome c (16 µM or 80 µM, respectively). Initial velocities were used for calculation. Incubation conditions for $O_2^{\cdot-}$ determination were 5 µg XO + 0.2 µmol hypoxanthine / ml phosphate buffer 66 mM (pH 7.5); 0.4 mg microsomal protein + 0.166 µmol NADPH / ml phosphate buffer 66 mM (pH 7.5); and 10^6 cells / ml Hepes buffer pH 7.6 containing adequate salt supplementation. $OH\cdot$ was determined by the formation of ethylene from methional[10]. Incubation conditions for $OH\cdot$ determination were 0.5 mg XO + 2 µmol hypoxanthine or 1.5 mg of microsomal protein + 0.5 µmoles NADPH / ml phosphate buffer 66 mM (pH 7.5). TBQ was synthetized from TBHQ by oxidation with MnO_2. The identity of the product was verified by mass spectrometry and NMR spectroscopy.

RESULTS AND DISCUSSION

The scavenger activity of two synthetic antioxidants, a gallic acid ester and a monophenol sufficiently water soluble to be studied in an aqueous medium, was tested in an $O_2^{\cdot-}$-producing aqueous model system, the XO system (Fig.1). The gallic acid ester PG decreased the concentration of $O_2^{\cdot-}$ available for cytochrome c reduction at concentrations between 5 and 500 µM. In addition, the availability of $OH\cdot$ radicals or an oxidant possessing oxidizing properties resembling those of $OH\cdot$ radicals (produced in the XO system by a basal rate of 1.5 nmoles x ml^{-1} x min^{-1}) for the formation of ethylene from methional was also decreased at 500 µM PG. The monophenolic antioxidant BHA was at least as active as PG towards $O_2^{\cdot-}$ and $OH\cdot$ at 500 µM. It is concluded that both PG and BHA are ROS scavengers in the XO system. Similar results were also obtained with other ROS detection methods and in other aqueous systems.

The results obtained in rat liver microsomes were different from those obtained in the XO system. Still, PG was a reliable inhibitor of $O_2^{\cdot-}$-mediated cytochrome c reduction

(Fig.2). Similarly, the more lipophilic gallic acid ester antioxidant OG effectively suppressed the reaction. However, BHA did no longer display net scavenger activity towards O_2^{-} but in contrast induced a massive increase of O_2^{-} production by about one order of magnitude at a concentration of 500 µM. Even at 5 µM BHA, O_2^{-} production by rat liver microsomes was significantly increased. The more lipophilic monophenol BHT was also tested and found to be only slightly active.

BHA has previously been shown to increase hydrogen peroxide formation[6,11] and NADPH consumption[11] in rat liver microsomes. Endogenous H_2O_2 formation in microsomes is assumed to result from O_2^{-} leakage from the monooxygenase reaction[12] and is increased by a number of monooxygenase substrates[13,14]. BHA is metabolized in microsomes and may thus act by accelerating the monooxygenase cycle. On the other hand, it has been proposed[15] that the extra production of H_2O_2 may be due to the presence of TBHQ. BHA is oxidatively demethylated to TBHQ in the endoplasmic reticulum of various species including rat and man[15-17]. TBHQ may be autoxidized to yield TBQ. In the presence of enzymes capable to mediate one electron reduction of TBQ, redox cycling accompanied by a burst of O_2^{-} production may therefore be induced by TBHQ and TBQ. Fig.3 suggests that this is indeed the case in microsomes. At 500 µM, TBHQ is considerably more active than BHA, leading to a 30fold stimulation of O_2^{-}-dependent cytochrome c reduction. TBQ is, in turn, much more active than TBHQ as shown by the data obtained at a concentration of 5 µM. Higher TBQ concentrations could not be tested because they inactivate SOD (S.Weinke, unpublished experiments). TBHQ and TBQ can also utilize an NADH-dependent activity for redox cycling Fig.3). BHA is inactive with NADH indicating that metabolic

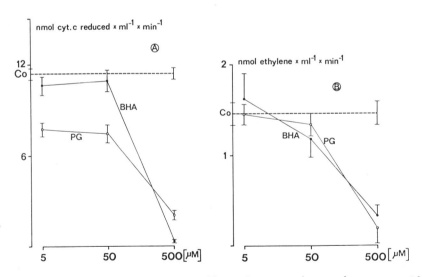

Fig.1 Suppression of ROS-mediated reactions in a xanthine oxidase system by propyl gallate (PG) and butylated hydroxyanisole (BHA). Co: control. Values are means ± S.E.(n=3). A. O_2^{-}-mediated cytochrome c reduction, n=3-12, p < 0.05 or less vs. control at 5, 50 and 500 µM PG and at 500 µM BHA. B. OH^{-}-mediated formation of ethylene from methional. n=3, p < 0.05 or less at 500 µM PG and BHA.

Tab.1   Stimulation of $O_2^{\bar{\cdot}}$ production in a xanthine oxidase system by tert-butyl hydroquinone
(TBHQ) and tert-butyl quinone (TBQ)

nmoles cyt.c reduced x ml^{-1} x min^{-1}

Control	10.2 ± 0.7
BHA 500 µM	0.3 ± 0.1*
TBHQ 500 µM	26.9 ± 2.4*
TBQ 5 µM	21.2 ± 1.4*

Means ± S.E. (n=5-13); * p < 0.05 or less vs. control

activation to TBHQ depending on the presence of NADPH is required for BHA-mediated $O_2^{\bar{\cdot}}$ production. Moreover, BHA-mediated $O_2^{\bar{\cdot}}$ production is suppressed by the monooxygenase inhibitor, metyrapone (data not shown).

One electron reduction of quinones can also be mediated
by XO.   Therefore, TBHQ and TBQ can initiate redox cycling in
the XO system (Tab.1), while BHA is inactive. Rather, BHA
will scavenge $O_2^{\bar{\cdot}}$ as already shown in Fig.1A.

In freshly isolated rat hepatocytes, the ability of BHA
and TBHQ to initiate O_2^{-} production could also be detected.
Tab.2 demonstrates an increase of SOD-inhibitable cytochrome
c reduction in the medium indicating release of O_2^{-} from the
cells. No cytochrome c reduction was observed in the medium
in the absence of cells.

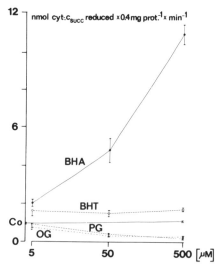

Fig.2   Influence of antioxidants on $O_2^{\bar{\cdot}}$-dependent cytochrome
c reduction in rat liver microsomes. Co: control. Values are
means ± S.E. (n=3-17). p < 0.05 or less vs. control at 5, 50
and 500 µM   BHA (butylated hydroxyanisole), 50 and 500 µM
BHT (butylated hydroxytoluene), 50 and 500 µM PG (propyl
gallate), 50 and 500 µM OG (octyl gallate).

Tab.2 Stimulation of $O_2^{\cdot}$ production in rat hepato-
cyte suspensions by butylated hydroxyanisole (BHA)
and tert-butyl hydroquinone (TBHQ)

	nmoles cyt.c_{succ}. reduced in the medium x 10^6 cells^{-1} x min^{-1}
Control	0.07 ± 0.01
BHA 250 µM	0.17 ± 0.06*
TBHQ 250 µM	0.43 ± 0.10*

Means ± S.E. (n=4-8); * p < 0.05 or less vs. control

Quinone toxicity is assumed to be related to univalent
reduction. The toxic effect may, on the one hand, be due to
adduct formation of the reactive semiquinones to critical
cellular constituents, or may, on the other hand, be the consequence of an oxidative stress imposed on the cell by the
formation of increased amounts of ROS due to redox cycling.
Data on the toxicity of TBHQ and TBQ are rare but a large
body of evidence exists indicating that BHA when administered
at high concentrations to rodents will initiate hyperplasia
and carcinomas in the forestomach (reviewed in[18]). We have
detected extra production of $O_2^{\cdot}$ by the addition of BHA, TBHQ
and TBQ to rat forestomach homogenate (S.Weinke, unpublished
results); however, the formation of TBHQ from BHA could not
be shown. Nevertheless, it is conceivable that oxidative
stress is involved in BHA toxicity in the forestomach.

The $O_2^{\cdot-}$ produced by redox cycling does not possess much

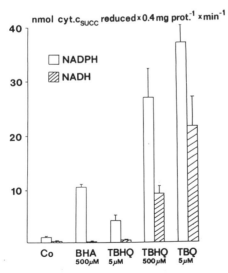

Fig.3 Stimulation of $O_2^{\cdot-}$-dependent cytochrome c reduction by
butylated hydroxyanisole (BHA), tert-butyl hydroquinone
(TBHQ) and tert-butyl quinone (TBQ) in rat liver microsomes
consuming NADPH or NADH. Co: control. Values are means ± S.E.
(n=3-9). NADPH experiments: p < 0.05 or less vs. control at
500 µM BHA, 5 and 500 µM TBHQ and 5 µM TBQ. NADH experiments:
p < 0.05 or less at 500 µM TBHQ and 5 µM TBQ.

cytotoxic potential. Rather, the highly reactive OH· formed from $O_2^{\overline{\cdot}}$ and H_2O_2 in the presence of a source of reduced transition metal is commonly assumed to be the ultimate toxic species. From Fig.1B it is obvious that BHA does, at high concentrations, possess a considerable OH· scavenging potential. Therefore, one might argue that even in the presence of a redox cycling metabolite, BHA will eventually not induce an oxidative stress since it is able to clear the OH· which may have been formed due to the extra production of O_2^-. However, our data suggest that in microsomes the scavenger activity of BHA towards OH· radicals as observed in the aqueous model system is compromised by its prooxidant activity. Fig.4 shows that PG is again a reliable radical scavenger similarly to what was seen in the XO system. However, BHA will no longer suppress OH· mediated reactions in microsomes. In contrast, a tendency to increase ethylene formation was found. The oxidative stress induced by TBHQ is visualized by an increase of OH·-mediated oxidation of methional by a factor of 2.

In the intact cell, oxidative stress induced by quinones will be less prominent than in subcellular fractions since endogenous antioxidant defense systems are operative. Moreover, quinones are subject to divalent reduction by the cytosolic quinone reductase so that less semiquinone radical will be formed. However, when cellular defense is overwhelmed due to preexisting oxidative stress, an antioxidant forming redox active metabolites may contribute to the pathologic state and may thus be unsuitable for use in antioxidant therapy of ROS-mediated disease. BHA is a prototype but BHT may be similar since we have observed an extra production of hydrogen peroxide with this antioxidant in rat liver micro-

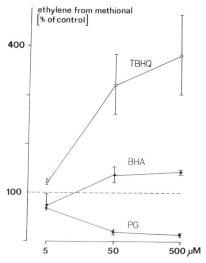

Fig.4 Influence of propyl gallate (PG), butylated hydroxyanisole (BHA) and tert-butyl hydroquinone (TBHQ) on hydroxyl radical-dependent formation of ethylene from methional in rat liver microsomes. Values are means ± S.E. (n=3-6). Absolute values of control were 2.00 ± 0.50 and 0.64 ± 0.07 nmoles x mg prot.-1 x min-1 in the PG and BHA experiments and in the TBHQ experiments, respectively. p < 0.05 or less vs. respective control at 50 and 500 µM and TBHQ. Co:control.

somes and, more extensively, in rat lung microsomes[8]. This is of interest because liver and, more so, lung are target organs of BHT toxicity[19,20]. In the present study, induction of O_2^- formation in liver microsomes by BHT was, however, marginal (Fig.2).

In summary, the gallic acid ester antioxidants PG and OG were shown to exhibit reliable scavenger activity toward superoxide and hydroxyl radicals both in aqueous model systems and in biological material. BHT was virtually inactive. BHA was moderately active in aqueous model systems. However, its scavenger activity was compromised in microsomes by the formation of the BHA metabolite TBHQ which induces redox cycling and massive superoxide production.

BHA ⟶ TBHQ

TBQ·

TBQ

O_2
O_2^-
O_2
O_2^-

This effect is probably counteracted efficiently by the antioxidative defense systems present in the intact cell. However, when these defense systems are overwhelmed in the presence of oxidative stress, compounds capable to exert prooxidant effects in addition to their antioxidative action may be less useful for therapy than antioxidants devoid of such prooxidant activity. Therefore, testing for potential prooxidant action should be performed with every antioxidant considered for therapeutic use.

Acknowledgement: This study was performed with the financial support of the Deutsche Forschungsgemeinschaft, Bonn, FRG. The able technical assistance of C.Engeholm and A.Kastien is gratefully acknowledged.

REFERENCES

1. K.Fukuzawa and J.M.Gebicki, Oxidation of α-tocopherol in micelles and liposomes by the hydroxyl, perhydroxyl, and superoxide free radicals, Arch.Biochem.Biophys. 226: 242 (1983).
2. K.Tajima, M.Sakamoto, K.Okada, K.Mukai, K.Ishizu, H.Sakurai and H.Mori, Reaction of biological phenolic antioxidants with superoxide generated by cytochrome P-450 model system, Biochem.biophys.Res.Commun.115:1002 (1983).
3. T.Ozawa and A.Hanaki, Spectroscopic studies on the reaction of superoxide ion with tocopherol model compound, 6-hydroxy-2-2-5-6-7-pentamethylchromane, Biochem.biophys. Res. Commun.126:873 (1985).
4. M.G.Simic and E.P.L.Hunter, Interaction of free radicals and antioxidants, in: "Radioprotectors and Anticarcinogens", O.F.Nygaard, M.G. Simic and J.N. Hauber, eds., Academic Press, New York (1983), p.449.
5. W.Bors, C.Michel and M.Saran, Inhibition of the bleach-

ing of the carotinoid crocin. A rapid test for quantifying antioxidant activity, Biochim.biophys.Acta 796:312 (1984).

6. D.Rössing, R.Kahl and A.G.Hildebrandt, Effect of synthetic antioxidants on hydrogen peroxide formation, oxyferro cytochrome P-450 concentration and oxygen consumption in liver microsomes, Toxicology 34:67 (1985).

7. S.Weinke, R.Kahl and H.Kappus, Effect of four synthetic antioxidants on the formation of ethylene from methional in rat liver microsomes, Toxicol.Lett. 35:247 (1987).

8. R.Kahl, A.Weimann, S.Weinke and A.G.Hildebrandt, Detection of oxygen activation and determination of the activity of antioxidants towards reactive oxygen species by use of the chemiluminigenic probes luminol and lucigenin, Arch.Toxicol. 60:158 (1987).

9. R.Kahl, Synthetic antioxidants: Biochemical actions and interference with radiation, toxic compounds, chemical mutagens and chemical carcinogens, Toxicology 33:185 (1984).

10. I.Mahmutoglu and H.Kappus, Oxy radical formation during redox cycling of the bleomycin-iron (III) complex by NADPH-cytochrome P-450 reductase, Biochem. Pharmacol.34: 3091 (1985).

11. S.W.Cummings and R.A.Prough, Butylated hydroxyanisole-stimulated NADPH oxidase activity in rat liver microsomal fractions, J.biol.Chem. 258:12315 (1983).

12. V.Ullrich and H.Kuthan, Autoxidation and uncoupling in microsomal monooxygenation, in: "Biochemistry, Biophysics and Regulation of Cytochrome P-450", J.Gustafsson, J. Carlstedt-Duke, A.Mode and T.Rafter, eds., Elsevier, Amsterdam (1980), p.267.

13. R.W.Estabrook and J.Werringloer, Cytochrome P-450 – Its role in oxygen activation for drug metabolism, in: "Drug Metabolism Concepts", D.M.Jerina, ed., ACS Symposium Series, 44:1 (1977).

14. A.G.Hildebrandt, G.Heinemeyer and I.Roots, Stoichiometric cooperation of NADPH and hexobarbital in hepatic microsomes during the catalysis of hydrogen peroxide formation, Arch.Biochem.Biophys.216:455 (1982).

15. S.W.Cummings, G.A.S.Ansari, F.P.Guengerich, L.S.Crouch and R.A.Prough, Metabolism of 3-tert-butyl-4-hydroxyanisole by microsomal fractions and isolated hepatocytes, Cancer Res.45:5617 (1985).

16. A.Rahimtula, In vitro metabolism of 3-t-butyl-4-hydroxyanisole and its irreversible binding to proteins, Chem.-biol.Interact.45:125 (1983).

17. R.El-Rashidy and S.Niazi, A new metabolite of butylated hydroxyanisole in man, Biopharm.Drug Dispos.4:389 (1983).

18. R.Kahl, The dual role of antioxidants in the modification of chemical carcinogenesis, J.environ.Sci.Health C4: 47 (1986).

19. Y.Nakagawa, K.Tayama, T.Nakao and K.Hiraga, On the mechanism of butylated hydroxytoluene-induced hepatic toxicity in rats, Biochem.Pharmacol. 33:2669 (1984).

20. A.A.Marino and J.T.Mitchell, Lung damage in mice following intraperitoneal injection of butylated hydroxytoluene, Proc.Soc.exp.Biol.Med. 40:122 (1972)

ANTIOXIDANT EFFECTS IN RADIOPROTECTION

P. BIENVENU, F. HERODIN, M. FATOME and
J.F. KERGONOU

Centre de Recherches du Service de Santé des
Armées
24, Avenue des Maquis du Grésivaudan - BP87
38700 LA TRONCHE/GRENOBLE (FRANCE)

I. OXIDATIVE EFFECTS IN RADIATION INJURY

Reactive Oxygen Intermediates and Oxygen Metabolism. Reactive Oxygen intermediates such as O_2^-, H_2O_2 and $\cdot OH$ might not be generated only during the initial radiolytic phase of radiation injury, but also in a later metabolic phase, involving oxygen. The consumption of oxygen and simultaneous oxidative damage in vivo have been considered as well documented hypotheses. Therefore oxygen supply and utilization seems to share a pivotal role in radiation effects, as well as in other domains where oxidative injury may happen[1,2,3]. As Koppenol wrote: Thus... "Oxygen not only makes the formation of an oxidizing radical possible, it also exacerbates damage through chain reactions"...

$\cdot OH$ + biomolecules ⟶ R$\cdot$ (rate constants close to the diffusion limit).

R$\cdot$ + O$_2$ $\xrightarrow{\text{rapidly}}$ ROO$\cdot$, itself an oxidizing species[4].

However the possible relationship of oxygen damage with its metabolism is not completely understood. The Latimer diagram presented in Figure 1 shows the various steps of oxygen reduction, leading to water generation. The four electron reduction exerted in the last step of respiration by cytochrome oxydase, a Cu/Fe/Zn/Mg enzyme has been omitted in this scheme. As molecular triplet oxygen is relatively unreactive, it must be activated, either by metal ions or by metalloenzymes.

Metals and Irradiation. Then the question to be addressed is the involvement of metals and/or metalloenzymes in biological systems after irradiation. Ionizing radiations have actually been shown to modify the serum content of some metals: iron, initially decreased and then increased, copper often much increased, and zinc which is often decreased[5,6,7].

Antioxidants in Therapy and Preventive Medicine
Edited by I. Emerit *et al.*
Plenum Press, New York, 1990

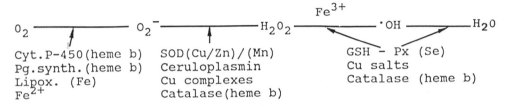

Figure I. Latimer diagram for oxygen reduction steps, and respective metal ions or metalloenzyme catalysts.

Serum copper concentrations have also been recently proposed as an index of lung injury in rats exposed to hemithorax irradiation, in which pulmonary prostacyclin was also increased[8]. Some metals in the free state, particularly copper and iron, are not only able to activate the initial step of oxygen metabolism, generating O_2^-, but they may also catalyze Fenton-type reactions, giving rise to the most detrimental hydroxyl radicals.

Metalloenzymes and Irradiation. Although numerous enzymes are inactivated by irradiation[9], some of them may be involved in oxygen activation. As far as free arachidonic acid and hydroperoxide may be generated during irradiation the heme-dependent[12] prostaglandin synthetase (=Pg.synth.) can be activated, as well as other related enzymes such as the cytochrome-P-450 dependent thromboxane synthetase[13] and prostacyclin synthetase[14], and the iron-dependent 5-lipoxygenase[15] (=Lipox.). Prostaglandin synthetase seems to be able to process some foreign substances[16] such as the lipid peroxides generated during irradiation. This enzyme also further generates O_2^-[17], as is the case for cytochrome P-450 monooxygenases (=Cyt.P450) which may even produce H_2O_2[18]. The synthesis of these monooxygenases is induced by xenobiotic substances[42], in a process requiring lipids[19], and they may catalyze various biotransformations[20,18]. Cytochrome P-450 is directly involved in malondialdehyde, as well as in hydrocarbon generation from lipid hydroperoxides[21].

Lipid Peroxidation. Whole-body gamma irradiation in rats causes lipid peroxidation most often evaluated by malonic dialdehyde (MDA) determination, and particularly increased in the most radiosensitive tissues[10,11]. The radiation-induced formation of lipid peroxides has been evidenced for many years and their toxic effects were proposed as an explanation for the lethality of radiation because linoleic acid autoxidation, without vitamin E, gave rise to a similar amount of toxic peroxides as was produced by X-rays in mice tissues[22]. Although vitamin E deficiency increased lipoperoxidation, it only slightly decreased (6.75vs 7.00Gy) the LD_{50} value of gamma rays[23]. Lipid peroxidation is generally accompanied by a gradual loss of intracellular tocopherol, despite sufficient glutathione supply[24].

To conclude about oxidative effects involved in radiation injury, we tentatively suggest that the initial step is primarily dependent on water radiolysis. This step is presumably associated with organic peroxide generation and some metalloprotein alteration, either liberating or taking up metal ions. Some of these processes, e.g. substrate

peroxidation, may then trigger a secondary processing consuming oxygen, further generating superoxide anion or other
reactive intermediates, and possibly catalyzed by metalloenzymes such as cytochrome P-450 oxygenases or prostaglandin
synthetase.

II. RADIOPROTECTANTS AS ANTIOXIDANTS

Radioprotection: Definition and Scope. As radioprotective drugs are tested in mammalian lethality, most often on
mice survival after 30 days following irradiation, they may
be considered as beneficial and even life-saving towards immediate and delayed harmful effects of irradiation, provided
they are administered before it happens. Radioprotectants are
not only useful in protecting healthy tissues during human
radiotherapy. They should also be beneficial to personnel engaged during peace time, in "cleanup of radioactive accident
area, and possibly in future spaceflights", as was recently
written by Giambarresi and Jacobs[25].

Mode of Action, of Aminothiols. WR-2721 (=S-2-(3-
aminopropylamino) ethyl phosphorothidic acid) is the most efficient radioprotective drug available now. In previous
experiments[26] we had hypothesized a possible involvement of
superoxide-decreasing prostaglandins and administered WR-2721
to mice, simultaneously with the cyclooxygenase inhibitor
aspirin, given "per os". Except for high radiation dose,
(12Gy), aspirin failed to antagonize, and even increased the
protection afforded by the aminothiol. An other experiment
showed that fatty acid deficiency accompanying WR-2721 administration did not protect mice as well as if they had received a standard pellet diet[27]. These data are not unequivoval,
as far as lipids may not only act as precursors of
prostaglandins, but also as activators of enzyme synthesis.
Furthermore, we tried to protect mice against the toxicity of
the radiomimetic mustard chloromethine. Twenty four hours after a 3mg per mice chloromethine dose had been given, the medulla cell count was 8, 450^+300 after 300mg/kg WR-2721 treatment versus $4,987^+598$, whereas the respective values after 72
hours were: $4,460^+455$ vs $1,610^+246$, and the initial control
count had been $16,340^+2,048$ (cells numbers x10^{-3}; femur medulla diluted in 1ml phosphate -buffered saline; (unpublished
results). These experiments confirmed the chemoprotective potency of aminothiols[28] which could be at least partly based,
as well as their radioprotective effect on their metal chelating potency. First of all their common thiol function, potentially interacting with oxidases such as the Cytochrome P-
450 mixed function oxidases, thereby stimulating oxygen
consumption[29]; secondly all protective aminothiols share a
common structural feature: they possess two amine functions
separated by a trimethylene group, and thus may potentially
chelate copper ions[30], giving compounds aften possessing superoxide dismutase (SOD)-like efficiency. Therefore WR-2721
may share similar antioxidant properties "in vivo" as happens
on erythrocyte membranes "in vitro"[31].

Clinical Use of WR-2721 and its Drawbacks. The efficiency of WR-2721 protection during radiotherapy has been
underlined[33], after several phase I clinical trials[34] where

some side effects such as hypotension, nausea, vomiting drowsiness, and sneezing have been described. However, selenium pretreatment enhances its radioprotective effect and reduces its lethal toxicity[35,26]. Besides its side-effects, the main drawback encountered with WR-2721 might be that its efficiency is greater in well oxygenated cells such as bone marrow cells[36,29]. Some authors have succeeded in increasing the biovailability of WR-2721, by using new galenic forms[37] such as liposomes, which have been studied by the EPR technique [74,39].

Arachidonic Acid Derivatives. Dimethylprostaglandin E_2 [40] and leukotriene C_4 [41] have recently proved good radioprotectors. This last substance, as a vasoconstrictor, might decrease oxygen supply, whereas E-and I-prostaglandins may decrease superoxide anion production[38] and show untowards side effects: nausea, vomiting, diarrhea, hypotension. Many other radioprotective drugs and mechanisms have been described in reviews[43,44,45,25,46,39] and hypoxic effects of radioprotectors have often been clained, in which an antioxidant component might not have been uncommon.

III. ANTIOXIDANTS AS RADIOPROTECTORS

Metal Chelators, metal therapy and immunostimulative drugs. As soon as 1953, Z.M. Bacq et al., claiming the metal chelating effects of the radioprotectors cysteamine and cyanide, showed that the copper chelator diethyldithiocarbamate was an efficient radioprotector and implicitly discovered that metal chelation was a radioprotective mechanism[47]. Later, Laser stated that the reduction in the electron transport systems is important, rather than the oxygen tension[48]. The scavenging of copper and iron ions was therefore proposed as a mechanism able to interrupt cellular oxidations initiated by radiations[43]. Among immune stimulators, diethyldithiocarbamate = Imuthiol [R]=DDC the most active radioprotector, is also a well-known copper chelator exerting antioxidant effects which may counteract oxygen toxicity[49,50]. DDC inhibits the oxygenase component of prostaglandin synthetase[51] probably by preventing the interaction of oxygen with the enzyme, and inactivates superoxide dismutase[52] as well as cytochrome P-450[53]. DDC may also exert peroxidase effects[54] which are additive with GSHPx activity, and might arise from iron oxidation into Fe^{3+}, causing an inactivation of heme-containing superoxide generating enzymes. Some stimulators such as endotoxins, polysaccharides[82,83] and IL-1 stimulate prostaglandin synthesis[55,56] and are radioprotective[25]. D-Penicillamine, which is also a copper and zinc chelator presents a limited radioprotective efficiency[58].
Zinc inhibits xanthine-xanthine oxidase-ferric iron-induced lipid peroxidation[59] and may also protect rats against gastric damage, by stimulating mucosal production of prostaglandin E_2 [60], which, in turn, might chelate zinc ions[61]. Zinc therapy is radioprotective, particularly on bone medulla and it may synergize thiols[61,62].

Radical Scavengers. Also involving metals, superoxide dismutase[81] (=SOD), as well as catalase are radioprotective[63] and chemioprotective enzymes; the copper deficiency - induced SOD decrease also decreases prostacyclin synthesis[64]. Copper complexes, possessing SOD-like properties are also radioprotective[65].

Peroxide Scavengers and peroxidative Chain Breakers. Selenium-dependent glutathione peroxidase seems effective against lipid peroxides[66,67], which are toxic and may be increased in tumor cells, as well as tocopherol[68]. Therefore, selenium derivatives are radioprotectors[26].

Vitamin E, being a powerful antioxidant[69,70], also shares radioprotective properties[25,71] and increases hematopoietic stem cell survival in mice after irradiation. Vitamin E may exert powerful effects on arachidonic acid metabolism[72] such as PgI_2 increased synthesis[73] and on the metabolism of xenobiotics[74]. Some interrelationships between prostaglandins, vitamin E, and metals have been described. Tocopherol deficiency may decrease prostacyclin synthesis and increase P_gE_2 production [75]. Conversely vitamin E supplementation decreases P_gE_2 synthesis[76] and increases immune responses in animals. Zinc deficiency also decreases P_gE_2[61] generation and possibly the suppressive effect of zinc on neutrophils might be explained by excess P_gE_2 synthesis[77]. A striking similarity of stabilizing effects has been shown with vitamin E, zinc and selenium on leukocyte membranes after burn injury[78]. As far a free radical interaction exists between vitamin E and vitamin C[79], ascorbic acid should be useful for tocopherol regeneration.Vitamin A - related retinoids decrease superoxide generation[80] and should therefore, be useful in radioprotection.

We tentatively classify radioprotectants in three groups acting either on oxygen availability, or oxygen reduction and peroxide disposable. Among substances acting on oxygen availability, two groups may be distinguished. The first one comprises compounds decreasing oxygen supply, usually not considered as antioxidants, and include vasoactive substances such as serotonin or histamine; peptidoleukotriene-like drugs or compounds stimulating their generation; also included in this group are substances inactivating hemoglobin fonction, such as CO, cyanides, etc. The second group stimulates oxygen consumptium and may share antioxidant properties; they may interact with oxygenases like Cyto. P-450 oxidases, and include thiols and aminothiols. Radioprotectants acting on oxygen reduction also present antioxidant potency and include metal chelators or metal antagonists, e.g. diethyldithiocarbamate and zinc derivatives respectively; radical scavengers e.g. SOD and copper complexes; (E- and I-type) prostaglandins and analogues and compounds stimulating their generation, e.g. endotoxins, yeast polyglucan or polymannan, glycoproteins such as Biostim(R) and Ribomunyl(R). The peroxide scavengers are typical antioxidants possibly sharing some radioprotective potency. They include vitamins E and A, possibly vitamin C, and also peroxidases such as glutathione peroxidase and catalase.

Some drugs belonging to different classes might be associated, so as to synergize their effects. For example WR2721 and P_gE_2 or metals   have been associated. Furthermore natural antioxidants, either enzymes or not are potential enhancers of chemical radioprotection.

REFERENCES

1. B. Halliwell and M. C. Gutteridge, Oxygen toxicity, oxygen radicals, transition metals and disease, Biochem. J. 219:1 (1984).
2. H. Sies, "Oxidative stress," Academic Press, New-York (1985).
3. J. Fehér, C. Csomos and A. Vereckei, "Free radical reactions in medicine," Springer, Berlin (1987).
4. W. H. Koppenol, The paradox of oxygen: thermodynamics versus toxicity, in: "Oxidases and related redox systems," T. King, H. S. Mason, M. Morrison, eds., A. R. Liss, New York (1988).
5. H. Altmann, Primary damage in cells by ionizing radiation, Biophysik 1:329 (1964)
6. A. H. Woods, P. R. O'Bar and S. L. Write, Trace metal behaviour in rabbits after whole body irradiation, Intern. J. Appl. Radiat. Isotopes 21:389 (1970).
7. A. De Bruin, "Biochemical toxicology of environmental agents," Elsevier, Amsterdam (1976).
8. W. F. Ward, A. Molteni, E. J. Fitzsimmons and J. Hinz, Serum coper concentrations as an index of lung injury in rats exposed to hemithorax irradiation, Radiat. Res. 114:613 (1988).
9. C. Von Sonntag, "The chemical basis of radiation biology", Taylor & Francis, London (1987).
10. J. F. Kergonou, P. Bernard, M. Braquet and G. Rocquet, Effect of whole-body gamma irradiation on lipid peroxidation in rat tissues, Biochimie 63:555 (1981).
11. J. M. Zajac and P. Bernard, Effects of whole body irradiation on the microsomal enzyme system and on cytochrome P-450 of rat liver, Enzymes, 27:19 (1982).
12. L. J. Marnett, Y. N. P. Chen, K. R. Maddipat, P. Ple and R. Labeque, Functional differentiation of cyclooxygenase and peroxidase. Activities of prostaglandin synthase by trypsin treatment. Possible location of a prosthetic heme-binding site, J. Biol. Chem. 263:6532 (1988).
13. M. Haurand and V. Ullrich. Isolation and characterization of thromboxane synthase as a cytochrome P-450 enzyme. Hoppe Seyler's Z. Naturforsch 363, (1982).
14. H. Graf, H.H. Ruf and V. Ullrich. Prostacyclin synthase, a cytochrome P-450 enzyme, Angew. Chemie. Int. Ed. 22: 487 (1983).
15. G. A. Veldink and F. G. Vliegenthart, Lipoxygenases : nonheme iron-containing enzymes, Adv. Inorg. Biochem 6:139 (1984).
16. M. B. Roberfroid, H. G. Viehe and J. Remacle, Free radicals in drug research, Adv. Drug. Res. 16:1 (1987).
17. R. F. Del Maestro, H. H. Thaw, J. Bjork, M. Planker and K.-E. Arfors, Free radicals as mediators of tissue injury, Acta Physiol. Scand., suppl. 492:43 (1980).
18. P. R. Ortiz de Montellano, Cytochrome P-450, Plenum Press, New York, (1986).
19. E. Hietanen, M. Laitinen and O. Hänninen, "Cytochrome P-450, Biochemistry, Biophysics and Environmental implications," Elsevier, Amsterdam (1982).

20. E. Bresnick, Metabolic activation, in: "Mechanism of toxicity of chemical carcinogens and mutagens", W.G. Flamm and R.I. Lorentzen, eds. Princeton Scientific Co, Princeton (1985).

21. M. J. Coon and A. D. N. Vaz, Role of cytochrome P-450 in hydrocarbon formation from xenobiotic and lipid hydroperoxides in: "Oxidases and related redox systems," T.E. King, H.S. Mason and M. Morrison, eds., A.R. Liss, New York (1988).

22. V. J. Horgan, J. S. L. Philpot, B. W. Porter and D. B. Roodyn, Toxicity of autoxidized squalene and linoleic acid and of simpler peroxides, in relation to toxicity of radiation, Biochem. J. 67:551 (1957).

23. A. W. T. Konings and E. B. Drijver, Radiation effects on membranes, Radiat. Res. 80:494 (1979).

24. M. S. Sandy, D. Dimonte and M. T. Smith, Relationships between intracellular vitamin E, lipid peroxidation and chemical toxicity in hepatocytes, Toxicol. Appl. Pharmacol. 93:288 (1988).

25. L. Giambarresi and A. J. Jacobs, Radioprotectants, in: "Military Radiobiology," J. J. Conklin and R. I. Walker,eds., Academic Press, New York, (1987).

26. P. Bienvenu, F. Herodin, M. Fatome and J.F. Kergonou, Prostanoids may mediate the radioprotective effects of selenium and WR 2721, in: J. Favier and J. Neve eds, Selenium in Medicine and Biology, de Gruyter, Berlin (1988).

27. P. Bienvenu, M. Fatome, J.F. Kergonou and R. Ducousso, Acides gras essentiels et radioprotection. Comptes-rendus Trav.S.S.A. 8:27 (1987).

28. J. M. Yuhas, Differential protection of normal and malignant tissues against the cytotoxic effects of mechlorethamine, Cancer Treat. Rep., 63:971 (1979).

29. E. L. Travis, The oxygen dependance of protection by aminothiols: implications for normal tissues and solid tumors, Int. J. Radiat. Oncol. Biol. Phys. 10:1495 (1984).

30. H. P. A. Illing, Interaction of thiol-containing compounds with cytochrome P-450, Biochem. Soc. Trans. 6:89 (1978).

31. D. P. Mellor, Chemistry of chelation and chelating agents, in:"The chelation of heavy metals, A. Catsch, A-E. Harmuth-Hoene and D.P. Mellor eds, Pergamon Press, Oxford (1979).

32. K. S. Kumar, A.M. Sancho, Y.N. Vaishnay and J.F. Weiss, Antioxidant properties of the radioprotector WR-2721 in red cell membrane model systems, Abstracts of International Conference on prostaglandin and lipid metabolism in radiation injury, AFRRI, Bethesda (1986).

33. M. Guichard, Bilan des essais cliniques en radiothérapie, in: "Actualités sur les substances radioprotectrices," E. D. F. Comite de radioprotection, Paris (1988).

34. A. T. Turrisi, D. J. Glover, S. Hurwitz, J. Glick, A. L. Norfleet, C. Weiler, J. M. Yuhas and M. M. Kligerman, Final report of the phase I trial of simple dose WR 2721 (S-2-(3-aminopropylamino) ethylphosphorothioic acid), Cancer. Treat. Rep. 70:1384 (1986).

35. J. F. Weiss, R. L. Hoover and K. S. Kumar, Selenium pretreatment enhances the radioprotective effect and reduces the lethal toxicity of WR 2721, Free Rad. Res. Comms. 3:33 (1987).

36. T. L. Phillips, Rationale for initial clinical trials and future development of radioprotectors, Cancer Clinical Trials 3:165 (1980).

37. M. Fatome, F. Courteille, J.D. Laval and V. Roman, Radioprotective activity of ethylcellulose microspheres containing WR2721 after oval administration, Int. J. Radiat. Biol., 52:21 (1987).

38. J. C. Fantone, W. A. Marasco, L. J. Elgas and P. A. Ward, Stimulus specificity of prostaglandin inhibition of rabbit polymorphonuclear leukocyte lysosomal enzyme release and superoxide anion production, Am. J. Pathol. 115:9 (1984).

39. V. Roman and F. Berleur. Les mécanismes d'action des radioprotecteurs chimiques, in: "Actualités sur les substances radioprotectrices", Comité de Radioprotection, EDF, Paris (1988).

40. T. L. Walden Jr, M. Patchen and S. L. Snyder, 16,16-dimethylprostaglandin E2 increases survival in mice following irradiation, Radiat. Res., 109:440 (1987).

41. T. L. Walden Jr, Leukotriene C4 induces radioprotection in part by a hypoxic mechanism, "Abstract 2nd annual program review," AFRRI, Bethesda (1988).

42. J. W. Finley and D.E. Schwass. Xenobiotic metabolism: nutritional effects, A.C.S., Washington (1985).

43. Z. M. Bacq, "Chemical protection against ionizing radiation," Thomas, Springfield (1965).

44. M. Fatome, La radioprotection chimique, Radioprotection 16:113 (1981).

45. O. F. Nygaard, and M.G. Simic. Radioprotectors and anticarcinogens, Academic Press, New-York, (1983).

46. M. Fatome. Les substances radioprotectrices. in: Actualités sur les substances radioprotectrices, Comité de Radioprotection, EDF, Paris (1988).

47. Z. M. Bacq, A. Herve and P. Fischer, Rayons X et agents de chelation, Bull. Acad. Roy. Med. Belge 18:226 (1953).

48. H. Laser, The "oxygen effect" in ionizing radiation, Nature 174:753 (1954).

49. L. Frank, D.L. Wood, R.J. Roberts, Effects of diethyl-dithiocarbamate on oxygen toxicity and lung enzyme activity in immature and adult rats, Biochem. Pharm., 27: 251 (1978).

50. J. J. Pocidalo, H. Mansour, P. Lacombe, M. Levacher, B. Rouveix and M.A. Gougerot, Immunomodulation par le diethyldithiocarbamate (Immuthiol): mise en évidence de propriétés immunoprotectrices anti-oxydantes, in: "Immunomodulateurs et thérapeutique antiinfectieuse", J.J. Pocidalo, J.P. Couland, F. Vachon and J.L. Vilde eds., Arnette, Paris (1986).

51. L. H. Rome, W.E.M. Lands, G.J. Roth and P.M. Majerus, Aspirin as a quantitative acetylating reagent for the fatty acid oxygenase that forms prostaglandins, Prostaglandins., 11:23 (1976).

52. R. E. Heikkila, Inactivation of superoxide dismutase by diethyldithiocarbamate, in: "Handbook of methods for oxygen radical research", R.A. Greenwald, ed., CRC Press Bocca Raton, (1985).

53. G. E. Miller, M. A. Zemaitis and F. E. Greene, Mechanisms of diethyldithiocarbamate induced loss of cytochrome P-450 from rat liver, Biochem. Pharm. 32:2433 (1983).

54. K. S. Kumar, A.M. Sancho and J.F. Weiss, Novel interaction of diethyldithiocarbamate with the glutathione/glutathione peroxidase system, Int. J. Radiat. Oncol. Biol. Phys., 12:1463 (1986).

55. B. Robertson, L. Gahring, R. Newton and R. Daynes, In vivo administration of interleukin 1 to normal mice depresses their capacity to elicit contact hypersensitivity responses: prostaglandins are involved in this modification of immune function, J. Invest. Dermatol. 88:380 (1987).

56. D. N. Tatakis, N. Weinfeld and R. Ziak, Interleukin-1 stimulates PgE$_2$ synthesis by osteoblastic cells- regulatory factors, J. Dent. Res. 65:291 (1986).

57. R. Neta, S. Douches and J. J. Oppenheim, Interleukin 1 is a radioprotector, J. Immunol. 136:2483 (1986).

58. W. F. Ward, A. Shih-Hoellwarth, P.M. Johnson, Survival of penicillamine-treated mice following Whole-body irradiation, Radiat. Res., 81:131 (1980)

59. A. W. Girotti, J.P. Thomas and J.E. Jordan, Xanthine Oxidase-Catalyzed crosslinking of cell membrane proteins, Arch. Biochem. Biophys., 251:639 (1988).

60. C. Navaro, G. Esolar, J.E. Banos, L. Casanovas and O. Bulbena, Effects of zinc acexamate in gastric mucosal production of prostaglandin E$_2$ in normal and stressed rats, Prostagl. Leuko. Ess. Fatty Ac. 33:75 (1988).

61. G. W. Evans, and P.E. Johnson, Defective prostaglandin synthesis in acrodermatitis enteropathica, Lancet 1:52 (1977).

62. G. L. Floersheim, N. Chiodetti and A. Bieri, Differential radioprotection of bone marrow and tumour cells by zinc aspartate, Brit. J. Radiol. 61:501 (1988).

63. A. Petkau, W.S. Chelack, S.D. Pleskach, B. Meeker and C.M. Brady, Radiation protection of mice by superoxide dismutase, Biochem.Biophys.Res.Commun 65:886 (1975).

64. L. L. Mitchell, K. G. D. Allen, and M. M. Mathias, Copper deficiency depresses rat aortal superoxide dismutase activity and prostacyclin synthesis, Prostaglandins 35:977 (1988).

65. J. R. J. Sorenson, Bis (3,5-diisopropyl-salicylato) copper (II), a potent radioprotectant with superoxide dismutase mimetic-activity, J. Med. Chem. 27:174 (1984).

66. M. J. Parnham and E. Graf, Seleno-organic compounds and the therapy of hydroperoxide - linked pathological conditions, Biochem. Pharm. 36:3095 (1987).

67. G. Batist, A.Reynaud A., A.G. Katki, E.L. Travis, M.C. Shoemaker, R.F. Greene et C.E. Myers. Enzymatic defense against radiation damage in mice. Effect of Selenium and Vitamin E depletion, Biochem. Pharmacol., 35:601 (1986).

68. G. W. Burton, K. H. Cheeseman, K. V. Ingold and T. F. Slater, Lipid antioxidants and products of lipid peroxidation as potential tumor protective agents, Biochem. Soc. Trans. 11:262 (1983).

69. M. A. Brown, Resistance of human Erythrocytes containing elevated levels of vitamin E to radiation - induced hemolysis. <u>Radiat. Res.</u>, 95 (2), 303-31 (1983).

70. A. Raleigh and F.Y. Shum, Radioprotection in model lipid membranes by hydroxyl radical scavengers: supplementary role for alpha-tocopherol in scavenging secondary peroxy radicals. <u>in</u>: "Radioprotectors anticarcinogens, O.F. Nygaard and M.G. Simic, eds. <u>Academic. Press</u>, New-York (1983).

71. R. M. Roy, M.A. Malick and G.M. Clark, Increased hematopoietic stem cell survival in mice injected with tocopherol after X-irradiation, <u>Strahlenther.</u> 58:312 (1982).

72. C. C. Reddy, C. E. Thomas and R.W. Scholz, Inadequate vitamin E and selenium nutrition, <u>in</u>: "Xenobiotic metabolism, nutritional effects," J. W. Finley and D. E. Schwass, eds., <u>ACS</u>, Washington (1985).

73. R. V. Panagamala, D.G. Cornwell. The effects of Vitamin E on arachidonic acid metabolism. <u>Ann. N.Y. Acad. Sci.</u>, 393:376 (1982).

74. M. Fatome, Les Aminothiols et leurs dérivés, in: Actualités sur les substances radioprotectrices, Comité de Radioprotection, EDF, Paris (1988).

75. W. E. M. Lands, R. J. Kulmacz and P. J. Marshall, Lipid peroxide actions in the regulation of prostaglandin biosynthesis, <u>in</u>: "Free radical in biology VI", W. A. Pryor, ed., <u>Academic Press</u>, New-York, (1984).

76. S. N. Meydani, M. Meydani, C.P. Verdon, A.A. Shapiro, J.B. Blumberg and K.C. Hayes, Vitamin E supplementation supresses prostaglandin E2 synthesis and enhances the immune response of aged mice, <u>Mech. Ageing Devel.</u> 34:191 (1986).

77. J. Yatsuyanagi, K. Iwai and T. Ogiso, Suppressive effect of zinc on some functions of neutrophils: studies with carrageenan-induced inflammation in rats, <u>Chem. Pharm. Bull.</u> 35:699 (1987).

78. M. Haberal, V. Mavi and G. Oner, The stabilizing effect of vitamin E, selenium and zinc on leucocyte membrane permeability: a study in vitro, <u>Burns</u> 13:118 (1987).

79. J. E. Packer, T.F. Slater, R.L. Willson, Direct observation of a free radical interaction between Vitamin E and Vitamin C, <u>Nature</u>, 278:737 (1979).

80. C. Camisa, B. Eisenstat, A. Ragaz and G. Weismann, The effects of retinoids on neutrophil functions in vivo, <u>J. Am. Acad. Dermatol.</u>, 6:620 (1982).

81. A. M. Michelson, K. Puget and P. Durosay. La superoxyde dismutase et la pathologie des radicaux libres, <u>C.R. Soc.Biol.</u>, 179:429 (1985).

82. J. R. Maisin, A. Kondi-Tamba, G. Mattelin, Polysaccharides induce radioprotection of murine hemopoietic stem cells and increase the LD 50/30 days, <u>Radiat. Res.</u> 105:276 (1986).

83. M. L. Patchen and T. J. Mac Vittie, Comparative effects of soluble and particulate glucan on survival in irradiated mice, <u>J. Biol. Resp. Mod.</u> 5:45 (1986).

84. Y. Matsubara, T. Shida, K. Ishioka, S. Egawa, T. Inada and K. Machida, Protective effect of zinc against lethality in irradiated mice. <u>Environ. Res.</u> 41:558 (1986).

NUCLEAR GLUTATHIONE TRANSFERASES WHICH DETOXIFY

IRRADIATED DNA

BRIAN KETTERER, GILLIAN FRASER AND DAVID J MEYER

CANCER RESEARCH CAMPAIGN MOLECULAR TOXICOLOGY
RESEARCH GROUP, UNIVERSITY COLLEGE AND MIDDLESEX
SCHOOL OF MEDICINE, LONDON W1P 6DB

HYDROPEROXIDES IN DNA

Free radicals have a number of toxic effects on the cell including the initiation of lipid peroxidation (Pryor, 1976) and the induction of potentially mutagenic lesions in DNA (Ames and Saul, 1986). Current knowledge suggests that some of the lesions in DNA may be peroxides. Irradiation of free thymidine gives a substantial yield of thymine hydroperoxides (Schulte-Frohlinde and von Sonntag, 1985) and the incubation of irradiated DNA with GSH and tissue extracts containing GSH peroxidase activity brings about the catalytic oxidation of GSH to GSSG which is assumed to be due to the reduction of hydroperoxide moieties in DNA by GSH (Christopherson, 1969).

Direct evidence for thymine hydroperoxide moieties in DNA, which has been exposed to radical damage, has yet to be found, but the thymine glycols and 5-hydroxymethyl uracil, which would result from reduction of these hydroperoxides, can be separated from digests of DNA which have either been irradiated (Frenkel et al., 1981; Frenkel et al., 1985) or exposed to the "oxygen burst" of phorbol ester stimulated polymorphonuclear leukocytes (Frenkel and Chrzan, 1987). The reduction products are also normal components of the urine and their levels are related to the rate of oxygen consumption of the species concerned (see Table 1; Ames and Saul, 1986).

The most reactive free radical produced in these systems is the hydroxyl radical. It reacts with thymine either by adding to the 5, 6 double bond or abstracting hydrogen from the 5-methyl group. According to the position of the attack either 5- or 6-ring or 5-methylene carbon radicals are formed. If sufficient hydroxyl radicals were present thymine glycols and 5-hydroxylmethyl uracil would be formed directly, but under most in vivo conditions the concentration of O_2 is much greater and it reacts rapidly with the carbon radicals to give the corresponding peroxy radicals. In order to arrive at the thymine glycol or 5-hydroxymethyl residues, peroxy radicals presumably undergo two reduction steps. The first step is

Antioxidants in Therapy and Preventive Medicine
Edited by I. Emerit *et al.*
Plenum Press, New York, 1990

301

Table 1   Levels of Thymine Glycol, Thymidine Glycol,
          Hydroxy Methyl Uracil and Hydroxymethyldeoxy-
          uridine in Rat and Human Urine

compound	molecules/cell/day human	rat
Thymine glycol	270	3800
Thymidine glycol	70	1200
Hydroxymethyl uracil	620	6000
Hydroxymethyldeoxyuridine	trace	trace

After Ames and Saul, 1986

either reduction to peroxide ions by, for example, $HO_2^-/O_2^-\cdot$ or GS^-; or reduction to hydroperoxides by hydrogen abstraction (von Sonntag, 1987). The second step is reduction of the hydroperoxides by the GSH peroxidase activity mentioned above to give the thymine glycol and 5-hydroxymethyluracil residues.

The importance of glutathione peroxidases is to remove the hydroperoxy moieties which have the potential to cause further radical damage. However, the resulting hydroxy derivatives, although stable, may inhibit DNA replication (Waschke et al., 1975) and are removed by enzymes such as thymine glycol DNA glycosylase and 5-hydroxymethyl uracil DNA glycosylase (Cathcart et al., 1984; Hollstein et al., 1984) leaving the DNA to undergo repair.

The excretion of the resulting thymine glycols and 5-hydroxymethyluracil under normal conditions of life suggest that oxygen-centred free radical damage is an everyday hazard to DNA and that a mechanism for its repair is always present and active.

In vivo, in addition to the action of stimulated phagocytes or possible high energy irradiation, oxygen-centred free radicals are also thought to arise by incomplete reduction of molecular oxygen during mitochondrial respiration and the operation of P450 cytochromes in the endoplasmic reticulum. Radicals formed this way may react with either mitochondrial or nuclear DNA according to their origins.

Radicals also cause lipid peroxidation. We and others have shown that in vitro lipid peroxidation can be inhibited by the co-operative actions of phospholipase A2 and Se-dependent or Se-independent GSH peroxidase activity (Sevanian et al., 1983; Grossman and Wendel, 1983; Tan et al., 1984). Free rather than esterified fatty acid hydroperoxides are substrates for these enzymes, hence the requirement for phospholipase which appears to be activated during peroxidation. Lipid hydroperoxides which escape reduction undergo metal-catalysed decomposition to alkyl and alkoxy radicals (Slater, 1984) which if they diffuse into the nucleus intact might contribute to DNA damage.

The reduction of DNA hydroperoxides by Se-dependent and Se-independent GSH peroxidases is our current interest and the main theme of this paper. We are particularly concerned with the Se-independent GSH peroxidases which are members of the GSH S-transferase (EC.2.5.1.18) gene superfamily.

GSH TRANSFERASES AND GSH PEROXIDASES

GSH transferase activity usually refers to the catalysis of the conjugation of carbon-centred electrophiles with GSH. The result is usually detoxication and excretion. Two examples from everyday life are catalysis of the reaction of GSH with N-acetyl benzoquinone imine (NAPQI) (the toxic metabolite of the common analgesic drug, paracetamol) in the liver to give 3-(glutathion-S-yl) paracetamol (Coles et al., 1988) and benzo(a)pyrene-7,8-diol-9,10-oxide (the carcinogenic metabolite of the common environmental pollutant benzo(a)pyrene) to give 9,hydroxy,10-glutathion-S-yl benzo(a) pyrene-7,8-diol (Jernström et al., 1985).

GSH peroxidase activity is also the attack of GSH on an electrophile, but in this case it is the electrophilic oxygen of hydroperoxide. The products of this reaction are the corresponding alcohol and GSSG. Se-dependent GSH peroxidase is a tetramer of mol. wt. 85,000 and has selenocysteine at the active centre. This enzyme can reduce both hydrogen peroxide and a range of organic hydroperoxides (Flohé, 1982). The GSH transferase isoenzymes which have Se-independent GSH peroxidase activity require substrates with a hydrophobic moiety, and none so far has exhibited activity towards hydrogen peroxide (Prohaska, 1980). The extent to which organic peroxide reducing activity is borne by Se-independent GSH peroxidases varies with the species. In human liver, the Se-independent enzymes predominate, but their relative contribution is less in rat liver (Lawrence and Burke, 1978).

The GSH transferases include both soluble and membrane-bound forms. The soluble enzymes are dimers in which the subunits behave independently of each other. These subunits are members of a gene superfamily composed of at least 4 families. The nomenclature is numerical, enzymes being numbered in the order of their characterization. The well established gene families are named alpha (subunits 1, 2, 8, 10), mu (subunits 3, 4, 6, 9, 11) and pi (subunit 7) (see Mannervik & Danielson, 1988; Ketterer et al., 1988; Kispert et al., 1989). Recent structural studies suggest that subunit 5 belongs to a new gene family which is tentatively called theta (D.J. Meyer, B. Coles & B. Ketterer, personal communication) (Table 2).

Membrane-bound enzymes are represented by microsomal GSH transferase which bears virtually no structural resemblance to the soluble forms, and other apparently very specific enzymes of unknown structure such as leukotriene C4 synthase (Morgenstern & DePierre, 1988; Ketterer et al., 1988).

Table 3 shows the activity of the GSH transferase isoenzymes towards a number of substrates including endogenous hydroperoxides, carbon-centred electrophiles either of endogenous origin or derived from drugs,

Table 2. GSH Transferase Subunits in the Rat

Nomenclature	Some physical characteristics		
	app mol wt x 10^{-3}	isoelectric point	retention time on RP-hplc (min)
alpha family			
1	25	10	48
2	28	9.8	33
8	24.5	6.0	54
10	25.5	9.6	33
mu family			
3	26.5	8.5	24
4	26.5	6.9	27
6	26	5.8	37
9	24	5.8	28
11	27	5.2	44
pi family			
7	24	7.0	31
theta family			
5	26.5	7.3	42

carcinogens and industrial chemicals. It shows that, from the information at present available, the ability of subunits to reduce hydroperoxides is as follows. For fatty acid hydroperoxides subunit 5 > 1 > 2 > 7 > 3 = 4 > 8 and for DNA hydroperoxide subunit 5 >> 4 > 3 > 7. Activity towards these hydroperoxides ranges from 0.01 - 3.5 µmol/min/mg. In the case of carbon-centred electrophiles present data shows that certain α,β unsaturated ketones e.g. NAPQI, 4-hydroxy-non-2-enal (HNE) and oxo-linoleate have substrate activity which far outstrips that for hydroperoxide reduction. On the other hand, where attack on epoxides and displacement of halides is concerned, activities tend to be much lower. HNE is a genotoxic and cytotoxic decomposition product of lipid peroxides (Marnett et al., 1985; Cadenas et al., 1983; Benedetti et al., 1986). Subunit 8 is a very powerful catalyst of HNE conjugation, and therefore a very important component of the systems involved in the detoxication of lipid peroxidation in vivo (Ketterer et al., 1988).

Where hydroperoxides are concerned the most interesting subunit is 5 which is active with both DNA and lipid hydroperoxides. The hydroperoxide moiety and the 5,6 carbons of the thymine are in the major groove and therefore it is the area of the major groove of DNA around the thymine hydroperoxide residue rather than the thymine hydroperoxide residue alone, which interacts with the enzyme active site. DNA is a poly-anion which may be an important consideration in substrate binding.

Thus, subunit 5 is able to detoxify oxygen-centred free radical damage at two key sites in the cell, namely at the membrane and in DNA. It will be of interest to see to what extent this subunit is also effective with genotoxic carbon-centred electrophiles.

Table 3 GSH Transferase Activity (μmol/min/mg) towards some Substrates of Biological Interest (after Ketterer et al., 1988)

| Substrate | Enzymes by Class | | | | | | | | |
| | Alpha | | | Mu | | | | Pi | Theta |
	1-1	2-2	8-8	3-3	4-4	6-6	11-11	7-7	5-5
Linoleic acid hydroperoxide(R)	3.0	1.6	0.1	0.2	0.2	nd	nd	1.5	5.3
DNA Hydroperoxide[1](R)	nil	nil	nd	0.02	0.03	0.01	nd	0.01	3.5
4-Hydroxy-non-2-enal(C)	2.6	0.7	170	2.7	6.9	nd	2.1	0.1	nd
Oxo-linoleate(C)	0.6	0.7	35.6	0.1	0.6	nd	nd	0.6	nd
N-Acetyl-p-benzoquinone imine(C)	24.0	48.0	nd	6.0	3.0	nd	nd	60.0	nd
Aflatoxin B1-8,9-oxide(C)	0.001	0.001	nd	nil	nd	nd	nd	nil	nd
Benzo(a)pyrene-7,8-diol-9,10-oxide(C)	0.1	0.08	nd	0.03	0.33	nd	nd	0.33	nd
1-Nitropyrene-4,5-oxide(C)	0.01	0.03	nd	0.30	0.30	nd	nd	0.02	nil
Ethylene dibromide(C)	0.011	0.117	nd	0.07	0.021	nd	nd	nd	nd

(R) = reduction reaction; (C) = conjugation reaction; [1]irradiated DNA lyophilized to remove H_2O_2 produced during irradiation; nd = not determined.

ISOLATION AND PROPERTIES OF GSH TRANSFERASE 5-5

GSH-agarose affinity columns which are so convenient for the selection of all other GSH transferases in one step cannot be used with GSH transferase 5-5. Early preparations of this enzyme were said to have a Km for GSH of 3 mM (Fjellstedt et al., 1973). It is possible that the inability of GSH transferase 5-5 to bind to GSH-agarose is related to its high Km for GSH. Present day separations of GSH transferase 5-5 depend on its gradient elution from sequential dye-ligand-, adsorption-, ion exchange- and hydrophobic interaction-chromatography. GSH transferase 5-5 has eluded purification in the past because of its lability and the use of β-mercaptoethanol and glycerol, particularly the former, are essential in order to obtain high levels of activity.

GSH transferase 5-5 is unique among mammalian isoenzymes so far studied, in not utilizing 1-chloro-2,4-dinitrobenzene as a substrate. Prior to the discovery of its ability to reduce certain hydroperoxides, the commonly used substrate was 1,2-epoxy-3(p-nitrophenoxy)propane (ENPP) with which its activity may be as high as 200 μmole/min/mg. The proof that GSH transferase 5-5 belongs to a new GSH transferase gene family comes, not only from its unusual substrate specificity, but also from partial amino acid sequence which shows little homology with members of the alpha, mu, pi and microsomal families.

THE INTRACELLULAR DISTRIBUTION OF GSH PEROXIDASES

It is important that the intracellular distribution of Se-dependent peroxidase and the GSH transferases which have Se-independent peroxidase activity should enable their access to the important substrates discussed above. The two most easily separated fractions, namely the soluble supernatant (cytosol) and the microsomes, have been studied in much detail as shown above and they have isoenzymes appropriate for the detoxication of lipid peroxidation. Preliminary work indicates the presence of Se-dependent GSH peroxidase (Flohé and Schlegel, 1971) and GSH transferases in the mitochondria (Kraus, 1980). However, it is not known to what extent GSH transferases in the mitochondria are related to those in in the cytosol and microsomes and might detoxify both lipid and DNA peroxidation in this organelle. In the context of oxygen-centred free radical damage the mitochondria are very important and the missing information is required. With respect to the nucleus, our present concern, work is bedevilled by uncertainty concerning the validity of methods of isolating this organelle. We have chosen to use the citric acid method developed by Taylor et al. (1973) and Bennett & Yeoman (1987). However, since the nuclear pore allows the passage of smaller proteins, all aqueous nuclear separations are suspect since some elution of soluble nuclear proteins may occur. The fact that these nuclei retain readily extractable GSH transferases suggest that they are the most appropriate for these separations.

The citric acid method has already been used to isolate

a DNA-binding protein, referred to as BA, which co-localizes with small nuclear ribonucleoprotein particles within interchromatinic regions of the cell nucleus. In the same extracts proteins have been detected which are apparently mu family GSH transferases (Bennett et al., 1986). When this nuclear fraction from rat liver is radiolabelled and micro-injected into Walker 256 carcinoma cells the apparent GSH transferase mu family proteins migrate into the nucleus (Bennett & Yeoman, 1987) which is further proof of their nuclear location.

In our own laboratory we have shown that GSH transferase subunits 1, 2, 3, 4 and 5 are extracted from citric acid-isolated nuclei with saline-EDTA. However, while GSH transferase subunit 5 is present in the same concentration in the nucleus as in the cytosol, subunits 1, 2, 3 and 4 occur at a 50 to 100 fold lower concentration. Homogenizing the nuclear residue with 8M urea gives an extract which, if rapidly concentrated in order to maximise the association of subunits and then dialysed in the presence of GSH in order to encourage the formation of a native conformation, gives enzymically active GSH transferases containing subunits 3, 4 and 5 only. This appears to demonstrate that subunits 3, 4 and 5 are more tightly bound to chromatin than subunits 1 and 2 (G. Fraser, D.J. Meyer & B. Ketterer, unpublished information).

The known enzymic activity of these subunits would suggest that, whereas subunits 3 and 4 may have some activity towards DNA hydroperoxides, they would be more effective in dealing with carbon-centred electrophiles. Subunit 5 on the other hand would be particularly effective in dealing with DNA hydroperoxide.

The Se-dependent GSH peroxidase activity has also been detected in both the saline-EDTA and urea extracts of liver nuclei (G. Fraser. D.J. Meyer and B. Ketterer, unpublished information). Like GSH transferase subunits 1, 2, 3 and 4, it is present in much lower amounts in the nucleus than the cytosol. Even so it contributes significantly to the overall activity towards DNA hydroperoxide reducing activity seen in the nucleus of the rat, the only species which has so far been studied. Thus, there may be a dual defence against DNA hydroperoxides, one involving Se-dependent GSH peroxidase and the other GSH transferase subunits 3, 4 and 5, the major burden among the transferases being borne by GSH transferase subunit 5. The role of Se-independent GSH peroxidases in the nucleus of human livers should be greater than in the nuclei from the rat.

Free radicals might be expected to affect sites other than thymine residues in DNA. In isolated DNA, attack on other bases and the deoxyribose moiety have been observed. Whether damage is so widespread in nuclear DNA in vivo is not known. Work on chromatin in the solid state indicates that the thymine moiety is the most electroaffinic and therefore the most likely site of free radical activity (Cullis et al., 1987) and it has been suggested that DNA strand breakage is the result of a secondary attack of a thymine radical on a deoxyribose moiety. 8-Hydroxyguanine has been detected and is also believed to be a product of oxygen-centred free radical attack on DNA in vivo however, the mode of its formation is not yet known (Kasai et al., 1986; Floyd et al., 1986).

CONCLUSIONS

Free radical damage DNA in the presence of oxygen is believed to generate thymine hydroperoxide residues. These residues which are potentially genotoxic may be reduced to their corresponding thymine glycol and 5-hydroperoxymethyl uracil residues by Se-dependent and Se-independent GSH peroxidases and analysis of fractions from the rat liver suggests that both types of GSH peroxidase reside in the nucleus and are therefore available to carry out this reaction. Of particular interest is GSH transferase 5-5 which has high levels of DNA hydroperoxide reducing activity and seems to be relatively firmly bound to chromatin.

REFERENCES

Ames, B. N. and Saul, R. L., 1985, Oxidative DNA damage as related to cancer and ageing, in: "Genetic Toxicology of Environmental Chemicals," C. Ramil, B. Lambert and J. Magnusson, eds., Alan Liss, New York.
Benedetti, A., Pompella, A., Fulceri, R., Romani, A. and Comporti, M., 1986, Detection of 4-hydroxynonenal and other lipid peroxidation products in the liver of bromobenzene poisoned mice. Biochim. Biophys. Acta, 876:658.
Bennett, C. F. and Yeoman, L. C., 1987, Microinjected glutathione S-transferase Yb subunits translocate to the cell nucleus, Biochem. J., 247:109.
Bennett, F. C., Spector, D. L. and Yeoman, L. C., 1986, Non-histone protein BA is a glutathione S-transferase localized to interchromatinic regions of the cell nucleus, J. Cell Biol., 102:600.
Cadenas, E., Müller, A., Brigelius, R., Esterbauer, H. and Sies, H., 1983, Effects of 4-hydroxynonenal on isolated hepatocytes, Biochem. J., 214:479.
Cathcart, R., Schwiers, E., Saul, R.L. and Ames, B.N., 1984, Thymine glycol and thymidine glycol in human and rat urine: A possible assay for DNA damage, Proc. Natl. Acad. Sci. USA, 81:5633.
Christopherson, B. O., 1969, Reduction of X-ray-induced DNA and thymine hydroperoxide by rat liver glutathione peroxidase, Biochem. Biophys. Acta, 186:387.
Coles, B., Wilson, I., Wardman, P., Hinson, J. A., Nelson, S. D. and Ketterer, B., 1988, The spontaneous and enzymatic reaction of N-acetyl-p-benzoquinoneimine with glutathione: a stopped flow study, Arch. Biochem. Biophys., 264:253.
Cullis, P. M., Jones, G. D. D., Symons, M. C. R. and Lea, J. S., 1987, Electron transfer from protein to DNA in irradiated chromatin, Nature. 330:773.
Fjellstedt, T. A. Allen, R. H., Duncan, B. K. and Jakoby, W.B., 1973, Enzymatic conjugation of epoxides with glutathione, J. Biol. Chem., 248:3702.
Flohé, L., 1982, Glutathione peroxidase brought into focus, in: "Free Radicals in Biology, vol V," W.A. Pryor, ed., Academic Press, New York.

Flohé, L. and Schlegel, W., 1971, Interzelluläre verteilung des glutathione-peroxidase-systems in der rattenleber, Z. Physiol. Chem., 352:1401.

Floyd, R. A., Watson, J. J. and Wong, P. K. 1986, Hydroxyl free radical adduct of deoxyguanosine: Sensitive detection and mechanisms of formation, Free Rad. Res. Commun., 1:163.

Frenkel, K., Goldstein, M. S., Duker, N. and Teebor, G. W., 1981, Identification of the cis-thymine glycol moiety in chemically oxidized and gamma-irradiated deoxyribolnucleic acid, Biochemistry, 20:750.

Frenkel, K., Cummings, A., Solomon, J., Cadet, J., Steinberg, J.J. and Teebor, G.W., 1985, Quantitative determination of the 5-(hydroperoxy methyl) uracil moiety in the DNA of gamma-irradiated cells, Biochemistry, 24:4527.

Frenkel. K. and Chrzan, K., 1987, Hydrogen peroxide formation and DNA base modification by tumor promoter-activated polymorphonuclear leukocytes, Carcinogenesis, 8:455.

Grossman, A. and Wendel. A., 1983, Activity of selenoenzyme glutathione peroxidase with enzymically peroxidized phospholipids, Eur. J. Biochem., 195:549.

Hollstein, M. C., Brooks, P., Linn, S. and Ames, B. N., 1984, Hydroxymethyluracil DNA glycolsylase in mammalian cells, Proc. Natl. Acad. Sci. USA, 81:4003.

Jernström, B., Martinez, M., Meyer, D. J. and Ketterer, B., 1985, Glutathione conjugation of the carcinogenic and mutagenic electrophile (±)-7β,8α-dihydroxy-9α, 10α-oxy-7,8,9,10-tetrahydrobenzo(a)pyrene catalysed by rat liver glutathione transferases, Carcinogenesis, 6:85.

Kasai, A.-H., Crain, P. F., Kuchino, Y., Nishimura, K., Oofsuyama, A and Tanooka, H. 1986, Formation of 8-hydroxyguanine moiety in cellular DNA by agents producing oxygen radicals and evidence for its repair, Carcinogenesis, 7:1849.

Ketterer, B., Meyer, D. J. and Clark, A. G., 1988, Soluble glutathione transferase isoenzymes, in: "Glutathione Conjugation: Mechanisms and Biological Significance," H. Sies and B. Ketterer, eds., Academic Press, London.

Kispert, A., Meyer, D. J., Lalor, E., Coles, B. and Ketterer, B., 1989, Purification and characterization of a labile rat glutathione transferase of the mu class, Biochem. J., 260:789.

Kraus, P., 1980, Resolution, purification and some properties of three glutathione transferases from rat liver mitochondria, Hoppe-Seylers 2. Physiol Chem., 361:9.

Lawrence, R. A. and Burke, R. F., 1978, Species, Tissue and subcellular distribution of non-Se-dependent glutathione peroxidase activity, J. Nutr., 108:211.

Mannervik, B. and Danielson, U.H. 198 8, Glutathione transferases - structure and catalytic activity, CRC Critical Reviews in Biochemistry, 23:283.

Marnett, L. J., Hurd, H. K., Hollstein, M. C., Levin, D. E., Esterbauer, H. and Ames, B. N., 1985, Naturally occurring carbonyl compounds are mutagens in Salmonella tester strain TA 104, Mutation Res., 148:25.

Morgenstern, R. and dePierre, J. 1988, Membrane-bound
    glutathione transferases isoenzymes, _in_:
    "Glutathione Conjugation: Mechanisms and Biological
    Significance," H. Sies, and B. Ketterer, eds.,
    Academic Press, London.
Pryor, W. A., 1976, The role of free radical reactions in
    biological systems, _in_: "Free Radicals in Biology,
    vol I", W.A. Pryor, ed., Academic Press, New
    York.
Prohaska, J. A. 1980, The glutathione peroxidase activity of
    glutathione S-transferases, _Biochem. Biophys. Acta_,
    611:87.
Sevanian, A. Muakassah-Kelly, S. F. and Montestruque, S.,
    1983, The influence of phospholipase A2 and
    glutathione peroxidase on the elimination of
    membrane lipid peroxides, _Arch. Biochem. Biophys._
    233:441.
Schulte-Frohlinde, P. and von Sonntag, C., 1985, Radiolysis
    of DNA and model systems in the presence of oxygen,
    _in_: "Oxidative Stress," H. Sies, ed., Academic
    Press, New York.
Slater, T. F., 1984, Free-radical mechansisms in tissue
    injury. _Biochem. J._, 222:1.
Tan, K. H., Meyer, D. J., Belin, J. and Ketterer, B., 1984,
    Inhibition of microsomal lipid peroxidation by GSH
    transferase B and AA: Role of phospholipase A2,
    Biochem. J., 220:243.
Taylor, C. W., Yeoman, L. C., Daskaly, I. and Busch, H.,
    1973, Two dimensional electrophoresis of proteins of
    citric acid prepared nuclei prepared with the aid of
    a Tissuemizer, _Exptl. Cell. Res._, 82:215.
von Sonntag, C. 1987, _in_: "The Chemical Basis of Radiation
    Biology," Taylor and Francis, London.
Waschke, S., Rufschläger, J., Bärwolff, D. and Langen, P.,
    1975, 5-Hydroxymethyl-2'-deoxyuridine, a normal DNA
    constituent in certain _Bacillus_ _subtilus_ phages is
    cytostatic for mammalian cells, _Nature_, 255:629.

URINARY BIOMARKERS IN RADIATION THERAPY OF CANCER

David S. Bergtold, Christine D. Berg* and Michael G. Simic

National Institute of Standards and Technology
Gaithersburg, MD 20899

*Department of Radiation Medicine, Georgetown University
Medical School, Washington, D.C. 20007

INTRODUCTION

The ability to monitor noninvasively the biological effect of a radiation dose in humans is potentially beneficial for screening cancer patients during the course of radiotherapy. Yet no universal approach exists for this purpose other than routine examination of patients and visible tumors, although in limited situations specific parameters are measured after treatment is completed (e.g., determinations of antibody levels after breast cancer therapy). In all cases, no appreciable indicators of the effects of the irradiation are apparent during the early part of the therapy regimen.

However, we recently demonstrated that both background levels of endogenous damage[1] and radiation-induced damage to DNA[2,3] may be followed noninvasively by measuring diverse urinary biomarkers, such as hydroxylated DNA bases (e.g., thymine glycol, 8-hydroxyguanine) and their nucleoside moieties. Based on the results of these studies, we have examined the urinary excretion of thymidine glycol and 8-hydroxy-2'-deoxyguanosine in cancer radiotherapy patients and, for the first time, have demonstrated an elevation in the levels of these compounds excreted after a single radiotherapeutic dose (1.8 Gy). We conclude, therefore, that urinary biomarkers resulting from radiation damage to DNA may be useful for monitoring biological effects during cancer radiotherapy as early as after the first dose fraction.

BACKGROUND

Radiation Therapy of Cancer

Radiation therapy of cancer is based on radiation-induced killing of cells, primarily via the induction of damage to DNA. The dose employed must be sufficient to eliminate all cancer cells. Consequently, to a certain extent, the surrounding normal tissue is damaged as well. One goal of radiation therapy, to reduce the damage to normal tissue, is achieved by fractionation of the therapy[4] so that about 2 Gy is delivered on a daily basis until all tumor cells are killed. A period of 24 hours between irradiations allows normal cells to repair themselves, thereby reducing the undesirable consequences of exposure.

Antioxidants in Therapy and Preventive Medicine
Edited by I. Emerit *et al.*
Plenum Press, New York, 1990

311

Mechanisms of DNA Damage

The killing of cells by ionizing radiations is a consequence of a series of chemical and biological processes triggered by the initial interaction of radiations with DNA and molecules around it, via ionization of these molecules and subcomponents of DNA.[5] The sequence of events is shown in Scheme 1.[6]

$$DNA \rightsquigarrow DNA^+ \rightarrow DNA\text{-}R\cdot \rightarrow DNA\text{-}P \rightarrow \quad Effects \rightarrow Consequences \quad (1a)$$

$$\uparrow DNA$$

$$RH \rightsquigarrow RH^+ \rightarrow \cdot R + H^+ \tag{1b}$$

Scheme 1. Sequence of events induced by ionizing radiations from the initial DNA and substrate molecule RH, their ions DNA^+ and RH^+, free radicals $DNA\text{-}R\cdot$ and $\cdot R$, products DNA-P, effects (e.g., DNA strand breaks, chromosome aberrations, crosslinks), to biological consequences (e.g., mutation, loss of white blood cells, cell inactivation, death).

Process (1a) is commonly referred to as direct action and process (1b) as indirect action of radiation on DNA. DNA-P resulting from indirect action and the products released from DNA-P are the biomarkers we propose as useful for monitoring radiation damage and the efficacy of radiotherapy. We have concentrated specifically on the DNA damage products originating from ionization of the surrounding water and the reaction of the resulting ·OH radicals with DNA.

Formation of DNA Products

Ionization of water, the major bioconstituent and the most abundant molecule around DNA, leads to formation of the primary water radicals,[5]

$$H_2O \rightsquigarrow H_2\overset{\bullet}{O}{}^+ + e^- \tag{2}$$

$$H_2\overset{\bullet}{O}{}^+ + H_2O \rightarrow \cdot OH + H_3O^+ \tag{3}$$

The resulting hydroxy radical, ·OH, generated in the immediate vicinity of DNA, reacts at diffusion controlled rates ($k \sim 10^{10}$ M^{-1} s^{-1}) with most subcomponents of DNA. The most abundant and distinguishable products of ·OH radical reactions with DNA are the pyrimidine and purine products. We have concentrated on investigations of the best characterized products of thymine and guanine, i.e., thymine glycol and 8-hydroxy-guanine,

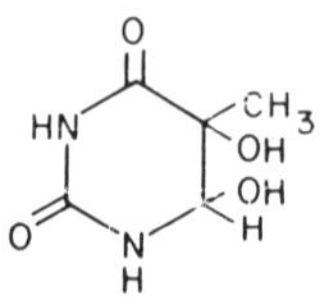

Thymine Glycol
TG

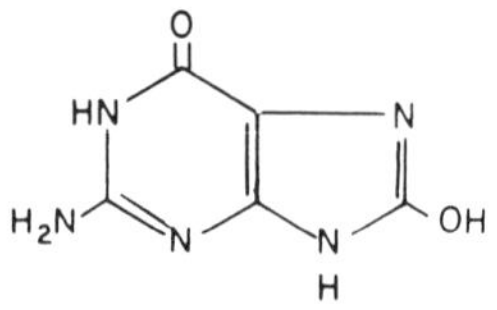

8-Hydroxy-Guanine
8-G-OH

by measuring their deoxynucleosides thymidine glycol (dRTG) and 8-hydroxy-2'-deoxyguanosine (dR-8-G-OH) (where dR stands for deoxyribose).

The reaction mechanisms that lead to the formation of these products are fairly well understood.[5] For example, reaction of •OH with thymine gives three radicals with distinguishable chemistry,[7]

$$\text{(4)}$$

All three radical forms react with oxygen to generate thymine glycol (TG), thymine hydrate, and 5-hydroxymethyluracil (5-HMU), respectively. For example (Ref. 5, pp. 141-147),

$$\text{(5)}$$

Reaction of •OH with guanine gives four distinguishable radicals. A series of reactions of the 8-hydroxyguanine radical, one of the four intermediates, leads to the formation of 8-hydroxyguanine moieties,[6] e.g.,

$$\text{(6)}$$

Conversion of DNA Products into Urinary Biomarkers

In this work we have restricted our approach to the measurement of products associated with the reaction of •OH with cellular DNA. The sequence of events that leads to radiolytic products, P, is shown in Scheme 1. The two processes not included in Scheme 1 are the chemical[8] and enzymatic repair.[9] Chemical repair takes place at the free radical stage (ions are too short-lived to consider) while enzymatic repair eliminates radiolytic products from DNA and replaces the excised components to restore the DNA to its original condition (Scheme 2).

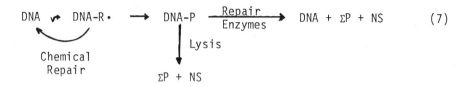

$$DNA \xrightarrow{} DNA\text{-}R \cdot \longrightarrow DNA\text{-}P \xrightarrow{\text{Repair Enzymes}} DNA + \Sigma P + NS \qquad (7)$$

Scheme 2. Chemical and enzymatic repair of damaged DNA. Mechanisms of elimination of radiolytic products, ΣP, and nuclear subcomponents, NS (DNA bases, nucleosides, nucleotides, and oligonucleotides).

The radiolytic products of interest are a variety of purine, pyrimidine, and deoxyribose products. Examples of these products are thymine glycol, TG, and thymidine glycol, dRTG. These two products may result from two types of elimination from DNA,[9]

$$DNA\text{-}dRTG \xrightarrow{\text{Excision Enzymes}} DNA' + dRTG \qquad (8a)$$
$$\xrightarrow{\text{TG Glycosylase}} DNA\text{-}dR + TG \qquad (8b)$$

To appear in the urine these products must enter the blood and then be removed to the urine by the kidneys. The fact that they appear in the urine as TG and dRTG supports the theory that these products are not metabolized completely. Partial metabolism, however, cannot be excluded at present. Lysis of dead cells would lead to an increased urinary level of both DNA and RNA bases and nucleosides, whereas repair functions would yield only DNA products.

EXPERIMENTAL

Irradiation of Cancer Patients

Patient 1 was a 59 year old female with a left breast mass and an abnormal chest x-ray. After radical mastectomy she received radiation therapy (55.8 Gy) to the mediastinum and right lung in 1.8 Gy fractions. First morning void urine samples were obtained just prior to the first treatment and again the following morning.

Patient 2 was a 57 year old female referred for postoperative radiation therapy after resection of adenocarcinoma of the descending colon. She received 45.0 Gy to the left upper quadrant of the abdomen in 1.8 Gy fractions. Urine (first void) was obtained on the morning of the first day of treatment and again the next day.

In both cases the patients were treated with 6 MV X rays from a Siemens 6740 linear accelerator.* Urine samples collected were refrigerated until analysis; other studies indicate that this treatment was unlikely to affect the results of this study.

----

*Commercial products are identified here only for technical purposes and are not meant to be an endorsement by the National Institute of Standards and Technology.

Measurement of Biomarkers in Urine

The general approach for measuring urinary oxidative DNA base damage biomarkers has been described[1,10,12] several times since such measurements were first introduced by Ames. Since the levels of the compounds measured were low and the samples were complex, the gas chromotography/mass spectrometry/selected ion monitoring (GC/MS/SIM)) techniques outlined by Bergtold, et al.[1] were used because they are highly sensitive and capable of reliably detecting compounds present in biological matrices. As described[1] multiple ions characteristic of each specific compound were monitored at the appropriate retention times to verify the identities of the compounds. In similar studies, the results obtained were tested by high performance liquid chromatography coupled with electrochemical detection.

RESULTS AND DISCUSSION

As an example of our data, measurements by GC/MS/SIM of two biomarkers in the urine of two representative radiation therapy patients are given in Table 1.

Table 1. Urinary excretion of radiolytic products by two cancer radiotherapy patients after the first dose fraction of 1.8 Gy.

	Baseline level (nmol/day)	Post-treatment level (nmol/day)
Patient 1		
thymidine glycol	10 ± 0.5*	37 ± 2
8-hydroxy-2'-deoxyguanosine	14 ± 0.7	40 ± 2
Patient 2		
thymidine glycol	8 ± 0.4	20 ± 1
8-hydroxy-2'-deoxyguanosine	8 ± 0.4	31 ± 1.5

*Error values are standard deviations reflecting variability in the results from three injections of each sample. Measured concentrations were converted to values of nmol/day by normalizing for daily creatinine excretion.[11]

It is evident from Table 1 that even irradiation of a partial body volume (5-10% of total body weight) at 1.8 Gy per treatment is sufficient to raise the levels of thymidine glycol (dRTG) and 8-hydroxy-2'-deoxyguanosine (8-dRG-OH) by significant factors. The factors were large enough to be significant because the levels of dR-TG and 8-dRG-OH in human urine before irradiation are fairly low. In small animals with high metabolic rates, such as mice and rats, the levels of these urinary biomarkers in unirradiated animals are about 15 times higher than in humans.[1,11] Consequently, measurement of radiation-induced increases in these biomarkers in mice and rats is achievable only at much higher doses.

It has been suggested that the baseline urinary levels of TG and 8-G-OH and their moieties are a consequence of metabolically generated •OH.[11] The evidence for such a process is still lacking, although our current experiments appear to indicate that TG and 8-G-OH levels are enhanced when •OH radicals are generated by radiation in the patient's body. Lean muscle tissue, for example, consists of about 70-80% water. Ionizing radiations, whether high energy electrons (4-10 MeV) or X- or

γ-rays, would ionize water in the tissue and generate •OH, reactions (2) and (3). The fraction of absorbed energy utilized for that process is equal to the mass fraction of the water in the tissue (in the first approximation). Consequently, the number of •OH radicals in irradiated tissue is quite high. Only a small fraction of those radicals, however, will react with DNA because of the high rate of reaction of •OH with subcomponents of other biomaterials, e.g., amino acids, sugars, fatty acids, and other organic constituents. Hence, only those •OH radicals generated in the immediate vicinity of DNA in the nucleus will contribute towards formation of TG and 8-G-OH as shown in reactions (4) to (6).

CONCLUSIONS

Radiolytic products resulting from •OH radical reaction with DNA in human patients treated by radiation therapy are excreted in the urine, as a result of enzymatic repair or lysis of damaged DNA. Measurements of the levels of these urinary biomarkers may be useful to monitor radiation damage in exposed tissue as early in therapy as after one day following the first radiotherapeutic dose (ca. 2 Gy).

ACKNOWLEDGMENT

Partial support for this work was provided by DNA.

REFERENCES

1. D. S. Bergtold, M. G. Simic, H. Alessio, and R. G. Cutler, Urine biomarkers for oxidative DNA damage, in: "Oxygen Radicals in Biology and Medicine," M. G. Simic, K. A. Taylor, J. F. Ward, and C. von Sonntag, eds., Plenum Press, New York (1988) pp. 483-490.
2. M. G. Simic and D. S. Bergtold, New approaches in biological dosimetry: urine biomarkers, in: "The Medical Basis for Radiation Accident Preparedness," R. C. Ricks, S. A. Fry, and C. C. Lushbaugh, eds., in press.
3. M. G. Simic, D. S. Bergtold, and L. R. Karam, Generation of oxy radicals in biosystems, Mutation Res., in press.
4. M. M. Elkind and H. Sutton, Radiation response of mammalian cells grown in culture. I. Repair of x-ray damage in surviving Chinese hamster cells, Radiat. Res. 13:556-593 (1960).
5. C. von Sonntag, "The Chemical Basis of Radiation Biology," Taylor and Francis, New York (1987).
6. M. G. Simic, S. V. Jovanovic, Free radical mechanisms of DNA base damage, in: "Mechanisms of DNA Damage and Repair," M. G. Simic, L. Grossman, A. C. Upton, eds., Plenum Press, New York (1986) pp. 39-50.
7. S. V. Jovanovic and M. G. Simic, Mechanisms of OH radical reaction with thymine and uracil derivatives, J. Am. Chem. Soc. 108:5968-5972 (1986).
8. M. G. Simic, Mechanisms of inhibition of free-radical processes in mutagenesis and carcinogenesis, Mutation Res. 202:377-386 (1989).
9. E. C. Friedberg, "DNA Repair," W. H. Freeman & Co., New York (1985).
10. M. Dizdaroglu and D. S. Bergtold, Characterization of free radical-induced base damage in DNA at biologically relevant levels, Anal. Biochem. 156:182-188 (1986).
11. R. Cathcart, E. Schwiers, R. L. Saul, and B. N. Ames, Thymine glycol and thymidine glycol in human and rat urine: A possible assay for oxidative DNA damage, Proc. Natl. Acad. Sci. (USA) 81:5633-5637 (1984).
12. K. C. Cundy, R. Kohen, and B. N. Ames, Determination of 8-hydroxydeoxy-guanosine in human urine: A possible assay for in vivo oxidative DNA damage, in: "Oxygen Radicals in Biology and Medicine," M. G. Simic, K. A. Taylor, J. F. Ward, and C. von Sonntag, eds., Plenum Press, New York (1988) pp. 479-482.

REDOX PARAMETERS ASSOCIATED TO CYTOTOXIC AND ANTITUMOR ACTIVITIES IN THE

SERIES OF ANTITUMOR DRUGS ELLIPTICINES AND DERIVATIVES

Christian Auclair

Laboratoire de Biochimie-Enzymologie, INSERM U 140, CNRS

UA 147, Institut Gustave Roussy, 94805 Villejuif Cedex
France

INTRODUCTION

Generation of virtual DNA breaks through the alteration of the catalytic activity of topoisomerase II are displayed by the most successful
antitumor drugs in clinical use including adriamycin, m-AMSA, hydroxy-
ellipticine derivatives, mitoxantrone and VP-16 (Nelson et al., 1984;
Tewey et al., 1984; Rowe et al., 1986). This feature is thought to be
responsible for the selective cytotoxic effect leading to the antitumor
activity of these drugs. comparative studies of the physicochemical properties of antitumor topoisomerase inhibitors clearly show that the single
parameter that they share in common is the capability to be oxidized to
reactive metabolites through one-electron transfer process (Auclair, 1987).
Along this line, most of them are able to undergo autoxidation generating
oxy-radicals (Dorowshow, 1983; Auclair et al., 1983a; Auclair, 1987;
Nakasawa et al., 1985; Kovacic et al., 1986) and are substrate for peroxidases (Auclair & Paoletti, 1981; Auclair et al., 1986; Auclair, 1987;
Trush, 1982; Shina et al., 1983, 1984; Reszha et al., 1986; Haim et al.,
1987). In view of rational design and screening of new antitumor topoisomerase modifiers, we have attempted to characterize significant para-
meter(s) in terms of redox properties associated to cytotoxic activity on
malignant cells. This study has been made possible in the series of
ellipticine in which are available, number of derivatives displaying
various cytotoxic activity associated to limited structural modifications.

RESULTS AND DISCUSSION

Structural requirement for cytotoxic activity in the series of ellipticine

In the series of ellipticine derivatives and related molecules (Table
1), drugs exhibiting the highest cytotoxic activity in vitro and efficient
antitumor property against experimental tumours in vivo are found among
those displaying a phenolic group at the nine position on the chromophore,
leading to the occurrence of a paraaminophenol structure (Paoletti et al.,
1979; Auclair et al., 1983b; Auclair, 1987).
Furthermore, the non-hydroxylated antitumor compounds of the series,
namely the natural alcaloid ellipticine and the 9-methoxy derivatives have
been found to be respectively hydroxylated or demethylated in vivo yielding
in both cases the active 9-hydroxy compounds. However, among 9-hydroxylated
compounds, only those having two methyl groups either at the 11 and 5 or at

Table 1. Structure of ellipticine derivatives

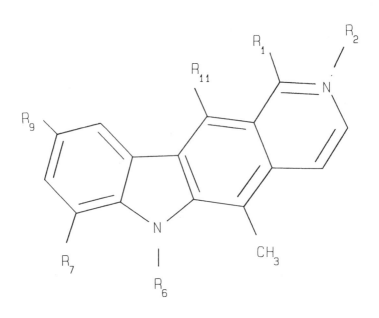

Names	R_1	R_2	R_6	R_7	R_9	R_{11}	Abbreviation
9-hydroxyellipticine	H	–	H	H	OH	CH_3	9-OH-E
9-hydroxyolivacine	CH_3	–	H	H	OH	H	9-OH-E
7-hydroxyellipticine	H	–	H	OH	H	CH_3	7-OH-E
6-N-Methyl-9-Hydroxyellipticine	H	–	CH_3	H	OH	CH_3	6-Me-9-OH-E
Demethyl-9-Hydroxyellipticine	H	–	CH_3	H	OH	H	DeMe-9-OH-E
2-N-Methyl-9-Hydroxyellipticine	H	CH_3	H	H	OH	CH_3	NMHE
1,2-Dihydro-2N-Methyl-9-Hydroxy-ellipticine	2H	CH_3	H	H	OH	CH_3	DhNMHE

the 1 and 5 positions and four aromatic rings display significant antitumor activity. For example, the suppression of the methyl group either at the 5 or at the 11 position results in a strong decrease of the cytotoxicity and in the disappearance of the antitumor activity. In a similar way, the partial or complete saturation of the pyridine ring yielding the dihydro and tetrahydro derivatives respectively, result in the suppression of the pharmacological activity. Despite these additional requirement the main structural feature which control the antitumor activity in the presence of the hydroxyl group at the nine position. It is of interest to notice that the antitumor topoisomerase inhibitors VP-16 and Daunorubicin display as well a phenolic group at the 4' and 11 position respectively. The absence or the methylation of these groups result in the disappearance of the anti-tumor activity.

Physicochemical and biological parameters resulting from the presence of the OH group on the ellipticine ring

Physicochemical

The main physicochemical feature resulting from the presence of the OH group on the ellipticine chromophore is the possible oxidation of the para-aminophenol structure to quinone imine product (Auclair & Paoletti, 1981). This oxidation has been found to occur through one-electron transfer process leading to the generation of phenoxy radical as intermediate. Both phenoxy radicals and quinone products have been demonstrated to be highly electrophilic and may undergo covalent binding with various biological nucleophiles. In addition, when molecular oxygen acts as electron acceptor, the oxidation of 9-hydroxy compounds results in the generation of oxy-radicals (Auclair et al., 1983). The ability to be oxidized to electro-philic species seems to be a common feature shared by antitumor topo-isomerase inhibitors. The epipodophyllotoxin derivative VP-16 (Shina & Myers, 1984a) mentioned above as well as the anilinoacridine derivative m-AMSA (Jurlina et al., 1987), have been found to be oxidized in biological conditions, to quinone imine and diimine respectively.

Biological

The mechanism by which antitumor ellipticines and related molecules exerted their cytotoxicity is thought to mainly involved the alteration of the catalytic activity of DNA topoisomerase II. This alteration results in the appearance of the so-called cleavable complex corresponding to virtual DNA breaks which can be revealed in cells using the alkaline elution technic. In the presence of DNA and purified topoisomerase II, the drug-induced formation of cleavable complex can be revealed by a denaturating treatment of the enzyme-DNA complex (Nelson et al., 1984). Structure acti-vity relationship study has allowed to demonstrate that oxidizable phenolic group was required to produce a significant amount of cleavable complex (J.M. Saucier, unpublished; Auclair, 1987; Auclair et al., 1988).

Relationship between redox potential and cytotoxic activity in the series of ellipticine

The redox potential of oxidizable ellipticines can be considered as a suitable parameter which characterize the chemical reactivity of the para-aminophenol structure of the drugs. The electrochemical studies of various ellipticines have been previously performed using cyclic voltametry (Moiroux & Ambruster, 1980; Meunier et al., 1987). In various experimental conditions, the anodic sweep always corresponds to the two-electron oxidation of the tested compounds. In acetonitrile/phosphate mixture, the oxidation peaks of hydroxylated drugs range from 0.035 to 0.42 V indicating that they can be easily oxidized. In most cases, because of the rapid poly-merization of the two electron oxidized product, the electrochemical process was either partially or not reversible. In these circumstances, the potential of the anodic signal was the only available electrochemical para-meter. Statistical treatment of experimental data published by Meunier et al. (1987) shows that there is no significant correlation (r=0.44) between the oxidation potential (anodic sweep) of height hydroxylated ellipticines and their cytotoxicity on L1210 cells as expressed in ID_{50}.

Relationship between antioxidant properties of ellipticines and their cyto-toxic activity

The capability of aromatic molecules, (essentially phenolic compounds) to undergo reversible redox reactions through one-electron transfer is often associated to antioxidant properties. The naturally occurring phenol-α-tocopherol (vitamin E) and butylated hydroxytoluene (BHT) are typical example of such molecules. Accordingly the phenolic antitumor drugs 9-OH-ellipticine derivatives (Searle et al., 1983; Rousseau-Richard et al., in press) and VP-16 (Shina et al., 1985) have been found to display anti-oxidant properties. Data summarized in Table II show that the inhibition of AIBN-induced methyl linoleate oxidation, strongly increase from 5-hydroxy-

Table 2. Mobility parameters of the hydrogen atom of phenolic groups

	r_o	D(O-H) (Kcal/mol)	ID_{50} (μM)
Phenol	–	88.2	
α-Tocopherol	0.10	78.2	
5-OH-I	0.28	–	
6-OH-C	0.12	–	
9-OH-E	0.05	79.6	0.015
9-OH-O	0.15	79.4	0.02
DeMe-9-OH-E	0.29	81.8	0.85
6-Me-9-OH-E	0.20	81.5	0.076
7-OH-E	0.34	86.3	4.26
NMHE	0.16	–	0.11
Dh-NMHE	0.36	–	36

r_o: ratio of the initial rate of methyl linolenate oxidation induced by the azo-bis-isobutyronitrile (AIBN) in the presence and in the absence of additives (After Rousseau-Richard, 1989a). D(O-H) Hydrogen bond dissociation energy of the phenolic group (After Rousseau-Richard et al., 1989b). The correlation coefficient r_o between r_o and log ID_{50}=0.94 and between D(O-H) and log ID_{50}= 0.92. ID_{50} is the concentration of drug which inhibits by 50 % the growth rate of L1210 cells in vitro after 48 H incubation. 5-OH-I indicates 5-Hydroxyindole and 6-OH-C indicates 6-Hydroxydimethylcarbazole.

indole to 9-Hydroxyellipticine, the latter drug displaying higher activity than α-tocopherol.

The suppression of the methyl group at the C-11 position of the ellipticine ring or the saturation of the 1-2 bond on the pyridinium of the quaternarized drug 2N-methyl-9-hydroxy-ellipticinium (NMHE) result in a marked decrease of the antioxidant activity. Further investigations have shown that the antioxidant efficiency was under the control of the hydrogen atom mobility of the phenolic group as quantified by the OH bond dissociation energy D(O-H) (Rousseau-Richard et al., 1989). Accordingly, a significant correlation (r=0.87) is observed between r_o and d(O-H). Both parameters were found in turn to be correlated with the cytotoxic activity of the drugs towards cultured L1210 leukemia cells.

CONCLUSION

In the series of ellipticines, the presence of a phenolic group at the 9 position on the chromophore is required for the expression of a high cytotoxic activity on malignant cells. For most hydroxylated drugs, the cytotoxic efficiency seems to be in part under the control of the hydrogen atom mobility of the phenolic group which control in turn the antioxidant activity and in general the ability to undergo one electron oxidation. It is of interest to notice that all antitumor topoisomerase inhibitors including hydroxyellipticines, m-AMSA, mitoxantrone, VP-16 and adriamycine display in their structure a labil hydrogen atom required for their activity. Most of them have been found to display antioxidant property. Antioxidant activity could be therefore a physicochemical parameter useful in the screening of new topoisomerase inhibitors.

REFERENCES

Auclair, C. and Paoletti, C., 1981, Bioactivation of the antitumor drug 9-Hydroxyellipticine and derivatives by a peroxidase-hydrogen peroxide system. J. Med. Chem., 24:289.

Auclair, C., Meunier, B. and Paoletti, C., 1983a, The generation of reactive molecular species during the oxidation of 9-hydroxyellipticine derivatives. Interest of prooxidant compounds in the design of anti-cancer drugs, In: The control of tumor growth and its biological bases, W. Davis, C. Maltoni and St. Tannenberger, Eds., Akademie-Verlag, Berlin.

Auclair, C., Hyland, K. and Paoletti, C., 1983b, Autoxidation of the anti-tumor drug 9-hydroxyellitpcine and its derivatives. J. Med. Chem., 26:1438.

Auclair, C., Dugue, B., Meunier, G., Meunier, B. and Paoletti, C., 1986, Peroxidase-catalyzed covalent binding of the antitumor drug N2-Methyl-9-Hydroxyellipticinium to DNA in vitro. Biochemistry, 25:1240.

Auclair, C., 1987, Multimodal action of antitumor agents on DNA: The ellipticine series. Arch. Biochem. Biophys., 258:1.

Auclair, C., Schwaller, M.A., René, B., Banoun, H., Saucier, J.M. and Larsen, A.K., 1988, Relationship between physicochemical and biological properties in a series of oxazolopyridocarbazole derivatives (OPCd); comparison with related antitumor agents. Anti-Cancer Drug Design, 3:133.

Bachur, N.R., Gordon, S.L. and Gee, M.V., 1978, A general mechanism for microsomal activation of quinone anticancer agents to free radicals. Cancer Res., 38:1745.

Dorowshow, J.H., 1983, Comparative cardiac oxygen radical production by anthracycline antibiotics, mitoxantrone, bisantrene, m-AMSA and neocarcinostatine. Clin. Res., 31:67.

Haim, N., Nemec, J., Roman, J. and Sinha, B.K., 1987, Peroxidase-catalized oxidation of etoposide (VP-16-213) and covalent binding of reactive intermediates to cellular macromolecules. Cancer Res., 15:5835.

Jurlina, J.L., Lindsay, A., Packer, J.E., Baguley, B.C. and Denny, W.A., 1987, Redox chemistry of the 9-anilinoacridine class of antitumor agents. J. Med. Chem., 30:473.

Kovacic, P., Ames, J.R., Lumme, P., Elo, H., Cox, O., Jackson, H., Rivera, L.A., Ramirez, L. and Ryan, M.D., 1986, Charge transfer-oxy radical mechanism for anticancer agents. Anti-Cancer Drug Design, 1:197.

Meunier, G., De Montauzon, D., Bernadou, J., Grassy, G., Bonnafous, M., Cros, S. and Meunier, B., 1987, The biooxidation of cytotoxic ellipticine derivatives: A key to structure-activity relationship studies. Mol. Pharmacol., 33:93.

Moiroux, J. and Ambruster, A.M., 1980, Electrochemical behaviour of ellipticine derivatives Part I. Oxidation of 9-Hydroxy-ellipticine. J. Electroanal. Chem., 114:139.

Nakasawa, H., Andrews, P.A., Callery, P.S. and Bachur, N.R., 1985, Superoxide radical reactions with anthracycline antibiotics. Biochem. Pharmacol., 34:481.

Nelson, E.M., Tewey, K.M. and Liu, L.F., 1984, Mechanism of antitumor drug action: Poisoning of mammalian topoisomerase II on DNA by 4'-(9-acridinylamino)-methanesulfon-m-anisidine. Proc. Natl. Acad. Sci. USA, 81:1361.

Paoletti, C., Cros, S., Dat-Xuong, N., Lecointe, P. and Moisand, A., 1979, Comparative cytotoxic and antitumoral effects of ellipticine derivatives on mouse L1210 leukemia. Chem. Biol. Inter., 25:45.

Rousseau-Richard, C., Auclair, C., Richard, C. and Martin, R., Free radical scavenging and cytotoxic properties in the ellipticine series. Free Rad. Biol. Med., in press.

Rousseau-Richard, C., Auclair, C., Richard, C. and Martin, R., 1989, Correlation between the OH bond dissociation energies of ellipticine hydroxylated derivatives and their cytotoxic properties. FEBS Lett., 252:58.

Reszka, K., Kolodziejczyk, P. and Lown, W.J., 1986, Horse radish peroxidase-catalized oxidation of mitoxantrone: spectrophotometric and electron paramagnetic resonance studies. J. Free Rad. Biol. Med., 2:25.

Rowe, C.T., Chen, G.L., Hsiang, Y.H. and Liu, L.F., 1986, DNA damage by antitumor acridines mediated by mammalian DNA topoisomerase II. Cancer Res., 46:2021.

Searle, A.J.F., Gee, C. and Wilson, R., 1983, Ellipticines and carbazole as antioxidants, In: Oxygen radicals in chemistry and biology, Proc. 3rd. International Conference, Neuherberg. W. Bors, M. Saran and D. Tait, Eds. Walter de Gruyter, Berlin.

Sinha, B.K., 1983, Irreversible binding of quinacrine to nucleic acids during horse radish peroxidase- and prostaglandin synthetase-catalized oxidation. Biochem. Pharmacol., 32:2604.

Sinha, B.K. and Myers, C.E., 1984a, Irreversible binding of etoposide (VP-16-213) to desoxyribonucleic acid and proteins. Biochem. Pharmacol., 33:3725.

Sinha, B.K., Trush, M.A., Kennedy, A. and Mimmaugh, E.G., 1984b, Enzymatic activation and binding of adriamycin to nuclear DNA. Cancer Res., 44:2892.

Sinha, B.K., Trush, M.A. and Kalyanaraman, B., 1985, Microsomal inter-actions and inhibition of lipid peroxidation by etoposide (VP-16-213) implications for mode of action. Biochem. Pharmacol., 34:2036.

Tewey, K.M., Rowe, T.C., Yang, L., Halligan, B.D. and Liu, L.F., 1984, Adriamycin-induced DNA damage mediated by mammalian topoisomerase. Science, 226:466.

Trush, M.A., Mimmaugh, E.G. and Gram, T.E., 1982, Activation of pharmaco-logic agents to radical intermediates. Implication for the role of free radicals in drug action and toxicity. Biochem. Pharmacol., 31:3335.

PEROXIDASE INDUCED METABOLISM AND LIPID PEROXIDE SCAVENGING BY ANTITUMOR

AGENTS

Pawel Kolodziejczyk and J. William Lown

Department of Chemistry, University of Alberta, Edmonton, Alberta, Canada
T6G 2G2

INTRODUCTION

Oxygen is often viewed as being synonymous with life. It is necessary for the living processes of mammals including humans. The arrest of breathing, apart from termination of other functions, is an evident sign of mammalian death. However, soon after death, the body begins to re-utilize oxygen and in a short period it consumes oxygen at much higher rate than in normal life processes. But this process no longer represents controlled biological oxidation. Rather it is uncontrolled autooxidation, peroxidation or free radical oxidation of polyunsaturated lipids. In recent years, interest has grown in the possible links between drug metabolic reactions and certain aspects of lipid biochemistry. Attention has been focused particularly on oxidative processes, namely those of lipid peroxidation and co-oxidation. Lipid peroxidation is now recognized as an important mediator of toxicity.[1] Reactive oxygen species (superoxide anion radical, hydroxyl radical, singlet oxygen and hydrogen peroxide) are produced during enzymatic oxidation of xenobiotics by cytochrome P-450 and other oxidoreductase systems.[2]

Additionally, it is now appreciated that under certain circumstances co-oxygenation involving oxidized lipid species may be important in drug oxidation. This has been shown with peroxide intermediates of prostaglandin biosynthesis and with peroxides and hydroperoxides of polyunsaturated fatty acids produced within microsomal membranes by the peroxidative action of cytochrome P-450.[3] Because of the instability and complexity of free radical oxidation products, their characterization and function is still far from being completely understood. However there is no longer any doubt that free radical processes play a key role in many vital functions of living organisms.

There are no simple answers to questions concerning the role of free radicals in biological systems. These species are both life-threatening to an organism as well as being essential for its normal function. They are extraordinarily powerful in a wide range of biological systems. The presence of free radicals has been established both in physiological and pathological states.[4] They are involved in fat metabolism, prostaglandin synthesis, and detoxication processes. Free radical activity is an essential part of phagocytosis, and free radical reactions may be a chemical link between cell damage and the inflammatory response. Free radicals are the reactive species used in destroying damaged cells from living organisms.

The fact that free radicals are not usually able to destroy normal cells is attributed to compartmentalization and maintenance of structural integrity. Many of the elements necessary to perform efficient free radical oxidation are present in the cells; and include: unsaturated lipids, oxygen, transition metals known to function as catalysts in the generation of chain-initiating radicals, and enzymes catalyzing oxidative processes. In an *in vitro* environment mixtures of these components are able to produce free radicals and promote oxidation reactions. It is because of the substantial potential energy of structural organization that living cells survive the oxidative stress. However, when this structural integrity is destroyed and the components are no longer prevented from interacting, then irreversible damage occurs. Therefore regulatory antioxidative mechanisms are essential for protecting cells from free radical oxidation.

Antioxidants in Therapy and Preventive Medicine
Edited by I. Emerit *et al.*
Plenum Press, New York, 1990

323

In addition to the structural integrity and compartmentalization elements of cells, enzymatic systems play protective roles against potentially damaging oxidants. Superoxide dismutases, widely distributed in living cells, act as scavengers of $O_2^{\cdot-}$, catalase protects against hydrogen peroxide, and glutathione peroxidase and glutathione S-transferase are also involved in antioxidative protection. Other protective agents include ceruloplasmin, transferrin, and α-tocopherol. These regulative systems function very efficiently under physiological conditions. However in pathological states antioxidant protection can fail for a variety of reasons, just as many different abnormalities can cause other homeostatic disturbances. Extensive oxidation processes are known in some tumor diseases where elevated levels of lipid peroxides have been observed.[5,6]

On the other hand, introduction into cells of anticancer drugs and other xenobiotics also affects homeostasis. These drugs may often be directly involved in the generation or capture of oxidizing free radicals. Thus certain drugs are able either to suppress or dramatically enhance free radical generation and therefore can play an important role in cancer chemotherapy. Moreover, it should be remembered that when prooxidants are introduced into living systems the normal prooxidant-antioxidant equilibrium of this system is altered. Therefore, existing protective mechanisms in tissues can be overwhelmed and the tissue may not be able to regenerate a sufficient quantity of antioxidant immediately. Such chains of events are now well established for anthracycline based antitumor therapy.[2]

There is a substantial amount of literature devoted to oxygen free radicals generated in mammalian systems as a result of chemotherapy. These radicals appear to be associated with base damage[7,8] and strand breakage of DNA. Chromosome damage and sister chromatid exchange by superoxide generating systems was demonstrated in human lymphocytes.[9]

The toxicity of oxygen radicals is generally linked to the initiation of lipid peroxidation chain reactions in membranes. Oxidized lipids can react directly with DNA or lead to further oxidative membrane damage resulting in stimulation of the arachidonic acid cascade and phospholipase A_2 activity. It is noteworthy that chromosomal breakage caused by phorbol myristate acetate, known to be a tumor promoter, could be prevented not only by antioxidants but also by inhibitors of arachidonic acid metabolism. This observation shed light on certain aspects of carcinogenesis and cancer chemotherapy. However this area of research is still subject to intensive study.[10] Nevertheless it is evident that relationships exist between the oxidative activation and biotransformation of anticancer agents, free radical formation and their involvement in lipid metabolism. The implications of these molecular processes for chemotherapy, especially for free radical generation and scavenging, and their role in cell enzymatic processes is discussed in this chapter. Examples of oxidative activation of xenobiotics by peroxidases and microsomal P-450 enzymatic systems are also presented.

ANTHRACYCLINE AND ANTHRACENEDIONE ORIGINATED FREE RADICALS

It is now well established that many xenobiotics are metabolized in living organisms. This not only affects the detoxication, but also may result in formation of reactive intermediates which subsequently manifest toxicity by interacting non-enzymatically with vital intracellular molecules, such as proteins, enzymes, and nucleic acids. Enzymatic activation of xenobiotics usually initiates this chain of events.

Anthracyclines, which are widely used as anticancer agents, have been extensively studied with respect to redox cycling.[2] In general, the drug is reduced in a one-electron step to a reactive intermediate, which is able to transfer one electron to molecular oxygen, resulting in formation of superoxide anion radical (Figure 1). The superoxide radical can undergo a variety of reactions. It spontaneously dismutates to hydrogen peroxide and dioxygen whereby singlet oxygen may be produced. On the other hand, the subsequently formed hydroxyl radical is the most reactive oxygen metabolite and is suggested to be responsible for some of the most serious damage occurring during redox cycling processes, e.g. peroxidation of membrane lipids, proteins and damage to DNA.[11,12] The enzymes involved in such activation are primarily membrane-bound flavoproteins like microsomal and mitochondrial NADPH, cytochrome P-450 reductase, mitochondrial NADH dehydrogenase, xanthine oxidase, and ferredoxin reductase. These enzymes have relatively low substrate selectivity and many other xenobiotics are able to maintain a redox cycle in their presence. Radical generating reactions have been demonstrated in all mammalian tissues examined thus far, including liver, heart, kidney, lung, spleen, and erythrocytes.[13]

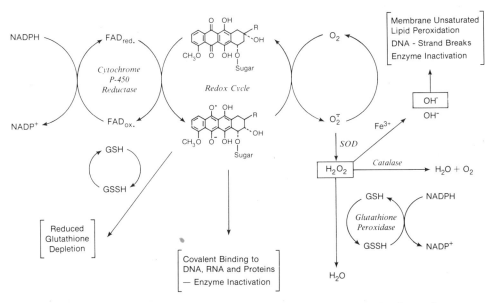

Figure 1. Scheme depicting free radical formation in redox-cycling of anthracyclines.

The anthracycline antibiotics, daunorubicin and doxorubicin (Adriamycin), are employed in the treatment of various solid and hematologic malignancies. However, their clinical use has been limited due to risk of cumulative and dose-dependent cardiotoxicity The molecular basis of antitumor action of the anthracyclines, as well as their cardiac toxicity, are commonly explained, at least in part, by reductive activation of the drug with cytochrome P-450 NADPH-reductase. Cardiac tissue, however, seems to be particularly susceptible to prolonged oxidative attack possibly because of the relatively low levels of antioxidants or protective enzymes in the heart. Thus, a connection between radical production and cardiotoxic effects of anthracyclines is indicated. The attack of reactive oxygen species on polyunsaturated fatty acids, essential constituents of biological membranes, has been shown to result in peroxidative damage of these lipids, leading to the formation of a number of peroxidative lipid breakdown products, including lipid hydroperoxides.[14]

Numerous anthracycline analogues have been prepared with the objective of separating the toxic side-effects from the cancer cytotoxicity.[14] Among these, 5-iminodaunorubicin has attracted attention because of its markedly lower cardiotoxicity when compared with the parent drug. Two related new classes of anticancer agents, namely anthracenediones and anthrapyrazoles, were also shown to exhibit lower cardiotoxic effects. Some of these less cardiotoxic agents are shown in Figure 2.

The lower cardiotoxicity of these compounds has been explained by the alteration of their anthraquinone system, resulting in structures that are more difficult to reduce and reoxidize. Indeed, electrochemical measurements demonstrate that mitoxantrone, ametantrone, 5-iminodaunorubicin, and anthrapyrazoles are more resistant to reduction than the corresponding anthracyclines, and this is reflected in the more negative polarographic reduction potentials for these compounds compared with daunorubicin.[12]

However our recent studies revealed another pathway common to these less cardiotoxic agents. We found that mitoxantrone, ametantrone, anthrapyrazoles, and 5-iminodaunorubicin undergo enzymatic oxidation, a process which is not observed for daunorubicin or doxorubicin. This potential for oxidative activation has not been recognized hitherto in the metabolism of anthracyclines or anthrapyrazoles. However, such enzymatic activation, accompanied by the generation of free radicals, was recently reported for the structurally related anticancer agent mitoxantrone.[15,16] Horseradish peroxidase and hydrogen peroxide have been used as the oxidizing system and in some cases EPR techniques have been employed to detect free radical metabolites.[17]

325

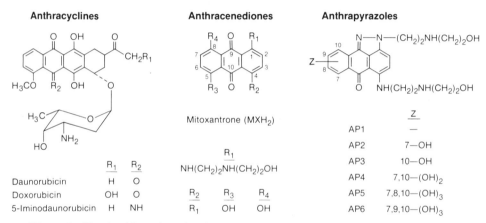

Figure 2. Structural formulae of anthracyclines, anthracenediones and anthrapyrazoles.

THE FORMATION OF N-CENTERED FREE RADICALS

Mitoxantrone is of clinical interest because it is believed to be less cardiotoxic than doxorubicin. Kharash and Novak have shown that mitoxantrone and other anthracenediones are poor substrates for microsomal cytochrome P-450 reductase.[18] These drugs do not increase NADPH utilization or $O_2^{\bullet-}$ formation in hepatic microsomes. Moreover mitoxantrone (and ametantrone) inhibit anthracycline-stimulated oxygen consumption and, what is even more remarkable, inhibit microsomal lipid peroxidation. On the other hand, mitoxantrone was also anticipated to undergo free radical formation, by analogy with daunorubicin, in a metabolic one-electron reduction. Indeed EPR signals of free radicals have been reported.[19] However, in contrast to daunomycin almost no basal NADPH consumption and $O_2^{\bullet-}$ production was observed. Therefore several questions arise concerning the mechanism of action of these drugs. Studies carried out in these laboratories in recent years brought about an alternative explanation, and revealed a new mechanism of free-radical formation for these drugs.[15-17,20] Horseradish peroxidase and hydrogen peroxide are often employed to mimic oxidative biological systems.[21] It was found that, in this medium, the characteristic absorption spectrum of mitoxantrone disappeared and a strong stable EPR free radical signal was observed. Additional studies revealed that mitoxantrone undergoes biotransformation resulting in the formation of a metabolite, MH2, hexahydronaphtho-[2,3-f]quinoxaline-6,12-dione in which one side chain has cyclized to the chromophore of MH2. The latter is also subject to redox cycling reactions, as shown in Figure 3, and concomitant formation of the action free radical species MH2+• was shown to be generated in this process.[16]

It was demonstrated that the N-centered free radical MH2+• may be formed in the direct electron transfer between the fully reduced and fully oxidized forms of the metabolite. The formation of the N-centered free radical MH2+•, which may be easily regenerated from both its reduced MH2 and oxidized MH2 2+ forms in a redox process, may be of significance for its mode of action. In particular, the strongly electrophilic and relatively long-lived free radical MH2+• may bind covalently to biomolecules, particularly proteins (including enzymes), thereby affecting their action(s) in a manner which is discussed below.

Another antitumor agent, 5-iminodaunorubicin (5-IMDR), was found to be substantially less cardiotoxic compared to its parent daunorubicin.[12] Apart from electrochemical evidence for a slightly higher one-electron reduction potential of 5-IMDR vs daunorubicin, there was no convincing rationale explaining such a substantial difference in their biological properties. Applying an approach similar to the one used for the mitoxantrone study, it was found that 5-IMDR also undergoes facile oxidation with concomitant formation of a N-centered free radical[17] (Figure 4). However, in contrast to mitoxantrone, the oxidation process of 5-IMDR is not reversible (or is only partially reversible), which may be explained by the lower stability or higher reactivity of the 5-IMDR-derived free radical. More detailed discussion on free radical formation, its structure and further transformation is given elsewhere.[17]

Enzymatic oxidation of some anthrapyrazoles was found to be a reversible process (e.g. AP 4 and AP 6, i.e. dihydroxy- and trihydroxy-substituted anthrapyrazoles); while mono-substituted

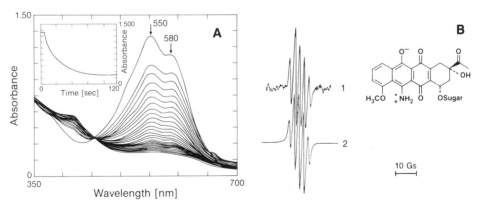

Figure 3. Schematic depiction of HRP-dependent oxidation and transformation of mitoxantrone, Mitoxantrone (MXH$_2$) is oxidized to a radical cation species MXH$_2$$^{\bullet+}$ followed by cyclization to radical cation MH$_2$$^{\bullet+}$ and establishing redox equilibria of the latter with isolated metabolite MH$_2$ and its unstable fully oxidized form MH$_2$$^{+2}$, Direct electron transfer between mitoxantrone metabolite and corresponding diimino form (MH$_2$$^{+2}$) in disproportionation-comproportionation reactions is indicated.

anthrapyrazole AP 2, undergoes irreversible oxidation.[20] There is an evident correspondence between our results and reported differences in inhibition of anthracycline-stimulated lipid peroxidation by anthrapyrazoles.[22] Anthrapyrazoles, with hydroxyl substituents in the 7 and 10 positions, effectively inhibit doxorubicin-stimulated lipid peroxidation. In contrast, a 7-monohydroxy anthrapyrazole, AP 2, was relatively ineffective in suppressing such peroxidation. Again, as in the case of mitoxantrone, it can be predicted that the reversible redox system may play an important role in the biological activity of anthrapyrazoles.

The object of our further study was to evaluate the evidence on the involvement of N-centered free radicals in scavenging primary or induced lipid peroxides and neutralization of these potentially damaging species by an oxidative pathway. It is conceivable, as we suggested earlier, that N-centered free radicals may bind covalently to proteins, specifically enzymes, and this may result in the inactivation of particular enzymatic systems by "suicidal inhibition", a process known to accompany oxidative metabolism of certain drugs.[16,20,23,24]

Figure 4. Enzymatic oxidation of 5-iminodaunorubicin: (a) The absorption spectrum changes reflect the oxidation caused by H$_2$O$_2$/HRP system, The incubation mixture contained 5-IMDR (140 μM), H$_2$O$_2$ (140 μM) and HRP (100 units) in phosphate buffer pH 6. Spectra were recorded every 3 seconds, arrows indicate direction of changes, The insert shows changes of the absorbance at 550 nm *vs* time, (b) EPR signal (1) of free radicals formed from 5-IMDR during oxidation and computer simulated signal (2).

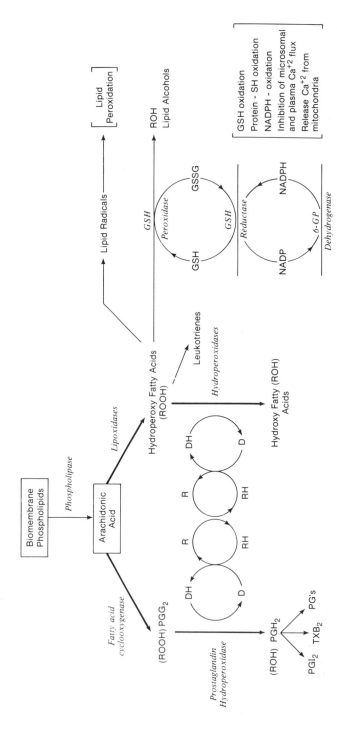

Figure 5. Proposed mechanism of alteration of lipid metabolism by drugs (D) undergoing oxidative activation. Arachidonic acid is enzymatically released from membrane phospholipids by phospholipases and initiates two biosynthesis pathways: (a) cyclooxygenase (function of PG synthase); (b) lipoxygenase. Accumulated hydroperoxy products: PGG$_2$ (pathway a) or hydroperoxyacids (pathway b) can cause cooxidation of drugs in reaction catalyzed by peroxidase function of enzymes involved in prostaglandin biosynthesis. Possible inhibition of enzymes by drug originated radicals may be reflected by alteration in prostaglandin and thromboxane production. On the other hand, drug-mediated removal of hydroperoxy fatty acids (pathway b) may affect the formation of lipid radicals and prevent lipid peroxidation processes. Regeneration of drugs, by oxidative cycling, can be afforded at the expense of the cell reductive systems (RH).

PEROXIDES IN TUMOR PROMOTION, CHEMOTHERAPY AND REGULATION

In addition to phorbol esters, which are known to be tumor promoters, various organic peroxides have also been found to effectively promote tumor formation.[10] Most inhibitors of the tumor promotion process, i.e. anti-inflammatory steroids, retinoids, protease inhibitors, prostaglandin synthase inhibitors, inhibitors of polyamine biosynthesis, and antioxidants have also been found to be effective inhibitors of tumor promotion.[25] Studies on the mechanism of tumor promoting agents confirmed the involvement of lipid peroxides and free radical intermediates in tumor promotion.[25] On the other hand, extensive lipid peroxidation and free radical injury have been observed during cancer chemotherapy. Oxygen free radical generation and its consequences such as lipid peroxidation play a major role in the pathogenesis of the injury occurring after ischemia and reperfusion. Reperfusion injury, appears to be due to a burst of free radical production during reoxidation, caused by autooxidation of reduced substances accumulated during ischemia.[26] Enzymatic activation of xenobiotics usually initiates a chain of events leading to free radical generation and lipid peroxidation.[26]

As has been mentioned above, anthracyclines have been extensively studied with respect to redox cycling - dependent production of lipid peroxides and free radicals. Metabolism of oxygen, which is considered to be relevant to cancer therapy is also directly related to cell death. The mechanism by which oxygen species kill cells is still obscure; however the view that oxygen toxicity is mediated directly by the superoxide radical has little support, owing to its low reactivity.[27]

It is evident that OH• radicals can destroy key molecular targets within the cell, such as DNA, if they are produced in close proximity to the target. It has been reported that anticancer agents like anthracyclines or anthracenediones[28] induce strand breakage in DNA via OH• radicals and that this may indeed be related to cell killing. These drugs also bind metals such as iron and copper. The latter undergo redox reactions, during the course of which OH• radicals are released, and the latter attack deoxyribose moieties which indirectly results in cleavage of the phosphate ester links in DNA. However, the high reactivity of OH radicals places great restrictions on their range of action (effectively the diffusion limit) and therefore OH• radicals produced within the cytoplasm have essentially no chance of reacting with DNA located in the nucleus or in mitochondria. Extracellularly produced OH• radicals cannot contribute significantly to cell killing because they will be scavenged before they reach a suitable target. However species like H_2O_2 and hydroxyperoxides (lipid hydroxyperoxides) which have longer life span are much better candidates for participating in cell killing. Hydroxyl radicals produced within the cell may be involved and it cannot be excluded that toxicity of H_2O_2 and lipid hydroperoxides may be mediated by OH• These active species may be responsible for the inactivation of important enzymatic systems (cytochrome P-450, detoxication enzymes, prostaglandin synthase oxidoreductases, etc.).[23,24] The biochemical changes within cells may alter their response to oxidizing agent.

There are recent indications that lipid peroxidation, apart from its cell damaging effect, is also related to the normal function of cells.[29] Furthermore, there is accumulating evidence that lipid peroxides, especially those of arachidonic acid, may be a key factor in an important regulatory mechanism in normal cell function. The important bioregulators, prostaglandins, and thromboxanes, are biosynthesized from polyunsaturated acids. In the first step, by the oxygenation process, a hydroperoxyendoperoxide prostaglandin G (PGG_2) is formed, which is then reduced to hydroxyendoperoxide PGH_2 (see Figure 5). Both reactions are catalyzed by an enzyme prostaglandin H synthase which exhibits both cyclooxygenase and peroxidase activities.[29] Those two activities are apparently in conflict, because the same enzyme is evidently responsible for both generating and removing a peroxide. In cyclooxygenase activity, which is hydroperoxide-initiated and comprises a free radical chain reaction resembling the non-enzymatic autooxidation of polyunsaturated acids, peroxide serves as a trigger for amplified production of peroxide. In the peroxidase activity of the enzyme, peroxides are reduced to the appropriate alcohols. It seems possible that many pathophysiological events are triggered by trace amounts of peroxides which in turn accelerate the production of lipid peroxides (or radicals formed therefrom) which might act directly on cell membranes and/or stimulate fatty acid oxygenases to produce specific eicosanoids.

In the second stage of prostaglandin biosynthesis the endoperoxide moiety of PGH_2 is transformed via metabolizing enzymes into functionalities characteristic of prostaglandin and thromboxanes. The metabolism of PGH_2 is determined by the levels of the metabolizing enzymes, and these are tissue dependent. For example, platelets make mainly thromboxane, whereas arterial endothelial cells make mainly prostacyclin. Regardless of which specific prostaglandin or thromboxane is biosynthesized, virtually all mammalian tissues display some prostaglandin H synthase activity. Substantial experimental evidence links prostanoids with cancer promotion, growth and metastasis.[30] In this regard, the anti-aggregatory PGI_2 and its endogenous antagonist, TXA_2, play an important role in tumor growth and spread. Their endogenous balance regulates the interaction between the circulating platelets and vascular endothelium.[30]

It has been shown that breast cancer tissue produces increased amounts of prostaglandins, especially the levels of PGI_2 and TXA_2 which were found to be three to six times higher than those in normal tissue.[31] Recent reports provide evidence for increased PGI_2 and TXA_2 production in ovarian malignancies. Interestingly, TXA_2 stimulation by human ovarian malignancy was almost three times larger than that of PGI_2, therefore a rise in prostanoid production is not a reflection of overall metabolic stimulation of the archidonic acid pathway but shows that a rather more selective process is involved.[32]

It was also shown that certain aromatic amines and substituted phenols can act as cyclooxygenase inhibitors.[33] These observations may be of great importance for understanding drug metabolism, because they imply that even when PGG_2 synthesis is blocked, some drugs could be metabolized in a similar way as occurs with other peroxidases (i.e. catalases or HRP discussed earlier). The precise reason for the inhibition of cyclooxygenase activity is not certain, but such inhibition may be caused by the covalent binding of oxidized or free radical forms of the drug to the enzyme. Such developments create quite a new situation in the enzymatic cellular environment. The fact that the first step of prostaglandin biosynthesis is "turned off" may affect not only further biosynthesis of prostaglandins by insufficient production of PGG_2, but may also affect other hydroxyperoxide utilizing enzymatic systems (e.g. glutathione peroxidase).

Peroxidase activity was found in several enzymatic systems including, of course, peroxidases but also catalases, ceruloplasmin, prostaglandin synthase and several other systems. But most remarkable is probably the peroxidative activity of cytochrome P-450 isoenzymes. In addition to catalyzing various O_2-dependent reactions, cytochrome P-450 also bring about the oxygenation of organic substrates at the expense of peroxy compounds. Several xenobiotics undergo reaction with a number of oxidants, such as alkylhydroperoxides, peroxyacids, iodobenzene, and N-oxides, which involve cytochrome P-450. Although the reaction of cytochrome P-450 isoenzymes with peroxides differs in several important respects from that of peroxidases (e.g. homolytic mechanism of oxygen-oxygen bond cleavage for P-450 *vs* the heterolytic process generally observed for peroxidases), the involvement of cytochrome P-450 in reactions similar to peroxidases can be expected.

DRUG BIOTRANSFORMATION *VIA* HYDROPEROXIDE-DEPENDENT OXIDATIONS INVOLVING ENZYMATIC PEROXIDATIVE FUNCTIONS

Drug biotransformation is frequently termed detoxication and detoxication enzymes are often used to denote the catalyst(s) involved in this process. These terms are, however, not entirely satisfactory, since the products formed are often biologically more active and toxic than the present compound. In fact, most, if not all, of the reactions generally referred to as detoxication reactions can result in generation of reactive and toxic metabolites. This is particularly important for cytochrome P-450 catalyzed reactions which are known to be critical events in the metabolic activation of a variety of drugs and chemical carcinogens.

The cytochrome P-450-linked monooxygenase system consists of families of inducible isoenzymes with broad and overlapping substrate specificities. In particular, the cytochrome P-450-linked system has been studied extensively in recent years and many of its properties are now known. Lipid peroxides can be substituted for NADPH and molecular oxygen in the cytochrome P-450 catalyzed oxidation of some substrates. It was shown that the mechanism of hydroperoxide-dependent oxidations is different from that of NADPH-mediated reactions.[34]

Interesting from the physiological point of view are reports describing cytochrome P-450 hydroperoxyeicosatetraenoic acid (15-HPETE) with rat liver microsomes in the absence of NADPH.[35] Under these conditions the only reactants were apparently the hydroperoxide and molecular oxygen. It was found that addition of 15-HPETE to liver microsomes prepared from phenobarbital treated rats results in rapid uptake of molecular oxygen. The stoichiometry of the reaction was shown to correspond to one equivalent of oxygen consumed for each equivalent of hydroperoxide acid added. In addition N,N-dimethylaniline, a substrate for cytochrome P-450, inhibited the oxygen uptake, undergoing N-demethylation in the process. Our observations pointed to evident differences between daunorubicin and mitoxantrone or anthrapyrazole (AP 4) in inhibition of oxygen uptake by microsomes, incubated in the presence of lipid hydroperoxides. The effect of antitumor agents daunorubicin, mitoxantrone, and anthrapyrazole AP4 on oxygen uptake by microsomes in the presence of 15-HPETE is shown in Figure 6.

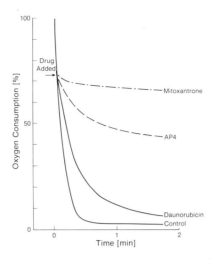

Figure 6. Inhibition of 15-HPETE metabolism in rat hepatic microsomes by antitumor agents, [Drug] = 100 μM, [15-HPETE] = 100μM, [cytochrome P-450] = 1.4 μM in 0.1 M TRIS buffer pH = 7.4, Ambient temperature.

Daunorubicin, which is known to be subject to reductive activation by NADPH-dependent cytochrome P-450 reductase, showed no apparent effect on oxygen consumption by microsomes in the presence of 15-HPETE. In contrast mitoxantrone and anthrapyrazole AP 4, which are known to undergo oxidative metabolism,[20] inhibit oxygen consumption and therefore inhibit cytochrome P-450 -dependent metabolism of 15-HPETE. These findings, indicating the ability of less cardiotoxic agents, to affect the metabolism of lipid hydroperoxides may be of significant value in understanding the differences between the mode(s) of action of these drugs. These observations confirm that antitumor drugs like mitoxantrone and anthrapyrazoles, which are subject to oxidative rather than reductive metabolism, may be involved in regulatory mechanisms, governing eicosanoid biosynthesis. In the light of the role of prostaglandins in regulation, and cancer promotion and its therapy, discussed above, these findings may open a promising avenue for further investigation.

Additional support for the hypothesis presented above, concerning the role of antitumor agents in the eicosanoid-based regulatory mechanisms, comes from our study on inhibition of lipoxygenase activity by these agents. Arachidonic acid is metabolized by lipoxygenases in addition to cyclooxygenase.[29] Soybean lipoxygenase (E.C.1.13.11.12) can be used for the assays of lipoxygenase activity, since it is a commercially available enzyme. In our preliminary experiments we have compared the inhibition of certain anticancer agents discussed above, with those of indomethacin, an inhibitor of lipoxygenase. The reaction of enzyme with arachidonic acid

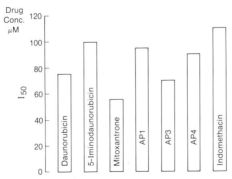

Figure 7. Effect of anticancer agents on arachidonic acid [AA] metabolism by soybean lipoxygenase [SL]. Bars represent drug concentration caused 50% inhibition (I_{50}) of diene formation, followed spectrophotometrically at λ = 234 nm, [AA] = 1 mM, [Drug] = 20-160 μM, [SL] = 10 unit/mL in 0.2 M borate buffer pH = 9.0.

was followed spectrophotometrically. The appearance of conjugated diene at 234 nm was recorded *vs* time for several drug concentrations. The results of these experiments are summarized in Figure 7 .Mitoxantrone was found to be the most potent inhibitor of soybean lipoxygenase (IC_{50}; 55 µM) and all drugs tested have shown inhibition effects greater than those of indomethacin, for which I_{50} was 110 µM. However it has to be born in mind that the results reported here can not be directly extrapolated to mammalian lipoxygenases. Their specific activities and the effects of inhibitors may vary. Nevertheless such an assay can be helpful for evaluation of lipoxygenase inhibition and has been applied in studies of non-steroidal antiinflammatory drugs.[36]

These studies demonstrate the role of xenobiotics in the regulation of cyclooxygenase activities, in addition to their involvement in peroxidative enzymatic functions established hitherto. Owing to our insufficient knowledge of the metabolic pathway of these drugs at present, one can only speculate on the reason for enzyme activation and try to find an answer by comparing analogous systems.

NITROGEN-CONTAINING INHIBITORS OF OXIDATIVE SYSTEMS

The biotransformation of drugs and other xenobiotics has important consequences for their biological activity. In many instances, biotransformation results in detoxication and facilitates elimination of potentially toxic agents. The reverse phenomenon, bioactivation of a substance to a more toxic species, is also well documented. In the case of many drugs, biotransformation becomes a limiting factor in the expression of pharmacological activity. The importance of biotransformation is not limited to response of the organism to xenobiotics. The capacity to activate an endogenous substance to a biologically effective form and/or to terminate its action by metabolic processes provides a sensitive and vital mechanism for the maintenance of homeostasis.

The cytochrome P-450 isoenzymes and related enzymes like plant peroxidases, and prostaglandin H synthase, represent important pathways for the biotransformation of both endogenous and exogenous chemicals.

Since most suicide inactivation processes involve alkylation of a nucleophilic group at the enzyme active site by an enzyme-generated electrophile, then nitrogen containing compounds are good candidates for the ultimate inactivating species.[37]

The action of cytochrome P-450 on secondary aliphatic amines usually results in α-hydroxylation and formation of a carbinolamine. The carbinolamine can either dissociate to a less substituted amine and a carbonyl compound or dehydrate to a Schiff base (imine). It was proved that electrophilic imines are indeed formed during the metabolism of nicotine and other nitrogenous compounds. Aromatic amines are even more prone to undergo oxidation with concomitant formation of imines. Electron exchange between amine and imine forms has often been observed to give a rise to monocationic radicals.[16] Iminocation radicals are well known as intermediates in chemical and electrochemical oxidations of amines. Recently more evidence has been obtained on peroxide-mediated oxidation of aromatic amines by various hemoproteins including cytochrome P-450.[38] Our studies proved the formation of such intermediates for mitoxantrone and 5-iminodaunorubicin.[20] The postulate that iminium radicals are intermediates in enzymatic control mechanisms has several virtues. As elaborated later, it provides a mechanism which can account for the inactivation of both cytochrome P-450 and prostaglandin synthase (including cyclooxygenase activity).

Interestingly, most of those N-centered radicals are formed by oxidative enzymatic systems: peroxidases, cytochrome P-450 or prostaglandin H synthase. Moreover, reaction of free-radical species with biomolecules has frequently been suggested and, in some cases, inhibitory effects on enzymes which generated them have been reported.[23,24,37]

COOXIDATION OF XENOBIOTICS

It is well documented that many xenobiotics undergo oxidation by enzymatic systems acting under normal physiological conditions. The most extensively studied systems include: peroxidases (lactoperoxidase, cytochrome peroxidase), cytochrome P-450 isoenzymes and prostaglandin synthase. The metabolism of xenobiotics by these systems is peroxidative in nature and often termed cooxidation. Many of the compounds which may serve as reducing cofactors for enzymes, especially those of pharamaceutical and toxicological interest, are metabolized to oxygenated products during cooxygenation. Such cooxidation of chemicals by enzymatic systems occurs by a free radical mechanism. Many amines undergo oxidation resulting in formation of N-centered free radical intermediates by direct one-electron oxidation or comproportionation-disproportionation reactions involving an imine cation. The resulting highly reactive radical species binds covalently to macromolecules. Such binding of nitrogen-containing species to cytochrome P-450 ultimately leads to denaturation of the enzyme.[23,34]

While the therapeutic consequences of such an inhibition are evident, however their significance remains to be fully assessed. As regards synergistic drug effects, in theory any drug which is a substrate of cytochrome P-450 can potentially inhibit the metabolism of other compounds. As a consequence of currently employed drug protocols, an enormous number of inhibitory interactions are possible in principle. However, unless some untoward factor prompts interest, it is unlikely that many cases will be investigated. In practical terms, the number of examples of inhibition of drug metabolism in living organisms caused by other drugs and xenobiotics in general is large and no attempt has been made to provide a comprehensive survey of the compounds involved. Rather, it is our intention in this chapter to provide illustrations of the more salient points referring to anthracycline and anthracenedione-based cancer therapy.

Interest in chemotherapeutic aspects of cytochrome P-450 inhibition centers largely on the possible consequences of altered effects of drugs. Indeed, these consequences may be either favorable (decreased toxicity) or detrimental (decreased transformation to a pharmacologically active metabolite and/or decreased detoxication). In other cases, monooxygenase inhibitors may act to cause a decrease in the pharmacological action of the anticancer drug. Thus, it is tempting to suggest that the inhibition of doxorubicin or daunorubicin metabolism by mitoxantrone or ametantrone prevents the formation of toxic metabolites responsible for cytotoxic activity (lipid oxidation, active oxygen species formation).[39]

Beneficial effects of cytochrome P-450 inhibition are gaining increasing interest and may be of potential clinical value. Apart from the previously mentioned protective role of mitoxantrone, anthrapyrazole-mediated inhibition of doxorubicin activation *in vitro* was observed.[39] Those effects may be dependent on the formation of the N-centered free radicals and their involvement in the action of enzymatic systems.[20]

Also of potential value is the considerable decrease in benzo[α]pyrene mutagenicity and carcinogenicity caused by ellipticine.[40] Since many cytochrome P-450 inhibitors act as inducers of this enzyme when used chronically, their protective uses are restricted to isolated administration. Long-term therapy would require non-inducing inhibitors, and mitoxantrone and anthrapyrazoles may be interesting candidates for such inhibitors.

INTERRELATIONSHIP BETWEEN XENOBIOTIC METABOLISM AND LIPID BIOSYNTHESIS

Over recent years there has been considerable interest in links between drug metabolism reactions and certain aspects of lipid biochemistry. Attention has focused particularly upon oxidative processes, those of lipid peroxidation and co-oxidation. Lipids which are crucial components of cell membrane structures also have diverse biological activities. Special attention has to be given to a group of acidic lipids, formed from polyunsaturated fatty acids, particularly those of arachidonic acids. Biosynthetic processes, originated from arachidonic acid, result in the formation of prostaglandins, thromboxanes and prostacyclin, the powerful regulatory substances.

The regulatory process is initiated by the release of arachidonic acid from phospholipids by lipases. This phase is activated by a variety of chemical, physiological and even mechanical stimuli. It is noteworthy that known carcinogenesis promotors, e.g. phorbol esters, are potent stimulating agents for arachidonic acid release in a number of cells and tissues.[41] Thrombin and collagen stimulate phospholipase activity, resulting in TXA_2 formation in platelets. Prostaglandin synthase, a cytochrome P-450 type enzyme, is responsible for biosynthesis of endoperoxide PGH_2 from arachidonic acid. This enzyme has dual cyclooxygenase and hydroperoxidase activity (see discussion in Introduction). The cyclooxygenase catalyzes the first stage, the formation of hydroperoxy endoperoxide PGG_2. In the second step, involving hydroperoxidase, PGG_2 is reduced to alcohol PGH_2 (Figure 5). Further enzymatic isomerization results in thromboxanes (TXA_2 or TXB_2) and prostacyclin PGI_2.

The fatty acid branch of biosynthesis results in hydroperoxy fatty acids, produced by lipoxygenases, and then hydroxy acids are formed upon reduction catalyzed by hydroperoxidases. There are known inhibitors affecting specific steps of biosynthesis. Steroids inhibit the formation of arachidonic acids, non-steroid anti-inflammatory agents like indomethacin, SFK 525 inhibit the cyclooxygenase step, but not the hydroperoxidase activity of prostaglandin synthase.

Co-oxidation of xenobiotics takes place in both prostaglandin and fatty acid branches, during reduction of the relevant hydroperoxide, employing appropriate hydroperoxidase activity. We may consider two possibilities: beneficial cycling of the xenobiotic, or alternatively suicidal inhibitor action. The former, a reversible process, is responsible for oxidation and consequent reduction of xenobiotic. This event requires that a sufficient level of reductant is available to perform transformation of the oxidized form of the drug (i.e. imine form) to its reduced form. Such a process may be beneficial to the cell (tissue) owing to removal of hydroperoxy fatty acid, or by suppressing further prostaglandin biosynthesis, if PGG_2 is removed.

In the latter scenario, a highly reactive N-centered free radical is formed, as a result of one-electron reduction. Alternatively due to insufficient level of reductant, this free-radical species is formed in a disproportionation - comproportionation reaction between imine and amine forms of drug.[16] Suicidal inhibition of hydroperoxidases may occur, affecting further prostaglandin or hydroxy fatty acid biosynthesis. The enhancement in the level of ROOH may be reflected in differences in leukotriene biosynthesis or in direct action of ROOH and ROOH-derived active oxygen species with cell components. Therefore co-oxidation of xenobiotics, accompanying prostaglandin biosynthesis, may have far reaching consequences for the regulation process in cells or tissues. Indeed, there exist data that support this concept. Possibly the most relevant information on the involvement of xenobiotics, including anticancer drugs like anthracyclines, anthracenediones and anthrapyrazoles, in prostaglandin biosynthesis and its effect on lipid peroxidation and platelet functions, comes from the work of Novak and collaborators.[39] Their studies showed that anthracenediones and anthrapyrazoles, in contrast to anthracyclines, have little effect on stimulation of NADH oxidation, superoxide production or H_2O_2 generation in the presence of NADPH-cytochrome P-450 reductase or in hepatic microsomes. It was shown that these agents inhibit oxidative microsomal drug metabolism and decrease anthracycline-stimulated lipid peroxidation.[22,39] They indicated that anthracenediones can act as very potent antioxidants, able to diminish linoleic acid oxidation at concentrations 20-fold lower than the most potent $OH^{\bullet}$ scavengers (thiourea or dimethylthiourea) and inhibit lipid peroxidation at concentrations 100-1000 fold lower than known inhibitors of lipid peroxidation.[39]

Perhaps even more significantly, they demonstrated that anthracenediones and anthrapyrazoles inhibit prostaglandin and thromboxane production in human platelets.[39] Thromboxane and prostacyclin (PGI_2) are important mediators of platelet physiology and function. It has been established that platelet aggregation plays a role in tumor cell metastasis and the balance between thromboxane (TXB_2) and prostacyclin (PGI_2) production is an important determinant in metastasis.[30] Platelet aggregation results from increased thromboxane production whereas prostacyclin is one of the most potent anti-aggregatory agents known so far. Both anthracenediones and anthrapyrazoles were found to produce a concentration-dependent decrease in platelet aggregation.[39] This observation may contribute to the understanding of the mode of action of these anticancer drugs, since the metastatic spread of tumors is linked to the association of circulating tumor cells with host platelets to form platelet-tumor cell aggregates which adhere to vascular intima. Further, angiogenesis, the formation of new capillary blood vessels, is necessary for the growth of solid tumors. This substantial body of evidence fully supports the hypothesis, presented above, on involvement of co-oxidation prone anticancer agents in the control mechanisms of prostaglandin synthesis and lipid peroxidation.

Additional support for this hypothesis comes from our studies on inhibition of prostaglandin synthase (RSV) with anticancer agents known to be a subject of oxidative metabolism. With arachidonic acid as the substrate, the inhibitory effect of drugs on the prostaglandin synthase activity was estimated following the oxygen consumption (oxygen electrode) in the enzymatic

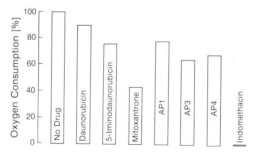

Figure 8. Inhibition of prostaglandin synthase (PS) by anticancer agents, Effect of various drugs on oxygen consumption during metabolism of arachidonic acid [AA] by PS, Bars represent relative initial velocity of oxygen consumption (no drug-100% consumption), [Drug] = 50 μM, [AA] = 200 μM, [PS] = 112 nM in 0.1 M TRIS buffer pH 7.4.

system. Again mitoxantrone was found to exhibit the most pronounced inhibition (the initial velocity of reaction was decreased to 43% with respect to system without drug). Daunorubicin was far less potent and caused only 11% decrease in initial velocity of the oxygen consumption (Figure 7). It is striking that the order of inhibition of metabolism of arachidonic acid in the prostaglandin synthase system by anticancer drugs is parallel to their platelet anti-aggregatory activity. These findings seem to support the hypothesis on possible co-oxidation of anticancer agents during prostaglandin biosynthesis (Figure 5).

CARDIOTOXICITY, LIPID PEROXIDATION AND CALCIUM MEMBRANE TRANSPORT

Calcium ions play an important role in the regulation of many enzyme systems, including those responsible for muscle contraction, nerve impulse transmission, blood clotting, neural activity, and modulation of hormone action. Specifically the intracellular Ca^{2+} concentration is modulated by binding to non-membraneous proteins, and by the operation of a membrane-bound transport system. Besides buffering the intracellular Ca^{2+}, the soluble proteins are also involved in signal processing and are therefore called Ca^{2+}-modulated proteins. Typical representatives are calmodulin, parvalbumin and troponin. However the most important role these proteins play is in transport of Ca^{2+} across the membranes. They are found in the plasma membrane, the endoplasmic (sarcoplasmic) reticulum and in mitochondria.[42]

Among their multiple functions in the cell, mitochondria act as safety devices against toxic increases of cytosolic Ca^{2+}; and once their ability to retain Ca^{2+} is compromised or lost the cell will die. There are two distinct intracellular Ca^{2+} pools: mitochnodrial and extra-mitochondrial. Different regulatory mechanisms seem to control these two Ca^{2+} pools. Metabolism of hydroperoxide causes release of Ca^{2+} transport via the redox state of mitochondrial pyridine nucleotides. Ca^{2+} uptake is favored by reduced, whereas its release is favored by oxidized, pyridine nucleotides. It was demonstrated that hydroperoxides are reduced in mitochondria by pyridine nucleotides with the aid of glutathione peroxidase, glutathione reductase and the energy linked transhydrogenase.[43] Hydroperoxides may induce Ca^{2+} release from intact cells. Recently it was shown that hydroperoxy-eicosatetraenoic acids (HPETEs) and less effectively, hydroxy-eicosatetraenoic acids (HETEs) stimulate Ca^{2+} release from rat liver mitochondria.[44] This supports the hypothesis that hydroperoxy-derivatives of arachidonic acid, which are the biosynthetic precursors of prostaglandins, thromboxanes, HETEs and leukotrienes, may represent a group of molecules which act as mediators of Ca^{2+} mobilization. This pathway of Ca^{2+} mobilization may be particularly important for the action of prooxidant xenobiotics and its relationship to cardiotoxicity of these agents is striking. Enzymatically-mediated redox cycling of anthracyclines resulting in prooxidants (i.e. lipid peroxides, H_2O_2, $^{\bullet}OH$) production is directly linked with anthracycline cardiotoxicity. It was recently suggested that toxicity of prooxidants is related to their action on plasma membranes, the endoplasmic (sarcoplasmic) reticulum and mitochondria, due to impairment of the ability of mitochondria to retain Ca^{2+}.[43] This feature is reflected in development of pathological states of the cell. The well known, dose-dependent risk of cardiotoxicity in administration of anthracyclines appears to be associated with cardiac lipid peroxidation. This peroxidative process may be suppressed by free radical scavengers like α-tocopherol or coenzyme Q_{10}.[45]

The structural evaluation of rat myocardium, following drug administration, revealed severe intramyocytic sarcotubular dilation, clearing of mitochondrial matrices, disorganization of contractile elements and accumulation of numerous lipofuscin pigment granules.[46] The ventricular tissue calcium content of adriamycin treated rats was significantly increased compared with controls. These increases in tissue calcium were associated with myocardial damage and preceded the onset of highest mortality rate. The observation of increased ventricular tissue calcium in rats following several dosages of doxorubicin suggests that the drug may produce an alteration of cell membrane permeability, accompanied by an influx of calcium and possible leakage of myocardial enzymes, both associated with fine structural cardiomyopathic alterations. Increased intracellular calcium is known to produce damaging effects on mitochondrial metabolism in heart muscle and anthracyclines have been shown to inhibit mitochondrial oxygen consumption in rat myocardium *in vivo*.[46]

Recent studies also stress the importance of GSH and protein thiols in maintenance of intracellular Ca^{2+} homeostasis in hepatocytes.[47] For example, plasma membrane Ca^{2+}-ATPase activity is inhibited by agents which complex or oxidize free thiol groups and is restored by subsequent exposure to reducing agents such as dithiothreitol. In addition, the glutathione content of mitochondria may affect the ability of this organelle to retain Ca^{2+}, either directly or *via* an effect on the redox status of intramitochondrial pyridine nucleotides.[43]

The consequence of the linkage between lipid peroxides and prostaglandin synthesis may influence the action of drugs capable of oxidative metabolism. When sufficient peroxide scavenging activity is present, these agents can be effective mediators in redox cycling processes. This leads to scavenging of hydroperoxides at the expense of reducing systems present in cells or they can be fairly effective inhibitors of prostaglandin biosynthesis. Drug-originated free radicals may play a role in inhibiting fatty acids oxygenases that produce eicosanoids. Therefore, the possible involvement of anticancer agents undergoing cooxidation, in disrupting or alterating eicosanoid formation may be more important than in direct interaction with DNA.

CONCLUDING REMARKS

Since the occurrence of metabolic interrelations between xenobiotics and lipids has only recently been recognized, it is not surprising that their functional significance remains a matter of speculation. However, the activation of antitumor agents by important enzymatic systems of lipid biochemistry and/or alteration of membrane function are clearly related to the metabolism of these xenobiotics.

Membrane-bound enzymes are particularly sensitive to lipid peroxidation since they need an integral lipid environment for their activity. Therefore the damage to membrane lipids alters the functions of membrane-bound proteins. As we have already discussed, anthracycline-related toxicity seems to be linked to their reductive activation and redox cycling, resulting in free radical production, disproportionate oxygen consumption, NADPH oxidation, and lipid peroxidation. The radicals react with oxygen forming active oxygen species, followed by oxidative stress and toxicity.

On the other hand, anthracenediones, anthrapyrazoles and 5-iminodaunorubicin, which are less cardiotoxic agents, undergo *in vitro* oxidative activation with concomitant formation of N-centered free radicals. Such activation was found for peroxidases, cytochrome P-450 and prostaglandin H synthase enzymes. Therefore, these chemotherapeutics may act as hydroperoxide (lipid hydroperoxide) scavengers in the presence of heme-containing enzymes. However much further work is required in order to elucidate the critical events involved in the mechanism of such scavenging. It should be emphasized that our evidence shows the ability to form N-centered free radicals and oxidative cycling *in vitro* but it has to be proved that this potential will be realized *in vivo*.

Although the interference of these drugs with prostaglandin and thromboxane synthesis and therefore platelets aggregation seems to support this hypothesis, more evidence is required from studies of changes in prostanoid metabolism profiles caused by action of anticancer drugs. Additionally it would be interesting to determine whether the above discussed anticancer agents are able to contribute to the maintenance of intracellular Ca^{2+} levels, which are known to be extremely sensitive to the oxidative stress caused by xenobiotics.

This work was presented in part at the NATO Advanced Study Institute; International School of Pharmacology: *Prostanoids and Drugs*, Erice, Italy, September 1988.

ABBREVIATIONS USED

HRP, horseradish peroxidase (E.C. 1.11.1.7); EPR, electron paramagnetic resonance; 5-IMDR, 5-iminodaunorubicin; MXH_2, mitoxantrone; MH_2, mitoxantrone cyclic metabolite; AP, anthrapyrazole(s); PG, prostaglandin(s); TX, thromboxane(s); PHS, prostaglandin H synthase; HPETE's, hydroperoxyeicosatetraenoic acids; HETE's, hydroxyeicosatetraenoic acids; AA, arachidonic acid; GSH, glutathione reduced form; RSV, ram seminal vesicles.

The skillful technical assistance of Miss Annabelle Wiseman is gratefully acknowledged.

REFERENCES

1. J. S. Bus and J. E. Gibson, Lipid peroxidation and its role in toxicology, *Rev. Biochem. Toxicol.* 1:125 (1979).
2. H. Kappus, Overview of enzyme systems involved in bioreduction of drugs and in redox cycling, *Biochem. Pharmac.* 35:1 (1986).
3. R. W. Estabrook, N. Chacos, C. Martin-Wixtrom, and J. Capdevila, Cytochrome P-450: A versatile vehicle of variable veracity, *in*: "Oxygenases and Oxygen Metabolism," M. Nozaki, S. Yamamoto, Y. Ishimura, M. J. Coon, L. Emster, R. W. Estabrook (eds.), Academic Press, New York (1982).

4. T. L. Dormandy, Free-radical oxidation and antioxidants, *Lancet* 647 (1978).
5. L. Masotti, E. Casali and T. Galeotti, Lipid peroxidation in tumor cells,*Free Radical Biol. Med.* 4:377 (1988).
6. P. A. Cerutti, Prooxidant states and tumor promotion, *Science* 227:375 (1985).
7. P. A. Cerutti, Repairable damage in DNA, *in*: "DNA Repair Mechanisms,"P. Phanawalt, E. Friedberg, C. Fox (eds.), Academic Press, New York (1978).
8. K. Brown and I. Fridovich, Superoxide radical and superoxide dismutase: Threat and defense, *Acta Physiol. Scand. Suppl.* 492:9 (1980).
9. I. Emerit, M. Keck, A. Levy, J. Feingold and A. M. Michelson, Activated oxygen species at the origin of chromosome breakage and sister-chromatid exchanges, *Mutat. Res.* 103:165 (1982).
10. I. Emerit, A. Levy and P. A. Cerutti, Suppression of tumor promoter phorbol-myristate-acetate induced chromosome breakage by antioxidants and inhibitors of arachidonic acid metabolism, *Mutat. Res.* 110:327 (1983).
11. L. Gianni, B. J. Corden and C. E. Meyers, The biochemical basis of anthracycline toxicity and antitumor action, *in*: "Reviews in Biochemical Toxicology, Vol. 5," E. Hodson, J. R. Bend and R. M. Philpot (eds.), Elsevier, Amsterdam (1983).
12. J. W. Lown, H-H. Chen, J. A. Plambeck and E. M. Acton, Further studies on the generation of reactive oxygen species from activated anthracyclines and the relationship to cytotoxic action and cardiotoxic effects, *Biochem. Pharmacol.* 31:575 (1982).
13. B. Chance, H. Sies and A. Boveris, Hydroperoxide metabolism in mammalian organs, *Physiol. Rev.* 59:527 (1979).
14. J. F. Mead, Free radical mechanism of lipid damage and consequences for cellular membranes, *in*: "Free Radicals in Biology," W. A. Pryor (ed.), Academic Press, New York (1976).
15. K. Reszka, P. Kolodziejczyk and J. W. Lown, Horseradish peroxidase-catalyzed oxidation of mitoxantrone, Spectrophotometric and electron paramagnetic resonance studies, *J. Free Rad. Biol. Med.* 2:25 (1986).
16. P. Kolodziejczyk, K. Reszka and J. W. Lown, Enzymatic oxidative activation and transformation of the antitumor agent mitoxantrone, *Free Radical Biol. Med.* 5:13 (1988).
17. P. Kolodziejczyk, K. Reszka and J. W. Lown, Oxidative activation of 5-iminodaunorubicin; action of horseradish peroxidase, Spectroscopic and electron paramagnetic resonance studies, *Biochem. Pharmacol.* in press (1989).
18. E. D. Kharash and R. F. Novak, Structural and mechanistic differences in quinone inhibition of microsomal drug metabolism; inhibition of NADPH-cytochrome P-450 reductase activity in cytochrome P-450, *Developments in Biochemistry* 23:623 (1983).
19. J. Basra, C. R. Wolf, J. R. Brown and L. H. Patterson, Evidence for human liver mediated Free-radical formation by doxorubicin and mitoxantrone, *Anticancer Drug Design* 1:45(1985).
20. P. Kolodziejczyk, K. Reszka and J. W. Lown, Alternative to the bioreductive activation of anthracyclines: Enzymatic oxidative metabolism of anthracenediones, 5-iminodaunorubicin and anthrapyrazoles, *in*: "Oxy-radicals in Molecular Biology and Pathology, Vol. 82," P. A. Cerutti, I. Fridovich, J. M. McCord (eds.), A. R. Liss, New York (1988).
21. B. Meunier, Horseradish peroxidase: A useful tool for modeling the extrahepatic biooxidation of exogens, *Biochimie* 69:3 (1987).
22. P. Frank and R. F. Novak, Effects of anthrapyrazole antineoplastic agents on lipid peroxidation, *Biochem. Biophys. Res. Commun.* 140:797 (1986).
23. B. Testa and P. Jenner, Inhibitors of cytochrome P-450s and their mechanism of action, *Drug Metabolism Reviews* 12:1 (1981).
24. R. A. Neal, T. Sawahata, J. Halpert and T. Kamataki, Chemically reactive metabolites as suicide enzyme inhibitors, *Drug Metabolism Reviews* 14:49 (1983).
25. R. K. Boutwell, Retinoids and prostaglandin synthesis inhibitors as protective agents against chemical carcinogenesis and tumor promotion, *in*: "Radioprotectors and Anticarcinogens," O. F. Nygaard, M. G. Simic (eds.), Academic Press, New York (1983).
26. J. M. McCord, Superoxide radical: A likely link between reperfusion injury and inflammation, *Adv. Free Rad. Biol. Med.* 2:325 (1986).
27. J. A. Fee, A comment on the hypothesis that oxygen toxicity is mediated by superoxide, *in*: Second BOC Priestley Conference Oxygen and Life, Royal Soc. of Chemistry, London (1981).
28. M. J. Waring, Overview of the interaction between chemotherapeutic agents and DNA, *Drugs Exptl. Clin. Res.* 12:441 (1986).
29. W. E. M. Lands, Interactions of lipid hydroperoxides with eicosanoid biosynthesis, *J. Free Rad. Biol. Med.* 1:97 (1985).
30. K. V. Honn, R. S. Bockman and L. J. Marnett, Prostaglandins and cancer: A review of tumor initiation through tumor metastasis, *Prostaglandins* 21:833 (1983).

337

31. A. Aitokallio-Tallbert, J. Karkkainen, P. Pantzar, T. Wahlstrom, and O. Ylikorkala, Prostaglandin and thromboxane in breast cancer: Relationship between steroid receptor status and medroxyprogesterone-acetate, *Br. J. Cancer* 5:671 (1985).

32. A. M. Aitokallio-Tallberg, L. V. Viinikka and R. O. Ylikorkala, Increased synthesis of prostacyclin and thromboxane in human ovarian malignancy, *Cancer Res.* 48:23968 (1988).

33. R. W. Egan, P. H. Gale, G. C. Beveridge, L. J. Marnett and F. A. Keuhl, Direct and indirect involvement of radical scavengers during prostaglandin biosynthesis, *Adv. Prostaglandin Thromboxane Res.* 6:153 (1980).

34. J. Capdevila, R. W. Estabrook and R. A. Prough, Differences in the mechanism of NADPH- and cumene hydroperoxide-supported reactions of cytochrome P-450, *Arch. Biochem. Biophys.* 200:186 (1980).

35. E. L. Wheeler, Cytochrome P-450 mediated interaction between hydroperoxide and molecular oxygen, Biochem. *Biophys. Res. Commun.* 110:646 (1983).

36. J. C. Sircar, C. F. Schwender and E. A. Johnson, Soybean lipoxygenase inhibition by nonsteroidal antiinflammatory drugs, *Prostaglandins* 25:393 (1983).

37. R. H. Tullman and R. P. Hanzlik, Inactivation of cytochrome P-450 and monoamine oxidase by cyclopropylamines, *Drug Metabolism Review* 15:11632 (1984).

38. C. Baarnhielm and G. Hansson, Oxidation of 1,4-dihydropyridines by prostaglandin synthase and the peroxic function of cytochrome P-450, Demonstration of a free radical intermediate, *Biochem. Pharmacol.* 35:14195 (1986).

39. R. F. Novak, E. D. Kharasch, P. Frank and M. Runge-Morris, Anthracyclines, anthracenediones and anthrapyrazoles: Comparison of redox cycling activity and effects on lipid peroxidation and prostaglandin production, *in*: "Anthracycline and Anthracenedione Anticancer Agents," J. W. Lown (ed.), Elsevier, Amsterdam (1989).

40. P. Lesca, P. Lecointe, C. Paoletti and D. Mansuy, Ellipticines as potent inhibitors of drug metabolism: Protective effect against chemical mutagenesis and carcinogenesis, *Biochemie* 60:1011 (1978).

41. L. Levine, Arachidonic acid transformation and tumor production, *Adv. Cancer Res.* 35:49(1981).

42. E. Carafoli, Intracellular calcium homeostasis, *Annu. Rev. Biochem.* 56:395 (1987).

43. C. Richter and B. Frei, Ca^{2+} release from mitochondria induced by prooxidants, *Free Rad. Biol. Med.* 4:365 (1988).

44. C. Richter, B. Frei and P. A. Cerutti, Mobilization of mitochondrial Ca^{2+} by hydroperoxyeicosatetraenoic acid, *Biochem. Biophys. Res. Commun.* 143:609 (1987).

45. N. Yamanaka, T. Kato, K. Nishida, T. Fujikawa, M. Fukushima and K. Ota, Elevation of serum lipid peroxide level associated with doxorubicin toxicity and its amelioration by d,l-alpha-tocopherol acetate or coenzyme Q_{10} in mouse, *Cancer Chemother. Pharmacol.* 3:223 (1979).

46. H. M. Olson and C. C. Capen, Chronic cardiotoxicity of doxorubicin (adriamycin) in the rat: Morphologic and biochemical investigations, *Toxicol. Appl. Pharmacol.* 44:605 (1978).

47. M. Moore, H. Thor, G. Moore, S. Nelson, P. Moldeus and S. Orrenius, The toxicity of acetaminophen and N-acetyl-p-benzoquinone imine in isolated hepatocytes is associated with thiol depletion and increased cytosolic Ca^{2+}, *J. Biol. Chem.* 24:13035 (1985).

REDUCTION OF TOXICITY AND INCREASE OF ANTITUMOR EFFECT OF ADRIAMYCIN

BY N-ACYL DEHYDROALANINES, A NEW FAMILY OF FREE RADICAL SCAVENGERS

P. Buc-Calderon[1], M. Praet[2], J.M. Ruysschaert[2], and
M. Roberfroid[1]

[1]Unité de Biochimie Cancérologique et Toxicologie, Dépt. de
Sciences Pharmaceutiques, Univ. Catholique de Louvain, 73
Avenue E. Mounier, B-1200; [2]Lab. de Chimie Physique des
Macromolécules aux Interfaces, Univ. Libre de Bruxelles
C.P. 206/2 Bd. du Triomphe, B-1050, Belgium

INTRODUCTION

Free radicals are highly reactive chemical species, they abstract
hydrogen atoms at sensitive sites, they add to olefinic C-atoms and they
are substrates of one-electron exchange reactions. Thus they react
easily with miscellaneous molecular targets. One of the main
characteristics of free radical reactions is that they produce new free
radicals, so that free radical processes propagate through a chain of
reactions which cause not only structural but mainly functional
disturbances at various cellular levels.

It has been proposed that carbon atoms bearing both an electron
donating and an electron withdrawing group (capto-dative substitution)
can be transformed to stabilized free radicals (1). Olefins having such
a capto-dative substituted geminal carbon atom have been shown to react
with free radicals to give stabilized free radical adducts, which often
disappear by dimerization or by reaction with another free radical. For
that reason they have been called radicophiles (2).

N-acyl dehydroalanines, indexed as AD compounds (Figure 1), are
original molecules which have been designed on the basis of the concept
of capto-dative substitution (3). In addition to the radicophilic site
(on the capto-dative substituted olefinic carbon), they have a methylene
group which has been called "proradical site". In fact, the cleavage of
the C-H bond yields a C-centered free radical which is stabilized by
capto-dative effect. Moreover, the aromatic ring may scavenge hydroxyl
radicals to yield hydroxylated aromatic derivatives. Therefore, AD
compounds may react with free radicals by offering 3 different reactive
sites, and whatever the type of reaction - radical addition to the
carbon-carbon double bond or hydrogen abstraction from the methylene
group - the resulting free radical adduct is stabilized by the capto-
dative effect. Thus they terminate the free radical chain reaction.

Considering the problem, in biological fluids, of trapping all
kinds of free radicals with different polarities and structures, the AD
compounds should present some interesting potentialities. Indeed, they

Antioxidants in Therapy and Preventive Medicine
Edited by I. Emerit *et al.*
Plenum Press, New York, 1990

339

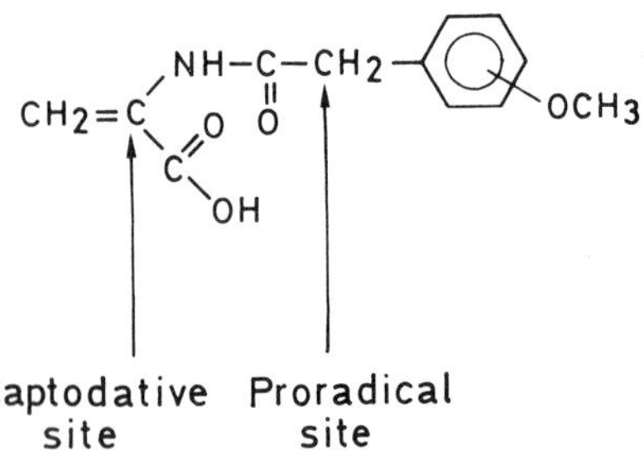

CH2=C、NH-C-CH2-〈◯〉-OCH3 ... structure with C=O, C-OH, Captodative site, Proradical site

Fig.1. Structure of AD compounds.
For the AD-20 derivative the methoxy group in ortho position.

have shown *in vitro* to react with and scavenge both superoxide anion and hydroxyl radical (4), inhibiting mostly in this way the *in vitro* and *in vivo* deleterious-mediated effects (4-7).

Adriamycin is an anticancer drug widely used against several types of malignancies. It displays antineoplastic activity towards a broad spectrum of human cancers such as carcinomas of the breast, lung, ovary, brain, and gastrointestinal tract (8). Nevertheless, its clinical usefulness is limited by a selective organ-toxicity (myelosuppression and gastrointestinal mucositis) and the risk of cardiomyopathy and congestive heart failure (9).

The impairment by adriamycin of the cardiac mitochondrial respiratory chain is mainly responsible for its cardiotoxicity, and evidence from several experimental studies suggests that it may be the consequence of the production of reactive oxygen species. Indeed, adriamycin is reduced to a semiquinone form at complex I of the mitochondrial electron transfer chain, and the redox cycling of the semiquinone radical (reaction with molecular oxygen to regenerate the parent drug) leads to a significant increase of superoxide anion, hydrogen peroxide, and hydroxyl radicals (10).

Since the antitumor activity of adriamycin can be dissociated from its toxic effect (thought to be mediated by free radical species), it was hypothesized that AD compounds by scavenging these species formed during adriamycin's metabolism, could decrease its toxicity thus enhancing its antitumor effect.

RESULTS AND DISCUSSION

Membrane toxicity induced by adriamycin

Both *in vitro* and *in vivo* experiments performed with adriamycin confirm its damaging effect on lipid membranes, as measured by lipid peroxidation of either rat liver microsomal suspensions or mouse heart mitochondrial membranes (table 1). Moreover, we have also reported changes in the membrane fluidity of heart mitochondria isolated from mice receiving adriamycin. This modification results from the increase in the rigidity of the membrane, indicating that the peroxidation of the membrane phospholipids may play a major role (7).

340

Table 1. Effect of AD-20 on adriamycin-induced lipid
         membrane damage

	Microsomal oxidation[a]	Mitochondrial oxidation[b]
No addition	0.3 ± 0.1	415 ± 20
Adriamycin	38.2 ± 1.5[c]	849 ± 57[c]
Adriamycin + AD-20	7.6 ± 0.4[d]	665 ± 63[d]

[a]Values are expressed as mean of nmol MDA-equivalents per mg of microsomal proteins ± S.D. from 3 separated experiments. Adriamycin and AD-20 were 0.05 mM and 5 mM, respectively.
[b]Values are expressed as mean of pmol MDA-equivalents per mg of mitochondrial proteins ± S.D. Mitochondria were isolated from mice heart two days after the onset of the treatment. Adriamycin and AD-20 were administered by intraperitoneal way at 20 mg/kg and 70 mg/kg, respectively. AD-20 was injected 15 min before adriamycin. Each group consisted of 6 mice.
[c]Significant statistically values (p less than 0.05) in relation to control values.
[d]Significant statistically values (p less than 0.05) in relation to adriamycin values.

Rat liver microsomal suspensions incubated _in vitro_ with adriamycin (at 0.05 mM) are rapidly peroxidized as seen by the increase in the MDA-equivalents (table 1). Reactive oxygen species have been involved in the initiation and/or propagation of microsomal lipid peroxidation. Since adriamycin belongs to the so-called redox-cycling drugs (generating superoxide anion), the resulting lipid peroxidation is thought to be the consequence of such a redox cycle. Supporting this view is the fact that the incubation of liver microsomes in the presence of both adriamycin and AD-20, resulted in a strong decrease of MDA production, as compared to the values obtained with adriamycin alone (table 1).

When mice received adriamycin at 20 mg/kg intraperitoneal way, the amount of MDA in mitochondrial membranes (isolated from mice heart 48 h after the injection of adriamycin) was higher as compared to membranes of untreated mice. The amount of MDA in heart mitochondrial membranes from mice receiving AD-20 in addition to adriamycin, remains still higher when compared to control values, but in a significant less extent than in animals receiving adriamycin alone.

Tables 2 and 3 show the toxicity which appears after the intraperitoneal administration of adriamycin to female NMRI mice. This toxicity has been evaluated either by recording the lethality induced by adriamycin administered by a single dose or a multiple dose protocol (table 2), or by measuring some parameters linked to a specific organ such as the serum transaminase levels, the heart mitochondrial enzymic activities, and the total peripheral leukocyte count (table 3).

Adriamycin administered at a single dose of 15 mg/kg produces a significant lethality with only 10% of mice surviving at the end of the

Table 2. Effect of AD-20 on lethality of mice induced by adriamycin

| | Acute[a] | | Chronic[b] | |
	MST	% S	MST	% S
Adriamycin	5.5	10	31.5	6
Adria + AD-20	56.0	50	40.0	20

[a]Adriamycin and AD-20 were administered at 15 mg/kg and 30 mg/kg, respectively. The survival rate (%S) was calculated at day 56. Each group consisted of 30 mice.
[b]Adriamycin and AD-20 were administered at 3 times (two weeks interval) at a dose of 3 x 5 mg/kg and 3 x 10 mg/kg, respectively. The survival rate (%S) was calculated at day 112. Each group consisted of 50 mice.
MST = medium survival time.

Table 3. Effect of AD-20 on some specific organ toxicity induced by adriamycin.

	Cardiac toxicity[a]	Hepatic toxicity[b]	bone marrow toxicity[c]
Control	328 ± 44	34 ± 3	85.5 ± 18.6
Adriamycin	184 ± 38[d]	260 ± 22	51.5 ± 6.1[d]
Adria + AD-20	286 ± 40[e]	150 ± 36[e]	83.7 ± 16.9[e]

[a]Values are expressed as pmol oxidized cytochrome c/mg protein per min ± S.D., representing the enzymatic activity of complex IV from heart mitochondria isolated from mice two days after the administration of adriamycin (30 mg/kg). AD-20 was at 70 mg/kg. Each group consisted of 6 mice.
[b]Values are expressed as UI/l ± S.D. of serum pyruvate-glutamic transaminase levels. Adriamycin and AD-20 were at 15 mg/kg and 30 mg/kg, respectively. Each group consisted of 12 mice.
[c]Values are expressed as peripheral white blood cells (WBC) x 100/cubic millimeter ± S.D. obtained nine days after the onset of the treatment. Adriamycin and AD-20 were 6 mg/kg and 50 mg/kg, respectively. Each group consisted of 5 mice.
[d]Statistically significant values (p less than 0.05) in relation to control values.
[e]Statistically significant values (p less than 0.05) in relation to adriamycin values.

experiment. AD-20 administered 15 min before adriamycin, enhances the survival time from 5.5 days to 56.0 days. In addition, the survival advantage, 8 weeks from the onset of the treatment, was 50% as compared to only 10% in the adriamycin-treated group.

Since the toxicity of adriamycin (especially to cardiac tissue) results mostly from its chronic administration, we have examined the effect of AD-20 in mice treated with multiple non-toxic doses of adriamycin. As shown in table 2, AD-20 not only delays the mortality but also enhances the survival time of mice. The survival rate, calculated at day 112, were 6% and 20% for animals receiving adriamycin alone and adriamycin plus AD-20, respectively. Thus, table 2 shows that AD-20 administration (15 min before adriamycin), results in a more elevated number of survivors than adriamycin alone, whatever the protocol used: a single acute dose (1 x 15 mg/kg) or multiple non-toxic doses (3 x 5 mg/kg).

The decrease in adriamycin's lethality by pretreatment with AD-20 may occur by different mechanisms. Among the principal tissue targets for the development of adriamycin's toxicity, the liver, the heart, and the bone marrow, appear to play a major role. We have hypothesized that AD-20 protects mice against the toxicity of adriamycin, by blocking its deleterious free radical-mediated effects which may take place in these tissues. Supporting this hypothesis, the pretreatment of mice with AD-20 (15 min before adriamycin), almost restored to normal values the heart mitochondrial enzymic activities, the serum transaminase levels, and the peripheral leukocyte count (table 3). Indeed, these 3 parameters were strongly modified in animals receiving adriamycin alone.

Antitumor effect of adriamycin

The antitumor effect of adriamycin alone or combined with AD-20, towards two types of murine ascitic leukaemia (L1210 and L311) is shown in table 4. In BDF1 mice bearing leukaemia L1210, the administration of AD-20 (at 25 mg/kg intraperitoneally 15 min before adriamycin at 7 mg/kg) increased from 15.3 days to 39.6 days the mean survival time of the animals. The enhancement of the life span was of 169%. On the other hand, the survival of LOU rats bearing leukaemia L311 was also increased, but this time from 28.4 days to 47.3 days (67%), when animals were administered with AD-20 at 100 mg/kg intraperitoneally 15 min before adriamycin. In this experimental condition adriamycin was administered intravenously at 5 mg/kg, in order to avoid the gastrointestinal mucositis which appears rapidly in rats following an intraperitoneal administration of adriamycin.

Since AD-20 has no antitumor effect by itself, and that the possibility of a local interaction between AD-20 and adriamycin in the peritoneal cavity, thus inactivating the drug, is excluded by the results obtained with the leukaemia L311, we conclude that AD-20 improves the chemotherapeutic effect of adriamycin.

CONCLUSIONS

Our results support the fact that the enhancement of the antitumor effect of adriamycin by AD-20, mostly results from the decrease of its toxicity. Indeed, we propose that AD-20, by its free radical scavenging abilities, blocks the radical chain reaction thus protecting mice against the deleterious effects of a adriamycin-induced free radical cascade.

Table 4. Effect of AD-20 on the antitumor activity of adriamycin.

| | L1210[a] | | L311[b] | |
	MST	ILS	MST	ILS
Control	10.4	-	16.1	-
Adriamycin	15.3	-	28.4	-
+ AD-20	39.6	169	47.3	67

[a]BDF1 mice bearing leukaemia L1210. Adriamycin and AD-20 were at 7 mg/kg and 25 mg/kg, respectively. Each group consisted of 20 mice.

[b]LOU rats bearing leukaemia L311. Adriamycin and AD-20 were at 5 mg/kg and 100 mg/kg, respectively. Each group consisted of 10 rats.

mst = mean survival time.

ILS = increase in life span.

REFERENCES

1. H.G. Viehe, Z. Janousek, R. Merényi, and L. Stella, The capto-dative effect, Acc. Chem. Res. 18: 148 (1985).
2. H.G. Viehe, R. Merényi, L. Stella, and Z. Janousek, Capto-dative substituent effects in synthesis with radicals and radicophiles, Angew. Chem. 18: 917 (1979).
3. M. Roberfroid, H.G. Viehe and J. Remacle, Free radicals in drug research in: "Advances in drug research", B. Testa, ed., Academic Press, London (1987).
4. P. Buc-Calderon and M. Roberfroid, Inhibition of superoxide anion and hydroxyl radical-mediated processes by a new class of free radical scavengers: the N-acyl dehydroalanines, Free Rad. Res. Commun. 5: 159 (1988).
5. P. Buc-Calderon, M. Praet, J.M. Ruysschaert and M. Roberfroid, Free radical modulation by N-substituted dehydroaglanines, a new way to improve therapeutic activity of anticancer drugs, Cancer Treat. Rev. 14: 379 (1987).
6. P. Buc-Calderon, M. Praet, J.M. Ruysschaert and M. Roberfroid, Increasing therapeutic effect and reducing toxicity of doxorubicin by N-acyl dehydroalanines, Eur. J. Cancer Cli. Oncol., in press (1989).
7. M. Praet, P. Buc-Calderon, G. Pollakis, M. Roberfroid and J.M. Ruysschaert, A new class of free radical scavengers reducing adriamycin mitochondrial toxicity, Biochem. Pharmacol. 37: 4617 (1988).
8. S.K. Carter, Adriamycin - a review, J. Natl. Cancer Inst. 55: 1265 (1975).
9. C. Praga et al., Adriamycin cardiotoxicity: a survey of 1273 patients, Cancer Treat. Rep. 63: 827 (1979).
10. N.R. Bachur, S.L. Gordon and M.V. Gee, A general mechanism for microsomal activation of quinone anticancer agents to free radical, Cancer Res. 38: 1745 (1978).

INHIBITION OF BLEOMYCIN-INDUCED TOXIC EFFECTS BY

ANTIOXIDANTS IN HUMAN MALIGNANT MELANOMA CELLS

H. Kappus and Ch. Reinhold

Department of Dermatology, Rudolf-Virchow-Clinic
Free University of Berlin, Augustenburger Platz 1
D-1000 Berlin 65, F.R.G.

INTRODUCTION

Bleomycin is an antibiotic drug used in tumor chemother-apy. It forms a complex with iron ions which can be reduced chemically or enzymatically. The bleomycin-Fe(II)-complex binds and activates oxygen and degrades cellular DNA. This activity of bleomycin is presumably responsible for the cytotoxic effects observed during therapy (for reviews see Burger et al., 1981; Hecht, 1986). Because reactive oxygen species are involved, we wondered whether antioxidants could inhibit bleomycin-induced cytotoxicity. For example, the seleno-organic compound Ebselen (2-phenyl-1,3-benzisoselena-zol-3(2H)one) has besides its glutathione peroxidase-like activities also antioxidant properties (Müller et al., 1984). It inhibits cell toxicity induced by adriamycin (Doroshow, 1986) and diquat (Cotgreave et al., 1987), both redox cycling compounds which can induce formation of reactive oxygen species. The other antioxidants examined were pyrogallol and propyl gallate which have also been shown to trap reactive oxygen species (Weinke et al., 1987).

METHODS

We used a malignant melanoma cell line which was estab-lished from a lymph node metastasis of a 40-year-old woman. Cells (10^5/ml) were grown in RPMI 1640 medium supplemented with 5 % fetal calf serum (Seromed, Berlin, F.R.G.) at 37 °C and 5 % CO_2. Bleomycin (Mack, Illertissen, F.R.G.) was added to the culture medium in concentrations up to 20 µg/ml and the antioxidants pyrogallol, propyl gallate (Sigma, Mün-chen, F.R.G.) and Ebselen (PZ51, kindley provided by Nattermann, Köln, F.R.G.) in concentrations up to 1×10^{-5} M. Cellular growth was determined 24 h after addition. Dead cells and cell debris were removed by washing with PBS and only cells firmly attached to the dishes were counted.

RESULTS

A significant increase in the number of cells was observable within 24 h of culture which decreased to 25 - 40 % of respective controls in the presence of 10 - 20 μg/ml bleomymycin (Table). When antioxidants were present at the same time, bleomycin-induced cytotoxicity was prevented to some extent, the optimal antioxidant concentrations being 10^{-7} - 10^{-8} M (Table). The most effective antioxidant was pyrogallol which completely compensated these cytotoxic effects of bleomycin, whereas propyl gallate and Ebselen lead to a lesser increase in the number of viable cells (75 % of respective controls) compared to cultures with bleomycin alone (Table).

DISCUSSION

The results demonstrate that the antioxidants pyrogallol, propyl gallate and Ebselen at low concentrations are able to prevent at least partially the cytotoxic effect of bleomycin in the malignant melanoma cells used, the concentration dependence curve reaching a plateau at 10^{-7}-10^{-8} M. It is also interesting that these antioxidants alone had no significant toxic effects to the cells when added in concentrations of 10^{-5}-10^{-9} M.

Our results indicate, that antioxidants may interfer with the molecular mechanism of bleomycin-induced cytotoxicity. They could trap reactive oxygen species which are presumably involved in DNA degradation by the reduced bleomycin-Fe-O_2-complex (Mahmutoglu and Kappus, 1985). Indeed, propyl gallate has been demonstrated to trap hydroxyl radicals (Weinke et al., 1987). On the other hand, DNA degradation most likely depends on redox cycling of the bleomycin-Fe-complex catalyzed by enzymes like NADPH-cytochrome P-450 reductase (Scheulen et al., 1981; Kilkuskie et al., 1984; Kappus et al., 1987) and NADH-cytochrome b_5 reductase (Mahmutoglu and Kappus, 1987; 1988). For example, Ebselen has been shown to efficiently inhibit NADPH-cytochrome P-450 reductase ($IC_{50} = 10^{-7}$ M, Wendel et al., 1986). This inhibition could be an alternative mechanism explaining the protective effect of the antioxidants used on bleomycin-induced cytotoxicity. But this has to be clarified in future experiments.

Table 1 Effect of antioxidants on bleomycin (BLM)-induced cytotoxicity in human malignant melanoma cells cultured for 24 h. Controls (100%) = number of cells without BLM and without antioxidants. Mean values $\pm$ SD (n=3) are shown.

% of control

Antioxidant conc. (M)	0	10^{-10}	10^{-9}	10^{-8}	10^{-7}	10^{-6}	10^{-5}
Pyrogallol	100±6	91±3	93±3	103±3	102±4	92±3	-
Pyrogallol BLM(10µg/ml)	43±7	61±4	66±3	100±2	98±6	69±1	-
Pyrogallol BLM(20µg/ml)	28±6	36±8	39±10	79±3	83±3	56±1	-
Propylgallate	100±9	101±5	105±3	100±12	100±14	96±10	-
Propylgallate BLM(10µg/ml)	40±3	40±2	54±11	78±7	77±4	63±6	-
Propylgallate BLM(20µg/ml)	26±2	34±3	42±11	69±13	73±12	47±9	-
Ebselen	100±7	-	101±2	103±11	100±8	99±9	94±9
Ebselen BLM(10µg/ml)	46±7	-	50±7	78±15	76±8	67±8	46±6
Ebselen BLM(20µg/ml)	31±6	-	39±15	66±3	70±5	56±8	59±12

ACKNOWLEDGEMENT

Ebselen (PZ51) was kindly provided by Nattermann, Köln, for which we are very grateful.

REFERENCES

Burger, R.M., Peisach, J., and Horwitz, S.B., 1981, Activated bleomycin. A transient complex of drug, iron, and oxygen that degrades DNA, J. Biol. Chem. 256:11636

Cotgreave, I.A., Sandy, M.S., Berggren, M., Moldeus, P.W., and Smith, M.T., 1987, N-Acetylcysteine and glutathione-dependent protective effect of PZ51 (ebselen) against diquat-induced cytotoxicity in isolated hepatocytes, Biochem. Pharmacol. 36:2899

Doroshow, J.H., 1986, Role of hydrogen peroxide and hydroxyl radical formation in the killing of Ehrlich tumor cells by anticancer quinones, Proc. Nat. Acad. Sci. USA 83:4514

Hecht, S.M., 1986, DNA strand scission by activated bleomycin group antibiotics, Fed. Proc. 45:2784

Kappus, H., Mahmutoglu, I., Kostrucha, J., and Scheulen, M.E., 1987, Liver nuclear NADPH-cytochrome P-450 reductase may be involved in redox cycling of bleomycin-Fe(III), oxy radical formation and DNA damage, Free Rad. Res. Comms. 2:271

Kilkuskie, R.E., MacDonald, T.L., and Hecht, S.M., 1984, Bleomycin may be activated for DNA cleavage by NADPH-cytochrome P-450 reductase, Biochemistry 23:6165

Mahmutoglu, I., and Kappus, H., 1985, Oxy radical formation during redox cycling of the bleomycin-iron (III) complex by NADPH-cytochrome P-450 reductase, Biochem. Pharmacol. 34:3091

Mahmutoglu, I., and Kappus, H., 1987, Redox cycling of bleomycin-Fe(III) by an NADH-dependent enzyme and DNA damage in isolated rat liver nuclei, Biochem. Pharmacol. 36:3677

Mahmutoglu, I., and Kappus, H., 1988, Redox cycling of bleomycin-Fe(III) and DNA degradation by isolated NADH-cytochrome b_5 reductase - Involvement of cytochrome b_5. Mol. Pharmacol. 34:578

Müller, A., Cadenas, E., Graf, P., and Sies, H., 1984, A novel biologically active seleno-organic compound - I. Glutathione peroxidase-like activity in vitro and antioxidant capacity of PZ51 (Ebselen). Biochem. Pharmacol. 33:3235

Scheulen, M.E., Kappus, H., Thyssen, D., and Schmidt, C.G., 1981, Redox cycling of Fe(III)-bleomycin by NADPH-cytochrome P-450 reductase, Biochem. Pharmacol. 30:3385

Wendel, A., Otter., R., and Tiegs, G., 1986, Inhibition by Ebselen of microsomal NADPH-cytochrome P-450 reductase in vitro but not in vivo, Biochem. Pharmacol. 35:2995

Weinke, S., Kahl, R., and Kappus, H., 1987, Effect of four synthetic antioxidants on the formation of ethylene from methional in rat liver microsomes, Toxicol. Letters 35:247

ARE OXYGEN RADICALS RESPONSIBLE FOR THE ACUTE CARDIOTOXICITY OF DOXORUBICIN?

F.Piccinini, E.Monti, L.Paracchini, G.Perletti

Istituto di Farmacologia Applicata, Universita' di Milano

Via Celoria, 26 - I-20133 Milano, ITALY

INTRODUCTION

Oxygen-derived free radicals have been implicated as mediators of the acute cardiotoxic effects induced by doxorubicin (DXR) treatment. Superoxide anions ($O_2^-\cdot$) may be generated by xanthine oxidase or by one-electron reduction of DXR to the corresponding semiquinone and subsequent electron transfer to molecular oxygen. ($O_2^-\cdot$) generation plays a central role in the sequence of reactions leading to the formation of the more active hydroxyl radical ($\cdot$OH) according to the Haber-Weiss reaction (figure 1) (1). An additional source of active oxygen species is represented *in vivo* by the leukocytic enzyme myeloperoxidase, since DXR has been shown to stimulate the production of oxidants by neutrophils (2). Free radical-mediated membrane lipoperoxidation and the subsequent impairment of ion transport processes would eventually account for DXR-induced cardiodepressant effects (3-5).

The present study was performed in order to test the effects produced by DXR on isolated rat atria in the presence of oxypurinol (OXY) and N-acetylcysteine (NAC). OXY was shown to act by multiple mechanisms, including xanthine oxidase inhibition and direct scavenging of $\cdot$OH (6) and of hypochlorous acid (HClO) (7), which is the main product of myeloperoxidase activity. NAC might exert a protective effect against membrane lipoperoxidation, either by a direct mechanism or by conversion to glutathione (8).

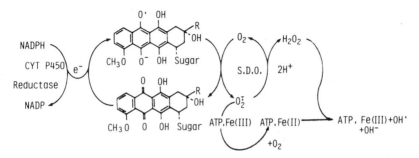

Figure 1 - Scheme for the enzymatic activation of the anthraquinone group and subsequent generation of superoxide anion, hydrogen peroxide and hydroxyl radicals.

Antioxidants in Therapy and Preventive Medicine
Edited by I. Emerit *et al.*
Plenum Press, New York, 1990

Based on the *in vitro* results, preliminary studies were performed *in vivo*, in order to check the ability of NAC and OXY to prevent the early cardiotoxic effects of DXR.

METHODS

In vitro studies

Groups of 4-6 spontaneously beating atria isolated from female CD rats (body weight 150 g) were allowed to equilibrate for 45 min in Tyrode's solution at 37 °C; either 0.2 mM OXY or 1.5 mM NAC were then added for 15 min, followed by a 60 min incubation in the presence of 0.2 mM DXR. Control preparations were maintained in Tyrode's solution throughout the experiment. The inotropic effects were evaluated by recording the maximal rate of tension development (dF/dt).

In vivo studies

Groups of 6 female CD rats (body weight 150 g) received i.v. a cardiotoxic dose of DXR (9.0 mg/Kg). OXY and NAC doses and administration schedules were chosen according to literature data (4,5): NAC (250 mg/Kg p.o.) one hour before and 2 hours after DXR administration, OXY (30.0 mg/Kg i.p.) one hour before DXR administration. ECG was recorded 1-2-7 days after treatment; QαT duration (calculated as the interval between the onset of the Q-wave and the apex of the T-wave) was assumed to be a reliable index of cardiac toxicity, according to the results of previous studies (9). The contractile performance of atria isolated from treated animals was also evaluated as described for the for the in vitro preparations after a 60 min incubation in Tyrode's solution.

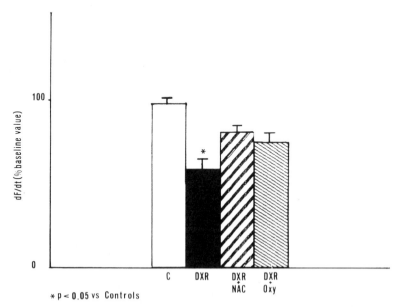

Figure 2 - Effect of doxorubicin (DXR, 0.2 mM), alone or in combination with oxypurinol (OXY, 0.2 mM) or N-acetylcyteine (NAC, 1.5 mM) on the contractile performance (dF/dt) of isolated rat atria. Mean ± S.E. of 4-6 preparations.

DXR *in vitro* produces a significant impairment of contractility in isolated rat atria (figure 2). This effect was significantly inhibited by NAC, which is *per se* devoid of pharmacological activity, thus suggesting that free radical production might be involved in this phenomenon. In contrast, OXY was found to exert a non significant effect on this parameter. This result seems to indicate that xanthine oxidase may not be involved in DXR-induced free radical generation; it is also possible that the concentration used in this study is too low to exert a direct scavenging action. However, this concentra- corresponds to the one usually adopted in vivo (6) and is the highest concen- tration devoid of intrinsic cardiotoxicity.

In vivo treatment with DXR produces a significant acute cardiotox- icity, which is documented by the progressive enlargement of QαT interval in the ECG. NAC was unable to affect this parameter (figure 3). This might be ex- plained by differences in pharmacokinetic behaviour: NAC levels in the heart might not be maintained as long as required in order to prevent the development of DXR-induced cardiotoxic effects. However, the possibility should be con- sidered that a "site-specific" free radical generation might occur very close to the bio logical targets, thus lessening the effectiveness of uniformly distri- buted scavengers. In contrast, OXY was found to inhibit the effect of DXR on QαT. The higher protective effect obtained *in vivo* with this agent suggests the possibility that production of active oxygen species by neutrophils is in- volved in DXR cardiotoxicity.

The contractile performance of atria isolated from treated rats was found to be unaffected by DXR. This observation seems to be in contrast with ECGraph- ic data; however, it has been shown that QαT is a very sensitive index of myo- cardial lesion in the rat (9,10), therefore this parameter may be altered by DXR administration before any gross functional damage becomes apparent. No differen- ces in contractile performance were observed with the two combined treatments.

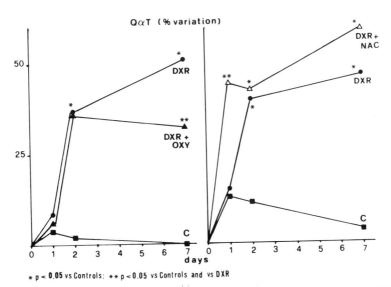

Figure 3 - Effect of doxorubicin (DXR, 9 mg/kg), alone or in combination with oxypurinol (OXY, 30 mg/kg; panel A) or N-acetylcysteine (NAC, 0.5 g/Kg; panel B) on QαT duration. Mean ± S.E. of 6 animals.

In conclusion, the results obtained in this study suggest that oxygen radicals probably play a causal role in the development of the acute phase of DXR-induced cardiotoxicity; the administration of free radical scavengers or antioxidants might prove useful in preventing DXR-induced cardiotoxicity, provided that adequate dosage and time schedules are adopted.

ACKNOWLEDGEMENT

Partially supported by C.N.R., Special Project "Oncology", grant n.87.01399.44

REFERENCES

1 - J. W. Lown, H.-H. Chen, Electron paramagnetic resonance characterization and conformation of daunorubicin semiquinone intermediate implicated in anthracycline metabolism, cardiotoxicity, and anticancer action, Can. J. Chem. 59: 3212 (1981).

2 - M. L. Schinetti, D. Rossini, A. Bertelli, Interaction of anthracycline antibiotics with human neutrophils: superoxide production, free radical formation and intracellular penetration., J. Cancer Res. Clin. Oncol. 113: 15 (1987).

3 - R. N. Harris, J. H. Doroshow, Effect of doxorubicin-enhanced hydrogen peroxide and hydroxyl radical formation on calcium sequestration by cardiac sarcoplasmic reticulum, Biochem. Biophys. Res. Commun. 130: 739 (1985).

4 - J. H. Doroshow, Role of reactive oxygen production in doxorubicin cardiac toxicity, in "Organ-Directed Toxicities of Anticancer Drugs", M. P. Hacker, J. S. Lazo, T. R. Tritton eds., Martinus Nijhoff Publ., Boston, p. 31, (1988).

5 - D. Ross, Glutathione, free radicals and chemotherapeutic agents, Pharmacol. Ther. 37: 231 (1988).

6 - P. C. Moorhouse, M. Grootveld, B. Halliwell, G. J. Quinlan, J. M. C. Gutteridge, Allopurinol and oxypurinol are hydroxyl radical scavengers, FEBS Lett. 213: 23 (1987).

7 - M. Grootveld, B. Halliwell, C. P. Moorhouse, Action of uric acid, allopurinol and oxypurinol on the myeloperoxidase derived oxidant hypochlorous acid, Free Rad. Res. Commun. 4: 69 (1988).

8 - J. H. Doroshow, G. Y. Locker, I. Ifrim, C. E. Myers, Prevention of doxorubicin cardiac toxicity in the mouse by N-acetylcysteine, J. Clin. Invest. 68: 1053 (1981).

9 - F. Villani, E. Monti, F. Piccinini, L. Favalli, E. Lanza, A. Rozza Dionigi, P. Poggi, Relationship between doxorubicin-induced ECG changes and myocardial alterations in rats, Tumori 72: 323 (1986).

10 - N. Bernardini, R. Danesi, M. C. Bernardini, M. Del Tacca, Fructose-1,6-diphosphate reduces acute ECG changes due to doxorubicin in isolated rat heart, Experientia 44: 1000 (1988).

INFLUENCE OF SELENIUM ON LIPID PEROXIDATION AND CARDIAC

FUNCTIONS IN CHRONICALLY ADRIAMYCIN-TREATED RATS

KOUKAY N*, MOUHIEDDINE S**, RICHARD M.J*, ARNAUD J*
De LEIRIS J**, FAVIER A*

* Labo Biochimie C, C.H.R.U.G., B.P 217 X
38043 GRENOBLE Cédex — FRANCE
** Labo. Physiologie Animale, URA C.N.R.S. 632, Université
J. FOURIER, 38000 Grenoble — FRANCE

INTRODUCTION

The Anthracycline antibiotic Adriamycin (Doxorubicin) is widely used in clinical medicine for cancer chemotherapy. Unfortunately, the clinical use of Adriamycin (ADR) is accompanied by side effects of which cardiotoxicity is the most serious. The cumulative dose-dependent cardiomyopathy severely limits the total dose of ADR that can be administered in the treatment of neoplastic diseases(1). The prevention of ADR cardiotoxicity without interference with the drug's antitumor activity would be of considerable advantage in achieving additional therapeutic benefit from this agent.

The identification of several biochemical activities of ADR in vitro has suggested different mechanisms for ADR cytotoxicity. These include direct interaction with nucleic acid, inhibition of coenzyme Q-dependent enzymes, direct cell surface interaction. However, the current leading hypothesis for the mechanism of ADR cardiotoxicity is that ADR produces reactive free radicals (2) that damage myocardial tissue by non-specific oxidation of membrane and cytosol molecules, ultimately leading to cell death.

In an effort to avoid the development of chronic cardiotoxicity of ADR, several potential cardioprotective compounds including selenium (Se), vitamin E, glutathione and other sulphydryl donors have been tested. There have been few studies in which the specific effect of Se deficiency on the toxicity of ADR has been investigated. DOROSHOW 1980 (3) found that Se deficiency increased the sensivity of mice to acute ADR toxicity. FACCHINETTI 1983 (4) showed that Se-supplemented rats had better resistance to chronic ADR treatment than Se-deficient rats.

In addition, mammalian cells are protected by several enzymatic systems e.g. superoxide dismutase, catalase and glutathione peroxidase. This latter enzyme is present in two different forms: selenium-independent glutathione peroxidase (GPx) and a selenium-dependent glutathione peroxidase (Se-GPx). The latter, which contains selenium, has a potential dual function by reducing hydrogen peroxide and already formed lipid hydroperoxides. Thus, exogenously administered selenium eliminates preferential oxygen centered free radicals thereby protecting the tissue from oxidative damage. This paper reports on the influence of selenium intake on endogenous antioxidant protective systems and therefore on cardiac functions in chronically adriamycin-treated rats.

Antioxidants in Therapy and Preventive Medicine
Edited by I. Emerit *et al.*
Plenum Press, New York, 1990

353

MATERIALS AND METHODS

Rats, diets and drug

Experiments were performed on male wistar rats of initial weight 180-200 g, supplied by IFFA CREDO, France. Sixty rats were allotted to six groups of 10 rats each, were housed in stainless steel cages and received granulated food and deionized distilled water ad libitum. Three groups were fed a Se-deficient basal diet obtained from U A R society (Villemoisson, France) with the following composition:

Beer yeast	30%
Sucrose	55,7%
Corn oil	8%
Salt mixture	5%
Vitamin mixture	1%
DL Methionine	0,3%
Selenium	0,058 ppm

The analysis of the basal diet showed that it contained 60 ug Se/Kg diet. The other three groups were fed the same basal diet supplemented with 1000 ug Se /kg diet as Na2SeO3. Rats were weighted weekly. Two weeks after beginning Se deficiency, two groups received intraperitoneal injections of ADR for 2,5 weeks (two injections of 4mg of ADR /kg body weight per week) for total doses of 20 mg ADR/kg body weight. ADR (Adriblastine), purchased from Roger Bellon, was reconstituted in deionized distilled water at a concentration of 2 mg/ml immediately before injection. Control rats (non drug treated) received an equivalent volume of sucrose solution at 10 mg/ml.

Two weeks after the last injection of ADR, rats were killed and hearts were quickly removed and mounted on the perfusion apparatus as described by LANGENDORFF 1895 (5). Hearts were perfused for 5 min with a constant perfusion pressure of 60 mm Hg through the coronary vessels, by a Krebs-Henseleit bicarbonate buffer. Ischaemia was performed at 37°C by a total stop of perfusion for 15 min. The reperfusion was allowed during 2 min after which hearts were immediately immersed in liquid nitrogen. During perfusion and reperfusion phases, coronary flow and cardiac frequence were recorded.

Biochemical analyses

After 18 hours of fasting, the rats were killed and tissues were removed and immersed in liquid nitrogen. Blood was collected from the chest cavity after heart had been excised. Se deficiency was verified by plasma Se determination (electrothermal AAS) and by measurement of Se-GPx in plasma, red blood cells (RBC) and in liver and heart homogenates by a modificaion of GUNZLER method 1984 (6). Plasma and tissue samples were analyzed for lipid peroxidation products: malondialdehyde (MDA) by YAGI method 1980 (7), dienes conjugated (D.C) as described by DORMANDY 1981 (8) and lipid hydroperoxides (LHP) by the enzymatic method of TAPPEL 1979 (9). Other antioxidant enzymatic and non enzymatic systems were also determined : Superoxide dismutase (SOD) was evaluated by MARKLUND method (10) glutathione reductase (GRx), glutathione transferase (GTx), catalase by the standard methods. The contents of reduced glutathione (GSH) and oxidized glutathione (Gs=sG) in the total blood and in tissues were determined as described by TIETZE 1969 (11). Alpha-tocopherol and xanthine oxidase activity were also evaluated.

STATISTICS

The results are expressed as means + standard deviation (S.D). Data were analysed statistically by using Student's t-test, and differences of P < 0,05 were considered statistically significant.

RESULTS

1- Selenium deficiency

Selenium deficiency was verified by plasma Se determination and by measurement of Se-GPx activities in blood, liver and heart homogenates. Deficient rats had 50% of plasma Se compared with the control rats. Se-GPx activities in red blood cells (RBC) and in the tissues confirmed Se deficiency realized on the appropriate experimental group . As shown in Table I, deficient rats exhibited a decreased Se and Se-GPx activities in all tested media.

Table I SELENIUM AND Se-GPX STATUS IN Se-DEFICIENT AND CONTROL RATS

DIET	PLASMA SELENIUM	PLASMA Se-GPX	RBC Se-GPX	HEPATIC Se-GPX	CARDIAC Se-GPX
	/umol/l	UI/ml	UI/g Hb	UI/g protein	UI/g protein
Selenium-adequate	4.42 ± 1.16	3.03 ± 0.62	643 ± 119	523 ± 77.5	256 ± 22.8
Selenium-deficient	2.36 ± 1.13 *	1.95 ± 0.35 *	281 ± 41.1 *	137 ± 42.0 *	208 ± 35.3 **

* p < 0.0001   ** p < 0.005

2- General toxicity

Thirty percent of animals receiving the drug died during the experiment. 75% of them developed ascites. They became generally unwell during the test and all of the ADR treated group suffered hair loss.

Selenium deficiency per se did not affect growth of rats. The ADR treatment completely depressed growth of Se-deficient and Se-supplemented rats (Fig. 1).

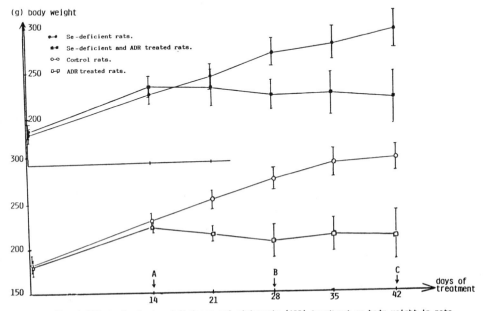

Fig. 1 Effect of selenium deficiency and adriamycin (ADR) treatment on body weight in rats. Beginning (A) and end (B) of ADR treatment; Analyses (C).

355

3- Effect of selenium and ADR treatment on cardiac functions :

i-effect of selenium

Perfusion phase

Selenium deficiency did not lead to any significant modifications. Coronary flow and heart rate measured were similar in the Se-deficient and Se-supplemented rats.

Reperfusion phase

Selenium deficiency did appear to affect the cardiac function after a period of ischaemia in vitro, especially at the level of coronary flow. The speed of cardiac recovery was faster in the Se-supplemented rats than it was in the Se-deficient rats.

ii-effect of ADR treatment

Perfusion phase

Chronic ADR treatment decreased the coronary flow but did not affect the heart rate in both groups studied, Se-deficient and Se-supplemented rats.

Reperfusion phase

Chronic treatment by ADR did not modify the speed of cardiac recovery after a period of ischaemia in vitro. Coronary flow and heart rate, measured after the ischaemia, were similar in ADR-treated or not ADR-treated rats.(Tab II)

Table II EFFECT OF SELENIUM AND ADR TREATMENT ON CARDIAC FUNCTIONS

DIET AND ADR TREATMENT		PERFUSION PHASE		REPERFUSION PHASE	
ADRIAMYCIN TREATMENT	DIET	CORONARY FLOW ml/min	HEART RATE pouls/min	CORONARY FLOW ml/min	HEART RATE pouls/min
ADR Treated rats	Se-deficient rats	6.6 ± 0.4	303 ± 7.2	5.3 ± 0.5	131 ± 28
	Se-suppl. rats	6.9 ± 1.59	313 ± 14.2	6.2 ± 1.1	182 ± 46
Non treated rats	Se-deficient rats	7.9 ± 1.25	305 ± 14.2	4.5 ± 1.26	136 ± 147
	Se-suppl. rats	8.1 ± 1.26	300 ± 35	5.8 ± 2.53	112 ± 146

These data demonstrate that ADR treatment decreased the coronary flow in all studied groups. Furthermore, Se supplementation increased the speed of cardiac recovery after the ischaemic phase, which is very important for the output of the heart and, therefore, on the total functions of the body.

4- Effect of selenium deficiency and chronic ADR treatment on lipid peroxides

The generation of malondialdehyde (MDA), as measured by the thiobarbituric acid (TBA) method, has been used by many investigators to determine the extent of lipid peroxidation in vivo. As can be seen in table III, Se deficiency or administration of ADR caused a marked increase in MDA content over that observed in the control rats. The estimations of conjugated dienes (D.C) and lipid hydroperoxides (LHP) have been used also. Both Se deficiency and chronic ADR treatment caused a significant increase in D.C levels as compared to the control rats.

Table III EFFECT OF SELENIUM DEFICIENCY AND CHRONIC MDA TREATMENT ON LIPID PEROXIDATION STATUS

ADRIAMYCIN TREATMENT	SELENIUM	PLASMA			HEPATIC TISSUE			CARDIAC TISSUE		
		MDA /uM/l	DC ABS	HP /uM/l	MDA nM/g prot	DC ABS	HP /uM/g prot	MDA nM/g prot	DC ABS	HP /uM/g prot
+	-	8.76+3.51	0.103+0.033	101+32.8	708+140	0.276+0.072	8.85+4.33	698+86.4	0.395+0.041	14.2+1.71
	+	7.61+2.51	0.092+0.028	118+20.9	865+315	0.290+0.031	7.39+0.81	806+63.6	0.425+0.092	12.9+2.26
	-	6.45+2.00	0.065+0.024	113+13.9	659+101	0.255+0.030	7.67+1.75	702+59.2	0.514+0.091	12.5+2.57
	+	5.79+2.71	0.065+0.017	123+19.8	592+79.8	0.244+0.040	8.06+1.72	601+58.9	0.423+0.058	10.2+1.88

5- Effect of selenium on the ischaemia-induced lipid peroxidation

We evaluated the ischaemia-induced lipoperoxide production with and without the ischaemic phase in Se-deficient and in control rats. Our results, table VI, demonstrated that the parameters studied (lipoperoxide production, enzymatic antioxidant systems) did not differ significantly from pre-ischaemia values. We think that 15 min of ischaemia is not long enough to obtain significant differences in the level of the mentioned parameters, and a longer period of ischaemia must be studied.

Table VI ISCHAEMIA - INDUCED LIPOPEROXIDE PRODUCTION

DIET	ISCHAEMIA	CARDIAC MDA nM/g prot	CARDIAC D.C. ABS	CARDIAC O.H.P. /uM/g prot	CARDIAC Se-GPX Ulg/g prot	Cardiac Vit E nM/g prot
SELENIUM	-	601 + 58.9	0.423 + 0.058	10.2 + 1.88	256 + 22.8	269 + 40.0
ADEQUATE	+	632 + 79.0	0.440 + 0.077	11.9 + 2.19	251 + 19.4	283 + 41.3
SELENIUM	-	702 + 59.2	0.514 + 0.091	12.45 + 2.57	208 + 35.3	326 + 54.0
DEFICIENT	+	640 + 72.9	0.450 + 0.075	13.3 + 1.55	199 + 34.6	239 + 27.3

DISCUSSION

Se deficiency obtained in our rats may be considered as satisfactory. Other investigators (CHEN X, 1986) (12), (FACCHINETTI T, 1983)(4) have used torula yeast as a protein source. This contains very little, if any, selenium (30 ug/Kg diet), and therefore leeds to profound Se deficiency. Our results demonstrated a less profound decrease in selenium level determined on the plasma blood and a corresponding decrease in Se-GPx measured in blood and tissue homogenates. This deficiency is closer to what we may expect to find in clinical situations.

Our results indicate that Se-deficient rats might be more sensitive to some of ADR toxic effects than are Se-supplemented rats. The magnitude of the growth depression seen in our Se-supplemented or Se-deficient rats treated with the ADR is similar to that reported by others. FACCHINETTI 1983 (4) found comparable weight losses in their Se-deficient and control rats after chronic ADR treatment. They used, as in the present study, adult rats that had been fed the experimental diet for two weeks before the drug treatment commenced.

The specificity of the TBA method for determining lipid peroxidation has drawn recent criticism. Nevertheless, many previous studies used MDA formation as an index of peroxidation to point out the effect of Se supplementation and chronic ADR treatment. MIMNAUGH 1981 (13) showed that ADR dramatically increased lipid peroxidation when added in vitro to cardiac microsomes from alpha-tocopherol-deficient rats. Later, these workers found that treatment of rats with ADR in vivo caused biochemical alterations in cardiac microsomal membranes that predisposed them to subsequent attack in vitro by anthracycine free radical-generated oxyradicals. MIMNAUGH 1985 (14) presented results supporting the concept that oxyradical-mediated membrane lipid peroxidation may also play a role in ADR-generation of mitochondrial dysfunction.

DOROSHOW 1980 (3) found that Se deficiency markedly increased the acute toxicity of ADR in mice and that acute doses of ADR depressed cardiac glutathione peroxidase activity. REVIS and MARUSIC 1987 (15) treated rabbits with 1,5 mg ADR/Kg body weight three times per week for three wks. They noted, as in our case, that both GPx activity and Se content were reduced in ADR-treated rats. They concluded that this increase in lipid peroxidation may be the result of decrease in GPx activity.

Our results indicate that ADR may exert some of its toxic effects by lipoperoxidative mechanisms. For example, plasma vitamin E levels were lowest in the Se-deficient and ADR-treated rats. Such depressed vitamin E levels could indicate increased destruction of alpha-tocopherol by the drug or by its radical intermediates. Administration of high doses of vitamin E alone had no substantial protective effect against ADR-induced cardiac toxicity, but little attention has been given to the possible effect of vitamin E and / or Se deficiency on the patient's tolerance to the drug. On this basis, we could conclude that Se-GPx, and therefore Se, play an important role in the prevention of free radical damage to the heart from ADR.

CONCLUSION

Our results are in accordance with previously reported data.: ADR produces its chronic cardiotoxicity through an oxidative mechanism, and Se deficiency potentiates this effect. We found lower GPx, Se, and vitamin E levels in the treated group.

Finally, selenium supplemented rats recover their coronary flow faster. So selenium and vitamin E are effective in preventing some of the cardiac lesions produced by ADR in experimental animals. This could be assumed to have some importance in humans, because patients treated by ADR may have low Se status due to their malnutrition, if they suffer from or have undergone surgery for intestinal cancer. Patients receiving parenteral nutrition often do not obtain sufficient Se supplementation. These problems are magnified in countries with very low Se status as in China or New Zealand. Many European countries (Finland, Italy, France..) also have a suboptimal status of Se. The protective effect of vitamin E and Se against ADR cardiotoxicity need to be evaluated, therefore, in clinical studies.

ACKNOWLEDGMENTS

The authors gratefully acknowledge the technical assistance of M. RUAL and J MEO, and to E Wheelwright for English correction..

This work was supported by LABCATAL laboratory 92120 MONTROUGE - FRANCE.

LITERATURE CITED

1-MINOW R.A., BENJAMIN R.S., & GOTTLIEB J.A., Adriamycin (NSC-123127) cardiomyopathy: An overview with determination of risk factors. *Cancer. Chemotherapy. Rep.* 1975, 3 **6** (2) : 195-201.

2-GOODMAN J., and HOCHSTEIN P., Generation of free radicals and lipid peroxidation by redox cycling of adriamycin and daunomycin. *Biochem. Biophys. Res. Commun.* 1977, **77** (2) : 797.

3-DOROSHOW J.H., LOCKER G.Y., and MYERS C.E., Enzymatic defenses of the mouse heart against reactive oxygen metabolites. Alterations produced by doxorubicin. *J. Clin. Invest.* 1980, **65** : 128.

4-FACCHINETTI T., DELAINI F., SALMONA M., DONATI M.B., FEUERSTEIN S., & WENDEL A., The influence of selenium intake on chronic adriamycin toxicity and lipid peroxidation in rat. *Toxicology. Lett.* 1983, **15** : 301-7.

5-LANGENDORFF O., Untersuchungen am überleben säugethierherzen. *Arch. Ges. Physiol.* 1895, **61** : 291.

6-GUNZLER W.A., KREMERS H., & FLOHE L., An improved coupled test procedure for glutathione peroxidase (EC 1.11.1.9) in blood. *Z. Klin. Chem. Klin. Biochem.* 1974, **10** : 444-448.

7-YAGI K., A simple fluorometric assay for lipoperoxide in blood plasma. *Biochem. Res.* 1976, **15** : 212-6.

8-WICKENS D., WILKINS M.H., LUNEC J., BALL G., & DORMANDY T.L., Free radical oxidation (peroxidation) products in plasma in normal and abnormal pregnancy. *Ann. Clin. Biochem.* 1981, **18** : 158-62.

9-HEATH R.L., & TAPPEL A.L., A new sensitive assay for the measurement of hydroperoxides. *Anal. Biochem.* 1976, **76** : 184-91.

10-MARKLUND S.L., Spectrophotometric study of spontaneous disproportionation of superoxide anion radical and sensitive direct assay for superoxide dismutase. J. *Biol. Chem.* 1976, **251** (23) : 7504-7.

11-TIETZE F., Enzymatic method for quantitative determination of nanogram amounts of total and oxidized glutathione: Applications to mammalian blood and other tissues. *Anal. Biochem.* 1969, **27** : 502-22.

12-CHEN X., XUE A., MORRIS V.C., FERRANS V.J., HERMAN E.H., EL-HAGE A., & LEVANDER O.A., Effect of selenium deficiency on the chronic toxicity of adriamycin in rats. J. *Nutr.* 1986, **116** (12) : 2453-65.

13-MIMNAUGH E.G., TRUSH M.A., & GRAM T.E., Stimulation by adriamycin of rat heart and liver microsomal NADPH-dependent lipid peroxidation. *Biochem. Pharmacol.* 1981, **30** (20) : 2797-804.

14-MIMNAUGH E.G., TRUSH M.A., BHATNAGAR M., & GRAM T.E., Enhancement of reactive oxygen-dependent mitochonrial membrane lipid peroxidation by anticancer drug adriamycin. *Biochem. Pharmacol.* 1985, **34** : 847-56.

15-REVIS N.W., & MARUSIC N., Glutathione peroxidase activity and selenium concentration in tne hearts of doxorubicin-treated rabbits. *J. Mol. Cell. Cardiol.* 1978, **10** : 945-51.

IMPROVEMENT OF ISCHEMIC AND POSTISCHEMIC MITOCHONDRIAL FUNCTION BY DEFER-

RIOXAMINE : THE ROLE OF IRON

H. van Jaarsveld, G.M. Potgieter, S.P. Barnard, and
S. Potgieter

Dept. Chemical Pathology, University of the Orange Free State
P.O. Box 339, Bloemfontein, Republic of South Africa

INTRODUCTION

Oxygen-derived free radicals have been implicated as a general mechanism of cell injury, including ischemia[1] and reperfusion[1,2]. Superoxide anions and other reactive products of oxygen metabolism can be formed from subsequent intracellular reduction of oxygen, including hydrogen peroxide and the hydroxyl radical. However, in the myocardium[3] and the mitochondria[4] there exist a series of defence mechanisms to protect the cell against cytotoxic oxygen metabolites, which may be overwhelmed by an overproduction of oxygen free radicals.

Xanthine oxidase, a generator of superoxide anions is an enzyme formed by conversion of xanthine dehydrogenase during reperfusion in the myocardial mitochondria[4]. During reperfusion oxygen is available and superoxide anions can therefore be formed. Iron again may play a key role by converting the relatively innocuous superoxide anion into the strongly oxidising hydroxyl radical via the iron-catalysed Haber-Weiss reaction[2]. Deferrioxamine (desferal) is an inhibitor of free radical production. It prevents the formation of OH by two conceptually different strategies: the first consists of inhibiting the Haber-Weiss reaction by reducing the availability of its substrates. The second consists of slowing the rate of the reaction by reducing the availability of its metal catalyst, ferri ions, through chelation[5].

Mitochondrial oxidative phosphorylation is responsible for most of the ATP needed for cellular integrity. Myocardial ischemia is always associated with defective mitochondrial oxidative phosphorylation as has been observed by the mitochondrial oxygen uptake in the presence of ADP[6]. It is therefore important to improve this defect. Mitochondria appear to be an important site of oxygen free radical production within the myocytes[3], and oxygen free radicals may play a role in mitochondrial damage during ischemia and reperfusion. The aim of this study was to determine whether desferal improves mitochondrial function and whether free iron plays a role.

Antioxidants in Therapy and Preventive Medicine
Edited by I. Emerit *et al.*
Plenum Press, New York, 1990

METHODS

Animals

Male Wistar rats (300 - 400 g) were used. The rats were allowed free access to food and water prior to experimentation.

Production of ischemia and reperfusion

The hearts were rapidly excised and arrested in ice-cold saline. The aorta was mounted on a cannula and perfused retrogradely at 65 mm Hg for a period of 10 min. During this time the left atrium was cannulated. At the end of this period the heart was perfused for 5 min according to the working heart technique of Neely et al[7]. Global ischemia was induced for different time-intervals by clamping both atrial inflow and aortic outflow. Reperfusion was performed by unclamping the aortic cannula, while the atrial output remained clamped. Retrograde perfusion of the heart was then performed for 20 min, followed by a further 10 min as working heart model[8].

Krebs Henseleit bicarbonate buffer[9] containing 2.5 mM calcium equilibrated with 95 % O_2 : 5 % CO_2 at 37 °C was used as perfusate. In some experiments desferal was added to a concentration of 0.89 mM.

Mitochondrial assays

Mitochondria were isolated in 0.18 mM KCl, 0.01 M EDTA (pH 7.4) as previously described[10]. Mitochondrial oxidative phoshorylation was determined polarographically using a Gilson oxygraph. Glutamate was used as substrate and state 3 respiration initiated by the addition of ± 500 nmoles ADP. The incubation medium contained the following: 0.25 M sucrose, 10 mM Tris-HCl (pH 7.4), 8.5 mM K_2HPO_4, 5 mM glutamate (Tris salt, pH 7.4), 1.5 mM EDTA (pH 7.4)[11]. In all experiments 3 ± 0.15 mg mitochondrial protein was added to the incubation medium.

Cytosolic iron

Hearts were homogenized in 0.25 M sucrose, 5 mM Tris (pH 7.4) and centrifuged 60 min at 105 000 g. A quarter volume of 20 % TCA was added to the supernatant and allowed to stand for 10 min, whereafter it was centrifuged for 10 min at 1 300 g. The supernatant was decanted and 500 µℓ of it was added to a tube containing 300 µℓ water, 50 µℓ 0.1 % o-phenanthroline, 50 µℓ saturated ammonium acetate. This solution was incubated 10 min at 37 °C for maximum color development. The absorbance was read at 510 nm against a blank containing all reagents except the o-phenanthroline. This was converted to iron concentration using a standard curve, and the results expressed as nmol iron / mg protein[12].

Protein content

The protein content was measured by the method of Lowry et al[13].

Statistical analysis

All values were expressed as mean ± SEM; p values were calculated by student's t-test, and values $p < 0.05$ were regarded as significant.

RESULTS AND DISCUSSION

Figure 1 shows the mitochondrial oxidative phosphorylation function as observed by the mitochondrial oxygen uptake in the presence of ADP. Ischemia alone impaired the mitochondrial function after 30 min. Reperfusion deteriorated the function further and a significant impairment was observed sooner. Desferal in the perfusate protected mitochondrial function to such an extent that it was not inhibited after 30 min ischemia or ischemia followed by reperfusion. This effect was however higher than that found without ischemia or reperfusion when desferal was absent from the perfusate. Desferal in the reperfusate improved the mitochondrial function statistically, but was not able to protect it completely. It therefore prevented the ischemic damage as well as the additional damage that occurred during reperfusion.

Unpublished results from our laboratory indicate that the mitochondrial iron content stays unaltered during ischemia as well as reperfusion. The iron-catalysed Haber-Weiss reaction was therefore not accelerated. It is however possible that the concentration of the substrates of the Haber-Weiss reaction increased and that enough iron was available to catalyse this reaction. This should accelerate hydroxyl radical production and therefore mitochondrial damage. Xanthine oxidase in the mithocondria was enhanced and more superoxide anions produced[4]. Unpublished data from our laboratory show that desferal does not inhibit this reaction. Xanthine oxidase activity is slow and some researchers failed to detect it in myocardial mitochondria[14]. The contribution of this reaction is therefore questionable.

Mitochondrial damage could also take place from outside the mitochondria. Table I shows that the cytosolic iron increased during

Table 1. Effect of ischemia and of reperfusion
on cytosolic iron content
(nmole / mg protein).

Min Ischemia	Without drugs	Desferal in perfusate	Desferal in reperfusate
0	0.452±0.067	0.414±0.04	---
10	0.715±0.059*	0.509±0.43°	---
+ Reperfusion	0.776±0.066*	0.518±0.081°	0.567±0.033°
20	1.084±0.051*	0.629±0.05°	---
+ Reperfusion	1.063±0.107*	0.614±0.026°	0.574+0.025°
30	1.089±0.073*	0.634±0.024°	---
+ Reperfusion	1.102±0.068*	0.652±0.04°	0.547±0.042°

* Versus control
° Desferal in (re)perfusate versus perfusion without drugs
Each value represents mean ± SEM of 6 - 8 samples

ischemia alone, but reperfusion did not alter it any further. The Haber-Weiss reaction could be accelerated and more of the strongly oxidising hydroxyl radicals formed. These radicals can attack the mitochondria from the outside. Desferal in the perfusate inhibited the rise in iron completely and also protected the mitochondrial function completely (figure 1). When deferrioxamine was added after ischemia in the reperfusate it chelated the iron to the value of hearts that were

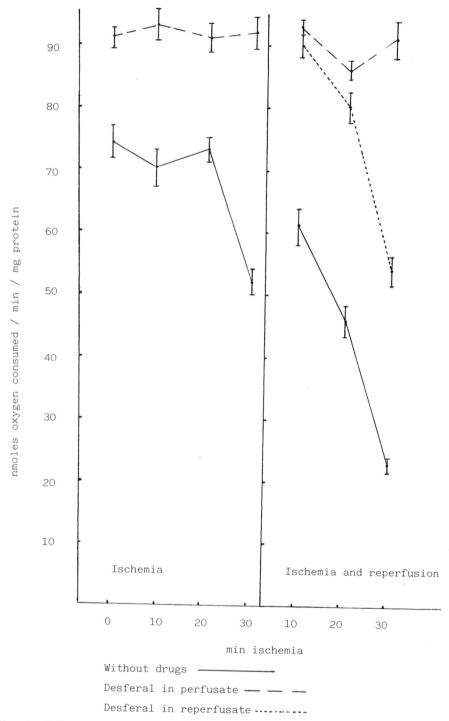

Fig. 1. Effect of ischemia and of reperfusion on mitochondrial function

not subjected to ischemia (table 1). The mitochondrial function was not protected completely, but the reperfusion damage was excluded (figure 1). Free radicals are produced even at low oxygen tensions and initiate lipid peroxidation[15]. This condition is present in ischemia.

Considering the high iron concentration, these results indicated that the Haber-Weiss reaction could take place to some extent during ischemia. Removal of the iron by chelation protected the mitochondria from ischemic damage and inhibited the impairment that occurred during reperfusion.

Iron therefore promoted the damage that occurred and if not available no impairment was observed. We came to the conclusion that the presence of cytosolic iron plays an important role in myocardial ischemia and reperfusion.

ACKNOWLEDGEMENTS

We thank the South African Research Council for financial aid and Miss H. Henning for typing the manuscript.

REFERENCES

1. H.A. Kontos, M.L. Hess. Oxygen radicals and vascular damage. Adv. Exp. Med. Biol.. 15: 269-78 (1983).
2. G. Ambrosio, J.L. Sweier, W.E. Jacobus, M.L. Weisfeldt, J.T. Flaherty. Improvement of postischemic myocardial function and metabolism by administration of deferoxamine at the time of reflow: the role of iron in the pathogenesis of reperfusion injury. Circulation 76 (4): 906-915 (1987).
3. R. Ferrari, C. Ceconi, S. Curella, A. Cargnoni, D. Medici. Oxygen free radicals and reperfusion on the cellular ability to neutrolise oxygen toxicity. J. Moll. Cell. Cardiol. 18 (Suppl. 4): 67-69 (1986).
4. H. van Jaarsveld, A.J. Groenewald, G.M. Potgieter, S.P. Barnard, W.J.H. Vermaak, H.C. Barnard. Effect of normothermic ischemic cardiac arrest and of reperfusion on the free oxygen radical scavenger enzymes and xanthine oxidase (a generator of superoxide anions). Enzyme 39: 8-16 (1988).
5. P. Menasche, C. Grousset, Y. Gauduel, C. Mouas, A. Piwnica. Prevention of hydroxyl formation: a critical concept for improving cardioplegia. Protective effect of deferoxamine. Circulation 76 (suppl. V): V180-185 (1987).
6. S. Sugiyana, Y. Miyazaki, K. Kotaka, T. Ozawa. On the mechanism of ischemia-induced mitochondrial dysfunction. Jap. Circ. J. 46: 296-302 (1982).
7. J.R. Neely, H. Liebermeister, E.J. Battersby, H.E. Morgan. Effect of pressure development on oxygen consumption by isolated rat heart. Am. J. Physiol. 212: 804-814 (1974).
8. A. Lochner, I. van Niekerk, J.C.N. Kotze. Normothermic ischaemic cardiac arrest of the isolated perfused rat heart: effects of trifluoperazine and lysolecithin on mechanical and metabolic recovery. Basic Res. Cardiol. 80: 363-376 (1985).
9. H.A. Krebs, K. Henseleit. Untersuchungen über die Harnstoff-Bildung im Tierkörper. Hoppe-Seylers. Z. Physiol. Chem. 210: 33-66 (1932).
10. H. van Jaarsveld, G.M. Potgieter, A. Lochner. Changes in NADH-ubiquinone reductase (complex I) with autolysis in the rat heart as experimental model. Enzyme 35: 206-214 (1986).

11. H. Kahles, G.G. Göring, H. Nordbeck, C.J. Preusse, P.G. Spiekermann. Functional behaviour of isolated heart muscle mitochondria of isolated heart muscle mitochondria after in situ ischemia. Polarographic analysis of mitochondrial oxidative phosphorylation. Basic Res. Cardiol. 72: 563-574 (1977).
12. G.S. Krause, K.M. Joyce, N.R. Nayini, C.L. Zonia, A.M. Garritano, T.J. Hoehner. Cardiac arrest and resuscitation: brain iron delocalization during reperfusion. Ann. Emerg. Med. 14: 1037-1043 (1985).
13. O.H. Lowry, N.J. Rosenbrough, A.L. Farr, R.J. Randall. Protein measurement with the Folin phenol reagent. J. Biol. Chem. 193: 265-275 (1951).
14. H. Nohl. A novel superoxide radical generator in heart mitochondria. FEBS Lett. 214: 269-273 (1987).
15. E. Röth, B. Török, T. Zsoldos, B. Matkovics. Lipid peroxidation and scavenger mechanism in experimentally induced heart infarcts. Basic Res. Cardiol. 80 : 530-536 (1985).

ALLOPURINOL IN ISCHEMIA - REPERFUSION INJURY OF HEART

Ingrid Emerit and Jean-Noel Fabiani

CNRS and University Paris VI,
15 rue de l'Ecole de Medicine
Hospital Broussais, 96 rue Didot, Paris

Allopurinol, chemically known as 4-hydroxypyrazolo-(3,4-d)pyrimidine, is an analogue of hypoxanthine, in which the positions of N7 and C8 are reversed. It is metabolized to oxypurinol, the 4,6-dihydroxy analogue. Both allopurinol and oxypurinol are inhibitors of xanthine oxidase. They also inhibit other enzymes of purine metabolism such as purine nucleoside phosphorylase and pyrimidine deoxyribosyl transferase. In ribonucleotide linkage, allopurinol inhibits early enzymes of both purine and pyrimidine biosynthesis. Extensive clinical experience attests to the usefulness of allopurinol as an antihyperuric agent in gout.[1]

In recent years, the drug has been considered to be protective in myocardial ischemia and reperfusion injury, but its mechanism of action in this context remains controversial. Dogs pretreated with allopurinol had better cardiac output and myocardial contractility after coronary artery ligation compared to untreated controls.[2] Reduction of infarct size was observed as well in the reperfused as in the permanently occluded canine heart, when allopurinol was applied.[3-5] The survival rate of dogs subjected to 10 min. of cardiac arrest after electrically induced fibrillation was increased with allopurinol.[6] A beneficial effect of allopurinol was also reported in the rat, as well for reperfusion induced arrhythmias as for reduction of infarct size.[7-8] The efficacy of allopurinol in preventing reperfusion-induced ventricular fibrillation could be observed in Japan in patients undergoing open heart surgery with cardiopulmonary bypass.[9] For ten years at the Broussais Hospital in Paris, in a large series of patients having open heart surgery, allopurinol improved the hemodynamic parameters, when added to the cardioplegic solution.[10]

However, failure of allopurinol to limit infarct size was also reported and it is possible that the controversial results may be due to differences in the experimental protocol which was discussed.[11,12] Allopurinol is less effective after prolonged ischemia and more effective, when given as a pretreatment. It has been claimed that pretreatment for at least 12-18 hours before ischemia is necessary for allopurinol to be of benefit. This may be because of the time required for accumulation of oxypurinol, its metabolite, which as a non competitive inhibitor of xanthine oxidase can more effectively antagonize the enzymatic activity of xanthine oxidase regardless of the substrate content. Acute treatment would be expected to be less effective.

Antioxidants in Therapy and Preventive Medicine
Edited by I. Emerit *et al.*
Plenum Press, New York, 1990

367

While the earlier reports related the beneficial effects of allopurinol to preservation of the pool of purine bases for the resynthesis of adenine nucleotides via the purine salvage pathway[2,6], much of the recent interest in allopurinol has focused on its ability to prevent the formation of oxygen free radicals.[13-15] The source of these free radicals is still subject of debate. It was proposed that oxygen radicals are generated during reoxygenation of the ischemic myocardium by the action of xanthine oxidase on hypoxanthine.[16] Adenosine triphosphate is progressively degraded during ischemia to yield hypoxanthine, which then acts as a substrate for xanthine oxidase. This enzyme is thought to be derived from xanthine dehydrogenase by the action of a protease. On reperfusion, molecular oxygen becomes available, and all participants are present for the reaction producing superoxide and hydrogen peroxide. Both EPR and spin trapping techniques have demonstrated a burst of oxygen radicals after post-ischemic reperfusion of the heart.[17-20] Also the detection of an ischemia-induced xanthine oxidase that is structurally distinct from the xanthine dehydrogenase found in normal tissues, comes in confirmation of the proposed hypothesis.[21] Endothelial cells, which contain xanthine dehydrogenase or oxidase,[22] produce superoxide and hydroxyl radical.[23] This was demonstrated on endothelial cells in culture with EPR techniques using the spin trap DMPO, Oxyradical formation was markedly decreased by oxypurinol.

However, the absence of xanthine oxidase in the rabbit and the pig heart, where allopurinol was protective, indicated that the mechanism of allopurinol cardioprotection cannot be universally ascribed to its inhibition of xanthine oxidase.[24,25,26] Other hypotheses were proposed. Since both allopurinol and oxypurinol significantly inhibited free radical signals generated by activated neutrophils, they were said to act as free radical scavengers.[27] Examination of the chemical/physical structure of allopurinol suggested that its effectiveness may indeed be due to an ability to scavenge hydrated electrons and hydrogen atoms that would react with oxygen to form superoxide.[28]

Work of our laboratory indicates that allopurinol can prevent the formation of clastogenic factors (CF) which are released by cells under conditions of oxidative stress. CF are observed after irradiation[29], in chronic inflammatory diseases[30], in cancer prone disorders such as ataxia telangiectasia[31] and Bloom's syndrome[32]. They may be induced in vitro by exposure of cells to a source of superoxide, e.g. the hypoxanthine xanthine oxidase system[33,34]. Therefore, it seemed interesting to search for CF in the plasma of patients undergoing coronary bypass surgery and to examine the protection of allopurinol against CF formation. In a blind study of 14 randomized patients, those operated without allopurinol treatment had clastogenic material in their plasma.[35] This was documented by the presence of increased chromosome damage in lymphocyte cultures set up with patients' blood and in test cultures set up with blood from healthy persons, to which ultrafiltrates from patients' plasma had been added. The clastogenic effect was not found in the cultures exposed to ultrafiltrates from patients treated with allopurinol. Also the cell cultures of treated patients did not exhibit chromosome damage. The blood was taken in the coronary sinus before and 20 min. after cross-clamping of the aorta. None of the preischemic samples contained CF. Two or three bypasses were performed with an ischemia time of 50 min. approximately.

Allopurinol was given twice: 100 mg were injected in 1 l of cold cardioplegic solution immediately after clamping of the aorta, the second injection of 100 mg was given in a warm blood cardioplegia solution before the clamp was released. The number of patients was too small for the study of correlations between CF formation and clinical or hemodynamic parameters.

When the kinetics of CF formation were studied, no CF was found in the samples collected immediately after the release of the aortic clamp. Samples taken 5 min. after reperfusion yielded discordant results, while all samples taken after 10 min. were clastogenic. For this reason, this sampling time is now used for further investigations. CF could be detected in venous blood 3 to 7 days after the operation. Later times have not been studied yet.

Since previous work had shown that the chromosome damaging effect of CF can be prevented by addition of superoxide dismutase to the culture medium, the specific substrate of this enzyme, the superoxide anion radical appears to be implicated not only in CF formation but also in CF action. We therefore examined whether CF containing ultrafiltrates stimulate O_2^- production by meeting neutrophils. SOD inhibited superoxide production was measured with the cytochrome C assay after exposure of cells to CF during 90 min. When the increases in superoxide production induced by pre- and post-ischemic samples were compared between 5 untreated and 6 allopurinol-treated patients, the mean values in the untreated group were significantly higher than in the treated group: 4.8 nmoles O_2^-/2.5 x 10^6 PMN/90 min. compared to 0.8 nmoles. Four of the six treated patients did not show any difference between the measurements done with pre- and postischemic samples. The other two showed moderate increases of 2.5 and 2.0 nmoles. Three of the untreated patients yielded values between 6 and 8 nmoles, while the other two were in the range of the treated group, 2.1 and 0.9 nmoles respectively.

Similar results had been observed with ultrafiltrates from xanthine oxidase-treated cell culture media and with plasma ultrafiltrates from patients with progressive systemic sclerosis.[36] The chemical nature of the clastogenic and superoxide stimulating components of the plasma ultrafiltrates is not known at present. From the pore size of the ultrafiltration membranes, it can be concluded that the material has a molecular weight of less than 10,000 daltons, as the other previously described CF. Hydrogen peroxide could be excluded as a candidate. Lipid peroxidation products, as they were found in the CF generated in vitro by a source of superoxide[34], may also be present in CF samples from patients' plasma. Certain of these products, in particular aldehydic breakdown products, have genotoxic properties. The aldehyde 4-hydroxynonenal has been shown to induce DNA fragmentation and sister chromatid exchanges.[37] It was found in 50% of CF samples from xanthine oxidase-treated cultures, and the synthetic product was indeed clastogenic (Emerit et al., in prep.). We were not able to confirm the presence of this clastogenic aldehyde in the CF samples from patients after ischemia-reperfusion. This may be due to insufficient sensitivity of the method. Further analysis of the material by HPLC fractionation and mass spectrometry is in progress. The chemical identification of CF will probably contribute to the elucidation of the mechanisms involved in ischemia-reperfusion injury and the protective effect of allopurinol. The observation of a superoxide stimulating effect of CF on competent cells may be quite relevant to neutrophil and monocyte induced endothelial cell damage and ultimate infarct size.

References

1. J.B. Wyngaarden and W.N. Kelley. Disorders of purine and pyrimidine metabolism. In "The metabolic basis of inherited disease". J.B. Stanbury, D.S. Fredrickson,J.B. Wyngaarden, J.L. Goldstein and M.S. Brown eds. McGraw-Hill Book Comp. New York, 1983, pages 1102-1105.
2. R.A. DeWall, K.A. Vasko, E.L. Stanley and P. Kezdi. Responses of the ischemia myocardium to allopurinol. Am. Heart J. 82:362-370 (1986).
3. D.E. Chambers, D.A. Parks, G. Patterson, R. Roy, J.M. McCord, S. Yoshida, L.F. Parmley and J.M. Downey. Xanthine oxidase as a source of free radical damage in myocardial ischemia. J. Mol. Cell Cardiol. 17:145-152 (1985).
4. S.W. Werns, M.J. Shea and B.R Lucchesi. Free radicals and myocardial injury: pharmacologic implications. Circulation 74:1-5 (1986).
5. S. Akizuki, S. Yoshida, D.E. Chambers, L.J. Eddy, L.F. Parmley, D.M. Yellon and J.M. Downey. Infarct size limitation by the xanthine oxidase inhibitor allopurinol in closed chest dogs. Cardiovasc. Res. 19:686-692 (1985).
6. J.C. Parker and E.E. Smith. Effects of xanthine oxidase inhibition in cardiac arrest. Surgery 71:339-344 (1972).
7. A.S. Manning. Reperfusion-induced arrhythmias: do free radicals play a critical role? Free Radical Biol. Med. 4:305-316 (1988).
8. J.N. Fabiani. The no-reflow phenomenon following early reperfusion of myocardial infarction and its prevention by various drugs. Heart Bull. (La Haye) 7:134-136 (1976).
9. H. Adachi, K. Motomatsu, I. Yara. Effect of allopurinol on patients undergoing open heart surgery. Japanese Circ. J. 43:3935-401 (1979).
10. P. Perier, J.N. Fabiani, M. Bocher. Protection myocardique pendant l'arret cardiaque ischemique; etude hemodynamique des effets de l'allopurinol dans une solution cardioplegique. Arch. Mal. Coeur 73:713-717 (1980).
11. K.A. Reimer and R.B. Failure of the xanthine oxidase inhibitor allopurinol to limit infarct size after ischemia and reperfusion in dogs. Circulation 71:1069-1075 (1985).
12. P.J. Simpson, J.K. Mickelson and B.R. Lucchesi. Free radical scavengers in myocardial ischemia. Federation Proc. 46:2413-2421 (1987).
13. D.E. Chambers, D.A. Parks, G. Patterson. Role of oxygen-derived free radicals in myocardial ischemia. Fed. Proc. 47:1093-1099 (1983).
14. J.R. Stewart, S.L. Crute, M.L. Loughlin, M.L. Hess and L.J. Greenfield. Prevention of free-radical-induced myocardial reperfusion injury with allopurinol. J. Thorac. Cardiovasc. Surg. 90:68-72 (1985).
15. D.J. Hearse, A.S. Manning, J.M. Downey and D.M. Yellon. Xanthine oxidase: a critical mediator of myocardial injury during ischemia and reperfusion? Acta. Physio. Scand. 126:65-78 (1986).
16. J.M. McCord. Oxygen-derived free radicals in postischemic tissue injury. N. Engl. J. Med. 312:159-160 (1985).
17. J.L. Zweier, J.T. Flaherty, M.L. Weisfeldt. Direct measurement of free radical generation following reperfusion of ischemic myodardium. Proc. Natl. Acad. Sci. (USA) 84:1404-1407 (1987).

18. P.B. Garlick, J.M. Davies, T.S. Slater, D.J. Hearse. Detection of free radical production in the isolated rat heart using a spin trap agent and electron spin resonance. Circ. Res. 61::757-760 (1987).

19. C.M. Arroyo, J.H. Kramer, R.H. Leibof, G.W. Mergner, B.f. Dickens and W.B. Weglicki. Spin trapping of oxygen and carbon-centered free radicals in ischemic canine myocardium. Free Rad. Biol. Med. 3:313-316 (1987).

20. R. Bolli, B.S. Patel, M.O. Jeroudi, E.K. Lai and P.B. McCay. Demonstration of free radical generation in "stunned" myocardium of intact dogs with the use of the spin trap alpha-phenyl N-tert-butyl- nitone. J. Clin. Invest. 82:476-485 (1988).

21. T.D. Engerson, T.G. McKelvey, D.G. Rhyne, E.B. Boggio, S.J. Snyder and H.P. Jones. Conversion of xanthine dehydrgenase to oxidase in ischemic rat tissues. J. Clin. Invest. 79:156401570 (1988).

22. E.D. Jarasch, C. Grund, G. Bruder, H.W. Heid, T.W. Keenan and W.W. Franke. Localization of xanthine oxidase in capillary endothelial cells. Acta. Phsyiol. Scand. Suppl. 548:39-46 (1986).

23. J.L. Zweier, P. Kuppusamy and G. A. Lutty. Measurement of endothelial cell free radical generation: Evidence for a central mechanism of free radical injury in postischemic tissues. Proc. Natl. Acad. Sci. (USA) 85:4046-4050 (1988).

24. D.V. Godin and S. Bhimji. Effects of allopurinol on myocardial ischemic injury induced by coronary artery ligation and reperfusion. Biochem. Pharmacol. 36:2101-2107 (1987).

25. C.M. Grum, R.A. Ragsdale, L.H. Ketai and M. Shlafer. Absence of xanthine oxidase or xanthine dehydrogenase in the rabbit myocardium. Biochem. Biophys. Res. Commun. 141:1104-1108 (1986).

26. D.K. Das, R.M. Engelmann, R. Clement, H. Otani, M.R. Prasad and P.S. Rao. Role of xanthine oxidase inhibitor as free radical scavenger: a novel mechanism of action of allopurinol and oxypurinol in myocardial salvage. Biochem. Biophys. Res. Comm. 148:314-319 (1987).

27. D.A. Peterson, B. Kelly and J.M. Gerrard. Allopurinol can act as an electron transfer agent. Biochem. Biophys. Res. Commun. 137:76-79 (1986).

28. L.G. Cockerham, C.M. Arroyo and J.D. Hampton. Effect of 4-hydroxypyrazolo (3,4-d) pyrimidine (allopurinol) on postirradiation celebral blood flow: implications of free radical involvement. Free Rad. Biol. Med. 4:279-284 (1988).

29. G.B. Faguet, S.M. Reichard, D.A. Welter. Radiation-induced clastogenic factors. Cancer Genet. Cytogenet. 12:73-80 (1984).

30. I. Emerit. Chromosome breakage factors: origin and possible significance. Prog. Mutat. Res. 4:61-74 (1982).

31. M. Shaham, Y. Becker and M.M. Cohen. A diffusable clastogenic factor in ataxia telangiectasia. Cytogenet. Cell. Genet. 27:155-161 (1980).

32. I. Emerit and P. Cerutti. Clastogenic activity from Bloom syndrome fibroblast cultures. Proc. Natl. Acad. Sci. (USA) 78:1868-1872 (1981).

33. I. Emerit, M. Keck. A. Levy, J. Feingold and A.M. Michelson. Activated oxygen species at the origin of chromosome breakage and sister chromatid exchanges. Mutat. Res. 103:165-172 (1982).

34. I. Emerit, S.H. Khan and P. Cerutti. Treatment of lymphocyte cultures with a hypoxanthine-xanthine oxidase system induces the formation of transferable clastogenic material. Free Rad. Biol. Med. 1:51-57 (1985).

35. I. Emerit, J.N. Fabiani, O. Ponzio, A. Murphy, F. Lunel and A. Carpentier. Clastogenic factor in ischemia-reperfusion injury during open heart surgery: protective effect of allopurinol. Ann. Thorac. Surg. 46:619-624 (1988).

36. I. Emerit, A. Autor, C. Auclair, A. Levy, S. Ensworth, M. Lahoud-Maghani, H. Stein and J.P. Camus. Role of oxygen free radicals in progressive systemic sclerosis (scleroderma). Proc. SFRR Biannual Meeting, Kyoto (1988), in press.

37. G. Brambilla, L. Sciaba, P. Faggin, A. Maura, U.M. Marinai, U.M. Ferro, and H. Esterbauer. Cytotoxicity, DNA fragmentation and sister chromatid exchange in Chinese hamster ovary cells exposed to the lipid peroxidation product 4-hydroxynonenal and homologous aldehydes. Mutat. Res. 171:169-176 (1986).

DIRECT MEASUREMENT OF FREE RADICAL GENERATION IN ISOLATED RAT HEART BY ELECTRON PARAMAGNETIC RESONANCE SPECTROSCOPY : EFFECT OF TRIMETAZIDINE

Véronique Maupoil[1], Luc Rochette[1], Alain Tabard[2]
Pascale Clauser[3], Catherine Harpey[3]

1 Laboratoire de Pharmacodynamie, Faculté de Pharmacie Dijon (France)
2 Laboratoire de Chimie des Hétérocycles, Faculté des Sciences, Dijon (France)
3 IRIS Neuilly sur Seine (France)

INTRODUCTION

Free radicals species have been implicated as important agents involved in myocardium ischemia and reperfusion injuries (Mac Cord, 1985). The major cytotoxic effect of oxygen-derivated free radicals may be peroxidation of lipid components of cellular membranes which may be an important process for cell damage during myocardial infarction. In this study, Electron Paramagnetic Resonance Spectroscopy (EPR) was used to directly mesure free radical generation in isolated perfused rat hearts (Maupoil and Rochette, 1988). We also investigated the formation of thiobarbituric acid reactants as an index of lipid peroxidation. The effect of TMZ in the postischemic heart was determined.

METHODS

Male Sprague-Dawley rats (300-350 g) were used for all studies. Isolated hearts were canulated via the aorta and perfused by the method of Langendorff at a constant flow rate of 12 ml/min and at 37°C. The perfusion fluid was a Krebs-Henseleit buffer consisting of (mM) : NaCl 118, NaHCO$_3$ 25, KH$_2$PO$_4$ 1.2, KCl 1:8, CaCl$_2$ 3, MgSO$_4$ 1.2, glucose 5.5, EDTA 0.027 which was gased with 95 % - 5 % O$_2$/CO$_2$.

Four groups (n = 8) were performed :
Control : hearts were perfused for a 40 min period.
Ischemic : hearts were perfused for 30 min and then subjected to 10 min of normothermic global ischemia with 1.2 ml/min coronary flow.
Reperfused : ischemia was followed by 20 sec of reflow.
TMZ-Reperfused : hearts were perfused for 15 min with normal perfusate and then TMZ 10^{-5} M was added to perfusate for 15 min of perfusion followed by 10 min of normothermic global ischemia and 20 sec of reflow.

Antioxidants in Therapy and Preventive Medicine
Edited by I. Emerit *et al.*
Plenum Press, New York, 1990

Free radical determination

The hearts were freeze clamped at 77 K, ground to fine powder under liquid nitrogen, and the powder was transferred to precision EPR tubes. EPR spectra were then recorded at a temperature of 100 K using a ESP 300 BRUKER spectrometer. Quantitation of the signal was performed by comparison of the double integrated signal aera with that of a known concentration of diphenylpicrylhydrazyl standard in identical conditions. Care was taken to perform these measurements with non-saturating microwave power (2.4 mW).

Lipid peroxidation evaluated by thiobarbituric acid (TBA) reactive substances expressed as malonedialdehyde

Lipid peroxidation was carried out on the residual liquid nitrogen frozen powder by the thiobarbituric acid reaction according to a modified method of Buege and Aust (1978).

RESULTS

The usual convention is to show the first derivative absorption function. These signals were defined by a magnetic field dependant g value which serves to identify chemical species.
The EPR spectrum recorded at 100 K (Fig 1, trace A) shows several lines and consists of different components. Temperature studies were performed to separate three different signals. Each of the component signals has different temperature stability and can be isolated by gradual warming of the sample.

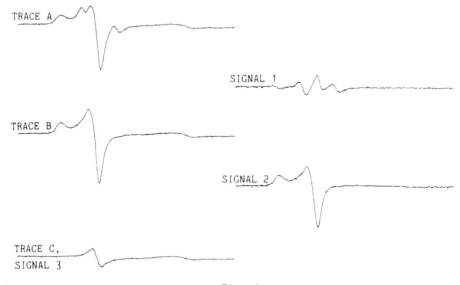

Fig. 1.

On warming the sample 1 min to 190 K, the spectrum shown in Fig 1, trace B is obtained. The difference spectrum A - B, shows that a signal, the signal 1, disappears. This could be a triplet signal g = 2.001, a_N = 25 G suggesting a nitrogen-centered radical.

On further warming to 240 K for 1 min, a second component disapears leaving only a remaining component. The difference spectrum B - C, reveals a signal 2 with axial symetry $g_{//}$ = 2.034 ang $g_\perp$ = 2.007 identical to signals previously reported for oxygen-centered alkylperoxyl free radical.
The remaining signal (Trace C), the signal 3 with g = 2.004 is similar to those of a carbon-centered ubiquinone free radical.
Similar spectra have been reported by Zweier et al. (1987) in isolated rabbit heart but the evolution of these three signals was different during ischemia and reperfusion.

The intensity of free radical signals in the perfused control group corresponds to a total free radical concentration of 222.18 ± 17.48 nmoles/g of dry tissue. This signal intensity was increased by 44 % (P < 0.05) after 10 min of normothermic global ischemia with a 10 % residual flow and by 31 % after 20 sec of reflow with oxygenated perfusate (P < 0.05). Compared with the reperfused group, TMZ 10^{-5} M decreased the free radical concentration from 291.3 ± 24.5 to 232.1 ± 20.9 nmoles/g of dry tissue, this reduction (- 20 %) was not significant (P = 0.087) (Fig 2A). With TMZ administration, the concentration of the oxygen-centered free radical, signal 2, was 24 % reduced (Fig 2B) and the concentration of ubiquinone free radical, signal 3, was not modified.

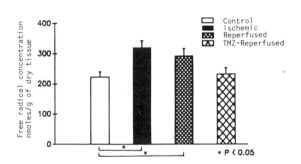

Fig. 2A. Total myocardial free radical concentration

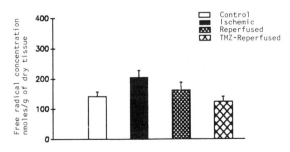

Fig. 2B. Myocardial free radical concentration of signal 2

However, lipid peroxides expressed by malonedialdehyde level remained unchanged in these different group (0.22 ± 0.02 nmoles MDA/mg of protein). In our experimental model, the duration of ischemia and reperfusion is probably too short to observe modifications of thiobarbituric acid reactive material. It is also possible that lipid peroxidation could not occur because protective systems were also effective or the production of TBA substances could be delayed in regard to the free radical production.

DISCUSSION

In our study formation of free radicals was measured directly during normothermic global ischemia and reperfusion. Similar EPR spectra have been reported by Zweier et al. (1987) in isolated rabbit hearts and by Rao et al. (1983). Production of free radicals takes place during ischemia (+ 44 % after 10 min of non-total global ischemia) and during reperfusion. In our experimental conditions (non total global ischemia with 1.2 ml/min coronary flow), the levels of free redicals did not increase above the levels measured at the end of ischemia. In return an exacerbation of free radical production during reperfusion after ischemia wad demonstrated in perfused rat heart (Maupoil and Rochette, 1988).

TMZ (10^{-5} M) decreased the free radical concentration (- 20 %) and the level of oxygen-centered free radical defined by signal 2 which was identical to signals previously reported for alkylperoxyl free radical. This decrease, although not statistically significant was important. The value in treated hearts did not differ from controls. This activity could partially explain the protective effect of this drug on free radical-induced human red cells membrane damage (Maridonneau-Parini and Harpey, 1985) and on cardiac cells (Kiyosue et al., 1986).

In conclusion, this study demonstrated that free radicals are generated in isolated rat heart during a short period of global ischemia and reperfusion and suggests that TMZ could have a beneficial effect due to an action on oxygen-centered alkyl peroxyl free radical. The variations of free radical production in myocardium were not associated with the concomittant formation of thiobarbituric acid reacting substances.

REFERENCES

Buege J.A. and Aust S.D., 1978, Methods Enzymol., 52 : 302-310.
Kiyosue T., Nakamura S. and Arita M.J., 1986, J. Mol. Cell Cardiol., 18 : 1301-1311.
Mac Cord J.M., 1985, N. Engl. J. Med., 312 : 159-163.
Maupoil V. and Rochette L., 1988, Cardiovas. Drugs and Ther., 2 : 615-621.
Maridonneau-Parini I. and Harpey C., 1985, Br. J. Clin. Pharmac., 20 : 148-151
Rao P.S., Cohen M.V. and Mueller H.S., 1983, J. Mol. Cell Cardiol., 15 : 713-716.
Zweier J.L., Flaherty J.T. and Weisfeldt M.L., 1987, Proc. Natl. Acad. USA, 84 : 1404-1407.

EFFECT OF A 5 DAY TRIMETAZIDINE PRETREATMENT IN A MODEL OF ISCHEMIC AND

REPERFUSED ISOLATED RAT HEART : SPIN TRAPPING EXPERIMENTS

Vincent Charlon, François Boucher, Pascale Clauser*
Catherine Harpey*, Alain Favier, Nizameddine Koukay
Sahar Mouhieddine and Joël de Leiris

Laboratoire de physiologie cellulaire cardiaque
URA CNRS 56, Université Joseph Fourier, Grenoble, France
* IRIS Neuilly-sur-Seine, France

INTRODUCTION

The purpose of the present study was to estimate the effects of a 5 day pretreatment with trimetazidine on the production, the nature and the fate of oxygen free radicals in isolated perfused rat hearts, submitted to an ischemia-reperfusion procedure, using PBN as a spin-trap.

MATERIAL AND METHODS

The experiment was performed on two groups of 10 Wistar rats weighing 220 to 250 g. Animals from the treated group received a daily intraperitoneal injection of trimetazidine (40 mg/kg ; 1 ml/kg) for 5 days, the control group received saline according to the same protocol. At the[1] end of the treatment, hearts were excised and perfused via the aorta, for 15 minutes with a Krebs Henseleit solution. At the end of this stabilization period, N-tert-butyl alpha phenylnitrone (PBN, 56 mM, Sigma Chemical, St. Louis, USA) was added to the perfusion fluid for 30 seconds. Hearts were then submitted to a 10 minute global and normothermic ischemia (37°C), and reperfused for 30 seconds with PBN-free perfusion fluid and frozen in liquid nitrogen.
Each heart[3] was used for biochemical[4] determinations of catalase,[2] superoxide dismutase,[3] glutathion peroxidase,[4] glutathion reductase,[5] xanthine oxidase,[6] total glutathion, malondialdehydes, conjugated dienes and organic hydroperoxides, and for electron spin resonance (ESR) studies. ESR signals were assessed on lyophilized tissue, before and after addition of 100 microliters of dimethylsulfoxide (DMSO, Sigma Chemical, St. Louis USA), at room temperature.

RESULTS

ESR studies

At least three signals were detected in lyophilized tissue samples from the two experimental groups (figure 1). The first one (A : g = 2.006) was detected in all lyophilized myocardial tissue samples of this study and

Antioxidants in Therapy and Preventive Medicine   377
Edited by I. Emerit *et al.*
Plenum Press, New York, 1990

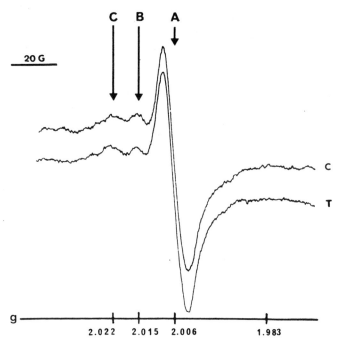

Figure 1. ESR signals recorded on lyophilized tissue samples
C : signal recorded with a lyophilized tissue sample from the control group.
T : signal recorded with a lyophilized tissue sample from the trimetazidine treated group.
Record settings : microwave power, 10 mW
                microwave frequency, 9.2 to 9.4 GHz
                sweep time, 8mn
                time constant, 250 ms
                modulation frequency, 100 kHz
                modulation amplitude, 4 G

of previous studies from our group,[7]. It has been reported in the literature that such a signal could well result from the lyophilization process,[8]. The origin of the signal "B" ($g = 2.022$) remains unknown but its intensity was equivalent in the two groups (control : 2.7 ± 2.2 vs trimetazidine : 4.3 ± 1.2 ; ns). The signal "C" ($g = 2.015$) seems to correspond to a typical nitroxide signal in an anisotropic medium as previously reported in another study,[7]. This signal could represent the signal of PBN-spin adducts. Its intensity, measured on the $g = 2.022$ signal (figure 2) was not significantly modified by the treatment (control : 3.1 ± 2.4 vs trimetazidine : 4.1 ± 2.0 ; ns).
These preliminary results did not show any difference between the two experimental groups and seem to demonstrate that the amount of PBN-spin adducts was not modified by the treatment.
In order to identify trapped radical species, it appeared necessary to solubilize spin-adducts contained in lyophilized samples. After various

378

tests with other solvents, we chose dimethylsulfoxide (DMSO). Three different signals have been observed in these conditions (see table I and figure 3) : i) Signal "1" was present in most of the samples of the control group (5/8) and in 6/10 hearts of the trimetazidine treated group. This signal appeared immediately after DMSO addition and could be due to alkoxy trapped radicals (LO°) according to splitting constants described in the literature,[9] . It should be noted that signal "1" represents 60 to

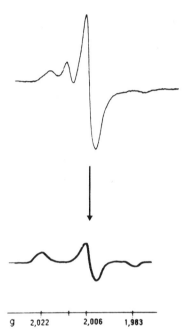

| g | 2,022 | 2,006 | 1,983 |

Figure 2. Plot of signal C extracted from ESR signal, corresponding to a typical nitroxide signal.
Upper spectrum : lyophilized sample from a control heart.
Lower spectrum : schematic representation of a nitroxide signal.

90% of the total nitroxide signal in the control group but only 10 to 30% in the trimetazidine treated group. ii) signal "2" appeared approximately 1 minute after DMSO addition and was clearly detected only in 2 hearts of the control group. Though its origin remains unknown, it may be postulated that this signal is due to DMSO-derived radicals,[7] . iii) signal "3" was detected in 8 of the 10 samples of the treated group immediately after DMSO addition but never in the control group. This signal represents 70 to 90% of the nitroxide signal in the treated group.

379

Table I. Characteristics of ESR nitroxide signals detected in lyophilized samples after addition of 100 microliters of DMSO.

Signal	Splitting constants	Number of hearts in which the signal was detected	
		Control	Trimetazidine
Signal 1	aN = 13.85 $\pm$ 0.03 aH = 2.31$\pm$ 0.01	5/8	6/10
Signal 2	aN = 15.35 $\pm$ 0.04 aH = 3.25 $\pm$ 0.04	2/8	0/10
Signal 3	aN = 14,98 $\pm$ 0.04 aH = 3.45 $\pm$ 0.04	0/8	8/10

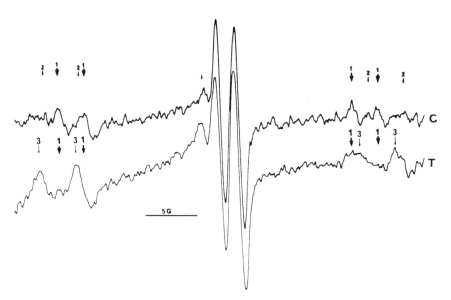

Figure 3. Signals recorded with lyophilized tissue samples + 100 microliters DMSO

C : signal recorded with a tissue sample from the control group.
T : signal recorded with a tissue sample from the trimetazidine treated group.
Record settings : microwave power, 7mW
microwave frequency, 9.2 to 9.4 GHz
sweep time, 8 mn
time constant, 1000 ms
modulation frequency, 100 KHz
modulation amplitude, 0.63 G

Biochemical measurements

In our experimental conditions, the treatment with trimetazidine did not significantly modify enzymatic activities or biochemical parameters measured in this study, except for malondialdehyde myocardial content which appeared significantly decreased in the treated group as shown on table II.

Table II.

Biochemical measurements	Control (n=10)	Trimetazidine (n=10)	p
Catalase (UI/mg proteins)	8.6 ± 0.4	8.5 ± 0.6	NS
Glutathion peroxidase	241 ± 8	247 ± 8	NS
Total superoxide dismutase (SODtt)	16.5 ± 0.8	17.2 ± 0.7	NS
Mitochondrial superoxide dismutase (SODMn)	7.8 ± 0.4	8.1 ± 1.6	NS
SODMn (%) / SODtt	47.1 ± 1.6	47.0 ± 1.6	NS
Xanthine oxidase mUI/mg proteins	0.453 ± 0.02	0.415 ± 0.03	NS
Malondialdehyde nmol/mg proteins	0.175 ± 0.01	0.145 ± 0.06	<0.001
Conjugated dienes OD/mg proteins	3.66 ± 0.5	4.06 ± 0.6	NS
Organic hydroperoxides nmol/mg proteins	14.7 ± 1.1	14.9 ± 1.2	NS
Glutathion reductase mUI/mg proteins	8.2 ± 0.5	8.5 ± 0.6	NS
Total glutathion (GSHtt) nmol/mg proteins	11.6 ± 0.8	11.9 ± 0.4	NS
Oxidized glutathion (GS-SG) nmol/mg proteins	4.2 ± 0.3	4.7 ± 0.2	NS
GS-SG (%) / GSHtt	36.6 ± 0.8	39.0 ± 0.8	NS
Glutathion transferase mUI/mg proteins	33.3 ± 1.2	32.4 ± 1.3	NS

CONCLUSIONS

Our results obtained on lyophilized samples + 100 microliters DMSO, indicate that, in our experimental conditions, trimetazidine or one of its metabolites could interact with some free radicals generated in ischemic-reperfused myocardium.
On the other hand, the treatment with trimetazidine did not significantly modify enzymatic activities or biochemical parameters assayed in this study, except for malondialdehyde myocardial content which appeared reduced in the treated group.

REFERENCES

1. O. Langendorff, Untersuchungen am überlebenden Saügertierherzen, Pflügers Arch. 61:291 (1895).
2. R. F. Beer, I. W. Sizer, A spectrophotometric method for measuring the breakdown of hydrogen peroxide by catalase, J. Biol. Chem. 195:133 (1952).
3. S. L. Marklund, Spectrophotometric study of spontaneous disproportionation of superoxide dismutase, J. Biol. Chem. 251:7504 (1976).
4. W. A. Gunzler, H. Kremers, L. Flohe, Improved coupled test procedure for glutathione peroxidase in blood, Z. Klin. Chem. Klin. Biochem. 10:444 (1974).
5. I. Carlberg, B. Mannervik, Glutathione reductase, Meth. Enzymol. 113:484 (1985).
6. M. Sugiura, K. Kato, T. Adachi, K. Hirano, A new method for the assay of xanthine oxidase activity, Chem. Pharm. Bull. 29:430 (1981).
7. J. de Leiris, V. Charlon, F. Boucher, Toxicité des radicaux libres de l'oxygène au cours de la reperfusion. Mythe ou réalité ?, Arch. Mal. Coeur. 81:35 (1988).
8. R. J. Heckly, Free radicals in dry biological systems, in : "Free radicals in biology," W. A. Pryor, Academic press (1976)
9. A. N. Saprin, L. H. Piette, Spin trapping and its application in the study of lipid peroxidation and free radicals production with liver microsomes. Arch. Biochem. Biophys. 480 (1977).

ANTILIPOPEROXYDANT EFFECT OF TRIMETAZIDINE IN POST ISCHAEMIC ACUTE RENAL FAILURE IN THE RAT

Philippe Catroux*, Jean Cambar*, Nabil Benchekroun°
Jacques Robert°, Pascale Clauser+ and Catherine Harpey+

* G.E.P.P.R., Faculté de Pharmacie de Bordeaux (France)
° Laboratoire de Biochimie, Faculté Médecine de Bordeaux
(France)
+ IRIS Neuilly sur Seine (France)

Trimetazidine (TMZ) is an antianginal drug largely used in clinical ischaemic disorders. Whereas classical antianginal drugs act by correcting the imbalance between myocardial vascular supply and demand for oxygen, TMZ has been reported to exert a direct cytoprotective effect inside the real ischaemic area[7,11]. Several mechanisms of action seem to be involved. These include more particularly the reduction of the ischaemia induced intracellular acidosis and a more rapid recovery of phosphorylation processes during the early phase of blood reperfusion[6,12,23]. It has also been reported that TMZ exerts a potent antioxidant activity, which could explain its cardioprotective role during ischaemic and reperfusion phases[10,15]. Oxygen free radicals are now well known to contribute to kidney damages induced by a temporary warm or cold ischaemia[13,21]. Inhibitors of xanthine oxidase by preventing superoxyde radical production, superoxydismutases by removing superoxyde radicals and scavengers of hydroxyl radicals reduce renal injury following ischaemia-reperfusion[9,13,14]. Moreover, several studies indirectly evidenced the production of free radicals following reflow in both warm and cold ischaemia by the presence of lipid peroxidation by products[6,18,19,20].

We have previously reported[4,5] that TMZ presents in rats a protective effect on renal damages induced by a warm renal ischaemia of 45 or 75 minutes duration. The reduction of the impairment in the whole kidney function after ischaemia and reflow is associated with a cytoprotective effect of trimetazidine on post ischaemic tubular damages, especially on convoluted and straight portions of the proximal tubule.

This paper will summarize some recent data evidencing that trimetazidine activity results in a decreased formation of lipid peroxydation products ; it is also reported that it preserves some aspects in the antioxidant status of the renal tissue during ischaemia-reperfusion.

Antioxidants in Therapy and Preventive Medicine
Edited by I. Emerit *et al.*
Plenum Press, New York, 1990

MATERIAL AND METHODS

The experiments were performed on male Sprague Dawley rats (200-250 g) orally treated with trimetazidine (2.5 mg/kg) or vehicle alone (NaCl) for 5 days before operative procedure. The treated group was also given 2.5 mg of trimetazidine intravenously just before renal pedicle occlusion. Rats were anesthetized with 30 mg/kg of I.V. pentobarbital.

The following groups were considered :
- group 1 (NaCl) and group 2 (TMZ) are used as control and were just sham-operated ;
- group 3 (NaCl) and group 4 (TMZ) were subjected to 60 minutes of bilateral renal pedicle occlusion. Kidneys were harvested after 2 hours of blood reperfusion ;
- group 5 (NaCl) and group 6 (TMZ) were subjected to 60 minutes of bilateral renal pedicle occlusion. Kidneys were reperfused with blood for 12 hours, then harvested ;
- group 7 (NaCl) and group 8 (TMZ) were subjected to 60 minutes of bilateral renal pedicle occlusion and allowed to reperfuse with blood for 24 hours.

Determination of lipid peroxidation and antioxidant status of the cell were performed on cortex and outer stripe of outer medulla.

RESULTS

Determination of by-products of lipid peroxydation

TBA reactive products. Tissue extracts were deproteinized in trichloroacetic acid (TCA) and the supernatant assayed for malonaldehyde (MDA) concentration according to OKHAWA et al.[17]. Tissular MDA patterns in control and ischaemic kidneys from TMZ-treated and untreated rats are shown in figure 1. Tissues harvested during blood reperfusion after 60 minutes of renal ischaemia displayed a higher MDA level than control tissues. At any time of blood reperfusion, increase in tissular TBA reactive substances was significantly lessened in TMZ treated rats.

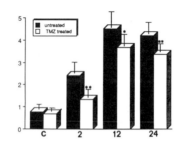

Fig. 1. Malonaldehyde (nmoles/mg protein) in rat kidneys during control (C) and post ischemic (2,12,24 h) periods.
$* p < 0.05 ; ** p < 0.01$

Diene conjugates. Diene conjugates were assayed in lipid extracts obtained by chloroforme methanol extraction. They were then measured for absorbance at 240 nm and expressed as change in optical density per mg of protein[22]. Our results showed that TMZ decreases the formation of diene conjugates in the early phase of blood reperfusion (figure 2).

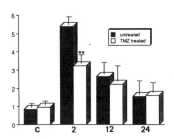 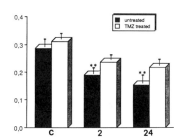

Diene conjugates (optical density/mg protein x 10)

Total glutathione (GSH+GSSG) (nmoles/mg) protein)

Fig. 2.   In rat kidneys during control(C) and post ischemic periods (2,12,24 h) ; ∗∗ p < 0.01.

Determination of antioxydant status

Determination of total, reduced and oxydized glutathione. Tissue fragments for glutathione determination were excised in situ, weighted and rapidely homogeneized in 5 % sulfosalicylic acid. Total and oxydized glutathione were determined using the recirculating assay describe by TIETZE et al.[24]. The ratio of GSSG/total glutathione is determined and then used as a relative index of redox activity. The results are summarized in figure 2 and 3. Our data indicated that TMZ lessens both the loss in glutathione and the increase in GSSG/total glutathione ratio during renal ischaemia reperfusion.

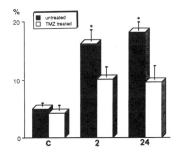

Fig. 3.   Glutathione redox activity (GSSG/total glutathione) in rat kidneys during control (C) and post ischemic (2,24 h) periods.
    ∗ p < 0.05

Superoxyde dismutase activity. Total superoxide dismutase activity in tissue extracts was measured by the pyrogallol method described by MARKLUND[16]. Table 1 shows that renal ischaemia reperfusion induced a loss of tissular superoxide dismutase activity, which was not significantly different between treated and untreated rats.

Catalase activity. Catalase activity was monitored by a spectrophotometry method described by BEERS and SIZER[2]. Decrease in catalase activity was noted during post ischaemic blood reflow. No significant differences were noted between treated and untreated groups (table 1).

Table 1. Superoxyde dismutase and catalase activities in rat kidneys during control (C) and post ischemic (2, 12, 24 h) periods.

Groups	1	2	3	4	5	6	7	8
SOD U × 10^3 / mg protein	0.25 ± 0.06	0.24 ± 0.04	0.2 ± 0.02	0.19 ± 0.04	0.15 ± 0.03	0.16 ± 0.01	0.14 ± 0.03	0.14 ± 0.03
Catalase U / mg protein	201 ± 13	201 ± 13	143 ± 27	162 ± 13	125 ± 12	130 ± 16	97 ± 8	88 ± 12

DISCUSSION

Renal ischaemia-reperfusion results in various and overlapped vascular and tubular changes involving pre- and/or post-glomerular vasoconstriction, medullary hyperemia, tubular leakage of glomerular filtrate and tubular obstruction. One of the early changes following ischemia is the loss of proximal tubular integrity, which is revealed by brush border microvilli and intracellular membranes degradation.

Previous studies in our laboratory have shown that TMZ, when administered preventively in rat, significantly reduced the impairment of the whole kidney function during the post ischaemic period[4,5]. Moreover, these studies have demonstrated that TMZ reduces post ischaemic tubular damages, as assessed by urinalysis of some renal originating enzymes. TMZ could exert its antiischaemic action either by a direct cytoprotective effect as reported in hypoxic or ischaemic heart[11,12], or indirectly by a possible vasoactive effect at renal level. However, our observations of an in vitro unresponsiveness of renal vascular structure to high concentrations of TMZ and the unability of TMZ to modify the vasoactive effect of endogenous mediators (as angiotensin II or arachidonate) led us to support the first hypothesis.

Numerous biochemical alterations have been proposed to explain the post-ischaemic renal tissue injury, including depletion of adenine nucleotide, loss of electrolyte homeostasis, alteration in phospholipid metabolism, intracellular acidosis and cell calcium influx after reperfusion[3]. Recently, the role of free radicals in post ischaemic renal injury has received considerable attention. In our knowledge, there has not been any report of the direct measurement of free radicals in post ischaemic renal tissue. The presence of free radicals after reflow has been documented by measuring lipid peroxidation[18,19], or by evidencing the protective effect of drugs known to interfere with free radical generation[1,8,9].

The results of the present study indicated that reperfusion after 60 minutes of ischaemia causes an increase in lipid peroxidation and an oxydative stress in rat kidney. Trimetazidine, administered just before renal occlusion, significantly lessened the increase in both MDA and diene conjugates. Renal ischaemia reperfusion was also evidenced by a decrease in reduced glutathione and concomitantly, by an increase in the oxidized form. TMZ significantly reduced the loss in tissular glutathione. Moreover, the shift in the glutathione redox ratio to a more oxidized value during reperfusion was significantly reduced in TMZ treated rats. In contrast, TMZ had no effect on the loss in both superoxide dismutase and catalase activities seen after ischaemia.

Our study demonstrated that TMZ reduces the toxic effect of oxygen free radicals during renal ischaemia reperfusion in the rat. The mechanism by which TMZ exerts this antilipoperoxidant effect is still under investigation. Recently, the results of various ex vivo[15] and in vitro[10] experiments have strongly suggested that TMZ or one of its metabolites presents a free radical scavenger activity. Their conclusions are discussed in this book (see P. Clauser, C. Harpey). It is not certain from our experiments, that TMZ exerts a direct effect on free radical generation. It must be pointed out that TMZ has been reported to reduce the intracellular acidosis during hypoxia[12] and to allow a more rapid recovery of adenine nucleotides in the early phase of reperfusion when tested in the hypoxic heart model. Such activities can lead to indirect interference with free radical generation and consequently to lessen their subsequent deleterious effects. Further studies in our laboratory are performed to evaluate precisely the relative part of these activities in the cytoprotective effect of trimetazidine.

REFERENCES

1. G.L. Backer, R.J. Corry and A.P. Autor, Oxygen free radical induced damage in kidneys subjected to warm ischemia and reperfusion : protective effect of superoxyde dismutase, Ann. Surg. 202 : 628-641 (1985).
2. R.F. Beers and I.W. Sizer, A spectrophotometric method for measuring the breakdown of hydrogen peroxyde by catalase, J. Biol. Chem. 195 : 133 (1952).
3. J.V. Bonventre, Cellular response to ischemia, in : "Acute renal failure", A. Whelton and K. Solez ed., Publisher New York, M. Dekker Inc.(1984).
4. P. Catroux, C. Dorian, C. Harpey and J. Cambar, Mise en évidence de l'effet protecteur de la trimétazidine vis-à-vis de l'enzymurie induite par clampage du pédicule rénal chez le rat, Néphrologie 7 : 124 (1986).
5. P. Catroux, C. Dorian, C. Harpey and J. Cambar, Une autre approche de l'ischémie pour la mise en évidence de l'effet protecteur de la trimétazidine, Conc. Méd. 36S : 3451-3454 (1987).
6. C. Cruz, A. Zaoui, I. Ayoub, C. Harpey, P. Goupit and A. Younes, Altérations des myocytes isolés des ventricules de coeur de rat adulte : protection par la trimétazidine, Conc. Méd. 36S : 3470-3475 (1987).
7. D. Garnier and M.J. Roulet, Vasoactivity of trimetazidine on guinea-pig isolated ductus arterious, Br. J. Pharmac. 84 : 517-524 (1985).

8. C.J. Green, G. Healing, S. Simpkin, J. Lunec and B.J. Fuller, Desferioxamine reduces susceptibility to lipid peroxidation in rabbit kidneys subjected to warm ischemia and reperfusion, <u>Comp. Biochem. Physiol.</u> 85 : 113-117 (1986).

9. R. Hansson, O. Jonsson, S. Lundstam, S. Petterson, T. Schersten and J. Waldenstrom, Effects of free radical scavengers on renal circulation after ischemia in the rabbit, <u>Clin. Sci.</u> 65 : 605-610 (1983).

10. C. Harpey, C. Labrid, L. Baud, B. Housset, I. Maridonneau-Parini, F. Piccini and P. Goupit, Evidence for antioxidant properties of trimetazidine, Xth International Congress of Pharmacology - Sydney (1987).

11. E. Honoré, M.M. Adamantidis, C.E. Challice and B.A. Chapuis, Cardioprotection by calcium antagonists, piridoxilate and trimetazidine, <u>Ircs, Med. Sci.</u> 14 : 938-939 (1986).

12. N. Lavanchy, J. Martin and A Rossi, Antiischemic effects of trimetazidine : P 31 NMR spectroscopy in the isolated rat heart, <u>Arch. Int. Pharmacodyn. Ther.</u> 286 : 97-110 (1987).

13. J.M. Mc Cord, Oxygen derived free radicals in post ischemic tissue injury, <u>N. Engl. J. Med.</u> 312 : 159-163 (1985).

14. R.N. Mc Coy, K.E. Hill, M.A. Ayon, J.H. Stein and R.F. Burk, Oxidant stress following renal ischemia : changes in the glutathione redox ratio, <u>Kidney Int.</u> 35 : 812-817 (1988).

15. I. Maridonneau-Parini and C. Harpey, Effect of trimetazidine on membrane damage induced by oxygen free radicals in human red cells, <u>Br. J. Clin. Pharmacol.</u> 20 : 148-151 (1985).

16. S. Marklund and G. Marklund, Involvement of superoxyde anion radical in the autoxidation of pyrogallol and a convenient assay for superoxyde dismutase, <u>Eur. J. Biochem.</u> 47 : 469 (1974).

17. H. Okhawa, N. Ohishi and K. Yagi, Assay for lipid peroxydes in animal tissues by thiobarbituric acid reaction, <u>Anal Biochem.</u> 95 : 351-358 (1979).

18. M.S. Paller, J.R. Hoidal and T.F. Ferris, Oxygen free radicals in ischaemic acute renal failure in the rat, <u>J. Clin. Invest.</u> 74 : 1156-1164 (1984).

19. M.S. Paller and R.P. Hebbel, Ethane production as a mesure of lipid peroxidation after renal ischemia, <u>Am. J. Physiol.</u> 251 : F839-F843 (1986).

20. M.S. Paller, Hypothyroidism protects against free radical damage in ischemic acute renal failure, <u>Kidney Int.</u> 29 : 1162-1166 (1986).

21. R.E. Ratych and G.B. Bulkley, Free radical-mediated post ischemic reperfusion injury in the kidney, <u>J. Free Rad. Biol. Med.</u> 2 : 311-319 (1986).

22. R.O. Recknagel and E.A. Glende, Spectrophotometric detection of lipid conjugated dienes, <u>Method. Enzymol.</u> 105 : 331-337 (1984).

23. J.F. Renaud, Internal pH, Na^+, and Ca^{2+} regulation by trimetazidine during cardiac cell acidosis, <u>Cardiovasc. Drugs and Ther.</u> 1 : 677-686 (1988).

24. F. Tietze, Enzymatic method for quantitative determination of nanogram amounts of total and oxidized glutathione : applications to mammalian blood and other tissues, <u>Anal Biochem.</u> 27 :502-522 (1969).

REPERFUSION INJURY AND RENAL METABOLISM : THE TEMPORAL RELATIONSHIP BETWEEN OXIDATIVE STRESS AND FUNCTIONAL CHANGE

B. J. Fuller*, J. Gower, L. Cotterill, G. Healing, S. Simpkin and C. J. Green

*Academic Department of Surgery, Royal Free Hospital & School of Medicine, London NW3 2QG and Section of Surgical Research MRC Clinical Research Centre, Northwick Park Hospital Harrow, UK

INTRODUCTION

A large number of studies have implicated oxygen-derived free radical (OFR) stress in the pathology of the so-called 'reperfusion injury' in a number of organs and tissues including brain (1), liver (2), intestine (3), skin (4), heart (5) and kidney (6). This is the type of injury sustained when an organ has been deprived of blood supply for a period of time, and is then subjected to a sudden restoration of blood supply bringing in high concentrations of oxygen; such events occur in many surgical interventions e.g. during repair of tissues after trauma, and particularly in organ transplantation (7). There is growing evidence that OFR scavengers, when introduced under such circumstances, can beneficially influence the outcome of reperfusion (8). However, there is still much debate about the time relationship between OFR events in a tissue and subsequent functional changes, which in turn will influence the timing of administration of any anti-OFR therapy. We have been particularly concerned with reperfusion injury in kidneys (9,10), and the present studies were undertaken to assess renal tissue metabolism (by gluconeogenesis) in rabbit kidneys after ischaemia / reperfusion. Renal cortical tissue gluconeogenesis was chosen as functional test since this is an active process of cortical tubular cells and these cells show characteristic early signs of ischaemia / reperfusion damage as expressed by tubular necrosis.

MATERIALS AND METHODS

Ischaemia / reperfusion in the rabbit kidney

Details of the surgical model have been published previously (9). In brief, adult male New Zealand White rabbits (2-3 Kg) were anesthetized by injection of fentanyl-fluanisone and diazepam. The abdomen was opened by a midline incision and the left kidney was completely skeletonized. The left renal artery, vein and ureter were clamped for the desired ischaemic period and the abdomen was temporarily closed. Kidneys were removed for assay immediately after 1 or 2 hours of ischaemia, or in other groups after additional blood reperfusion for 1 or 24 hours. Untreated kidneys were used for control assays.

Assays for indices of lipd peroxidation

For assessment of OFR stress we have used an in vitro incubation assay,

Antioxidants in Therapy and Preventive Medicine
Edited by I. Emerit *et al.*
Plenum Press, New York, 1990

which has been shown to correlate with degree of ischaemic damage in kidney tissue (9). In this assay kidneys were sliced and the majority of renal tissue (about 8g) was homogenised in buffer to give a 10% w/v concentration, and incubated in open vessels with shaking at 37 C. Production of fluorescent Schiff bases and thiobarbituric acid reactive substances were measured over the time course of incubation (90 minutes) in extracted lipid fractions as described previously (9).

Assay for gluconeogenesis

From the remaining (approximately 2g) tissue, thecortex was dissected from the medulla, and the cortical fragments used to produce tissue slices (0.3 mm thick) on a McIlwain tissue slicer. The slices were weighed and incubated in Krebs Ringer bicarbonate at 37 C, with or without added pyruvate as gluconeogenic substrate as described previously (11). The incubated super-natants were assayed for glucose production, and from this the pyruvate-stimulated glucose synthesis was assessed.

RESULTS AND DISCUSSION

In Table 1 are shown the results for assays of OFR-induced lipid perox-idation products and the tissue metabolism as assessed by gluconeogenesis. It can be seen that ischaemia for 1 and 2 hours showed a rising trend in Schiff base production (groups II and V). Reperfusion tended to increase still further the marker of lipid peroxidation, and this became a statisticall significant increase on reperfusion for 2 hours ischaemia (group VI). A similar trend was seen for TBA-reactive compounds measured at 2 hours ischaemi and with reperfusion.

When gluconeogenesis was examined, it can be seen that ischaemia for 1 hour (group II) caused a significant drop in activity, whilst activity was almost eradicated by 2 hours ischaemia (group V). After reperfusion for 1 hour (groups III and IV) there was an improvement in gluconeogenic activ-ity. This is in spite of the enhanced markers of oxidative stress seen on reperfusion. Only when reperfusion was continued for 24 hours did gluconeo-genesis decrease again compared to that seen in the short (1 hour) reperf-usion group (group III versus group IV).

These data show that there is no simple relationship between OFR stress in reperfusion injury and changes in a particular metabolic process within the ischaemic organ. Particularly for the kidneys subjected to 2 hours ischaemia, gluconeogenesis was virtually completely absent, yet a short period of reperfusion allowed a significant recovery of this active process. This occurred in spite of the increased evidence of oxidative stress resultin from reperfusion. Only when reperfusion was carried out for prolonged periods (24 hours) did gluconeogenic activity again decrease. Since in all cases the kidney slices were supplied with substrate and oxygen sufficient to allow gluconeogenesis to proceed, the changes could not be attributed to depletion of any of these factors. The recovery of metabolism on reperfusion in the 2 hour group (V) must have resulted from recovery of some organelle function, required co-factor or alleviation of inhibition by waste products which were brought about by the restored blood flow. The secondary decline in gluconeogenic activity during 24 hour reperfusion would then have resulted from some expression of damage which may have been completely unrelated to the primary loss of function at the end of ischaemia.

A similar relationship between recovery of metabolic activity and time of reperfusion has been demonstrated in a model of myocardial ischaemia (12). Here, the cardiac muscle ATP content recovered immediately on reperfusion after ischaemia, but then declined again as reperfusion continued. Taken

Table 1. Production of Schiff bases, TBA-reactive material and gluconeogenesis after ischaemia / reperfusion.

Group n=6 in each	Schiff bases*	TBA-react.* substances	Glucose** production
I	1.05 ± 1.04	0.85 ± 0.96	36.8 ± 6.2
II	2.53 ± 0.77	ND	14.4 ± 2.8
III	2.93 ± 0.68	ND	21.6 ± 2.9 [a]
IV	4.93 ± 1.27 [a]	ND	11.2 ± 4.3
V	4.00 ± 0.83	4.32 ± 1.88	2.8 ± 1.4
VI	5.67 ± 0.75	6.28 ± 0.98 [a]	16.1 ± 4.8 [a]

Experimental groups : I - fresh kidney; II - 1 hour ischaemia; III - 1 hour ischaemia + 1 hour reperfusion ; IV - 1 hour ischaemia + 24 hour reperfusion; V - 2 hour ischaemia; VI - 2 hour ischaemia + 1 hour reflow.
* - fluorescence intensity units / hr / mg protein; ** umol glucose / mg / inncubation.
[a] denotes statistically significant difference from ischaemia alone (Student's T test)

together with our data the results suggest a complex relationship between ischaemia / reperfusion and tissue damage with possible primary, secondary and even later events all playing a role. Such interactions need to be considered when undertaking studies of reperfusion damage and any preventive therapy.

ACKNOWLEDGEMENTS

Parts of this study were funded by grants from the National Kidney Research Fund and the Peter Samuel Fund of the Royal Free Hospital awarded to BJF.

REFERENCES

1. Siesjo B, Rehncrona S & Smith D. Neuronal cell damage in the brain : possible involvemnet of oxidative mechanisms. Acta Physiol. Scand. Suppl. 492, 121-128 (1980).
2. Arthur M J. Reactive oxygen species and liver injury. J Hepatol. 6, 125-131 (1988).
3. Whadwa S & Perry M. Gastric injury induced by hemorrhage, local ischaemia and oxygen radical generation. Am. J Physiol. 253, G129-133 (1987).
4. Green C J, Dhami L, Prasad S, Healing G, and Shurey C. The effect of desferrioxamine on lipid peroxidation and survival of ischaemic skin flaps in rats. Med. Sci. Res. 16, 1045-1046 (1988).
5. Burton K, McCord J & Ghai G. Myocardial alterations due to free radical generation. Am. J. Physiol. 246, H776-782 (1984).
6. Lunec J, Fuller B J, Healing G, Simpkin S & Green C J. Lipid peroxidation and survival of ischaemic rabbit kidneys. IRCS Med Sci. 13, 818-819 (1985).

7. Fuller B J, Gower J, & Green C J. Free radical damage and organ preservation – fact or fiction? Cryobiology 25, 317–331 (1988).

8. Shlafer M, Kane P F and Kirsch M M. Superoxide dismutase plus catalase enhaces efficacy of hypothermic cardioplegia to protect the globally-ischaem reperfused heart. J. Thorac. Cardiovasc. Surg. 83, 830–839 (1982).

9. Green C J, Healing G, Simpkin S, Lunec J & Fuller B J. Increased susceptibility to lipid peroxidation in rabbit kidneys subjected to warm ischaemia and reperfusion. Comp. Biochem. Physiol. 83B, 603–606 (1986).

10. Fuller B J, Lunec J, Healing G, Simpkin S & Green C J. Reduction in susceptibility to lipid peroxidation by desferrioxamine in rabbit kidneys subjected to 24 hour cold ischaemia and reperfusion. Transplantation 43, 604–606 (1987).

11. Fuller B J, Marley S & Green C J. Gluconeogenesis in stored kidneys from normal and cold-acclimated rats. Cryo-Letters 6, 91–98 (1985).

12. Lewandowski E, Devous M & Nunally R. High energy phosphates and function in the isolated working rabbit heart. Am. J. Physiol. 253, H1215–1223 (1987)

IRON REDISTRIBUTION AND LIPID PEROXIDATION IN THE COLD ISCHAEMIC KIDNEY

Jon D Gower, Guy Healing, Barry J Fuller* and Colin J Green

Section of Surgical Research, Clinical Research Centre,
Watford Road, Harrow, Middlesex, HA1 3UJ and *Academic
Department of Surgery, Royal Free Hospital School of
Medicine, Pond Street, Hampstead, London, NW3 2QG

INTRODUCTION

Oxygen-derived free radicals may play an important role in the damage which occurs to organs subjected to extended periods of cold storage followed by reperfusion with oxygenated blood upon transplantation into the recipient[1]. One damaging radical-mediated process is the peroxidation of membrane-bound polyunsaturated fatty acids and we have previously demonstrated significant elevations in lipid peroxidation markers in kidneys subjected to cold storage and autotransplantation[2]. Iron is required for the initiation of lipid peroxidation[3] which may be the consequence of highly reactive $OH^\bullet$ radical formation from O_2^- and H_2O_2 via the Haber-Weiss reaction or direct attack on polyunsaturated fatty acids by iron complexes with oxygen[4]. In addition, iron also catalyses the decomposition of lipid hydroperoxides (LOOH) to alkoxy ($LO^\bullet$) and peroxy ($LOO^\bullet$) radicals which stimulate the chain reaction of lipid peroxidation[4]. In order to minimize the likelihood of these damaging reactions, iron is transported and stored in specific proteins. However, a small pool of iron exists in the cell as low molecular weight chelates[5] which are able to exert catalytic activity.

In this study, we have measured the level of iron which is available for chelation by desferrioxamine (DFX) in supernatants of rabbit kidneys. We demonstrate that a period of 24 hours cold storage results in a redistribution of intracellular iron to more available forms and this may be an important factor responsible for the increased level of lipid peroxidation observed when the organs are subsequently subjected to reperfusion with oxygen[2].

METHODS

Kidneys were removed from anaesthetised New Zealand White rabbits and flushed with 30ml hypertonic citrate solution (HCA)[2]. Organs were either assayed immediately (control) or stored in HCA at $0^\circ C$ for 24 hours. Following storage, the organs were either assayed or perfused at $37^\circ C$ for 5,15,30 or 60 mins with a perfusate gassed with 95% O_2:5% CO_2 and containing 25mM $NaHCO_3$, 0.6mM $MgCl_2$, 0.3mM KH_2PO_4, 0.465mM Na_2HPO_4, 5mM glucose, 5mM hydroxybutyric acid, 5mM hypoxanthine, haemaccel to provide colloidal pressure and carbenicillin for antibacterial protection[6]. Kidneys were

Antioxidants in Therapy and Preventive Medicine
Edited by I. Emerit *et al.*
Plenum Press, New York, 1990

393

divided into cortex and medulla, homogenised in 0.1M Tris-HCl buffer pH7.4 and centrifuged at 10,000 gav for 15 minutes. Triplicate aliquots of the supernatants were incubated at 37°C with desferrioxamine (DFX) (2mM) for 1 hour and the parent drug and its iron-bound form, ferrioxamine, were extracted using Bond-Elut C18 cartridges[7]. Quantitation of the two forms of the drug was achieved using reversed-phase hplc with u.v. detection and the (ferrioxamine)/(desferrioxamine) ratio calculated[7]. The amount of chelatable iron in each sample was determined from a standard curve obtained from triplicate standards which contained 0, 10, and 25nmoles Fe^{3+}/ml and were subjected to the same procedure. The total iron content of the samples was determined by atomic absorption spectroscopy and protein contents by the method of Lowry et al[8]. Statistical analysis was performed using Student's t-test.

RESULTS AND DISCUSSION

There was a detectable pool of DFX-chelatable iron in homogenates of fresh kidneys which increased significantly ($P < 0.005$) in both the cortex and medulla after 24 hours cold ischaemia (Table 1). These increased levels fell rapidly upon reperfusion, returning to control levels after only 5 minutes of reperfusion (Figure 1). There was, however, no difference in the total iron content of the tissues; hence a redistribution of iron to more available forms had occurred as a result of 24 hours cold ischaemia. It is likely that this extra available iron was derived from ferritin. The release of ferritin iron involves the reduction of ferric ions to the ferrous state and may therefore be stimulated during ischaemia due to the build up of reducing species including a decreased pH.

It cannot be stated with certainty that the iron measured in this study is in a form which is able to catalyse the formation of OH$\cdot$ radicals by the Haber-Weiss reaction and the initiation of lipid peroxidation. However, the availability of iron for chelation by desferrioxamine is highly suggestive of its availability to catalyse these damaging reactions. The increased amount of chelatable iron in ischaemic renal cells will therefore contribute to the level of oxidative stress posed to the tissue at the point of reperfusion with oxygen and may therefore cause the increased rates of lipid peroxidation which we have previously observed[2]. This series of damaging reactions, having been initiated at the start of the reperfusion period,

Table 1. The Effect of Cold Ischaemia on the level of Desferrioxamine-Available Iron in the Rabbit Kidney

	Cortex	Medulla
Control	296 ± 100	345 ± 177
24h C.I.	604 ± 118*	620 ± 139*

Desferrioxamine-available iron (pmol/mg protein) was determined by hplc. Values represent the mean ± S.D. of 6 determinations. *Significantly ($P < 0.005$) different from control.

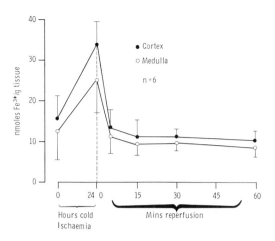

Fig. 1. Levels of desferrioxamine-chelatable iron in kidneys subjected to 24 hours cold ischaemia and reperfusion for up to 60 minutes.

then continues as a self-perpetuating chain reaction for an extended period of at least 60 minutes[2] which results in extensive reperfusion damage to the organ.

We have previously shown that DFX is effective at inhibiting lipid peroxidation in rabbit kidneys subjected to cold ischaemia[9]. The new data presented here further supports the hypothesis that redistribution of intracellular iron is important in the pathogenesis of ischaemic tissue damage and suggests that iron chelation may be a worthwhile therapeutic approach to this problem.

REFERENCES

1. B.J. Fuller, J.D. Gower, and C.J. Green, Free radical damage and organ preservation: fact or fiction?, Cryobiology 25:377 (1988).
2. C.J. Green, G. Healing, J. Lunec, B.J. Fuller, and S. Simpkin, Evidence of free radical-induced damage in rabbit kidneys after simple hypothermic preservation and autotransplantation, Transplantation 41:161 (1986).
3. E.D. Wills, Mechanisms of lipid peroxide formation in animal tissues, Biochem. J. 99:667 (1966).
4. J.M.C. Gutteridge, B. Halliwell, and D.A. Rowley, Catalytic iron complexes in biological material: A potential for oxygen radical damage, Life Chem. Rep. 1, Suppl 2:15 (1981).
5. D.L. Bakkeren, C.M.H. de Jeu-Jaspars, C. van der Heul, and H.G. van Eijk, Analysis of iron-binding components in the low molecular weight fraction of rat reticulocyte cytosol, Int. J. Biochem. 17:925 (1985).
6. G. Healing, J.D. Gower, B.J. Fuller, and C.J. Green, Release of chelatable iron and free radical damage to rabbit kidneys subjected to ischaemia and reperfusion, Med. Sci. Res. 17:67 (1989).
7. J. Gower, G. Healing, and C. Green, Measurement by hplc of desferrioxamine-available iron in rabbit kidneys to assess the effect of ischaemia on the distribution of iron within the total pool, Free Rad. Res. Comms. 5:291 (1989).
8. O.H. Lowry, N.J. Rosebrough, A.L. Farr, and R.J. Randall, Protein measurement with the folin phenol reagent, J. Biol. Chem. 192:265 (1951).

9. B.J. Fuller, J. Lunec, G. Healing, S. Simpkin, and C.J. Green, Reduction of susceptibility to lipid peroxidation by desferrioxamine in rabbit kidneys subjected to 24-hr cold ischaemia and reperfusion, <u>Transplantation</u> 43:604 (1987).

EVIDENCE THAT CALCIUM MEDIATES FREE RADICAL DAMAGE THROUGH ACTIVATION OF

PHOSPHOLIPASE A$_2$ DURING COLD STORAGE OF THE RABBIT KIDNEY

Lisa A Cotterill, Jon D Gower, Barry J Fuller* and
Colin J Green

Section of Surgical Research, Clinical Research Centre
Harrow, Middlesex, HA1 3UJ, U.K. and *Academic Department
of Surgery, Royal Free Hospital School of Medicine
London, NW3 1QG, U.K.

INTRODUCTION

Renal transplantation involves periods of warm and cold ischaemia which often result in poor renal function and death of tubular cells soon after engraftment. There is growing evidence that oxygen derived free radicals are responsible, at least, in part, for reperfusion injury and may be involved in ischaemic damage[1]. We have shown in previous studies that markers of lipid peroxidation, resulting from oxidative damage, increase in rabbit kidneys following storage[2]. In addition, we have recently reported evidence for a relationship between altered calcium homeostasis and oxidative damage to the rabbit kidney following cold storage[3].

One possible mechanism by which calcium ions may potentiate oxidative damage is through the activation of phospholipases. Free fatty acids (FFA), the products of phospholipase action, may accumulate during periods of cold ischaemia and such accumulation would be detrimental to the cell due to the detergent-like properties of these molecules[4]. In addition, the disruption of the membrane by lysophosphatides, the residual products of phospholipid hydrolysis[5], may render the membrane more susceptible to attack by free radicals.

We have now investigated the free fatty acid status of rabbit kidneys following cold storage in hypertonic citrate (HCA) using gas liquid chromatography. In addition, we have examined the effects of adding the calcium ionophore A23187, and the phospholipase inhibitor dibucaine on rabbit kidneys during the storage period.

MATERIALS AND METHODS

Groups of New Zealand White rabbits (average weight 3kg) were anaesthetised, the abdomen opened by a mid-line incision, and the kidneys completely skeletalised and removed[2]. The renal artery of each kidney was cannulated and the organs flushed with either 30ml sterile isotonic saline solution or 30ml HCA already cooled to 4°C. 10µM A23187 and 250µM dibucaine were added to storage solutions where appropriate. The kidneys were placed in sterile beakers containing 60ml of the identical flush solution and

Antioxidants in Therapy and Preventive Medicine
Edited by I. Emerit *et al.*
Plenum Press, New York, 1990

397

stored in ice at 0°C for either 72 hours or 24 hours. Kidneys were dissected into cortex and medulla and immediately frozen in liquid nitrogen. Samples were ground to a powder using a mortar and pestle while still frozen. Free fatty acids were extracted and methylated with methyl iodide according to the method of Allen[6]. 10μg of the internal standard pentadecanoic acid was added to the samples prior to extraction. Fatty acid methyl esters were analysed by gas-liquid chromatography using a glass column packed with 10% DEGS on chromosorb WAW 80-100 mesh. Fatty acid methyl esters were identified on the basis of their retention times compared to known standards. The rate of malonaldehyde formation was taken as a marker of lipid peroxidation[3]. Statistical analysis was performed using the students paired t-test.

RESULTS AND DISCUSSION

Following storage of rabbit kidneys for 72 hours in HCA, there were significant rises in all observed unsaturated fatty acids oleic acid (18:1), linoleic acid (18:2) and arachidonic acid (20:4) in both cortex and medulla of kidneys (Table 1). No such increases were observed in the saturated fatty acids, stearic acid (16:0) and palmitic acid (18:0) (Table 1). The most conspicuous increase observed in cortex and medulla was the rise in arachidonic acid.

Addition of the phospholipase A_2 inhibitor dibucaine led to a reduction in the accumulation of free arachidonic acid in both cortex and medulla of rabbit kidneys following 72 hours storage in HCA (Table 1). It is likely, therefore, that calcium activation of phospholipase A2 is responsible at least in part for augmenting the release of arachidonic acid during storage. However dibucaine did not completely abolish arachidonic acid accumulation

Table 1. Free Fatty Acid (FFA) analysis in cortex and medulla of rabbit kidneys ± 250μM dibucaine before and after storage in HCA for 72 hours.

	Ratio of 15:0 Standard: FFA			
	CORTEX		MEDULLA	
FFA	Non-stored	Stored	Non-stored	Stored
16:0	4.76 ± 1.99	7.42 ± 3.06	6.22 ± 2.73	6.71 ± 4.02
18:0	6.35 ± 2.69	8.16 ± 3.77	9.32 ± 3.58	7.55 ± 4.85
18:1	0.21 ± 0.32	1.69 ± 0.87*	0.00 ± 0.00	1.06 ± 0.49*
18:2	0.38 ± 0.47	1.85 ± 1.66*	0.28 ± 0.15	0.78 ± 0.39*
20:4	0.13 ± 0.22	1.66 ± 0.59*	0.10 ± 0.06	0.41 ± 0.16*
20:4 250μM Dibucaine		0.44 ± 0.20*		0.21 ± 0.08*

Free fatty acids (FFA) were measured by GLC. Values represent mean ± S.D. of 8 determinations. *Significantly (P < 0.05) different from control.

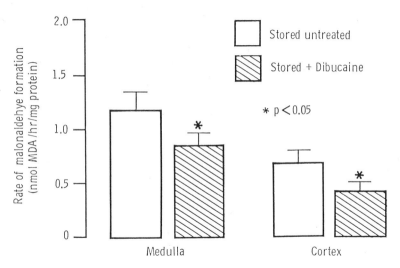

Figure 1. Rate of malonaldehyde formation in cortex and medulla
of rabbit kidneys stored for 24 hours in saline
containing 250μM dibucaine.

Values represent mean ± S.D. of 8 determinations. *Significantly
(P<0.05) different from control.

during storage. Thus it may be postulated that blocked β-oxidation, due to
shortage of ATP during ischaemia is a contributory factor in producing FFA
accumulation.

In similar studies dibucaine reduced the level of markers of lipid
peroxidation (Figure 1), indicating the involvement of phospholipase A_2 in
potentiating lipid peroxidation. Disruption of the lipid bilayer through
increased phospholipase activity may render the membrane more susceptible to
lipid peroxidation. In addition, free fatty acids, without the protective
environment of the membrane, form excellent targets for free radical attack.

Increases in the accumulation of free arachidonic acid through
phospholipase A_2 activation may lead to a burst in prostaglandin synthesis
via cyclooxygenase. Upsetting the delicate eicosanoid balance may have
important consequences in the vascular bed of the organ due to production of
potent vasoconstrictors[7].

It would appear therefore that an increase in arachidonic acid
accumulation may be potentiated by calcium-dependent activation of
phospholipase A_2. Such effects are likely to mediate oxygen free radical
damage. Exact mechanisms are yet to be elucidated, but there may be a role
for phospholipase inhibitors in the improvement of organ preservation.

REFERENCES

1. R.E. Raych, and G.B. Bulkley, Free-radical-mediated postischaemic
     reperfusion injury in the kidney, J. Free Radic. Biol. Med. 1:311
     (1986).
2. C.J. Green, G. Healing, J. Lunec, B.J. Fuller, and S. Simpkin, Evidence
     of free-radical induced damage in rabbit kidneys after simple
     hypothermic preservation and autotransplantation, Transplantation
     41:161 (1986).

3. L.A. Cotterill, J.D. Gower, B.J. Fuller, and C.J. Green, A role for calcium in cold ischaemic damage to the rabbit kidney? in: "Free Radicals: Chemistry, pathology and medicine," C. Rice-Evans, and T. Dormandy, Eds., Richelieu Press, London, 1988.

4. H.M. Piper, and A. Das, Detrimental actions of endogenous fatty acids and their derivatives. A study of ischaemic mitochondrial injury, in: "Lipid metabolism in the normoxic and ischaemic heart," H. Stam, and G.J. van der Vussue. Eds.. Springer-Verlag. New York. (1986).

5. N.A. Shaikh, and E. Downar, Time course of changes in porcine myocardial phospholipid levels during ischaemia, CRC. Res, 49:316 (1981).

6. K.G. Allen, A new procedure to analyse free fatty acids. Application to 20mg brain tissue samples, J. Chromat., 309:33 (1984).

7. J.D. Gower, G. Healing, B.J. Fuller, S. Simpkin, and C.J. Green, Protection against oxidative damage in cold-stored rabbit kidneys by desferrioxamine and indomethacin, Cryobiology. In press.

ISCHEMIA-REPERFUSION INJURY AND FREE RADICAL INVOLVEMENT IN GASTRIC MUCOSAL

DISORDERS

T.YOSHIKAWA, Y.NAITO, S.UEDA, S.TAKAHASHI, H.OYAMADA, T.YONETA, S.SUGINO, and M.KONDO

First Department of Medicine, Kyoto Prefectural University of Medicine, Kamigyo-ku, Kyoto 602, Japan

INTRODUCTION

There has been great interest in a possible role of oxygen radical species in ischemia-reperfusion injury in the heart,[1] small intestine,[2] liver,[3] pancreas,[4] kidney,[5] brain,[6] skin,[7] and organs being prepared for transplantation.[8] The ischemia itself causes tissue damage and eventual death, but further injuries can occur while oxygen is reintroduced to the tissue. Much evidence suggests that free radical and active oxygen species derived from molecular oxygen, such as superoxide, hydrogen peroxide, hydroxyl radical, and singlet oxygen, contribute to the tissue injury. At least five possible sources are under investigation for the production of the active oxygen species : 1)the hypoxanthine-xanthine oxidase system ; 2)the activated polymorphonuclear leukocytes; 3)The disrupted mitochondrial electron transport system; 4)the metabolism of arachidonate via the lipoxygenase pathway, and 5)vascular endothelial cell.

The object of this investigation was to study the chronological changes in gastric mucosal injury and in TBA-reactive substances in the gastric mucosa after ischemia or ischemia-reperfusion, and also to evaluate the anti-oxidative effect of several radical scavengers. In addition, to clarify the source of oxygen radicals, the effect of the treatment with allopurinol or with anti-neutrophil serum on acute gastric mucosal injury was investigated.

MATERIALS AND METHODS

Experimental model of ischemia-reperfusion injury

Male Sprague-Dawley rats each weighing 190-210 g, from Keari Co., Ltd., Osaka, were used. The animals were not fed for 18 h prior to the experiment, but were allowed free access to water. Ischemia was created under intraperitoneal pentobarbital anesthesia (25 mg/kg) by applying small clamps to the celiac artery, and reoxygenation was produced by removal of the clamps.

Determination of gastric mucosal blood flow

The microcirculatory blood flow in the gastric mucosa was measured by the laser Doppler flowmeter (ALF 2100, Advance Co., Ltd.).

Evaluation of gastric mucosal lesions

After ischemia or ischemia-reperfusion, rats were killed by

Antioxidants in Therapy and Preventive Medicine
Edited by I. Emerit *et al.*
Plenum Press, New York, 1990

401

exsanguination via the abdominal aorta. The gastric mucosa was carefully examined macroscopically and microscopically, and the extent of the gastric mucosal lesion was indicated by the total area of the erosions.

Biochemical assay

Thiobarbituric acid (TBA)-reactive substances, an index of lipid peroxidation, were measured in serum by the method of Yagi,[9] and in tissue by that of Ohkawa et al.[10] The level of TBA reactants was indicated as nmol of malodialdehyde. TBA (BDH Chemical, Poole, England) and 1,1,3,3-tetramethoxy propane (Tokyo Kasei Co., Tokyo) were used for TBA assay and all other chemicals were of reagent grade. The serum α-tocopherol content was determined by the method of Abe et al.,[11] using a high-speed LC-6A liquid chromatograph (Shimazu Co., Kyoto). The serum cholesterol level was assayed according to the method of Richmond.[12]

Treatment with radical scavengers

To assess the effect of superoxide dismutase (SOD) and catalase, human SOD (Nippon Kayaku Co., Ltd., Tokyo) at a dose of 50,000 U/kg or/and catalase from bovine liver (Sigma Chemical Co., St Louis, MO) at a dose of 90,000 U/kg were injected subcutaneously 1 h before ischemia, and 10,000 U/kg of SOD was intravenously injected just before the reoxygenation. The control rats were treated with physiological saline in the same manner.

Allopurinol (Sigma Chemical Co.), an inhibitor of xanthine oxidase, dissolved in distilled water (pH 10.8) was orally administrated to rats at a dose of 100 mg/kg 48, 24 h before experiment. For the control, distilled water adjusted to pH 10.8 by 0.5 N NaOH was administrated in the same manner.

Anti-rat polymorphonuclear leukocytes(PMN) antibody obtained from the immunized rabbits according to the modified method of Ward et al.,[13] was intraperitoneally injected 18 h before experiments. The control rats were injected normal rabbits serum in the same manner.

RESULTS

Blood flow of the gastric mucosa

Clamping of celiac artery could reduce the gastric blood flow to 10% of that measured before the clamping. In the cases of less than thirty min of clamping, the blood flow in the gastric mucosa could recover from ischemic status promptly and completely just after the removal of the clamps (Fig.1). However, over 45 min of clamping inhibited prompt recovery from ischemic status. The decrease of blood blow after clamping or the prompt recovery following removal of clamps were not influenced by the treatment with SOD and/or catalase (Fig.2).

The chronological changes in the total area of the erosion

The total area of the erosions(TAE), a morphological index of gastric injury, did not increase after 30 min ischemia. TAE significantly increased after 60 min ischemia ($p < 0.001$), and 30 min ischemia with 30 min reperfusion ($p < 0.001$) ; however, the increase in TAE in the former was significantly ($p < 0.05$) much higher than that in the latter. Between the cases of 90 min ischemia and 30 min ischemia with 60 min reperfusion, there was little difference in TAE(Fig.3).

The chronological changes in the TBA-reactive substances in the gastric mucosa

The TBA reactants in the gastric mucosa, an index of lipid peroxidation, did not increase after 30 min ischemia. The TBA reactants in the gastric mucosa significantly increased after 60 or 90 min ischemia, and after 30 min ischemia with 30 or 60 min reperfusion; however, the increase in TBA reactants was similar between the ischemic injury group and the ischemia-reperfusion injury group (Fig.4).

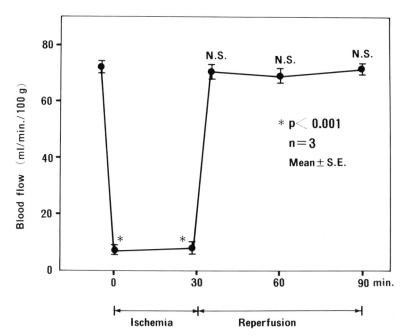

Fig. 1 Gastric mucosal blood flow in ischemia-reperfusion model measured by laser doppler blood flowmeter.

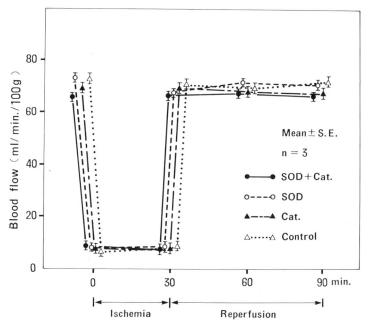

Fig. 2 Effects of SOD or/and catalase on the gastric mucosal blood flow in ischemia-reperfusion.

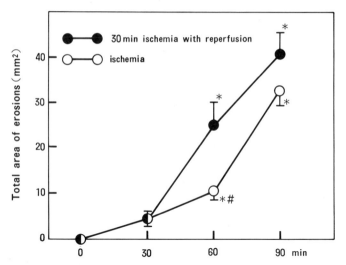

Fig. 3 The chronological changes in the total area of the gastric erosions
induced by ischemia or ischemia-reperfusion.
Each value indicates mean±SE of 7-10 experiments. *p<0.001 when compared
with the value before ischemia, #p<0.05 when compared with the value after
30 min ischemia with 30 min reperfusion.

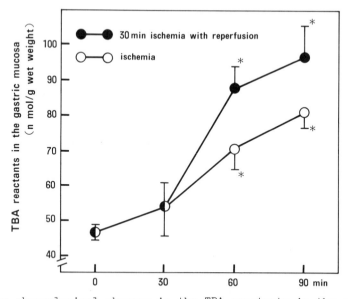

Fig. 4 The chronological changes in the TBA reactants in the gastric mucosa
after ischemia or ischemia-reperfusion.
Each value indicates mean± of 6-19 experiments. *p<0.001 when compared with
the value before ischemia.

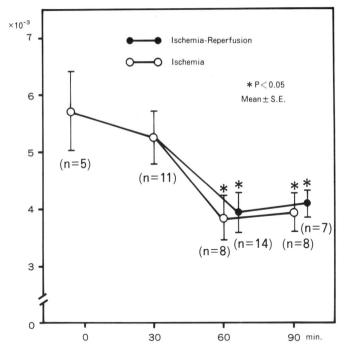

Fig. 5 The chronological changes in the α-tocopherol/cholesterol ratio in serum. *p<0.05 when compared with the value before ischemia.

The chronological changes in the α-tocopherol/cholesterol ratio in serum

The α-tocopherol/cholesterol ratio in serum was significantly reduced after 60 or 90 min ischemia, and after 30 min ischemia with 30 or 60 min reperfusion. The decrease in the α-tocopherol/cholesterol ratio was in the same degree between the ischemia group and ischemia-reperfusion group (Fig.5).

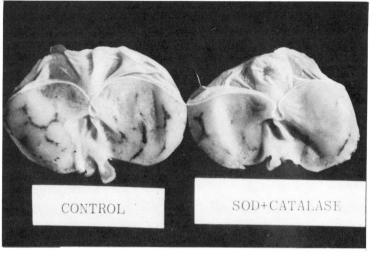

Fig. 6 Effect of SOD + catalase on the gastric erosions induced by ischemia-reperfusion in rats.

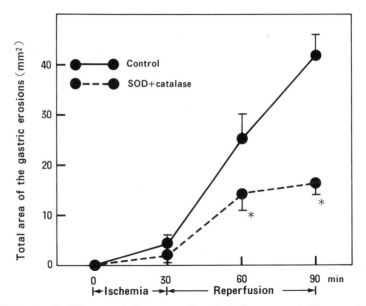

Fig. 7 Effect of SOD + catalase on the total area of the gastric erosions induced by ischemia-reperfusion.
Each value indicates mean±SE of 5-12 experiments. *p<0.01 when compared with the value of control.

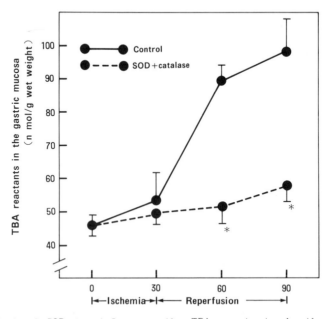

Fig. 8 Effect of SOD + catalase on the TBA reactants in the gastric mucosa after ischemia-reperfusion.
Each value indicates mean±SE of 5-14 experiments. *p<0.01 when compared with the value of control.

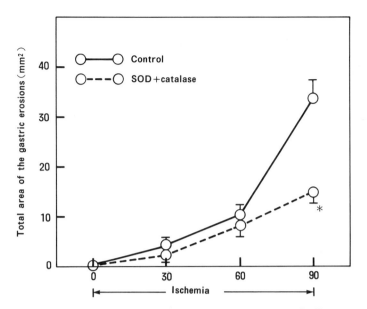

Fig. 9   Effect of SOD + catalase on the total area of the gastric erosions after ischemia.
Each value indicates mean±SE of 5-10 experiments.   *p<0.01 when compared with the value of control.

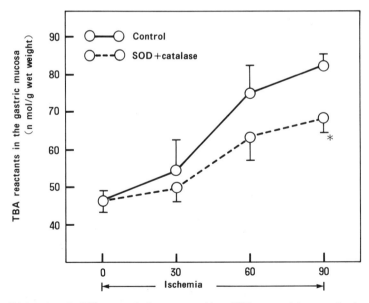

Fig. 10   Effect of SOD + catalase on the TBA-reactive substances in the gastric mucosa after ischemia.
Each value indicates mean±SE of 5-10 experiments.   *p<0.01 when compared with the value of control.

The effect of radical scavengers on ischemic or ischemia-reperfusion injury

TAE in the gastric mucosa induced by 30 min ischemia with 30 or 60 min reperfusion significantly decreased by the treatment with SOD, SOD+catalase, or catalase (Fig.6, 7). TAE induced by 90 min ischemia also significantly decreased by the treatment with SOD+catalase (Fig.8). The increase in TBA reactants in the gastric mucosa induced by 30 min ischemia with 30 min or 60 min reperfusion significantly reduced by the treatment with SOD+catalase (Fig.9). The increase in TBA reactants induced by 90 min ischemia was also significantly inhibited by the treatment with SOD+catalase (Fig.10).

Treatment with allopurinol could attenuate the gastric mucosal injury induced by 30 min ischemia with 60 min reperfusion, but the increase in TBA reactants was not significantly inhibited (Fig.11). The treatment with anti-PMN antibody did not show significant inhibition against the increase in TAE and TBA in the gastric mucosa induced by ischemia-reperfusion.

DISCUSSION

These studies demonstrated that relatively little injury was produced during the 30 min ischemia, whereas most of the gastric mucosal injury resulting from 30 min of ischemia and 30 min of reperfusion occurred during the following stage of reperfusion. TBA reactants in the gastric mucosa, which are important and damaging products of free radical-mediated lipid peroxidation, did not increase during 30 min ischemia, but increased dramatically and significantly during the following stage of reperfusion. These results suggest that lipid peroxidation or lipid peroxides may play an important role in the formation of reperfusion injury. The time course study confirmed that more than 60 min incomplete ischemia itself produced

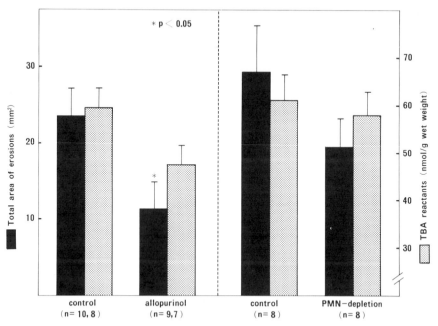

Fig. 11 Effects of allopurinol or PMN depletion on the total area of the erosions and on the increase of TBA reactants induced by ischemia-reperfusion.
Each value indicates mean±SE. *p<0.05 when compared with the value of control group.

gastric mucosal injury with increased lipid peroxidation, while the reperfusion injury associated with the 30 min ischemia greatly exceeded the injury produced by 60 min of ischemia without reperfusion.

In addition, SOD and catalase inhibited the aggravation of gastric mucosal injury induced by ischemia-reperfusion as well as by ischemia itself without recovering the reduced gastric mucosal blood flow. These results indicate that the effectiveness of these enzymes possibly appear not through the effect on gastric wall microcirculation, but through the catalyzation of oxygen radicals, especially superoxide radical.

The mechanism for generation of active oxygen metabolites in ischemic and ischemia-reperfusion tissue was firstly proposed by Granger et al.,[2] for the small intestine, who presented that pretreatment with xanthine oxidase inhibitor (allopurinol, folic acid, pterinaldehyde) or inactivation of xanthine oxidase by placing animals on a tungsten-supplemented molybdenum-deficient diet, resulted in dramatic protection of the intestinal mucosa from ischemia-reperfusion injury.[14] Consequently, they have proposed the following mechanism : during ischemia, ATP is catabolized to hypoxanthine which accumulates in the tissue. Concurrently, the NAD^+-reducing enzyme xanthine dehydrogenase is converted to xanthine oxidase by the action of a protease or oxidation of the enzyme. Introduction of molecular oxygen by reoxygenation of ischemic intestine results in the xanthine oxidase-catalyzed production of superoxide radical and hydrogen peroxide. Subsequently, superoxide and hydrogen peroxide react in the presence of transitional metals or their chelates to form the highly reactive and cytotoxic hydroxyl radical.

In the gastric injury model induced by hemorrhagic shock, Itoh and Guth[15] reported that gastric lesions were reduced by the treatment with SOD, or allopurinol. In our ischemia-reperfusion injury model in the stomach, pretreatment with allopurinol largely prevented the gastric injury, indicating that xanthine oxidase is one of the major sources of reactive oxygen metabolites in the gastric injury induced by ischemia-reperfusion.

Another source of oxygen radicals is the polymorphonuclear leukocytes (PMNs). When PMNs are stimulated by particle or specific soluble inflammatory mediators, they produce superoxide radicals via activation of their NADPH oxidase in the cell membrane.[16] Most hydrogen peroxide released during the stimulation of phagocytic cells appears to be directly derived from the dismutation of superoxide. Evidence regarding a role for PMNs in ischemic or ischemia-reperfusion injury to the gastric or intestinal mucosa is derived from the studies using tissue-associated myeloperoxidase (MPO) activity,[17] anti-neutrophil serum (ANS), or monoclonal antibody.[18] Using MPO assay, Grisham and Granger[19] have demonstrated a document increase in neutrophil infiltration into the gastric, intestinal, and colonic mucosa during ischemia and reperfusion. To clarify the role of PMNs, we produced PMN depletion rats by administration of the ANS from the immunized rabbits. By the treatment with ANS, the gastric mucosal injury induced by ischemia-reperfusion could not be inhibited. As compared with the hypoxanthine-xanthine oxidase system catalyzed by xanthine oxidase, PMNs seem to play a relatively small part in the formation of gastric injury induced by ischemia-reperfusion.

Many the other sources of active oxygen species should be considered. When mitochondria was injured by ischemia, much electrons from the electron transport chain may be leaked. Oxygen radicals released from endothelial cells, platelets, macrophages, and smooth muscle cell may contribute to the ischemia-reperfusion injury. Broader knowledge about the sources of free radicals, mechanisms of injury initiation, and biochemical regulation of antioxidant defense mechanism should provide insight into new therapeutic strategies for the modulation of many disorders induced by ischemia-reperfusion.

REFERENCES

1)T.J.Gardner, J.R.Stewart, A.S.Casale, et al: Reduction of myocardial ischemic injury with oxygen-derived free radical. Surgery, 94:423 (1983).

2)D.N.Granger, G.Rutili, and J.M.McCord: Superoxide radicals in feline intestinal ischemia. Gastroenterol, 81:22 (1981).

3)M.J.P.Arthur, I.S.Bentley, A.R.Tanner, et al: Oxygen-derived free radicals promote hepatic injury in the rat. Gastroenterol., 89:1114 (1985).

4)H.Sanfey, G.B.Bulkley, and J.L.Cameron: The role of oxygen-derived free radicals in the pathogenesis of acute pancreatitis. Ann.Surg., 200:405 (1984).

5)G.L.Baker, R.J.Corry, and A.P.Autor : Oxygen free radical induced damage in kidneys subjected to warm ischemia and reperfusion. Ann Surg., 202:628 (1985).

6)E.S.Flamm, H.B.Demopoulos, M.L.Seligman, R.G.Poser, and J.Ransohoff: Free radicals in cerebral ischemia. Stroke, 9:445 (1978).

7)P.N.Manson, R.N.Anthenelli, M.J.Im, G.B.Bulkley, and J.E.Roopes : The role of oxygen-free radicals in ischemic tissue injury in island skin flaps. Ann.Surg., 198:87 (1983).

8)I.Koyama, G.B.Bulkley, G.M.Williams, and M.J.Im : The role of oxygen free radicals in mediating the reperfusion injury of cold-preserved ischemia kidneys. Transplantation, 40:590 (1985).

9)K.Yagi : A simple fluorometric assay for lipid peroxides in blood plasma. Biochem.Med., 15:212 (1976).

10)H.Ohkawa, N.Ohishi, and K.Yagi : Assay for lipid peroxides for animal tissues by thiobarbituric acid reaction. Anal.Biochem., 95:351 (1979).

11)K.Abe, Y.Yuguchi, and G.Katui: Quantitative determination of tocopherols by high-speed liquid chromatography. J.Nutr.Sci.Vitaminol., 21:183 (1975).

12)W.Richmond: Preparation and properties of a cholesterol oxidase from nocardia sp. and its application to the enzymatic assay of total cholesterol in serum. Clin.Chem., 19:1350 (1973).

13)P.A.Ward, C.G.Cochrance : Bound complement and immunologic injury of blood vessels. J.Exp.Med., 121:215 (1965).

14)D.N.Granger, M.Hallwarth, and P.A.Parks : Ischemia-reperfusion injury, role of oxygen-derived free radicals. Acta.Physiol.Scand.Suppl, 548:47 (1986).

15)M.Itoh and P.H.Guth : Role of oxygen-derived free radicals in hemorrhagic shock-induced gastric lesion in rats. Gastroenterol., 88:1162 (1985).

16)B.M.Babior, R.S.Kipnes, and J.T.Curnutte : Biological defense mechanisms. The production by leukocytes of superoxide, a potential bactericidal agent. J.Clin.Invest, 52: 741 (1973).

17)K.Suzuki, H.Ota, S.Sasagawa, T.Sakatani, and T.Fujikura: Assay method for myeloperoxidase in human polymorphonuclear leukocytes. Anal.Biochem., 132:345 (1983).

18)L.A.Hernandez, M.B.Grisham, B.Twohig, K.E.Arfos, J.M.Harlan, and D.N.Granger: Role of neutrophils in ischemia-reperfusion-induced microvascular injury. Am.J.Physiol., 253:H699 (1987).

19)M.B.Grisham and D.N.Granger: Neutrophil-mediated mucosal injury; Role of reactive oxygen metabolites. Dig.Dis.Sci.Suppl., 33:6s (1988).

ZINC-CARNOSINE CHELATE COMPOUND (Z-103) ATTENUATES ACUTE GASTRIC MUCOSAL

INJURY INDUCED BY ISCHEMIA-REPERFUSION IN RATS

Y.NAITO, T.YOSHIKAWA, T.TANIGAWA, T.YONETA, S.UEDA, H.OYAMADA
T.TAKEMURA, S.SUGINO, and M.KONDO

First Department of Medicine, Kyoto Prefectural University of
Medicine, Kamigyo-ku, Kyoto 602, Japan

INTRODUCTION

A novel synthesized agent, zinc N-(3-aminopropionyl)-L-histidine (Z-103, Fig. 1), is a chelate compound that consists of zinc iron and L-carnosine. Carnosine was discovered in 1900 by Gulewitsch and Amiradzibi[1] from meat extract, and is reportedly present in the range of 1-20 mM in the skeletal muscle and brain of many animals and humans[2]. Recently, several reports have described the antioxidative activity of carnosine, such as efficient singlet oxygen scavengers[3], peroxy radical scavengers[4], efficient chelating agent for copper and other transitional metals[4], and superoxide scavenging activity in the presence of copper or zinc[5].

Zinc compounds have been reported to enhance the rate of healing human gastric ulcers and to protect against various kinds of experimental ulcers[6,7]. Although it was considered that their anti-ulcerative effects was due to membrane stabilizing activity of zinc ions especially on mast cell and on lysosomes[8], zinc has also been reported to mediate a free radical reaction including lipid peroxidation. Willson showed that zinc may play a vital role in masking labile S sites so reducing damage from radical processes at critical times during the life of the cell[9]. Girotti has demonstrated that Zn strongly inhibited lipid peroxidation in a natural membrane exposed to xanthine-xanthine oxidase and iron, and concluded that Zn interfered with lipid peroxidation at the membrane level, possibly by altering or preventing iron binding[10].

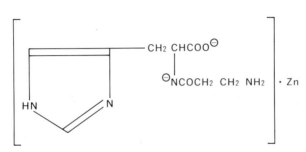

Fig. 1   Zinc N-(3-aminopropionyl)-L-histidine (Z-103)

Antioxidants in Therapy and Preventive Medicine
Edited by I. Emerit *et al.*
Plenum Press, New York, 1990

411

Recently, much interest has been focused on the cytotoxic effects or reactive oxygen metabolites in the ischemia-reperfusion tissue injury. Lipid peroxidation and active oxygen metabolites have been suggested to play an important role in the pathogenesis of acute gastric or intestinal mucosal injury induced by ischemia-reperfusion in rats and cats[11,12]. The object of this study was to determine whether Z-103 prevents or attenuates acute gastric mucosal injury induced by ischemia-reperfusion in rats.

MATERIALS AND METHODS

Experimental model of ischemia-reperfusion injury

Male Sprague-Dawley rats each weighing 190-210g, from Keari Co.,Ltd., Osaka, were used. The animals were not fed for 18 hr. prior to the experiment, but were allowed free access to water. Ischemia was created under intraperitoneal pentobarbital anesthesia (25 mg/kg) by applying small clamps to the celiac artery for 30 min and reoxygenation was produced by removal of the clamps for 30 min[11]. The agent Z-103 was dissolved in 0.5 % carboxymethyl cellulose sodium solution, and given to rats by gastric intubation 1 h before ischemia.

Determination of gastric mucosal blood flow

The microcirculatory blood flow in the gastric wall was measured by the laser Doppler flowmeter (ALF 2100, Advance Co., Ltd.). The effect of Z-103 at a dose of 30 mg/kg on the blood flow of the gastric wall was evaluated at before, just and 30 min after ischemia, and just and 30 min after reoxygenation.

Evaluation of gastric mucosal lesions

After ischemia-reperfusion, rats were killed by exsanguination via the abdominal aorta. The gastric mucosa was carefully examined macroscopically and microscopically, and the extent of gastric mucosal lesions was expressed by the total area of the erosions.

Biochemical assay

Thiobarbituric acid (TBA)-reactive substances, an index of lipid peroxidation, were measured in serum by the method of Yagi[13], and in tissue by that of Ohkawa et al[14]. The level of TBA reactants was expressed as nmol of malondialdehyde. TBA (BDH Chemicals, Poole, England) and 1,1,3,3-tetramethoxy propane (Tokyo Kasei Co., Tokyo) were used for TBA assay and all other chemicals were of reagent grade. Total protein in tissue homogenates was measured by the method of Lowry[15].

RESULTS

Blood flow of the gastric wall

Clamping of celiac artery could decrease the gastric blood flow to 10 % of that measured before the clamping. Just after the removal of the clamps, the blood flow in the gastric wall recovered from ischemic status completely. The decrease of blood flow after ischemia was not inhibited by the treatment with Z-103 at a dose of 30 mg/kg.

Gastric mucosal lesions

The total area of the gastric mucosal erosions induced by ischemia-reperfusion was significantly decreased by the treatment with Z-103 at doses of 10, 30, and 100 mg/kg (Fig. 2).

TBA reactive substances in serum and in the gastric mucosa

The increase in TBA-reactive substances in the gastric mucosa after ischemia-reperfusion was significantly inhibited by the treatment with Z-103 at doses of 30 and 100 mg/kg.

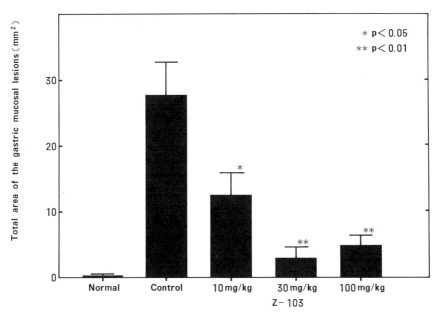

Fig. 2 Effect of Z-103 on the total area of the gastric erosions induced by ischemia-reperfusion.   Each value indicates mean ± SE of 7 experiments. *p<0.05, **p<0.01 when compared with disease control group.

Serum TBA reactants of the disease control group and groups treated with Z-103 did not show any significant changes after ischemia-reperfusion compared with that of the sham-operated rats (normal control).

DISCUSSION

In the present study, Z-103 inhibited the increase of gastric mucosal lesion and the increase of TBA reactants in the gastric mucosa induced ischemia-reperfusion without recovering the reduced microcirculatory blood flow in the gastric wall.   In this ischemia-reperfusion injury model of the stomach in rats, we have presented some evidences that tissue lipid peroxidation may play an important role in the formation of gastric mucosal injury, and SOD, catalase, and allopurinol, a competitive inhibitor of xanthine oxidase, inhibit the aggravation of gastric injury.

TABLE 1

Effect of Z-103 on the increase of TBA-reactive substances in the gastric mucosa after ischemia-reperfusion

		TBA reactants	
Pretreatment (n)		gastric mucosa (nmol/mg protein)	serum (nmol/ml)
Normal control (7)		0.34 ± 0.12	2.37 ± 0.18
Disease control (7)		0.62 ± 0.18	2.57 ± 0.28
Z-103 10 mg/kg (7)		0.48 ± 0.17 *	2.51 ± 0.31
30 mg/kg (7)		0.38 ± 0.11 *	2.49 ± 0.23
100 mg/kg (7)		0.41 ± 0.11 *	2.59 ± 0.37

Each value indicates mean ± SD. *p<0.05 when compared with the disease control group.

In addition to several reports described the antioxidative action of carnosine or zinc ion, Z-103 showed more efficient superoxide scavenging activity compared with carnosine in the presence of zinc ion[5]. The increase of lipid peroxide produced in rat brain homogenate and in rat hepatic microsome was both significantly inhibited by the presence of Z-103 (Data was not shown).

These results suggest that the protective effect of Z-103 against the aggravation of gastric mucosal injury induced by ischemia-reperfusion may be due to its antioxidative action.

REFERENCES

1. W.Gulewitsch and S.Amiradzibi : Uber das Carnosine, eine neue organische Base des Fleischextraktes. Ber.J.G., 33:1902 (1900).
2. C.R.Scriver, T.L.Perry, and W.Nutzenadel : Disorder of beta-alanine, carnosine, and homocarnosine metabolism. in "The metabolic basis of inherited disease", J.B.Stanbury et al eds. mcGraw-Hill, New York (1983).
3. P.E.Hartman : Interception of toxic agents/mutagens/carcinogens: some of nature's novel strategies. in "Antimutagenesis and anticarcinogenesis mechanism", D.M.Shankel et al eds., Plenum press, New York and London (1988).
4. R.Kohen, Y.Yamamoto, K.C.Cundy, and B.N.Ames : Antioxidant activity of carnosine, homocarnosine, and anserine present in muscle and brain. Proc.Natl.Acad.Sci. USA, 85:3175 (1988).
5. T.Yoshikawa, Y.Naito, T.Tanigawa, T.Yoneta, H.Oyamada, S.Ueda, T.Takemura, S.Sugino, and M.Kondo : Effect of zinc-carnosine chelate Z-103 on the generation of superoxide anion radicals from hypoxanthine-xanthine oxidase system. J.Clin.Exp.Med., 147:927(1988).
6. C.H.Cho, C.W.Ogle, and S.Dai: Effect of zinc chloride on gastric secretion and ulcer formation in pylous-occluded rats. Eur.J.Pharmacol., 38:337 (1976).
7. S.H.Wong, C.H.Cho, and C.W.Ogle : Protection by zinc sulfate against ethanol-induced ulceration ; Preservation of the gastric mucosal barrier. Pharmacology, 33:94 (1986).
8. W.J.Bettger and B.L.O'Dell : A critical physiological role of zinc in the structure and function of biomembrane. Life Science, 28:1425 (1981).
9. R.L.Willson : Vitamin, selenium, zinc and copper interaction in free radical protection against ill-placed iron. Proceed.Nut.Soc., 46:27 (1987).
10. A.W.Girotti, J.P.Thomas, and J.E.Jordan :Inhibitory effect of Zn(II) on free radical lipid peroxidation in erythrocyte membrane. J.Free.Rad.Biol. Med., 1:395 (1985).
11. T.Yoshikawa : Pathophysiology of ischemia-reperfusion injury. in "Medical, biochemical and chemical aspects of free radicals", O.Hayaishi et al eds., Elsevier Sci., Amsterdam.New York.Oxford (1988).
12. D.N.Granger, M.E.Hollwarth, and D.A.Parks: Ischemia-reperfusion injury: role of oxygen-derived free radicals. Acta.Physiol.Scand., Suppl.548:47 (1986).
13. K.Yagi: A simple fluorometric assay for lipid peroxides in blood plasma. Biochem.Med.,15:212 (1976).
14. H.Ohkawa, N.Ohishi, and K.Yagi: Assay for lipid peroxides in animal tissues by thiobarbituric acid reaction. Anal.Biochem, 95:351 (1979).
15. O.H.Lowry, N.J.Roserough, A.L.Farr, et al: Protein measurement with the folin phenol reagent. J.Biol.Chem., 193:265 (1951).

ROLE OF XANTHINE-XANTHINE OXIDASE SYSTEMS AND POLYMORPHONUCLEAR LEUKOCYTES

IN LIPID PEROXIDATION IN COMPOUND 48/80-INDUCED GASTRIC MUCOSAL INJURY IN RATS

Toshiki Takemura, Toshikazu Yoshikawa, Yuji Naito
Shigenobu Ueda, Toru Tanigawa, Shigeru Sugino, Motoharu Kondo

First Department of Medicine
Kyoto Prefectural University of Medicine
Kawaramachi, Hirokoji, Kamigyo-ku, Kyoto 602, Japan

INTRODUCTION

Recent studies have demonstrated that active oxygen species and lipid peroxidation might be involved in the formation of gastric mucosal lesions[1-3]. Compound 48/80 is a mast cell degranulator to release histamine and serotonin. Repeated administration of a small dose of this agent can induce gastric mucosal injury in rats with a low mortality rate [4]. This study was designed to evaluate a role of active oxygen species and lipid peroxidation in the pathogenesis of gastric mucosal injury induced by compound 48/80 in rats. Furthermore, the role of xanthine-xanthine oxidase systems and polymorphonuclear leukocytes (PMN) for the source of active oxygen species was examined.

MATERIALS AND METHODS

Male Donryu rats weighing 180-220g were fed a normal rat chow and tap water ad libitum during the entire experimental period.

Induction of Gastric Mucosal Injury by Compound 48/80

Compound 48/80 (Sigma, USA), dissolved in distilled water, was given intraperitoneally once daily for either 4 days at a dose of 0.75mg/kg[4]. The rats were killed with blood sampling, and their stomachs were removed 24 h after the final administration of compound 48/80. The total area of the gastric erosions were measured macroscopically. Thiobarbituric acid (TBA) reactants in the gastric mucosa were measured by the method of Ohkawa[5]. Serum TBA reactants and serum α-tocopherol/total cholesterol ratio were determined by the method of Yagi and HPLC-fluorescence method, respectively. Protein concentration in the gastric mucosa was measured by Lowry's method.

Effects of Various Agents on Compound 48/80-induced Gastric Injury

Polyethylene glycol-superoxide dismutase (PEG-SOD, 150,000 U/kg, i.v., Takeda Chemical Industry, Japan), which is a manganese SOD from serratia marcescens[6], and/or bovine catalase (90,000 U/kg, s.c., Sigma, USA), dissolved in physiological saline, were injected once daily for either 4 days with the injection of compound 48/80. Allopurinol dissolved in NaOH solution (pH 10.8), an inhibitor of xanthine oxidase, was orally administered at a dose of 50mg/kg 48 and 24 h before the initiation of the compound 48/80 treatment and once daily for either 4 days with the injection of compound 48/80. Anti-rat PMN an-

Antioxidants in Therapy and Preventive Medicine
Edited by I. Emerit *et al.*
Plenum Press, New York, 1990

415

tiboby obtained from the immunized rabbits was intraperitoneally injected at a dose of 10 ml/kg 18 h before the initiation of the compound 48/80 treatment and once daily for either 4 days with the injection of compound 48/80[7,8].

Determination of Arterial Blood Pressure and Gastric Mucosal Blood Flow

Systolic blood pressure of tail artery and gastric mucosal blood flow were determined by a programmable sphygmomanometer (PS-100, Rick, Japan) and by a laser doppler flowmeter (ALF 2100, Advance, Japan), respectively, before the compound 48/80 treatment and 1/6, 1/2, 1, 3, 6, 24 h after the single administration of 0.75mg/kg of compound 48/80.

Determination of Superoxide Generation from Peritoneal PMN

Superoxide generation from peritoneal PMN stimulated by phorbol myristate acetate (PMA) and opsonized zymosan was determined by 2-methyl-6(p-methoxyphenyl)3,7-dihydroimidazo(1,2-a)pyrazin-3-one (MCLA)-dependent chemiluminescence[9] 24 h after the single administration of 0.75mg/kg of compound 48/80.

The data were statistically analyzed by Student's t-test, and values of $p < 0.05$ were regarded as significant.

RESULTS

Gastric Mucosal Injury Induced by Compound 48/80

Gastric mucosal injury such as marked erosions and edema were observed in the glandular stomach after the repeated administration of compound 48/80. Histological findings showed an apparent loss of the superficial layer of the gastric mucosa and marked submucosal edema with the infiltration of PMN and eosinophiles.

Changes in TBA Reactants in the Serum and in the Gastric Mucosa, and Serum α-Tocopherol/Total Cholesterol Ratio

TBA reactants in the gastric mucosa were significantly increased (Table 1), and serum α-tocopherol/total cholesterol ratio was significantly decreased after the compound 48/80 treatment (Table 2). There were no changes in serum TBA reactants.

Effects of Various Agents on the Gastric Injury Induced by Compound 48/80

The total area of the gastric erosions and the increase in TBA reactants in the gastric mucosa were significantly reduced by the pretreatment with SOD, catalase, SOD plus catalase, allopurinol, and anti-PMN antibody (Table 1).

Changes in Arterial Blood Pressure and Gastric Mucosal Blood Flow (Table 3)

Systolic blood pressure was slightly decreased only 10 min after the injection of compound 48/80. Gastric mucosal blood flow was rapidly decreased after the injection of compound 48/80, and recovered to the normal range 24 h after the treatment. The pretreatment with SOD, catalase, SOD plus catalase, and allopurinol could not affect the gastric mucosal blood flow.

Superoxide Generation from Peritoneal PMN (Table 4)

In MCLA-dependent chemiluminescence study, superoxide generation from peritoneal PMN stimulated by PMA was significantly increased after the compound 48/80 treatment. There were no changes in superoxide generation stimulated by opsonized zymosan.

416

Table 1. Effects of Various Agents on Total Area of the Gastric Erosions and TBA Reactants in the Gastric Mucosa.

Groups	Total Area of the Gastric Erosions (mm^2)				TBA Reactants in the Gastric Mucosa (nmol/mg protein)			
	Mean	S.D.	n		Mean	S.D.	n	
Normal Conrol	–	–	–		0.372	0.071	10	***
Disease Control	73.0	31.4	10		0.512	0.104	10	
SOD	38.5	25.6	10	**	0.405	0.059	10	**
Catalase	46.0	19.6	10	*	0.427	0.070	10	*
SOD+Catalase	28.8	29.0	10	***	0.398	0.087	10	**
Allopurinol	22.5	17.5	8	*	0.392	0.055	8	***
Anti-PMN Ab.	11.2	24.4	16	***	0.371	0.088	10	***

$*p<0.05$, $**p<0.02$, $***p<0.01$

Table 2. Changes in Serum α-Tocopherol/Total Cholesterol Ratio after the Compound 48/80 Treatment.

Groups	n	α-Tocopherol/Total Cholesterol Ratio ($\times 10^{-2}$)
Normal Control	7	1.95 ± 0.29
Compound 48/80	11	1.14 ± 0.23 *

Values are mean $\pm$ S.D.. $*p<0.001$

Table 3. Changes in Gastric Mucosal Blood Flow after the Compound 48/80 Treatment

Time (hr)	Systolic Blood Pressure (mmHg)		Gastric Mucosal Blood Flow (ml/100g/min)	
before	117 ± 21		54.0 ± 4.9	
1/6	102 ± 27	*	27.9 ± 7.0	**
1/2	102 ± 26		23.7 ± 7.7	**
1	112 ± 13		22.0 ± 6.2	**
3	113 ± 9		21.0 ± 5.7	**
6	109 ± 10		22.5 ± 5.0	**
24	110 ± 13		54.3 ± 6.0	

Value are mean $\pm$ S.D. from 6 rats. $*p<0.05$, $**p<0.001$

Table 4. Superoxide Generation Detected by MCLA-dependent Chemiluminescence from Rat Peritoneal PMN Stimulated by PMA and Opsonized Zymosan.

Groups	Maximum Light Intensity ($\times 10^{-2}$ counts/min)	
	PMA	Opsonized Zymosan
Normal Control	9.1 ± 2.5	4.0 ± 1.3
Compound 48/80	15.3 ± 2.2 *	4.7 ± 1.1

Values are mean $\pm$ S.D. from 5 rats per group. $*p<0.01$

DISCUSSION

After the repeated treatment of compound 48/80, TBA reactants in the gastric mucosa was increased, and serum level of α-tocopherol, an important antioxidant in vivo, was decreased. The total area of the gastric erosions and the increase in TBA reactants in the gastric mucosa were reduced by the pretreatment with SOD, catalase, SOD plus catalase, allopurinol, and anti-PMN antiboby. Gastric mocosal blood flow was rapidly decreased for several hours after the injection of compound 48/80, and recovered 24 h later. This fact suggests that compound 48/80-induced gastric mucosal injury may be a kind of ischemia-reperfusion injury. Furthermore, superoxide generation from PMN stimulated by PMA was increased after the compound 48/80 treatment.

These results suggests that active oxygen species and lipid peroxidation may play an important role in the pathogenesis of the gastric mucosal injury induced by compound 48/80, and both xanthine-xanthine oxidase system and PMN may be important for the sources of active oxygen species in this gastric injury.

REFERENCES

1. T. Yoshikawa, H. Miyagawa, N. Yoshida, T. Takemura, T. Tanigawa, S. Sugino, and M. Kondo, Increase in lipid peroxidation in rat gastric mucosal lesions induced by water-immersion restraint stress, J. Clin. Biochem. Nutr. 1:271 (1985)
2. T. Takemura, T. Yoshikawa, K. Tainaka, Y. Morita, K. Itani, O. Seto, S. Sugino, and M. Kondo, Role of lipid peroxidation in acute gastric mucosal lesions experimentally induced by radio-frequency hyperthermia in rats, J. Clin. Exp. Med. 143: 545 (1987)
3. T. Yoshikawa, N. Yoshida, H. Miyagawa, T. Takemura, T. Tanigawa, S. Sugino, and M. Kondo, Role of lipid peroxidation in gastric mucosal lesions induced by burn shock in rats, J. Clin. Biochem. Nutr. 2: 163 (1987)
4. K. Takeuchi, H. Ohtsuki, and S. Okabe, Pathogenesis of compound 48/80 -induced gastric lesions in rats, Digestive Disease and Science 31: 392 (1986)
5. H. Ohkawa, N. Ohnishi, and K. Yagi, Assay for lipid peroxide in animal tissue by thiobarbituric acid reaction, Anal. Biochem. 95: 351 (1976)
6. K. Maejima, K. Miyata, and K. Tomoda, A manganese superoxide dismutase from serratia marcescens, Agric. Biol. Chem. 47:1537 (1983)
7. P. A. Ward and C. G. Cochrane, Bound compliment and immunologic injury of blood vessels, J. Exp. Med. 121:215 (1964)
8. T. Takemura, T. Yoshikawa, O. Seto, T. Tanigawa, K. Tainaka, Y. Morita, N. Yoshida, S. Sugino, and M. Kondo, Role of lipid peroxidation and polymorphonuclear leukocyte-derived oxygen radicals in acute gastric lesions induced by hyperthermic treatment, Jpn. J. Gastroenterol. 2:(1989, submitted)
9. A. Nishida, K. Sugioka, M. Nakano, and K. Yagi, Use of 2-methyl-6-phenyl -3,7-dihydroimidazo(1,2-a)pyrazin-3-one for assay of superoxide anions in phagocytizing granulocytes, J. Clin. Biochem. Nutr. 1: 5 (1986)

EFFECT OF THE PLATELET-ACTIVATING FACTOR (PAF) ANTAGONIST, BN 52021, ON

FREE RADICAL-INDUCED INTESTINAL ISCHEMIA-REPERFUSION DAMAGE IN THE RAT

Marie-Thérèse Droy-Lefaix, Yvette Drouet, Gérard Géraud
Pierre Braquet

I.H.B./IPSEN Research Laboratories
17 avenue Descartes
92350 Le Plessis Robinson, France

INTRODUCTION

During ischemia-reperfusion, oxygen-derived free radicals mediate tissue injury.[1-21] The sources of oxygen-free radicals might include mucosal oxidases and activated polymorphonuclear leukocytes (PMNLs).[7,11,20]

It has been demonstrated that attraction and activation of neutrophils may be initiated by a variety of compounds, such as bacterial formyl peptides (FMLP) and activated complement components (C5a).[9,10]

In this regard, a phospholipid mediator of inflammatory reactions, platelet-activating factor (PAF),[12] synthetized and released from various cells types (platelets, polymorphonuclear neutrophils, macrophages mastocytes, endothelial cells) might also play a physiopathologic role.[12,19] There is evidence that at picomolar concentrations, PAF dramatically potentiates the generation of free radicals in PMNLs stimulated with f-methionylleucylphenylalanine.[12,13] Furthermore gastrointestinal mucosal damage has been described after PAF administration.[14,15]

Thus in intestinal ischemia-reperfusion, the involvement of PAF was studied using the ginkgolide BN 52021, a potent and specific PAF receptor antagonist. Evidence of oxygenated-free radical release was provided by superoxide dismutase (SOD), dimethyl-sulfoxide (DMSO) and allopurinol.

METHODS

Rats were anaesthetised with sodium pentobarbital (50 mg/kg/ip). After a midline abdominal incision, mesenteric artery occlusion was maintained for 1 hour and followed by 24 h reperfusion. Rats were then sacrificed after an overnight fasting. Jejunum, ileon and colon were removed and fixed with Bouin liquid for microscopic observation. Mucosa alterations were quantified using a score 0-5.[4]

A total of fifty six Sprague-Dawley female rats (190-200 g) were used. Eight were untreated controls, eight were pretreated with allopurinol (50 mg/ kg/po/2d) and BN 52021 (20 mg/kg/45 min/ip) before clamping and eight

Antioxidants in Therapy and Preventive Medicine
Edited by I. Emerit *et al.*
Plenum Press, New York, 1990

419

were treated with SOD (15.000 U/kg), DMSO (50 mg/kg), BN 52021 (20 mg/kg) given intravenously after declamping.

Data were analysed using an analysis of variance by contrasts.

RESULTS

This investigation showed that ischemia-reperfusion promoted severe damage in the intestinal mucosa (Figure 1).

Histologically, lesions in the jejunum and ileon (Figure 2) were characterized by marked destruction of the villi and crypts epithelium with hemorrhagic necrosis, oedema and polymorphonuclear neutrophil infiltration. In the colonic mucosa, necrosis, hemorrhage and inflammation were extended to the muscularis mucosa.

Jejunal, ileal (Figure 1) and colonic damage scores were significantly reduced by SOD (15.000 U/kg/iv), allopurinol (50 mg/kg/po), DMSO (50 mg/kg/iv), BN 52021 (20 mg/kg/iv) and BN 52021 (20 mg/kg/ip).

DISCUSSION

As reported by others, the results suggest that clamping of the mesenteric artery followed by reperfusion induces marked deep intestinal injury with infiltrating polymorphonuclear neutrophils.[5,8,11,16,21]

The protection afforded by superoxide dismutase, allopurinol and dimethyl-sulfoxide indicated that release of superoxide radicals during reperfusion are primary mediators of tissue lesions.

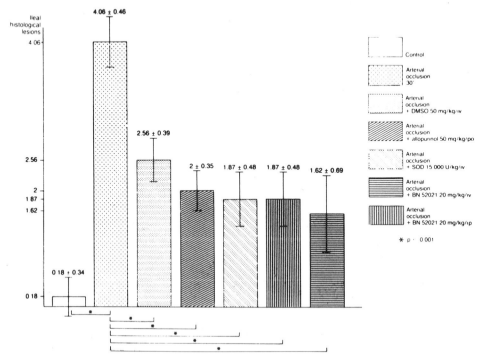

Figure 1.   In the rat ischemia-reperfusion induced severe damage of ileal mucosa. Pretreatments with SOD, allopurinol and BN 52021 iv and ip significantly reduced histological lesions.

420

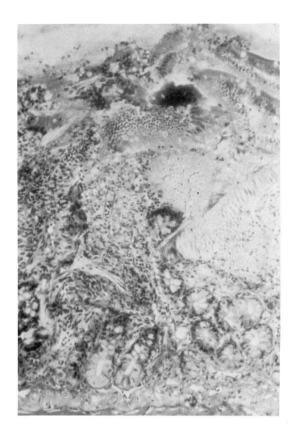

Figure 2.   Optical micrography of rat ileal mucosa after ischemia-
             reperfusion : marked hemorrhagic and necrotic erosion with
             polymorphonuclear neutrophil infiltration (X 250).

     The fact that intestine mucosal damage could be ameliorated by
iv or ip administration of BN 52021, a PAF-antagonist, suggests that
PAF,[12] a potent chemotactic agent and mediator of inflammation,[22] plays
an important role in ischemia-reperfusion induced damage in the rat
intestine, which appears to be mediated by free radicals produced by
infiltrating PMNLs.

REFERENCES

1.   M. T. Droy-Lefaix, Y. Drouet, G. Geraud and B. Schatz, Radicaux
     libres et tube digestif, Cah. Nutr. Diet. XXII:44 (1987).
2.   M. T. Droy-Lefaix, Y. Drouet, G. Geraud and P. Braquet, The amplificative
     role of PAF-acether in the oxidative stress following reperfusion
     of ischemic stomach, in: Oxygen radicals in biology and medicine,
     M. G. Simic and K. A. Taylor, eds., Plenum Press, New York
     (1988) (in press).
3.   M. T. Droy-Lefaix, Y. Drouet, G. Geraud and P. Braquet, Involvement
     of platelet-activating factor in rat ischemia-reperfusion gastric
     damage, in: Ginkgolides - chemistry, biology, pharmacology
     and clinical perspectives, P. Braquet, ed., J.R. Prous, Barcelona
     (1988).
4.   D. N. Granger, M. E. Hollwarth and D. A. Parks, Ischemia-reperfusion
     injury : role of oxygen-derived free radicals, Acta Physiol.
     Scand. 548:47 (1986).
5.   M. B. Grisham, L. A. Hernandez and D. N. Granger, Xanthine oxidase
     and neutrophil infiltration in intestinal ischemia, Am. J.
     Physiol. 251:G567 (1986).

6. M. T. Droy-Lefaix, Y. Drouet, G. Geraud and P. Braquet, Platelet-
    activating factor : a potent agent of the oxidative stress
    in post-ischemic gastrointestinal lesions, in: Free radicals
    in digestive diseases, M. Tsuchiya et al, eds., Elsevier, Amsterdam
    (1988).
7. D. N. Granger, L. A. Hernandez and M. B. Grisham, Reactive oxygen
    metabolites : mediators of cell injury in the digestive system,
    Viewp. Dig. Dis. 18:13 (1986).
8. D. A. Parks and D. N. Granger, Contributions of ischemia and reperfusion
    to mucosal lesion formation, Am. J. Physiol. 250:G749 (1986).
9. D. N. Granger, B. J. Zimmerman, E. Sekizuka and M. B. Grisham,
    Intestinal microvascular exchange in the rat during luminal
    perfusion with formylmethionyl-leucyl-phenylalanine, Gastroenterology
    94:673 (1988).
10. C. Von Ritter, E. Sekizuka, M. B. Grisham and D. N. Granger, The
    chemotactic peptide N-formyl-methionyl-leucyl-phenylalanine
    increases mucosal permeability in the destal ileum of the rat,
    Gastroenterology 95:651 (1988).
11. L. A. Hernandez, M. B. Grisham, B. Twohig, K. E. Arfors, J. M.
    Harlan and D. N. Granger, Role of neutrophils in ischemia-reperfusion
    induced microvascular injury, Am. J. Physiol. 253:H699 (1987).
12. P. Braquet, L. Touqui, T. Y. Shen and B. B. Vargaftig, Perspectives
    in platelet-activating factor research, Pharmacol. Rev. 39:97
    (1987).
13. M. Baggiolini, B. Dewald and M. Thelen, Effects of PAF on neutrophils
    and mononuclear phagocytes, Progr. Biochem. Pharmacol. 22:90
    (1988).
14. F. Clostre, A. Etienne, J. M. Mencia-Huerta and P. Braquet, Prevention
    of the platelet-activating factor induced gastrointestinal
    damages by BN 52021 and BN 52063, in: Ginkgolides - chemistry,
    biology, pharmacology and clinical perspectives, P. Braquet,
    ed., J. R. Prous, Barcelona (1988).
15. J. L. Wallace, PAF and endotoxin induced gastrointestinal ulceration :
    inhibitory effects of BN 52021, in: Ginkgolides - chemistry,
    biology, pharmacology and clinical perspectives, P. Braquet,
    ed., J. P. Prous, Barcelona (1988).
16. C. Tagesson, L. Lindahl and T. Otamiri, BN 52021 ameliorates mucosal
    damage associated with small intestinal ischaemia in rats,
    in: Ginkgolides chemistry, biology, pharmacology and clinical
    perspectives, P. Braquet, ed., J. R. Prous, Barcelona (1988).
17. D. A. Parks, G. B. Bulkley, D. N. Granger, S. R. Hamilton and
    J. M. Mc Cord, Ischemic injury in the cat small intestine role
    of superoxide radicals, Gastroenterology 82:9 (1986).
18. S. P. Dunn, K. R. Gross, M. Dalsing, R. Hon and J. L. Grosfeld,
    Superoxide : a critical oxygen-free radical in ischemic bowel
    injury, J. Pediatr. Surg. 19:740 (1984).
19. J. M. Lynch, G. Z. Lotner, S. J. Betz and P. M. Henson, The release
    of a platelet-activating factor by stimulated rabbi neutrophils,
    J. Immunol. 123:1219 (1979).
20. R. T. Briggs, J. M. Robinson, M. L. Karnovsky and M. J. Karnovsky,
    Superoxide production by polymorphonuclear leukocytes - a cytochemica
    approach, Histochemistry 84:371 (1986).
21. M. H. Schoenberg, B. B. Fredholm, U. Haglund, H. Jung, D. Sellin,
    M. Younes and F. W. Schildberg, Studies on the oxygen radical
    mechanism involved in the small intestinal reperfusion damage,
    Acta Physiol. Scand. 124:581 (1985).
22. J. L. Wallace, P. Braquet, G.C. Ibbotson, W. K. Mac Naughton and
    G. Cirino, Assessment of the role of platelet-activating factor
    in an animal model of inflammatory bowel disease, J. Lipid.
    Mediators (pre issue) (1988).

422

ANTIOXIDANT THERAPY IN

HEMATOLOGICAL DISORDERS

Michael R. Clemens

Eberhard-Karls-Universität Tübingen
Medizinische Klinik und Poliklinik
D-7400 Tübingen 1, Federal Republic of Germany

INTRODUCTION

Numerous investigations indicate the involvement of free
radical reactions in the pathogenesis of hemolytic diseases;
i.e. in vitamin E deficiency, abnormal glutathione metabolism,
decreased NADPH production (e.g. glucose-6-phosphate dehydrogenase deficiency), catalase deficiency, sickle cell disease,
thalassemia, and paroxysmal nocturnal hemoglobinuria (1-3).
Other conditions leading to free radical mediated anemia may
be the anemia of chronic renal failure and Fanconi's anemia.
Many drugs induce oxidative hemolysis in both normal individuals
(4) and in patients with a defect in red cell antioxidant capacity, such as glucose-6-phosphate dehydrogenase (G-6-PD) deficiency and vitamin E deficiency. The majority of these drugs
are aromatic compounds containing amino, nitro, or hydroxy
groups. Clinical studies with antioxidants treating free radical-mediated hemolysis are quite numerous. However, reports
with valuable trials are rare. Therefore, only a few studies
are presented here with results, obtained with patients suffering from G-6-PD deficiency and anemia of chronic renal failure
who were treated with either vitamin E or deferoxamine.

The treatment of leukemias with antitumor drugs depleting
glutathione or acting via free radicals in addition to total
body irradiation (in the case of bone marrow transplantation)
is another condition in which essential antioxidants like alpha-
tocopherol and beta-carotene may have a therapeutic potential
in hematological disorders. Therefore, our recently obtained
data is presented from patients, who have been treated with a
high dose chemotherapy and often additionally with total body
irradiation preceding bone marrow transplantation.

Rational Basis for the Treatment of Hemolytic Disorders with
Antioxidants

G-6-PD Deficiency. Deficiency of G-6-PD is the most
common inborn metabolic disorder of red blood cells. The clinical expression of G-6-PD variants encompasses a continuous

Antioxidants in Therapy and Preventive Medicine
Edited by I. Emerit *et al.*
Plenum Press, New York, 1990

423

spectrum of hemolytic syndromes. With the most prevalent G-6-PD variants - the Mediterranean type - hemolytic crisis are induced by certain drugs,bacterial or viral infection, stress and following ingestion of fava beans. Under most of these conditions, red cell glutathione (GSH) oxidation to the disulfide (GSSG) is increased due to the failure of NADPH regenration in the deficient cells. The biochemical basis of the onset of hemolytic crisis in G-6-PD deficient subjects is still unknown.

We have recently shown that younger cells, in particular reticulocytes, are much more susceptible to lipid peroxidation than older red blood cells (5). Analyzing the main substrate for peroxidation, i.e. the polyunsaturated fatty acids in the membrane, we found that cells from reticulocyte-rich suspensions contain higher amounts of arachidonic and docosahexaenoic acid than older cells. Corresponding changes were found in subjects with G-6-PD deficiency: the red cell sensitivity to lipid peroxidation was higher than in normal subjects, whereas the proportion of arachidonic acid showed a clear increase (2). The increased susceptibility to lipid peroxidation might be a consequence of both the failure to regenerate NADPH and the lipid abnormalities. It formed the basis for clinical studies with antioxidants like vitamin E and deferoxamine.

Results with Antioxidant Treatment in G-6-PD Deficiency

Vitamin E. The observation of Spielberg et al. (6) that high-dose oral vitamin E supplementation (800 IU per day) improved red cell survival in a few patients suffering from G-6-PD deficiency and glutathione synthetase deficiency prompted a similar trial in 23 patients with Mediterranean G-6-PD deficiency (7). Three months of vitamin E administration resulted in decreased chronic hemolysis as evidenced by improved red cell life span, with an improved red cell half-life from 22.9 + 0.7 days to 25.1 + 0.6 days, increased hemoglobin concentration (p < 0.001), and decreased reticulocytosis (p < 0.001) as compared with base-line values. Some of these patients were followed for a full year, with increased improvement and normalization of all hematologic values. During the study, none of the subjects had an acute hemolytic crisis, but since this study was designed to observe only mild chronic hemolysis, this aspect remains open to speculation. The authors postulate that under conditions in which normal cellular antioxidant defenses are not operative, vitamin E may assume a more prominent role in cellular protection. Newman et al. (8) provided supporting evidence for this concept with in vitro studies that showed some improved preservation of reduced glutathione levels under oxidant stress when G-6-PD deficient cells were preincubated with vitamin E.

Johnson et al. (9) studied three men with various variants of G-6-PD deficiency and found that high-dose vitamin E therapy (2000 - 2400 IU per day) given for four weeks does not decrease rapid chronic hemolysis secondary to G-6-PD deficiency.

Recently, the antioxidant effect of high-dose vitamin E alone and in combination with selenium in patients with G-6-PD deficiency with mild chronic hemolysis was studied (10). 36 male children with such manifestations were enrolled consecutively into two equal groups. Group 1 received 800 IU vitamin E

daily, and group 2 received 800 IU vitamin E in combination
with 25 μg selenium. The hematologic status before and 2 months
after treatment was evaluated. After treatment, there was a
significant change toward normal in both groups. The mean red
cell half-life increased in group 1 from 16.9 to 22.8 days
(p < 0.01), and in group 2 from 15.6 to 24.3 days (p < 0.01). A
comparison of the mean difference of paired values in the two
groups revealed a more significant increase in hemoglobin
(0.9 + 0.1 g/dl vs. 1.2 + 0.2 g/dl, p < 0.05), hematocrit (2.4%
+ 0.4% vs. 3.8% + 0.3%, p < 0.05), and red cell half-life (5.9
+ 3.0 days vs. 9.1 + 4.4 days, p < 0.01), and more significant
reduction in reticulocytes (-0.7% + 0.2% vs. -1.5% + 0.4%,
p < 0.01) in group 2.

Deferoxamine. A preliminary observation was reported
suggesting the arrest of hemolysis in favism by a single injection of deferoxamine (11). The patient was a six-year-old
Greek child who had ingested a meal of fava beans 24 hours
before admission. The child was treated with a single dose of
500 mg deferoxamine (20 mg per kilogram of body weight). Seven-
teen hours after deferoxamine had been given, hemolysis ceased.
It was suggested to investigate, if a single injection of deferoxamine is effective in acute hemolytic crisis in G-6-PD
deficiency.

This single observation prompted Meloni et al. (12) to
treat three G-6-PD deficient children with deferoxamine (30 mg
per kilogram of body weight) in an acute hemolytic crisis
following ingestion of fava beans. In there hands, deferoxamine
was unsuccessful in the treatment of favism; the above mentioned single observation may probably be interpreted as spontaneous arrest of hemolysis.

Discussion of Antioxidant Treatment in G-6-PD Deficiency

Vitamin E. Studies with antioxidants in the treatment of
G-6-PD deficiency had different designs, depending on the drug
used. Vitamin E was applicated to improve chronic hemolysis,
which is in general not a severe manifestation of G-6-PD deficiency. In contrast, deferoxamine was used to treat hemolysis
in an acute crisis. The distinct indications for both drugs are
primarily a result of differences in the pharmacology. Only
20-40% of orally ingested tocopherol and/or its esters are ab-
sorped. As the dose increases, the percentage of absorption decreases. When high doses of all-rac-alpha-tocopheryl acetate
are given to normal humans, blood levels increase proportio-
nately to the dose. Thus, if higher tissue or red cell membrane
levels afford some advantage, there is at least a potential
for benefit from administration of high levels of the vitamin.
However, as shown in Fig. 1, even to double the plasma tocopherol concentration with a 800 IU-daily intake needs 21 days.
Theoretically, an alternative may be provided by an acute,
single, and high-dose administration of tocopherol intravenously. Since there is no high-dose preparation of free alpha-tocopherol available for an intravenous application, an acute intervention in hemolytic crisis of G-6-PD deficiency is not feasi-
ble. The metabolism of intravenously given alpha-tocopheryl
acetate is still obscure, so those preparations are not recommendable. This situation elucidates the need for a free toco-

pherol or a related compound such as Trolox for the intravenous administration, which is of interest not only in acute
hemolytic states, but under acute conditions like reperfusion
injury and adult respiratory distress syndrome as well.

In conclusion, vitamin E can not be considered as a
treatment of choice in acute hemolytic crisis in G-6-PD deficiency. The therapeutic potential in the treatment of chronic
hemolysis is not yet clear. Further studies are required in
order to evaluate the role of vitamin E for the improvement
of chronic hemolysis and for the prevention of acute hemolytic
crisis in G-6-PD deficient subjects.

Deferoxamine. The possibility that autoxidation of oxy-
hemoglobin to methemoglobin involves the generation of a superoxide radical had been previously shown. Additionally, the formation of superoxide radicals during the autoxidation of divicine, an unstable aglycone involved in the hemolytic anemia
occuring in favism, has been demonstrated by EPR (13). The enzymatic and spontaneous dismutation of superoxide radicals
yields hydrogen peroxide. The presence of both $\cdot O_2^-$ and H_2O_2
creates the possibility for the generation of hydroxyl radicals
(OH$\cdot$). The latter reaction can be catalyzed by iron or other
transition metals and allows the reaction to proceed at a much
faster rate. At this point the availibility of reducing equivalents is critical,since hydroxyl radicals have the potential
to initiate lipid peroxidation. A recent study showed that
iron bound to hemoglobin is unlikely to produce hydroxyl radicals in free solution, but that peroxides can cause a release
of iron from the protein. This released iron, which can be
bound by transferrin or deferoxamine, is likely to be the true
promotor of hydroxyl radical generation in red cells and contribute to the peroxidation in hemoglobin-containing systems
(14). Thus, hemoglobin may act as an iron supplying agent but
not as a promotor of red cell lipid peroxidation per se. This
has been confirmed for red cells by our experiments performed
with hemolysates and deferoxamine (5), showing that deferoxamine
can almost completely block the hydrogen peroxide-induced
lipid peroxidation and fatty acid breakdown. The results of
these in vitro studies impicate that deferoxamine may be effective in the treatment of acute hemolysis of G-6-PD deficient
red cells. However, the clinical studies presented so far do
not support the effectiveness of deferoxamine in vivo. Because
the numbers of patients investigated are small and because
another schedule of deferoxamine administration may improve
the drug's acute action further studies are needed to evaluate
the role of deferoxamine in acute hemolytic crisis of G-6-PD
deficient subjects. A treatment of chronic hemolysis with
deferoxiamine is not recommendable due to the side effects
occuring with chronic administration, which may overwhelm the
benefit of an improvement of chronic hemolysis.

Antioxidant Treatment of Other Hematological Disorders

Vitamin E. In homozygous beta-thalassemia low serum
levels of alpha-tocopherol have been found. The administration
of high doses of the vitamin increased serum levels, decreased
lipid peroxidation, and, in some cases, prolonged red blood
cell survival; no significant change in transfusion requirement was obtained. Miniero et al. (16) studied 131 patients
aged 1 to 72 years with heterozygous thalassemic trait and

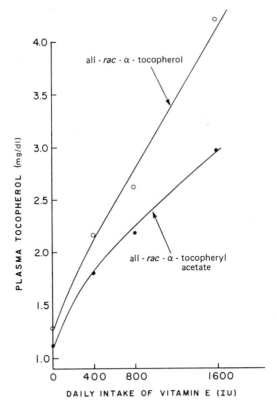

Figure 1. Relationship of intake of all-rac-alpha-tocopherol
or all-rac-alpha-tocopheryl acetate to plasma levels of tocopherol in adult humans after 21 days. "Zero" intake was the
value prior to supplementation. Each point represents mean of
three male and three female sujects (from 15).

and 218 age-matched controls. Serum levels of alpha-tocopherol
was statistically lower in patients. The effect of vitamin E
treatment (400 - 600 mg per day) has been studied in 10 of
these patients. Hematological values were unchanged after
therapy in all patients. However, in half of them biochemical
examinations showed a reduced lipid peroxidation and an increased red cell survival.

Giardini et al. (17) examined the effect of vitamin E
administration (300 mg orally and intramuscularly, respectively) in patients with homozygous beta-thalassemia. The treatment appears to be effective to reduce red cell oxidative
damage as well as to improve hemoglobin concentration, when
given i.m.

Deferoxamine. In the hemoglobinopathies, iron overload
may result from repeated blood transfusions or from excessive
absorption of iron from the diet. Iron chelation therapy
emerged in an unusual context about 15 to 2o years ago. Deferoxamine , previously used primarily for acute iron intoxication, was introduced for patients with chronic iron overload. The primary goal of iron chelation therapy is prolonged
survival for patients who require regular red all trans-

427

fusions. Data from two overlapping British studies and one American study strongly suggest that survival is, in fact, prolonged (for review, see 18). To summarize the results of these complex statistical analyses, the ratio of observed to expected deaths in patients with thalassemia major in Great Britain was reduced when the average dose of deferoxamine was 2 g or more per week for at least 5 years. The patients in these centers received most of their chelation therapy by intramuscular injection. The American study compared patients who since 1977 were treated with subcutaneous infusions of deferoxamine more than 3 days per week and maintained at a minimum hemoglobin level of 11 mg per dl with historical controls who were not chelated and who were maintained at a minimum hemoglobin level of 8 gm per dl. Median survival increased from 17 to 3o years with the change in treatment. The current use of daily subcutaneous or intravenous infusions of deferoxamine usually yields a much larger weekly dose so that the effect of chelation therapy on survival may be even more striking in the coming years. At the same time , the probable dose-dependent effect of deferoxamine on survival emphasizes that halfhearted attempts at chelation therapy may not be better than no chelation therapy at all.

ANEMIA OF CHRONIC RENAL FAILURE

Rational Basis for the Treatment of Anemia of Chronic Renal Failure with Antioxidants

Erythropoietin deficiency is probably the primary etiological factor in the anemia of renal disease. Inhibitors of erythropoiesis which include inhibitors of heme synthesis in bone marrow have been demonstrated in plasma and urine of anemic patients (19). A shortened red cell life span due to "the uremic toxins" is also considered to be an important factor in the mechanism of the anemia of renal disease. Some authors mentioned that uremic serum contains substances that impair the function of pentose phosphate shunt enzymes in red cells (20), which may result in an impairment of the red cell's antioxidant mechanisms (3). Indeed, there is some evidence for an increased sensitivity of red cells to lipid peroxidation and for oxidative damage of red cells in uremia (21-23). However, the cause of anemia in renal failure appears to be complex and may not be related solely to uremic toxins, e.g. in chronic dialysis patients, aluminium may be involved in oxidative damage of red cells (24).

Lipid abnormalities, in particular hyperlipidemia, are a common feature of the nephrotic syndrome and uremia. Enhanced hepatic synthesis of lipoprotein lipids may be stimulated by a decreased plasma albumin concentration or oncotic pressure. The resulting serum lipid abnormalities may influence the membrane lipid composition of red cells, which on the other hand may result in an increased or decreased susceptibility of red cells to lipid peroxidation (2,3). Recently, we have demonstrated the importance of the membrane lipid composition for red cell lipid peroxidation (25). A present study demonstrates that red cells in nephrotic syndrome are abnormal susceptible to peroxides, even in a patient without renal failure (M.R. Clemens, unpublished data). It is suggested that these findings may be caused

428

by abnormalities in the lipid composition, i.e. by a decreased
content of cholesterol in the membrane, which in turn may be a
result of an elevated activity of LCAT in contrast to changes
observed in LCAT deficiency. Other factors like abnormalities
in activities of enzymes of the pentose phosphate shunt and of
the content of vitamin E in red cell membranes may contribute
to the increased sensitivity of red cells to lipid peroxidation
as well. An increased sensitivity of red cells to lipid peroxidation will become of significant relevance when they are exposed to oxygen-delivered free radicals, e.g. during hemodialysis by trace amounts of aluminium and iron, by iron overload
due to a transfusion therapy and by an inflammation resulting
in the release of oxygen radicals from leucocytes' respiratory
burst. However, even without the latter conditions the enhanced
sensitivity of red cells to lipid peroxidation in nephrotic
syndrome might be involved in the pathogenesis of anemia in
renal disease, particularly since our results demonstrate that
it may precede terminal renal failure and that it is unrelated
to hemodialysis. Thus, antioxidant therapy of patients with
nephrotic syndrome and/or uremia is expected to be successful,
preventing or treating anemia of chronic renal failure.

Results with Antioxidant Treatment of Anemia of Chronic Renal
Failure

Vitamin E. Sinsakul et al. (26) conducted a prospective,
randomized, double-blind therapeutic trial of vitamin E in a
group of patients with chronic renal failure who were undergoing
hemodialysis. Sixteen patients received 400 IU of vitamin E
(d-alpha tocopheryl acetate) per os twice daily and 19 patients
received a placebo twice daily for 20 weeks. For the treated
group the initial hematocrit was 24.8 + 3.0 and the final hematocrit 25.8 + 3.8. For the placebo group the initial hematocrit
was 24.9 + 3.0 and the final hematocrit was 23.5 + 2.7. The
treated group received a total of 40 blood transfusions, and
the placebo group received a total of 35 blood transfusions.
Thus, vitamin E had no effect on the anemia or transfusion requirements of patients undergoing chronic hemodialysis for
chronic renal failure.

Ono (27) investigated the effect of vitamin E treatment on anemia in 30 regular dialysis patients, who were treated with
600 mg daily for 30 days. Hematocrit increased from 26.1 + 1.0
to 28.1 + 1.2% (p 0.05). Two years later Lillo-Ferez et al. (28)
reported in the same journal that 600 mg vitamin E per day did
not improve anemia in 10 hemodialysis patients.

Deferoxamine.Since the first description of the so-called
"Dialysis dementia syndrome" and the demonstration of the role
of aluminium overload in its pathogenesis, aluminium intoxication
has become one of the most preoccupying problems in the long-
term management of patients with end-stage renal failure (29).
Besides the now classical neurologic manifestations and vitamin
D-resistant osteomalacia, aluminium intoxication can be responsible for a microcytic anemia. Aluminium salts greatly accelerate the peroxidation induced by iron, in particular the peroxidation observed when human erythrocytes are treated with hydrogen peroxide at pH 7.4 (30). Deferoxamine decreases the peroxidation (30), which is of interest for the treatment of anemia

in renal failure. Indeed, Tielemans et al. (29) found that deferoxamine treatment (4 months period, 50 mg per kg body weight
given at the end of each dialytic procedure, which was performed 9 to 12 hours a week) of dialysis patients with aluminium
bone disease can markedly improve their anemia, even in the absence of recent aggravation, microcytosis and hypochromia. In
1988, 2 investigators (24,31) showed that deferoxamine may benefit the anemia of chronic hemodialysis patients through improvement of erythropoiesis. The effect seems not to be related
to chelation of a heavy aluminium overload. In one study (24)
16 chronic hemodialysis patients (group I), with non-microcytic
anemia (mean hemoglobin 7.2 g/dl, SD 1.0, range 5.8-9.8), moderate aluminium overload (serum aluminium 44 ug/l; SD 16, range
21-74), and normal or high iron stores (ferritin 800 µg/l; SD
464, range 34 - 2013) were treated with intravenous deferoxamine
1g at the end of each dialysis for six months. 8 patients with
similar characteristics served as controls (group II). After
six months group I showed a rise in hemoglobin to 9.1 (SD 2.5)
g/dl and a decrease in blood transfusion requirements, both significant, whereas group II showed no changes.

In the other study (31) 15 patients whose exposure to aluminium
had been low were treated for three months with deferoxamine,
30 mg/kg intravenously at the end of each dialysis session.
After one month of deferoxamine, serum aluminium had risen from
54.6 (SEM 11.2) to 167.0 (27.5) µg/l; and after three months
hemoglobin had risen from 8.46 (0.70) to 10.43 (0.80) g/l. Mean
cell volume and mean cell hemoglobin concentration also increased significantly. The maximum rise in hemoglobin correlated with
the patients' aluminium burden as estimated by mean serum aluminium concentration after one month of deferoxamine therapy
($p=0.85$).

Discussion of Antioxidant Treatment in Anemia of Chronic Renal
Failure

        Data concerning the benefit of vitamin E treatment in
patients suffering from anemia of chronic renal failure are conflicting. However, deferoxamine treatment did improve anemia in
chronic hemodialysis patients in three different studies. Results
obtained with a combined modality therapy (vitamin E plus deferoxamine) are not yet reported. The mechanism of the beneficial
effect of deferoxamine on anemia of chronic renal failure is not
completely clear and requires further investigations.

ESSENTIAL ANTIOXIDANTS IN PATIENTS UNDERGOING HIGH DOSE RADIO-
CHEMOTHERAPY

        Treatment of leukemia and aplastic anemia includes bone
marrow transplantation, which follows a high-dose chemotherapy,
generally in combination with total body irradiation for antitumor cell and immunosuppressive treatment. This conditioning
therapy approaches the limit of tolerance for several tissues.
Severe acute and delayed toxicity may occur involving the mucous
membranes, gastrointestinal tract, liver, lung, bladder, CNS
and, rarely, other tissues. The acute toxicity demands parentera
nutrition. Cyclophosphamide, frequently used in conditioning

chemotherapy, depletes hepatic glutathione, thus potentially initiating peroxidative processes. Additionally, lipid peroxidation has been suggested as one of the main causes of ionizing radiation damage. Thus, toxicity of conditioning regimens may results in part from radical-mediated tissue damage.

In a recently performed study, we investigated the blood from 19 patients having bone marrow transplants for essential antioxidants alpha-tocopherol and beta-carotene before, at and after bone marrow transplantation. After conditioning therapy the plasma levels of absolute and lipid standardized alpha-tocopherol and beta-carotene decreased significantly, presumably as a results of an enhanced breakdown of these antioxidants (32). Our results clearly demonstrate that alpha-tocopherol and beta-carotene have not been maintained on the initial basis of concentrations as determined prior to the conditioning therapy. The intravenously applicated amount of alpha-tocopherol (about 9 mg daily) covers the recommended dietary allowance (RDA: women 8 mg, men 10 mg alpha-tocopherol equivalents). Diplock suggested that the present RDA will prove to be too low and that there are cogent reasons for the recognition of carotene as a vitamin in its own right. Our result support this suggestion in respect of special situations appearing with an enhanced requirement for antioxidants. A particular recommendation for the supplementation of antioxidants in patients undergoing highly toxic cancer treatment is required, since the deleterious effects of chemotherapy, radiotherapy and total parenteral nutrition with often a high content of polyunsaturated fatty acids accumulate.

REFERENCES

1. D. Chiu, B. Lubin, and S.D. Shohet, Peroxidative reactions in red cell biology, in: "Free Radicals in Biology, Vol.V", W.A. Pryor, ed., Academic Press, New York, p. 115 (1982).
2. M.R. Clemens, H. Einsele, and H.D. Waller, The fatty acid composition of red cells deficient in glucose-6-phosphate dehydrogenase and their susceptibility to lipid peroxidation, Klin. Wochenschr., 63:578 (1985).
3. M.R. Clemens, and H.D. Waller, Lipid peroxidation in erythrocytes, Chem. Phys. Lipids, 45:251 (1987).
4. M.R. Clemens, H. Remmer, and H.D. Waller, Phenylhydrazine-induced lipid peroxidation of red blood cells in vitro and in vivo: monitoring by the production of volatile hydrocarbons, Biochem. Pharmacol., 33:1715 (1984).
5. M.R. Clemens, H. Einsele, C. Ladner, and H.D. Waller, Some new aspects on free radical reactions in red cell pathology, in: "Free Radicals: Chemistry, Pathology and Medicine", C. Rice-Evans and T. Domandy, eds., Richelieu Press, London, p. 383 (1988).
6. S.P. Spielberg, L.A. Boxer, L.M. Corash, and J.D. Schulman, Improved erythrocyte survival with high-dose vitamin E in chronic hemolyzing G-6-PD and glutathione synthetase deficiencies, Ann. Intern. Med., 90:53 (1979).
7. L. Corash, S. Spielberg, C. Bartsocas, L. Boxer, R. Steinherz, M. Scheetz, M. Egan, J. Schlessleman, and J.D. Schulman, Reduced chronic hemolysis during high-dose vitamin E administration in Mediterranian-type glucose-6-phosphate dehydrogenase deficiency, N. Engl. J. Med., 303:416 (1980).

8. J.G. Newman, T.B. Newman, L.J. Bowie, and J. Mendelsohn, An examination of the roll of vitamin E in glucose-6-phosphate dehydrogenase deficiency, Clin. Biochem., 12:149 (1979).

9. G.J. Johnson, G.T. Vatassery, B. Finkel, and D.W. Allen, High-dose vitamin E does not decrease the rate of chronic hemolysis in glucose-6-phosphate dehydrogenase deficiency, N. Engl. J. Med., 308:1014 (1983).

10. M. Hafez, E.S. Amar, M. Zedan, H. Hammad, A.H. Sorour, E.S.A. El-Desonky, and N. Gamil, Improved erythrocyte survival with combined vitamin E and selenium therapy in children with glucose-6-phosphate dehydrogenase deficiency and mild chronic hemolysis, J. Pediatr., 108:558 (1986).

11. H. Ekert, and I. Rawlinson, Deferoxamine and favism, N. Engl. J. Med., 312:1260 (1985).

12. T. Meloni, G. Forteleoni, and G.F. Gaetani, Desferrioxamine and favism, Br. J. Haematol., 63:394 (1986).

13. G. Musci, I. Mavelli, and G. Rotilio, Evidence for superoxide generation from the autoxidation of the favism-inducing aglycone divicine, Biochim. Biophys. Acta, 926:369 (1987).

14. J.M.C. Gutteridge, Iron promoters of the Fenton reaction and lipid peroxidation can be released from haemoglobin by peroxides, FEBS Lett., 201:291 (1986).

15. L.J. Machlin, Vitamin E, in: "Handbook of Vitamins", L.J. Machlin, ed., Marcel Dekker, New York, p. 99 (1984).

16. R. Miniero, E. Canducci, D. Ghigo, P. Saracco, and C. Vullo, Vitamin E in beta-thalassemia, Acta Vitaminol. Enzymol., 4:21 (1982).

17. O. Giardini, A. Cantani, A. Donfrancesco, F. Martino, O. Mannarino, P. D'Eufemia, C. Miano, U. Ruberto, and R. Lubrano, Biochemical and clinical effects of vitamin E administration in homozygous beta-thalassemia, Acta Vitaminol. Enzymol., 7:55 (1985).

18. A. Cohen, Management of iron overload in the pediatric patient, Hematol. Oncol. Clin. North. Amer., 1:521 (1987).

19. J.W. Fisher, Mechanism of the anemia of chronic renal failure, Nephron, 25:106 (1980).

20. Y. Yawata, R. Howe, and H.S. Jacob, Abnormal red cell metabolism causing hemolysis in uremia, Ann. Intern. Med., 79:362 (1973).

21. D.P. Cauhan, P.H. Gupta, M.R.N. Nampoothiri, P.C. Singhal, K.S. Chugh, and C.R. Nair, Clin. Chim. Acta, 123:153 (1982).

22. M. Taccone-Galucci, O. Giardini, R. Lubrano, D. Bandino, V. Mazzarella, O. Mannarino, C. Meloni, M. Morosetti, M. Elli, C. Tozzo, L. Strolighi, and C.U. Casciani, Red blood cell lipid peroxidation in predialysis chronic renal failure, Clin. Nephrol., 27:238 (1987).

23. A. Miguel, A. Miguel, M. Linares, A. Perez, R. Moll, J. Sanchis, J.M. Escobedo, and J.M. Miguel-Borja, Evidence of an increased susceptibility to lipid peroxidation in red blood cells of chronic renal failure patients, Nephron, 50:64 (1988).

24. F.J. De la Serna, M. Praga, F. Gilsanz, J.L. Rodicio, L.M. Ruilope, and J.M. Alcazar, Improvement in the erythropoiesis of chronic haemodialysis patients with desferrioxamine, Lancet, I: 1009 (1988).

25. M.R. Clemens, M. Ruess, Z. Bursa, and H.D. Waller, The relationship between lipid composition of red blood cells and their susceptibility to lipid peroxidation, Free Rad. Res. Comms, 3:265 (1987).

26. V. Sinsakul, J.R. Drake, J.N. Leavitt, B.R. Harrison, and C.D. Fitch, Lack of effect of vitamin E therapy on the anemia of patients receiving hemodialysis, Am. J. Clin. Nutr., 39:223 (1984).
27. K. Ono, Effect of large dose vitamin E supplementation on anemia in hemodialysis patients, Nephron, 40:440 (1985).
28. M. Lillo-Ferez, B. Allain, C. Dupommereulle, P. Prieur, and M. Petrover, Inefficacy of vitamin E supplementation on anemia in hemodialysis patients, Nephron, 45:79 (1987).
29. C. Tielemans, F. Collart, R. Wens, J. Smeyers - Verbeeke, J. van Hooff, M. Dratwa, and D. Verbeelen, Improvement of anemia with deferoxamine in hemodialysis patients with aluminium-induced bone disease, Clin. Nephrol., 24:237 (1985).
30. J.M.C. Gutteridge, G.J. Quinlan, J. Clark, and B. Halliwell, Aluminium salts accelerate peroxidation of membrane lipids stimulated by iron salts, Biochim. Biophys. Acta, 835:441 (1985).
31. P. Altmann, D. Plowman, F. Marsh, and J. Cunningham, Aluminium chelation therapy in dialysis patients: evidence for inhibition of hemoglobin synthesis by low levels of aluminium, Lancet, I: 1012 (1988).
32. M.R. Clemens, C. Ladner, G. Ehninger, H. Einsele, W. Renn, E. Bühler, H.D. Waller, and K.F. Gey, Vitamin E and beta-carotene decreased during radiochemo-therapy preceding bone marrow transplantation, Am. J. Clin. Nutr., in press

THE ADULT RESPIRATORY DISTRESS SYNDROME (ARDS) AND OXIDATIVE

STRESS: THERAPEUTIC IMPLICATIONS

Carroll E. Cross, Balz Frei*, and Samuel Louie

Department of Medicine, University of California
Davis, CA 95616 and *Department of Biochemistry
University of California, Berkeley, CA 94720

ARDS is an important acute clinical disorder of the
lung, the   pathogenesis of which is still incompletely
understood.   Evidence   from   in vitro studies,   from
experimental animal models, and from some clinical studies,
suggest that oxidants may play a contributory role, possibly
by mediating direct cellular injury or initiating or
pertubating interactive inflammatory processes which lead to
cell injury and leakage of blood components into the lung
interstitial space and alveoli.   Further investigations to
define the contributions of oxidants and the efficacy of
antioxidant administration will be needed before antioxidant
therapies are routinely used to bolster what has been largely
supportive care for patients with ARDS.

INTRODUCTION

ARDS is a common end-point in response of the lung, and
to varying degrees some other organs, to injury of various
causes. Two decades after its first definitive description
(1), and in spite of a burgeoning research activity, its
pathogenesis   remains   controversial.   The   disease   is
characterized by: (i) a predisposing condition, most
frequently sepsis, shock, trauma, multiple transfusions,
severe pneumonia, lung contusion, aspiration or inhalation
injury; (ii) diffuse pulmonary infiltrates on chest x-ray
not explained by heart failure or fluid overload; and (iii)
severe hypoxemia, necessitating high concentrations of O_2
delivered by mechanical ventilators in an intensive care unit
(ICU).

ARDS remains a perennial problem in ICUs because of its
frequency and low survival of less than 50% (2-5).   It is
estimated that there are 150,000 cases/yr in the United
States (6), although the incidence appears to be lower by an

Antioxidants in Therapy and Preventive Medicine
Edited by I. Emerit *et al.*
Plenum Press, New York, 1990

order of magnitude in England (7). Because of the non-uniform entry criteria used to make the diagnosis, the severity of the ARDS, and the varying gravity of the underlying condition(s), the true incidence and mortality is controversial (8, 9). As patients considered at high risk for developing ARDS may manifest subclinical abnormalities of lung dysfunction early in their course, the disorder may actually represent a very wide spectrum of disease expression.

Although there are many primary inciting etiologies, the histologic picture of diffuse lung damage, inflammation and edema, with varying degrees of fibrosis in later stages, is remarkably constant. Particularly perplexing has been an understanding of the interacting roles played by the various injury and inflammatory effector systems involved, how they initiate the disorder, perpetuate it, and finally inactivate it. In this paper we provide a short synopsis of known operative pathobiologic mechanisms in ARDS, both non-oxidant and oxidant-related. In this context, potential but largely unproven newer therapeutic approaches are discussed.

PATHOBIOLOGICAL CONSIDERATIONS IN ARDS

There is now abundant evidence that most forms of ARDS, whatever the precipitating insult, are accompanied by activation of the inflammatory system. Reviews of the many inflammatory system pathways which appear to be involved in the pathobiology of ARDS are listed in Table 1. It is no surprise that it has been difficult to unravel the vast array of seemingly redundant and interactive effector systems responsible for ARDS, for like the situation in most other clinical disorders involving inflammation, the precise pathobiology, including specific signal transduction processes, interactive dependencies and temporal sequences, remain both complex and elusive.

Table 1. Pathobiologic Considerations in ARDS

cellular systems	non-cellular systems
Polymorphonuclear leukocytes	Complement, coagulation
Monocytes and macrophages	and fibrinolytic cascades
Platelets	Proteolytic processes and
Red blood cells	antiproteolytic systems
Endothelial cells	Arachidonic acid metabolites
Epithelial cells	Platelet activating factor
Connective tissue cells	Vasoactive agents (e.g. kinins
(e.g. fibroblasts,	Cytokines (e.g. TNF, IL-1)
mast cells)	Growth factors
Intracellular antioxidant	Extracellular antioxidant
systems	systems

In the last decade attention has been directed toward the important roles of complement (10, 11) and PMN activation (12-15) in ARDS. The conventional pathogenetic sequence is as follows: the precipitating injury, most notably sepsis or trauma, activates complement, which in turn "activates" PMNs; the activated PMNs, when sequestered in the lung, seem well equipped to act both as initiators and potentiators of tissue damage via their arsenal of proteolytic, oxidative, and active lipid mediator generation processes. Although sepsis and trauma, the leading causes of ARDS, are associated with marked changes in circulating PMNs and their state of activation (14-16), most patients with sepsis and/or trauma do not develop ARDS. In this regard, the levels of complement activation in sepsis do not appear to reflect a particular susceptibility to develop ARDS (17).

As in the pathology of many allergic and inflammatory disorders, arachidonic acid metabolites of both cyclooxygenase and lipoxygenase pathways have been strongly implicated in ARDS. In experimental models of ARDS, blood concentrations of archidonic acid metabolites are markedly increased and modulators of cyclooxygenase and lipoxygenase pathways appear to influence manifestations of ARDS (18, 19). The observations of increased levels of metabolites in the bronchoalveolar fluid of ARDS patients (20) buttresses the possibility that eicosinoids may be involved in the pathobiology of the condition.

More recently, interest has focused on the role of cytokines and macrophages in mediating endogenous responses to inflammation (21-24), and this has been applied to considerations of ARDS. For example, alveolar macrophages derived from ARDS patients have been shown to be activated and to produce cytokines (25) which induce PMN degranulation and "prime" both their oxidative burst capabilities and their cytotoxicity (21, 22). These and other observations indicate a dynamic role for alveolar macrophages, and possibly intravascular macrophages in the pulmonary circulation (26), in acute lung injury. Changes in their states of activation may play a significant orchestrating role in both the pathobiology and resolution of ARDS. Prostaglandins are capable of modulating cytokine production (27), demonstrating the complex interactions between the various effector systems in vivo and underscoring the difficulties in establishing an exact role for any one system.

Pulmonary vascular endothelial cells, strategically positioned to influence interactions between circulating constituents such as PMNs, toxins, complement and various cytokines, and the lung parenchymal cells, are undoubtedly more active participants in inflammatory reactions than previously assumed (28). For example, they possess a vigorous metabolism of biological active lipids, and they produce and/or respond to such substances as endotoxin, TNF, IL-1, platelet derived growth factor and PMN-activating factors (29, 30). In addition, they can produce oxidants themselves (31). It is very likely that the nature and degree of interaction between phagocytes, such as PMNs and monocyte-macrophages, and endothelial cells, play a complex but pivotal role in the initiation and progression of ARDS. Likewise, although less studied, it is probable that the parenchymal epithelial cells play a more active role in lung injury mechanisms in ARDS than formerly recognized.

Sorting out the roles of various multifacted, interacting, cascading processes involving inflammatory cells, mediators and cytokines, the factors responsible for their activation and deactivation processes, and their mechanisms of damaging lung cells in ARDS will continue to be a research challenge for a number of years.

EXPERIMENTAL EVIDENCE OF A ROLE OF OXIDANTS IN ARDS

The experimental evidences suggesting a role for oxidants in ARDS consist of identification of the presence of oxidants, or the specific products of their reactions, and the ability of scavengers of oxidants to inhibit the development of ARDS in appropriate model systems. Interpretations are often complicated because after tissue injury of any kind, it is likely that the injured cells will undergo oxidative damage more readily than normal cells (32). Thus, even if products of oxidative damage are verified, it is difficult to ascertain whether this is cause or consequence. An argument could be made for antioxidant therapy strategies in either case.

As listed in Table 2, it is likely that the local release of oxidants by phagocytes represents an important mechanism by which oxidants are involved in ARDS. Although endothelial cells, like other cells, have a significant capability to protect themselves from oxidants in a well-defined range of environmental oxidant concentrations (33), they have repeatedly been shown to be particularly sensitive to injury by activated PMNs (33-36) and by reactive O_2 species (33, 35, 37), presumably when exposed to oxidants at rates exceeding their oxidative degradative capabilities. That PMN oxidants alone, rather than PMN granular enzyme proteases, can be responsible for the injury is confirmed by the finding that PMN cytoplasts devoid of proteolytic activities can injure endothelium (38). It is tempting to relate the extra-pulmonary organ failure seen in ARDS to structural and/or functional alterations occurring as a result of phagocyte-related effects on endothelial cells in other organs, a hypothesis not without some experimental evidence to support it. (39).

Table 2.   Potential Sources of Oxidant Production in ARDS

Phagocytic cell oxidant generating systems
   (PMNs, monocytes, macrophages, ?endothelial cells)
Arachidonic acid metabolism
   (cyclooxygenase and lipoxygenase pathways)
Ischemia-reperfusion phenominae including xanthine oxidase
   oxidant generation
   (in areas of microthrombi and microemboli)
Secondary oxidations due to cell injury
   (e.g., decompartmentalization of iron, pro-oxidants and antioxidants)
Disordered lung cell P-450 and mitochondria oxidant systems
Antioxidant vitamin deficiencies
High therapeutic oxygen administration

A number of lipoxygenase inhibitors are known to exert their action via antioxidant mechanisms, presumably because a minimal level of hydroperoxide ("peroxide tone") is needed for initiation of the reaction (40, 41). This, and the fact that these reactions generate oxidants and lipid hydroperoxides themselves (41), indicate that eicosinoid biosynthesis must be considered to play a role in determining oxidant-related tissue injuries. This argument is supported by the fact that certain lipoxygenase products are capable of activating and recruiting additional phagocytes via production of chemotactic leukotrienes, but the argument is complicated by the abilities of some cyclooxygenase products to down-regulate inflammatory processes (19, 23).

ARDS is accompanied by widespread pulmonary microthrombosis (42) and by considerable evidence of inhomogeneous lung parenchymal involvement (43), suggesting that hypoxia-reoxygenation phenominae (44), known to occur in the lung (45, 46), may be operative in ARDS. Oxidant production from xanthine oxidase (or closely related aldehyde or sulfite oxidases) have been shown to play a role in the mediation of lung injury after rat skin burns (47). Furthermore, increased levels of plasma xanthine oxidase activity and its hypoxanthine substrate have been reported in human patients with ARDS (48), demonstrating the potential for this mechanism of oxidant production in ARDS.

A variety of antioxidant interventions have been shown to attenuate experimental animal models of ARDS (49-51). More supporting evidence for a role of oxidants in ARDS include the recovery of myeloperoxidase and oxidized $\alpha-1$ antitrypsin from bronchoalveolar lavage fluid (12, 15, 52) and the presence of increased amounts of hydrogen peroxide in the expired breath of patients with ARDS (52). Although indirect evidences for circulating lipid hydroperoxides have been reported in some animal models (54-56) and in patients with sepsis, shock and ARDS (57, 58), we have not been able to verify this in man (59). However, the finding of reduced amounts of plasma antioxidants (59, 60) and lipid hydroperoxides in bronchoalveolar secretions (60), provides some additional support that oxidant stress exists in these patients.

TREATMENT STRATEGIES IN ARDS

Therapeutic measures for patients with ARDS focus on four aspects: (i) attention to the underlying condition(s) which lead to ARDS; (ii) prevention of the development and progression of ARDS; (iii) supportive physiologic manipulations during ARDS; and (iv) strategies to reverse the mechanistic processes underlying ARDS.

The first two measures are interrelated and in fact invoke provocative strategies. Obviously, if such inciting etiologies as sepsis or shock are prevented, ARDS is averted; if present, the earlier it is treated, the less the inciting stimulus triggering the ARDS may be. More problematic areas are anticipatory strategies directed toward incompletely understood pathobiologic mechanisms.

These strategies often represent double-edged swords. For example, corticosteroid administration ameliorates the activation and cytokine production by macrophages incubated with endotoxin (61), and appear effective in various animal models of ARDS (49, 62). However, their predominant action is most effective only if they are administered before endotoxin activations have been initiated, a situation not strictly analogous to the ICU patient with ARDS. Clinical trials have in fact shown no benefit of steroid administration, while they increased secondary infections in patients with sepsis and shock (63).

The reader is referred elsewhere for details concerning the complex physiologic manipulations directed towards support of O_2 delivery systems used until reparative processes in the lungs and various other organs can effectively dominate the clinical condition (2, 64, 65). Therapies that would inhibit or block the activation of various operative pathophysiologies would be expected to be beneficial in so far as these mechanisms are providing an unfavorable amplification of injury sequences rather than facilitating reparative processes. Problems abound and include: (i) the need to take into account "window effects", i.e., the mediator of the injury mechanism may only be operative for a short time; (ii) the "site" effect, i.e., therapies must be effective at site of where injury mechanism is operative; and (iii) the need for a variety of intervention strategies to meet the multiplicity of synergising, redundant injurious oxidants. Strategies designed to get systemically administered therapeutic agents into interfaces between PMNs and endothelium, into endothelium, into interstitium, and into parenchymal cells will be a particularly challenging problem.

Patients with ARDS commonly develop intra-alveolar fibrosis that can play a role in the progressive respiratory failure. In these patients, administration of agents designed to inhibit fibroblast proliferation (such as growth factor agonists) or modulating transcription or degradation rates of cellular procollagen mRNA or extracellular collagen metabolism could be considered (66).

Table 3.   Experimental Approaches in ARDS

Phospholipase A_2 inhibitors
Cyclooxygenase, thromboxane and lipoxygenase inhibitors
Prostanoids
PAF inhibitors
Anti-endotoxin antibodies
Anti-cachectin/TNF antibodies
Anti-adherence promoting factor antibodies
   (phagocytes; endothelial cells)
Down-regulation of phagocyte activation
   (e.g., inhibitors of O_2 burst)
Other modulators of signal transduction
   (e.g., calcium channel blockers)
Aggressive measures to combat secondary infection
Artificial surfactant administration
Agents designed to inhibit fibrogenesis
Antiproteases
Antioxidant approaches (see Table 4)

Table 3 summarized some of the numerous approaches that have been proposed to ameliorate the hypothesized inflammatory mediatory and effector processes believed to contribute to the injury processes. Future emphasis will almost surely be given to protective approaches designed to inhibit more than one pathobiologic mechanism (e.g., antiproteases and antioxidants).

ANTIOXIDANT THERAPEUTIC APPROACHES IN ARDS

Some simple therapeutic strategies to limit oxidant damage in ARDS are already being employed. For instance, although increased inspired O_2 tensions are necessary to maintain tissue oxygenation, this is kept minimal by interventions such as positive end-expiratory pressure during mechanical ventilation and careful body fluid management (2, 64). This is important in ARDS not only because the increased lung O_2 tensions give rise to toxic O_2 species, but also because PMN oxidant production appears increased at elevated O_2 tensions (67). A variety of more complex antioxidant therapeutic strategies which have been considered are shown in Table 4.

Table 4. Antioxidant Therapeutic Approaches in ARDS

Minimize increased therapeutic O_2 administrations
Maintain or enhance intracellular and extracellular antioxidant levels
Decrease phagocyte activations and/or recruitments to the lung
Xanthine oxidase inhibitors
Antioxidant enzyme administrations
Free radical scavengers
Iron chelators

Although there has been a vast amount of basic and animal research interest in oxidant damage to the lung (68-70), and a clear-cut rationale for the therapeutic use of antioxidant agents exists, few of the essential clinical trials have been done. This is primarily due to considerations of safety and concern for side-effects. Substances which deactivate PMN activation processes may predispose patients to infection, as has been implicated for the case of corticosteroid administration in ARDS (63). As sepsis represents a major predisposing condition, and a major cause of death in ARDS (2, 4, 5), this is an important and frustrating consideration. This concern is further highlighted by the finding that some antioxidants clearly decrease PMN bactericidal functions (71). Another problem for the clinician to contend with is that some antioxidants, especially at high concentrations, may exert paradoxical pro-oxidant effects, rendering them less useful for therapeutic consideration.

Given the prominence of PMN-related injury mechanisms in the recent literature, and regarding their potential for causing endothelial cell injury (34-36, 39), it is not surprising that interference with PMN activation processes have been proposed. For example, antibodies to endotoxin (72) are expected to lead to diminutions in the large number of inflammatory endotoxin-induced activation processes, among which are complement and PMN, giving rise to increased oxidant production.

Phagocyte activation and adherence to endothelium is an important first step in the development of acute inflammatory reactions. Thus, antibodies against complement (73) or the major PMN adhesive glycoprotein (74), appear to represent promising ways to decrease PMN injuries directed against lung endothelial cells. Likewise, antibodies against the endothelial PMN adhesion glycoprotein, by interfering with endothelial cell adhesiveness specific for PMNS, may be protective. TNF and IL-1 cause increased adhesive surface glycoprotein expression in both PMNs and endothelium (75-77), and as both these cytokines along with IL-2 appear capable of eliciting lung injury similar to ARDS (23, 78-80) antibodies directed against them could be expected to be protective, in part by helping to reduce the untoward effects of PMN and endothelial activation and their interaction (76, 78, 79). The intricacies of interdependence is illustrated by the fact that endotoxin induces TNF formation (81).

These concepts have both pathobiological and therapeutic implications. Although PMNs are believed to play a critical role in ARDS (33, 36, 68), ARDS is seen in patients with severe leukopenia and sepsis (82). This can perhaps be explained by direct effects of endotoxin to injure cells and by indirect effects such as complement activation, cytokine production and the "priming" effects of both endotoxin and various cytokines on oxidant generation by remaining phagocytes. Since lymphokines themselves are capable of causing pulmonary edema typical of ARDS (23, 78-80), possibly in part due to their stimulating effects on O_2 radical production by phagocytes, agents which decrease cytokine production for a short period of time may have an efficacy in ARDS. With the recent newer understandings of the molecular mechanisms of the PMN oxidant generating systems, it can be expected that the development of novel modulators of PMN respiratory burst activity will become available.

Small quantities of endotoxin (83) or even TNF and IL-1 (84) confer protection against oxidant lung injury, presumably by increasing lung antioxidant protective systems and possibly by increasing lung cell stress (heat shock) protein levels. In this regard, studies designed to take a critical look at signals which drive the regulation of antioxidant protective systems in animal cells are still needed. This would include further studies designed to delineate the role of the heat shock proteins in oxidant stress responses in man (85).

Enzymes such as superoxide dismutase and catalase (86) clearly afford protection in animal models of ARDS, but have not yet been administered to ARDS patients. Protein engineering approaches, such as conjugating antioxidant enzymes (such as SOD) to molecules which reversibly bind to albumin, thereby generating longer-lived species permeant to wherever albumin goes, would appear to be one particularly enterprising approach. These newer techniques will be especially interesting when added to further derivatizations designed to target organ delivery (personal communication, Professor Masayasu Inoue, Kumamoto University, Japan). Likewise, protective interventions utilizing xanthine oxidase inhibitors, oxidant scavengers and iron chelators have yet to be used in controlled studies in ARDS patients. Although ARDS patients have reduced levels of antioxidant micronutrients such as vitamins C and perhaps E (59), and replenishment of vitamin E may diminish lung injury in some models of ARDS (87), there has been no systematic study of their use in these patients.

It should be emphasized that just as the various injurious oxidant species may synergize with one another, so may the antioxidant protective systems, as has already been noted for the case of vitamins C and E (88). The recent demonstration that pulmonary oxidant toxicity in dogs can be ameliorated by the antioxidant N-acetylcysteine (89) is of special interest, and in fact this agent is currently undergoing clinical trials in ARDS (van Asbeck, personal communication).

The outcome of ARDS has not improved over the two decades since the syndrome was first recognized, in spite of the improved and costly techniques for ICU monitoring and supportive care. A variety of self-destructive mechanisms, some related to oxidant pathways, can contribute to cause acute lung injury in ARDS and thus provide targets for therapeutic antioxidant strategies. At the present time, it remains to be seen whether antioxidant therapeutic approaches will lead to substantive improvements in the outlook for these critically ill patients in the 1990's.

REFERENCES

1. Asbaugh D G, Bigelow D B, Petty T L et al. Acute respiratory distress in adults. Lancet, 2:319-323, 1967.
2. Rinaldo J E, Rogers R M. Adult respiratory distress syndrome: changing concepts of lung injury and repair. N Engl J Med, 306:900-909, 1982.
3. Fein A, Weiner-Kronish J P, Niederman M et al. Pathophysiology of the adult respiratory distress syndrome. What have we learned from human studies? Crit Care Clin 2:429-453, 1986.
4. Fowler, A A, Hamman R F, Zerbe G O et al. Adult respiratory distress syndrome: prognosis after onset. Am Rev Respir Dis 132:472-78, 1985.
5. Montgomery, A B, Stager, M A, Carrico et al. Causes of mortality in patients with the adult respiratory distress syndrome. Am Rev Respir Dis 132:485-89, 1985.
6. American Lung Program. Respiratory Diseases. Task Force Report on Problems, Research Approaches, Needs. The Lung Program. National Heart and Lung Institute. DHEW Publ No (NIH) 73-432: 165-80, 1972.
7. Webster N R, Cohen A T, Nunn J F. Adult respiratory

distress syndrome...how many cases in the UK? Anesthesia, 43:923-926, 1988.

8. Rinaldo J E. The prognosis of the adult respiratory distress syndrome; inappropriate pessimism. Chest, 90:470-471, 1986.

9. Murray J F, Matthay M A, Luce J M et al. Pulmonary perspectives: an expanded definition of the adult respiratory distress syndrome. Am Rev Respir Dis, 138:720-723, 1988.

10. Till G O, Johnson K J, Kunkel R et al. Intravascular activation of complement and acute lung injury. J Clin Invest 69:1126-1135, 1982.

11. Ward P A, Till G O, Kunkel R et al. Evidence for role of hydroxyl radical in complement and neutrophil-dependent tissue injury. J Clin Invest 72:789-801, 1983.

12. Fantone J C, Feltner D E, Brieland J K. Phagocytic cell-derived inflammatory mediators and lung disease. Chest 91:428-434, 1987.

13. Tate R M, Repine J E. Neutrophils and the adult respiratory distress syndrome. Am Rev Respir Dis 125:552-559, 1983.

14. Warschawski F J, Sibbald W J, Driedger A A et al. Abnormal neutrophil-pulmonary interaction in the adult respiratory distress syndrome. Am Rev Respir Dis 133:797-804, 1988.

15. Weiland J E, Davis W B, Holter J F et al. Lung neutrophils in the adult respiratory distress syndrome. clinical and pathophysiological significance. Am Rev Respir Dis 133:218-225, 1986.

16. Fletcher M P, Vassar M J, Holcroft J W. Patients with ARDS demonstrate in vivo neutrophil activation associated with diminished binding of neutrophil-specific monoclonal antibody 31D8. Inflamm 12:455-473, 1988.

17. Weinberg P F, Matthay M A, Webster R O et al. Biologically active products of complement and acute lung injury in patients with the sepsis syndrome. Am Rev Respir Dis 130:791-796, 1984.

18. Malik A B, Perlam M B, Cooper J A et al. Pulmonary microvascular effects of arachidonic acid metabolites and their role in lung vascular injury. Federation Proc 44:36-42, 1985.

19. Perlman M B, Johnson A, Jubiz W et al. Lipoxygenase products induce neutrophil activation and increase endothelial permeability after thrombin-induced pulmonary microembolism. Circ Res 64:62-73, 1989.

20. Stephenson A H, Lonigro A J, Hyers T M et al. Increased concentration of leukotrienes in bronchoalveolar lavage fluid of patients with ARDS or at risk of ARDS. Am Rev Respir Dis 138:714-719, 1989.

21. Le J, Vilcek J. Tumor necrosis factor and interleukin-1: cytokines with multiple overlapping biological activities. Lab Invest 56:234-248, 1987.

22. Sherry B, Cerami A. Cachectin/tumor necrosis factor exerts endocrine, paracrine, and autocrine control of inflammatory responses. J Cell Biol 107:1269-1277, 1988.

23. Remick D G, Kunkel S L. Toxic effects of cytokines in vivo. Lab Invest 60:317-319, 1989.

24. Rich E A, Panuska J R, Wallis R S et al. Dyscoordinate expression of tumor necrosis factor-alpha by human blood monocytes and alveolar macrophages. Am Rev Respir Dis 139:1010-1016, 1989.

25. Jacobs R F, Tabor D R, Campbell G D. Interleukin-1 release by alveolar macrophages from adult respiratory distress syndrome. Clin Res 37:73A, 1989.

26. Dehring D J, Wismar B L. Intravascular macrophages in pulmonary capillaries of humans. Am Rev Respir Dis 139:1027-1029, 1989.

27. Kunkel S L, Spengler M, May M et al. Prostaglandin E regulates macrophage-derived tumor necrosis factor gene expression. J Biol Chem 263:5380, 1988.

28. Cotran R S. New roles for the endothelium inflammation and immunity. Amer J Path 129:407-413, 1987.

29. Strieter R M, Kunkel S L, Showell H J et al. Endothelial cell gene expression of a neutrophil chemotactic factor by TNFα, LPS and IL-1β. Science 243:1467-1469, 1989.

30. Schroder J M, Christophers E. Secretion of novel and homologous neutrophil-activating peptides by LPS-stimulated human endothelial cells. J Immunol 142:244-251, 1989.

31. Gorog P, Pearson J D, Kakkar V V. Generation of reactive oxygen metabolites by phagocytosing endothelial cells. Arteriosclerosis 72:19-27, 1988.

32. Halliwell B, Gutteridge J M C. Lipid peroxidation, oxygen radicals, cell damage, and antioxidant therapy. Lancet 1:1396-1397, 1984.

33. Dobrina A, Patriarca P. Neutrophil-endothelial cell interaction. J Clin Invest 78:462-471, 1986.

34. Fantone J C, Ward P A. Role of oxygen-derived free radicals and metabolites in leukocyte-dependent inflammatory reactions. Am J Pathol 107:397-418, 1982.

35. Varani J, Fligiel S E G, Till G O et al. Pulmonary endothelial cell killing by human neutrophils. Lab Invest 53:656-663, 1985.

36. Lewis R E, Granger H J. Neutrophil-dependent mediation of microvascular permeability. Fed Proc 45:109-113, 1986.

37. Ryan U S, Vann J M. Endothelial cells: a source and target of oxidant damage. In: Oxygen Radicals in Biology and Medicine. Simic M G, Taylor K A, Ward A F, eds. New York: Plenum Publishing 963-968, 1987.

38. Antony V B, Owen C L, English D. Polymorphonuclear leukocyte cytoplasts mediate acute lung injury. J Appl Physiol 65:706-713, 1988.

39. Mizer L A, Weisbrode S E, Dorinsky P M. Neutrophil accumulation and structural changes in nonpulmonary organs after lung injury induced by phorbol myristate acetate. Am Rev Respir Dis 139:1017-1026, 1989.

40. Hemler M E, Lands W E M. Evidence for a peroxide-initiated free radical mechanism of prostaglandin biosynthesis. J Biol Chem 255:6253-6261, 1980.

41. Cleland L G. Oxy-radicals, "peroxide tone" and inflammation. J Rheum 11:725-726, 1984.

42. Vesconi S, Rossi G P, Pesenti A et al. Pulmonary microthrombosis in severe adult respiratory distress syndrome. Crit Care Med 16:111-113, 1988.

43. Maunder R J, Shuman W P, McHugh J W et al. Preservation of normal lung regions in the adult respiratory distress syndrome: analysis by computed tomography. JAMA 255:2463-2465, 1986.

44. McCord J M. Oxygen-derived free radicals in postischemic tissue injury. N Engl J Med. 312:159-161, 1985.

45. Koyama I, Toung T J K, Rogers M C et al. O_2 radicals mediate reperfusion lung injury in ischemic O_2-ventilated canine pulmonary lobe. J Appl Physiol 63:111-115, 1987.

46. Lynch M J, Grum C M, Gallagher K P et al. Xanthine oxidase inhibition attenuates ischemic-reperfusion lung injury. J Surg Res 44:538-544, 1988.

47. Burton L K, Patt A, Velasco S E et al. Toxic O_2 metabolites from oxidase systems cause rapid serum complement activation, RBC fragility and acute edematous oxidative lung injury in rats subjected to skin burn. Clin Res 36:194A, 1988.

48. Grum C M, Ragsdale R A, Ketai L H et al. Plasma xanthine oxidase activity in patients with adult respiratory distress syndrome. J Critical Care 2:22-26, 1987.

49. Flick M R. Mechanisms of acute lung injury. What have we learned from experimental animal models. Crit Care Clin 2:455-70, 1986.

50. Taylor A E, Martin, D J, Townsley M I. Oxygen radicals and pulmonary edema. In The Pulmonary Circulation and Acute Lung Injury, Said S I, ed. Furtana Pub. Co. Inc., Mt. Kisco, N.Y., 1985, pp 307-320.

51. Johnson A, Perlman M B, Blumenstock F A et al. Superoxide dismutase prevents the thrombin-induced increase in lung vascular permeability: role of superoxide in mediating the alterations in lung fluid balance. Circ Res 59:405-415, 1986.

52. Cochrane C G, Spragg R G, Revak S D et al. The presence of neutrophil elastase and evidence of oxidation activity in bronchoalveolar lavage fluid of patients with adult respiratory distress syndrome. Am Rev Respir Dis 127:525-527, 1983.

53. Baldwin S R, Gun C M, Boxer L A et al. Oxidant activity in expired breath of patients with adult respiratory distress syndrome. Lancet 1:11-14, 1986.

54. Takeda K Y, Shimada T, Okada M et al. Lipid peroxidation in experimental septic rats. Crit Care Med 14:719-723, 1986.

55. Ward P A, Till G O, Hatherill J R et al. Systemic complement activation, lung injury, and products of lipid peroxidation. J Clin Invest 76:517-527, 1985.

56. Demling R H, Lalonde C, Ryan P et al. Endotoxemia produces an increase in arterial but not venous lipid peroxides in sheep. J Appl Physiol 64:592-598, 1988.

57. Takeda, K Y, Shimada M. Plasma lipid peroxides and alpha-tocopherol in critically ill patients. Crit Care Med 12:957-959, 1984.

58. Richard C F, Lemonnier M. Lipoperoxidation and vitamin E consumption during adult respiratory distress syndrome. Am Rev Respir Dis 135:425a, 1987.

59. Frei B, Yamamoto Y, Niclas D et al. Analysis of oxidants and antioxidants in human plasma of healthy subjects and of patients with adult respiratory distress syndrome. In: Free Radicals: Methodology and Concepts, Rice Evans C, Halliwell B, eds. Richlieu Press, London 1988, pp 349-368.

60. Cross C E, Frei B, Stocker R et al. Evidence for oxidative stress in ARDS: bronchoalveolar vs plasma compartments. Am Rev Respir Dis 139:A221, 1989.

61. Movat H Z, Cybulsky M I, Colditz I G et al. Acute inflammation in gram negative infection: endotoxin, interleukin I, tumor necrosis factor, and neutrophils. Fed Proc 46:97-104, 1987.

62. Brigham K L, Bowers R E, McKeen C R. Methylprednisolone prevention of increased lung vascular permeability following endotoxemia in sheep. J Clin Invest 67:1103-1110, 1981.

63. Bone R C, Fisher C J Jr, Clemmer T P et al. Methylprednisolone severe sepsis study group. A controlled clinical trial of high-dose methylprednisolone in the treatment of severe sepsis and septic shock. N Engl J Med 317:653-658, 1987.

64. Broaddus V C, Berthiaume Y, Biondi J W et al. Hemodynamic management of the adult respiratory distress syndrome. J Intensive Care Med 2:190-213, 1987.

65. Simmons R S, Berdine G G, Seidenfeld J et al. Fluid balance and the adult respiratory distress syndrome. Am Rev Respir Dis 135:924-929, 1987.

66. Salvador R A, Fiedler-Nagy C, Coffey J W. Biochemical basis for drug therapy to prevent pulmonary fibrosis in ARDS. In: Acute respiratory failure, Zapol W, ed. New York: Dekker, 1985, pp 477-506.

67. Krieger, B P, Loomis W H, Spragg R G. Minireview. Granulocytes and hyperoxia act synergistically in causing acute lung injury. Exper Lung Res 7:77-83, 1984.

68. Brigham K L. Role of free radicals in lung injury. Chest 89:859-863, 1986.

69. Fridovich I, Freeman B. Antioxidant defences in the lung. Ann Rev Physiol 48:703-720, 1986.

70. Jamieson D, Chance B, Cadenas E et al. The relation of free radical production to hyperoxia. Ann Rev Physiol 48:703-720, 1986.

71. Jackson J H, Berger E M, Repine J E. Thiourea and dimethylthiourea decrease human neutrophil bactericidal function in vitro. Inflamm 12:515-524, 1988.

72. Ziegler E J. Perspective: Protective antibody to endotoxin core: the emperor's new clothes. J Infect Dis 158:286-290, 1988.

73. Stevens J H, O'Hanley P, Shapiro J M et al. Effects of anti-C5a antibodies in the adults respiratory distress syndrome in septic primates. J Clin Invest 77:1812-1816, 1986.

74. Vedder N B, Winn R K, Chi E Y et al. A monoclonal antibody to the adherence-promoting leukocyte glycoprotein, CD18, reduces organ injury and improves survival from hemorrhagic shock and resuscitation in rabbits. J Clin Invest 81:939-944, 1988.

75. Belivacqua M P, Stengelia S, Gimbrone M A et al. Endothelial leukocyte adhesion molecule 1: an inducible receptor for neutrophils related to complement regulatory proteins and lectins. Science 243:1160-1165, 1989.

76. Jutila M A, Berg E L, Amento E P et al. Effect of systemic cytokine administration on leukocyte/endothelial cell interactions during inflammation. FASEB J 3:A319, 1989.

77. Smith C W, Rothlein R, Hughes B J et al. Recognition of an endothelial determinant for CD18-dependent human neutrophil adherence and transendothelial migration. J Clin Invest 82:1746-1756, 1988.

78. Tracey K J, Fong Y, Hesse D G. Anti-cachectin/TNF monoclonal antibodies prevent septic shock during lethal bacteraemia. Nature 330:662-664, 1987.

79. Stevens K E, Ishizaka A, Wu Z et al. Granulocyte depletion prevents tumor necrosis factor-mediated acute lung injury in guinea pigs. Am Rev Respir Dis 138:1300-1307, 1988.

80. Rosenstein M, Ettinghausen S E, Rosenberg S A. Extravasation of intravascular fluid mediated by the systemic administration of recombinant interleukin 2. J Immunol 137:1735-1742, 1986.

81. Beutler B A, Milsark I W, Cerami A. Cachectin/tumor necrosis factor: Production, distribution and matabolic fate in vivo. J Immunol 135:3972-7, 1985.

82. Ognibene F P, Martin S E, Parker M M et al. Adult respiratory distress syndrome in patients with severe neutropenia. N Engl J Med 315:547-511, 1986.

83. Frank L, Summerville F L, Massaro D. Protection from oxygen toxicity with endotoxin: role of the endogenous antioxidant enzymes of the lung. J Clin Invest 65:1104-1110, 1980.

84. White C W, Ghezzi P, Dinarello C A et al. Recombinant TNF/cachectin and IL-1 pretreatment decreases lung oxidized glutathione accumulation, lung injury, and mortality in rats exposed to hyperoxia. J Clin Invest 79:1868-1873, 1987.

85. Hass, M A, Massaro D. Regulation of the synthesis of superoxide dismutases in rat lungs during oxidant and hyperthermic stresses. J Clin Invert 263:776-781, 1988.

86. Turrens J F, Crapo J D, Freeman B A. Protection from oxygen toxicity by intravenous injection of liposome-entrapped catalase and SOD. J Clin Invest 73:87-97, 1984.

87. Bucher J R, Roberts R J. Effects of α-tocopherol treatment on newborn rat lung development and injury in hyperoxia. Ped Pharmacol 2:1-9, 1982.

88. Niki E. Lipid antioxidants. How they may act in biological systems. Br J Cancer 55:Suppl VIII, 153-157, 1987.

89. Wagner P D, Mathieu-Costello O, Debout D E et al. Protection against pulmonary O_2 toxicity by N-acetylcysteine. Eur Respir J 2:116-126, 1989.

ANTIOXIDANT ACTIVITY IN FETAL AND NEONATAL LUNG

Mary McElroy, Tony Postle and Frank Kelly

Departments of Human Nutrition and Child Health
University of Southampton, Southampton SO9 3TU

CLINICAL SITUATION

    Preterm infants, especially those born 8-15 weeks prematurely, have
a high risk of developing respiratory distress syndrome [RDS]. RDS is
caused by a relative lack of pulmonary surfactant and is characterized by
areas of lung collapse and the deposition of protein containing hyaline
membranes within the alveolus (Battenburg, 1982). RDS is treated by
positive pressure ventilation together with supplemental oxygen; the
success of this therapy can be seen by the greater than 50% survival rate
of infants delivered as early as 26 weeks of gestation in many centres.

    While some infants experience complications such as patent ductus
arteriosus and interventricular haemorrhage during the acute phase of RDS
(O'Brodovich and Mellins, 1985), at least one quarter of the smallest
infants go on to develop severe chronic lung disease (CLD). Baurotrauma
associated with positive pressure ventilation and prolonged exposure to
hyperoxia have been implicated as primary aetiological factors in the
pathogenesis of CLD (Phillips, 1975) and it has been suggested that
bronchiolar damage with subsequent fibrosis leads to the long term
sequelae in oxygen-treated infants who survive RDS.

    The potential role that hyperoxia may play in the development of CLD
has received considerable attention in recent years (Massaro, 1986). It is
now generally appreciated that exposure to high concentrations of oxygen
causes alterations throughout the respiratory tract in humans and other
animals (Smith, 1899; Kapanci et al, 1969, Gould et al., 1972). Such
damage is not thought to be caused directly by molecular oxygen, but to
be mediated by the production of partially reduced oxygen species such as
superoxide ions (O_2^-), hydrogen peroxide (H_2O_2) and the hydroxyl radical
($OH^\cdot$). Cellular changes that can be induced by these very toxic moeities
include inactivation of important enzymes, destruction of cell membranes
by lipid peroxidation and damage to DNA (Slator, 1984; Halliwell and
Gutteridge, 1984).

    Oxygen free radicals are continually produced as a consequence of
aerobic metabolism. The ordered transport of electrons from reducing
substances to oxidizing species within the respiratory chain is believed

Antioxidants in Therapy and Preventive Medicine   449
Edited by I. Emerit *et al.*
Plenum Press, New York, 1990

to be one such site of oxygen radical formation. As hyperoxic conditions are thought to markedly increase the generation of these toxic species (Freeman and Crapo,1981; Freeman et al,1982), it is a logical hypothesis that such an increased production of oxygen free radicals during hyperoxia may be central to the subsequent tissue damage.

Under normal circumstances the lungs are protected from the basal level of oxygen free radical production by an elaborate and extensive array of antioxidant defences. These biochemical defences against oxygen free radicals are based on enzymatic and non-enzymatic defences. Enzymatic defences include both Cu/Zn and Mn forms of superoxide dismutase (SOD), catalase (CAT) and glutathione peroxidase (GSH-Px). Non-enzymatic antioxidants include the membrane lipid constituent vitamin E and the tripeptide glutathione. It is now generally accepted that successful aerobic cell function is only possible due to a balance between the anti-oxidant defences of that cell and the oxygen free radical flux within it.

In utero, fetal lungs are exposed to a relatively low oxygen tension ($\approx$ 25-30 mmHg). Following birth however, as the infant takes its first breath expelling the fluid from its lungs and replacing it with air, the oxygen tension within the alveoli rises dramatically (> 100 mmHg). The work of Freeman and colleagues (1981, 1982) suggests that a direct increase in oxygen free radical production follows this event. Birth represents a relatively hyperoxic experience for the fetus and to make the transition succesfully fetal lungs must be prepared for this increased oxidative insult. Over recent years a number of investigators have demonstrated substantial increases in antioxidant enzyme activity over the final 10-15% of gestation. (Tanswell and Freeman, 1984; Gerdin et al., 1985; Frank and Sosenko, 1987; Hass and Massaro, 1987). Human studies have been few due to the considerable difficulties inherent in obtaining suitable samples.

If the development of antioxidant enzyme activity in human lung follows a course similar to that seen in other species, then infants born prematurely will be at risk from oxidative-induced cell damage at birth. In addition, those infants who are born with a deficiency of pulmonary surfactant, develop RDS and are then treated with supplemental oxygen will be at an even greater risk from the free radical/antioxidant imbalance. It is therefore of considerable importance to establish the antioxidant status of the human fetal lung. At this meeting we report our preliminary finding regarding this question.

EXPERIMENTAL

Sample Collection

Lung and liver tissue was obtained within 24h of death, either from fetuses therapeutically aborted between 8 and 20 weeks gestation or from infant mortalities (26w gestation to 3 months post-term). In order to minimize any possible intra tissue variation the same section of tissue was taken each time ie. the top right hand lobe of the lung and middle section from the right side to the liver. All samples were rinsed in saline, blotted dry and frozen in liquid N_2 prior to storage at -70°C.

Tissue Preparation

Tissue samples ($\approx$ 100mg) were homogenized in 10 volumes of ice cold 0.05M phosphate buffer (pH 7.4) with a Janke and Kunkel Ultra-Turrax tissue grinder (2 x 15 sec bursts). The homogenates were then sonicated

(30 x 1 sec bursts, MSE Soniprep) and a 100μl aliquot removed for DNA (Sterzel et al., 1985) and protein analyis (Lowry et al., 1951) The remaining homogenate was centrifuged (10,000 rpm, 5 min) and the resultant supernatant divided into three aliquots for subsequent enzyme analysis. Supernatant fractions were stored at $-20°C$ and assayed within 10 days.

Enzyme Analysis

Total Cu/Zn and Mn SOD was determined by the pyrogallol autoxidation method of Marklund (1985). Pyrogallol autoxidation was monitored at 420nm with a reaction rate spectrophotometer (LKB Mk II Kinetic Analyser) at $25°C$ over 2 minutes. A standard curve was constructed using standard Cu/Zn SOD (Sigma).

GSH-Px activity was determined by the method of Beutler (1979). In this linked enzyme reaction oxidized glutathione formed by the action of GSH-Px on lipid hydroperoxide is converted back to its reduced form in the presence of glutathione reductase and NADPH. Glutathione is thus maintained at a constant concentration and the reaction followed by the oxidation of NADPH. The reaction was monitored for 2 min at 320nm using an LKB reaction rate kinetic analyzer at $37°C$.

CAT activity was determined using the method of Aebi (1984) in which the initial rate of hydrogen peroxide decomposition is determined. The reaction was followed at 240nm using a LKB Ultraspec II spectrophotometer. Units of CAT are defined in terms of initial velocity of hydrogen peroxide decomposition per minute.

As the collection and assay of human tissue samples was performed over a 15 month period a quality control system was included to ensure the identification of any interassay variation which developed over this period. Enzyme activities were expressed relative to unit DNA and results grouped over 10 week intervals.

RESULTS

All three lung antioxidant enzyme activities increased with gestational age (Table 1). This was greatest for CAT which increased 8-fold between the period 11-20 weeks and 31-40 weeks gestation and a further 2-3 fold following birth. Both SOD and GSH-Px approximately doubled between the gestation periods 11-20 and 21-30w and then remained constant until birth. After term, SOD increased further by 51-60w, while GSH-Px exhibited a transient increase between 41-50w (Table 1).

Liver catalase activity increased throughout gestation, a trend which also continued into the early neonatal period (Table 2). Likewise, both total SOD and GSH-Px activities in the liver increased during gestation. Superoxide dismutase activity increased about 15 fold between the earliest point considered in gestation (11-20w) and 51-60 weeks post conception, while liver GSH-Px activity increased 2-3 fold over the same period (Table 2).

Antioxidant enzyme activities were always higher in liver than lung, although the magnitude of difference varied considerably with time. For instance, liver CAT activity was 6-fold greater than lung CAT at 11-20 weeks gestation, but by birth both activities were approximately the same. Total SOD activity was marginally higher in liver than lung during gestation , while following birth the increase in liver SOD activity far exceeded the change in lung SOD activity. GSH-Px activity was found to be always 2-3 fold higher in liver than in lung.

Table 1. Human fetal and early neonatal lung antioxidant
enzyme activities

	Post conceptual age in weeks				
	11-20	21-30	31-40	41-50	51-60
CAT IU/mg DNA	200 ± 17	847 ±177	1624 ±160	4159 ±563	2797 ±213
Total SOD U/mg DNA	30.2 ±1.62	50.5 ±5.7	53.0 ±15.1	46.6 ±1.4	89.4 ±11.4
GSH-Px U/mg DNA	0.30 ±0.02	0.52 ±0.06	0.58 ±0.12	0.96 ±0.08	0.64 ±0.04
n	20	11	3	5	7

Results are expressed as means ± SEM

Table 2. Human fetal and early neonatal liver antioxidant
enzyme activities

	Post conceptual age in weeks				
	11-20	21-30	31-40	41-50	51-60
CAT IU/mg DNA	1282 ± 96	1724 ± 246	2248 ±166	5433 ± 751	6178 + 888
Total SOD U/mg DNA	50.7 ± 2.8	86.8 ±13.4	178.1 ±38.7	559.4 ±120.3	789.3 ± 66.9
GSH-Px U/mg DNA	0.75 ± 0.03	1.10 ±1.14	1.90 ±0.75	2.20 ±0.23	1.83 ±0.18
n	21	11	4	5	10

Results are expressed as means ± SEM

DISCUSSION

The preliminary data presented here generally support the hypothesis that pulmonary antioxidant enzyme activities is low in early gestation and increases as gestation proceeds. This is most dramatically illustrated by the change in CAT, which increased 8-fold between the earliest and latest points considered in gestation. Unfortunately, due to the small number of samples which have been obtained for study at the critical period of 31-40w in gestation (3 out of a total of 46) it is not yet possible to determine if there are even more dramatic increases in antioxidant activities over the final 10-15% gestation as has been observed in several animal species (Tanswell and Freeman, 1984; Gerdin et al., 1985; Frank and Sosenko, 1987; Hass and Massaro, 1987). In the lung higher CAT and GSH-Px activities were recorded in tissue obtained from 40-51w neonates, which may tend to support this hypothesis.

Of the three antioxidant enzymes considered SOD showed the smallest change in activity with time. Similar observations have been reported by Strange et al. (1988) This group also reported that they were unable to detect any change in GHS-Px activity during gestation when expressed per unit protein (Fryer et al., 1986). In respect to this study we feel that it is important to qualify these results as all the samples after 24 weeks gestation were obtained at post-mortem. Thus it is not presently possible to assess whether the antioxidant activity in the lungs of these infants who died is reflective of the normal control situation.

The majority of studies on the ontogeny of the antioxidant enzymes, SOD, CAT and GSH-Px have been conducted in the lung (Tanswell and Freeman, 1984; Gerdin et al., 1985; Frank and Sosenko, 1987; Hass and Massaro, 1987). These studies have focussed on the lung due to the direct and dramatic increase in oxygen exposure this tissue undergoes at birth. In essence however, many other tissues also experience increased oxidative stress at birth as their metabolic rate and hence free radical generation increases at this time. To determine if extra pulmonary tissues also undergo a pre-birth surge in antioxidant activity we also determined CAT, SOD and GSH-Px activity changes in the liver throughout gestation. Study of this tissue revealed two main characteristics. First, antioxidant activity increased throughout gestation and into early neonatal life. Second, the absolute activity of all three antioxidants was always greater on a cellular basis in the liver than lung. These results suggest that the lung is not unique in its increasing antioxidant requirement following birth and that these requirements are likely to be determined by the metabolic rate of the particular tissue under consideration.

In summary, although there are considerable problems associated with antioxidant measurements in human tissue such as the inevitable post-mortem delays and limited availability of appropriate time points, such studies are essential for the advancement of normal non-pathological samples and our understanding of diseases associated with premature delivery. The toxic effects of hyperoxic exposure may in preterm infants be compounded by inadequate antioxidant defence mechanisms in several tissues. Unfortunately at present, while efforts are being made to reduce the level of oxygen exposure, treatment with elevated oxygen remains an essential component of the therapy for infants with RDS.

ACKNOWLEDGEMENTS

This study was supported in part by funds provided by the British Lung Foundation and the Sir Halley Stewart Trust.

REFERENCES

Aebi, H., 1984, Catalase in vitro, <u>Meth. Enzymol.</u>, 105, 121-126.
Battenburg, J. J., 1982, <u>in</u> "Lung Development, Clinical Perspectives",
    Vol 1, pp. 359-390, P. M. Farrel ed. , Academic Press
Beutler, E., 1979, Glutathione peroxidase, <u>in</u>: "Red Cell Metabolism. A
    manual of Biochemical Methods" 2nd edition, pp. 71-73, Grune and
    Stratton, New York.
Frank, L. and Sosenko, I. R. S., 1987, Prenatal development of lung
    antioxidant enzymes in four species, <u>J. Pediatr.</u>, 110:106-110.
Freeman, B. A. and Crapo, J. P., 1981, Hyperoxia increases oxygen radical
    production in rat lungs and lung mitochondria, <u>J. Biol. Chem.</u>, 256,
    10986-10992.
Freeman, B. A., Topolsky, M. K. and Crapo, J. P., 1982, Hyperoxia
    increases oxygen radical production in rat lung homogenates,
    <u>Biochem. Biophys.</u>, 216:477-484.
Fryer, A. A., Hume, K. and Strange, R.C. , 1986, The development of
    glutathione S-transferase and glutathione peroxidase activities in
    human lung, <u>BBA</u>, 883:448-453
Gerdin, E., Tyden, O. and Eriksson, U. J., 1985, The development of
    antioxidant enzymatic defence in the perinatal rat lung, <u>Pediatr.
    Res.</u>, 198:687-691.
Gould, V. E., Tosco, R., Wheelis, R. G., Gould, N. S. and Kapanci, Y.,
    1972, Oxygen pneumonitis in man: Ultrastructural observations on
    the development of alveolar lesions, <u>Lab. Invest.</u>, 26:499-508.
Halliwell, B. and Gutteridge, J. M. C., 1984, Oxygen toxicity, oxygen
    radicals, transition metals and disease, <u>Biochem. J.</u>, 219:1-14.
Hass, M.A. and Massaro, D., 1987, Developmental regulation of rat lung
    Cu/Zn superoxide dismutase, <u>Biochem. J.</u>, 246:697-703.
Kapanci, Y., Weibel, E. R., Kaplan, H. P. and Robinson, F. R., 1969,
    Pathogenesis and reversibility of the pulmonary lesions of oxygen
    toxicity in monkeys II. Ultrastructural and morphometric studies,
    <u>Lab Invest.</u>, 20:101-17
Lowry, O. H., Rosenbrough, N. J., Farr, A. L. and Randall, 1951, Protein
    measurement with the Folin reagent, <u>J. Biol. Chem.</u>, 193:265-275.
Marklund, S. L., 1985, Pyrogallol autoxidation, <u>in</u>: "CRC Handbook of
    Methods for Oxygen Radical Research", pp. 243-247.
Massaro, D., 1986, Oxygen: Toxicity and Tolerance, <u>Hospital Practice</u>, July
    95-101.
O'Brodovich, H. M. and Mellins, R.B., 1985, Bronchopulmonary Dysplasia,
    <u>Am. Rev. Respir. Dis.</u>, 132:694-709.
Phillips, A. G. S., 1975, <u>Pediatrics</u>, 55:44-50.
Smith, J. L., 1899, The pathological effects due to increase of oxygen
    tension in the air breathed, <u>J. Physiol.</u>, 24:19-35.
Slater, T. F., 1984, Free radical mechanisms in tissue injury, <u>Biochem. J</u>,
    22:1-15.
Sterzel, W., Bedford, P. and Eisenbrand, G., 1987, Automated determination
    of DNA using the Fluorochrome Heochrit 33258, <u>Anal. Biochem.</u>,
    147:462-567.
Strange, R. C., Cotton, W., Fryer, A. A., Drew, R., Bradwell, A. R.
    Marshall, T., Collins, M.F., Bell, J., Hume, R. , 1988, Studies on
    the expression of Cu,Zn superoxide dismutase in human tissues during
    development, <u>BBA</u>, 883:448-453.
Tanswell, A. K. and Freeman, B. A., 1984, Pulmonary antioxidant enzyme
    maturation in the fetal and neonatal rat, <u>Pediatr. Res.</u>, 18:584-587.

PRO- AND ANTI-OXIDANT FACTORS IN RAT LUNG CYTOSOL

Cornelis J.A. Doelman and Aalt Bast

Department of Pharmacochemistry, Faculty of Chemistry, Vrije Universiteit
De Boelelaan 1083, 1081 HV Amsterdam, The Netherlands

INTRODUCTION

During the process of inflammation oxygen radicals are produced by phagocytic cells like neutrophils, eosinophils, monocytes and macrophages. Little is known about the effects of oxygen radicals on lung tissue in inflammatory diseases such as asthma[1]. Activated phagocytes release superoxide anions, synthesized by NADPH oxidase which is located in the cellular membrane. Superoxide anions dismutate to hydrogen peroxide enzymatically (superoxide dismutase) or non-enzymatically. These superoxide anions and hydrogen peroxide together form very reactive oxygen species, namely the hydroxyl radical and singlet oxygen. In these reactions iron plays a catalytic role[2]. Oxygen metabolites cause DNA strand breaks, protein destruction (particularly sulfhydryl groups are sensitive to oxidative stress) and lipid peroxidation. The lipid peroxidation reaction products are very toxic too and cause DNA and protein damage as well[3,4].

Receptor responses, like the lung beta-adrenergic and lung muscarinic receptor responses, are also influenced by oxidative stress. The receptor proteins can be destroyed but also other proteins which are part of the receptor signalling system in the membrane are susceptible to reactive oxygen species. Especially when these proteins contain an essential sulfhydryl group. We recently reported that hydrogen peroxide causes a disbalance between lung beta-adrenergic and muscarinic receptor responses[5]. This disbalance was explained by the dissimilarity in receptor structure and membrane located signal transfer. The so-called G-protein (coupling factor between the beta-adrenergic receptor complex and adenylate cyclase) contains an essential sulfhydryl moiety and the activity of this protein is determined by the composition of the surrounding lipids[6]. During the process of lipid peroxidation the composition of the lipid bilayer changes, and the bilayer becomes more rigid. This may affect the receptor-responses-coupling directly. Moreover the breakdown of the lipid bilayer initiates the arachidonic acid cascade which results in formation of prostaglandins, leukotrienes, prostacyclins and lipoxins[7].

Antioxidants in Therapy and Preventive Medicine
Edited by I. Emerit *et al.*
Plenum Press, New York, 1990

Vitamin E (a lipid soluble anti-oxidant) and selenium (an essential element for the activity of the cytosolic selenium-dependent glutathione peroxidase) preserve against hydrogen peroxide in the rat lung[8]. Ascorbic acid, a non-enzymatic lung cytosolic anti-oxidant, has been described as neutralizer of reactive oxidants released by hyperactive phagocytes[9]. The protective action of ascorbic acid is proposed to be linked to the reduction of glutathione[10]. Glutathione is able to reduce free radicals, including the vitamin E radical. Glutathione may also act as substrate for the glutathione peroxidase system, by which peroxides are transformed to alcohols.

In this study we determined several pro- and anti-oxidant factors in the rat lung and tested some of them on their pro- or anti-oxidant activity towards lipid peroxidation in lung microsomes, initiated by 100 µM cumene hydroperoxide.

MATERIALS AND METHODS

Male Wistar rats (200 - 300 gram) were killed by cervical dislocation and the lungs were rapidly removed. The lungs were homogenized in an ice-cold buffer (50 mM sodium phosphate/ 0.1 mM EDTA, pH=7.4) using a Polytron homogenizer and a Teflon Potter homogenizer. Mitochondrial and nuclear fractions were separated by ultracentrifugation at 10.000 x g (2 x 20 min.). The supernatant was centrifuged for 60 min. at 105.000 x g. The pellet (microsomes) was resuspended in the same buffer (1:1 lung weight/volume). Cytosolic fraction and microsomes were stored at -80°C until use.

Ascorbic acid, alpha-tocopherol and glutathione were determined according to the methods of Ramell[11], Jagota and Dani[12] and Griffith[13]. The content of iron was measured according to the method of Brumby and Massey[14]. The Biorad assay was used to determine the protein content in both microsomes and cytosol with bovine serum albumine as standard[15]. The activity of the following enzymes: glutathione peroxidase (Se-dependent and Se-independent), glutathione reductase and superoxide dismutase, were determined according to the methods of Wendel[16], Worthington and Rosemeyer[17] and McCord and Fridovich[18].

Before use the microsomes were washed as previously described by Haenen and Bast[19]. These washed lung microsomes (approximately 0.5 mg protein in a total incubation volume of 4 ml) were incubated in a Tris buffer solution (50 mM Tris-HCl, 150 mM KCl, pH=7.4) with 100 µM cumene hydroperoxide at 37°C. To the incubation medium were added either 0.3 mM $FeSO_4$, 57 µM ascorbic acid or 100 µM glutathione.

The cytosol which was added to the incubation medium, was dialysed overnight in distilled water at 4°C. Of the dialysed cytosol (diluted to 3.5 mg protein per ml) 1 ml was added to the incubation medium (total volume of 4 ml) either with or without supplementation of ascorbate and/or glutathione. Lipid peroxidation was measured as described by Haenen and Bast[19], determining the amount of thiobarbituric acid reactive material.

RESULTS

Table 1 shows the content of several factors present in rat lung cytosol and in rat lung

Table 1

Content of several pro- and anti-oxidant factors in the rat lung:

Cytosol:	per mg protein
Fe (II):	340 ± 17 nmoles
Fe (III):	n.d.
Ascorbate:	65.2 ± 2.6 nmoles
Glutathione (reduced):	110 ± 15 nmoles
Glutathione peroxidase activity	
Se-dependent:	8.6 ± 0.14 nmoles NADPH/ min.
Se-independent:	10.1 ± 0.07 nmoles NADPH/ min.
Glutathione reductase activity:	41.3 ± 5.4 nmoles NADPH/min.
Superoxide dismutase activity:	12.5 ± 2.0 U
α-tocopherol:	0.34 ± 0.003 nmoles
Microsomes:	
α-tocopherol:	5.25 ± 0.08 nmoles

n.d.= not detectable.

microsomes. No Fe(III) was detectable according to the method of Brumby and Massay[14] in the cytosolic fraction.

As shown in figure 1, microsomes and 100 μM cumene hydroperoxide (the control) gave no lipid peroxidation after 50 min. incubation time. Also glutathione did not initiate or stimulate lipid peroxidation in lung microsomes in the presence of 100 μM cumene hydroperoxide. When either ascorbic acid or $FeSO_4$ were added to the incubation medium, containing lung microsomes and 100 μM cumene hydroperoxide, lipid peroxidation was found. So ascorbic acid and $FeSO_4$ stimulate lipid peroxidation in lung microsomes in the presence of 100 μM cumene hydroperoxide compared to the control.

When control cytosol was added to the incubation medium, also no lipid peroxidation was detectable. Notably addition of dialysed cytosol to the reaction mixture resulted in formation of thiobarbituric acid reactive material, as shown in figure 2. Subsequent supplementation of either ascorbic acid or glutathione resulted in a diminished amount of lipid peroxidation compared to the incubation in which only dialysed cytosol was added (figure 2). In the presence of dialysed cytosol, ascorbic acid had anti-oxidant properties.

DISCUSSION

Lungs must be highly protected against the damaging effects of radicals because not only in inflammatory processes these radicals are formed, but also oxidant gasses from the air can damage the lungs via radical processes[20,21,22].

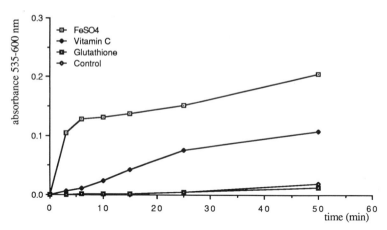

Figure 1. Effects of different cytosolic components (FeSO$_4$, vitamin C and glutathione) on the amount of lipid peroxidation in rat lung microsomes initiated by 100 µM cumene hydroperoxide.

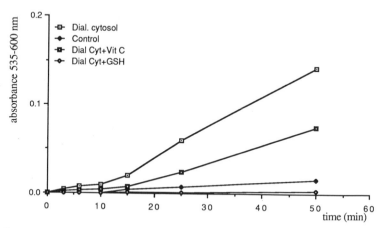

Figure 2. Effects of different cytosolic components (vitamin C and glutathione) on the amount of lipid peroxidation, induced by 100 µM cumene hydroperoxide in rat lung microsomes in the presence of dialysed cytosolic fraction.

The physiological concentration of Fe(II) stimulates lipid peroxidation in rat lung microsomes, induced by 100 μM cumene hydroperoxide. Also Ascorbic acid stimulated lipid peroxidation whereas glutathione had no effect. For establishing lipid peroxidation an optimal ratio Fe(II)/Fe(III) must be present in the incubation medium[23,24]. Fe(II) or Fe(III) alone do not initiate and/or promote lipid peroxidation. In our incubation Fe(II) is oxidized to Fe(III) during lipid peroxidation. Ascorbate is able to reduce Fe(III) to Fe(II) non-enzymatically and in this way may produce an optimal ratio which results in further promotion of lipid peroxidation.

Addition of control cytosol does not stimulate lipid peroxidation in the presence of cumene hydroperoxide.This might be due to a too high level of reducing equivalents, which prevents an optimal Fe(II)/Fe(III) ratio being formed. This is consistent with our observation that Fe(III) is not detectable in rat lung cytosol. Dialysis of cytosol results in reduction of ascorbic acid to 16% and glutathione to 12% of the control concentration. No diminished Fe(II) concentration was observed after dialysis, as determined by iron complexation by 1,10-phenanthroline compared to the method of Brumby and Massey[14]. May be iron is bound to several proteins present in the cytosol, which makes it not dialysable but apparently still measurable with the high affinity complexator 1,10-phenenthroline. Dialysed cytosol promoted lipid peroxidation. This is probably explained by the diminished reducing capacity of dialysed cytosol, compared to the control cytosol. Dialysed cytosol may reduce Fe(III) to Fe(II) in such a rate, that an optimal ratio of Fe(II)/Fe(III) for lipid peroxidation is created. Addition of either ascorbate or glutathione (supplementation to the physiological concentration) enhanced the reducing capacity of the dialysed cytosol. This may be the reason for a diminished extent of lipid peroxidation as we observed.

Our results indicate that ascorbate and glutathione play a role in the protection of the rat lung against lipid peroxyradicals. These two reductors in the cytosol are probably involved in maintaining cytosolic iron in its Fe(II) redox state.

ACKNOWLEDGEMENT

This study was financially supported by the Dutch Asthma Foundation (grant nr. 87-30).

REFERENCES

1. P.J. Barnes, K.F. Chung, and C.P. Page, Inflammatory mediators and asthma, Pharmacol. Rev. 40:49 (1988).
2. D. Roos, and R.S. Weening, Defects of oxidative killing of micro-organisms by phagocytic leukocytes, in: "Oxygen Free Radicals and Tissue Damage", Ciba Foundation Symposium 65, Exerpta Medica, Amsterdam (1979).
3. C.E. Vaca, J. Wilhelm, and M. Harms-Ringdahl, Interaction of lipid peroxidation products with DNA, Mutation Res. 195:137 (1988).
4. R. Leurs, B. Rademaker, K. Kramer, H. Timmerman, and A. Bast, The effects of 4-hydroxy-2,3-trans-nonenal on β-adrenoceptors of rat lung membranes, Chem.-Biol. Interactions 59:211 (1986).

5. K. Kramer, C.J.A. Doelman, H. Timmerman, and A. Bast, A disbalance between beta-adrenergic and muscarinic responses caused by hydrogen peroxide in rat airways in vitro, <u>Biochem. Biophys. Res. Comm.</u> 145:337 (1987).

6. N. Ben-Arie, C. Gileadi, and M. Schramm, Interaction of the beta-adrenergic receptor with Gs following delipidation, <u>Eur. J. Biochem.</u> 176:649 (1988).

7. A. Achari, D. Scott, P. Barlow <u>et al</u>., Facing up to membranes: structural function relationships in phospholipases, <u>in</u> Cold Spring Harbor Symposia on Quantitive Biology, Vol LII (1987).

8. C.J.A. Doelman, K. Kramer, H. Timmerman, and A. Bast, Vitamin E and selenium regulate balance between β-adrenergic and muscarinic responses in rat lungs, <u>FEBS Lett.</u> 233:427 (1988).

9. R. Anderson, A.J. Theron, and G.J. Ras, Ascorbic acid neutralizes reactive oxidants released by hyperactive phagocytes from cigarette smokers, <u>Lung</u> 166:149 (1988).

10. H. Wefers, and H. Sies, The protection by ascorbate and glutathione against microsomal lipid peroxidation is dependent on vitamin E, <u>Eur. J. Biochem.</u> 174:353 (1988).

11. C.G. Ramell, B. Cunliffe, and A.J. Kieboom, Determination of alpha-tocopherol in biological specimens by high-performance liquid chromatography, <u>J. Liq. Chromatogr.</u> 6:1123 (1983).

12. S.K. Jagota, and H.M. Dani, A new colorimetric technique for the estimation of vitamin C using folin phenol reagent, <u>Anal. Biochem.</u> 127:178 (1982).

13. O.W. Griffith, Determination of glutathione and glutathione disulfide using glutathione reductase and 2-vinylpyridine, <u>Anal. Biochem.</u> 106:207 (1988).

14. P. E. Brumby, and V. Massey, Determination of nonheme iron, total iron and copper, <u>J. Biol. Chem.</u> 229:763 (1957).

15. M. Bradford, A rapid and sensitive method for the quantitation of microgram quantities of protein utilizing the principle of protein-dye binding, <u>Anal. Biochem.</u> 72:248 (1976).

16. A. Wendel, Glutathione peroxidase, <u>Methods in Enzymol.</u> 77:325 (1981).

17. D.J. Worthington, and M.A. Rosemeyer, Human glutathione reductase: purification of the crystalline enzyme from erythrocytes, <u>Eur. J. Biochem.</u> 48:167 (1974).

18. J.M. McCord, and I. Fridovich, Superoxide dismutase, <u>J. Biol. Chem.</u> 244:6049 (1969).

19. G.R.M.M. Haenen, and A. Bast, Protection against lipid peroxidation by a microsomal glutathione-dependent labile factor, <u>FEBS Lett.</u> 159:24 (1983).

20. M.G. Mustafa, and D.F. Tierney, Biochemical and metabolic changes in the lung with oxygen, ozone, and nitrogen dioxide toxicity, <u>Am. Rev. Resp. Dis.</u> 118:1061 (1978).

21. J.R. Wright, H.D. Colby, and P.R. Miles, Cytosolic factors which affect microsomal lipid peroxidation in lung and liver, <u>Arch. Biochem. Biophys.</u> 206:296 (1981).

22. J.R. Wright, H.D. Colby, and P.R. Miles, Lipid peroxidation in guinea pig lung microsomes, <u>Biochim. Biophys. Acta</u> 619:374 (1980).

23. G. Minotti, and S.D. Aust, The requirement for iron (III) in the initiation of lipid peroxidation by iron (II) and hydrogen peroxide, J. Biol. Chem. 262:1098 (1987).

24. J.M. Braughler, R.L. Chase, and J.F. Pregenzer, Stimulation and inhibition of iron-dependent lipid peroxidation by desferrioxamine, Biochem. Biophys. Res. Comm. 153:933 (1988).

OXIDANTS, JOINT INFLAMMATION AND ANTI-INFLAMMATORY STRATEGIES

Ewa J Dowling, Vivienne R Winrow, Peter·Merry and
David R Blake

The Inflammation Group, The London Hospital Medical College
London E1 2AD   U.K.

INTRODUCTION

Biological reduction of molecular oxygen in cells is accompanied by the
production of dangerously reactive free radical and non-radical oxygen
species. Because of the ubiquity of molecular oxygen and its ability to
accept electrons, their production is readily associated with cellular
damage[1].

The superoxide radical (O_2^-) is the first univalent reduction product
of oxygen, which by dismutation via the enzyme superoxide dismutase (SOD) is
transformed to hydrogen peroxide (H_2O_2). H_2O_2 can easily penetrate the
membranes of surrounding cells, whereas O_2^- usually cannot. However, in the
presence of ions of a suitable transition metal (usually iron), H_2O_2 can
interact with the reduced form of the metal ion to form several highly
oxidising species, the most important of which is probably the hydroxyl
radical ($OH^·$). The hydroxyl radical is so highly reactive that it will
combine with whatever molecules are present at or close to its site of
formation[2]. Indeed it has been well documented that reactive oxygen
species (ROS) when stimulated in the environment of critical biomolecules
such as DNA, lipids, proteins and carbohydrates, promote oxidative damage[3].
It is hardly surprising, therefore, that there is currently considerable
evidence implicating ROS as mediators of tissue damage in inflammatory
processes[4], and more importantly in such diseases as rheumatoid arthritis[5].

ROS and Synovial Inflammation

Reactive oxygen metabolites, in vitro, are able to damage tissue
components essential for joint function[6]. For example, they are able to
damage cartilage-based proteoglycans[7] making them more susceptible to

Antioxidants in Therapy and Preventive Medicine
Edited by I. Emerit *et al.*
Plenum Press, New York, 1990

proteolytic attack and also depolymerise glycosaminoglycans (hyaluronic acid) altering synovial fluid viscosity[8]. Indeed in isolated systems almost all biomolecules are susceptible to radical damage[6,9] - exposure of many mammalian cells to H_2O_2 causes DNA damage; it has been suggested[10] that this damage occurs because H_2O_2 reacts with transition metal ions bound to the DNA to produce OH·, which immediately fragments the DNA molecule. Exposure of proteins to ROS will lead to denaturation, loss of function, cross-linking, aggregation and fragmentation. Again OH· generation at sites of transitional metal binding will produce localised site-specific damage. Methionine residues on ∝1-antiproteinase (the major circulating inhibitor of serine proteases such as elastase and collagenase) are oxidised by ROS, primarily hypochlorous acid (HOCl)[11,12], rendering the molecule biologically inactive.

Polyunsaturated fatty acids, which are integral structural components of cell membranes are particularly susceptible to radical and transition metal promoted peroxidation[13]. Lipid peroxidation, an autocatalytic chain reaction, may lead to decreased membrane fluidity and hence changes in cell function. Cell death can result from prolonged exposure to ROS and these cells, including phagocytic cells, will then release their inflammatory mediators into the extracellular space evoking a broad spectrum of inflammatory effects. Perhaps more significant in RA, exposure of human IgG to free radical generating systems, e.g. UV radiation, copper/H_2O_2, peroxidising lipids or activated neutrophils results in several marked changes to the immunoglobulins. They develop a characteristic autofluorescence and form monomeric and polymeric complexes capable of stimulating further radical production[14].

The involvement of ROS in promoting inflammatory processes _in vivo_ are evident by:

a) the measurement of products by peroxidative decomposition of lipid (principally malondialdehyde as thiobarbituric acid (TBA) reactive material) at the site of injury, plasma and liver; and also by the detection of ethane and pentane, a non-invasive indicator of _in vivo_ lipid peroxidation in exhaled air[15].

b) the detection of modified protein characteristic of radical damage[16].

c) the measurement of high levels of free radical activity, as intense luminol-amplified chemiluminescence (LAC) directly from the inflamed site in an experimental model of foot pad oedema[17]. Certainly in patients with inflammatory joint disease, lipid peroxidation products (diene conjugates and TBA-reactive material) were found to be elevated in both their synovial fluid and sera[18,19]. Similarly levels of protein consistent with being radical-altered IgG were also significantly raised in both synovial fluid and sera[14]. Interestingly, Humad et al[20] found increased levels of pentane expired by patients with RA, and the amount of pentane exhaled was directly proprotional to the activity of the disease.

A notable illustration of the potential pro-inflammatory action of ROS in the context of arthritis is provided by the condition alcaptonuria, caused by a defect of the enzyme homogentisate oxidase which is involved in tyrosine metabolism. Alcaptonuric patients accumulate homogentisic acid, show pigmentation of cartilage and connective tissue, and may gradually develop severe inflammatory arthritis. Martin and Batkoff[21] showed that oxidation of homogentisic acid generates O_2^- and H_2O_2 as well as $OH^{\cdot}$ if iron ions are present. They therefore proposed that the arthritis of alcaptonuria is induced by increased radical production.

HYPOXIC REPERFUSION INJURY AND INFLAMMATORY JOINT DISEASE

There seems to be little doubt that oxidant damage occurs in the inflamed rheumatoid joint. However, oxidative damage requires O_2 and the fact that the oxygen tension in inflammatory synovial fluid is low[22] might suggest that cells will have a depressed rather than an increased ability to generate ROS[23]. To address the paradox that oxidative damage seems to arise in a hypoxic environment, we have speculated that inflammatory synovitis is an example of hypoxic reperfusion injury[24].

Hypoxic-reperfusion injury is a known mechanism of ROS production. In brief this process described in detail by McCord[25] suggests that on the restoration of blood supply following a transient ischaemic event, ROS are generated by the uncoupling of a variety of intracellular redox systems.

It has been shown that transient ischaemic injury to the small bowel of cats results in increased intestinal capillary permeability and albumin clearance[26]. However, the increased intestinal capillary permeability that is observed during reperfusion can be blocked in these animals by pre-dosing with the radical scavenging enzymes SOD or catalase, or allopurinol, a xanthine oxidase inhibitor[27]. In this animal model, tissue injury is thought to be related to the generation of O_2^-, H_2O_2 and $OH^{\cdot}$. The effect of ischaemia on myocardial function in various animals has been modified by treatment with SOD, catalase, and mannitol, all putative scavengers of ROS[28].

The mechanism for the production of O_2^- in ischaemic tissues appears to be effected by changes in purine metabolism within ischaemic cells. During temporary ischaemia low oxygen concentrations halt mitochondrial oxidative phosphorylation, and cellular ATP production becomes dependent on anaerobic glycolysis. This is an inefficient means of ATP production from glucose, and leads to raised levels of adenosine and of its breakdown products, including hypoxanthine and xanthine which are the substrates for the xanthine oxido-reductase enzyme system[29].

Xanthine oxidoreductase is a cytosolic enzyme normally oxidising hypo-xanthine and xanthine to uric acid. It is a well described oxygen free radical generating system and is known to be instrumental in models of hypoxic reperfusion injury[30].

Hypoxic-reperfusion injury has been applied to many disease states, including, transient coronary or cerebral ischaemia and ischaemic acute renal failure. Animal models support the view that similar processes should occur in man, but direct evidence has been lacking. We have recently developed a human model that demonstrates hypoxic-reperfusion injury occurring within the inflamed human joint[33].

Hypoxic-reperfusion Injury and Inflammatory Synovitis

Several physiological and biochemical features present within the inflamed joint suggest that exercise will provide the potential environment for hypoxic-reperfusion injury.

The intra-articular pressure (IAP) in the normal knee joint of both humans and animals is at or slightly below atmospheric pressure[34]. Jayson and Dixon[35] demonstrated that in the normal joint, quadriceps contraction produces a subatmospheric pressure. In contrast, patients with rheumatoid arthritis (RA) had significantly higher resting pressures than control subjects with a simulated effusion of the same volume[31]. On quadriceps setting RA patients produced IAPs as high as 200mmHg, well in excess of the synovial capillary perfusion pressure of 30-60mmHg[35]. We have recently supported these findings, and also shown a dynamic inverse relationship between SF oxygen tension and IAP in inflammatory synovitis[33].

The pO2 in inflamed joints has been measured by several research groups, and it is agreed that inflammatory effusions have lower oxygen tensions than non-inflammatory effusions[22,36,37]. Also, the severity of the hypoxia correlates with the synovial histological findings of synovial cell proliferation, focal necrosis and focal obliterative microangiopathy.[38]

One determinant of SF oxygen tension is the blood supply to the synovium. This should be related to the IAP, as the synovial membrane and joint capsule form a closed environment in which the IAP is transmitted directly to the synovial membrane vasculature. We have recently studied microvascular perfusion dynamics within the synovium of the knee during exercise using laser doppler flowmetry. In the normal knee there was a negligible reduction of capillary perfusion during exercise. In contrast, exercise of the inflamed knee produced occlusion of the synovial capillary bed for the duration of the exercise period. Reperfusion of the synovial membrane occurred on cessation of exercise[33].

A recent study has shown that endothelial cells subjected to anoxia/ reoxygenation cycles generated superoxide-derived hydroxyl radicals leading to endothelial cell injury[39]. In this study the radicals were generated by xanthine oxidoreductase, an enzyme system we have previously demonstrated to be present in both normal and diseased synovial tissue[40].

Thus exercise of the inflamed human knee joint provides the potential pathophysiological environment for the promotion of hypoxic-reperfusion injury. We have recently verified this by demonstrating exercise induced oxidative damage to lipids and IgG within the knee joint of patients with inflammatory synovitis[33].

There is, therefore, strong evidence to support the hypothesis that the peculiar persistence of synovial inflammation is a consequence of exercise induced, radical promoted hypoxic-reperfusion injury[24].

ANTIOXIDANTS AND INFLAMMATORY JOINT DISEASE

To circumvent the destructive damage caused by the production of ROS from numerous enzymatic and non-enzymatic biological processes, multiple defence systems have arisen. The cytochrome oxidase system localised on the inner mitochondrial membrane tetravalently reduces the major portion of oxygen produced by aerobic cells. The second line of defence is provided by enzymes that catalytically scavenge the intermediates of O_2 reduction. O_2^- is eliminated by superoxide dismutase (SOD). In mammals two types of SOD have been described, one cytosolic (CuZn SOD) and the other mitochondrial (Mn SOD). Extracellular levels of SOD are very low although Marklund[41] has described the presence of a glycoprotein CuZn enzyme called 'Extra-cellular SOD' (EC - SOD) in human plasma. It has been proposed that EC - SOD is normally present attached to the surfaces of endothelial cells, offering localised protection[42]. This ability of EC - SOD to bind to cell surfaces does suggest that it might act as an effective antioxidant in protecting against reperfusion injury[42].

The subsequent formation of H_2O_2 by SOD is removed intracellularly by catalases and peroxidases to form water. Finally these enzymic systems are supplemented by a variety of biochemical defence systems, such as vitamin E, ascorbic acid, glutathione and caeruloplasmin which provide further protection from uncontrolled free radical attack. Simplistically, from the data, one might expect antioxidants which act as free radical scavengers to be effective in the treatment of inflammation. Indeed, numerous reports have shown that free radical scavengers and antioxidants alleviated symptoms of both clinical[43-45] and experimental inflammation[46-48].

Purified bovine copper or zinc SOD has been formulated for clinical use as Orgotein. The suppression of the inflammatory response by exogenous SOD has been demonstrated in certain animal models[49]. The administration of Orgotein has also produced clinical improvements in certain arthritic conditions[50] implying a pathogenetic role for O_2^-; but this data has been criticised[51]. Progress in this area has recently been extensively reviewed by Flohe[45]. More recently human SOD in combination with human catalase, Epurox, has been reported to give encouraging results in preliminary trials[52]. Other clinical studies have reported that in patients with inflammatory joint disease, high doses of vitamin E provided increased protection of tissues against chronic high levels of ROS[53].

Several observations suggest a role for iron-dependent radical reactions in promoting inflammatory joint disease. Mild nutritional iron deficiency significantly reduces the severity of adjuvant joint inflammation in rats[54], whilst the iron chelator, desferrioxamine, has been shown to be anti-inflammatory in both Glynn-Dumonde synovitis in guinea pigs, and adjuvant arthritis in rats[55]. In RA, infusion of iron complexes are known to exacerbate synovitis[56], and 'bleomycin iron' is detectable in synovial fluid, the concentration correlating loosely with clinical and laboratory parameters of disease activity[5]. Desferrioxamine is used clinically in the treatment of iron overload. Preliminary clinical trials of the compound in RA were curtailed by toxicity[57].

Non-steroidal anti-inflammatory drugs (NSAID) are the drugs most commonly used for the symptomatic relief of rheumatoid arthritis. Many reports, however, have provided substantial evidence to suggest that several of these drugs act as effective free radical scavengers, e.g. $OH^\cdot$ by salicylate[58], and this inhibition of O_2^- derived radicals may contribute to the anti-inflammatory action of these agents[59,60].

D-penicillamine, a disease modifying drug, has been used as an effective treatment for patients with RA. A number of mechanisms underlying its beneficial effects have been attributed to a number of its actions, such as its anti-inflammatory action[61], interference with leucocyte chemotaxis[62], and a SOD-like action[63]. More recently it has been reported that this agent may synthetise H_2O_2 in the presence of copper; and this generation of H_2O_2 has been shown to inhibit T cell function[64] and have a pronounced anti-angiogenic effect through inhibition of endothelial cell proliferation[65]. These authors have proposed that the reduction of rheumatoid synovitis by

D-penicillamine may be due to the reduction of small blood vessels available for the emigration of chronic inflammatory cells, and hence the proliferation of synovial tissue.

Stress Proteins and Inflammatory Joint Disease

Without doubt, aerobic organisms are continually exposed to many potentially damaging oxidative reactants and clearly numerous defence mechanisms exist. More recently, numerous reports have suggested a major family of proteins, namely the heat shock (stress) proteins, to have a protective role in inflammation and more specifically in diseases such as rheumatoid arthritis[66,67]. The heat shock protein response was first observed in the larvae of the fruit fly Drosophila busckii[68]. Under conditions of elevated temperature, chromosome 'puffs' were observed, general protein synthesis was switched off and a set of specific mRNAs were translated to produce the heat shock proteins (hsps). It is now well established that hsps occur in all living organisms and are produced follow- ing not only heat shock but also other conditions of physiological stress[69]. Indeed, it is now known that mammalian cells when subjected to a number of stressors including hypoxia, ROS and cytokines, generate stress proteins of variable molecular weight; known functions include the removal of incorrectly folded or glycosylated (i.e. damaged) proteins and a protective effect in aiding recovery of the cell from physiological stress[66,70].

Collaborators in Israel and Holland have shown that adjuvant arthritis induced in rats is associated with the 65kD hsp of mycobacteria, the major component of Freund's adjuvant[71]. We have shown that antibodies to myco- bacterial 65kD proteins bind to synovial membrane intimal cells and vessel endothelial cells. More importantly, a human 72kD homologue, which is specifically induced following physiological stress is also present in these cells[72]. As discussed earlier, normal movement of an inflamed joint gener- ates sufficient pressure that transient synovial capillary ischaemia inevitably follows. As the pressure falls, reperfusion occurs, leading to the release of oxidants (including H_2O_2 and $OH^{\cdot}$), which have the capacity to denature proteins[14]. Transient anaerobic exposure leads to the induc- tion of glucose regulated proteins, a group of stress proteins induced by glucose deprivation. Reoxygenation induces the major heat shock proteins, which in humans will include the 72kD protein[73]. Thus, both H_2O_2 and oxidatively modified proteins will serve as stress signals and trigger the activation of heat shock genes. Since H_2O_2 specifically induces the production of the 32kD stress protein haeme oxygenase[74], further damage from haeme iron catalysed hydroxyl radical formation could be abrogated. It is interesting to note that penicillamine acts by generating low levels of H_2O_2[65].

Other evidence includes the observation that chondrocytes from patients with osteoarthritis produce 70kD and 90kD proteins[75] and sera from patients with rheumatoid arthritis recognise 70kD and 28kD antigens in extracts of synoviocytes[76]. In addition raised IgG and IgA antibodies have been detected to human 70kD and mycobacterial 65kD hsps[77].

SUMMARY

In summary, we support the hypothesis that the rheumatoid knee is subjected to repeated hypoxic reperfusion injury. Moreover, we speculate that oxidative damage induced as a consequence of hypoxic reperfusion injury leads to the synthesis of intracellular stress proteins. In certain individuals this triggers an autoimmune response which might explain the persistance of the rheumatoid disease process[67]. This gives scope for novel therapeutic approaches in the future development of anti-inflammatory agents.

REFERENCES

1. B. A. Freeman and J. D. Crapo, Biology of disease: free radicals and tissue injury, Lab. Invest. 47: 412-426 (1982).
2. B. Halliwell and J. M. C. Gutteridge, Iron as a biological pro-oxidant, ISI Atlas Sci. Biochem. 1: 48-52 (1988).
3. J. M. C. Gutteridge, Free radical damage to lipids, amino acids, carbohydrates and nucleic acids determined by TBA reactivity, Int. J. Biochem. 14: 649-653 (1982).
4. T. F. Slater, Free radical mechanisms in tissue injury, Biochem. J. 222; 1-15 (1984).
5. D. Rowley, J. M. C. Gutteridge, D. R. Blake and B. Halliwell, Lipid peroxidation in rheumatoid arthritis: TBA activity and catalytic iron salts in synovial fluid from RA. Clin. Sci. 66: 691-695 (1984).
6. D. R. Blake, R. E. Allen and J. Lunec, Free radicals in biological systems - A review orientated to inflammatory processes. Br. Med. Bull. 43: 371-385 (1987).
7. H. Burkhardt, M. Schwingel, H. Menninger, H. W. Macartney and H. Tschesche, Oxygen radicals are effectors of cartilage destruction, Arth.Rheum. 29: 379-387 (1986).
8. R. A. Greenwald and W. W. Moy, Effect of oxygen derived free radicals on hyaluronic acid, Arth.Rheum. 23: 455-463 (1980).
9. B. Halliwell, J. M. C. Gutteridge and D. R. Blake, Metal ions and oxygen radical reactions in human inflammatory joint disease. Phil. Trans. R. Soc. London, 311: 659-671 (1985).
10. M. Larramendy, A. C. Mello-Filho, E. A. L. Martins and R. Meneghini, Iron-mediated induction of sister-chromatid exhanges by hydrogen peroxide and superoxide anion. Mutat. Res. 178: 57-63 (1987).
11. H. Carp and A Janoff, In vitro suppression of serum elastase inhibitory capacity by reactive oxygen species generated by phagocytosing poly-morphonuclear leucocytes, J. Clin. Invest. 63: 793-797 (1979).
12. B. Halliwell, J. R. Hoult and D. R. Blake, Oxidants, inflammation and anti-inflammatory drugs, FASEB J. 2: 2867-2873 (1988).
13. J. P. Clavel, J. Emerit and A. Thullier, Lipid peroxidation and free radicals. Role in cellular biology and in pathology, Path. Biol. 33: 61-69 (1985).

470

14. J. Lunec, D. R. Blake, S. J. McCleary, S. Brailsford and P. A. Bacon, Self-perpetuating mechanisms of immunoglobulin G aggregation in rheumatoid inflammation, J. Clin. Invest. 76: 2084-2090 (1985).

15. E. J. Dowling, The role of lipid peroxidation in inflammation, PhD Thesis, University of Surrey, Guildford, U.K. (1985).

16. J. Lunec, A. Wakefield, S. Brailsford and D. R. Blake, Free radical altered IgG and its interaction with rheumatoid factor, in: Free radicals, cell damage and disease, C. Rice-Evans, ed., Richelieu Press, London, 241-261 (1986).

17. E. J. Dowling, A. M. Symons and D. V. Parke, Free radical production at site of an acute inflammatory reaction as measured by chemiluminescence, Agents Actions, 19: 203-207 (1986).

18. G. Rowley, J. M. C. Gutteridge, D. R. Blake, M. Farr and B. Halliwell, Lipid peroxidation in RA, thiobarbituric acid reactive material and catalytic iron salts in synovial fluid from rheumatoid patients. Clin. Sci. 66: 691-695 (1984).

19. J. Lunec, S. P. Halloran, A. G. White and T. L. Dormandy, Free radical oxidation (peroxidation) products in serum and synovial fluid in RA, J. Rheum. 8: 233-245 (1981).

20. S. Humad, E. Zarling, M. Clapper and J. L. Skosey, Breath pentane excretion as a marker of disease activity in rheumatoid arthritis, Free Rad. Res. Comm. 5: 101-6 (1988).

21. J. P. Martin and B. Batkoff, Homogentisic acid autoxidation and oxygen radical generation: implications for the aetiology of alcaptonuric arthritis, Free Rad. Biol. Med. 3: 241-250 (1987).

22. K. Lund-Oleson, Oxygen tension in synovial fluids, Arth.Rheum. 13: 769-776 (1970).

23. S. W. Edwards, M. B. Hallet and A. K. Campbell, Oxygen radical production may be limited by oxygen concentration, Biochem. J. 217: 851-854 (1984).

24. T. Woodruff, D. R. Blake, J. Freeman, F. J. Andrews, P. Salt and J. Lunec, Is chronic synovitis an example of reperfusion injury? Ann. Rheum. Dis. 45: 608-611 (1986).

25. J. M. McCord, Oxygen-derived free radicals in postischaemic tissue injury, New Eng. J. Med. 312: 159-163 (1985).

26. D. N. Granger, M. Sennet et al, Effect of local arterial hypotension on cat intestinal permeability, Gastroenterology, 79: 474-480 (1980).

27. D. N. Granger, G. Rutili and J. M. McCord, Superoxide radicals in feline intestinal ischaemia, Gastroenterology, 81: 22-29 (1981).

28. J. R. Stewart, W. H. Blackwell, S. L. Crute, V. Loughlin et al, Prevention of myocardial ischaemia reperfusion injury with oxygen free radical scavengers, Surg. Forum, 33: 317-320 (1982).

29. R. B. Jennings, K. A. Reimer, M. A. Hill and S. E. Mayer, Total ischaemia in dogs hearts I. A comparison of high energy phosphate production, utilisation and depletion and of adenine nucleotide catabolism in total ischaemic in vitro, Circ. Res. 49: 892-900 (1981).

30. E. Della Corte and F. Stirpe, The regulation of rat liver xanthine oxidase: involvement of thiol groups in the conversion of the enzyme activity from the dehydrogenase (type D) to the oxidase (type O) and purification of the enzyme, Biochem. J. 126: 739-743 (1972).

31. W. Cao, J. M. Carney, A. Duchon, R. A. Floyd and M Chevion, Oxygen free radical involvement in ischaemia and reperfusion injury to brain, Neuroscience Letters, 88: 233-238 (1988).

32. C. Canavese, P. Stratta and A Vercellone, The case for oxygen free radicals in the pathogenesis of ischaemic acute renal failure, Nephron. 49: 9-15 (1988).

33. D. R. Blake, P. Merry, J. Unsworth, B. Kidd, J. M. Outhwaite, R. Ballard, C. J. Morris, L. Gray and J. Lunec, Hypoxic-reperfusion injury in the inflamed human joint, Lancet i: 289-293 (1989).

34. M. I. V. Jayson, A. St.J. Dixon, Intra-articular pressure in rheumatoid arthritis of the knee I. Pressure changes during passive joint distension, Ann. Rheum. Dis. 29: 261-265 (1970).

35. M. I. V. Jayson and A. St.J. Dixon, Intra-articular pressure in rheumatoid arthritis of the knee II. Pressure changes during joint use. Ann. Rheum. Dis. 29: 401-408 (1970).

36. P. S. Treuhaft and D. J. McCarty, Synovial fluid pH, lactate, oxygen and carbon dioxide partial pressure in various joint diseases. Arthritis Rheum. 14: 475-484 (1971).

37. A. I. Richman, E. Y. Su and G. Ho, Reciprocal relationship of synovial fluid volume and oxygen tension, Arthritis Rheum. 24: 701-705 (1981).

38. K. H. Falchuk, E. J. Goetzl and N. P. Kulka, Respiratory gases of synovial fluids, Am. J. Med. 49: 223-231 (1970).

39. J. L. Zweier, P. Kuppusamy and G. A. Lutty, Measurement of endothelial cell free radical generation. Evidence for a central mechanism of free radical injury in postischaemic tissues, Proc. Natl. Acad. Sci. U.S.A. 85: 4046-4050 (1988).

40. R. E. Allen, J. M. Outhwaite, C. J. Morris and D. R. Blake, Xanthine oxidoreductase is present in human synovium, Ann. Rheum. Dis. 46: 843-845 (1987).

41. S. L. Marklund, E. Holme and L. Hellner, Superoxide dismutase in extra-cellular fluids, Clin. Chim. Acta, 126: 41-51 (1982).

42. K. Karlsson and S. L. Marklund, Heparin-induced release of extracellu-lar superoxide dismutase to human blood plasma, Biochem. J. 242: 55-59 (1987).

43. S. L. Marklund, SOD, catalase and GSH peroxidase in degenerative disease, Bull. Europ. Physiopath. Resp.17: 259 (1981).

44. E. Munthe, E. Kass, E. Jellum, Evidence for enhanced radical scaveng-ing prior to drug response in RA, Adv. Inflam. Res. 3: 211-235 (1982).

45. L. Flohe, Superoxide dismutase for therapeutic use: clinical experience, dead ends and hopes, Molec. Cell Biochem, 84: 123-131 (1988).

46. J. M. McCord, Free radicals and inflammation protection of synovial fluid by SOD, Science, 185: 529-531 (1974).

47. M. U. Dianzani, M. V. Torrielli, L. Paradisi and J. S. Franzone, Influence of antioxidants and oxygen scavengers on experimental inflammatory processes, Eur. J. Rheum. Inflamm. 1: 187-196 (1978).

48. K. Hirschelmann and H. Bekemeier, Effects of catalase, peroxidase, SOD and 10 scavengers of O_2 radicals in carrageenin oedema and in adjuvant arthritis in rats, Experientia, 37: 1313 (1981).

49. A. M. Michelson, K. Puget, G. Jadot, Anti-inflammatory activity of superoxide dismutase: comparison of enzymes from different sources in different models in rats; mechanism of action, Free Radical Res. Commun. 2: 43-56 (1986).

50. W. Huber, Orgotein - (bovine Cu-Zn, SOD) an anti-inflammatory protein drug: discovery, toxicology and pharmacology, Europ. J. Rheum. Inflamm. 4: 173-192 (1981).

51. R. A. Greenwald, Therapeutic benefits of oxygen radical scavenger treat-ments remain unproven, J. Free Rad. Biol. 1: 173-177 (1985).

52. G. Szegli, A Herold, E. Negut et al, Clinical efficacy of a new anti-inflammatory drug with free radical scavenging properties: superoxide dismutase (SOD) and catalase of human origin, Arch. Roum. Pathol. Exp. Microbiol. 45: 75-89 (1986).

53. K. H. Schmidt, Efficacy of vitamin E as a drug in inflammatory joint diseases, SFRR Abstracts, Paris Meeting 1988.

54. F. J. Andrews, Effect of nutritional iron deficiency on acute and chronic inflammation, Ann. Rheum. Dis. 46: 859-865 (1987).

55. F. J. Andrews, C. J. Morris, G. Kondratowicz and D. R. Blake, Effect of iron chelation on inflammatory joint disease, Ann. Rheum. Dis. 46: 327-333 (1987).

56. P. G. Winyard, D. R. Blake, S. Chirico, J. M. C. Gutteridge and J. Lunec, Mechanism of exacerbation of rheumatoid synovitis by total-dose iron-dextran infusion: in vivo demonstration of iron-promoted oxidative stress, Lancet i: 69-72 (1987).

57. D. R. Blake, P. Winyard, J. Lunec et al, Cerebral and ocular toxicity induced by desferrioxamine, Quart. J. Med. 56: 345-355 (1985).

58. M. Grootveld and B. Halliwell, Aromatic hydroxylation as a potential measure of hydroxyl radical formation in vivo. Identification of hydroxylated derivatives of salicylate in human body fluids, Biochem. J. 237: 499-504 (1986).

59. E. J. Dowling, A. M. Symons and M. K. Jasani, the ex-vivo measurement of malondialdehyde and chemiluminescence as possible indices for anti-inflammatory drug evaluation, Int. J. Tiss. Reac. IX(5): 385-391 (1987).

60. B. Halliwell, J. R. Hoult and D. R. Blake, Oxidants, inflammation and anti-inflammatory drugs, FASEB J. 2: 2867-2873 (1988).

61. J. R. J. Sorenson, Copper chelates as possible active forms of the anti-arthritic agents, J. Med. Chem. 19: 135-148 (1976).

62. H. Chwalinska-Sadowska and J. Bacon, The effect of D-penicillamine on polymorphonuclear function, J. Clin. Invest. 58: 871-879 (1976).

63. E. Longfelder and E. F. Elstner, Determination of the superoxide dismutating activity of D-penicillamine copper, Hoppe-Seyler's J. Physiol. Chem. 359: 751-757 (1978).

64. P. E. Lipsky, Immunosuppression by D-penicillamine in vitro. Inhibition of human T lymphocyte proliferation by copper - or caeruloplasmin - independent generation of hydrogen peroxide and protection by monocytes, J. Clin. Invest, 73: 53-65 (1984).

65. T. Matsubara, R. Saura, K. Hirohata and M. Ziff, Inhibition of human endothelial cell proliferation in vitro and neovascularisation in vivo by D-penicillamine, J. Clin. Invest. 83: 158-167 (1989).

66. B. S. Polla, A Role for heat shock proteins in inflammation, Immunology Today, 9: 134-137 (1988).

67. V. R. Winrow, I. L. McLean, C. J. Morris and D. R. Blake, The heat shock protein response and its role in inflammatory disease, Ann. Rheum. Dis. 1989 (in press).

68. F. Ritossa, A new puffing pattern induced by temperature shock and DNP in Drosophila, Experientia, 18: 571-3 (1962).

69. S. Lindquist, The heat shock response, Ann. Dev. Biochem. 55: 1151-1159 (1986).

70. W. J. Welch, The mammalian heat shock (or stress) response: A cellular defence mechanism, Adv. Exp. Mod. Biol. 228: 287-304 (1987).

71. W. van Eden, J. E. R. Thole, R van der Zee et al, Cloning of the myco-bacterial epitope recognised by T lymphocytes in adjuvant arthritis, Nature, 331: 171-173 (1988).

72. I. L. McLean, V. R. Winrow, P. I. Mapp, A. H. Cherrie, J. R. Archer and D. R. Blake, Synovial fluid T cells and 65kD heat-shock protein, Lancet ii: 856-857 (1988).

73. J. J. Sciandra, J. R. Subjeck and C. S. Hughes, Induction of glucose-regulated proteins during anaerobic exposure and of heat-shock proteins after reoxygenation, Proc. Natl. Acad. Sci. (USA) 81: 4843-7 (1984).

74. S. M. Keyse and R. M. Tyrrell, Haeme oxygenase is the major 32kD stress protein induced in human skin fibroblasts by UVA radiation, hydrogen peroxide and sodium arsenite, Proc. Natl. Acad. Sci. (USA) 86: 99-103 (1989).

75. T. Kubo, C. A. Towle, H. J. Makin and B. V. Tradwell, Stress-induced proteins in chrondrocytes from patients with osteoarthritis, Arthritis Rheum. 28: 1140-5 (1985).

76. B. Perez-Maceda, C. Bernaben, J. P. Lopez-Bote, A. Marquet and V. Larraga, Autoantibodies from rheumatoid arthritis patients recognise antigens on the synoviocyte surface, Scand. J. Immunol. 27: 295-304 (1988).

77. G. Tsoulfa, G. A. W. Rook and J. D. A. van Embden et al, Raised serum IgG and IgA antibodies to mycobacterial antigens in rheumatoid arthritis, Ann. Rheum. Dis. 48: 118-23 (1988).

ANTIOXIDANT THERAPY IN NEUROLOGICAL DISORDERS

D. P. R. MULLER

Department of Child Health
Institute of Child Health
30 Guilford Street
LONDON   WC1N 1EH, U.K.

INTRODUCTION

Oxygen derived free radical species have been implicated in an increasing number of disease processes (Halliwell and Gutteridge 1985a) including neurological disorders (Halliwell and Gutteridge 1985b), and as a result antioxidant therapy has been suggested for a number of diseases affecting the nervous system. In vivo antioxidant defenses can be divided into 2 categories. Firstly preventative antioxidants such as catalase and glutathione peroxidase which prevent the formation of free radical species, and secondly chain breaking antioxidants such as superoxide dismutase, ascorbate, urate and vitamin E (alpha-tocopherol) which trap oxygen derived free radicals and halt the chain reaction. This paper will concentrate on the therapeutic role of vitamin E in neurological disorders, as in practice the administration of this fat soluble vitamin is a simple and safe way of altering antioxidant status in vivo. Alpha-tocopherol is also potentially important as it is the only well recognised lipid soluble chain breaking antioxidant in vivo (Burton et al 1983) and may, therefore, be expected to play an important role in such highly lipid structures as the brain, spinal cord and peripheral nerves.

There are three situations where serum vitamin E concentrations are reduced and where supplements might be expected to be beneficial in preventing deficiency syndromes including neurological disease. These are patients with a) chronic fat malabsorption and b) a selective vitamin E deficiency and c) the newborn, in particular the premature infant. It has also been suggested that antioxidant therapy may be beneficial in certain neurological disorders (e.g. tardive dyskinesia, Parkinson's disease and the neuronal ceroid lipofuscinoses) where serum vitamin E concentrations are not reduced but where free radical mediated processes

Antioxidants in Therapy and Preventive Medicine
Edited by I. Emerit *et al.*
Plenum Press, New York, 1990

have been implicated. The role of vitamin E therapy in these conditions and situations will be discussed in turn.

CHRONIC FAT MALABSORPTIVE STATES

As vitamin E is a fat soluble vitamin, concentrations would be expected to be reduced in chronic fat malabsorptive states. In a study of groups of children with specific defects affecting fat absorption it was found that they all had mean serum vitamin E concentrations which were significantly reduced below normal (Muller et al., 1974). All the patients with abetalipoproteinaemia (an inborn error of lipoprotein metabolism) had undetectable serum concentrations of vitamin E from birth (Muller et al., 1974), and they, therefore, provide an ideal model for the study of the role of vitamin E in human nutrition. Amongst the clinical features of abetalipoproteinaemia are an ataxic neuropathy and pigmentary retinapathy which typicallly develop during the second decade of life. These features are progressive leading eventually to crippling and blindness and have been described by Herbert et al (1978) as "devastating". No cases of spontaneous improvement have been reported. The neurological syndrome is characteristic and involves the central and peripheral nervous system, the retina and muscles. It comprises areflexia, cerebellar ataxia, distal loss of position sense, loss of vibration sense, pes cavus, scoliosis, abnormalities of eye movements, a pigmentary retinopathy and generalised muscle weakness (Muller et al., 1983). Because of the absence of vitamin E in this condition and the fact that a deficiency of this vitamin in experimental animals can cause similar neurological sequelae (Nelson et al., 1981), we decided to treat our patients with abetalipoproteinaemia with large oral doses of vitamin E. A dose of approximately 100 mg/kg/day (normal requirements 10-30 mg/day) was required to achieve an adequate vitamine E status and we (Muller et al., 1977) and others (Azizi et al., 1978; Herbert et al., 1978) have shown that if such therapy is given sufficiently early, the development of the neurological features can be prevented, or if already present, progression can be halted and in some cases reversed.

Identical neurological findings have also been reported in patients with very low serum vitamin E concentrations and fat malabsorption as a result of liver disease, intestinal resection and cystic fibrosis (Muller et al., 1983). If appropriate vitamin E supplementation is given (see later), the response is similar to that in abetalipoproteinaemia. Evidence for a causal relationship between vitamin E deficiency and the neurological sequelae was provided by a study by Sokol et al (1985) who followed the effect of vitamin E supplementation in children with cholestatic liver disease. Two young patients (less than 3 years of age) remained normal after 18 months of therapy. Neurological function in 3 symptomatic children also less than 3 years normalised after 18 - 32 months of treament and in a group of older symptomatic children (5-17½ years) there was improvement (using a clinical scoring system) after 18-48 months of treatment.

SELECTIVE VITAMIN E DEFICIENCY

In recent years a few patients have been reported with a severe deficiency of vitamin E but without generalised fat malabsorption (Harding et al., 1985; Sokol et al., 1988). They have the same neurological syndrome as described above and responded in a similar manner to vitamin E supplements. As there does not appear to be any evidence for a deficiency of any other nutrient, these patients provide further evidence for a causal relationship between a deficiency of vitamin E and the neurological sequelae.

PREMATURE INFANT

Many studies have documented reduced serum concentrations of vitamin E in the newborn and particularly the premature infant compared to both adults in general and also the maternal concentration (see Muller 1987). Initially it was thought that this resulted from an impaired permeability of the placenta to alpha-tocopherol but it is now thought to be a function of a reduced transport capacity in the blood of the neonate. Thus Haga et al (1982) reported that the ratio of serum concentrations of alpha-tocopherol to low density lipoprotein (a major carrier of alpha-tocopherol) is constant and similar in cord and maternal blood. As vitamin E is a lipid soluble antioxidant it may be more relevant to express concentrations per lipid. When total body tocopherol is related to total body lipid, the ratio is similar for the foetus, newborn and adult (Bell and Filer 1981). It is, therefore, arguable whether the "functional" vitamin E status is reduced in the newborn.

Prophylactic, pharmacological doses of vitamin E have been suggested for the possible prevention of two neurological disorders (periventricular haemorrhage and retinopathy of prematurity) commonly seen in very premature infants, and the use of such antioxidant therapy will now be discussed.

Periventricular Haemorrhage

Periventricular haemorrhage is the most common finding in the brains of premature babies who die during the first week of life and is probably one of the most important lesions responsible for handicap in surviving infants. In a preliminary pilot study, Chiswick et al (1983) reported that intramuscular vitamin E reduced the incidence of periventricular haemorrhage in premature infants under 32 weeks of gestation. This observation was confirmed in a subsequent randomised controlled trial of intramuscular vitamin E in a total of 231 babies where Chiswick and colleagues (Sinha et al., 1987) reported that the supplemented babies had a significantly lower frequency of periventricular haemorrhage than controls (10.8 v 40.7% respectively, p<0.001).

Two studies designed to assess the role of vitamin E in the prevention of retinopathy of prematurity have, however, given conflicting results with regard to the incidental

protection against periventricular haemorrage. Thus Speer
et al (1984) observed a positive effect, whereas Phelps et
al (1987) found that patients given intravenous vitamin E
had a higher frequency of severe grades of periventricular
haemorrage. The route and dose of vitamin E in the latter
study has been criticised (Johnson et al., 1988), but more
randomised controlled trials are clearly required to prove
the efficacy of vitamin E in protecting against this
condition.

The mechanism underlying the periventricular
haemorrhage is not understood. Sinha et al (1987) have,
however, speculated that it results from ischaemia and
subsequent reperfusion of the subependymal region of the
brain which is particularly vulnerable to ischaemia as a
result of systemic hypotension, a common finding in ill
preterm babies. The oxygen derived free radicals which
would normally be produced following the subsequent
reperfusion phase would be scavenged by the vitamin E and
thus tissue damage and the extent of the haemorrhage may be
limited.

Retinopathy of Prematurity (Retrolental Fibroplasia)

This retinal disease of the preterm infant was first
described by Terry (1942) and subsequent studies indicated
that the condition resulted principally from excessive use
of oxygen (Kinsey 1956) although other factors undoubtedly
contribute. Following a reduction in the use of oxygen, the
number of cases of retinopathy of prematurity decreased
dramatically in the 1960s but this was associated with an
increased mortality and morbidity particularly from
respiratory distress syndrome. In recent years an increase
in the survival of low-birth-weight infants has again led to
an increased incidence of the condition. Owens & Owens
(1949) were the first to suggest that prophylactic vitamin E
supplementation could reduce the incidence and severity of
the retinopathy but at that time other workers were unable
to repeat their findings. More recently, Johnson et al
(1974), in a controlled but non-random study, reported a
significant decrease in severity of the disease in infants
receiving vitamin E. This prompted a number of independent,
double-blind, randomised controlled trials which have
recently been reviewed (Aranda et al., 1986). A number of
these studies came to broadly similar conclusions; that
administration of vitamin E may reduce the severity but not
the overall incidence of retinopathy of prematurity.

Cost Benifit Relationship of Vitamin E Therapy in the
Premature Infant

Phelps (1982) has estimated that in the US, 22,000 of
the 37,000 infants born weighing less than 1500g survive,
and that of these, 2000 (9.5%) would have severe retinopathy
of prematurity and 500 (2.5%) would be blind. If,
therefore, prophylactic vitamin E is given to all premature
infants, a very large percentage would receive the vitamin
who would not develop the severe form of the condition.
This is acceptable if it is certain that such
supplementation is completely safe or that the benefits

strongly outweigh the risks. Reports have, however,
appeared which suggest that there may be risks associated
with the prophylactic administration of vitamin E to the
premature infant. Thus an increased incidence of
necrotising enterocolitis (Finer et al., 1984; Johnson et
al., 1985) and sepsis (Johnson et al., 1985) have both been
reported in premature infants receiving vitamin E
supplements. An unusual syndrome comprising pulmonary
deterioration, thrombocytopaenia, ascites, liver and renal
failure which in some cases resulted in death has been
associated with the use of one particular intravenous
preparation of tocopheryl acetate marketed as E-Ferol (Lorch
et al., 1985). It is, however, highly probable that the
carrier (polysorbate 80) was the cause of the problem rather
than the tocopheryl acetate itself (Alade et al., 1986).
Because of the risks associated with the prophylactic use of
vitamin E supplementation in the preterm infant I would
agree with the recommendation of an American Committee on
Fetus and Newborn (1985) which concluded "At this time,
however, the Committee regards prophylactic use of
pharmacological vitamin E as experimental and cannot
recommend that high doses of vitamin E be given routinely to
infants weighing less than 1500g even if such use is limited
to infants who require supplemental oxygen."

ANTIOXIDANT THERAPY IN OTHER NEUROLOGICAL DISORDERS

It has recently been suggested that oxygen derived free
radical species may be involved in neurological conditions
such as tardive dyskinesia, Parkinson's disease and the
neuronal ceroid lipofuscinoses, and that vitamin E may,
therefore, have a role in the treatment of these conditions.

Tardive Dyskinesia

Tardive dyskinesia is a major complication of the long
term use of neuroleptic drugs for the control of acute
psychotic behaviour. The condition is a complex motor
syndrome characterised by abnormal mouth, trunk and limb
movements. It has been suggested that free radicals
produced as a result of increased metabolism of
catecholamines resulting from the use of the neuroleptic
drugs may be involved (Cadet et al., 1986). On the basis of
this hypothesis Lohr et al (1987) reported a study in which
patients with tardive dyskinesia were treated with either
alpha-tocopherol or placebo in a randomized cross-over study
on two separate occasions. The patients were clinically
evaluated by an abnormal involuntary movement scale and
showed a significant improvement whilst receiving the
vitamin. The results of this single study are, therefore,
encouraging but need to be confirmed by other centres.

Parkinson's Diseaese

The cause of the degeneration of the dopamine
containing neurons in the substantia nigra of patients with
Parkinson's disease is unknown. It has, however, been
suggested that excessive free radical production, as a
consequence of the catabolism of the monoamines, may be

involved (Dexter et al., 1989). In favour of a free radical hypothesis, it has been shown that the activities of protective enzymes such as catalase and glutathione peroxidase are reduced in the substantia nigra of Parkinsonian patients, as are concentrations of glutathione. It has, therefore, been suggested that additional antioxidant therapy including vitamins E and C may be helpful in slowing the progression of the condition. Fahn (in press) has reported a preliminary uncontrolled study in which 14 patients diagnosed as having Parkinson's disease but not yet receiving L-dopa therapy were recommended to take vitamins E and C. He observed that in this group the antioxidant therapy appeared to increase the time before L-dopa therapy was required, compared to a similar group of patients from another centre not receiving antioxidant therapy. On the basis of these preliminary findings a major double blind study is currently being undertaken to examine the efficacy of antioxidant therapy in Parkinson's disease.

Neuronal Ceroid Lipofuscinoses

In this group of conditions there is a characteristic accumulation of ceroid and lipofuscin lipopigments in both neuronal and non-neuronal tissues. The pathogenesis of the neuronal ceroid lipofuscinoses is unknown but it is probable that the accumulation of the ceroid and lipofuscin, which are presumed to be products of peroxidation, is a secondary phenomenon. In Finland antioxidant therapy has been used in such patients from an early age (Santavuori et al., 1988), and although it does not prevent or correct the condition it may improve some of the symptoms in some of the patients for variable periods of time. It is, therefore, possible that antioxidant therapy may be of benefit by preventing some of the secondary consequences of the disease.

CONCLUSIONS

Appropriate vitamin E replacement therapy in patients with a severe and chronic deficiency of this fat soluble vitamin can, if given sufficiently early, prevent the development of the characteristic neurological features, or if they are already present halt and sometimes reverse the deterioration. A deficiency of vitamin E is, therefore, an important diagnosis to make and a recent editorial emphasised the importance of vitamin E deficiency as a cause of neurological disease and concluded by stating that "the possibility of vitamin E deficiency should be thought of in any condition associated with lipid malabsorption, and also in spinocerebellar syndromes of unknown cause" (Anonymous 1986).

If vitamin E therapy is given it is essential that it should be administered appropriately, and this will vary with different patient groups. In patients with abetalipoproteinaemia it is necessary, as discussed above, to give very large oral doses of about 100 mg/kg/day of the regular fat soluble tablets if an adequate vitamin E status is to be achieved. In other fat malabsorptive conditions such as patients with pancreatic insufficiency resulting from cystic fibrosis a much smaller oral dose of 10 mg/kg/day for one month followed by a maintenance dose of 200 mg/day is usually adequate to maintain normal serum concentrations (Stead et al., 1986). The defect in patients with selective vitamin E deficiency can also be overcome by modest oral doses (10-20 mg/kg/day) of the vitamin (Harding et al., 1985; Sokol et al., 1988). However, in patients with cholestatic liver disease and greatly reduced luminal bile salt concentrations it is unlikely that significant quantities of the regular oral preparations will be absorbed. In general, therefore, the vitamin has to be given by intramuscular injection (Harries and Muller 1971; Guggenheim et al., 1982). Before commencing long term regular intramuscular treatment it is our practice to assess intestinal absorption by giving a large oral load (1-2g) of tocopheryl acetate and monitoring serum concentrations of the vitamin at regular intervals over the following 24 hours. If there is evidence of absorption, as judged by a significant increase in serum concentrations, large oral doses are given and serum concentrations carefully monitored to ensure that normal levels are reached and maintained. If no absorption of the oral load can be documented intramuscular therapy is given.

There have been reports that children with cholestatic liver disease are able to absorb tocopheryl polyethylene glycol succinate, which is an oral water soluble preparation of vitamin E (Sokol et al., 1987). Further clinical studies will, however, be required to confirm the efficacy of this form of treatment. In addition hydrolysis of this tocopherol ester releases polyethylene glycol, a small proportion of which is absorbed. The long term effects of polyethylene glycol are not known and, therefore, careful monitoring will be required to document the long term safety of this preparation.

The role of routine prophylactic vitamin E supplements in the premature infant remains controversial. The results of the only randomised controlled study carried out to date, regarding the use of vitamin E supplements in periventricular haemorrhage look encouraging but need to be confirmed by other centres. The role of vitamin E in retinopathy of prematurity is still not entirely clear although the evidence to date suggests it does not reduce the incidence but may reduce the severity of the condition. However, the risk/benefit relationship of prophylactic vitamin E therapy in the premature infant still needs to be clarified. The trials of vitamin E and other antioxidant therapy in neurological disorders such as tardive dyskinesia, Parkinson's disease and the neuronal ceroid lipofuscinoses are still experimental and more studies are required.

With the increasing awareness of the importance of free radicals in disease processes including neurological disorders, there is likely to be an increasing potential for antioxidant related therapy. Properly controlled trials will, however, be essential for the proper evaluation of the efficacy of such treatment regimes.

REFERENCES

Alade,S.L., Brown,R.E., and Paquet,A., 1986, Polysorbate 80
    and E-Ferol toxicity, Pediatrics, 77:593.
Anonymous, 1986, Vitamin E deficiency, Lancet, 1:423.
Aranda, J.V., Chemtob, S., Laudignon, N., and Sasyniuk,
    B.I., 1986, Furosemide and vitamin E: two problem
    drugs in neonatology, Ped Clin N Amer., 33:583.
Azizi, E., Zaidman, J. L., Eshchar, J., and Szeinberg, A.,
    1978, Abetalipoproteinemia treated with parenteral and
    oral vitamins A and E, and with medium chain
    triglycerides, Acta Paediatr Scand., 67:797.
Bell, F.E., and Filer, L.J., 1981, The role of vitamin E in
    the nutrition of premature infants, Am J Clin Nutr.,
    34:414.
Burton, G.W., Joyce, A., and Ingold, K.U., 1983, Is vitamin
    E the only lipid-soluble, chain-breaking antioxidant in
    human blood plasma and erythrocyte membranes?, Arch
    Biochem Biophys., 221:281.
Cadet, J.L., Lohr, J.B., and Jeste, D.V., 1986, Free
    radicals and tardive dyskinesia, Trends Neurosci.,
    9:107.
Chiswick,M.L., Johnson, M., Woodhall, C., Gowland,M.,
    Davies, J., Toner,N., and Sims,D.G., 1983, Protective
    effect of vitamin E (DL-alpha-tocopherol) against
    intraventricular haemorrhage in premature babies, Br
    Med J., 287:81.
Committe on Fetus and Newborn, 1985, Vitamin E and the
    prevention of retinopathy of prematurity, Pediatrics.,
    76:313.
Dexter, D.T., Carter, C.J., Wells, F.R., Javoy-Agid, F.,
    Agid,Y., Lees,A., Jenner,P., and Marsden,C.D., 1989,
    Basal lipid peroxidation in substantia nigra is
    increased in Parkinson's disease, J Neurochem., 52:381.
Fahn, S., High dosage antioxidants in early Parkinson's
    disease, Ann NY Acad Sci., in press.
Finer, N.N., Peters, K,L., Hayek, Z., and Merkel C.L., 1984,
    Vitamin E and necrotizing enterocolitis, Pediatrics,
    73:3.
Guggenheim, M.A., Ringel, S.P., Silverman, A., and Grabert,
    B.E., 1982, Progressive neuromuscular disease in
    children with chronic cholestasis and vitamin E
    deficiency: diagnosis and treatment with alpha
    tocopherol, Pediatrics, 100:51.
Haga, P., Ek, J., and Kran, S., 1982, Plasma tocopherol
    levels and vitamin E/B-lipoprotein relationships during
    pregnancy and in cord blood, Am J Clin Nutr., 36:1200.
Halliwell, B., and Gutteridge, J.M.C., 1985a, Oxygen
    radicals and the nervous system, Trends Neurosci.,
    8:22.
Halliwell, B., and Gutteridge, J.M.C., (1985b) Free Radicals
    in Biology and Medicine, Oxford University Press,
    Oxford .
Harding, A.E., Matthews, S., Jones,S., Ellis, C.J.K.,
    Booth,I.W., and Muller, D.P.R., 1985, Spinocerebellar
    degeneration associated with a selective defect of
    vitamin E absorption, N Engl J Med., 313:32.
Harries, J.T., and Muller, D.P.R., 1971, Absorption of
    vitamin E in children with biliary obstruction, Gut,
    12:579.

Herbert, P.N., Gotto, A.M., and Fredrickson, D.S., 1978, Familial lipoprotein deficiency, in: The Metabolic Basis of Inherited Diseases, Stanbury, J.B., Wyngaarden, J.B., Fredrickson, D.S., eds, McGraw Hill, New York.

Johnson, L., Schaffer, D., and Goggs, T.R., 1974, The premature infant, vitamin E deficiency and retrolental fibroplasia, Am J Clin Nutr., 25:1158.

Johnson, L., Bowen F.W., Abbasi,S., Herrmann,N., Weston, M., Sacks,L., Porat,R., Stahl,G., Peckham, G., Delivoria-Papadopoulos, M., Quinn,G., and Schaffer, D., 1985, Relationship of prolonged pharmacologic serum levels of vitamin E to incidence of sepsis and necrotizing enterocolitis in infants with birth weight 1,500 grams or less, Pediatrics, 75:619.

Johnson, L., Abbasi,S., Quinn, G.E., Otis, C., and Bowen, F.W., 1988, Vitamin E and retinopathy of prematurity, Pediatrics, 81:329.

Kinsey, V.E., 1956, Retrolental fibroplasia: cooperative study of retrolental fibroplasia and the use of oxygen, Arch Ophthalmol., 56:481.

Lohr, J.B., Cadet J.L., Lohr, M.A., Jeste, D.V., and Wyatt, R.J., 1987, Alpha-tocohperol in tardive dyskinesia, Lancet, 1:913.

Lorch,B., Murphy, D., Hoersten, L.R., Harris,E., Fitzgerald, J., and Sinha, S.N., 1985, Unusual syndrome among premature infants: association with a new intravenous vitamin E product, Pediatrics, 75:598.

Muller, D.P.R., Harries, J.T., and Lloyd, J.K., 1974, The relative importance of the factors involved in the absorption of vitamin E in children, Gut, 15:966.

Muller, D.P.R., Lloyd, J.K., and Bird, A.C., 1977, Long-term management of abetalipoproteinaemia, possible role for vitamin E, Arch Dis Childh., 52:209.

Muller, D.P.R., Lloyd, J.K., and Wolff, O.H., 1983, Vitamin E and neurological function, Lancet, 1:225.

Muller, D.P.R., 1987, Free radical problems of the newborn. Proc Nutr Soc., 46:69.

Nelson, J.S., Fitch, C.D., Fischer, V.W., Broun, G.O., and Chou, A.C., 1981, Progressive neuropathologic lesions in vitamin E-deficient rhesus monkeys, J Neuropath Exp Neurol., 40:166.

Owens, W.C., and Owens, E.U., 1941, Retrolental fibroplasia in premature infants. II. Studies on the prophylaxis of the disease. The use of alphatocopheryl acetate, Am J Ophthalmol., 32:1631.

Phelps, D.L., 1982, Vitamin E and retrolental fibroplasia in 1982, Pediatrics, 70:420.

Phelps, D.L., Rosenbaum, A.L., Isenberg, S.J., Leake, R.D., and Dorey, F.J., 1987, Tocopherol efficacy and safety for preventing retinopathy of prematurity: a randomized, controlled, double-masked trial, Pediatrics, 79:489.

Santavuori, P., Heiskala, H., Westermarck, T., Sainio, K., and Moren, R., 1988, Experience over 17 years with antioxidant treatment in Spielmeyer-Sjogren disease, Am J Med Gen Suppl., 5:265.

Sinha, S., Davies, J., Toner, N., Bogle, S., and Chiswick, M., 1987, Vitamin E supplementation reduces frequency of periventricular haemorrhage in very preterm babies, Lancet., 1:466.

Sokol, R.J., Guggenheim, M., Iannaccone, S.T., Barkhaus, P.E., Miller, C., Silverman, A., Balistreri, W.F., and Heubi, J.E., 1985, Improved neurologic function after long-term correction of vitamin E deficiency in children with chronic cholestasis, N Engl J Med., 313:1580.

Sokol, R.J., Kayden, H.J., Bettis, D.B., Traber, M.G., Neville, H., Ringel, S., Wilson, W.B., and Stumpf D.A., 1988, Isolated vitamin E deficiency in the absence of fat malabsorption – familial and sporadic cases: characterization and investigation of causes, J Lab Clin Med., 111:548.

Sokol, R.J., Butler-Simon, N.A., Bettis, D., Smith, D.J., and Silverman, A., 1987, Tocopheryl polyethylene glycol 1000 succinate therapy for vitamin E deficiency during chronic childhood cholestasis: neurological outcome, J Peds., 111:830.

Speer, M.E., Blifeld, C., Rudolph, A.J., Chadda, P., Holbein, M.E.B., and Hittner, H.M., 1984, Intraventricular hemorrhage and vitamin E in the very low-birth-weight infant: evidence for efficacy of early intramuscular vitamin E administration, Pediatrics, 74:1107.

Stead, R.J., Muller, D.P.R., Matthews, S., Hodson, M.E., and Batten, J.C., 1986, Effect of abnormal liver function on vitamin E status and supplementation in adults with cystic fibrosis, Gut, 27:714.

Terry, T.L., 1942, Extreme prematurity and fibroplastic overgrowth of persistent vascular sheath behind each crysalline lens, Am J Ophthalmol., 25:203.

FREE RADICALS AND TRIALS OF ANTIOXIDANT THERAPY IN MUSCLE DISEASES

M.J. Jackson and R.H.T. Edwards

Department of Medicine
University of Liverpool
P.O. Box 147
Liverpool, L69 3BX, U.K.

INTRODUCTION

It has been recognised for a number of years that the severest forms of the degenerative muscular dystrophies share many of the characteristics of vitamin E or selenium deficiency myopathy of animals and indeed the animal disorders have been incorrectly termed 'nutritional muscular dystrophies' [Bradley and Fell, 1980]. The recognition that the major functions of vitamin E and selenium within the body are likely to be as antioxidants acting to inhibit the toxic effects of free radicals has prompted investigations of the possible role of free radicals in some of the human disorders, particularly myotonic and Duchenne muscular dystrophies.

Involvement of Free Radicals in Muscle Diseases

The evidence for an involvement of free radical-mediated processes in the muscular dystrophies is largely based on extrapolation from animal work, studies of the antioxidant content of tissues and body fluids and measurement of free radical reaction products in blood or tissues of patients. The evidence is summarised in table 1.

It can be seen that there is little concensus concerning the content of dietary antioxidants in these patients or in the activity of the antioxidant enzymes. There appears to be an undisputed elevation of muscle thiobarbituric acid-reactive substances in Duchenne dystrophy patients although the lack of specificity of this parameter is well known [Halliwell and Gutteridge, 1985].

In addition, we have also been studying an alternative approach by which therapies for these disorders might be devised. This is to try to reduce the amount of damage to the muscle fibres which results from the basic biochemical defect. Patients with muscular dystrophy have histological changes in their muscle indicative of an increased rate of loss of muscle fibres and have plasma activities of muscle - derived

Antioxidants in Therapy and Preventive Medicine
Edited by I. Emerit *et al.*
Plenum Press, New York, 1990

Table 1

Evidence for an involvement of free radicals in muscular dystrophy

A. Duchenne muscular dystrophy

Antioxidant status	Content cf. controls	References
Vitamin E (Blood)	Unchanged	Jackson et al. 1984b
	Reduced	Hunter and Mohamed, 1986
Selenium (Blood)	Decreased	Westermark, 1977
	Normal	Jackson et al. 1989
Superoxide dismutase (Blood)	Decreased	Burri et al. 1980
	Unchanged	Hunter et al. 1981
	Unchanged	Mechler et al. 1984
Superoxide dismutase (Muscle)	Increased	Kar and Pearson, 1979
	Increased	Burr et al. 1989
Catalase (Muscle)	Increased	Kar and Pearson, 1979
	Unchanged	Burr et al. 1987
Glutathione peroxidase (Blood)	Unchanged	Jackson et al. 1989
Glutathione peroxidase (Muscle)	Increased	Kar and Pearson, 1979
	Increased	Burr et al. 1987
	Unchanged	Jackson et al. In press

Free radical reaction products

Thiobarbituric acid - reactive materials (Blood)	Increased	Hunter and Mohamed, 1986
Thiobarbituric acid - reactive materials (Muscle)	Increased	Kar and Pearson, 1979
	Increased	Jackson et al. 1984b
	Increased	Mechler et al. 1984
Diene conjugates (Blood)	Increased	Hunter and Mohamed, 1986
Lipofuscin pigments (Blood)	Increased	Hunter and Mohamed, 1986

B. Myotonic dystrophy

Selenium (Blood)	Decreased	Orndahl et al. 1982

enzymes [e.g. creatine kinase (CK)] up to 3 orders of magnitude greater than normal. However, any attempt to interfere with the process of muscle cell damage requires a knowledge of the biochemical mechanisms underlying this process. To this end we have been studying these

mechanisms using both 'in vitro' and 'in vivo' preparations. These studies have suggested that free radical processes may play some role in the more fundamental general process of muscle damage and hence may be relevant to muscular dystrophy, whatever the fundamental underlying defect in this disorder. [Jackson et al., 1984a, 1985, 1987, 1988: Phoenix et al., 1989].

Therapeutic Trials of Antioxidants

There have been a number of therapeutic trials of antioxidants in patients with Duchenne muscular dystrophy, some of which have claimed benefit to the patients (table 2). Controlled therapeutic trials in patients with these disorders are difficult to undertake because of the heterogeneity of the clinical course of the disease, but it appears that all well-controlled trials of antioxidants in Duchenne dystrophy have shown no beneficial effects [Edwards et al., 1984].

The only situation where the published data supports the case of antioxidant therapy is in myotonic dystrophy but the same group of workers have published the 2 (uncontrolled) trials apparently demonstrating beneficial effects of selenium plus vitamin E therapy. [Orndahl et al., 1983, 1986]. It is imperative that these studies are repeated in a double-blind, controlled manner.

During our studies of zinc [Jackson and Edwards, 1986] or selenium supplementation [Jackson et al., 1989] in Duchenne muscular dystrophy it became apparent that even at relatively high oral doses these substances may not exert the required biological effect at the tissue of interest. This is because many naturally occuring nutrients are controlled by efficient homeostatic mechanisms to prevent cellular toxicity of the nutrient. Thus in the case of zinc we have demonstrated [Jackson et al., 1984] that an elevation of the oral zinc intake via dietary supplementation only leads to a transitory positive balance for zinc with homeostatic adaptation occuring to prevent tissue accumulation of the element. This adaptation occurs at the gastrointestinal level reducing the proportion of the zinc absorbed from an increased dietary intake and increasing the amount of zinc excreted into the gut (see table 3). The implication of this is that, in zinc replete subjects, supplements do not lead to any sustained tissue retention of zinc and hence would be unlikely to lead to any tissue specific pharmacological effect.

The situation for selenium appears to be similar with selenium supplemented subjects rapidly re-establishing zero balance for the element [Jackson et al., 1989], although in this case the adaptations appear to be achieved both by a reduction in the fractional gastrointestinal absorption of the dietary selenium and an increased urinary excretion of the element. In addition, in the case of selenium, it is necessary for the element to 'stimulate' the synthesis of, and be incorporated into glutathione peroxidase in order to exert an antioxidant effect. Studies by Gebre-Medhin et al., [1985] demonstrated that large oral doses of selenium did not result in an increased red cell glutathione peroxidase activity and we have found that muscle glutathione peroxidase activities were unaffected by selenium therapy in patients with Duchenne and other muscular dystrophies (table 4).

Table 2

Antioxidant Agents Tested in Patients with Muscular Dystrophy

A. Duchenne muscular dystrophy

Agent	Stated result	Reference
Vitamin E	Benefit	Bicknell, 1940
	No benefit	Fitzgerald & McArdle, 1941
	Benefit	Robinovitch et al. 1951
	No benefit	Berneske et al. 1960
	No benefit	Edwards et al. 1984
Vitamin E and selenium	No benefit	Gamstorp et al. 1986
Selenium	No benefit	Jackson et al. 1989
Superoxide dismutase ('Orgotein')	No benefit	Stern et al. 1982
Allopurinol	Benfit	Thompson and Smith, 1982
	No benefit	Mendell and Wichen, 1979
	No benefit	Bertorini et al. 1985
	No benefit	Griffiths et al. 1985
Penicillamine	No benefit	Bradley et al. 1977
	No benefit	Roelfs et al. 1979
Zinc	No benefit	Jackson and Edwards, 1986

B. Myotonic dystrophy

Selenium and vitamin E	Benefit	Orndahl et al. 1983
	Benefit	Orndahl et al. 1986

It is therefore apparent that much work remains to be undertaken on the possible roles of free radicals in muscle diseases and the place of antioxidants in their therapy. More fundamental studies are required of the ways in which antioxidant defences can be increased in nutritionally replete patients. Pharmacological agents having antioxidant properties

Table 3

Adaptation of a normal subject to increased zinc intake

Period	Dietary Zn (uM/day)	Absorbed Zn (uM/day)	GI Zn excretion (uM/day)	Balance (uM/day)
1.	111	65	58	-5
2.	109	51	46	-6
3.	231	104	69	+27
4.	224	73	71	-7
5.	473	142	97	+36
6.	472	98	90	-3

Table 4

Lack of effect of 6 months selenium supplementation on red cell and muscle glutathione peroxidase activities

	Post-placebo therapy	Post-selenium therapy
Red cell glutathione peroxidase activity (Units/g Hb)	33 $\pm$ 15	36 $\pm$ 15
Muscle glutathione peroxidase activity (Units/g soluble protein)	25 $\pm$ 13	19 $\pm$ 8

Data derived from Jackson et al. 1989.

may be useful in this respect because of the ability to 'target' them to tissues and to be active regardless of nutritional status.

The subject was infused with 20mg ^{67} zinc and the abundance measured in plasma and fecal samples during the course of a balance study with 4 day balance periods. The dietary zinc intake was doubled after every second period. The rates of gastrointestinal absorption and excretion were calculated from the enrichments of 70 zinc. (Data derived from Jackson et al. 1984c).

There appears to be little progress in attempts to find the basic gene or protein defect responsible for myotonic dystrophy, but in the study of Duchenne dystrophy progress in elucidating the basic genotypic and phenotypic defects has recently been rapid. The gene which is defective in patients with Duchenne dystrophy has recently been described [Monaco et al., 1986] and the protein for which it normally codes partially characterized and named 'dystrophin' [Hoffmann et al., 1987]. It remains to be seen whether the lack of 'dystrophin' in patients with Duchenne dystrophy does indeed lead to an activation of free radical-mediated degenerative processes.

Acknowledgements

The authors would like to thank the Muscular Dystrophy Group of Great Britain and Northern Ireland for their continued financial support.

References

Berneske G.M., Burton A.R.C., Gould E.N. and Levy D. (1960) Neurology 35: 61-65.

Bertorini T.E., Palmieri G.M.A., Griffin J., Chasney C., Pifer D. et al. (1985) Neurology 35: 61-65.

Bicknell F. (1940) Lancet i: 10-13.

Bradley R. and Fell B.F. (1980). In: Disorders of Voluntary Muscle Ed. J.N. Walton. Pub. Churchill, London pp 824-872.

Bradley W.G., Enomoto A and Gardner-Medwin D. (1977). Proc. Roy. Soc. Med. 70: 94.

Burr I.M., Asayama K. and Fenichel G.M. (1987) Muscle and Nerve, 10: 150-154.

Burri B.J., Chan S.G., Gerry A.J. and Yarnell S.K. (1980) Clin. Chim. Acta. 105: 249-255.

Edwards R.H.T., Jones D.A. and Jackson M.J. (1984) Med. Biol. 62: 143-147.

Fitzgerald G. and McArdle B. (1941) Brain 64: 19-42.

Gamstorp I., Gustavson K.H., Hellstron O. and Nordgren B. (1986) Child Neurol 1: 211-214.

Gebre-Medhin M., Gustavson K.H., Gamstorp I., and Plantin L.O. (1985) Acta. Paed. Scand. 7: 886-890.

Griffiths R.D., Cady E.B., Edwards R.H.T. and Wilkie D.R. (1985) Muscle and Nerve 8:760-767.

Halliwell B. and Gutteridge J.M.C. (1985) 'Free Radicals in Biology and Medicine', Oxford Science.

Hoffmann EP, Brown RH and Kunkel LM (1987) Cell 51: 919-928.

Hunter M.I.S., Brzeski M.S. and de Vane P.J. (1981) Clin. Chim. Acta. 115: 93-98.

Hunter M.I.S. and Mohamed J.B. (1986) Clin. Chim. Acta. 155: 123-132.

Jackson M.J. and Edwards R.H,T. (1986) Cardiomyology 5: 31-38.

Jackson M.J., Jones D.A. and Edwards R.H.T. (1984a) Europ. J. Clin. Invest. 14: 369-374.

Jackson M.J., Jones D.A. and Edwards R.H.T. (1984b) Med. Biol. 62:135-138.

Jackson M.J., Jones D.A., Edwards R.H.T., Swainbank I.G. and Coleman M.L. (1984c) Br. J. Nutr. 51: 199-208.

Jackson M.J., Jones D.A, and Edwards R.H.T. (1985) Clin. Chim. Acta. 147: 215-221.

Jackson M.J., Wagenmakers A.J.M. and Edwards R.H.T. (1987) Biochem. J. 241: 403-407.

Jackson M.J., Roberts J. and Edwards R.H.T. (1988) Br. J. Nutr. 60: 217-224.

Jackson M.J., Coakley J., Stokes M., Edwards R.H.T. and Oster O. (1989) Neurology (In press).

Kar N.C. and Pearson C.M. (1979) Clin. Chim. Acta. 94: 277-280.

Mechler F., Imre S. and Dieszeghy P. (1984) J. Neurol. Sci. 63: 279-283.

Mendell J.R. and Wichers D.O. (1979) Muscle and Nerve 2: 53-56.

Monaco AP, Neve RL, Colletti-Faener et al (1986) Nature 323: 646-650.

Orndahl G., Rindby A. and Selin E. (1982) Acta. Med. Scand 211: 493-499.

Orndahl G., Rindby A. and Selin E. (1983) Acta. Med. Scand. 213: 237-239.

Orndahl G., Seliden U., Hallin S., Wetterqvist H., Rindby A. and Selin E. (1986) Acta. Med. Scand. 219: 407-414.

Phoenix J., Edwards R.H.T. and Jackson M.J. (1989) Biochem. J. 287:207-213.

Roelfs R.I., De Arango G.S., Law P.K. et al. (1979) Arch. Neurol. Neurosurg. Psych. 14: 95-100.

Stern L.Z., Ringel S.P., Ziter F.A. et al. (1982) Arch. Neurol. 39: 342-346.

Thompson W.H.S. and Smith I. (1978) Metabolism 27: 151-163.
Westermark T. (1977) Acta. Pharmacol. et. Toxicol. 41: 121-128.

PARKINSON'S DISEASE AND ALZHEIMER'S DISEASE :NEURODEGENERATIVE DISORDERS DUE TO BRAIN ANTIOXIDANT SYSTEM DEFICIENCY ?

Irène Ceballos, France Javoy-Agid, André Delacourte
André Defossez, Annie Nicole and Pierre-Marie Sinet

Laboratoire de Biochimie Génétique, CNRS URA 1335
Hopital Necker, 149 rue de Sèvres, Paris, FRANCE

INTRODUCTION

Parkinson disease (PD) and Alzheimer's disease (AD), the two most common types of adult chronic degenerative disorders of the central nervous system are characterized by degeneration of certain populations of neurons with relative sparing of other groups of nerve cells (1). Dysfunction of at-risk neurons is associated with several types of cytoskeletal pathology including neurofibrillary tangles (NFT), granulovacuolar degeneration, senile plaques, in AD and Lewy bodies in PD. Dysfunction and death of neurons lead to the clinical syndromes of PD and AD. The bradykinesia and rigidity of PD are associated with lesions in the nigrostriatal dopaminergic systems, whereas the dementia of AD is attributed to abnormalities of neurons in monoaminergic brainstem nuclei,cholinergic basal forebrain , and neuronal populations within amygdala, hippocampus and neocortex (1).

The etiology of these diseases is still unknown. It has been suggested that hydrogen peroxide (H_2O_2) and related oxy-radicals are involved in the degeneration of dopamine neurons in PD (2-4) and pyramidal neurons in AD (5). These reactive oxygen species are all able to disrupt a variety of biomolecules including proteins (6) nucleic acids(7) and the phospholipids in cell membranes (8).

Normally, a variety of scavenger systems prevent the excessive accumulation of these potentially destructive oxygen species. These natural defence mechanisms include copper-zinc superoxide dismutase (SOD-1) and catalase, which inactivate superoxide radicals (O_2^-) and H_2O_2 respectively, as well as the glutathione peroxidase system, in which enzymatic oxidation of reduced glutathione (GSH) to oxidized glutathione (GSSG) is coupled to detoxification of H_2O_2 or other peroxides, notably those derived from the oxidation of membrane phospholipids. Levels of GSH are believed to be limiting in this process, requiring the regeneration of GSH and maintenance of a high GSH/GSSG ratio. This is achieved by glutathione reductase which utilizes NADPH as a reducing cofactor derived from the hexose monophosphate shunt(9,10) . In brain tissue, the major protection against both H_2O_2 and lipid peroxides is conferred

Antioxidants in Therapy and Preventive Medicine
Edited by I. Emerit *et al.*
Plenum Press, New York, 1990

493

by GSH-PX (2,11). In addition to these enzymatic mechanisms, antioxidants such as ascorbate and α -tocopherol also react as radicals quenchers in vivo (11). However, any regional impairment of these defence mechanisms arising, for example from inadequate levels of GSH (12), or from decreased and/or increased activities of enzymes involved in GSH metabolism (9,13,14) could initiate a process of progressive neuronal degeneration.

PARKINSON DISEASE

PD is a slowly progressive neurodegenerative disorder which is characterized by the loss of dopaminergic neurons from the pigmented substantia nigra, zona compacta . The etiology of PD is unknown and the disease does not appear to have strong genetic basis (15). One hypothetical mechanism implicate free radicals and/or environmental toxins (3,16-18).

It is interesting that the neuronal cells primarily affected in PD are highly pigmented. The melanin, a pigment resulting from the oxidation and polymerization of the dopamine (DA) precursor, dopa, may in fact provide an index of the relative inability of DA cells to prevent the accumulation of oxydized by-products of DA metabolism (19). Oxidative metabolism of DA in brain homogenates is associated with the production of H_2O_2 and the conversion of GSH to GSSG (20). Pigmentation of the substantia nigra increases between childhood and early adult life. It is also known that normal human aging is associated with a progressive loss of nigrostriatal neurons (21). These observations, taken together with evidence that aging may be associated with an increased succepibility of cells to oxidative damage (22), suggest that oxidative events peculiar to the substantia nigra may cause the progressive loss of dopaminergic neurons in PD. One suggestion is that the final insult is a consequence of excessive lipid peroxidation provoked by free radicals(4) . Nigral GSH and GSH-PX, protective mechanisms against such insult are reduced in PD (12,13,23). A marker of lipid peroxidation, the malonedialdehyde (MDA) is increased in parkinsonian nigral tissue (4) and increased iron content in parkinsonian nigra may contribute to increased lipid peroxidation in this brain region (24).
Morever, hydroxylated analogues of DA namely 6-hydroxydopamine are able to cause selective degeneration of adrenergic neurons in experimental animals (18). Conversion of DA to 6-hydroxy-dopamine might occur in vivo by spontaneous autoxidation, possibly mediated by $O_2^{\cdot-}$ (2). Manganese,another specific DA neurotoxin may also operates through the formation of free radicals (25). Normally, optimal functioning of various radical scavenging processes would prevent the accumulation of reactive-oxygen species. However, increase of SOD-1 activity in the substantia nigra of PD (3) and decrease in GSH-PX activity in the same brain region (13,23) could lead to an increased production of H_2O_2 and may reflect the involvement of radical-induced cell damage in the pathogenesis of PD. Recently, we have specifically localized the mRNA of SOD-1 in melanized-dopaminergic neurons by using in situ hybridization.

Interest in the area has been stimulated by the finding that the neurotoxin 1-methyl-4-phenyl - 1,2,3,6 tetrahydropyridine (MPTP) via its active metabolite MPP+ may kill nigral dopamine-containing cells by free radical-induced lipid peroxidation (26). In vitro studies show $O_2^{\cdot-}$ -formation during the conversion of MPTP to MPP+(27).This step catalized by MAO-B probably takes place in glial cells,but there is evidence of further oxidation of MPP+ inside DA neurons which could result in intraneuronal O_2^- formation. The toxicity of MPP+ after MPTP administration has been ascribed to the production of O_2^- (17), and to the presence of transition metals(28). In addition, there is a chemical similarity between MPP+ and the herbicide Paraquat, a toxic which diverts electron flow from the water-producing cytochrome pathway to O_2^- -producing route and this causes cellular death (29). If MPTP induces DA neuronal damage by such mechanism, systemic administration of antioxidants such as ascorbate or α- tocopherol may prevent its toxic effects. Depletion of GSH content in the substantia nigra of mice caused by MPTP is prevented by antioxidant pretreatment with α-tocopherol and β-carotene (30). Moreover, pretreatment with anyone of four antioxidants α-tocopherol, β-carotene, ascorbate or N-acetylcysteine significantly decreased MPTP-induced striatal DA loss and neuronal loss in the substantia nigra of the mice (31). If idiopathic PD is caused by damage to membranes of dopaminergic nigrostriatal neurons by free radicals generated from exogenous and/or endogenous neurotoxin - like MPTP, patients with PD might be treated at the onset of symptoms with antioxidant drugs such as ascorbate or α-tocopherol, or with sulfhydryl compounds like N-acetylcysteine, penicillamine, or cysteamine which could enter brain cells and substitute for GSH in glutathione- S-transferase conjugation reactions. Epidemiological studies have shown that PD is less common among persons who have smoked cigarettes for many years than among non smokers (32). Heavy smokers have chronically elevated blood levels of carboxyhemoglobin, and presumably increased partial pressures of carbon monoxide in brain cells. This might provide a reducing environment which could partially protect nigrostriatal neurons from oxidant damage. Antioxidants such as ascorbate and α-tocopherol, in doses designed to maximize their antioxidant properties are probably harmless enough for prophylactic use in asymptomatic older persons.

ALZHEIMER'S DISEASE

The nature of biochemical events which lead to the loss of specific pyramidal neurons is not understood, but oxy-radicals may play a significant causal role in the etiology and the pathogenis of AD.Altered activity of cellular antioxidant systems have been implicated in the neuronal cell death that is associated with AD (5,14).Of particular pertinence to the injurious effects of oxidant stress in relation to brain damage in AD is the proposed role of oxy-radicals in the aging process(22) and their reactive capacity to change synaptic transmission(33) and degrade brain membrane phospholipids(8). The development of clinical features in AD is linked to the amount of deposition of amyloid in the limbic areas and cerebral cortex.Moreover,amyloid formation may arise as a consequence of membrane damage may be due to lipid peroxidation(34). Neurofibrillary tangles are primary neuropathological features of AD which appear as paired helical filaments (PHF) on electron microscopic examination (1). About 6% of PHF is comprised of the amino- acid, hydroxyproline. This amino-acid is not a constituent of cytoplasmic protein in normal brain and the abundance of hydroxyproline in cytoplasmic PHF involves non-enzymatic hydroxylation of proline residues (37) probably by hydroxyl free radicals. This free radical hypothesis of PHF formation suggests that AD is an acceleration of the normal aging process in affected brain regions .

In brain, ample levels of the scavenging enzyme, SOD-1 have been reported in AD patients with modest increases in SOD-1 activities in some brain regions(36). Moreover, recent studies using immunohistochemistry have demonstrated that pyramidal neurons which are susceptible to degenerative processes in AD have the property to contain higher amount of SOD-1 protein than other hippocampal cells (5).The higher levels of the SOD-1 protein and mRNA expressed in the pyramidal neurons of the human hippocampus(37) could suggest that biochemical pathways leading to O_2^- generation are particularly active in these neurons, requiring a high SOD-1 content to eliminate these radicals. Alternatively, a high SOD-1 content might be noxious for neurons(9,38). This hypothesis is motivated by two observations. First, in human the gene coding for SOD-1 is located on chromosome 21 and in Down's patients increased SOD-1 activity reflects a gene dosage effect in various tissues (9) including brain (39). Second, patients with DS develop an accelerated aging and the neuropathology of the disease is reminiscent to that of AD (40).In order to test this hypothesis we have developed neuronal cell lines(41) and transgenic mice which mimic SOD-1 overexpression after human SOD-1 gene transfer. Furthermore, the high content of polyunsaturated fatty acids in brain tissue, and the low level of the endogenous antioxidants GSH-PX (42) and vit E (43) render the brain particularly succeptible to oxidant stress. Thus antioxydants such as vit E and ascorbate might offer protection against loss of pyramidal neurons in AD brains.

If more data in support of this hypothesis are forthcoming,to determine the role of antioxidants in preventing or delaying the neurodegenerative processes in AD should become an area of active research.

REFERENCES

1. D.L.Price, P.J.Whitehouse and R.G.Struble, cellular pathology in Alzheimer's and Parkinson's disease, TINS, 29 (1986).
2. G.Cohen, the pathobiology of Parkinson's disease: biochemical aspects of dopamine neuron senescence, J.Neural Transm., 19:89 (1983).
3. R.J.Marttila, H.Lorentz and U.K.Rinne, oxygen toxicity protecting enzymes in Parkinson's disease, J.Neurol.Sci., 86:321 (1988).
4. D.Dexter, C.Carter, F.Agid, Y.Agid, A.J.Lees, P.Jenner and C.D. Marsden, lipid peroxidation as cause of nigral cell death in Parkinson's disease, Lancet, 639 (1986).
5. A.Delacourte, A.Defossez, I.Ceballos, A.Nicole and P.M Sinet, preferential localisation of copper- zinc superoxide dismutase in the vulnerable cortical neurons in Alzheimer's disease, Neurosci.Lett., 92:247 (1988).
6. J.M.C.Gutteridge, free-radicals damage to lipids, amino-acids,carbohydrates and nucleic acids determined by thiobarbituric acid reactivity, Int.J.Biochem., 14:649 (1982).
7. J.A.Imlay, S.M.Chin and S.Linn, toxic DNA damage by hydrogen peroxide through the Fenton reaction in vivo and in vitro. Science, 240:640 (1988).
8. P.H.Chan, M.Yurko and R.A. Fishman, phospholipids degradation and cellular edema by free radicals in brain cortical slices. J.Neurochem., 38:525 (1982).
9. P.M.Sinet, metabolism of oxygen derivatives in Down's

syndrome. Ann.N.Y.Acad.Sci.USA, 396:83 (1982).

10. I.Fridovich, superoxide dismutases, In: Advances in enzymology and related areas of modern biology, A.Meister, ed., New York (1986).

11.B.Halliwell and M.C.Gutteridge, oxygen radicals and the nervous system.Trends Neurosci., 8:22 (1985).

12.T.L.Perry, D.V.Godin and S.Hansen, Parkinson's disease: a disorder due to nigral glutathione deficiency?, Neurosci. Lett., 33:305(1982).

13.S.J.Kish, C.Morito and O.Hornykiewics, glutathione peroxidase activity in Parkinson's disease brain, Neurosci.Lett., 58:343 (1985).

14.R.N.Martins, C.G.Harper, G.B.Stokes and C.L.Masters, increased glucose-6-phosphate dehydrogenase activity in Alzheimer's disease may reflect oxydative stress, J.Neurochem., 46:1042 (1986).

15.D.B.Calne and J.W Langston, the etiology of Parkinson's disease, Lancet, 2:1457 (1983).

16.T.L.Perry and V.W.Yong, Idiopathic Parkinson disease, progressive supranuclear palsy and glutathione metabolism in the substantia nigra, Neurosci.Lett., 67:269 (1986).

17.B.K.Sinha, Y.Singh and G.Krishna, formation of superoxide and hydroxyl radicals from 1-methyl-4-phenylpyridinium ion (MPP+): reductive activation by NADPH cytochrome P450 reductase, Biochem.Biophys.Res.Comun., 135:583 (1986).

18.G.Cohen, R.E.Heikkila and D.Mac Namee, the generation of hydrogen peroxide, superoxide radical and hydroxyl radical by 6-hydroxydopamine, dialuric acid and related cytotoxic agents.J.Biol.Chem., 249:2447 (1974).

19.M.J.Bannon, M.Goedert and B.Williams, the possible relation of glutathione, melanin and 1-methyl-4-phenyl-1,2,5,6-tetrahydropyridine (MPTP) to Parkinson's disease, Biochem.Pharmacol., 33:2697, (1984).

20.H.S.Maker, C.Weiss, D.G.Silides and G.Cohen, coupling of dopamine oxidation to glutathione oxidation via the generation of hydrogen peroxide in rat brain homogenates, J.Neurochem., 36:589 (1981).

21.P.L.Mc.Geer, E.G.Mc.Geer and J.S.Suzuki, aging and extrapyramidal function, Arch.Neurol., 34:33 (1977).

22.D.Harman, The aging process, Proc.Natl.Acad.Sci.USA, 78:7124 (1981).

23.L.M.Ambani, M.H.Van Waert and S.Murphy, Brain peroxidase and catalase in Parkinson's disease, Arch.Neurol., 32:114 (1975)

24.D.T.Dexter, F.R.Wells, F.Agid, Y.Agid, A.J.Lees, P.Jenner and C.D.Marsden, Increased nigral iron content in postmortem parkinsonian brains, Lancet, 2:1219 (1987).

25.J.Mc Gregor and F.Labella, Manganese neurotoxicity : a model of free radical induced neurodegeneration ?, Can.J.Physiol. Pharmacol., 60:1398 (1982).

26.C.Rios and R.Tapia, Changes in lipid peroxidation induced by 1-methyl-4-phenyl-1,2,3,6-tetrahydropyridine and 1-methyl-4-phenyl-pyridinium in mouse brain homogenates, Neurosci.Lett., 77:321 (1987).

27.J.Poirier and A.Barbeau, a catalyst function for MPTP in superoxide formation, Biochem.Biophys.Res.Commun., 131:1284 (1985).

28.J.Poirier, J.Donaldson and A.Barbeau, The specific vulnerability of the substantia nigra to MPTP is related to the presence of transition metals, Biochem. Biophys. Res.Commun., 128:25 (1985).

29.A.Barbeau, L.Dallaire, N.T.Buu, J.Poirier and E.Runcinska,

Comparative behavioral, biochemical and pigmentary effects of MPTP, MPP+ and paraquat in rana pipiens, Life sciences, 37:1529 (1986).

30.V.W.Yong, T.L.Perry and A.A.Krisman, Depletion of glutathione in brainstem of mice caused by N-Methyl-4-phenyl-1,2,3,6-tetrahydro-pyridine is prevented by antioxydant pretreatment, Neurosci. Lett., 63:56 (1986).

31.T.L.Perry, V.W.Yong, R.M.Clavier, K.Jones, J.M.Wright, J.G.Foulks and R.A.Wall, Partial protection from the dopaminergic neurotoxin N-methyl-4-phenyl-1,2,3,6-tetrahydropyridine by four different antioxidants in the mouse, Neurosci. Lett., 60:109 (1985).

32.M.D.Nefzger, F.A.Quadfosel and V.C.Karl, A retrospective study of smoking and Parkinson's disease, Amer.J.Epidemiol., 88:149 (1968).

33.C.A.Calton, F.S.Calton and D.L.Gilbert, Changes in synaptic transmission produced by hydrogen peroxide. J.Free Radic.Biol. Med., 2:141 (1986).

34.T.Dyrks, A.Weidemann, G.Multhaup, J.M. Salbaum, H.G.Lemaire, J.Kong, B.Muller-Hill, C.L.Masters and K.Beyreuther, Identification, transmembrane orientation and biogenesis of the amyloid A4 precursor of Alzheimer's disease, EMBO J, 7:949, (1988).

35.F.P.Zemlan, O.J.Thienshaus and H.B.Bosmann, Superoxide dismutase activity in Alzheimer's disease : possible mechanism for paired helical filament formation, Brain Res. (1988), in press.

36.S.L.Marklund, R.Adolfson, C.Gottfries and B.Winblad, Superoxide dismutase isoenzymes in normal brains and in brains from patients with dementia of Alzheimer's type, J.Neurol.Sci., 67:319 (1985).

37.I.Ceballos, F.Agid, A.Delacourte, E.Hirsch, P.M.Sinet and Y.Agid. Preferential localization of copperzinc superoxide dismutase in the injured hippocampal neurons in Alzheimer's disease, Neurosci.Abst., 437.1, 1083 (1988).

38.O.Elroy-Stein and Y.Groner, Impaired neurotransmitter uptake in PC12 cells overexpressing human CuZn SOD - Implication for gene dosage effects in Down's Syndrome. Cell, 52:259 (1988).

39.B.W.L. Brooksbank and R.Balazs. Superoxide dismutase, glutathione peroxidase and lipoperoxidation in Down's Syndrome fetal brain, Develop.Brain.Res., 16:37 (1984).

40.K.E.Wisniewski, H.M.Wisniewski and G.Y.Wen, Occurence of neuropathological changes and dementia of Alzheimer's disease in Down's Syndrome, Ann. Neurol., 17:278 (1985).

41.I.Ceballos, J.M.Delabar, A.Nicole, R.E.Lynch, R.A. Hallewell, P.Kamoun and P.M.Sinet, Expression of transfected human CuZn superoxide dismutase gene in mouse L cells and NS20Y neuroblastoma cells induces enhancement of glutathione peroxidase activity, Biochim.Biophys.Acta, 58:59 (1988).

42.O.de Marchena, M.Guarnieri and G.Mc Khann, Glutathione peroxidase levels in brain, J.Neurochem, 22:773 (1974).

43.Y.Nadia, P.L.Mc Geer and E.G. Mc Geer, Lipid peroxides in brain during aging and vitamin E deficiency, Neurobiol.Aging, 3:173 (1982).

OXIDATIVE STRESS, POLY(ADP)RIBOSYLATION AND AGING: IN VITRO STUDIES ON
LYMPHOCYTES FROM NORMAL AND DOWN'S SYNDROME SUBJECTS OF DIFFERENT AGE
AND FROM PATIENTS WITH ALZHEIMER'S DEMENTIA

Franceschi C., Monti D., Cossarizza A., Tomasi A., *Sola P.
#Zannotti M

Istituto di Patologia generale, via Campi 287, 41100 Modena
*Clinica Neurologica, Universitá di Modena
#Istituto di Istologia ed Embriologia generale Universitá di
Bologna, Italy

INTRODUCTION

Free radicals are formed in the body as a consequence of aerobic
metabolism. Cells have developed a variety of antioxidant systems, that
include classical antioxidant enzymes (superoxide dismutase, glutathione
peroxidase, and catalase) as well as nonenzymatic oxy-radicals
scavengers (vitamin E, urea, β-carotene and some more recently described
substances such as carnosine) (1). However, a certain fraction of active
oxygen species escapes the cellular defence and may cause transient or
permanent damage to cellular components. According to one of the most
interesting theory of aging is the "free radical theory of aging",
proposed by D. Harman (2) more than thirty years ago, where oxidative
damage has been suggested as a major cause of aging. One of the
prediction of this theory is an age-related decrease of the efficiency
of antioxidant defence mechanisms.

Recently, other enzymes such as poly(ADP-ribosyl) transferases (ADPRT)
appear to be involved in the cellular response to oxygen free radicals
(OFR) (3). The physiological role of ADPRT is still poorly understood,
even if several data suggest that it may play a role in cell
proliferation and differentiation and DNA repair (4). ADPRT is activated
by DNA single strand breaks directly or indirectly (endonuclease ?)
produced by OFR. Once activated ADPRT utilizes NAD^+ as substrate to add
chains of poly (ADP-ribose) to several nuclear proteins. A great number
of breaks may lead to a marked depletion of intracellular NAD^+ pool and
eventually to cell death (5).

It has been shown that 3-amino benzamide (3-ABA), an inhibitor of
ADPRT, protects human lymphocytes damaged by an oxygen radical producing
system, i.e. xanthine oxidase (XOD) - hypoxanthine (HYP), which mimics
conditions occurring in vivo (6). This observation is at variance with
many other data where 3-ABA increases cytotoxicity by alkylating agents
and other toxic agents (4). Using this interesting experimental systems
we have tested the following hypothesis: i) are peripheral blood
lymphocytes (PBL) from healthy aged people, from subjects with Down's
syndrome (DS) - a classical syndrome of accelerated aging where OFR are
supposed to play an important role (7-8) - and from Alzheimer's disease
(AD) patients more sensitive to OFR in comparison with PBL from young
subject ? ii) are PBL from aged DS and AD subjects protected by 3-ABA as
PBL from young subjects ? iii) is the protective effect of 3-ABA due to
a possible OFR scavenger activity ?

Antioxidants in Therapy and Preventive Medicine
Edited by I. Emerit *et al.*
Plenum Press, New York, 1990

MATERIALS AND METHODS

Subjects. Eleven young subjects (20-25 years old), 9 healthy old donors (80-85 years old), 17 DS children, 4 DS subjects of relatively old age (40-55 years) and 9 AD patients (50-60) were studied.

Lymphocyte cultures and OFR damage. Resting human PBL, suspended in medium containing 10% human inactivated AB serum (3×10^6 cells/ml), were damaged by 1 hour exposure to a fixed concentration of XOD (0.5 mU/ml) plus graded doses of HYP (1-100 μM) in the presence or in the absence of 3-ABA (5 mM) washed, plated in microtiter wells (10^5 cells/well in 0.1 ml) stimulated for 96 hours with an optimal dose of phytohemagglutinin (PHA-P Difco, USA; 1μl/ml) and pulsed with 0.5 μCi tritiated thymidine (^{3}H-TdR, spec. act. 5 C/mmol, Amersham, U.K.) during the last 6 hours, as previously described (9). All other chemicals were purchased from Sigma (USA).

Spin trapping competition experiments The spin trapping 5,5-dimethyl-1-pyrroline-1-oxide (DMPO) was purchased from Sigma and redistilled under vacuum before use. The electron spin resonance (ESR) spectrometer used was a Bruker 200 SRD spectrometer, operated with the usual instruments settings: modulation amplitude 1 G, microwave power 10 mW, scan range 100 G. All recordings were performed at room temperature using a flat cells for liquid sampling. A typical sample used in the spin trapping experiments contained RPMI 1640, 100 μM hypoxanthine, 0.5 U/ml of Xanthine oxidase, and 50 mM DMPO, final volume 200 μl. When indicated desferal 1 mM and 3-ABA 5 mM was added to the solution. The ESR signal was recorded at consecutive time intervals, up to 300 seconds.

RESULTS AND DISCUSSION

The data obtained show that: i) the survival of the lymphocytes from old donors and DS subjects of relatively old age was markedly decreased by low concentration of OFR which had no effects on lymphocytes from young subjects and DS children; ii) higher concentrations of HYP, i.e. a higher concentration of OFR, decrease PBL survival in every groups, being such a damaged more evident in PBL from aged and old DS subjects in comparison with young donors and in DS children; iii) the presence of 3-ABA during the damaging period was able to fully protect cells from young controls and from DS children; in PBL from old people and relatively old DS subjects, 3-ABA was able to protect cells exposed to a high concentration of OFR, but not those damaged by low concentration of OFR; representative experiments are summarized in table 1. iv) the sensitivity to OFR and to 3-ABA of PBL from AD patients was intermediate between that of young and old donors. v) The protective effect of 3-ABA on the survival of PBL damaged by OFR in the XOD-HYP system was likely not due to the OFR scavenger capability of the compound. The potential scavenger effect of 3-ABA towards OFR produced during the HYP-XOD reaction was tested through a competition experiment: a solution containing, RPMI 1640, the enzymatic system, 50 mM DMPO was placed in a ESR flat cell. The developing of the signal due to the trapping of oxygen-derived free radicals was followed in time with or without addition of 3-ABA. A similar experiment was carried out in the presence of 1 mM desferal with the aim of preventing any influence due to the presence of metals.

The scavenging effect of 3-ABA was weak and only evident in serum-free medium (Fig. 1).

These data suggest that PBL from normal healthy subjects of old age and form relatively old DS patients have shown a higher sensitivity to OFR. The major difference between cells of the above groups and those from young subjects and DS children appears to be ascribed to an abnormal sensitivity to damages by low concentration of OFR, which likely do not involved ADPRT activation and NAD^+ depletion. On the whole our data fit

Table 1

Survival of PBL exposed to 0.5 mU/ml XOD and to an increasing concentration of HYP; values express the percentage of thymidine incorporation in comparison to control PBL cultures.

| | hypoxanthine (μM) | | | | | |
	1	5	10	25	50	100
young subject	143	134	133	109	13	9
+ 3-ABA	128	164	117	116	104	76
old subject	44	48	46	25	30	5
+ 3-ABA	76	76	67	38	65	57

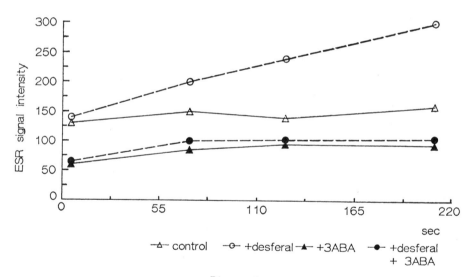

Figure 1

A solution containing RPMI 1640, 100 μM hypoxanthine, 0.5 U/ml of Xanthine oxidase, and 50 mM DMPO, final volume 200 μl, was inserted in the ESR cavity and the OFR-spin adduct formed was measured. Desferal, 1 mM was added in order to suppress any interference due to metals The addition of 3-ABA (5 mM) caused a decrease in the rate of OFR trapping.

with the free radical theory of aging and confirm the hypothesis that the precocious aging observed in Down's syndrome may involve a prooxidative state as suggested by the fact that critical region of chromosome 21, responsible for the syndrome, contains the superoxide dismutase-1 gene (7-8). The protective mechanism exerted by 3-ABA can not be ascribed to its OFR scavenging activity, which was not apparent in experiments performed in the presence of serum and weak when the same experiments were performed in serum free medium. The protective role exerted by inhibitors of ADPRT, such as 3-ABA on OFR-induced damages appear to vary according to the age of the donor, suggesting a role of ADPRT in the aging process at cellular level.

ACKNOWLEDGEMENTS

This work has been supported by grants of the MPI (Italy Rome) Progetti 40% and 60% to CF., by a grant given by SIGMA TAU (Rome). AT acknowledges the financial help given by AICR (St. Andrews, U.K.)

REFERENCES

1) R. Kohen, Y. Yamamoto, K.C. Cundy, and B.N. Ames, Antioxidant activity of carnosine, homocarnosine, and anserine present in muscle and brain. Proc. Natl. Acad. Sci. (USA) 85: 3175 (1988).

2) D. Harman, A theory based on free radicals and radiation chemistry. J. Gerontol. 11: 298 (1956).

3) D.R. Spitz, W.C. Dewey, and C.G. Li, Hydrogen peroxide or heat shock induced resistance to hydrogen peroxide in Chinese hamster fibroblasts, J. Cell Physiol. 131: 364 (1987).

4) S. Shall, ADP-ribosylation of proteins: a ubiquitous cellular control mechanism. In: Advances in post-translational modifications of proteins and aging, V. Zappia, P.Galletti, R.Porta and F. Wold eds, Plenum Press, NY 1988.

5) J.E. Cleaver, C.Borek, K. Milam, and W.F. Morgan, The role of poly (ADP-ribose) synthesis ion toxicity and repair of DNA damage, Pharmacol. Ther. 31: 269 (1987).

6) D.A., Carson, S. Seto, and B. Wasson, Lymphocyte disfunction after DNA damage by toxic oxygen species. A model for immunodeficiency. J. Exp. Med. 163: 746 (1986).

7) C. Franceschi, F. Licastro, M. Chiricolo, M. Zannotti, and M. Masi, Premature senility in Down's syndrome: a model for and an approach to the molecular genetic of the ageing process. In: Immunoregulation in Aging, A. Facchini, J.J. Harman, and G. Labó, Eurage, Rijswijk (1986).

8) J. Kedziora and G. Bartosz, Down's syndrome: a pathology involving the lack of balance of reactive oxygen species, Free Rad. Biol. Med. 4: 317 (1988).

9) P. Sola, E. Merelli, P. Faglioni, D. Monti, A Cossarizza, and C. Franceschi, DNA repair, sensitivity to gamma radiation and to heat shock in lymphocytes from acute, untreated multiple sclerosis patients, J. Neuroimmunol. 21: 23 (1989).

PROTEIN OXIDATION AND PROTEOLYTIC DEGRADATION

GENERAL ASPECTS AND RELATIONSHIP TO CATARACT FORMATION

Kelvin J. A. Davies

Institute for Toxicology & Department of Biochemistry
The University of Southern California
Los Angeles, California, U. S. A.

GENERAL ASPECTS OF INTRACELLULAR PROTEIN DEGRADATION

The past few years have seen an explosion of knowledge on the subject of intracellular protein degradation. The view that lysosomes (or rather intra-lysosomal proteases) are primarily responsible for degrading intracellular proteins has now been discredited, and a large number of cytoplasmic proteolytic enzymes have been discovered. It now appears that lysosomes are mostly responsible for the degradation of cellular organelles, whereas most cytoplasmic proteins are degraded by soluble, cytoplasmic proteinases, proteases, and peptidases.[1]

Because the terminology in the proteolysis field is somewhat confusing it is worth taking a little space to define a few key terms. A proteinase is an initial recognition and cleavage enzyme. A protease can degrade the large peptides generated by proteinases and further cleave them to small peptides. A peptidase can degrade the small peptides generated by proteases to produce free amino acids. The terms "protein degradation," "proteolysis," and "protein cleavage" are used interchangeably to denote the peptide bond hydrolysis by proteolytic enzymes. In contrast, the term "protein fragmentation" refers to the non-enzymatic, direct chemical breakdown of proteins. Protein fragmentation can be caused by oxidants such as the hydroxyl radical,[2-7] and may occur at sites other than peptide bonds.[2-6]

A major stimulus to the study of intracellular proteolysis has been the discovery of ATP- and ubiquitin-stimulated proteolytic systems in eucaryotes.[8-10] In procaryotes ATP-stimulated proteolytic systems operate without the need for ubiquitin.[9,10] In eucaryotes ATP appears to be required for the conjugation of ubiquitin (a small peptide) to certain abnormal proteins. ATP is again required to stimulate the activity of a proteinase which recognizes protein-ubiquitin conjugates.[8-10] In procaryotes (which have no ubiquitin) ATP appears to directly stimulate the activity of two cytoplasmic proteinases, named "La" and "Ti."[11,12]

Several abnormal proteins, truncated polypeptides, and products of nonsense/missense mutations are thought to be degraded by ATP- or ATP-ubiquitin-stimulated proteolytic pathways.[7-12] In the past few years, however, it has become clear that oxidatively modified proteins are degraded by ATP- and ubiquitin-independent proteolytic pathways in both eucaryotes and procaryotes.[2-6,13-33] These newly discovered, non-lysosomal and ATP/ubiquitin-independent, proteolytic pathways form the main subject of this review.

Antioxidants in Therapy and Preventive Medicine
Edited by I. Emerit *et al.*
Plenum Press, New York, 1990

INTRACELLULAR DEGRADATION OF OXIDATIVELY MODIFIED PROTEINS

Mechanisms of Protein Oxidation

Although lipids have long been accepted as major targets for intracellular oxidation, the proof that proteins are readily oxidized *in vivo* has only come in the past few years. Two major mechanisms for intracellular protein oxidation have been demonstrated: 1) Oxidation by free radicals or other activated species, and 2) Oxidation catalyzed by mixed-function oxidases.[33,37-44] Oxidation by free radical (often oxygen radical) species and other activated (often oxygen based) species may be considered as a somehwat random process of cytotoxicity. Oxidation by mixed-function oxidase enzymes may be used as a directed mechanism for "marking" proteins to be degraded. Although these two mechanisms may appear to be quite distinct, they actually share many common features such as oxygen activation, transition-metal catalysis, and product profiles. Finally, the products of both free radical oxidation and mixed-function oxidase oxidation exhibit similar increases in proteolytic susceptibility.[2-7,13-44]

Degradation of Oxidatively Modified Proteins

Following exposure to oxidative stress, both eucarotes and procaryotes exhibit large increases in protein degradation.[2-7,13-44] When purified proteins are oxidatively modified by exposure to oxygen radicals or hydrogen peroxide,[2-6,13-32] or by exposure to mixed-function oxidation systems,[33,37-44] the modified proteins are rapidly and selectively degraded if they are subsequently incubated with cell-free extracts (which retain the proteolytic systems of the original intact cells). Such experiments have been conducted with various cells and extracts, including erythrocytes, reticulocytes, skeletal muscles, cardiac muscles, hepatocytes, *Escherichia coli* and *Salmonella* strains.

As has been observed for the degradation of other proteins,[1,9-12] oxidatively modified proteins are completley degraded to free amino acids in both intact cells and cell-free extracts.[2-6,13-44] This observation implies a close cooperation between proteinases, proteases, and peptidases; and may, further, indicate that proteinase activity (initial recognition and cleavage) is rate-limiting whereas peptidase activity (final cleavage of small peptides to amino acids) is in excess of all other components.

Although many proteins are degraded by an ATP- or ATP-ubiquitin-stimulated proteolytic pathway,[1,8-12] the degradation of oxidatively modified proteins is neither stimulated by, nor dependent upon, ATP or ubiquitin.[2-6,13-33] In many cases ATP actually inhibits the degradation of oxidatively modified proteins in cell extracts and further purified fractions. Furthermore, *E. coli lon⁻* mutants which lack the ATP-stimulated protease La still degrade oxidatively modified proteins at the same rate as wild-type strains.[17] From such results we came to the early conclusion that oxidatively modified proteins must be degraded by a proteolytic pathway or system distinct from the ATP-(ubiquitin)-stimulated system.

Macroxyproteinase (M.O.P.) and Other Proteolytic Enzymes

Mammalian Cells. In erythrocytes some 80-90% of the proteolytic activity against oxidatively modified protein is expressed in a single "high molecular weight proteinase complex."[18,29,30] This proteinase complex has a molecular size of 600,000-700,000 daltons in various mammalian cells, and is composed of 8-12 discrete polypeptides. In naming the complex we chose the prefix "macro-" (previously suggested by Rechsteiner *et al.*[45]) because of its large size. "Oxy" denotes the preference of the complex for oxidatively modified protein substrates, and the suffix "-proteinase" was chosen as a generic and easily identifiable functional title. Finally, we abbreviated "Macro-oxy-proteinase" to "Macroxyproteinase" for convenience. The resultant acronym "M.O.P." helps us to remember that the complex "mops-up" oxidatively damaged proteins.

A high molecular weight proteinase complex with properties similar or identical to M.O.P. has been reported to exist in all mammalian cell-types studied; however, the function of this complex was previously unknown and there were conflicting reports of inhibitor profiles in the literature (see refs. 45 and 46 for reviews). We have now shown that the red cell M.O.P. complex contains multiple proteolytic sites including sulfhydryl-enzymes, metallo-enzymes, and serine-enzymes.[18,29,30] These studies have used a variety of proteolytic substrates including oxidatively modified red-cell hemoglobin and superoxide dismutase, "foreign proteins," and several fluorogenic peptides. We find that the relative importance of each of the three active sites (sulfhydryl, metallo, and serine) varies with the proteolytic substrate, as does the effectiveness of various inhibitors.

As indicated above, our work and literature evaluations suggest that the 600,000-700,000 dalton proteinase complex from many mammalian cells represents different forms of M.O.P. Furthermore, it appears likely (to us) that the 300,000 dalton proteinase complex which Rivett[41,46] partially purified from liver, on the basis of its activity against oxidatively modified glutamine synthetase, represents a fragment (approximately one-half) of M.O.P. In our hands M.O.P. exhibits no ATP or ubiquitin stimulation (in fact, ATP is typically inhibitory) which agrees well with the lack of ATP stimulation reported for the 600,000-700,000 proteinase complex by other investigators.[45,46]

The larger, approximately 1,500,000 dalton, proteinase complex of reticulocytes (and probably other mammalian cells) appears to account for most of the ATP-ubiquitin-stimulated proteolysis in red cells.[47,48] Names such as Megapain and Megasin have been suggested for this proteinase compelx.[45] In our hands the 1,500,000 proteinase complex can account for no more than 10-15% of the degradation of oxidatively damaged red cell proteins such as hemoglobin.[13,29,30] Furthermore, the compelx exhibits no ATP or ubiquitin stimulation in the degradation of oxidatively modified proteins, yet the degradation of casein is stimulated 3-4 fold by ATP and ubiquitin (plus "conjugating fraction").[18,29,30]

Bacterial Cells. Bacteria such as *E. coli* possess active proteolytic systems which preferentially degrade the oxidatively modified forms of proteins.[2,3,6,16,17,21,23,27,28,33,37-40,42-44] Both endogenous proteins (oxidatively modified *in vivo*) and exogenous proteins (added to cell-free extracts) become excellent proteolytic substrates following oxidative modification. As reported above for mammalian cells, bacteria exhibit no ATP stimulation in the degradation of oxidatively modified proteins and the ATP-stimulated proteases "La" and "Ti"[11,12] do not preferentially degrade oxidized protein substrates.[2,16,17,21] The ATP-independent proteolytic enzymes, "Do," "Re," "Mi," "Fa," and "So" preferentially degrade several oxidatively damaged proteins,[2,16,17,21] and Levine *et al.* have reported an (ATP-independent) enzyme which has selective activity against an oxidatively modified form of glutamine synthetase. These studies indicate that bacteria use several proteolytic enzymes to degrade oxidatively modified proteins. The relative importance of each proteinase or protease appears to vary with the particular protein substrate. It is tempting to hypothesize that this series of bacterial proteinases and proteases may represent an evolutionary forerunner for the mammalian M.O.P. complex of enzymes.

Denaturation and Hydrophobicity as Proteolytic Signals

In our studies of both eucaryotic and procaryotic protein degradation we observed a relationship between protein unfolding (denaturation) or hydrophobicity, and proteolytic susceptibility.[2-6,13-32] Our results indicated that both denaturation and hydrophobicity increase with mild oxidative modification. More severe protein oxidation (especially in the absence of oxygen) results in the formation of covalent cross-links (both intra- and inter-molecular) which prevent further denaturation. Proteolytic susceptibility increases linearly with mild denaturation/hydrophobicity, exhibiting correlation coefficients (linear regression) of 0.98 for erythrocytes[6] and 0.96-0.98 for *E. coli*,[16] until the point at which covalent cross-linking prevents further denaturation. Samples which are oxidatively modified in the presence of high oxygen concentrations, however, exhibit fragmentation rather than covalent cross-linking, and both denaturation/hydrophobicity and proteolytic susceptibility continue to increase even with severe oxidative damage.[2-6,13-32]

The above results led us to first propose that denaturation and hydrophobicity could represent a "signal" for the intracellular proteolysis of oxidatively modified proteins.[2,6] Recently, Cervera and Levine[49] published work which confirms our original hypothesis. Our further work with M.O.P.,[18,29,30] and evaluations of the literature reports of M.O.P.-like enzyme complexes[45-48] have further confirmed our view that denaturation/hydrophobicity is the mechanism by which oxidative modification converts proteins into proteolytic substrates. Both Hough, Pratt, and Rechsteiner's[48] studies (of the red-cell 600,000-700,000 dalton proteinase complex) and our own studies[13,29,30] (of M.O.P.) have shown a distinct preference for hydrophobic and bulky (aromatic) amino acid sequences. We have, therefore, proposed that the exposure of hydrophobic and bulky amino acid residues, which occurs during oxidative denaturation, provides new proteolytic cleavage sites for M.O.P. In other words, we propose that M.O.P. exists in an active form in mammalian cells. Undamaged proteins "hide" their hydrophobic and bulky residues within their tertiary (and quaternary) structures and are, therefore, poor substrates for M.O.P. During oxidative modification, however, partial unfolding (only a minor degree of unfolding is necessary) exposes these hydrophobic/bulky amino acid residues resulting in an overall increase in hydrophobicity.[2-6,15-32] The newly exposed hydrophobic/bulky amino acid residues provide peptide bonds which are highly susceptible to hydrolysis by M.O.P., and the initial cleavage of oxidatively damaged proteins (the rate-limiting step in complete degradation) begins.

Macroxyproteinase has been described as a cysteine proteinase, a serine proteinase, a multicatalytic proteinase, and a multisubunit proteinase.[45-48] Our recent work has confirmed each of these titles, and has revealed a metallo-proteinase activity (inhibited by transition metal chelators).[18,29,30] Furthernmore, the products of protein digestion by the M.O.P. complex appear to be free amino acids rather than peptides.[18,29,30] We, thus, propose that M.O.P. is a proteolytic complex of proteinase, protease, and peptidase activities which cooperate to degrade oxidatively denatured protein substrates by a processive mechanism. Work by Arrigo et al.[50] indicates that M.O.P. is shaped rather like a doughnut with a central hole or cavity. It is tempting to speculate that hydrophobic/bulky amino acid residues may bind across, or within, the M.O.P. cavity and then allow the whole protein to be "threaded" through the M.O.P. complex. Such speculation implies an unfoldase or unwindase activity for M.O.P.; in addition to the cooperative cysteine, serine, and metallo active sites.

PROTEIN OXIDATION, PROTEOLYSIS, AND CATARACT FORMATION IN THE EYE LENS

Lens Structure and Function

The function of the eye lens is to collect and focus light on the retina. In order to do so it must remain clear through life. The lens is composed of approximately 60% water and 38% (w/w) proteins; primarily a set of long-lived, organ-specific gene products called the crystallins.

Although it appears totally homogenous, the lens is a highly organized structure. formed during embryogenesis the eye lens develops from invaginated ectodermal tissue. After establishment of polarity, epithelial cell division is observed at the anterior equatorial region. Certain epithelial cells migrate to the equatorial region where they begin to express the lens specific major gene products, the crystallins. Continued crystallin elaboration and perfect close packing of these molecules result in an elongated fiber cell. As new epithelial cells are produced, the more mature fiber cells are compressed toward the interior regions. When crystallin synthesis is largely complete, most, if not all, of the subcellular organelles (such as lysosomes, mitochondria, and nuclei) are lost. thus, the lens core or nucleus is composed of cells and proteins elaborated during fetal stages of development; it contains the oldest tissue.

The epithelium or youngest tissue is made of cells which contain all their organelles but which have yet to express their major gene products and elongate. The cortex contains cells at various developmental stages between thesee two extremes. The ability to segregate and examine differentially aged or maturing lens tissue makes it possible to perform cross-sectional analysis and obtain revealing information about biochemical events in the lens during development, maturation and aging. A collagenous capsule surrounds the lens.

Cataract Formation

Extensive postsynthetic modification of lens proteins occurs during aging and it has been estimated that over 90% of the major crystallins may be damaged in the aged lens. One of the major causes of crystallin damage is photooxidative insult.

Protein turnover in the lens is extremely slow. It appears that much of the protein produced during fetal development is retained throughout life.[51-54] During this prolonged time in diurnal animals, lens proteins are exposed to ultraviolet light and a variety of active oxygen species.[53-62] Exposure to these high-energy species in cell-free systems and in vivo has been correlated with oxidation and aggregation of many lens proteins.[51,53-64] Thus, in the aged lens there is enhanced formation of disulfide and bityrosine bonds, deamidation, aggregation, dehydration, and, in the extreme, protine precipitation or cataract. Even cataract induction by abnormal levels of hexoses has been related to oxidative stress.[65] It should, however, be noted that not all forms of cataract are related to photooxidative stress.

The interplay of several factors may actually determine the ultimate fate of lens proteins. These include the degree of photooxidative stress to which lens proteins are exposed, the adequacy and efficiency of antioxidant enzymes and compounds, and the effective operation of lens proteolytic enzymes in degrading oxidatively damaged lens proteins. High photooxidative stress, diminished antioxidant capacity, or decreased proteolytic capacity could all contribute to increased lens protein damage and aggregation (i.e., cataract formation). The extent of the problem from a public health perspective can be appreciated if one considers that cataract is one of the major causes of blindness worldwide and that over 50,000,000 people suffer from this debility.

<u>Degradation of Oxidized Proteins in the Lens</u>

Lens cells contain both endo-proteinase activities and exo-peptidase activities. A high molecular weight endo-proteinase (approximately 600,000 daltons) appears to be the major endo-protoelytic activity in the lens: recent, collaborative, studies indicate that this endo-proteinase is extremely similar if not identical to the red-cell M.O.P. complex.[31,32] Leucine aminopeptidase appears to be the major exo-peptidase activity in lens.

In recent, collaborative, work we have demonstrated that purified M.O.P. from lens epithelium (the youngest lens tissue) selectively degrades the oxidatively denatured form of lens α-crystallin (oxidatively modified by exposure to the hydroxyl radical).[31,32] The increased susceptibility of α-crystallin to degradation by lens M.O.P. is observed with mild oxidative denaturation (2-3 mol of hydroxyl radical per mol of crystallin). With greater oxidative stress extensive cross-linking of α-crystallin occurs and the ability of lens M.O.P. to degrade the damaged crystallin declines rapidly. Since erythrocytes and lens cells share many common features it is, perhaps, not surprising that M.O.P. may perform a similar function in both cell types; the degradation of oxidatively denatured proteins.

Exposure to photooxidative stress by UV light treatment has been used to model cataract formation *in vivo*.[63,64] Such treatment causes extensive cross-linking of lens crystallins and severe opacity of the lens. Blondin *et al.*[64] have also demonstrated a clear decline in lens exopeptidase activity with exposure to UV light. Thus proteolytic activity (or capacity) may also be compromised by photooxidative stress.

<u>Diminished Proteolytic Capacity in Cataract and Aging</u>

As stated above, photooxidative stress (of the kind associated with lens opacity or cataract) can result in diminished proteolytic capacity in the lens. Cataract is primarily a disease associated with aging and so it is instructive to monitor age-related changes in lens tissue. Several reports in the literature indicate that lens proteolytic capacity declines with the age of individual lens cells, and with the age of the whole lens. Proteolytic enzymes such as calpain I and II, the neutral endoproteinase (which we have called M.O.P.), cathepsin B, and the aminopeptidases (including leucine aminopeptidase) all exhibit clear age or maturation related declines in activity.[66-72]

<u>Antioxidant Protection Against Cataract</u>

The activities of several lens enzymes involved in antioxidant defenses are reported to decline with age.[15,73] A growing literature suggests that ascorbic acid (or ascorbate, or vitamin C) can protect lens components against photooxidative stress. Studies performed *in vitro* indicate that ascorbate can delay UV light induced alterations to lens proteins, proteases, lipids, and the sodium/potassium pump.[55,57,63] Scorbutic guinea pigs reportedly experience a significantly greater incidence of cataract formation than do non-scorbutic animals.[74,75] Supporting arguments for a protective role of ascorbate have been gleaned from the observations that human lens contains 30 times more ascorbate than does blood,[76,77] and that cataractous lenses have lower ascorbate contents than do cataract-free lenses.[78-80]

It would be misleading to suggest that human cataract can now be prevented by antioxidant therapy. Nevertheless, some interesting and potentially important discoveries have been made using the physiological antioxidant ascorbic acid. In studies in which guinea pigs were maintained on either low or high dietary ascorbate regimens, Blondin *et al.*[63] found that lenses from the animals fed a low level of ascorbate in the diet were more susceptible to photooxidative stress than lenses from animals fed a high level of dietary ascorbate. Both protein cross-linking, and loss of exopeptidase activity were more pronounced in the lenses from animals given a diet which was low in ascorbate than in lenses from guinea pigs which were fed high levels of dietary ascorbate.[63] Lenses from animals on the high ascorbate diet were found to contain more than three times the concentration of ascorbate found in lenses from guinea pigs on the low ascorbate diet. Furthermore, in a subsequent study, it was found that levels of both total ascorbate and reduced ascorbate increased in the lens in proportion to the dietary intake, up to a daily intake of approximately 11 mg/animal (this intake is some ten times higher than that required to prevent scurvy).[81]

Summary and Conclusions

1) Intracellular proteins are subject to oxidative and photooxidative denaturation.

2) Proteolytic systems recognize and selectively degrade oxidatively denatured, and photooxidatively denatured proteins. By degrading mildly denatured proteins these proteolytic systems prevent further oxidative/photooxidative damage which could otherwise result in the formation of cross-linked (undigestible) proteins, or protein fragments with toxic biological activities. Proteolytic systems also provide amino acids for the synthesis of new (replacement) proteins.

3) A 700,000 dalton neutral endoproteinase, which we have called macroxyproteinase or M.O.P., appears to be mostly responsible for the degradation of oxidatively denatured proteins. M.O.P. has been shown to function in red blood cells and in the eye lens, and appears to also exist in many other mammalian cell types.

4) Cataract is a disease associated with aging, and with photooxidative denaturation (and cross-linking) of lens crystallins and other proteins.

5) Both cataract and aging of lens cells are associated with declining proteolytic capacity and diminished antioxidant protection.

6) Lens aging and *in vivo* photooxidative stress can cause opacity ("cataract"), cross-linking of crystallins, and diminished proteolytic capacity.

7) High levels of dietary ascorbate increase ascorbate concentrations in lens tissue, and are associated with greater resistance of lens proteins and lens proteolytic enzymes to oxidative/photooxidative stress *in vitro*.

Over 95% of the dry mass of the eye lens consists of specialized proteins called crystallins. Aged lenses are subject to cataract formation, in which damage, cross-linking, and precipitation of crystallins contribute to a loss of lens clarity. Cataract is one of the major causes of blindness, and it is estimated that over 50,000,000 people suffer from this disability. Damage to lens crystallins appears to be largely attributable to the effects of UV radiation and/or various active oxygen species. Photooxidative damage to lens crystallins is normally retarded by a series of antioxidant enzymes and compounds. Crystallins which experience mild oxidative damage are rapidly degraded by a system of lenticular proteases, particularly M.O.P. Extensive oxidation and cross-linking, however, severely decrease the proteolytic susceptibility of lens crystallins. Thus, in the young lens the combination of antioxidants and proteases serves to prevent crystallin damage and precipitation in cataract formation. The aged lens, however, exhibits diminished antioxidant capacity and decreased proteolytic capabilities. The loss of proteolytic activity may actually be partially attributable to oxidative damage which proteases (like any other protein) can sustain. We propose that the rate of crystallin damage increases as antioxidant capacity declines with age. The lower protease activity of aged lens cells may be insufficient to cope with such rates of crystallin damage, and denatured crystallins may begin to accumulate. As the concentration of oxidatively denatured crystallins rises, cross-linking reactions may produce insoluble aggregates which are refractive to protease digestion. Such a scheme could explain many events which are known to contribute to cataract formation, as well as several which have appeared to be unrelated. This hypothesis is also open to experimental verification and intervention. Recent studies further indicatre that ascorbate may have promise as a cataract retarding agent *in vivo*, when administered by dietary supplementation.

ACKNOWLEDGEMENTS

My thanks to Dr. Allen Taylor, without whose collaborative efforts (in previous work[15,31,32]) this chapter would not have been possible. The studies reported in this chapter were made possible by grant support from the National Institutes of Health/National Institute of Environmental Health Sciences (ES 03598 and ES 03785).

REFERENCES

1. J. S. Bond and P. E. Butler, Intracellular proteases, <u>Annu. Rev. Biochem.</u> 56:333 (1987).
2. K. J. A. Davies, Intracellular proteolytic systems may function as secondary antioxidant defenses: An hypothesis, <u>J. Free Radicals Biol. Med.</u> 2:155 (1986).
3. K. J. A. Davies, Protein damage and degradation by oxygen radicals. I. General aspects, <u>J. Biol. Chem.</u> 262:9895 (1987).
4. K. J. A. Davies, M. E. Delsignore, and S. W. Lin, Protein damage and degradation by oxygen radicals. II. Primary structure, <u>J. Biol. Chem.</u> 262:9902 (1987).
5. K. J. A. Davies and M. E. Delsignore, Protein damage and degradation by oxygen radicals. III. Secondary and tertiary structure, <u>J. Biol. Chem.</u> 262:9908 (1987).
6. K. J. A. Davies, S. W. Lin, and R. Pacifici, Protein damage and degradation by oxygen radicals. IV. Degradation of denatured protein, <u>J. Biol. Chem.</u> 262:9914 (1987).
7. S. P. Wolff, A. Garner, and R. T. Dean, Free radicals, lipids, and protein degradation, <u>Trends Biochem. Sci. (TIBS)</u> 11:27 (1986).
8. A. Hershko and A. Ciechanover, Mechanisms of intracellular protein breakdown, <u>Annu. Rev. Biochem.</u> 51:335 (1982).
9. A. L. Goldberg and F. J. Dice, Intracellular protein degradation in mammalian and bacterial cells, <u>Annu. Rev. Biochem.</u> 43:835 (1974).
10. A. L. Goldberg and A. C. St. John, Intracellular protein degradation in mammalian and bacterial cells. II, <u>Annu. Rev. Biochem.</u> 45:747 (1976).
11. A. L. Goldberg and L. Waxman, The role of ATP hydrolysis in the breakdown of proteins and peptides by protease La from *Escherichia coli*, <u>J. Biol. Chem.</u> 260:12029 (1985).
12. B. J. Hwang, W. J. Park, C. H. Chung, and A. L. Goldberg, *Escherichia coli* contains a soluble ATP-dependent protease (Ti) distinct from protease La, <u>Proc. Natl. Acad. Sci. U. S. A.</u> 84:5550 (1987).
13. K. J. A. Davies and A.L. Goldberg, Oxygen radicals stimulate intracellular proteolysis and lipid peroxidation by independent mechanisms in erythrocytes, <u>J. Biol. Chem.</u> 262:8220 (1987).
14. K. J. A. Davies and A. L. Goldberg, Proteins damaged by oxygen radicals are rapidly degraded in extracts of red blood cells, <u>J. Biol. Chem.</u> 262:8227 (1987).
15. A. Taylor and K. J. A. Davies, Protein oxidation and diminished proteolytic capacity in cataract formation during aging, <u>Free Radical Biol. Med.</u> 3:371 (1987).
16. K. J. A. Davies and S. W. Lin, Degradation of oxidatively denatured proteins in <u>Escherichia coli, Free Radical Biol. Med.</u> 5:215 (1988).
17. K. J. A. Davies and S. W. Lin, Oxidatively denatured proteins are degraded by an ATP-independent pathway in *Escherichia coli*, <u>Free Radical Biol. Med.</u> 5:225 (1988).
18. D. C. Salo, S. W. Lin, R. E. Pacifici, and K. J. A. Davies, Superoxide dismutase is preferentially degraded by a proteolytic system from red blood cells following oxidative modification by hydrogen peroxide, <u>Free Radical Biol. Med.</u> 5:335 (1988).
19. O. Marcillat, Y. Zhang, S. W. Lin, and K. J. A. Davies, Mitochondria contain a proteolytic system which can recognize and degrade oxidatively denatured proteins, <u>Biochem. J.</u> 254:677 (1988).
20. O. Marcillat, Y. Zhang, and K. J. A. Davies, Oxidative and non-oxidative mechanisms in the inactivation of cardiac mitochondrial electron transport chain components by doxorubicin, <u>Biochem. J.</u> 254:677 (1988).
21. K. J. A. Davies, Proteolytic systems as secondary antioxidant defenses, <u>in</u>: "Cellular Antioxidant Defense Mechanisms," C. K. Chow, ed., Vol. 2, p. 25, CRC Press, Boca Raton (1988).
22. K. J. A. Davies, Free radicals and protein degradation in human red blood cells, <u>in</u>: "Cellular and Molecular Aspects of Aging: The Red Cell as a Model," J. W. Eaton, D. K. Konzen, and J. G. White, eds., p. 15, Alan R. Liss, New York (1985).
23. K. J. A. Davies, The role of intracellular proteolytic systems in antioxidant defenses, <u>in</u>: "Superoxide and Superoxide Dismutase in Chemistry, Biology, and Medicine," G. Rotillio, ed., p. 443, Elsevier, Amsterdam (1986).
24. K. J. A. Davies, Protein oxidation, protein cross-linking, and proteolysis in the formation of lipofuscin: Rationale and methods for the measurement of protein degradation, <u>in</u>: "Lipofuscin-1987: State of the Art," I. Zs.-Nagy, ed., p. 109, Elsevier Science, Amsteram (1988).
25. K. J. A. Davies, Oxidative stress causes protein degradation and lipid peroxidation by different mechanisms in red blood cells, <u>in</u>: "Lipid Peroxidation in Biological Systems," A. Sevanian, ed., p. 100, American Oil Chemists Society, Champaign, Ilinois (1988).

26. K. J. A. Davies, Possible importance of proteolytic systems as secondary antioxidant defenses during ischemia-reperfusion injury, in: "The Role of Oxygen Radicals in Cardiovascular Diseases," A. L'Abbate and F. Ursini, eds., p. 143, Kluwer Academic Publishers, Dortrecht (1988).

27. R. E. Pacifici, S. W. Lin, and K. J. A. Davies, The measurement of protein degradation in response to oxidative stress, in: "Oxygen Radicals in Biology and Medicine," M. G. Simic and K. A. Taylor, eds., p. 531, Plenum Press, New York (1988).

28. K. J. A. Davies, Intracellular proteolytic systems as secondary antioxidant defenses, in: "Oxygen Radicals in Biology and Medicine," M. G. Simic and K. A. Taylor, eds., p. 575, Plenum Press, New York (1988).

29. R. E. Pacifici and K. J. A. Davies, A 700-kDa proteinase which selectively degrades oxidatively denatured hemoglobin, FASEB J. 2:A1007 (1988).

30. K. J. A. Davies, S. W. Lin, and R. E. Pacifici, The degradation of oxidatively denatured proteins: A housekeeping function of M. O. P., International Committee on Proteolysis (I. C. O. P.) Newsletter, p. 3, August (1988).

31. A. Taylor, B. Blondin, K. J. A. Davies, and K. Murakami, Relationships between ascorbate levels, accumulation of damaged proteins, and proteolytic capabilities in the presence and absence of photooxidative stress to the guinea pig eye lens, Abstracts of the fourth International Congress on Oxygen Radicals, W-20 (1987).

32. K. Murakami, J. H. Jahngen, S. W. Lin, K. J. A. Davies, and A. Taylor, A lens protease which shows enhanced rates of degradation of oxidatively modified alpha-crystallin, Free Radical Biol. Med. (1989, in press).

33. R. L. Levine, C. N. Oliver, R. M. Fulks, and E. R. Stadtman, Turnover of bacterial glutamine synthetase: oxidative inactivation precedes proteolysis, Proc. natl. Acad. Sci. U. S. A. 78:2120 (1981).

34. R. T. Dean and J. K. Pollak, Endogenous free radical generation may influence proteolysis in mitochondria, Biochem. Biophys. Res. Commun. 126:1082 (1985).

35. R. T. Dean, S. M. Thomas and A. Garner, Free-radical-mediated fragmentation of monoamine oxidase in the mitochondrial membrane, Biochem. J. 240:489 (1986).

36. S. P. Wolff and R. T. Dean, Fragmentation of proteins by free radicals and its effect on their susceptibility to enzymatic hydrolysis, Biochem. J. 234:399 (1986).

37. L. Fucci, C. N. Oliver, M. J. Coon, and E. R. Stadtman, Inactivation of key metabolic enzymes by mixed-function oxidation reactions: Possible implication in protein turnover and aging, Proc. Natl. Acad. Sci. U. S. A. 80:1521 (1983).

38. R. L. Levine, Oxidative inactivation of glutamine synthetase: I. Inactivation is due to loss of one histidine residue, J. Biol. Chem. 258:11823 (1983).

39. R. L. Levine, Oxidative modification of glutamine synthetase: II. Characterization of the ascorbate model system, J. Biol. Chem. 258:11828 (1983).

40. K. Nakamura and E. R. Stadtman, Oxidative inactivation of glutamine synthetase subunits, Proc. Natl. Acad. Sci. U. S. A. 81:2011 (1984).

41. A.J. Rivett, Preferential degradation of the oxidatively modified form of glutamine synthetase by intracellular mammalian proteases, J. Biol. Chem. 260:300 (1985).

42. E. R. Stadtman and M. E. Wittenberger, Inactivation of Escherichia coli glutamine synthetase by xanthine oxidase, nicotinate hydroxylase, horseradish peroxidase, or glucose oxidase: effects of ferredoxin, putidaredoxin and manadione, Arch. Biochm. Biophys. 239:379 (1985).

44. J. E. Roseman and R. L. Levine, Purification of a protease from Escherichia coli with specificity for oxidized glutamine synthetase, J. Biol. Chem. 252:2101 (1987).

45. R. Hough, G. Pratt, M. Rechsteiner, J. S. Bond, and M. Orlowski, 'A rose by any other name'--or opinions of naming enzymes, International Committee on Proteolysis (I. C. O. P.) Newsletter, p. 3, January (1988).

46. A. J. Rivett, The multicatalytic proteinase of mammalina cells, Arch. Biochem. Biophys. 268:1 (1989).

47. L. Waxman, J. M. Fagan, and A. L. Goldberg, Demonstration of two distinct high molecular weight proteases in rabbit reticulocytes, one of which degrades ubiquitin conjugates, J. Biol. Chem. 262:2451 (1987).

48. R. Hough, G. Pratt, and M. Rechsteiner, Purification of two high molecular weight proteases from rabbit reticulocyte lysates, J. Biol. Chem. 262:8303 (1987).

49. J. Cervera and R. L. Levine, Modulation of the hydrophobicity of glutamine synthetase by mixed-function oxidation, FASEB J. 2:2591 (1988).

50. A.-P. Arrigo, K. Tanaka, A. L. Goldberg, and W. J. Welch, Identity of the 19S 'prosome' particle with the large multifunctional protease complex of mammalian cells (the proteasome), Nature 331:192 (1988).

51. J. J. Harding and M. J. C. Crabbe, The lens: development, proteins, metabolism and cataract, in: "The Eye," M. Davson, ed., Vol. 1B, p. 207, Academic Press, New York (1984).

52. A. M. J. Blow, Proteolyses in the lens, in: "Proteinase in Mammalian Cells and Tissues," A. J. Barrett, ed., North Holland Publishing Co., New York, p. 501 (1979).

53. H. J. Hoenders and H. Bloemendal, Aging of lens proteins, in: Molecular and Cellular Biology of the Lens," H. Bloemendal, ed., p. 279, John Wiley and Sons (1981).

54. J. J. Harding, Changes in lens proteins in cataract, in: "Molecular and Cellular Biology of the Lens," H. Bloemendal, ed., John Wiley and Sons, New York, p. 327 (1981).

55. S. D. Varma, D. Chand, Y. R. Sharma, J. R. Kuck, Jr., and R. D. Richards, Oxidative stress on lens and cataract formation: role of light and oxygen, Curr. Eye Res. 3:35 (1984).

56. J. S. Zigler and J. D. Goosey, Singlet oxygen as a possible factor in human senile nuclear cataract development, Curr. Eye Res. 3:59 (1984).

57. J. S. Zigler, H. M. Jernigan, N. S. Perlmutter, nd J. H. Kinoshita, Photodynamic cross-linking of polypeptides in intact rat lens, Exp. Eye Res. 35:239 (1982).

58. S. D. Varma, S. Kumar, and R. D. Richards, Light induced damage to ocular lens cation pump-prevention by vitamin C, Proc. Natl. Acad. Sci. 76:3501 (1979.

59. M. H. Garner and A. Spector, Selective oxidation of systeine and methionine in normal and senile cataractous lenses, Proc. Natl. Acad. Sci. U. S. A. 77:1274 (1980).

60. O. Roy, J. Dillon, W. Wada, W. Chaney, and A. Spector, Nondisulfide polymerization of gamma and beta crystallin in the humn lens, Proc. Natl. Acad. Sci. U. S. A. 81:2878 (1984).

61. J. S. Zigler, Jr. and H. H. Hess, Cataracts in the Royal College of Surgeon rats: evidence for initiation by oipid peroxidation products, E.p. Eye Res. 41:67 (1985).

62. S. Zigman, The role of sunlight in human cataract formation, Survey of Cataract Formation 27:317 (1983).

63. J. Blondin, V. Baragi, E. Schwartz, J. A. Sadowski, and A. Taylor, Delay of UV-induced eye lens protein damage in guinea pigs by dietary ascorbate, J. Free Radicals Biol. Med. 2:275 (1986).

64. J. Blondin and A. Taylor, Measures of leucine aminopeptidase can be used to anticipate UV-induced age-related damage to lens proteins, Mech. Aging Develop. 41:39 (1987).

65. N. H. Ansari, A. Schulter, and S. K. Srivastava, Antioxidant (BHT) significantly delays galactose cataract formation, Invest. Ophthalmol. Vis. Sci. 28:192 (1987).

66. K. K. Sharma and B. J. Ortwerth, Isolation and characterization of a new aminopeptidase from bovine lens, J. Biol. Chem. 261:4295 (1986).

67. H. Yoshida, T. Murachi, and I. Tsukahara, Distribution of calpain I, calpain II, and calpastatin in bovine lens, Invest. Ophthalmol. Vis. Sci. 25:953 (1985).

68. D. A. Eisenhauer and A. Taylor, protease activities in cultured bovine lens epithelial cells of various passage, Invest. Ophthalmol. Vis. Sci. 28:384 (1987).

69. A. Taylor, Leucine aminopeptidase activity is diminished in aged hog, beef and human lens, in: "Intracellular Protein Catabolism," D. Kharallah, J. S. Bond, and J. W. C. Bird, eds., p. 299, Alan R. Liss, New York (1985).

70. G. R. McCarty and A. Taylor, comparison of aminopeptidase sensitivity of Mn^{2+} and bestatin in bovine, human and rabbit lens, Int. Soc. Eye Res. 148 (1984).

71. G. R. McCarty and A. Taylor, Resolution and partial purification of new aminopeptidase activities in beef lens, Fed. Proc. 44:878 (1985).

72. K. R. Fleshman, J. W. Margolis, S. C. J. Fu, and B. J. Wagner, Age changes in bovine lens endopeptidase activity, Mech. of Ageing and Develop. 31:37 (1985).

73. C. Ohrloff, O. Hichwin, R. Olson, and S. Dickman, Glutathione peroxidase, glutathione reductase and superoxide dismutase in the aging lens, Curr. Eye Res. 3:109 (1984).

74. A. Ferrara, Respirazione e glicolisi del cristallino di cavie sottoposte a diéta scorbutigena, Annali D. Ottalmolog. E. Clin. Oculistica 68:529 (1940).

75. N. K. Monjukowa and M. J. Fradkin, Neue experimentelle Befunde über die pathogenese der katarakt, Archiv für Ophthalmoligie (Albrecht von Graefes) 133:329 (1935).

76. R. Hill and C. F. Mills, Chemical composition of blood, in: "The Biochemists Handbook," C. Long, ed., p. 839, Van Nostrand, Princeton (1968).

77. R. Heyninger, The component parts of the eye, in: "The Biochemists Handbook," C. Long, ed., p. 706, Van Nostrand, Princeton (1968).

78. J. Bellows, Biochemistry of the lens VII, Some studies on vitamin C and the lens, Arch. Ophthalmol. (Chicago) 16:58 (1936).

79. B. Nakamura and O. Nakamura, über das vitamin C in der linse und dern kammerwasser der menschlichen katarakte, Graefes Arch. Clin. Ophthalmol. 134:197 (1935).

80. H. K. Muller and W. Buschke, Vitamin C in linse, kammerwasser und blut bei normalen und pathologischem liasenstoffwechsel, Arch. Augenheilkd 108:368 (1934).

81. J. Berger, D. Shephard, F. Morrow, J. Sadowski, T. Haüe, and A. Taylor, Reduced and total ascorbate in guinea pig eye tissues in response to dietary intake, Curr. Eye Res. 7:681 (1988).

FREE RADICALS AND ANTIOXIDANTS IN THE PATHOGENESIS OF EYE DISEASES

G.E. Marak*, Y. de Kozak** and J.P. Faure**

*Center for Sight, Georgetown University
Washington, DC, USA; **Unité de Recherche
d'Ophtalmologie, INSERM U 86, Paris, France

INTRODUCTION

The eye is uniquely exposed to, susceptible to and protected against the damaging effects of free radicals. The lens absorbs increasing amounts of ultra-violet light with aging.[1] The eye is exposed to the photogeneration of free radicals by the photosensitization of reducing substances in the retina and the interaction of photosensitizers such as retinal with oxygen to generate singlet oxygen.[2-4] The vertebrate retina consumes 5-10 times more oxygen per mg than any other tested tissue.[5] Photoreceptors contain high concentrations of polyunsaturated fatty acids: for example compare the 50% concentration of docosahexanoic acid of retinal rod phospholipids to the 20% concentration found in vertebrate brain tissue.[6,7].

The eye also has high concentrations of antioxidant enzymes, free radical scavengers and inhibitors of superoxide production.[8-12] Immunohistochemical studies indicate that superoxide dismutase (SOD), catalase and glutathione peroxidase (Gpx) have a similar distribution.[13-15] These enzyme activities are concentrated at the corneal epithelium, corneal endothelium, lens and ciliary epithelia, photoreceptor inner segments and retinal pigment epithelium. These are the same locations where free radicals appear to be generated in catalase inactivated, normal ocular tissues as demonstrated by deposits of cerium perhydroxide (indicative of hydrogen peroxide production).[16] Additional antiioxidants include ascorbate, carotenoids, vitamin E and quercetin.[17-23]

The most extensive evaluation of free radical participation in ocular disease is in the development of cataracts, photic damage, retinal degeneration and uveitis. Free radical injury has also been implicated in retrolental fibroplasia, vitamin deficiencies, reperfusion injury and other rare eye diseases. These less thoroughly studied entities will not be reviewed.

Antioxidants in Therapy and Preventive Medicine
Edited by I. Emerit *et al.*
Plenum Press, New York, 1990

The role of free radicals in ocular disease has been most thoroughly studied in human senile cataracts as well as chemically-induced and congenital cataracts in experimental animals.

Free radicals have been detected by electron spin resonance in lens tissue.[24] Significant increases in the levels of hydrogen peroxide (H_2O_2) have been observed in both the aqueous and lenses of cataract patients when compared to age matched controls.[25] In rabbits treated with 3-amino triazole, aqueous H_2O_2 levels were 2-3 times above normal controls.[11]

Physiologic activities of the lens such as the Na-K pump activity are damaged by increased H_2O_2.[26] Photochemically generated toxic oxygen metabolites also damage electrolyte transport and are accompanied by an increase in products of lipid peroxidation.

The photosensitized oxidation of lens proteins has been demonstrated in vivo.[29] There is protein aggregation with formation of disulfide bonds.[30] Oxidation products of methionine and cysteine are increased in cataracts.[31] A variety of proteolytic enzymes are decreased in aging or with U-V light exposure.[32] There are increased levels of lipid peroxidation products in human cataracts as well as in congenital and experimentally induced cataracts. Antioxidant enzymes appear to be decreased in cataracts when compared to age matched controls. SOD is consistently decreased.[33] Catalase changes are equivocal. The glutathione redox system has also generally been found to have reduced activity. The glutathione complex activity is linked to the pentose phosphate pathway.[34] It has been shown that glucose-6-phosphate dehydrogenase deficiency is associated with cataract.[35] Catalase depletion with 3-amino triazole is cataractogenic. Glutathione depletion with dinitrochlorobenzene does not produce cataracts but does increase the susceptibility of the lens to oxidant stress.[36] The damage is reduced by treatment with antioxidant enzymes and hydroxyl radical scavengers.[27,28]

Antioxidants are protective against the development of experimental cataracts and reduce levels of oxidized proteins and lipid peroxidation products. The antioxidant butylated hydroxytoluene was found to significantly delay the onset of galactose cataracts.[37] A variety of antioxidants have been demonstrated to inhibit aminotriazole induced cataracts.

Agents that produce free radicals are also cataractogenic. Ultra-violet light, X-rays, hyperbaric oxygen and photosensitizing chemicals have all been implicated in the induction of cataracts.[38-42]

Epidemiologic studies associate cataracts with sunlight and U-V light (280-315 μm) exposure. Similar observations were obtained in Australia, Nepal and the United States.[43-45]

FREE RADICALS, PHOTIC RETINOPATHY AND RETINAL DEGENERATION.

The observation that ophtalmoscopes and operating microscopes could produce retinal damage[46-49] and similarities between senile macular degeneration (SMD) and photic damage[50,51] have stimulated some interest in the role of free radicals in the pathogenesis of retinal disease.

Epidemiologic studies suggest that SMD could be associated with light exposure. There appears to be less SMD with darkly pigmented irises compared to lightly pigmented irises.[52,53] Uncontrolled series suggest that the development of cataracts is protective against the development of SMD.[54,55]

The retina is sensitive to light damage either by direct effects on the photoreceptors and retinal pigment epithelium and by thermal damage and the accompanying inflammation.[56-59] The disorganization of retinal rod outer segment structure upon light exposure was first described by Wolken.[60] Most attention, however, is given to Noell's subsequent in vivo studies of this phenomenom in albino rats.[61] Photic damage to the retina is associated with a decrease in unsaturated lipids and an increase in products of lipid peroxidation.[62] Some antioxidants provide significant protection against photic damage; however, vitamin E deficiency does not enhance photic damage.[64,65]

The results of vitamin E deficiency studies have been somewhat confusing. Vitamin E deficiency results in degeneration of retinal rod outer segments, increased fluorescence and lipofuchsin content. The damage is more severe in the central compared to the peripheral retina with loss of some photoreceptor cell nuclei.[66] Combined vitamin E and selenium deficiency has had variable effects[66] but it must be remembered that most of the retinal GPx activity is derived from non-selenium containing peroxidases. Patients with severe senile macular degeneration have been reported to have reduced serum GPx levels.[67] Vitamin C supplementation was found to protect against photic injury. Photic injury was associated with increased levels of reduced ascorbate in the retina and retinal pigment epithelium.[68-69] β-carotene treatment was protective against photic damage in experiments involving increased oxygen tensions.[70]

There is no direct demonstration of free radical products in the retina although the cerium perhydroxide studies implicate the normal retina and the retinal pigment epithelium in the production of H_2O_2. The rhodopsin chromophore, retinal, has been demonstrated to be a photosensitizer capable of generating singlet oxygen. Intraocular injection of xanthine oxidase or lipid hydroperoxides has been observed to produce retinal degenerative changes.[71,72] Antioxidant treatment prevented the damage induced by lineolic acid hydroperoxide.[72]

Hyberbaric oxygen has produced retinal damage in rabbits[73] and dogs[74] but may kill other species before structural retinal damage can be observed.[75] Reversible functional defects such as electroretinographic changes or visual fields defects have been observed with hyperbaric oxygen in humans.[76] Hyberbaric oxygen enhances photic damage. The damaging effects of hyperbaric oxygen are enhanced by photosensitizing agents.[77] Retained intraocular iron foreign bodies produce severe retinal damage.[78,79] In experimental studies iron toxicity has been associated with increased levels of saturated lipids.[80]

Photosensitizing drugs such as hematoporphyrin derivative, chloroquine and methylpsoralen have been found to enhance retinal photic damage.[81-84]

The retina contains photosensitizing agents and has been shown to generate H_2O_2. There are high levels of polyunsaturated fatty acids that are susceptible to free radical damage. There is an experimental model, photic retinopathy that has some similarities to human SMD and may involve free radical damage. This system has received much less attention than experimental cataracts. Certainly there are all of the indications that free radical mechanisms may be involved in the pathogenesis of photic damage and senile macular degeneration but this speculation cannot be directly verified by the available evidence.

FREE RADICALS AND UVEITIS

Oxygen derived free radicals have been implicated as important mediators of cytotoxicity in acute inflammation[85-88] as well as in bacterial and tumor cell killing.[89-90] The superoxide anion is the principal product associated with the oxygen burst of stimulated inflammatory cells.[91-94] Many of the cytotoxic effects of the superoxide radical are believed to result from more highly reactive metabolites of superoxide such as H_2O_2 and hydroxyl radicals.

The production of free radicals by stimulated inflammatory cells in the eye has been adequately demonstrated. The participation of reactive oxygen metabolites in inflammatory cytolysis has also been thoroughly established. Toxic oxygen metabolites are produced in ocular tissues.[16] Free radical generating systems produce severe ocular tissue damage and enhance inflammation.[95-98] The involvement of free radical damage in ocular as in other acute inflammations is well established. The potential for the treatment of acute ocular inflammation with antioxidants and inhibitors of superoxide production is now being evaluated.

Studies on the therapy of acute ocular inflammation have principally employed experimental models of uveitis although there has recently also been some interest in optic neuritis.[99] The two principal models of uveitis are experimental allergic uveitis (EAU), a disease that has many similarities to human sympathetic ophthalmia[100], and experimental phacoanaphylactic endophthalmitis (EPE) which is virtually identical with the human disease that results after lens injury or extra-capsular cataract extraction.[101]

EAU displays a spectrum of severity dependent upon the dose of sensitizing antigen.[102] EAU develops 9-11 days after a

Table 1. Antioxidant treatment of experimental uveitis

TREATMENT	% REDUCTION OF INFLAMMATION		
	EAU		EPE
	Guinea pig	Rat	Rat
Antioxidant enzymes			
Superoxide dismutase (2000 Units)	63%	80%	70%
Catalase (2400 Units/kg)	73%	ND	70%
Glutathione peroxidase (5 Units)	ND	80%	46%
Iron chelators			
Deferoxamine (osmotic pump 70 mg/day)	ND	69%	ND
2,3 dihydroxybenzoic acid (100 mg/kg)	ND	ND	32%
Scavengers			
Na-benzoate (100 mg/kg)	49%	ND	NS
Dimethylsulfoxide (1.5 ml/kg)	ND	ND	39%
Vitamin E (diet + 25 mg/kg/day 9 weeks)	ND	ND	50%
Dimethylthiourea (500 mg/kg)	ND	ND	62%

ND : Not done
NS : Not significant

single intradermal injection of retinal "S" antigen and is principally a cell mediated immune disease.[103] The severe form of EAU induced by 50 μg of retinal "S" antigen is complement dependent and accompanied by a number of acute inflammatory cells.[104] This severe form of the disease has been employed in studies of antioxidants and agents which suppress free radical generation.

EPE may also have a varying severity of expression depending upon the sensitization protocol.[105,106] EPE is an immune complex disease that develops after lens injury in animals that have received a series of subcutaneous sensitizing injections.[107] A severe Arthus reaction develops within 24 hours after lens injury in Lewis rats that have been sensitized by a series of 4 inocula of 10 mg rabbit lens protein in complete Freund's adjuvant given at 2 week intervals.[108]

Severity of inflammation in these forms of uveitis was measured by morphometric analysis of choroidal thickness. The morphometry was highly correlated with neutrophil numbers or myeloperoxidase levels.[109].

The dose of antioxidant employed was selected on the basis of the maximal effective dose in cobra venom induced pulmonary disease or the maximal in vitro effective concentration adjusted to total extracellular fluid volume. Deferoxamine was administered continuously by an osmotic pump. The other agents were given intraperitoneally 1-2 times per day. In EPE the test agents were given at the time of lens injury and the eyes removed 24 hours later. In EAU the agents were given daily from day 9-10 post immunization to day 16-21, the day of enucleation.All of the tested agents were effective in inhibiting experimental uveitis[110-119] (Table 1).

Suppressing superoxide production in activated inflammatory cells is another potential approach to the therapy of acute inflammation. There are a number of points at which activation and signal transmission of neutrophils may be modulated by agonists and antagonists : agents that alter the activities of receptors, guanine nucleotide binding proteins, phospholipase C and phosphoinositides, calcium channels and intracellular calcium concentrations, phospholipid/calcium dependent protein kinase C, the hexose monophosphate shunt or the plasmalemma NADPH oxidase and electron transport chain complex.[120]

Agents which interfere with the activation of neutrophils at the receptor level have been shown to have an anti-inflammatory effect in vivo. Prostaglandin E and adenosine were effective in reducing pulmonary inflammation.[120-122] Adenosine has an antiphlogistic effect in EAU in rats.[123] Although adenosine inhibits superoxide production by human neutrophils[124], this effect does not occur with rat neutrophils[125] so that antiphlogistic mechanism other than inhibition of superoxide production must be involved.

When a ligand binds to a receptor it produces conformational changes that are transmitted to an associated G-protein. The G-protein is altered so that guanosine triphosphate binding is enhanced and in turn promotes the activation of an associated phospholipase-C.[126] Pertussis toxin inactivates the G-protein and disrupts signal transmission and superoxide production.[127-131] Pertussis toxin treatment at the onset of the effector response completely inhibited the development of EAU.[132] Pertussis toxin given at the time of lens injury markedly reduced the expected inflammation in EPE.[133] In the case of EPE the inhibition of superoxide production may be an important antiphlogistic mechanism. In EAU pertussis toxin presumably interfered with lymphocyte homing.[134,135] In the absence of any significant inflammation it is difficult to involve a free radical mechanism for the anti-inflammatory effect.

The hexose monophosphate shunt is the source of the electrons employed by the plasmalemma NADPH oxidase for the reduction of oxygen in superoxide production by stimulated acute inflammatory cells.[132] An antagonist of the pentose phosphate pathway is 6-aminonicotinamide, an agent which inhibits superoxide generation.[137] This agent has a significant antiphlogistic effect in EAU.[138] The effects of 6-amino-nicotinamide on inflammation have not been thoroughly evaluated but this drug also appears to depress lymphocyte functions.[139]

Quercetin[140-142] and imidazole[143,144] are known to interfere with superoxide production. Both of these agents have considerable antiphlogistic activity in experimental uveitis.[144,145] It is tempting to speculate that these agents may be antagonists of the oxidase-transport chain complex but both compounds are also effective hydroxyl radical scavengers.[146,147]

The pharmacologic probes employed for in vitro testing of neutrophil activation are unfortunately not sufficiently specific and many have a number of immunomodulatory effects unrelated to neutrophil stimulation. There are significant species differences in free radical metabolism that are not well understood. The possibility that agonists and antagonists of stimulus response coupling reactions may be applicable to modulating clinical inflammation merits more detailed evaluation as many of these agents are safe and have been useful in other situations.

SUMMARY

There is fairly convincing evidence that free radical mechanisms are involved in the pathogenesis of cataracts and uveitis and that antioxidants may be protective. Studies on retinal degeneration are almost entirely limited to dietary manipulation of vitamins C and D. Unfortunately, antioxidant properties are not easily isolated from other metabolic effects of vitamins. Cataracts, uveitis, and retinal degeneration cause nearly one-third of all blindness.[148] The evidence that free radical mechanisms are important in the pathogenesis of these diseases is compelling incentive to encourage more extensive and detailed investigation.

REFERENCES

1. S. Lerman, Biophysical aspects of corneal and lenticular transparency, *Curr. Eye Res.* 3, 3-14 (1984).
2. M. Delmelle, Generation of singlet oxygen by retinal, *Photochem, Photobiol.* 27, 731-734 (1978).
3. A.A. Shvedova, O. Alekseeva, I.Y.Kuliev, K.O. Muranov, Y.P. Kozlov and V.E. Kagan, Damage of photoreceptor membrane lipids and proteins induced by photosensitized generation of singlet oxygen, *Curr. Eye Res.* 2, 683-689 (1983).
4. J.L. Calkins and B.F. Hochheimer, Retinal light exposure from ophtalmoscopes, slit lamps and overhead surgical lamps, *Invest. Ophthalmol. Vis. Sci.* 19, 1009-1015 (1980).
5. W. Sickel, Retinal metabolism in dark and light, *in*:

"The Handbook of Sensory Physiology", 7, M. Fuortes, ed., Springer Verlag, Berlin, 667-727 (1972).

6. W.L. Stone, C.C. Farnsworth and E.A. Dratz, A reinvestigation of the fatty acid content of bovine, rat and frog retinal rod outer segments, *Exp. Eye Res.*, 28, 387-397 (1979).

7. J. Tinoco, Dietary requirements and function of linolenic acid in animals, *Prog. Lipid Res.*, 21, 1-45 (1982).

8. K.C. Bhuyan and D.K. Bhuyan, Molecular mechanism of cataractogenesis: Toxic metabolites of oxygen as initiators of lipid peroxidation and cataract, *Curr. Eye Res.*, 3, 67-81 (1984).

9. R. Fried and P. Mandel, Superoxide dismutase of bovine and frog rod outer segments, *J. Neurochem.* 24, 433-438 (1975).

10. D. Armstrong, B. Santangelo and F. Connole, The distribution of peroxide regulatory enzymes in the canine eye, *Curr. Eye Res.*, 1, 225-242 (1981).

11. K.C. Bhuyan and D.C. Bhuyan, Regulation of hydrogen peroxide in eye humors. Effect of 3-amino 1-H, 1, 2, 4 triazole in catalase and glutathione peroxidase of rabbit eye, *Biochem. Biophys. Acta* 497, 641-651 (1977).

12. W.L. Stone and E.A. Dratz, Selenium and non-selenium glutathione peroxidase activities in selected ocular and non-ocular rat tissues, *Exp. Eye Res.* 35, 405-412 (1982).

13. N.A. Rao, L.G.Thaete, J.M. Delmage and A. Sevanian, Superoxide dismutase in ocular structures, *Invest.Ophthalmol. Vis. Sci.*, 26, 1778-1781 (1985).

14. F. Attala, M.A. Fernandez and N.A. Rao, Immunohistochemical localization of catalase in ocular tissue, *Curr. Eye Res.*, 6, 1181-1187(1987).

15. L.R. Atalla, A. Sevanian and N.A. Rao, Immunohistochemical localization of glutathione peroxidase in ocular tissue, *Curr. Eye Res.*, 7, 1023-1028 (1988).

16. L.R. Atalla, A. Sevanian and N.A. Rao, Hydrogen peroxide localization in ocular tissue, an electron microscopic cytochemical study, *Curr. Eye Res.*, 7, 931-936 (1988).

17. H. Heath, The distribution and possible functions of ascorbic acid in the eye, *Exp. Eye Res.*, 1, 362-367 (1962).

18. M.O.M. Tso, A.J. Woodford and K.W. Lam, The distribution of ascorbate in the normal primate retina and after photic injury: A biochemical morphologic correlated study, *Curr. Eye Res.*, 3, 181-191 (1984).

19. S.D. Varma, D. Chand, Y.R. Sharma, J.F. Kuch and R.D. Richards, Oxidative stress on the lens and cataract formation : Role of light and oxygen, *Curr. Eye Res.*, 3, 35-57 (1984).

20. K. Kirschfeld, Carotenoid pigments : Their possible role in protection against photooxidation in eyes and photoreceptor cells, *Proc. Roy Soc., London B.*, 216, 71-85 (1982).

21. J. Nishiyama, E.C. Ellison, G.R. Mizuno and J.R. Chipault, Microdetermination of α tocopherol in tissue lipids, *J. Nutr. Sci. Vitaminol.* 21, 355-361 (1975).

22. D.F. Hunt, D.T. Organisciak, H.M. Wang and R.L. Wu, α tocopherol in the developing rat retina : A high pressure liquid chromatographic analysis, *Curr. Eye Res.*, 3, 1281-1288 (1984).

23. E.L. Paulter, J.A. Maga and A. Tengerdy, A pharmacologically potent natural product in the bovine retina, *Exp. Eye Res.*, 42, 285-288 (1986).

24. J.J. Weiter and S. Subramanian, Free radicals produced in human lenses by a biphotonic process, *Invest. Ophthalmol. Vis. Sci.*, 17, 869-873 (1978).

25. A. Spector and W.H. Garner, Hydrogen peroxide and human cataract, *Exp. Eye Res.*, 33, 673-681 (1981).

26. W.H. Garner, M.H. Garner and A. Spector, H_2O_2 induced uncoupling of bovine lens Na+, K+ ATPase, *Proc. Natl. Acad. Sci. U.S.A.* 80, 2044 (1983).

27. S.D. Varma and J.M. Mooney, Photodamage to the lens in vitro: Implications of the Haber-Weiss reaction, *Free Radical Biol. Med.*, 2, 57-62 (1986).

28. S.D. Varma, V.K. Srivastava and R.D. Richards, Photoperoxidation in lens and cataract formation; Preventive role of superoxide dismutase, catalase and vitamin C, *Ophthalmic Res.* 14, 167-175 (1982).

29. S. Lerman, M. Jacoy and R.F. Borkman, Photosensitization of the lens by 8-methoxysporalen, *Invest. Ophthalmol. Vis. Sci.* 16, 1065-1068 (1977).

30. A. Spector and D. Roy, Disulfide linked high molecular weight protein associated with human cataract, *Proc. Natl. Acad. Sci U.S.A.* 75, 3244-3248 (1978).

31. A. Spector, The search for a solution to senile cataracts. *Invest.Ophthalmol. Vis. Sci.* 25, 130-146 (1984).

32. A Taylor and K.V.A. Davies, Protein oxidation and loss of protease activity may lead to cataract formation in the aged lens, *Free Radical Biol. Med.* 3, 371-377 (1987).

33. C.Ohrloff, O.Hockwin, R.O. Olsen and S. Dickman, Glutathione peroxidase, glutathione reductase and superoxide dismutase in the aging lens, *Curr. Eye Res.* 3, 109-116 (1984).

34. F. Giblin and J. McCready. The effect of inhibition of glutathione reductase on detoxification of H_2O_2 by rabbit lens, *Invest. Ophthalmol. Vis. Sci.* 24, 113-118 (1983).

35. N. Orzalesi, R.Sorcinelli and G. Guiso, Increased incidence of cataracts in male subjects deficient in glucose-6-phosphate dehydrogenase, *Arch. Ophthalmol.* 99, 69-71 (1981).

36. S.K. Srivastava, H.H. Ansari and Y.C. Awasthi, Lens glutathione depletion of 1-chloro-2,4 dinitrobenzene and oxidative stress, *Curr. Eye Res.* 3, 112-119 (1984).

37. H.H. Ansari, A. Schulter and S.K.Srivastava, Antioxidant (BHT) significantly delays galactose cataract formation in rats, *Invest. Ophthalmol. Vis. Sci.* 28 (Suppl.), 192 (1987).

38. S.Zigman and W.Vaughn, Near U-V light effects on the lenses and retinas of mice, *Invest. Ophthalmol.* 13, 462-465 (1974).

39. S. Lerman, Human U-V radiation cataracts, *Ophthalmic Res.* 12, 303-314 (1980).

40. S.Duke-Elder and P.A. MacFaul, "System of Ophthalmology" Vol.9, Kimpton, London, 985-1001(1972).

41. S.S. Schocket, J.Esterson, B. Bradford, M. Michaelis and R.D. Richards, Induction of cataracts in mice by exposure to oxygen, *Isr. J. Med. Sci.* 8, 1596-1601 (1972).

42. A. Pirie and J.N. Rees, Diquat cataract in the rat, *Exp. Eye Res.* 9, 198-203 (1970).

43. F. Hollows and D. Moran, Cataract - The ultraviolet risk

factor, *Lancet* 2, 1249-1250 (1981).

44. L.B. Brilliant, N.C. Grassett, R.P. Pokhrel, A. Kolstad, J.M. Lepkowski, G.E. Brilliant, W.N. Hawkes and R. Parajosegaram, Associations among cataract prevalence, sunlight and altitude in the Himalayas, *Am. J. Epidemiol.* 118, 250-264 (1983).

45. R. Hiller, R.D. Spurduto and F. Ederer, Epidemiologic associations with cataract in the 1971-1972 National Health and Nutrition Survey, *Am. J. Epidemiol.* 118, 239-249 (1983).

46. J.L. Calkins, B.F. Hockheimer and S.A. D'Anna, Potential hazards from specific ophtalmic devices, *Vision Res.* 20, 2039-2053 (1980).

47. J.L. Calkins and B.F. Hockheimer, Retinal light exposure from operating microscopes, *Arch. Ophthalmol.* 97, 2363-2367 (1979).

48. H.R. McDonald and A.R. Irvine, Light induced maculopathy from the operating microscope in extracapsular cataract extraction and intraocular lens implantation, *Ophthalmology* 90, 945-951, (1983).

49. D.M. Robertson and R.B. Feldman, Photic retinopathy from the operating microscope, *Am. J. Ophthalmol.* 101, 561-569 (1986).

50. M.O.M. Tso, "Retinal Diseases", Lippincott, Philadelphia, 187-214 (1988).

51. M.O.M. Tso, Pathogenic factors of aging macular degeneration, *Ophthalmology* 92, 628-635 (1985).

52. F.L. Ferris, Senile macular degeneration : Review of epidemiologic features, *Am. J. Epidemiol.*, 118, 132-1 51(1983).

53. L.G. Hayman, Senile macular degeneration : A case control study, *Am. J. Epidemiol.* 118, 213-227 (1983).

54. J. Von der Hoeve, Eye lesions produced by light rich in ultraviolet rays, senile cataract, senile degeneration of the macula, *Am. J. Ophthalmol.*3, 178-194 (1920).

55. H.G.A. Gjessing, Is there an antagonism between senile cataract and senile macular degeneration?, *Acta Ophthalmol.* 31, 401-421 (1953).

56. W.K. Noell, There are different kinds of retinal light damage in the rat, *in*: "The Effects of Constant Light on the Visual Process", T. Williams and B. Baker eds., Plenum Press, New-York, 3-28 (1980).

57. W.T. Ham, Jr., J.J. Ruffolo, H.A. Mueller, A.M.Clark and M.E. Moore, Histologic analysis of photochemical lesions produced in rhesus retina by short-wavelength light, *Invest. Ophthalmol. Vis. Sci.* 17, 1029-1035 (1978).

58. W.T. Ham, Jr., J.J. Ruffolo, H.A. Mueller, D. Guerry, The nature of retinal light damage. Dependence on wave length, power level and exposure time, *Vision Res.* 20, 1105-1111 (1980).

59. W.T. Ham, H.A. Mueller, Retinal sensitivity to damage from short wave length light, *Nature,* 253-255 (1976).

60. J.J. Wolken, Biophysics and biochemistry of the retinal photoreceptors, *in*: "Vision" : Thomas, Springfield, 52-61 (1966).

61. W.K. Noell, V.S. Walker, B.S. Kang and S. Berman, Retinal damage by light in rats, *Invest. Ophthalmol.* 5, 450-463 (1966).

62. R.J. Wiegland, N.M. Giusto, L.M.Rapp and R.E. Anderson, Evidence for rod outer segment lipid peroxidation

following constant illumination of the rat retina, *Invest. Ophthalmol. Vis. Sci.* 24, 1433-1435 (1983).

63. V.E. Kagan, "Lipid Peroxidation in Biomembranes", CRC Press, Boca Raton, 139-140 (1988).

64. W.L. Stone, M.L. Katz, M. Lurie, M.M. Marmor and E.A. Dratz, Effects of dietary vitamin E and selenium on light damage to the rat retina, *Photochem. Photobiol.* 29, 725-730 (1979).

65. V.E. Kagan, I.Y. Kuliev, V.B.. Spiricht , A.D.Shvedova and Y.P. Kozlov, Accumulation of lipid peroxidation products and depression of electrical activity in vitamin E deficient rats exposed to high intensity light, *Bull. Exp. Biol. Med.* 91, 144-147.

66. G.J. Handleman and E.A. Dratz, The role of antioxidants in the retina and retinal pigment epithelium and the nature of prooxidant-induced damage, *Free Radical Biol. Med.* 2, 1-90 (1986).

67. J. Weiter, E. Dratz, K. Fitch and G. Handleman, Role of selenium nutrition in senile macular degeneration, *Invest. Ophthalmol. Vis. Sci.* 26, (suppl.) 58 (1985).

68. D.T. Organisciak, H.M. Wang, Z.Y. Li and M.O.M. Tso, The protective effect of ascorbate in retinal light damage of rats, *Invest. Ophthalmol. Vis. Sci.* 26, 2580-2588 (1985).

69. Z.Y. Li, M.O.M. Tso, H.M. Wang and D.T. Organisciak, The amelioration of photic injury in rat retina by ascorbic acid, a histopathologic study, *Invest. Ophthalmol. Vis. Sci.* 26, 1589-1598 (1985).

70. W.T. Ham, H.A. Mueller, J.J. Ruffolo, Jr, J.E. Millen, S.F. Cleary, R.K. Guerry and D. Guerry, III, Basic mechanism underlying the production of photochemical lesions in the mammalian retina, *Curr. Eye Res.* 3, 165-174 (1984).

71. T.W. Sery and R. Petrillo, Superoxide anion radical as an indirect mediator in ocular inflammatory disease, *Curr. Eye Res.* 3, 243-252 (1984).

72. D. Armstrong, T. Hiramitsu, J. Gutteridge and S.E Nilsson, Studies on experimentally induced retinal degeneration; Effects of lipid peroxides on electroretinographic activity in the albino rabbit, *Exp. Eye Res.* 35, 157-171 (1982).

73. W.K. Noell, Metabolic injuries of the visual cell, *Am.J. Ophthalmol.* 40, 60-68 (1955).

74. C.C. Beehler, N.L. Newton, J.F. Culver and T.J. Tredici, Retinal detachment in dogs resulting from oxygen toxicity, *Arch. Ophthalmol.* 71, 665-670 (1964).

75. W.K. Noell, Studies on visual cell viability and differenciation, *Ann. N.Y. Acad. Sci.* 74, 337-361 (1958).

76. C.W. Nichols and C.J. Lambertson, Effects of high oxygen pressures on the eye, *N. Eng. J. Med.* 281, 25-30 (1969).

77. C.C. Beehler and W. Roberts, Experimental retinal detachment induced by oxygen and phenothiazines, *Arch. Ophthalmol.* 79, 759-762 (1968).

78. P.A. Libis and T. Yamashita, Experimental aspects of ocular siderosis, *Am. J. Ophthalmol.* 48, 465-479 (1959).

79. A.M. Roth, R.Y. Foos, Ocular pathologic changes in primary hemochromatosis, *Arch. Ophthalmol.* 87, 507-514 (1972).

80. R.E. Anderson, L.M. Rapp and R.D. Wiegand, Lipid peroxidation and retinal degeneration, *Curr. Eye Res.* 3, 223-228 (1984).

81. J. Legros, I.Rosner and C.Berger, Influence du niveau

d'éclairement ambiant sur les modifications oculaires induites par l'hydroxychloroquine chez le rat, *Arch. Ophtalmol. (Paris)* 33, 417-424 (1973).

82. S.Lerman, K. Megaw, Y.U. Gardner, Y.Tadei, Y.Franks and Gammon, Photobinding of ^{3}H 8-methoxypsoralen to monkey intracular tissues, *Invest. Ophthalmol. Vis. Sci.* 25, 1267-1274 (1984).

83. J.Winther and N. Ehlers, The combined effect of hematoporphyrin derivative and light on the normal mouse retina, *Arch. Ophthalmol.* 62, 112-122 (1984).

84. C.J. Gomer, D.R. Dorion, L. While, J.V. Jester, S. Dunn, B.C. Szirth, J.J. Razum and A.L. Murphee, Hematoporphyrin derivative photoradiation induced damage to normal and tumor tissue of the pigmented rabbit eye, *Curr. Eye Res.* 3, 229-237 (1984).

85. S.J. Weiss and P.A. Ward, Immune complex induced generation of oxygen metabolites by human neutrophils. *J. Immunol.* 129, 209-213 (1982).

86. K.J. Johnson and P.A. Ward, Role of oxygen metabolites in immune complex injury of the lung, J. *Immunol.* 126, 2365-2369 (1981).

87. A. Rehan, K.J. Johnson, R.C. Wiggins, R.G. Kunkel, P.A. Ward, Evidence for the role of oxygen metabolites in acute nephrotoxic nephritis, *Lab. Invest.* 51, 396-403 (1984).

88. G.E. Marak, N.A. Rao, J.M. Scott, R. Duque and P.A. Ward, Free radicals and phacoanaphylaxis, *in:* "Advances in Immunology and Immunopathology of the Eye",G. R. O'Connor and J. Chandler, eds., Masson, New York, 144-145 (1985).

89. B.M. Babior, Oxygen dependent microbial killing by phagocytes, *New England J. Med.* 298, 659-668 (1978).

90. C. Nathan and Z. Cohn, Role of oxygen dependent mechanisms in antibody induced lysis of tumor cells by activated macrophages, *J. Exp. Med.* 152,198-208 (1980).

91. I. Fridovich, Oxygen radicals, hydrogen peroxide and oxygen toxicity, *in:* "Free Radicals in Biology", W. Pryor, ed., Academic Press, New York, 259-277 (1976).

92. S.J. Klebanoff, Oxygen dependent cytotoxic mechanisms of phagocytes, *in:* "Advances in Host Defense Mechanisms", R. Galin and A. Fauci, eds. Vol 1. Phagocytic cells, 111-162, Raven Press, New York (1982).

93. B.A. Freeman and V.S. Crapo, Free radicals and tissue injury, *Lab.Invest.* 47, 412-426 (1982).

94. S.J. Weiss and A.F. LoBuglio, Phagocyte-generated oxygen metabolites and cellular injury, *Lab. Invest.* 44, 5-18 (1982).

95. T.W. Mittag, Role of free radicals in ocular inflammation and cellular damage, *Exp. Eye Res.* 39, 759-767 (1984).

96. T.W. Mittag, B.R. Hammond, K.E. Eakins and P. Bhattacherjee, Ocular responses to superoxide generated by intraocular injection of xanthine oxidase, *Exp. Eye Res.* 40, 411-419 (1985).

97. L. Feeney and E.R. Berman, Oxygen toxicity : Membrane damage by free radicals, *Invest. Ophthalmol.* 15, 789-792 (1976).

98. T.W. Sery, A.W. Vogel, R. Folberg and R. Petrillo, Oxygen free radicals in ocular inflammatory disease, *in:* "Uveitis Update", K.M. Saari, ed., Elsevier Science Publishers, Amsterdam, 39-45 (1984).

99. J. Guy, E.A. Ellis, G.M. Hope and N.A. Rao, Influence of

anti-oxidant enzymes in reduction of optic disc edema in experimental optic neuritis, *Free Radical Biol. Med.* 2, 349-351 (1986).

100. W.B. Wacker, N.A. Rao and G.E. Marak, Experimental sympathetic ophthalmia, *in:* "Immunology and Immunopathology of the Eye", A.M. Silverstein and G.R. O'Connor, eds, Masson, New York, 121-126 (1979).

101. G.E. Marak, R.L. Font, L.N. Czawlytko and F.P. Alepa, Experimental lens induced granulomatous endophtalmitis : Preliminary histopathologic observations, *Exp. Eye Res.* 19, 311-16 (1974).

102. N.A. Rao, W.B. Wacker and G.E. Marak, Experimental allergic uveitis : Clinicopathologic features associated with varying doses of S-antigen, *Arch. Ophthalmol.* 97, 1954-1958 (1979).

103. J.P. Faure, Autoimmunity and the retina, *Curr. Topics Eye Res.* 2, 215-302 (1980).

104. G.E. Marak, W.B. Wacker, N.A. Rao, R. Jack, and P.A. Ward, Effect of complement depletion on experimental allergic uveitis, *Ophthalmic Res.* 11, 97-107 (1979).

105. G.E. Marak, R.L. Font, and N.A. Rao, Strain differences in autoimmunity to lens protein, *Ophthalmic Res.* 13, 320-329 (1981).

106. G.E. Marak, N.A. Rao, G. Antonakou and A. Slewinski, Experimental lens-induced granulomatous endophthalmitis in common laboratory animals, *Ophthalmic Res.* 14, 292-297 (1982).

107. G.E. Marak, Abrogation of tolerance to lens protein, *in:* "New Directions in Ophtalmic Research", M. Sears, ed., Yale Univ. Press, New Haven, 47-61 (1981).

108. G.E. Marak, R.L. Font and F.P. Alepa, Arthus-type panophtalmitis in rats sensitized to heterologous lens protein, *Ophthalmic Res.*, 162-170 (1977).

109. G.E. Marak, N.A. Rao, (unpublished observations).

110. N.A. Rao, A.J. Calandra, A. Sevanian, B. Bowe, J.A. Delmage and G.E. Marak, Modulation of lens induced uveitis by superoxide dismutase, *Ophthalmic Res.* 18, 41-46 (1986).

111. N.A. Rao, M.A. Fernandez, A. Sevanian, G.O. Till and G.E. Marak, Antiphlogistic effect of catalase on experimental phocoanaphylactic endophthalmitis, *Ophthalmic Res.* 18, 185-191 (1986).

112. N.A. Rao, B.E. Bowe, A. Sevanian, G.O. Till and G.E. Marak, Modulation of lens induced uveitis by dimethyl sulfoxide, *Ophthalmic Res.* 18, 193-198 (1986).

113. N.A. Rao, J.L. Romero, M.A. Fernandez, A. Sevanian and G.E. Marak, Effect of iron chelation on severity of ocular inflammation in an animal model, *Arch. Ophthalmol.* 104, 1369-1371 (1986).

114. N.A. Rao, A. Sevanian, M.A. Fernandez, J.L. Romero, J.P. Faure, Y. de Kozak, G.O. Till and G.E. Marak, Role of oxygen radicals in experimental allergic uveitis, *Invest. Ophthalmol. Vis Sci.* 28, 886-892 (1987).

115. N.A. Rao, J.L. Romero, A. Sevanian, M.A. Fernandez, C. Wang, P.A. Ward and G.E. Marak, Anti-inflammatory effect of glutathione peroxidase on experimental lens induced uveitis, *Ophtalmic Res.* 20, 106-11 (1988).

116. N.A. Rao, M.A. Fernandez, A. Sevanian, J.L. Romero, G.O. Toll and G.E. Marak, Treatment of experimental lens-induced uveitis by dimethylthiourea, *Ophthalmic Res.* 20, 106-111 (1988).

117. G.E. Marak, N.A. Rao, J.M. Scott, R. Duque and P.A. Ward, Free radicals and phocoanaphylaxis, *in:* "Advances in Immunology and Immunopathology of the Eye", G.R. O'Connor and J. Chandler, eds, Masson, New York, 144-145 (1985).

118. Y. de Kozak, J.P. Nordmann, J.P. Faure, N.A. Rao and G.E. Marak, The effect of anti-oxidant enzymes on experimental uveitis in rats, *Ophthalmic Res.* (in press).

119. G.E. Marak, N.A. Rao, A. Sevanian, V. Zdravkovich, G.O. Till and P.A. Ward, Modulation of experimental phacoanaphylactic endophthalmitis with the anti-oxidants sodium benzoate and 2,3,dihydroxybenzoic acid, *Ophthalmic Res.* 19, 120-128 (1987).

120. J.A. Badway and J.L. Karnovsky, Production of superoxide by phagocytic leukocytes: A paradigm for stimulus response phenomena, *Curr. Topics in Cell Regulation* 28, 183-208 (1986).

121. S.L. Kunkel, J.C. Fantone, P.A. Ward and R.B. Zurier, Modulation of inflammatory reactions by prostaglandins, *Prog. Lipid Res.* 20, 633-640 (1981).

122. G.O. Till and P.A. Ward (unpublished observations).

123. G.E. Marak, Y. de Kozak, J.P. Faure, N.A. Rao, J.L. Romero, P.A. Ward and G.O. Till, Pharmacological modulation of acute ocular inflammation I. Adenosine, *Ophthalmic Res.* 20, 220-226 (1988).

124. B.N. Cronstein, S.G. Kramer, B. Weissman, R. Hirschorn, Adenosine : A physiologic modulation of superoxide anion generation of human neutrophils, *J. Exp. Med.* 158, 1160-1177 (1983).

125. G.O. Till and P.A. Ward, (unpublished observations).

126. S. Cockcroft and B.D. Gambers, The role of guanine nuceotide binding protein in the activation of phosphoinositide phosphodiesterase, *Nature*, 314, 534-536 (1985).

127. A.G. Gilman, G-proteins and dual control of adenylate cyclase, *Cell* 36, 557-579 (1984).

128. K. Kraus, W. Schlagel, C. Wollheim, T. Anderson, F. Waldvogel and P. Lew, Chemotactic peptide activation of human neutrophils and HL-60 cells, *J. Clin. Invest.* 76, 1348-1354 (1985).

129. T. Matsumoto, T. Molski, C. Valpi, Y. Pelz, Y. Kanako, E.L. Becker, M. Feinstein, P. Naccache and R. Saafi, Treatment of rabbit neutrophils with phorbol esters results in increased ADP ribosylation catalyzed by pertussis toxin and inhibition of the GTPase stimulated by F-met-leu-phe, *FEBS Lett.* 198, 295-300 (1986).

130. T. Molski, P. Naccache, M. Marsh, J. Germode, E. Becker and R. Saafi, Pertussis toxin inhibits the rise in the intracellular concentration of free calcium that is induced by chemotactic factors in rabbit neutrophils : Possible role of the G-proteins in calcium mobilization, *Biochem. Biophys. Res. Comm.* 121, 644-650 (1984).

131. M.W. Verghese, C.D. Smith and R. Snyderman, Potential role for a guanine nucleotide regulatory protein in chemoattractant receptor mediated polyphosphoinositide metabolism, Ca++ mobilization and cellular response by leukocytes, *Biochem. Biophys. Res. Comm.* 127, 450-457 (1985).

132. G.E. Marak et al (unpublished observations).

133. G.E. Marak, P.A. Ward and G.O. Till (unpublished observations).

134. S.T. Tamara, Y. Nakanisi, A. Kojima, M. Otokawa,
     N. Uchida, H. Sato and Y. Sato, Effect of pertussis toxin
     (Pt) on T cell populations sensitized for delayed-type
     hypersensitivity in mice, *Cell. Immunol.* 85, 351-363
     (1984).
135. G.J. Spangrude, B.A. Araneo and R.A. Daynes, Site
     selected homing of antigen primed lymphocyte population
     can play a critical role in the efferent limb of cell
     mediated immune responses in vivo. *J. Immunol.* 134, 2900-
     2907 (1985).
136. B.D. Cheson, R.L. Christenson and R. Sperling, The origin
     of the chemiluminescence of phagocytosing granulocytes,
     J. Clin. Invest. 58, 789-796 (1976).
137. J.G. Bender and D.E. Van Epps, Inhibition of human
     neutrophil function by-6-aminonicotinamide : The role of
     the hexose monophosphate shunt in cell activation,
     Immunopharmacol. 10, 191-199 (1985).
138. G.E. Marak et al. (unpublished observations).
139. S.J. Berger, I. Manory, D.C. Sudar, D. Krothapalli and
     D. Berger, Pyridine nucleotide analog interference with
     metabolic processes in mitogen stimulated human
     T lymphocytes, *Exp. Cell. Res.* 173, 389-387 (1987).
140. C. Pagonis, A.I. Tauber, N. Pavlotsky and E.R. Simons,
     Flavinoid impairment of neutrophil response, *Biochem.*
     Pharmacol. 35, 237-245 (1985).
141. A.I. Tauber, J.R. Fay, M.A. Marietta, Flavinoid
     inhibition of the human neutrophil NADPH-oxidase,
     Biochem. Pharmacol.,33, 1367-1369 (1984).
142. G. Berton, C. Schneider and D. Romeo, Inhibition by
     quercetin of activation of polymorphonuclear leukocyte
     functions : Stimulus-specific effects, *Biochem. Biophys.*
     Acta. 595, 47-56 (1980).
143. T. Iizuka, S. Kanegasaki, R. Makino, T. Tanaka and Y.
     Ishimura, Pyridine and imidazole reversibly inhibit the
     respiratory burst of porcine and human neutrophils,
     Biochem. Biophys. Res. Comm. 30, 621-626 (1985).
144. J. Romero, G.E. Marak and N.A. Rao, Pharmacologic
     modulation of acute ocular inflammation with quercetin,
     Ophthalmic Res. (in press).
145. G.E. Marak et al. (unpublished observations).
146. W.W. Busse, D.E. Kopp, E. Middleton, Flavinoid modulation
     of human neutrophil function, *J. Allerg. Clin. Immunol.*
     73, 801-09 (1984)
147. G.O. Till and P.A. Ward (unpublished observations).
148. H.A. Kahn and H.B. Moorehead, Statistics on blindness in
     the model reporting area 1969-1970, DHEQ Publication No
     (NIH) 73-427.

SINGLET OXYGEN-INDUCED DAMAGE TO RAT LENSES IN VITRO:

PROTECTION BY ANISYLDITHIOLTHIONE

Gérard TISSIE, Elisabeth LATOUR, Claude COQUELET and Claude BONNE*

Centre de Recherches Laboratoire Chauvin, B.P. 1174, 34009 Montpellier
cedex, France.
* Laboratoire de Physiologie Cellulaire, Université Montpellier I, 15 avenue
Charles Flahault, 34060 Montpellier cedex, France

Oxidation is now recognized as being the major process involved in lens changes associated with cataract development (1). Visible light and near UV (300-400 nm) could be an important source of oxygen reactive species, in particular singlet oxygen. Photosensitizers such as riboflavin and N-formylkynurenine have been shown to be present in the lens (2). Singlet oxygen provokes in vitro the same alterations in lens cristallins as those observed in aging and cataracts (3, 4). This reactive species also produces an impairment of transport systems in the rat lens (5). Furthermore, it is a powerful inducer of lipid peroxidation considered as an initiating event in cataract formation (6).

Antioxidative defenses, enzymes (superoxide dismutase, catalase, glutathione peroxidase), as well as endogenous scavengers have been shown to be lowered in cataracts (1, 7). An effective means of protecting the lens could be by compensating the loss of these endogenous scavengers by antioxidants. We have undertaken to screen drugs that could scavenge oxygen-derived species and enter the lens, thus protecting it from peroxidative damage (8). This paper illustrates our approach and describes the activity of anisyldithiolthione (ADT).

MATERIALS and METHODS

Reactivity of Drugs with Singlet Oxygen

We used the method described by Ames et al. (9), with minor modifications. Briefly, incubates containing sodium phosphate buffer (50 mM, pH 7.8), EDTA (0.1 mM), uric acid (50 uM), rose bengal (250 nM) and various concentrations of the drugs were irradiated under tungsten light and the disappearance of uric acid was monitored at 292 nm.

Lipoperoxidation Inhibition

Peroxidation of egg yolk phosphatidylcholine liposomes was induced by singlet oxygen and quantified by measuring malonaldehyde formation by the thiobarbituric acid reaction (10). Briefly, chloroform solution of phospholipids (400 µg/assay) and drugs at various concentrations was evaporated to dryness under N_2. The residue was resuspended in 0.9 ml Tris-HCl (50 mM, pH 7.0), NaCl (50 mM) by sonication. The suspensions were incubated at 37°C for 1hr under

Antioxidants in Therapy and Preventive Medicine
Edited by I. Emerit *et al.*
Plenum Press, New York, 1990

529

white light irradiation after addition of rose bengal (100nM). Thiobarbituric acid reagent was then added and the reaction mixture boiled for 15 min to develop coloration. After centrifugation, the optical density of the clear supernatant was recorded at 532 nm.

Lens Culture and Photooxidation

Lenses from 100-125 g male Sprague Dawley rats were cultured as described (11) in 2 ml medium containing rose bengal (100 nM). Lenses were incubated at 37°C in a modular incubator chamber (Flow Laboratories) under 5 % CO_2 / 95 % air atmospher and illuminated by a tungsten light (100 W) at 40 cm. [^{3}H] choline transport (1 μ Ci/ml) was measured during 3 hrs. Lenses were homogenized, treated with trichloracetic acid (TCA) and the ratio [^{3}H] choline lens / [^{3}H] choline medium was calculated. Total glutathione (GSH) concentration was also determined in the TCA supernatant (12).

RESULTS and DISCUSSION

Inhibition of Singlet Oxygen-Induced Oxidation

As reported in Figure 1 (A), ADT was the most efficient compound for inhibiting singlet oxygen degradation of uric acid. However, the tocopherol activity was probably underestimated due to its low water solubility. For the same reason it was impossible to test the reference compound, β-carotene, in these experiments.

When compounds were tested in the phosphatidylcholine liposome system, β-carotene appeared to be the most potent inhibitor of singlet oxygen-induced lipid peroxidation. ADT presented a lower but significant inhibiting effect (Figure 1 (B)).

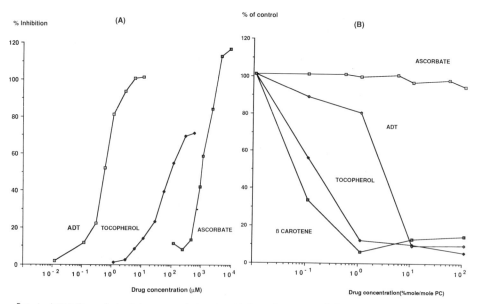

Fig. 1   Inhibition of singlet oxygen-induced oxidation of uric acid (A) and phosphatidylcholine liposomes (B).

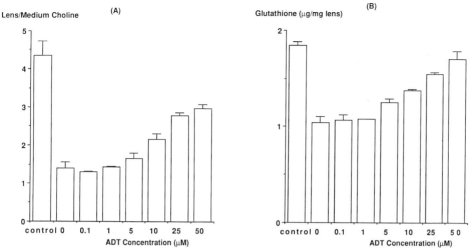

Fig. 2. Inhibition of singlet oxygen-induced lens damage in vitro as measured by choline transport (A) and GSH level (B)
Controls were incubated without rose bengal.

<u>Inhibition of Singlet Oxygen-Induced Lens Damage</u>

Lens damage was evaluated by inhibition of the choline transport and by decrease in GSH level induced by oxidation in vitro. When the lens were treated with ADT before the generation of singlet oxygen, they were protected against oxidative stress as reported in Figure 2.

ADT has been previously described as a potent inhibitor of rat liver microsome peroxidation (13). In the present study, we have shown that ADT was an efficient quencher of singlet oxygen, able to enter the lens in culture to protect transport systems and maintain the GSH level in this organ when subjected to oxidative stress. Such a drug deserves to be evaluated for cataract treatment in clinical trials.

REFERENCES

1. S.D. Varma, D. Chand, Y.R. Sharma, J.F. Kuck Jr., and R.D. Richards, Oxidative stress on lens and cataract formation : role of light and oxygen, Current Eye Res. 3:35 (1984).

2. J.S. Zigler Jr. and J.D. Goosey, Photosensitized oxidation in the ocular lens : Evidence for photosensitizers endogenous to the human lens, Photochem. Photobiol. 33:869 (1981).

3. J.S. Zigler Jr. and J.D. Goosey, Singlet oxygen as a possible factor in human senile cataract development, Current Eye Res. 3: 59 (1984).

4. K. Mandal, M. Kono, S.K. Bose, J. Thomson, and B. Chakrabarti, Structure and stability of gamma-cristallins. IV. Aggregation and structural destabilization in photosensitized reactions, Photochem. Photobiol. 47:583 (1988).

5. M.H. Jernigan Jr. and A.S. Vallari, Effects of photo-oxidation on transport systems in lens, Lens Research. 2:159 (1985).

6. K.C. Bhuvan and D.K. Bhuvan, Molecular mechanism of cataractogenesis : III. Toxic metabolites of oxygen as initiators of lipid peroxidation and cataract, Current Eye Res. 3: 67 (1984).

7. R.S. Dwivedi and V.B. Pratap, Alteration in glutathione metabolism during cataract progression, Ophthalmic Res. 19:41 (1987).

8. G. Tissié, V. Guillermet, E. Latour, C. Coquelet, and C. Bonne, Oxidative stress and lens opacity: an overall approach to screening anticataractous drugs, Ophthalmic Res. 20:27 (1988).

9. B.N. Ames, R. Cathcart, E. Schwiers, and P. Hochstein, Uric acid provides an antioxidant defense in humans against oxidants and radical-caused aging and cancer : a hypothesis, Proc. Natn. Acad. Sci. USA, 78:6858 (1981).

10. D.C. Liebler, D.S. Kling, and D.S. Reed, Antioxidant protection of phospholipids bilayers by α-tocopherol, J. Biol. Chem. 261:12114 (1986).

11. H.M. Jernigan Jr, H.N. Fukui, J.D. Goosey, and J.H. Kinoshita, Photodynamic effects of rose bengal or riboflavin on carrier-mediated transport systems in rat lens, Exp. Eye Res. 32: 461 (1981).

12. F. Tietze, Enzymic method for quantitative determination of nanogram amounts of total and oxidized glutathione : Applications to mammalian blood and other tissues, Anal. Biochem. 27:502 (1969).

13. D. Mansuy, A. Sassi, P.M. Dansette, and M. Plat, A new potent inhibitor of lipid peroxidation in vitro and in vivo. The hepatoprotective drug anisyldithiolthione, Biochem. Biophys. Res. Commun., 135:1015 (1986).

Dermatologic Antioxidant Therapy may be Warranted to Prevent

Ultraviolet Induced Skin Damage

Jürgen Fuchs [1], Margaret Huflejt, Laurie Rothfuss, David Wilson,
Gerardo Carcamo and Lester Packer
Zentrum der Dermatologie und Venerologie, Abteilung II, Klinikum der J.W.
Goethe Universität Frankfurt, FRG[1], and Department of Physiology
University of California Berkeley, USA.

Introduction

Experimental and clinical evidence is increasing that free radical processes are implicated in various pathological conditions in skin, such as ultraviolet and ionizing irradiation damage, thermal trauma, phototoxicity and photoallergy, drug toxicity, skin aging, skin autoimmune disease and tumor promotion. Direct evidence for ultraviolet induced free radical formation in skin has been obtained by low temperature electron paramagnetic resonance spectroscopy (Norrins 1962, Pathak 1968). It was suggested that there is no direct evidence relating formation of reactive oxygen species with acute or chronic ultraviolet effects (Epstein 1977). However, a considerable body of circumstantial evidence has been ammassed that strongly inferes that reactive oxygen species or reactions initiated by them are responsible for at least some of the deleterious effects of ultraviolet upon skin (Black 1987). Exposure of human and animal skin to UVB (280-320 nm) and UVA (320-400 nm) irradiation results in a significant elevation of lipid peroxidation products. Lipid peroxides are toxic in mammalian skin; epidermal damage due to epicutaneous application of lipid peroxides could be mediated by propagation of lipid peroxidation in situ. Injection of an aqueous extract of ultraviolet irradiated linoleic acid into skin induces necrosis of the epidermis and dermis followed by inflammatory response and disruption of collagen and elastic fibres (Waravdekar 1965).

Antioxidative enzymes can be inactivated during oxidative exposure in-vitro and in-vivo. Superoxide dismutase and glutathione peroxidase can be inhibited by superoxide anion radicals. Photodestruction of catalase by visible light was demonstrated in rat hepatocytes (Cheng 1979). In UVB exposed skin inhibition of the enzymic skin antioxidant system (superoxide dismutase, glutathione peroxidase and glutathione reductase) is reported (Maisuradze 1987). A single exposure of mouse skin to UVB irradiation results in significant decrease of SOD activity 24 to 48 hours after exposure, returning to the normal level 72 hour after irradiation (Miyachi 1987). Furthermore, irradiation of skin of hairless mice with UVB results in a decrease in epidermal reduced glutathione to 60% of its control level within 10 minutes and an increase in oxidized glutathione. This returned back to control levels within 30

Antioxidants in Therapy and Preventive Medicine
Edited by I. Emerit *et al.*
Plenum Press, New York, 1990

533

minutes (Connor 1987). The objective of our investigation (Fuchs 1989a) was to obtain a functional analysis of the acute responses of the cutaneous antioxidant defense system immediately after a single exposure to a large fluence of ultraviolet light. Our study departs substantially from previous approaches towards investigating photooxidative damage of skin antioxidants, since we analyze lipophilic, hydrophilic and enzymic antioxidants as well as free radical scavenging activity of the epidermis under identical experimental conditions.

Materials and methods

Skin irradiation
To ensure homogeneous irradiation of the total body skin and to avoid any interference with systemic anesthetics, which could interfere with tissue antioxidants, skin excisates of female hairless mice, 10-12 weeks old, (Jackson Laboratory, Bar Harbor, MA) were irradiated. To simulate solar irradiation, skin excisates of hairless mice were illuminated with an Oriel 1000 Watt Xenon lamp. The collimated radiation was projected from a distance of 50 cm onto the skin in a uniform circular field. The light was filtered through three neutral density filters and a WG 305 long pass filter. The energy output of the collimated beam was 1.0 mW/cm^2 at 305 nm at a distance of 50 cm, the samples were irradiated with a single dose of 300 mJ/cm^2.

Assays for small molecular and enzymic antioxidants
Small molecular antioxidants were measured in total skin homogenates, enzymic antioxidants in supernatants. Alpha-tocopherol, ubiquinol, and ubiquinone contents were analyzed simultaneously by HPLC as described (Lang 1986) using in line electrochemical detection of tocopherol and ubiquinol 9, and UV detection of ubiquinone 9. For glutathione measurements, aliquots of homogenized samples were processed and analyzed by HPLC (Reed 1980). Ascorbate levels were measured according to a spectrophotometric method (Roe 1954). Enzymic antioxidants were assayed according to conventional methods: catalase (Del Rio 1977) glutathione reductase (Karni 1984) and (Akerboom 1981), glutathione peroxidase (Flohe 1984b) and superoxide dismutase (Flohe 1984a).

ESR spectroscopy
Free radical scavenging activity at the epidermal surface of skin biopsies was measured by electron spin resonance spectroscopy (ESR) using the nitroxide (2,2,6,6- tetramethyl -1-piperidinoxy -4- [2',4',6'- trimethyl] methylpyridinium perchlorate), (Fuchs 1989b). First derivative ESR spectra (100 kHz modulation) were recorded at ambient temperature (22° C) on a Bruker ER 200 D-SCR spectrometer (X-band).

Results

Tissue concentrations of the lipophilic antioxidants alpha-tocopherol, ubiquinol-9 and ubiquinone-9 decreased significantly in skin excisates of hairless mice immediately after a single exposure to ultraviolet light. Tocopherol is decreased to 80%, ubiquinol-9 to 38%, ubiquinone-9 to 57% and the total amount of ubiquinol/one-9 is reduced to 52% of its control value. Concentrations of ubiquinol-10, ubiquinone-10 and beta-carotene were below detectability level in controls. A significant loss of hydrophilic antioxidants was also observed immediately after irradiation. Reduced glutathione declined to 63%, glutathione oxidized increased from 27.5% to 44.5% and total glutathione content declined to 83% of its control value. Tissue concentration of total ascorbate did not change significantly. Catalase and glutathione reductase activity were significantly inhibited after irradiation to 68% and 82% of their control values, respectively. There was only a minor and insignificant decrease in glutathione peroxidase and superoxide dismutase activity. Free radical scavenging activity at the surface of skin biopsies was assayed by measuring reduction of a cationic, persistent nitroxide radical. A significant inhibition of nitroxide scavenging activity was measured in skin biopsies immediately after ultraviolet exposure.

Discussion

The study indicates that immediately after exposure to a large fluence of ultraviolet irradiation the enzymic and non enzymic antioxidant potential of skin decreases significantly. This could be the consequence of oxidative exposure. Photolysis of water by UVC light results in direct formation of superoxide anion radical. Direct generation of reactive oxygen species in water is however not observed during irradiation with UVB or UVA light. Several endogenous metabolites like NADPH, riboflavin and certain nucelosides can sensitize the transmission of photon energy from solar radiation to oxygen resulting in formation of superoxide anion radical (Cunningham 1985). Besides NADPH, flavins are potent photochemical sources of superoxide anion radicals and have been clearly indicated as endogenous photosensitizers in human diploid lung fibroblasts (Pereira 1976).

Skin is a biological interface with the environment and functions as the first line of defense against noxious stimuli. Therefore skin primarily provides protection of the bodies interior from external insults. This protective mechanism can break down under stresses caused by environmental extremes, e.g., by conditions of imbalanced prooxidant / antioxidant equilibrium in favor to the former. The involvement of free radicals and reactive oxygen species in environmental stress conditions such as hyperoxia, hypoxia and reperfusion, heat schock and metabolism of environmental pollutants and toxic chemicals is well documented. Skin is a potential target organ of oxidative injury because it is continuously exposed to high oxygen concentrations, visible and ultraviolet irradiation, contains a variety of oxidizable structures critical for maintenance of cellular homeostasis and possesses ceaseless mitotic activity. It is one of the largest body organs and serves as a major portal of entry for many airborne environmental pollutants, some of them are free radical generating agents.

Ultraviolet light has short and long term pathological effects in skin. Acute skin reaction is commonly known as sun burn and is a phototoxic effect. It is characterized by development of erythema and epidermal and dermal cell damage. The sun burn cell is a histologic characteristic of UV induced epidermal injury (Danno 1984). It is a type of individual cell death appearing in epidermis after exposure to UV radiation. Sun burn cells are formed if reactive oxygen species production overcomes the capacity of normally present quenchers in skin (Danno 1987). Chronic skin reactions comprise degenerative and neoplastic tissue changes such as elastin and collagen fiber alterations, changes in the mesenchymal ground substance and transformation of epidermal cells to autonomous growth and malignancy. Activity of photoprotective agents is frequently assessed by measuring inhibition of skin erythema, although inflammation and epidermal cytotoxicity are not necessarily related pathological events. Cyclooxygenase inhibitors, such as indomethacin, prevent erythema development if applied immediately after UV exposure on the skin but do not inhibit formation of sun burn cells (Snyder 1975, Kaidbey 1976). In contrast, antioxidants such as alpha-tocopherol prevent both formation of erythema and epidermal cytotoxicity (Msika 1988). It was suggested, that sunburn cell formation can be prevented in skin by scavenging reactive oxygen species (Miyachi 1983). Formation of ultraviolet light induced lipid peroxidation is also inhibited by alpha-tocopherol (Glavind 1967, Ohsawa 1984) and liposomal superoxide dismutase. The findings, that the skin antioxidant defense is impaired by ultraviolet light and that antioxidants prevent inflammatory and cytotoxic effects of UV warrant further clinical trials to evaluate their therapeutic potential in photo-dermatology.

References

Akerboom TPM, Sies H. Assay of glutathione, glutathione disulfide and glutathione mixed disulfides in biological samples. Methods Enzymol 77, 373-382, 1981.

Black HS. Potential involvement of free radical reactions in ultraviolet light mediated cutaneous damage. Photochem Photobiol 46, 213-221,1987.

Cheng LYL,Packer L. Photodamage to hepatocytes by visible light. FEBS Letters 97, 124-128, 1979.

Connor MJ, Wheeler LA. Depletion of cutaneous glutathione by ultraviolet radiation. Photochem Photobiol 45, 239-245, 1987.

Cunningham ML, Krinsky NI, Giovanazzi SM, Peak MJ. Superoxide anion is generated from cellular metabolites by solar radiation and its components. J Free Rad Biol Med 1, 381-385,1985.

Danno K, Horio T, Takigawa M, Imamura S. Role of oxygen intermediates in UV-induced epidermal cell injury. J Invest Dermatol 83, 166-168, 1984.

Danno K, Horio T: Sunburn cell: factors involved in its formation. Photochem Photobiol. 45, 683-690, 1987.

Del Rio LA, Ortega MG, Lopez AL, Gorge JL. A more sensitive modification of the catalase assay with the clark oxygen electrode. Methods Enzymol. 105, 409-415, 1977.

Epstein JH. The pathological effects of light on the skin. In:" Free radicals in biology" , Pryor WA. ed, Academic Press, New York, pp 219- 249, 1977.

Flohe L, Ötting F. Superoxide dismutase assays. Methods Enzymol 105, 93-105, 1984a.

Flohe L, Günzler W. Assays of glutathione peroxidase. Methods Enzymol 105, 114-121, 1984b.

Fuchs J, Huflejt ME, Rothfuss LM, Wilson DS, Carcamo G, Packer L. Impairment of enzymic and non-enzymic antioxidants in skin by UVB irradiation. To be published, 1989a.

Fuchs J, Mehlhorn RJ, Packer L. Free radical reduction mechanisms in mouse epidermis and skin homogenates. To be published, 1989b.

Glavind J, Christensen F. Influence of nutrition and light on the peroxide content of the skin surface lipids of rats. Acta Derm Venereol 47, 339-344, 1967.

Kaidbey KH, Kurban AK. The influence of corticosteroids and topical indomethacin on sunburn erythema. J Invest Dermatol 66, 153-156, 1976.

Karni L, Moss SJ, Tel-Or E. Glutathione reductase activity in heterocysts and vegetative cells of cyanobacterium Nostoc muscorum. Arch Microbiol 140, 215-217, 1984.

Lang JK, Gohil K, Packer L. Simultaneous determination of tocopherols, ubiquinols, and ubiquinones in blood, plasma, tissue homogenates, and subcellular fractions. Anal Biochem 157,106-116, 1986.

Maisuradze VN, Platonov AG, Gudz TI, Goncharenko EN, Kudriashov IUB. Effect of ultraviolet rays on lipid peroxidation and various factors of its regulation in the rat skin (in Russian). Biol Nauki 5, 31-35, 1987.

Miyachi Y, Horio T, Imamura S. Sunburn cell formation is prevented by scavenging oxygen intermediates. Clin Exp Dermatol 8, 305-310, 1983.

Miyachi Y, Imamura S, Niwa Y. Decreased skin superoxide dismutase activity by a single exposure of ultraviolet radiation is reduced by liposomal superoxide dismutase pretreatment. J Invest Dermatol 89, 111-112, 1987.

Msika P, Cesarini JP, Poelman MC. Antioxidants and aggressions from ultraviolet radiation in man. Society for Free Radical Research Winter Meeting on" Free Radicals in Medicine: Current status of antioxidant therapy", Paris, December 8-9, 1988.

Norrins AL. Free radical formation in the skin following exposure to ultraviolet light. J Invest Dermatol 39, 445-448, 1962.

Ohsawa K, Watanabe T, Matsukawa R, Yoshimura Y, Imaeda K. The possible role of squalene and its peroxide of the sebum in the ocurrence of sunburn and protection from the damage caused by UV iradiation. J Toxicol Sci 9, 151-159, 1984.

Pathak MA, Stratton K. Free radicals in human skin before and after exposure to light. Arch Biochem Biophys 123, 468-476, 1968.

Pereira O, Smith JR, Packer L. Photosensitization of human diploid cell cultures by intracellular flavins and protection by antioxidants. Photochem Photophysiol 24, 237-242, 1976.

Reed DJ, Babson JR, Beatty PW, Brodie AE, Ellis EE. High performance liquid chromatography of nanomole levels of glutathione, glutathione disulfide and related thiols and disulfides. Anal Biochem 106: 55-62, 1980.

Roe JH. Chemical determination of ascorbic, dehydroascorbic and diketogluonic acids. In: "Methods of biochemical analysis", Glick D. ed, Interscience, New York, Vol. 1, pp 115-139, 1954.

Snyder DS. Cutaneous effects of topical indomethacin, an inhibitor of prostaglandin synthesis, on UV-damaged skin. J Invest Dermatol 64, 322-325, 1975.

Waravdekar VS, Saslaw LD, Jones WA, Kuhns JG. Skin changes induced by UV-irradiated linoleic acid extract. Arch Pathol 80, 91-95, 1965.

Antioxidant and Prooxidant effects of the Antipsoriatic Compound

Anthralin in Skin and Subcellular Fractions

Jürgen Fuchs, Wolfgang Nitschmann[1] and Lester Packer[1]

Zentrum der Dermatologie und Venerologie, Abteilung II, Klinikum der J.W. Goethe Universität Frankfurt, FRG, and Department of Physiology University of California Berkeley, USA[1]

Introduction

Anthralin (1.8-dihydroxy-9-anthrone) is a potent antipsoriatic compound (Ashton 1983). External therapy of psoriasis with anthralin has been introduced by Galewsky in 1916 and structure activity relationship of anthralin derivatives was first studied by Unna. The minimal structure for antipsoriatic activity of anthrone compounds has been defined by Krebs and corresponds to 1- hydroxy -9- anthrone (Krebs 1969). It contains I) a hydroxyl group in position 1, II) a carbonyl group in position 9, and III) two hydrogen groups in position 10. Altering only one group results in loss of antipsoriatic activity. Anthralin is a strong reductant that is readily oxidized by light, trace concentrations of metal ions and oxygen. Reactive oxygen species (singlet oxygen, superoxide anion and the hydroxyl radical) are formed as reaction intermediates during oxidation of anthralin. The anthralin metabolites anthraquinone and anthralin dimer have been identified in human and rat skin. A significant portion of anthralin is converted in intact skin into an ether insoluble product (Cavey 1985) which is assigned to anthralin "dark structures" (Mustakallio 1984). Anthralin dark structures are not well characterized chemically, naphtodianthrones and other polycyclic aromatic hydrocarbons are constituents of this product. A persistent free radical was detected in pig skin treated with anthralin, and it was suggested that the 10-anthranyl radical is the species reported (Shroot 1986). Anthralin binds to the plasma membrane of keratinocytes and is subsequently accumulated in mitochondria. It inhibits oxygen consumption in keratinocytes, fibroblasts, yeast and ascites cells and in skin. Research into mechanism of action of anthralin has elucidated mainly four target sites: 1) alteration of mitochondrial functions, 2) inhibition of cellular key enzymes, 3) interaction with nucleic acid metabolism, and 4) modulation of neutrophil function and arachidonic acid metabolism. Mitochondria are target organelles of drug action (Morlier 1985), in keratinocytes respiration is inhibited at lower anthralin concentrations than is thymidine incorporation. Thus, the antirespiratory effect of anthralin was suggested as the main reason for its antipsoriatic activity (Reichert 1986).

In this report we summarize our recent results (Fuchs 1986, Fuchs 1989a, Fuchs 1989b) about redox interactions of anthralin with biomembranes and skin. We investigated anthralin derived free radical formation in skin, subcellular particles and chemical model systems using electron spin resonance spectroscopy (ESR). The interaction of anthralin with energy transducing membranes using mammalian mitochondria (rat liver mitochondria) and intact cells of the cyanobacterium (blue-green alga) Synechococcus was studied using the

Antioxidants in Therapy and Preventive Medicine
Edited by I. Emerit *et al.*
Plenum Press, New York, 1990

537

same method. Mitochondrial respiration, lipid peroxidation and membrane antioxidant status were assessed by oxypolarography, analyzing membrane alpha-tocopherol, ubiquinol/ubiquinone content and measuring formation of thiobarbituric acid reactive products.

Results

Anthralin derived free radical formation in skin, subcellular particles and model systems

Topical application of a single dose of anthralin (2% solution in chloroform) on skin of hairless mice results in the formation of products containing persistent free radicals. The radical concentration in skin reaches a maximum at about 12 hours after application and persists over the next 60 hours. About one molecule of radical is formed per 500 molecules of anthralin. The antioxidant alpha-tocopherol significantly inhibits formation of anthralin radicals in skin when applied at the same time or 5 hours after exposure of skin to anthralin. The electron spin resonance spectrum is characterized by an unresolvable singlet (g=2.0036, one resonance line, line width 6 Gauss), that saturates at 5 mW half saturation power. A similar paramagnetic product appears after addition of anthralin to a microsomal suspension or to well coupled mitochondria. The radical formation in mitochondria is unaffected by inhibitors of the electron transport chain, uncouplers of oxidative phosphorylation and NAD-dehydrogenase blocker. In an chemical system, alkaline oxidation of the anthralin anion results in the sequential formation of different free radical species. The primary radical, that is converted over a period of several hours into a secondary, unidentified radical, can be assigned to a paramagnetic dianthrone species. The uncharacterized secondary radical undergoes further conversion into a persistent tertiary radical with similar spectral characteristics as the radical detected in skin. Anthralin dark structures were synthesized as the ether insoluble oxidation product of anthralin dimer as described (Auterhof 1962), and found to contain similar paramagnetic products. Since the hyperfine structure and the g-value of the persistent skin radical are different from the values reported for the highly reactive 10-anthranyl radical (g=2.0028, 7 resonance lines) (Ducret 1985), we conclude that the skin radical is not identical with 10-anthranyl. The skin radical is assigned to paramagnetic products of anthralin dark structures.

Interaction of anthralin with energy transducing membranes

Conversely to an uncoupler of oxidative phosphorylation, anthralin increases the pH gradient across the mitochondrial membrane and enhances the proton motive force. Mitochondrial energization is therefore increased. In isolated cells, anthralin increases the proton gradients of dark anaerobic Synechococcus. However, both in mitochondria and in Synechococcus ATP production is decreased by anthralin. Anthralin is a strong inhibitor of mitochondrial respiration in metabolic state 3 (substrate and ADP present). Respiration in metabolic state 4 (substrate, no ADP) is only slightly stimulated in a cyanide insensitive manner. Stimulation of mitochondrial oxygen consumption by the uncoupler CCCP is abolished by anthralin, if given before or after addition of CCCP in presence of NAD and FAD dependent substrates. The redox state of mitochondrial ubiquinol/ubiquinone couples was investigated in the presence of anthralin. Anthralin caused a significant reduction of mitochondrial ubiquinone 9 and 10 to the corresponding ubiquinols. Membrane concentrations of the lipophilic antioxidant alpha-tocopherol remained unchanged and thiobarbituric acid reactive products, indicative of lipid peroxidation, were not formed under these conditions.

Discussion

Anthralin radicals

The antipsoriatic compound anthralin is converted in skin into several oxidized products, including persistent free radicals that are not well characterized. Anthralin oxidation was investigated by electron spin resonance spectroscopy in a biological system and in a chemical system. Free radical formation in the skin of hairless mice is reduced by the antioxidant alpha-tocopherol. The data indicate, that tocopherol acts by interfering with free radical formation rather than by scavenging persistent anthralin radicals directly. The skin radicals do not correspond to 10-anthranyl, the initial paramagnetic anthralin oxidation product. Similar radicals obtained in skin are formed by anthralin exposed to ultraviolet light or alkaline solution and by mitochondria and microsomes. The pertinent skin radical is attributed to products derived from anthralin dark structures, the final oxidation products of anthralin. It is suggested that resonance stabilized, paramagnetic polycyclic hydrocarbons are the compounds detected. Their stability and low reactivity indicate a low potential for cutaneous irritation and tumor promotion. Reactive oxygen species which have been reported to be formed concomitantly during oxidation of anthralin and the initially formed highly reactive 10-anthranyl radical are more potent candidates for mediating tumor promotion and inflammation.

Anthralin and energy transducing membranes

Bioenergetic parameters and redox properties of energy transducing membranes in rat liver mitochondria and cyanobacteria were investigated in the presence of anthralin. Transmembrane pH and electrical gradients were determined using electron paramagnetic resonance spectroscopy. In mitochondria ubiquinones 9/10 and other redox components of the electron transport chain are reduced by anthralin, the proton motive force is increased. In the absence of ADP anthralin slightly stimulates mitochondrial cyanide insensitive oxygen consumption. It is concluded that increased cyanide insensitive respiration is due to enhanced autoxidation of mitochondrial components and/or catalyzed oxidation of anthralin, suggesting formation of reactive oxygen species. In presence of ADP mitochondrial respiration is decreased and ATP synthesis is inhibited. Uncoupler induced mitochondrial respiration is also decreased by anthralin, indicating inhibition of the electron transport chain. In the cyanobacterium Synechococcus PCC 6311 anthralin increases the pH gradient and decreases ATP levels. Thus, anthralin acts as an electron donor to membrane associated redox components and inhibits ATP synthesis in two different biological systems. Since anthralin reacts with redox components in different biological membranes, alterations of subcellular/cellular redox status and energy metabolism might contribute significantly to its antipsoriatic activity.

General considerations

Anthralin has a complex redox behavior and may act in-vivo both as an anti- and as a prooxidant. Antioxidant activity could be mediated by its properties as a strong reductant, its strongly negative redox potential of $Eo' = -0.76$ V (Mustakallio 1984) allows reduction of e.g., ubiquinones to ubiquinols (table 3) and stimulating NADP reduction to NADPH (Kohen 1986). Epidermal NADPH quinone reductase is induced by anthralin (Merk 1988) which might prevent toxicity of redox cycling molecules. Anthralin inhibits in a dose dependent manner superoxide anion generation in stimulated neutrophils in-vitro (Schröder 1985).

Two mechanisms contribute to the establishment of a prooxidant cellular state by anthralin: a) inhibition of antioxidant defense mechanisms, and b) direct formation of reactive oxygen species and initiation of lipid peroxidation. A third mechanism may be stimulation of oxidant release of neutrophils. Although anthralin inhibits the oxygen burst of stimulated neutrophils (Schröder 1985), increased formation of reactive oxygen species in the early neutrophil response was found in another study (Anderson 1987). Further work is required to fully understand these results.

Anthralin inhibits epidermal superoxide dismutase and catalase activity (Solanki 1981). However, reduced glutathione content, glutathione reductase and glutathione peroxidase activity in skin are affected by anthralin only at very high, toxic concentrations (Kingston 1985).

The anthralin anion is a photosensitizer (Müller 1986a) and generates hydrogen peroxide (Unna 1916, Krebs 1969), singlet oxygen (Müller 1986b, Joshi 1984), superoxide anion radical (Joshi 1984) and the hydroxyl radical (Müller 1988). Unna already pointed out in 1916 that anthralin may induce lipid peroxide formation in skin (Unna 1916). Vice versa, lipid peroxides may also be initiators of anthralin free radical formation in skin (Ducret 1985). Anthralin increases ultraviolet induced formation of lipid peroxides in skin surface extracts of psoriatic patients (Meffert 1969) and skin microsomal fractions are peroxidized by anthralin (Finnen 1984). In-vivo, topical application of anthralin increases ethane exhalation in animals indicating increased lipid peroxidation in skin (Müller 1987).

Anthralin enhances the extracellular release of reactive oxidants by activated neutrophils in vitro. Intracellular generation of reactive oxygen species in activated neutrophils is inhibited by anthralin (Anderson 1987). Inhibition of the intracellular response is probably due to auto-inactivation of neutrophil respiratory burst by high concentrations of extracellular oxidants (Edwards 1986). Apparently, anthralin does not directly activate membrane associated oxidative metabolism of neutrophils but induces hyperreaction of these cells to receptor mediated stimuli. It was suggested that anthralin potentiates the formation of phagocyte derived immunosuppressive reactive oxidants which are the probable mediators of the therapeutic activity (Anderson 1987).

A common concept in reaction mechanism of antipsoriatic drugs such as anthralin, coal tar, pyrogallol, UVB light, PUVA therapy and hematoporphyrin phototherapy has been proposed. These topical antipsoriatic agents may act via the induction of reactive oxygen dependent chain reactions in skin (Gruner 1987). Since anthralin reacts with redox components in different biological membranes, alterations of the cellular redox status and energy metabolism, as well as impairment of neutrophil function might contribute significantly to its antipsoriatic activity. The scenario of reactions induced by anthralin is far from being completely understood. Since its redox interactions with subcellular/cellular membranes and intact skin are rather complex, sophisticated experiments are warranted to acquire more detailed knowledge about its antipsoriatic mechanism in-vivo. This could finally help in better understanding of the pathophysiology of psoriasis and in development of new antipsoriatic drugs with less side effects.

References

Anderson R, Lukey PT, Dippenaar U, Eftychis HA, Findlay GH, Wooten MW, Nel AE. Dithranol mediates pro-oxidative inhibition of polymorphonuclear leukocyte migration and lymphocyte proliferation. Brit J Dermatol, 117: 405-418, 1987.

Ashton RE, Andre P, Lowe NJ, Whitefield M. Anthralin: Historical and current perspectives. J Am Acad Dermatol 9, 173-191, 1983.

Auterhof H, Sachdev R. Das Verhalten von Hydroxydianthronderivaten in alkalischem Milieu. Arch Pharm 295, 850-852, 1962.

Cavey D, Dickinson RG, Shroot B, Schaefer H. The in vivo fate of topically applied dithranol in the skin of hairless rat. Arzneim Forsch/Drug Res 35, 605-609, 1985.

Ducret F, Lamblet P, Löliger J, Savoy MC. Antipsoriatic drug action of anthralin oxidation reactions with peroxidizing lipids. J Free Rad Biol Med 1, 301-306, 1985.

Edwards SW, Swan TF. Regulation of superoxide generation by myeloperoxidase during the respiratory burst of human neutrophils. Biochem J 237, 601-604, 1986.

Finnen MJ, Lawrence CM, Shuster S. Anthralin increases lipid peroxide formation in skin and free radical scavengers reduce anthralin irritancy. Brit J Dermatol 111, 717, 1984.

Fuchs J, Nitschmann WH, Packer L. The antipsoriatic compound anthralin influences bioenergetic parameters and redox properties of energy transducing membranes, to be published, 1989a.

Fuchs J, Packer L. Investigations on anthralin free radicals in model systems and in skin of hairless mice. J Invest Dermatol, in press, 1989b.

Fuchs J, Zimmer G, Wölbling RH, Milbradt R. On the interaction between anthralin and mitochondria: a revision. Arch Dermatol Res 279, 59-65, 1986.

Gruner S, Strunk D, Diezel W. Der Einfluß topischer Antipsoriatika auf ATP-ase-positive epidermale Langerhans Zellen. Tierexperimentelle Untersuchungen. Dermatol Monschr 173, 436-440, 1987.

Joshi PC, Pathak MA. The role of active oxygen ($^1O_2$ and O_2^-) induced by crude coal tar and its ingredients used in photochemotherapy of skin diseases. J Invest Dermatol 82,67-73, 1984.

Kingston TP, Connor MJ, Lowe NJ. Anthralin and cutaneous glutathione. Brit J Dermatol, 113, 780-781, 1985.

Kohen E, Kohen C, Morliere P, Santus R, Reyftmann JP, Dubertret L, Hirschberg JG, Coulomb B. A microspectrofluorometric study of the effect of anthralin, an antipsoriatic drug, on cellular structure and metabolism. Cell Biochem Funct 4, 157-168, 1986.

Krebs A. Untersuchungen zur Strukturspezifität der Psoriasisheilmittel Chrysarobin und Dithranol. Der Hautarzt 20, 204-209, 1969.

Meffert H, Reich P. Beeinflussung der Lipoperoxide der menschlichen Hautoberfläche durch ultraviolette Strahlung in vitro und in vivo. Dermatol Monschr 155, 948-954, 1969.

Merk HF, Khan WA, Bickers DR, Mukhtar H. Induction of epidermal and liver NAD(P)H: quinone reductase by anthralin. J Invest Dermatol 90, 248, 1988.

Morlier P, Dubertret L, Melo TSE, Salet C, Fosse M, Santus R. The effects of anthralin (dithranol) on mitochondria. Br J Dermatol 112, 509-515, 1985.

Müller K, Mayer KK, Wiegrebe W. $^1O_2$ - Oxidation of dithranol to chrysazin. Arch Pharm (Weinheim) 319, 1009-1018, 1986a.

Müller K, Eibler E, Mayer KK, Wiegrebe W. Dithranol, singlett oxygen and unsaturated fatty acids. Arch Pharm (Weinheim) 319, 2-9, 1986b.

Müller K, Wiegrebe W, Younes M. Dithranol, active oxygen species and lipid peroxidation in vivo. Arch Pharm (Weinheim) 320, 59-66, 1987.

Müller K, Kappus H. Hydroxyl radical formation by dithranol. Biochem Pharmacol 37, 4277-4280, 1988.

Mustakallio KK, Martinmaa J, Vilvala R, Halmekoski J. Free radicals and the treatment of psoriasis with special reference to dithranol. Medical Biology 62, 155-158, 1984.

Reichert U. Skin toxicity and celluar metabolism: in vitro models. Br J Dermatol 115, Suppl 31, 108-116, 1986.

Schröder JM, Kosfeld U, Christophers E. Multifunctional inhibition by anthralin in nonstimulated and chemotactic factor stimulated human neutrophils. J Invest Dermatol 85, 30-34, 1985.

Shroot B, Brown C. Free radicals in skin exposed to dithranol and its derivatives. Arzneim Forsch/Drug Res 36, 1253-1255, 1986.

Solanki V, Rana RS, Slaga TJ. Diminution of mouse epidermal superoxide dismutase and catalase activities by tumor promotors. Carcinogenesis 2: 1141-1146, 1981.

Unna PG. Cignolin als Heilmittel der Psoriasis. Dermatol Wochenschr, 7,150-163, 1916.

ANTIOXIDANT TREATMENT IN EXPERIMENTAL THERMAL INJURY

Gerd O. Till, Hans P. Friedl, and Peter A. Ward

Department of Pathology
University of Michigan Medical School
Ann Arbor, Michigan 48109-0602

INTRODUCTION

There is increasing experimental evidence to suggest that oxygen-derived free radicals play an important role in the pathophysiology of thermal injury. Following thermal injury of skin, oxygen radicals have been linked to the appearance of lipid peroxidation products (1-3) and the pathogenesis of burn shock (4). Experimental studies have shown that edema formation in the skin of rats following a partial-thickness burn is mainly mediated by oxidants derived from xanthine oxidase (5, 6). In addition, there is evidence that the development of intravascular hemolysis and acute lung injury secondary to thermal trauma can be linked to the generation of oxygen radicals from complement-activated blood neutrophils (7, 8). In these model systems of thermal injury, antioxidant interventions have demonstrated pronounced, protective effects supporting the concept that oxygen radicals are playing an important role in local as well as distant cell and tissue damage following thermal injury.

Evidence that oxygen radicals are participating in the pathogenesis of thermal injury has largely been derived from interventional measures including treatment with antioxidant enzymes, radical scavengers, iron chelators, and by elimination or inhibition of oxidant sources such as neutrophils and xanthine oxidase. By the same token, some of these compounds and means, that have been employed as tools or probes to elucidate the in vivo role of oxidants may potentially be used in the treatment of thermal injury-related cell and tissue damage. In the following, we will briefly describe the experimental models of thermal injury and interventions that have proven to be beneficial in the prevention of edema formation in thermally injured skin as well as acute lung injury secondary to thermal trauma.

EXPERIMENTAL THERMAL INJURY

The experimental models of thermal injury used to study the effects of interventional measures (treatment) on skin edema formation and secondary acute lung injury are well established in our laboratory (1, 5, 8). Both skin edema as well as lung injury were studied employing the same model of thermal injury. Briefly, specific pathogen-free Long Evans rats (350-450 g) were anesthetized with xylazine (13 mg/kg body weight) and ketamine hydrochloride (100 mg/kg). Following anesthesia, the shaved dorsal skin over the lumbosacral area was exposed to 70°C water for 30 seconds. This resulted in a deep second degree burn involving 28-30% of

Antioxidants in Therapy and Preventive Medicine
Edited by I. Emerit *et al.*
Plenum Press, New York, 1990

the total body surface area. Experimental animals were kept under anesthesia until sacrifice by exsanguination via the posterior vena cava and cervical dislocation. Control animals were treated the same way except that the shaved skin was exposed to water of 22^0C.

Edema Formation in Thermally Injured Skin

As a measure of microvascular leakage and edema formation in thermally injured skin, the extravasation of radiolabeled bovine serum albumin (^{125}I-BSA) into skin tissue was determined. ^{125}I-BSA was injected intravenously into experimental animals five minutes before thermal injury. At various time points after thermal injury, skin biopsies were taken and the radioactivity determined. A representative example of the time course of edema formation in thermally injured skin is depicted in Figure 1. As can be seen, maximal edema formation (leakage of ^{125}I-BSA) could be observed at 60 minutes after thermal trauma. These changes in skin radioactivity correlated well with changes in total water content of skin as determined by dry/wet weight determinations (5).

Lung Injury Secondary to Thermal Trauma

Lung injury secondary to skin burns was determined by the increase in lung vascular permeability, which was determined by the leakage of ^{125}I-BSA into the extravascular compartment of the lung. Five minutes before thermal injury, 0.5 ml phosphate-buffered saline containing ^{125}I-BSA was injected intravenously into experimental animals. They were killed at various time points post thermal trauma by exsanguination from the posterior vena cava followed by cervical dislocation. One ml of blood was collected for assessment of radioactivity. The lung vasculature was then perfused with 10 ml of physiologic saline to clear any radioactivity remaining within the vasculature. Lung permeability values were calculated by assessing the content of radioactivity in the lung and then expressed as a ratio to the amount of radioactivity present in 1 ml blood. A representative example of the time course of development of acute lung injury secondary to thermal trauma to skin is shown in Figure 2.

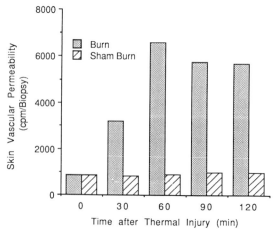

Fig. 1. Time course of edema formation in thermally injured skin of rats.

Pathomechanism of Edema Formation in Thermally Injured Skin

The pathomechanism of edema formation in thermally injured skin is only poorly understood. The list of mediators suggested to play a role in burn edema formation includes among others histamine (10), bradykinin (11), serotonin (12), prostaglandins (9), and leukotrienes (13, 14), which have either been demonstrated in the lymph coming from the burn area or whose role has been inferred by the results of pharmacologic interventions.

Recent evidence suggests that oxygen radicals may play a significant role in the pathogenesis of microvascular damage in thermally injured skin. It has been reported that oxygen radicals are involved in the appearance of lipid peroxidation products in burned skin (1-3). The ability of catalase and superoxide dismutase to attenuate burn edema formation in thermally injured rat paws (15) and the hamster cheek pouch (14) also suggest that oxygen radicals may play a role in the pathogenesis of burn edema.

Using the above described experimental model of thermal injury, we have now obtained data which indicate that oxygen radicals derived from xanthine oxidase are involved in edema formation in burned skin (5). The plasma of thermally injured rats showed increases in levels of xanthine oxidase activity, with peak values appearing at 15 minutes after thermal trauma. The failure of neutrophil depletion to protect against microvascular injury of burned skin and the protective effects of xanthine oxidase inhibitors and scavengers of hydroxyl radical support the concept that xanthine oxidase is the source of oxygen radicals (most likely hydroxyl radicals) involved in edema formation. Using the same experimental model system, we have obtained additional data which suggest that histamine plays an important role too (6). We have observed that in thermally injured rats levels of plasma histamine and xanthine oxidase rise in parallel. Furthermore, in vitro and in vivo studies have shown that histamine can enhance the activity of xanthine oxidase (6).Because prevention of histamine release in experimental animals resulted in protection from edema formation (6), it can be concluded that histamine is indirectly contributing to the development of oxygen radical-mediated microvascular injury in burned skin.

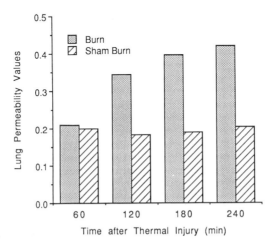

Fig. 2   Increase in lung vascular permeability following thermal injury of skin of rats.

Pathomechanism of Acute Lung Injury Secondary to Thermal Trauma

Oxygen radicals have also been suggested to play a role in the pathogenesis of pulmonary injury secondary to thermal trauma. We have shown that experimental thermal injury to the skin of rats can result in activation of the complement system leading to the appearance of C5-related chemotactic activity in plasma, transient neutropenia, sequestration of blood neutrophils in pulmonary capillaries, and oxygen radical-mediated vascular endothelial cell damage (8). Because antioxidant interventions or depletion of experimental animals of neutrophils or complement prior to thermal injury resulted in significant protection from lung injury, it is believed that oxygen-derived free radicals released from complement-activated neutrophils are instrumental in the induction of lung injury secondary to skin burns.

This assumption is further supported by the demonstration of lipid peroxidation products in plasma and lung tissues of thermally injured rats. Complement or neutrophil depletion of experimental animals prior to thermal injury, early removal of the burned skin or antioxidant treatment significantly diminished the appearance in plasma of lipid peroxidation products (1), suggesting a linkage between thermal trauma of skin, secondary injury of lung, and appearance in plasma and lung tissues of oxygen radical-derived lipid peroxidation products.

ANTIOXIDANT TREATMENT

Because oxygen radicals appear to be involved in the pathophysiology of both edema formation in the burned skin as well as in the development of acute pulmonary injury secondary to thermal skin burns, we have investigated the effects of antioxidants including specific enzymes, hydroxyl radical scavengers, and iron chelators. In addition, we have employed drugs like xanthine oxidase inhibitors and a mast cell stabilizer which will prevent, directly or indirectly, oxygen radical formation in vivo.

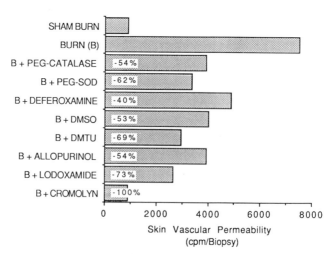

Fig. 3.   Protection by various compounds against edema formation in thermally injured skin of rats. Skin biopsies were taken at 60 minutes after thermal injury for radioactivity measurements.

Protection from Edema Formation in Skin

As outlined above, edema formation in thermally injured skin of rats is largely mediated by oxygen radicals (hydroxyl radicals) generated by xanthine oxidase. In order to investigate the protective effect of antioxidant treatment, experimental animals were pretreated (at five minutes before thermal injury) with one of the following compounds: polyethylene glycol-modified catalase (PEG-catalase, 1200 U/kg body weight, i.v.), polyethylene glycol-modified superoxide dismutase (PEG-SOD, 1200 U/kg, i.v.), the iron chelator deferoxamine (15 mg/kg, i.v.), the hydroxyl radical scavengers dimethyl sulfoxide (DMSO, 1.5 g/kg, i.p.) and dimethylthiourea (DMTU, 1g/kg, i.p.), the xanthine oxidase inhibitors allopurinol (50 mg/kg, i.p.) and lodoxamide tromethamine (5 mg/kg, i.v.), and the mast cell stabilizer cromolyn (20 mg/kg). Skin biopsies were taken at 60 minutes after thermal injury and the radioactivity determined. It should be noted that the concentrations for all of the compounds employed do not necessarily represent optimal doses since extensive dose/effect studies have not been performed.

The results of the treatment are depicted in Figure 3 which shows a representative set of data. Depending on the compound employed, the observed protection from burn edema formation ranged from 40% (for deferoxamine) to 100% (for cromolyn). These observations demonstrate that antioxidants or compounds which prevent oxidant formation are beneficial in the treatment of burn edema. In addition, they support the concept that xanthine oxidase-derived hydrogen peroxide and its iron-catalyzed conversion product, the hydroxyl radical, are involved in edema formation of skin following thermal injury. Somewhat surprising is the profound protection provided by cromolyn. One explanation would be that cromolyn prevents histamine release from mast cells, thus preventing enhancement of xanthine oxidase activity by histamine (6).

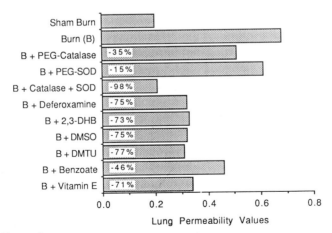

Fig. 4.   Protection by various compounds against lung microvascular injury of rats secondary to thermal trauma of skin. Pulmonary injury was determined at 3 hours after thermal trauma.

Attenuation of Acute Lung Injury

To evaluate antioxidant treatment on lung injury following skin burns, experimental animals were treated 10 minutes prior to thermal injury with one of the following compounds: PEG-catalase (1200 U/kg, i.v.), SOD (1200 U/kg, i.v.) or a mixture of the two compounds, the iron chelators deferoxamine (15 mg/kg, i.v.) and 2,3-dihydroxybenzoic acid (2,3-DHB, 100 mg/kg, i.v.), the hydroxyl radical scavengers DMSO (1.5 ml/kg, i.p.), DMTU (1g/kg, i.p.) and sodium benzoate (100 mg/kg), and vitamin E (25 U.S.P. units at 18 hours and 2 hours before thermal injury). The other compounds were administered at 10 minutes before thermal injury.

Figure 4 shows a typical result of the effects of antioxidant treatment. as can be seen, there was significant attenuation of acute pulmonary trauma in thermally injured rats For most of the compounds tested, the protection achieved was approximately 70%.

CONCLUSION

In a rat model system of thermal injury, antioxidant treatment has been shown to significantly reduce the development of skin edema as well as acute lung injury following skin burns. The use of long-lived polyethylene glycol derivatives of catalase and superoxide dismutase was particular effective, as were iron chelators, and hydroxyl radical scavengers. Because xanthine oxidase-derived oxygen radicals are playing a prominent role in edema formation in the burned skin, and because histamine has been shown to substantially enhance the activity of xanthine oxidase (6), it came as no surprise that inhibitors of xanthine oxidase or of histamine release demonstrated dramatic protection from edema formation. Based on these promising observations in an experimental model of thermal injury, it might be anticipated, that at least some of the antioxidant treatment procedures may also prove to be beneficial in the severely burned human patient.

REFERENCES

1. Till GO, Hatherill JR, Tourtellotte WW, Lutz MJ, Ward PA: Lipid peroxidation and acute lung injury after thermal trauma to skin. Evidence of a role for hydroxyl radical. Am J Pathol 1985, 119:376-384.
2. Nishigaki I, Hagihara M, Hiramatsu M, Izawa Y, Yagi K: Effect of thermal injury on lipid peroxidation levels of rat. Biochem Med 1980, 24:185-189.
3. Demling RH, Katz A, Lalonde C, Ryan P, Jin L-J: The immediate effect of burn wound excision on pulmonary function in sheep: The role of prostanoids, oxygen radicals, and chemoattractants. Surgery 1987, 101:44-55.
4. Saez JC, Ward PH, Gunther B, Vivaldi E: Superoxide radical involvement in the pathogenesis of burn shock. Circ Shock 1984, 12:229-239.
5. Till GO, Guilds LS, Mahrougui M, Friedl HP, Trentz O, Ward PA: Role of xanthine oxidase in thermal injury of skin. Am J Pathol 1989, in press.
6. Friedl HP, Till GO, Trentz O, Ward PA: Roles of histamine, complement and xanthine oxidase in thermal injury of skin. Am. J. Pathol. 1989, in press.
7. Hatherill JR, Till GO, Bruner LH, Ward PA: Thermal injury, intravascular hemolysis, and toxic oxygen products. J Clin Invest 1986, 78:629-636.
8. Till GO, Beauchamp C, Menapace D, Tourtellotte W, Kunkel R, Johnson KJ, Ward PA: Oxygen radical dependent lung damage following thermal injury of rat skin. J Trauma 1983, 23:269-277.
9. Anggard E, Jonsson CE: Efflux of prostaglandins in lymph from scalded tissue. Acta Physiol Scand 1971, 81:440-443.
10. Elrod PD, McCleery RS and Ball CT:. 1951. An experimental study of the effect of heparin on survival time following lethal burns. Surg Gynecol Obstet 1951, 92:35-42.

11. Rocha E, Silva M, Antonio A: Release of bradykinin and the mechanism of production of thermic edema ($45^{o}C$) in the rats paw. Med Exp 1960, 3:371-382.
12. Carvajal HF, Brouhard BH, Linares HA: Effect of antihistamine-antiserotonin and ganglionic blocking agents upon increased capillary permeability following burn edema. J Trauma 1975, 15:969-975.
13. Alexander F, Mathieson M, Teoh KH, Huval WV, Lelcuk S, Valeri CR, Shepro D, Hechtman HB: Arachidonic acid metabolites mediate early burn edema. J Trauma 1984, 24:709-712.
14. Hambrecht GS, Hilton JG: The effect of catalase, indomethacin and FPL 55712 on vascular permeability in the hamster cheek pouch following scald injury. Prostaglandins Leukotrienes Med 1984, 14:297-304.
15. Bjork J, Arturson G: Effect of cimetidine, hydrocortisone, superoxide dismutase and catalase on the development of edema after thermal injury. Burns Incl Therm Inj 1983, 9:249-256.

PRODUCTION OF SUPEROXIDE ANION BY GLYCATED PROTEINS:

INVOLVEMENT IN COMPLICATIONS OF DIÀBETES MELLITUS

Monboisse, J.C., Gillery, P., Maquart, F.X. and Borel, J.P.

Laboratoire de Biochimie, Centre Hospitalier Universitaire, Hôpital

R. Debré, Rue Alexis Carrel, F51092, Reims Cedex, France

INTRODUCTION

Oxygen free radicals have been implicated in the etiology of a number of disorders such as mutagenesis, radiation disease, aging, inflammation myocardial infarction, atherosclerosis, cataract or immunological disorders[1]. As regards diabetes mellitus, oxygen free radicals are considered likely to participate in the very early stages of the disease but they have not previously been thought to be involved in the long term diabetes complications. The results presented in this paper suggest that oxygen free radicals can be formed in increasing amounts in diabetes mellitus by an electron exchange occuring between the sugar moiety of glycated proteins and molecular oxygen.

PRODUCTION OF OXYGEN FREE RADICALS BY GLYCATED PROTEINS

During the course of diabetes mellitus, increased concentrations of glucose molecules bind to free amine radicals of proteins and they form aldimines which are rearranged (Amadori's rearrangement) into 1-desoxy-2-amino-2-ketose derivatives (the so-called "fructosamine" derivatives of glycated proteins)[2] :

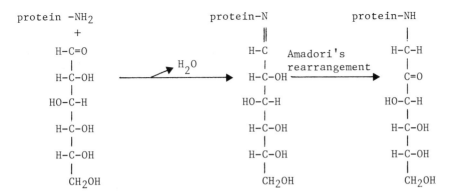

Antioxidants in Therapy and Preventive Medicine
Edited by I. Emerit *et al.*
Plenum Press, New York, 1990

The assay technique of glycated proteins has been largely used since 1982[3], and consists in the reduction of nitroblue tetrazolium at alkaline pH :

$$
\begin{array}{ccc}
\begin{array}{c}
\text{Protein-NH} \\
\mid \\
\text{H-C-H} \\
\mid \\
\text{C=O} \\
\mid \\
\text{HO-C-H} \\
\mid \\
\text{H-C-OH} \\
\mid \\
\text{H-C-OH} \\
\mid \\
\text{CH}_2\text{OH}
\end{array}
&
+ \text{ NBT} \longrightarrow
&
\begin{array}{c}
\text{protein-NH} \\
\mid \\
\text{H-C-H} \\
\mid \\
\text{C=O} \\
\mid \\
\text{C=O} \\
\mid \\
\text{H-C-OH} \\
\mid \\
\text{H-C-OH} \\
\mid \\
\text{CH}_2\text{OH}
\end{array}
\quad + \text{ NBTH}_2
\end{array}
$$

It has been demonstrated previously by other authors that organic substances containing a ketol radical are capable of reacting with dioxygen to form superoxyde anion O_2^- according to the reaction [4] :

$$
\begin{array}{ccc}
\begin{array}{c}
\text{R} \\
\mid \\
\text{C=O} \\
\mid \\
\text{HO-C-H} \\
\mid \\
\text{R'}
\end{array}
&
+ \ 2O_2 \longrightarrow
&
\begin{array}{c}
\text{R} \\
\mid \\
\text{C=O} \\
\mid \\
\text{C=O} \\
\mid \\
\text{R'}
\end{array}
\quad + \ 2O_2^- + 2H^+
\end{array}
$$

In addition, O_2^- ions are capable of reducing NBT to form diformazan.

We demonstrated in vitro that glycated proteins or 1-deoxy-2-morpho-lino-fructose (DMF), an analog of the "fructosamine" radical was capable of reducing NBT and ferricytochrome c at alkaline pH and, in addition, at pH 7.4. This reduction was inhibited by the addition of superoxide dismutase (SOD), suggesting that O_2^- ions were produced by these coumpounds. The levels of NBT and ferricytochrome c reduction induced by sera from diabetic patients were higher than that of controls. We purified the glycated proteins from diabetic sera by affi-nity chromatography on phenylboronic acid. These glycated proteins were capable of reducing ferricytochrome c, whereas the non-glycated ones were unable[5].

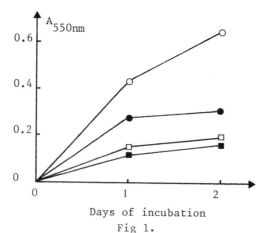

Days of incubation

Fig 1.

Reduction of ferricytochrome c at pH 7.4 in phosphate buffer by glycated (O , ●) or non glycated (□ , ■) proteins, in the absence (O , □) or presence (● , ■) of SOD (50 U/ml).

DIABETES MELLITUS AND OXYGEN FREE RADICALS

The effects of oxygen free radicals have been incriminated in various types of cell damages and in the pathogenesis of many diseases. The most pronounced toxic effets were lipid peroxidation, destruction of biomembranes, nucleic acid damage, enzyme inhibition and protein degradation[5]. Lipid peroxidation is initiated by substracting hydrogen atoms from unsaturated lipids, followed by free radical propagation and a final liberation of malondialdehyde. Malondialdehyde has been found in increased amounts in the erythrocytes from diabetic patients[6]. In addition, blood lipoprotein abnormalities have been described in diabetic patients, resulting either from peroxidation of lipid components by oxygen free radicals or from reaction of malondialdehyde with the apoproteins. Modified lipoproteins are known to be toxic to cultured endothelial cells and to attract phagocytic cells[7]. Oxygen free radicals are capable of degrading DNA, supporting the acceleration of aging. They also degrade proteins and connective tissue components such as collagen [8,9].

The defenses against oxygen free radicals seem to be decreased during the course of diabetes mellitus. The superoxide dismutase glycation leads to the inactivation of the enzyme[10] and glutathione peroxidase activity is decreased in erythrocytes from diabetic patients[11].

These effects prompt us to propose the testing of antioxidants in the therapy of diabetic patients. Vitamin E has been shown to inhibit the glycation of proteins and could be a candidate for the prevention of diabetic complications[12].

REFERENCES

1 - J.Z. Byczkowski and T. Gessner. Biological role of superoxide ion-radical. Int. J. Biochem. 20 : 569 (1988).
2 - D.A. Armbruster. Fructosamine : structure, analysis and clinical usefulness. Clin. Chem. 33 : 2153 (1987).
3 - R.N. Johnson, P.A. Metcalf and J.R. Baker. Fructosamine : a new approach to the estimation of serum glycosylprotein. An index of diabetic control. Clin. Chim. Acta 127 : 87 (1982).
4 - J. Garst, P. Stapleton and J. Johnston. Mutagenicity of alpha hydroxyketones may involve superoxide anion radical. In "Oxy radicals and their scavenger systems" Vol. II. R.A. Greenwald and G. Cohen, eds. Elsevier Biomedical New-York (1983) 125.
5 - P. Gillery, J.C. Monboisse, F.X. Maquart and J.P. Borel. Glycation of proteins as a source of superoxide. Diab. Metab. 14 : 25 (1988).
6 - R. Selvam and C.V. Anuradha. Lipid peroxidation and antiperoxidative enzyme changes in erythrocytes in diabetes mellitus. Ind. J. Biochem. Biophys. 25 : 268 (1988).
7 - P.A. Southorn. Free radicals in Medicine. II. Involvement in human disease. Mayo Clin. Proc. 63 : 390 (1988).
8 - S.P. Wolff, A. Gamer and R.T. Dean. Free radicals, lipids and protein degradation. Trends Biochem. Sci. 11 : 27 (1986)
9 - J.C. Monboisse, M. Gardes-Albert, A. Randoux, J.P. Borel and C. Ferradini. Collagen degradation by superoxide anion in pulse and gamma radiolysis. Biochim. Biophys. Acta. 965 : 29 (1988).

10 - K. Arai, S. Iizuka, Y. Taka, K. Oikawa and N. Taniguchi. Increase in the glucosylated form of erythrocyte Cu-Zn-superoxide dismutase in diabetes and close association of the nonenzymatic glucosylation with the enzyme activity. Biochim. Biophys. Acta 924 : 292 (1987).

11 - S.A. Wohaieb and D.V. Godin. Alterations in free radical tissue defense mechanisms in streptozotocin-induced diabetes in rat. Effect of insulin treatment. Diabetes 36 : 1014 (1987).

12 - A. Ceriello, D. Giugliano, A. Quatraro, P. Dello Russo and R. Torella. A preliminary note of inhibiting effect of α - tocopherol (vit. E) on protein glycation. Diab. Metab. 14 : 40 (1988).

RATIONALE FOR ANTIOXIDANT THERAPY IN PANCREATITIS AND CYSTIC FIBROSIS

S. Uden, D. Bilton, P.M. Guyan, P.M. Kay and J.M. Braganza

The Department of Gastroenterology
Manchester Royal Infirmary
Oxford Road
Manchester M13 9WL, U.K.

SUMMARY

The overlapping features of the acquired diseases acute pancreatitis and chronic pancreatitis on the one hand, and of chronic pancreatitis and pancreatic involvement in the congenital condition cystic fibrosis on the other, suggest that the basic mechanism of pancreatic injury may be the same in each illness. We propose that pancreatic oxidant stress is the common denominator and, furthermore, that this is facilitated by a shortfall of micronutrient antioxidants in the face of heightened free radical activity through different sources. If so antioxidant supplements should alleviate symptoms. This deduction was supported by an exploratory dose-seeking study that spanned five years in 20 patients with recurrent (non-gall stone) acute or chronic pancreatitis and confirmed by a 20-week double-blind placebo-controlled crossover trial of the successful combination (daily doses of 600 µg organic selenium, 0.54 g vitamin C, 9000 IU β-carotene, 270 IU vitamin E and 2 g methionine) in a further 20 cases. A randomised trial of glutathione precursors, given intravenously for 24 hours after admission in patients with a first attack of acute pancreatitis, is in progress. Long-term trials of oral antioxidant formulas are planned in patients with cystic fibrosis.

INTRODUCTION

Acute pancreatitis (AP) manifests as a severe attack of upper abdominal pain accompanied by elevated serum levels of pancreatic enzymes, including amylase, lipase and zymogens of proteases. The morphological counterpart is pancreatic interstitial oedema with stunting of acinar cells – the latter representing an early stage of their dedifferentiation into "tubular

Antioxidants in Therapy and Preventive Medicine
Edited by I. Emerit *et al.*
Plenum Press, New York, 1990

555

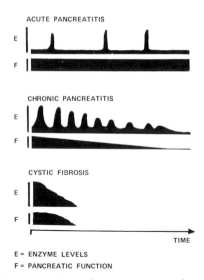

ACUTE PANCREATITIS

CHRONIC PANCREATITIS

CYSTIC FIBROSIS

TIME

E = ENZYME LEVELS
F = PANCREATIC FUNCTION

Fig 1. Schematic representation of
the chemical pathology of pancreatitis and cystic fibrosis

complexes"[1,2]. In the majority of cases, the process is self-limiting so
that by 72 hours enzyme levels return to normal and pain disappears. The
likelihood of further attacks is low if an aetiological factor, such as gall
stones, can be identified and removed; otherwise, the risk continues but
exocrine functional recovery is complete between attacks (Fig 1). However,
in up to 20% of patients admitted with a first attack, the gland undergoes
a metomorphosis to haemorrhagic pancreatic necrosis: death from shock lung,
diffuse intravascular coagulation and kidney failure is the usual outcome.
The transformation is traditionally blamed on premature intrapancreatic
activation of trypsinogen to trypsin[1] but excess levels of the zymogen, and
not its protease, is found in the bloodstream in the earliest stage of the
debacle[3,4] and, in any case, the presence of active trypsin in serum is not
a prerequisite for the multiple organ failure syndrome[4-6]. Treatment is
essentially supportive.

The typical features of chronic pancreatitis (CP) include recurrent
painful attacks, progressive loss of exocrine secretory tissue from increasing fibrosis within the gland (Fig 1), tubular complexes representing
dedifferentiated acini (Fig 2) and ductal protein plugs that tend to calcify
as time goes by. Evidence of permanent pancreatic damage between attacks is
thus the basis of the distinction from recurrent AP, but this may be impossible to obtain in the early stages of CP without access to a surgically

resected specimen of the gland for serial histology[1]. The chronic consumption of alcohol is an important aetiological factor for CP in developed countries but teetotalism does not guarantee freedom from attacks, whilst in our practice some 50% of cases do not fulfil any definition of alcoholism. The disease runs a particularly virulent course in underprivileged communities of several tropical zones: malnutrition, dietary cyanogens and recurrent infections have been incriminated[7]. There is increasing suspicion that CP, especially the calcific variant, predisposes to pancreatic cancer. The grim alternatives for the treatment of pancreatic pain in CP are near-total pancreatectomy or the use of narcotic analgesics[8].

Tubular complexes (Fig 2) are the pathological hallmark of pancreatic involvement in cystic fibrosis (CF). Other shared features with CP are ductal protein plugs and intense pancreatic fibrosis[9]. The principal differences are the accelerated course of exocrine destruction which is usually complete by the first decade of life, persistent hypertrypsinogenaemia before that time[10] (Fig 1), and the generally painless nature of pancreatic injury - although frank attacks of pancreatitis, which may be haemorrhagic, have been documented[11]. The gene for cystic fibrosis has been pinpointed to chromosome 7; yet, there is no plausible explanation for the cardinal features of the disease - exocrine pancreatic damage; hyper-

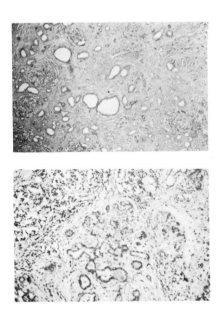

Fig 2. Pancreatic histology showing "tubular complexes" in chronic pancreatitis (top frame) and cystic fibrosis (bottom frame).

secretion of mucus; hyperpolarisation of membranes due to a defect in
chloride reabsorption; and bronchiectasis which is an acquired problem since
it is not present at birth. Steatorrhoea from pancreatic exocrine failure
is relatively easy to control with microencapsulated preparations of pancreatic enzymes. Longevity is determined by the rate of pulmonary destruction
and most patients succumb by the second or third decade. Recent observations
suggest that if they live long enough they run the risk of pancreatic
cancer[12,13].

OXIDANT STRESS: UNIFYING MECHANISM OF PANCREATIC INJURY?

The increased levels of pancreatic enzymes in the bloodstream in AP,
CP and CF (Fig 1) can be interpreted as indicating a reversal of secretory
polarity in the pancreatic acinar cell, so that a higher fraction of secretions than normal – including zymogens, not activated enzymes[3,4] – is dis-
charged directly into the interstitium and bloodstream. This is clearly
shown in animal models of oedematous AP[4], which also indicate that the
"exit block" is imposed by unquenched oxygen free radicals, ie oxidant
stress – since prior treatment with superoxide dismutase and catalase aborts
the problem[14]. Against this background, the sporadic nature of attacks in
AP may imply that pancreatic oxidant stress is intermittent and transient,
whilst the recurrent nature of attacks in CP may indicate persistent oxidant
stress in acinar cells, and the pattern in CF suggests that the fundamental
genetic defect facilitates "leakage" of oxygen free radicals from one or
more cellular systems (Fig 3). Dedifferentiation of acini into tubular
complexes – transient and reversible in oedematous AP[2], persistent in CP and

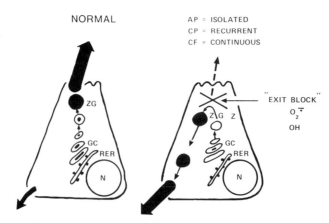

Fig 3. Oxidant stress-induced interference
with exocytosis, resulting in a reversal of
secretory polarity, would explain elevated
serum levels of enzymes in pancreatitis and
cystic fibrosis.

CF (Fig 2), is in keeping with this philosophy, because these complexes can be produced experimentally by reactive intermediates from certain chemicals[15].

It is now accepted that controlled production of oxygen free radicals in cells is essential for normal function and cell turnover[16], but it is equally certain that excessive free radical activity is potentially damaging. Those hazards which may be especially relevant in the context of impaired exocytosis (Fig 3) are the propensity of unquenched free radicals to induce microtubular dysfunction and to attack key amino acids, such as methionine, and/or sulphydryl groups in proteins and enzymes that are involved in stimulus-secretion coupling. Other potential problems would stem from the ability of free radicals to peroxidise polyunsaturated fatty acid (PUFA), to render lysosomal membranes leaky, to interfere with ionic pumps in cell membranes, to provoke fibrosis, to interact with critical cellular macromolecules and to cause chemotaxis[17-19].

As a possible source of excessive free radical activity, the enzyme xanthine oxidase would seem a likely candidate in oedematous AP from a migrating gall stone, hypertriglyceridaemia or ischaemia[14]; the cytochrome P450 complex in drug-related AP[20] and also in alcoholic, idiopathic[20] and tropical[21] CP; and aberrant activity of a constituent of microsomal[9,22] or mitochondrial[23-25] electron transport chain in CF. Excessive free radical activity is unlikely to be directly involved in the transformation from oedematous AP to haemorrhagic pancreatic necrosis, however, since the zymogens of human pancreatic proteases are not activated by exposure to high doses of the primary or secondary metabolites of oxygen, whilst pancreatic trypsin inhibitor is not destroyed under those circumstances[26]. This observation, and the failure of oxygen free radical scavengers to prevent the fatal transformation in suitable animal models[27,28], suggested that the carnage may be initiated by the "overt" secretions from activated leucocytes that are drawn into the pancreatic bed – perhaps by the entry of excessive amounts of pancreatic enzymes (? and free radical oxidation products) into the interstitium. This inadvertent response occurs when the primary chemotactic stimulant is overwhelming: cholesterol crystals, bile salts and other debris in the pancreatic duct seem to be especially provocative in this regard[4]. The chemical arsenal of activated leucocytes includes active elastase and phospholipase A_2; oxygen free radicals; cachectin and tumour necrosis factor; platelet and plasminogen-activating factors; cathepsins and collagenases; activators of the complement cascade; and myeloperoxidase. The last of these generates the powerful oxidant hypochlorous acid which attacks methionine residues in α_1 trypsin inhibitor and thus severely erodes defence against active elastase: this is seen as a critical component in the genesis of (non pancreatitis) shock lung syndrome[29] as well as bronchiectasis[30].

The only sure way to demonstrate excessive free radical activity is by electron spin resonance studies of tissues but this is clearly inapplicable in day to day clinical practice. Instead one must rely upon the measurement of validated "markers" of free radical activity in biological fluids and/or observations in patients on supplemental antioxidants. Most markers reflect a free radical attack on PUFA in cell membranes. The classical route of attack involves lipid peroxidation, generating hydroperoxides, endoperoxides, long-lived aldehydes (notably 4-hydroxynonenal) and the end-products malondialdehyde, ethane and pentane. The first two classes are inherently unstable, whilst the end-products are non specific in that malondialdehyde also reflects activity of the normal prostaglandin pathway whilst the concentration of short-chain volatile hydrocarbons in expired air is influenced by gut flora[31]. In recent years a non-peroxide pathway of a free radical attack on PUFA has been documented by exposing linoleic acid (9,12 LA) to ultraviolet irradiation in the presence of albumin. A stable non-peroxide isomer, octadeca 9 cis 11 trans dienoic acid, was generated. The concentration of this isomer, 9 11 LA', expressed as a percent molar ratio of the 9,12 LA concentration in body fluids and tissues, provides a convenient index of free radical activity along this pathway[32]. Certain anaerobic bacteria – for example lactobacillus, corynebacterium and butyrovibrio fibrosolvens (which colonises the gut of ruminants) can produce 9,11 LA' along the biohydrogenation route from 9,12 LA to stearic acid. However, feeding human beings with milk products, which contain the isomer, does not lead to increased serum levels[32]; whilst feeding a corn oil-enriched diet to hamsters results in a decrease in 9,11 LA' secretion into bile[33]. Nevertheless, because of the

TABLE I   FREE RADICAL MARKERS OF PUFA ATTACK IN DUODENAL JUICE[a]

	Isomerization pathway 9,11 LA' (umol)		Peroxidation pathway Ultraviolet fluorescence(units)	
	n	mean $\pm$ SE	n	mean $\pm$ SE
Controls	7	1.62 $\pm$ 0.31	11	536 $\pm$ 115
AP	11	4.26 $\pm$ 0.62[b]	9	1091 $\pm$ 173[c]
CP	5	6.14 $\pm$ 1.41[b]	12	1542 $\pm$ 299[b]

[a]outputs in 10 min after secretin

[b]$p<0.01$ cf controls

[c]$p<0.05$ cf controls

difficulties in interpreting changes in the level of individual free radical oxidation products, it is accepted that use of a panel of markers provides more reliable information[34].

We have adopted this approach in patients with AP, CP and CF. Duodenal juice, collected in the first 10 min after an intravenous injection of secretin (given as part of a test to detect exocrine functional impairment and so distinguish between AP and CP[1]), is an admixture of hepatic bile and pancreatic juice. We ascertained by endoscopic cannulation studies of bile or pancreatic ducts that the raised levels of free radical oxidation products, in patients who were studied six weeks after full clinical recovery (Table I), emanated in bile[35]. The 9,11 LA'/9,12 LA molar ratio in serum was measured in a different set of subjects (Fig 4). The median molar ratio in 20 controls was 2.09% (range 1.70-3.01) whilst ratios in 14 patients with AP and 17 patients with CP were higher (AP median 3.26%, range 1.98-6.89, p<0.05; CP median 3.85% , range 1.63-6.29, p<0.05) : very high values were found in two patients with chronic renal failure who had recently suffered an attack of AP, and in a patient with recurrent AP who had several calculi in the common bile duct. Three points can be made from the combined studies. Firstly, there was an overlap between values in controls and patients (Fig 4, Table I). Secondly, whilst the higher levels in patients with AP and CP - as a whole - suggest increased free radical activity, the results cannot be extrapolated to provide information on the duration of that problem. Thirdly, increased levels of a "marker" in serum cannot pinpoint the source of heightened free radical activity, but direct studies of bile, pancreatic and duodenal juice

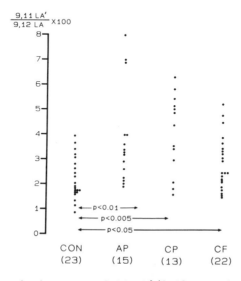

Fig 4.   Serum 9,11 LA'/9,12 LA molar ratios (%) in 23 controls (con) and 50 patients with exocrine pancreatic disease.

facilitate interpretation. The notion that in CP heightened free radical activity in both liver and pancreas is a persistent problem, whereas it is generally sporadic in AP, is supported by the morphological findings of excessive lipofuscin deposits (representing condensation of malondialdehyde in tissues) and microvesiculation (as accompanies disrupted microtubular traffic routes) in hepatocytes and acinar cells – despite increased levels of the natural antioxidants (apo)lactoferrin and mucin in pancreatic juice, and increased caeruloplasmin (ferroxidase I) levels in serum (reviewed in Refs 21 and 35).

The pancreas is an early target in CF and subclinical involvement of the liver is common[36]. Therefore, as in patients with pancreatitis, analysis of duodenal juice for a panel of free radical markers would provide useful information. However, the difficulties in intubating chronically breathless patients, as well as the viscid nature of duodenal secretions in CF makes this approach impractical. Analysis of serum from CF patients showed high levels of the 9 11 LA'/9,12 LA molar ratio (Fig 4) – 22 CF adults median 2.48%, range 1.60–5.24%, p = 0.035 versus values in controls. The difference was especially impressive when nasal epithelial cells – another consistent target in CF – were examined: %MR in 17 adults with CF, median 2.09%, range 1.70–3.01% versus 1.56, 0.92–2.23% in 20 controls, p = 0.0002[23]. We[23] and others[37] have not found raised levels of markers of the lipid peroxidation pathway in serum from CF patients but the notion that protracted oxidant stress is tied in with the pathophysiology of this disease is supported by the lack of correlation between molar ratios and conventional indices of disease severity – which suggests that the raised levels are a primary feature of CF. Furthermore, morphological studies show large amounts of lipofuscin in both pancreas and nasal epithelial cells of CF patients, whilst the mobilisation of natural antioxidants to counter oxidant stress would rationalise the finding of raised serum levels of lactoferrin as well as mucin hypersecretion by epithelia[9].

EVIDENCE FOR A SHORTFALL OF MICRONUTRIENT ANTIOXIDANTS

The higher level of free radical oxidation markers in biological material from groups of patients with AP, CP and CF compared with controls (Fig 4, Table I) gives no clue as to whether this reflects an absolute increase in the rate of free radical production from potential cellular sources as discussed earlier, or reflects an absolute deficiency of antioxidants so that the normal cellular quota of radicals cannot be mopped up, or an imbalance in favour of radicals although both production rates and antioxidant levels are within the broad limits of "normal". Cells contain an intricate antioxidant defence system, incorporating components that prevent radical-mediated damage

and those that intercept the chain process of lipid peroxidation. Many of
the raw ingredients are derived from the diet. Seven-day home inventories
were therefore conducted to obtain a global view of antioxidant supply in
patients with idiopathic CP. When the results were compared with information
from age/sex matched healthy volunteers, the patients - who had not changed
their diets between painful attacks - were found to ingest less selenium
($p = 0.001$), vitamin C ($p < 0.001$), vitamin E ($p = 0.03$) and riboflavin
($p = 0.03$) than controls: female patients also ingested less β-carotene
($p < 0.05$) and sulphur amino acids ($p < 0.01$) than did female controls[38]. When the
information from patients with CP was compared with that from an equally
enzyme-induced group of controls, ie patients on anticonvulsant drugs, using
computer-assisted discriminant analysis to examine all the dietary information and also theophylline clearance as an indicator of enzyme induction, the
lower intakes of vitamin C and methionine by the CP group emerged as the key
discriminators[39]. The essential aminoacid methionine is a precursor for
cysteine which is the rate-limiting component in the synthesis of glutathione
(GSH) - which has multifarious roles in removing reactive intermediates from
drugs and chemicals[40] - whilst vitamin C helps to refurbish GSH from GSSG.
Furthermore, the smooth incorporation of methionine into S-adenosylmethionine,
which then participates in numerous methylating and transulfurating reactions,
is essential for cellular integrity. Thus, these dietary studies hinted at
the involvement of non-biological reactive intermediates in the pathogenesis
of CP, and it is interesting that compounds which are known to undergo bioactivation via cytochromes P450 - including sodium valproate, steroids, oral
contraceptives, azathioprine, inhaled halogenated hydrocarbons - are suspected
aetiological factors in CP[1,41,42]. The notion that a deficit of dietary antioxidants facilitates oxidant stress in this disease is supported by blood

TABLE II   HABITUAL DAILY INTAKES OF ANTIOXIDANTS[a]

	n	selenium (ug)	Vit C (mg)	Vit E (mg)	β-carotene (mg)	Methionine (g)
controls	15	85 + 26	121 + 73	6.3 + 2.5	2.5 + 1.8	1.8 + 0.4
epilepsy controls	15	43 + 15	95 + 23	6.0 + 2.5	3.4 + 2.2	2.3 + 0.7
CP	15	38 + 22[b]	43 + 23[bc]	4.3 + 2.0[b]	2.4 + 1.3	1.6 + 0.9
AP	7	73 + 29[d]	56 + 28[b]	6.8 + 3.5[d]	2.7 + 1.3	2.5 + 1.1

[a]derived from 7-day home inventories, mean + SD
[b]$p < 0.05$ cf controls
[c]$p < 0.05$ cf epilepsy controls
[d]$p < 0.05$ cf CP

measurements[43,44] and it should be stressed that even in the case of lipid antioxidants, subnormal levels could not be simply ascribed to exocrine secretory impairment. Of particular relevance is a recent study indicating GSH consumption along with impaired transulfuration pathways in patients with alcohol-induced relapses of CP[45].

Home dietary studies in a small group of patients with AP showed that their antioxidant intakes were similar to those in controls, except for lower intakes of ascrobic acid ($p = 0.015$, Table II)[46]. A systematic investigation of blood antioxidant status has not as yet been done in this subgroup, but a pilot study of serum selenium showed lower levels at the time of admission[4] (Fig 5).

Pre-morbid dietary studies in adults with CF, ie before pancreatic involvement, are impossible considering the rapid time-course of the pancreatic problem (Fig 1). When the exocrine pancreas is destroyed, low blood levels of lipid antioxidants are only to be expected, but subnormal serum selenium levels[23] (Fig 5) and reports of taurine deficiency in children[47] are difficult to explain on that basis. Those observations could, however, be rationalised in terms of oxidant stress – since we have found that repetitive exposure to xenobiotics is accompanied by a reduction in serum selenium levels suggesting sequestration into tissues; whilst reduced taurine levels could reflect impaired transulfuration pathways due to a free radical

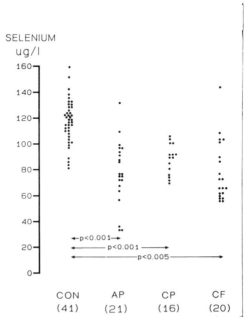

Fig 5. Serum selenium levels in 41 controls (con) and 57 patients with exocrine pancreatic diseases

attack on a relevant enzyme, eg those that are involved in the conversion of S-adenosyl methionine to S-adenosyl homocysteine and further metabolites.

As with serum "molar ratios" (Fig 4), so with serum selenium concentrations, there was an overlap between data from controls and patients with chronic pancreatic disease, ie CP and CF. This overlap was virtually abolished when the levels of those substances were viewed alongside, using computer assisted discriminant analysis (Fig 6). At any given level of free radical activity, as judged by the molar ratio, patients with CP or CF had substantially lower selenium concentrations than did controls. This is in keeping with the notion of oxidant stress. The identity in the discriminant line separating controls from CP patients, and controls from CF patients, suggests to us that selenium status may play a role in determining whether a free radical attack on lipid terminates with the formation of the non-peroxide isomer or whether it is propagated along the classical lipid peroxidation pathway. We speculate that suboptimal selenium availability facilitates the former route: furthermore, since there was no correlation between serum selenium and plasma GSH peroxidase activity in blood samples from the CP patients (data not shown), we suspect that this proposed function of selenium may be independent of its peroxidase activity. If this interpretation is validated, it could have important implications with regard to the development of pancreatic cancer in CP and CF, in that lower (pre-morbid) serum selenium levels[48] and/or high molar ratios[49] have been reported in other malignant states. The stability of 9,11 LA' in contrast to the extreme fragility of lipid peroxidation products — the concentrations of which are low in malignant cells[50] — would interfere with cellular turnover, as pointed out by Dormandy[16].

SUPPLEMENTAL ANTIOXIDANTS IN PANCREATITIS AND CYSTIC FIBROSIS

Our various studies in patients with CP (Figs 4-6; Tables I,II) could be dismissed as epiphenomena, or they could be interpreted as supporting the hypothesis that oxidant stress in pancreatic acinar cells — as a consequence of chronic induction of cytochromes P450 as well as suboptimal nutrition — initiates the disease. In the absence of a good animal model for CP it seemed legitimate to test the hypothesis by prescribing antioxidants — using a combination of items, since deficiency of a single micronutrient would be expected to jeopardise the smooth working of the whole antioxidant machinery in cells. Permutations and combinations were tried in a group of 20 patients with idiopathic CP over the course of five years. It is perhaps not surprising, considering the cited dietary data, that the consistently successful combination included organic selenium, β-carotene, vitamin C, vitamin E and methionine in daily doses up to 600 μg, 9000 IU, 0.54 g, 270 IU

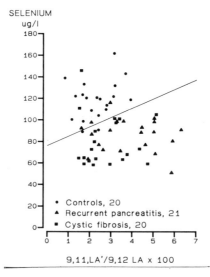

SELENIUM
ug/l

Fig 6. Computer assisted discriminant
analysis reveals different relationships
between free radical activity and selenium
status in health and exocrine pancreatic
disease.

and 4 g, respectively. Detailed reports on the youngest and oldest member
in this exploratory dose-seeking study have been published[51,52]. The preparation selenium ACE was a convenient source of the first three items: a
particular attraction was that it provided selenium in an organic form, so
increasing the therapeutic safety margin. Methionine had to be given as a
separate tablet.

The successful antioxidant combination (but with the dose of methionine
restricted to 2 g daily) was tested by a 20-week double-blind placebo-controlled switchover trial in a further group of patients with frequent attacks
of pancreatitis. 20 patients completed the trial, including five with
recurrent (non-gallstone) AP, seven with alcohol-related CP and eight with
idiopathic CP. Their baseline blood levels of selenium, plasma glutathione
peroxidase, β-carotene and vitamin E (expressed as μmol per mmol cholesterol)
are given in Table III. The levels of micronutrient antioxidants in the
patients were lower than in controls. However, the levels of plasma glutathione peroxidase generally lay within the reference range and, except in the
few patients with serum selenium $<90 \mu$.g/1, bore no relationship to selenium
concentrations. The blood levels of selenium and β-carotene normalised after
10 weeks on active treatment but the vitamin E/cholesterol ratios, although
higher than baseline values, remained less than in controls (n = 20,3.12 $\pm$
0.70, p <0.05). In the subgroup of eight patients who received placebo in
the second phase, the blood levels of each item had reverted to baseline -

thus endorsing the switchover design of the trial.

The beneficial effect of antioxidant supplementation was evidenced by absence of clear-cut attacks during active treatment (as opposed to three episodes necessitating admission during the placebo phase), along with a reduction in background pain as gauged both by analysis of visual analogue score sheets incorporating 36 appropriate descriptor words at the pretrial, crossover and completion stage; and daily diaries analysed by conventional and time-series methods. Furthermore, this improvement was accompanied by a reduction in free radical activity as assessed by a fall in serum 9,11 LA'/ 9,12 LA molar ratio. Full details of these studies will be reported elsewhere. We cannot tell which of the prescribed antioxidants ameliorated symptoms and it may be that this was different in each patient. Alternatively, the cocktail may have boosted cellular levels of a critical component that is involved in stimulus/secretion coupling within acinar cells (Fig 3), and our on-going trials are geared to testing this possibility.

As discussed in the introduction, most attacks of AP are self-limiting. However, up to 40% of patients admitted with a first attack go on to develop a complication (psuedocyst, abscess) whilst half of these die from pancreatic organ failure. There is no foolproof way to predict the outcome at the time of admission. Once the various enzyme cascades are activated - including coagulation, kinin, complement and eventually trypsinogen systems - therapeutic measures are heroic rather than practical or efficacious[4]. The available options include: surgical debridement of pancreatic slough followed by open drainage of the abdomen (laparostomy) and supervision on an intensive care unit so that positive pressure respiration can be instituted without delay (Fig 7); or repeated exchange transfusions/plasmapharesis to remove activated enzymes and toxic principles from the bloodstream. If the concept

TABLE III BASELINE BLOOD ANTIOXIDANT LEVELS IN PATIENTS WHO COMPLETED THE PLACEBO CONTROLLED TRIAL

		Serum selenium μg/l		Plasma GSH-Px i.u/l		Serum β-carotene mg/l		Serum vit E μmol/mol cholesterol
	n	m $\pm$ SD	n	m $\pm$ SD	n	m $\pm$ SD	n	m $\pm$ SD
controls	41	117 $\pm$ 18	14	137 $\pm$ 40	19	106 $\pm$ 61	17	4.7 $\pm$ 0.82
patients	20	37 $\pm$ 13	20	156 $\pm$ 47	17	37 $\pm$ 26	15	1.6 $\pm$ 0.66
		$p < 0.0005$		NS		*$p < 0.01$		$p < 0.0005$

*
non-parametric, as data skewed

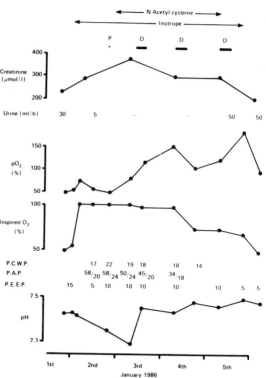

Fig 7. REproduced from ref. 53 by kind permission of editor of the Lancet.
PEEP = pulmonary end-expiratory pressure
PAP = pulmonary arterial pressure
PWWP - pulmonary capillary edge pressure
D = bicarbonate haemodialysis
P = bolus methylprednisolone

that oxidant stress plays a pivotal role in initiating pancreatitis is correct, then the early provision of antioxidants – including a GSH precursor such as N-acetylcysteine, to protect against primary radicals as well as those that may be released by activated leucocytes – should be helpful. This was indeed the case in a patient with gall stone-related AP who went into shock lung syndrome and anuric renal failure within hours of a laparotomy to drain a pancreatic abscess: administration of N-acetylcysteine was followed by a dramatic recovery (Fig 7)[53]. A randomised study of this agent, given along with S-adenosylmethionine – which should promote exocytosis by ensuring methylation of membrane phospholipids – is in progress and 20 patients have been recruited to date.

The suggestion that antioxidant therapy could ameliorate symptoms of CF was made by Wallach[54] a decade ago, but the premise on which that suggestion was based – ie that CF is not a congenital disease but a perinatal manifestation of selenium deficiency – is untenable (Fig 5). Our proposal is that the (as yet undiscovered) genetic defect facilitates "leakage" of oxygen free radicals in cells. If so, environmental influences that promote free radical activity – eg ultraviolet light, exposure to enzyme-inducing enzymes, drugs that undergo redox reactions to generate superoxide – would aggravate the problem, as would a deficiency of micronutrient antioxidants (maternal or in CF patient). In this scheme, epithelial chloride impermeability, persistent hypertrypsinogenaemia in neonates and mucin hypersecretion are envisaged as secondary phenomena reflecting oxidant stress or mobilization of natural defences to counter it. We view bronchiectasis in CF as an even more remote consequence, and suggest that lung damage is mediated by "frustrated phagocytosis", ie recurrent extracellular secretions from activated leucocytes that are drawn into the pulmonary interstitium by the entry of excessive amounts of secretions (? along with free radical oxidation products) from airways epithelial cells that are affected by the same exocytosis problem as seems to affect pancreatic acinar cells (Fig 3): Clara cells are a prime candidate[55] and their exposure to high concentrations of oxygen after birth compared with the levels before birth might rationalise the paradox that the control of progressive lung damage is a major therapeutic challenge in CF although the lungs are pristine in neonates. This concept of a reversal of secretory polarity as the key pathogenetic problem in CF is in keeping with recent in-vitro studies showing failure to induce secretion by jejunal epithelial cells - as monitored by changes in short circuit current - exposed to various secretagogues including acetylcholine, prostaglandin E_2, dibutryl cyclic AMP and a calcium ionophore[56]. Until genetic engineering becomes an everyday event, the prescription of antioxidants and/or sulphur aminoacid precursors to ensure adequate levels of key cellular ingredients such as sulphadenosylmethionine and GSH, seems logical - to mothers who are known to carry a CF foetus, and from the time of birth to CF patients. We have planned a 30 week double blind study in which young adults with CF will serve as their own controls with regard to changes in clinical and biochemical parameters, including measurement of 9,11 LA' in nasal epithelial cells; measurement of potential difference across respiratory epithelium; measurement of sweat osmolality; and conventional indices of respiratory and hepatic involvement. Three limbs of treatment are envisaged: optimal conventional treatment which will include the standard supplement of vitamin E along with intermittent antibiotic treatment; optimal conventional treatment plus selenium, β-carotene CE and S-adenosylmethionine; optimal conventional treatment plus S-adenosylmethionine but not the antioxidants. (Studies in conjunction with Dr K Webb of The Regional Adult Cystic Fibrosis Unit, Monsall Hospital, Manchester).

ACKNOWLEDGEMENTS

Dr S Uden and his predecessor Dr D W K Acheson were supported by the Wellcome Trust. Dr D Bilton is supported by the Cystic Fibrosis Trust, as

was her predecessor Dr B Salh. Dr Guyan has been funded by the North West
Regional Health Authority for the past five years. We thank these organisa-
tions and also the following companies for generous gifts of antioxidants or
placebos for previous and/or current studies: Wassen International (Leather-
head, U.K.); Evans Medical Ltd (Beaconsfield, U.K.); Bioresearch Spa (Milan,
Italy) and the Boots Company,(Nottingham, U.K.). We acknowledgement the
excellent secretarial assistance of Mrs J Hanbridge and thank the Department
of Medical Illustrations for the Figures.

REFERENCES

1. J. M. Braganza, The pancreas in: "Recent Advances in Gastroenterology,
     volume 6, "R Pounder, ed., Churchill Livingstone, London: 251 (1986).
2. G. Adler and H. F. Kern, Fine structural and biochemical studies in
     human acute pancreatitis, in: "Pancreatitis: concepts and classifica-
     tion", K. E. Gyr, M. V. Singer and H. Sarles, eds., Elsevier Science
     Publishers, Amsterdam: 37 (1984).
3. P. R. Durie, K. J. Gaskin, J. E. Ogilvie, C. R. Smith, G. G. Forstner,
     and C. Largman, Serial alterations in the forms of immuno-reactive
     pancreatic cationic trypsin in plasma from patients with acute
     pancreatitis, J Paediatr. Gastroenterol. 4: 199 (1985).
4. J. M. Braganza, Free radicals and pancreatitis, in: "Free radicals:
     chemistry, pathology and medicine". C. Rice-Evans and T. L. Dormandy,
     eds., Richelieu Press, London: 357 (1988).
5. H. Rinderknecht, Fatal pancreatitis, a consequence of excessive leucocyte
     stimulation? Int. J. Pancreatol. 3: 105 (1988).
6. J. M. Braganza and H. Rinderknecht, Free radicals and acute pancreatitis,
     Gastroenterology 94: 1111 (1988).
7. A. Abu-Bakare, G. V. Gill, R. Taylor and K. G. M. M. Alberti, Tropical
     or malnutrition-related diabetes: a real syndrome? Lancet 1:1135
     (1986).
8. A. R. Moossa, Surgical treatment of chronic pancreatitis: an overview,
     Br. J. Surg. 74: 66 (1987).
9. J. M. Braganza, Cystic Fibrosis: casualty of detoxification?, Med.
     Hypoth. 20: 233 (1986).
10. F. Pederzini, P. Armani, E. G. Barlocco, M. Canciani, D. Olivieri,
     P. Rizzotti, M. Zanchetta, and G. Mastella, Longitudinal study of
     blood pancreatic enzymes in CF from birth to 24 months, in: "Cystic
     fibrosis: horizons", D. Lawson, ed., John Wiley and Sons, Chichester:
     212 (1984).
11. H. Schwachmann, E. Lebenthal, and K. T. Khaw, Recurrent acute pancreat-
     itis in patients with cystic fibrosis with normal pancreatic enzymes,
     Pediatrics 55: 86 (1975).
12. R. J. Stead, A. N. Reddington, L. J. Hinks, B. E. Clayton, M. E. Hodson,
     and J. C. Batten, Selenium deficiency and possible increased risk of
     carcinoma in adults with cystic fibrosis, Lancet 2: 862 (1985).
13. J. McIntosh, R. A. Schoumacher, and R. E. Tiller, Pancreatic adeno-
     carcinoma in a patient with cystic fibrosis, Am. J. Med. 4: 592 (1988)
14. H. Sanfey, M. G. Sarles, G. B. Bulkley, and J. L. Cameron, Oxygen derived
     free radicals and acute pancreatitis: a review, Acta Physiol. Scand.
     126 (suppl. 54): 109 (1986).
15. D. E. Bockman, O. Black, L. R. Mills, and P. D. Webster, Origin of
     tubular complexes developing during induction of pancreatic adeno-
     carcinoma by 7, 12 - dimethylbenz(a)anthracene, Am. J. Pathol. 90:
     645 (1978).
16. T. L. Dormandy, In praise of peroxidation, Lancet 2: 1126 (1988).
17. T. L. Slater, Free-radical mechanisms in tissue injury, Biochem. J.
     222:1 (1984).

18. D. J. Jollow, I. J. Kocsis, R. Snyder, and H. Vainio, eds., "Biological reactive intermediates", Plenum Press, London (1977).

19. G. Weissman, C. Serhan, H. M. Korchak, and J. E. Smolen, Neutrophils: release of mediators of inflammation with special reference to rheumatoid arthritis, Ann. N.Y. Acad. Sci., 389:11 (1982).

20. D. W. K. Acheson, L. P. Hunt, P. Rose, J. B. Houston, and J. M. Braganza, Factors contributing to the accelerated clearance of theophylline and antipyrine in adults with exocrine pancreatic disease, Clin. Sci: in press.

21. J. M. Braganza, Cytochrome P450-mediated oxidant stress: common denominator in the pathogenesis of temperate-zone tropical chronic pancreatitis? in: "Proceedings of symposium on malnutrition related diabetes, London 1988", G. Alberti, ed., Oxford University Press, Oxford: in press.

22. D. C. Knoppert, M. Spino, R. Beck, J. J. Thiessen, and S. M. MacLeod, Cystic fibrosis: Enhanced theophylline metabolism may be linked to the disease, Clin. Pharmacol. Ther. 44: 254 (1988).

23. B. Salh, K. Webb, P. M. Guyan, J. P. Day, D. Wickens, J. Griffin, J. M. Braganza, and T. L. Dormandy, Aberrant free radical activity in cystic fibrosis, Clin. Chim. Acta: in press.

24. R. J. Feigal and B. L. Shapiro, Mitochondrial calcium uptake and oxygen consumption in cystic fibrosis, Nature 278: 276 (1979).

25. A. A. VonRuecker, R. Bertele, and H. K. Harms, Calcium metabolism and cystic fibrosis: mitochondrial abnormalities suggest a modification of the mitochondrial membrane, Paediatr. Res. 14: 594 (1984).

26. P. M. Guyan, J. M. Braganza, and J. Butler, The effect of oxygen metabolites on the zymogens of human pancreatic proteases, in: "Free radicals: chemistry, pathology and medicine", C. Rice-Evans and T. L. Dormandy, eds., Richelieu Press, London: 471 (1988).

27. A. Saluja, R. E. Powers, M. Saluja, P. Rutledge, and M. L. Steer, The role of oxygen derived free radicals in caerulein-induced pancreatitis, Gastroenterology 90: 1613 (1986).

28. S. Uden, A. L. Blower, T. V. Taylor, E. Bentow, R. T. F. McMahon, and J. M. Braganza, Exploration of a role for activated leucocytes in fatal pancreatitis, J. Clin. Path. 155: 351A (1988).

29. J. Travis, Oxidants and antioxidants in the lung, Am. Rev. Resp. Dis. 135: 773 (1983).

30. D. Burnett, A. Chambra, S. L. Hill, and R. A. Stockley, Neutrophils from subjects with chronic obstructive lung disease show enhanced chemotaxis and extracellular proteolysis, Lancet 2: 1043 (1987).

31. B. Halliwell and J. M. C. Gutteridge, "Free radicals in biology and medicine", Clarendon Press, London (1985).

32. D. G. Wickens and T. L. Dormandy, The possible origins of human octadeca 9,11-dienoic acid, in: "Free radicals: chemistry, pathology and medicine". C. Rice-Evans and T. L. Dormandy. eds., Richelieu Press, London: 237 (1988).

33. A. A. Ali, J. M. Braganza, P. M. Guyan, I. J. M. Jeffrey, and S. C. B. Rutishauser, Effects of corn-oil enriched diets on the biliary excretion of markers of free radical oxidation os lipids in Syrian golden hamsters, J. Physiol. 403:112 (1988).

34. C. V. Smith, and R. E. Anderson, Methods for determination of lipid peroxidation in biological samples, free radical biol. med. 3: 341 (1987).

35. J. M. Braganza, The role of the liver in exocrine pancreatic disease, Int. J. Pancreatol. 3: S19 (1988).

36. I. Benett, B. Salh, N. Y. Haboubi, and J. M. Braganza, Sclerosing cholangitis with microvesicular steatosis in cystic fibrosis and chronic pancreatitis, J. Clin. Path.: In press.

37. B. Matkovics, K. Gynrokovits, A. Laszlo, and L. Szabo. Altered peroxide metabolism in erythrocytes from children with cystic fibrosis, Clin. Chim. Acta 125: 59 (1982)

38. P. Rose, E. Fraine, L. P. Hunt, D. W. K. Acheson, and J. M. Braganza, Dietary antioxidants and chronic pancreatitis, Hum. Nutr. Clin. Nutr. 40C: 151 (1986).

39. S. Uden, D. W. K. Acheson, J. Reeves, H. V. Worthington, L. P. Hunt, S. Brown, and J. M. Braganza, Antioxidants, enzyme induction and chronic pancreatitis, Eur. J. Clin. Nutr. 42: 561 (1988).

40. S. Orrenius, and P. Mondens, The multiple roles of glutathione in drug metabolism, Trends Pharmacol. 5: 432 (1984).

41. J. M. Braganza, J. E. Jolley, and W. R. Lee, Occupational chemicals and pancreatitis: a link? Int. J. Pancreatol. 1:9 (1986).

42. M. Dossing, O. Jacobsen, and S. M. Rasmussen, Chronic pancreatitis possibly caused by occupational exposure to organic solvents, Human Toxicol. 4: 237 (1985).

43. J. M. Braganza, C. D. Hewitt, and J. P. Day, Serum selenium in patients with chronic pancreatitis: lowest values during painful exacerbations, Trace Elements in Medicine 5:79 (1988).

44. I. Kalvaria, D. Labadrios, G. S. Shephard, L. Vesser, and I. N. Marks, Biochemical vitamin E deficiency in chronic pancreatitis, Int. J. Pancreatol. 1: 119 (1986).

45. J. Martensson, and T. Bolin, Sulfur aminoacid metabolism in chronic relapsing pancreatitis. Am. J. Gastroent. 81: 1179 (1986).

46. J. M. Braganza, E. Fraine, P. Rose, and H. Martin, Dietary anti-oxidants, unsaturated fat, cytochromes P450, and pancreatitis, proceedings of Second Meeting of the International Association of Pancreatology, Sao Paulo, (1986).

47. G. N. Thompson, and F. M. Thomas, Protein metabolism in cystic fibrosis: responses to malnutrition and taurine supplementation, Am. J. Clin. Nutr. 46: 606 (1987).

48. W. C. Willett, J. S. Morris, S. Pressel, J. O. Taylor, B. F. Polk, M. J. Stamper, B. Rasner, R. Schneider, and C. G. Harries, Prediagnostic serum selenium and risk of cancer, Lancet 2: 130 (1983).

49. A. Singer, S. K. Tay, J. F. A. Griffin, D. G. Wickens, and T. L. Dormandy, Diagnosis of cervical neoplasia by the estimation of octadeca 9,11 dienoic acid, Lancet 1: 537 (1987).

50. C. Benedetto, A. Bocci, M. U. Dianzani, B. Ghiringhello, T. F. Slater, A. Tomasi, and V. Vannini, Electron spin resonance studies on normal human uterus and cervix and on benign and malignant uterine tumours, Cancer Res. 41: 2936 (1981).

51. J. M. Braganza, A. Thomas, and A. Robinson, Antioxidants to treat chronic pancreatitis in childhood? Int. J. Pancreatol. 3: 209 (1988).

52. J. M. Braganza, I. J. M. Jeffrey, J. Foster, and R. F. McCloy, Recalcitrant pancreatitis: eventual control by antioxidants, Pancreas 2: 489 (1987).

53. J. M. Braganza, A. M. Holmes, A. R. Morton, L. Stalley, R. Ku, and R. R. Kishen, Acetylcysteine to treat complications of pancreatitis, Lancet I: 914 (1986).

54. J. D. Wallach, and B. Germaise, Cystic fibrosis: a perinatal mani-festation of selenium deficiency in: "Trace substances in environmental health, volume 12", D. D. Hemphill, ed., University of Missouri Press, Missouri: 469 (1979).

55. J. G. Widdicombe and R. J. Packe, The Clara Cell, Eur. J. Respi. Dis. 63: 202 (1982).

56. C. J. Taylor, P. S. Baxter, J. Hardcastle, and P. T. Hardcastle, Failure to induce secretion in jejunal biopsies from children with cystic fibrosis, Gut 29: 957 (1988).

LIPID PEROXIDATION LEVELS AND ANTIOXIDANT ACTIVITIES OF BLOOD PLASMA

IN PARTURIENTS AND NEW-BORN INFANTS IMMEDIATLY AFTER NORMAL DELIVERY

P. Deligné, J.P Bonnardot, R. Couderc, S. Kerisit, J.F Périer
and P. Laruelle

Université Paris VI, Faculté de Médecine Saint-Antoine
Departments of Anesthesiology, Biochemistry, and Obstetrics
Gynecology, Hôpital Tenon, 4 rue de la Chine
F-75970 Paris Cedex 20, France

INTRODUCTION

Previously, personal surveys of free radicals in medicine and surgery
have emphasized some main clinical involvements and therapeutic possibilities (1,2,3). The purpose of the present study is to give evidence of
oxygen-derived free radicals generation in parturient women and new-born
infants immediately after normal delivery by the evaluation of lipid peroxidation levels and the evaluation of the plasmatic antioxidant activities.
    In such eutocial and supposed physiological conditions, devoid of
maternal risks, the mother-baby couple seems but slightly affected by these
deleterious and cytotoxic chemical species, as shown by the mother's
clinical data and by several clinical evaluations performed on the new-born
infants (Apgar score, Amiel-Tison test).
    Therefore, this search is a preliminary work to further biological
studies on obstetrics after complicated or high risk pregnancies, on dystocial deliveries, on caesarean sections for placenta abnormalities
(placenta praevia), maternal pathology (gravidic toxemia) or fetal distress.
In these cases, the pathological signs may be related to an overproduction
of oxygen-derived free radicals. This occurrence could be found particularly
in the case of fetal distress while the in-utero treatment of the fetus may
produce a true experimental ischemia-reperfusion syndrome, or in the respiratory failure of the new-born while the resuscitation might theoretically
lead to the same result.

MATERIAL AND METHODS

Subjects

    Both maternal and new-born levels of lipid peroxidation were studied
immediatly after 24 eutocial normal deliveries and compared with those of
25 fertile but non pregnant control women free from oral contraception.
    All the Apgar scores of the new-born infants were > 9 at the fifth
minute after birth.

Assay techniques

    The venous parturient and control blood samples and the venous and

Antioxidants in Therapy and Preventive Medicine
Edited by I. Emerit *et al.*
Plenum Press, New York, 1990

573

arterial cord blood of the new-born infants were collected on an ethylene-
diamine tetra-acetate solution, the anticoagulant EDTA (4).

The lipoperoxidation has been estimated by malondialdehyde (MDA) assay
according to the Satoh's colorimetric technique (5) with thiobarbituric
acid and standardization with 1,1,3,3 tetraethoxypropane.

The antioxidant plasmatic activity was simultaneously measured by
ceruleoplasmin and transferrin assays with laser immunonephelometry
(Behring) and by evaluation of superoxide dismutase (SOD) activity with
the Salin-Mc Cord method adapted to Cobas Bio -Roche- (6).

Statistical analysis

The statistical evaluations were made by the Student tests, and
correlations studied with the Spearman test.

RESULTS

They are collected in table 1 and expressed in nmol/ml for MDA, in
U/ml for SOD, in g/l for transferrin and ceruleoplasmin.

Malondialdehyde (MDA) :
-significant maternal increase of MDA plasmatic concentrations
compared with controls (p < 0,01),
-significant increase in the venous cord blood of the new-born infants
compared with mothers (p < 0,002),
-no significant difference between arterial and venous cord blood MDA
concentrations of the new-born infants, and
-positive correlation between these last values (figure 1a) and
between maternal MDA levels and venous cord blood MDA levels of the new-
born infants (figure 1b) (p < 0,01).

It must be emphasized that in our study, control MDA values are in
agreement with those of previous medical litterature, and maternal MDA
increase confirms former values found in pregnant women (7).

Superoxide dismutase (SOD) :
-clear-cut decrease of the SOD activity in mothers compared with

Table 1. Lipid peroxidation levels (MDA) and activities of superoxide
dismutase (SOD) and of antioxidant proteins (Transferrin and
Ceruleoplasmin) in controls, mothers, and new-borns.

	CONTROL (venous blood)	MOTHERS (venous blood)	NEW-BORNS	
			(venous cord-blood)	(arterial cord-blood)
MAD (nmol/ml)	0,58 ± 0,14	1,03 ± 0,75	2,23 ± 1,55	2,89 ± 2,04
SOD (U/ml)	0,30 ± 0,20	0,11 ± 0,11	0,20 ± 0,52	0,26 ± 0,32
TRANS-FERRIN (g/l)	3,00 ± 0,51	4,28 ± 0,69	2,09 ± 0,42	1,93 ± 0,39
CERULEO-PLASMIN (g/l)	0,28 ± 0,08	0,64 ± 0,17	0,13 ± 0,07	0,11 ± 0,06

controls (p < 0,001),
    -decrease of the SOD activity of venous and arterial cord blood samples
compared to control samples,
    -significant correlation (p < 0,01) between venous cord blood values
and maternal values, and also significant correlation (p < 0,01) between
SOD activity and MDA concentrations in controls.This last correlation was
neither found in the mothers nor in the new-born infants.

Transferrin :
    -significant increase in mothers compared with control samples
(p < 0,001) but
    -significant decrease in the venous cord blood compared with the
maternal blood (p < 0,001) and control blood samples (p < 0,001).

Ceruleoplasmin :
    -same features as for transferrin with a significant correlation in
all the cases (p < 0,001) between the blood concentrations of these two
protective antioxidant proteins.

DISCUSSION

    A plasmatic MDA increase had been previously described in pregnant
women (7). In the present study, MDA, transferrin, and ceruleoplasmin
assays, and SOD activity evaluation, have shown an increase in lipid peroxidation during normal delivery in mothers and new-born infants.
    However, a better appreciation might be expected with assays performed
on other degradation products such as arachidonic metabolites, diene
conjugates, and hydroperoxides.
    The interpretation of simultaneous changes in protective factors
(SOD, antioxidant proteins) is difficult and speculative.Nevertheless, the
decrease of SOD activity in mothers and new-born infants compared with
controls, and the low cord blood concentrations of transferrin and ceruleoplasmin suggest a particular susceptibility of these new-borns to the
nocuous action of oxygen-derived free radicals.

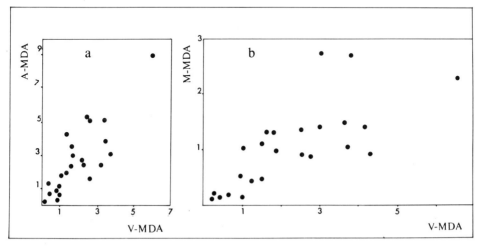

Fig.1. Correlations between arterial cord blood (A) and venous cord blood
       (V) malondialdehyde (MDA) concentrations (fig.1a) and between
       maternal (M) MDA concentrations and V-MDA (fig.1b).

These results must also be confrimed by further research works, in the same conditions, on the eventual changes of other antioxidant factors such as selenium or selenium-glutathione peroxidase in red blood cells.

CONCLUSIONS

This study shows a correlated increase of lipid peroxidation in parturient women and in new-born infants after a normal pregnancy and immediatly after an uneventful delivery. It also suggests a proper liability of fetus and new-born infants to the oxidant aggresive action of free radicals. However, we believe that these results are physiological in such conditions. They are inclined to show perhaps only a previous and continuing state of the lipoperoxidation chain which, from archidonic and eicosanoid metablites led, beyond MDA, to leufotrienes (by the cyclo-oxygenase pathways) (1). In fact, all these by-products of degradation are involved:

-in the adaptation of the fetus to birth,
-in the activation of the uterus smooth muscle contraction, and
-in the initiation of labor in women (8,9,10).
This work will be completed by further comparable studies in pathological obstetric conditions and dystocial labors.

REFERENCES

1. Deligné (P.).Les radicaux libres. I-Réductions de l'oxygène et lipoperoxidation membranaire. Aperçu des conséquences patholo-goques, Conv. Méd., 4 (1) : 49-59 (1985).
2. Deligné (P.). Les radicaux libres. II-Détoxication enzymatique, anti-oxidants et autres dépolluants. Aperçu des possibilitiés thérapeutiques, Conv. Méd., 4 (2) : 145-158 (1985).
3. Deligné (P.).Les radicaux libres. III-A propos du rôle antioxidant de quelques substances psychotropes utilisées en anesthésie-réanimation, Conv. Méd.,4 (6) : 445-452 (1985).
4. Knight (J.A), Smith (S.E),Kinder (V.E), Anstall (H.B).Reference inter-vals for plasma lipoperoxides : age-, sex-, and specimen-related variations,Clin.Chem., 33 (12) : 2289-2291 (1987).
5. Satoh (K.) .Serum lipid peroxide in cerebrovascular disorders determi-ned by a new colorimetric method,Clin.Chim.Acta, 90 (1) : 37-43 (1978).
6. Pasquier (C.),Laoussadi (S.), Sarfati (G.), Raichvarg (D.),Amor (B.). Superoxide dismutase in polymorphonuclear leukocytes from patients with ankylosing spondylitis or rheumatoid arthritis.Clin.Exp.Rheumatol.,3 (2) : 123-126 (1985).
7. Maseki (M.), Nishigaki (I.), Hagihora (M.), Tomada (Y.) Yagi (K.).Lipid peroxide levels and lipids content of serum lipoprotein fractions of pregnant subjects with or without pre-eclamp-sia.Clin.Chim.Acta, 115 (2) : 155-161 (1981).
8. Mitchell (M.D).Pathway of arachidonic acid metabolism with specific applications to the fetus and mother.Seminar in Peri-natalogy, 10 (4) : 224-254 (1986).
9. Casey (L.), Mac Donald (P.C). Regulation of phospholipid and arachidonic acid metabolism and of prostaglandin production.Seminar in Perinatalogy, 10 (4) : 270-275 (1986).
10.Riemer (R.K), Roberts (J.M).Activation of uterine smooth muscle contra-ction : implications for eicosanoid action and interactions. Seminar in Perinatalogy, 10 (4) : 276-287 (1986).

ATTENDEES

ABEDINZADEH Z.
Université Paris V - René Descartes
45, Rue des Saints Pères
Lab. Chimie Physique
75270 PARIS CEDEX 06 - FRANCE

ADOLPHE M.
Pharmacologie Cellulaire - EPHE -
Institut Biomédical des Cordeliers
15, rue de l'Ecole de Médecine
75006 PARIS - FRANCE

ALCINDOR L.
Fac. de Médecine Paris Ouest
45, rue des Saints Pères
75270 PARIS CEDEX 06 - FRANCE

ANTEBI H.
Fac. de Médecine Paris Ouest
Service de Biochimie
45, rue des Saints Pères
75270 PARIS CEDEX 06 - FRANCE

ANTILA E.
Dept. of Anatomy
University of HELSINKI
FINLAND N

ARSTILA A.
Dept. Cell. Biology
Univ. Jyvaskyla   FINLAND
VAPAUDENKATU 4 -40100 JYVASKYLA

ASMUS M. & Mme
Hahn Meitner Institut fur
Kernforsschung BERLIN Gmbh
post fach 39 01 28
D1000 BERLIN 39

ASS. RECHERCHE REANIMATION
MEDICALE IMMUNOLOGIE APPLIQUEE
Hôp. St Antoine
184, rue du Fg. St Antoine
75012 PARIS - FRANCE

AUCLAIR C.
Institut Gustave Roussy
94800 VILLEJUIF - FRANCE

AUTOR A.
Pulmonary Research Laboratory
St Paul's Hospital
1081 Burrard Street
VANCOUVER BC V6Z 1Y6 - CANADA

BARET A.
Lab. Trichereau S.A.
11, allée Duquesnes
44000 NANTES - FRANCE

BARIETY J.
Hôpital Broussais
96, rue Didot
75674 PARIS CEDEX 14 - FRANCE

BAST A.
Dept. of Pharmacology Chemistry
Faculty of Chemistry. VRIJE Universiteit
De Boelelaan 1083
1081 HV AMSTERDAM - NETHERLANDS

BEAUMONT M.
Research Division North East Wales Institute
DEESIDE - CLWYD CH5 4BR
WALES - UK

BECK I.
L'OREAL - 1 Av. E. Schueller
93600 AULNAY S/BOIS - FRANCE

BENATTAR C.
Lab. de Biochimie - Hôpital A. Béclère
157, rue de la Pte Trivaux
92141 CLAMART - FRANCE

BENSASSON R. V.
Lab. de Biophysique - INSERM. U. 201 -
Muséum National d'Histoire Naturelle
43 rue Cuvier
75231 PARIS CEDEX 05 - FRANCE

BENVENISTE J.
Centre de Recherche de l'INSERM. U. 200
Immunopharmacologie de l'Allergie et de
l'Inflammation
32, rue des Carnets
93140 CLAMART - FRANCE

BERGER H. M.
NEONATAL UNIT DEPT. PAEDIATRICS
UNIVERSITY HOSPITAL
RIJNSBURGERWEG 10
LEIDEN - NETHERLANDS

BERGMANN A.
HOECHST AKTIENGESELLSCHAFT
PHARMA-SYNTHESE
6230 FRANKFURT AM MAIN 80

BERNARDINI D.
30, RUE BONAPARTE
76006 PARIS - FRANCE

BERREBI A.
121, RUE DE L'OUEST
75014 PARIS - FRANCE

BERTIERE M. C.
LABO. PHARMA 2000
584, RUE FOURNY
78530 BUC - FRANCE

BESANCON D.
LAB. DELAGRANGE
46, BD. DE LA TOUR MAUBOURG
75340 PARIS 7ème - FRANCE

BETTS W. H.
RHEUMATOLOGY UNIT
THE QUEEN ELIZABETH HOSPITAL
WOODVILLE S.A. 5011
AUSTRALIA

BIENVENU P.
CRSSA
38702 LA TRONCHE - FRANCE

BILTON D.
DEPT. OF GASTROENTEROL.
MANCHESTER ROYAL INFIRMARY
MANCHESTER - ENGLAND

BLACK C.
RADIATION ONCLOGY - NATIONAL CANCER INSTITUE
NIH (BLDG 10 RM B3B69)
BETHESDA M.D. 20892 - U.S.A.

BLAKE R.
THE BONE A JOINT RESEARCH UNIT
LONDON HOSPITAL
LONDON E I Z A D - ENGLAND

BLAKEWAY J.
ROURE S.A.
55, VOIE DES BANS
95102 ARGENTEUIL - FRANCE

BLANDIN C.
CLINIQUE NÉPHROLOGIQUE U. 25
HÔPITAL NECKER
149, RUE DE SÈVRES
75015 PARIS - FRANCE

BLOUNT
DEPARTMENT OF BIOCHEMISTRY
SELLY OAK HOSPITAL
BIRMINGHAM B29 6 JD UR. - U.K.

BOLANN B.
LAB. OF CLINICAL BIOCHEMISTRY
UNIVERSITY OF BERGEN
HAUKELAND SYKEHUS
BERGEN - NORWAY

BONNE C.
LAB. DE PHYSIOLOGIE CELLULAIRE
UNIVERSITÉ MONTPELLIER I
34000 MONTPELLIER - FRANCE

BONTE F.
LVHM RECHERCHE
GROUPE MOET HENNESSY, LOUIS VUITTON
48 - 50, RUE DE SEINE
92704 COLOMBE CEDEX - FRANCE

BORS W.
INST. STRAHLENBIOLOGIE, GSF.
FORSCHUNGSZENTRUM, D8042
NEUHERBERG FRG.

BOUCHER F.
UNIVERSITÉ J. FOURIER
BP. 53X
38041 GRENOBLE CEDEX - FRANCE

BOUGHTON - SMITH
WELCOME RESEARCH LABS
BECKENHAM - KENT - U. K.

BOURRE J. M.
INSERM. U. 26
HÔPITAL FERNAND WIDAL
200, RUE DU FAUBOURG ST DENIS
75475 PARIS CEDEX 10 - FRANCE

BRACCO U.
NESTLÉ RESEARCH
P.O. Box 353
VEVEY - CH.

BRAQUET P. G.
IHB RESEARCH CENTRE
17, AV. DESCARTES
92350 LE PLESSIS ROBINSON - FRANCE

BRAVO CUELLAR A.
ICIG. HÔPITAL PAUL BROUSSE
14-16, AV. PAUL-VAILLANT COUTURIER
94800 VILLEJUIF - FRANCE

BREARD
ICIG. HÔPITAL PAUL BROUSSE
14-16, AV. PAUL VAILLANT COUTURIER
94800 VILLEJUIF - FRANCE

BRUNK
Dept. of Pathology
University of Linkoping
S581 85 LINKOPING - SWEDEN

BUARD A.
INSERM. U. 26
Hôpital Fernand Widal
200, rue du Fg. St Denis
75475 PARIS CEDEX 10 - FRANCE

BUC CALDERON P.
Unité de Biochimie Cancérologique et
Toxicologique, UCL 7369 -
Dept de Sciences Pharmaceutiques
LOUVAIN - B. 1200 -

BUCHMAN T.
600 North Wolf Street, Halsted 612
The John Hopkins Hospital
BALTIMORE, MD 21210 - USA

BUFFINTON G.
Inst. Vet. Virology - Univ. Berne
Langgass Strasse 122
CH. 3012 BERNE - SWITZERLAND

BURDON R.H.
Dept. Bioscience - Biotechnology
University of Strathclyde
GLASGOW G A O N R - SCOTLAND - U.K.

BUSI F.
Via Castagnoli
40126 BOLOGNA - ITALY

BUTCHER G.P.
University Dept. of Medecine and Walton
Hospital
LIVERPOOL - U. K.

CAMBO - GROS C.
U. 87 INSERM -
2, rue F. Magendie
31400 TOULOUSE - FRANCE

CAMUS J.P.
Hôpital de la Pitié - Bât. Benjamin Delessert
83, Bd. de l'Hôpital
75651 PARIS CEDEX 13 - FRANCE

CAND F.
Lab. de Physiologie Cellulaire
URA CNRS n° 56
Université Joseph Fourier - BP. 53 X -
38041 GRENOBLE CEDEX - FRANCE

CAPUL C.
INSERM U. 87
Lab. de Biochimie Appliquée
2, rue F. Magendie
31400 TOULOUSE - FRANCE

CARDONA F.
Anenstr. 6
7775 BERMATINGEN 2 - FRG

CARRERA G.
INSERM U. 87
2, rue François Magendie
31400 TOULOUSE - FRANCE

CARINI R.
Dipartimento di Medicina ed
oncologia sperimentale, Sezdi
patologia generale
Universita di TORINO - ITALY

CATROUX P.
Groupe d'étude de Physiologie
et Physiopathologie rénales
Faculté de Pharmacie
3, Place de la Victoire
33000 BORDEAUX - FRANCE

CEBALLOS I.
Lab. de Biochimie génétique
Pr. Sinet et Pr. Kamoun
Hôpital Necker
149, rue de Sèvres
75015 PARIS - FRANCE

CERBION
22, Av. du Doyen Lépine
69500 - BRON - FRANCE

CESBRON J.Y.
INSERM U176
Immunologie Parasitaire
Institut Pasteur Lille
Rue du Prof. Calmette
59019 LILLE CEDEX - FRANCE

CHARLON V.
IRIS
Université J. Fourier
B.P. 53 X
38041 GRENOBLE CEDEX - France

CHEESEMAN K.
Dept. Biochemistry
Brunel University
UXBRIDGE - U.K.

CHEVION M.
Hebrew University
Hadassah Medical School
JERUSALEM - ISRAEL

CHIMI H.
Lab. Botanique et Biologie cellulaire
UER. Médicament
35043 RENNES -FRANCE

CHIRICO S.
Academisch Ziekenhuis Utrech
University Hospital
Dept. of Haematology
UTRECH - NETHERLANDS

CHOPRA
Ninewelles Hospital
Dundee DDI 9 SY
SCOTLAND - U.K.

CRISTEN O.
Laboratoire de Thérapeutique Moderne
BP. 22
92151 SURESNES CEDEX - FRANCE

CIAVATTI M.
INSERM U. 63
22, av. du Doyen Lépine
69576 BRON - FRANCE

CILLARD J.
Laboratoire de Botanique et Biologie cellulaire
4UE. Médicament
Av. du Professeur Léon Bernard
35043 RENNES - FRANCE

CLAUSER P.
Institut de Recherches Internationales Servier
27, rue du Pont
92202 NEUILLY S/SEINE - FRANCE

CLEMENS M.R.
Medizinische Klinik
Otfried Müller Strabe 10
D - 7400 TUBINGEN

CLEMENT M.
INSERM U. 26
Hôpital F. Widal
200, rue du Fg St Denis
75010 PARIS - FRANCE

COMPAGNON
Laboratoire de Toxicologie
Faculté de Pharmacie
4, Av. de l'Observatoire
75006 PARIS - FRANCE

COMPORTI M.
Instituto di Patologia Generale
dell' Universita di Siena
53100 - SIENA - ITALY

CONTI
Unité de Recherche d'Hépatologie
Pédiatrique - INSERM U. 56
Hôpital de Bicêtre
78, rue du Général Leclerc
94270 - LE KREMLIN BICETRE - FRANCE

CONSTANTIN M.
R.L. C E R M.
63203 RIOM CEDEX - FRANCE

COTGREAVE I.
Dept. of Toxicology
Karolinska Institute
STOCKHOLM - SWEDEN

COTTERILL L.
Section of Surgical Research
Clinical Research Centre
Watford Road
HARROW - MIDDLESEX - U.K.

COUTURIER
Unité de Recherche d'Hépatologie
Pédiatrique - INSERM U. 56
Hôpital de Bicêtre
78, rue du Général Leclerc
94270 LE KREMLIN BICETRE - FRANCE

CROSS C.
University of California
Davis and Berkeley
CA - USA - 95616

DAMON
INSERM U. 56
60, rue de Navacelles
34090 MONTPELLIER - FRANCE

DANSETTE P.M.
Lab. de Chimie et Biochimie pharmacologiques
et toxicologiques - U A 400 CNRS
Université René Descartes
45, rue des Sts Pères
75270 PARIS 6ème - FRANCE

DARLEY USMAR V.
Wellcome Research Laboratories
Dept. Biochemistry
BECKENHAM - KENT - U.K.

DAVIES KELVIN J.A.
Institute for Toxicology and Dept. of Biochemistry
The University of Southern California
1985 Zonal Avenue
HSC PSC 614 - 616
LOS ANGELES - CA 90033 - USA

DAVIES R.
346 Alma Road
Enfield Middx
LONDON - ENGLAND

DAWSON M.
Dept. of Pharmacy
University of Strathclyde
204 George Street
GLASGOW 911 XW - SCOTLAND - U.K.

DECUYDER J.
Clinical Chemistry Lab.
Erasmus Hospital
808 route de Lennik
1070 BRUXELLES - BELGIQUE

DEFLANDRE A.
Laboratoires de Recherches L'Oréal
1, av. Eugène Schueller
93600 AULNAY S/BOIS - FRANCE

DE HAAN M.
Hoefstraat 39
2311 PP LEIDEN - NETHERLANDS

DELIGNE P.
Department d'Anesthésie Réanimation chirurgicale
Hôpital Tenon
4, rue de la Chine
75970 PARIS CEDEX 20 - FRANCE

DEMOPOULOS H.
New York University
Medical Center
550 First Avenue
NEW YORK N.Y. 10016 - USA

DERACHE P.
Laboratoire de Biochimie
Faculté des Sciences Pharmaceutiques
35, chemin des Maraîchers
31400 TOULOUSE - FRANCE

DEREU N.
A. Nattermann et Cie GMBH
Nattermannallee 1
D. 5000 KOLN - RFA

DEXTER D.
Department of Neurology
Institute of Psychiatry
Decrospgny Park
LONDON - ENGLAND

DIANZANI M.
Dipartmento di Medicina e oncologia
sperimentale
Sezione di patologia generale
Corso Rafaello 30
TORINO 10125 - ITALY

DIQUET H.
Laboratoires Hoechst
Tour Roussel Hoechst
1 Terrasse Bellini
92800 PUTEAUX - FRANCE

DILLINGERⱵU.
Wolgramestr. 8
1000 BERLIN 42 - ALLEMAGNE

DI SIMPLICIO P.
Dept. of Environmental Biology
Faculty of Science
University of Siena
SIENA - ITALY

DOELMAN C.
Department of Pharmacochemistry
Faculty of chemistry
Vrije Universiteit, de Boelelaan 1083,
1081 HV AMSTERDAM - NETHERLANDS

DONKOR M.R.
Biochemistry Lab.
Brunel University
UXBRIDGE - MIDDX UB8 3PH - U.K.

DORDONI P.
Dpt. of Anesthesiology
Catholic University
ROME - ITALIE

DORMANDY
Dept. of chemical Pathology
Whittington Hospital
Highgate
LONDON N 19 5 NF - ENGLAND

DORNAND J.
CNRS E. 228
Laboratoire de Biochimie des Membranes
Centre Paul Lamarque
34000 - MONTPELLIER - FRANCE

DOWLING E. J.
Dept. of Biochemistry
University of Surrey
Guildford
SURREY GU 2 5 X H - U.K.

DROY - LEFAIX M.T.
IHB - IPSEN
17, avenue Descartes
92000 LE PLESSIS ROBINSON - FRANCE

DUBERTRET L.
U. 312 - Service de Dermatologie
Hôpital Henri Mondor
94010 CRETEIL - FRANCE

DUNSTER C.A.
Brunel University
UXBRIDGE MIDDX UB8 3PH - ENGLAND

EMERIT I.
Groupe de Recherche Radicaux Libres
Institut Biomédical des Cordeliers
15, rue de l'Ecole de Médecine
75006 PARIS - FRANCE

ESTERBAUER H.
Institut fur Biochemie der Universitat Graz
Schubertstrasse 1
A8010 GRAZ - AUSTRIA

ETIENNE J.J.
Parfums Rochas
75, rue d'Aigremont
78300 POISSY - FRANCE

EVANS P.
MRC, Dun Nutrition Unit,
Milton Road
CAMBRIDGE - ENGLAND CB 41XJ

FABIANI
Chirurgie Cardiovasculaire
Hôpital Broussais
96, rue Didot
75014 PARIS - FRANCE

581

FAURE F.
S.H.D. - Vichy
6-10, avenue Ste Anne
92000 ASNIERES - FRANCE

FAVIER A.
Laboratoire de Biochimie C
C.H.R.U.G.
BP. 217 X
38043 GRENOBLE CEDEX - FRANCE

FERNANDEZ Y.
INSERM. U. 87
2, rue François Magendie
31400 TOULOUSE - FRANCE

FERRADINI
Laboratoire de Chimie Physique
Université René Descartes
45, rue des Sts Pères
75006 PARIS - FRANCE

FINET M.
Laboratoire Innothera
10, avenue Paul Vaillant-Couturier
94111 ARCUEIL - FRANCE

FLETCHER
Department of Biological Sciences
University of Salford
SALFORD M5 4WT - U.K.

FONTAGE J.
Institute of Pharmacology
Faculty of Medecine
15, rue de l'Ecole de Médecine
75006 PARIS - FRANCE

FONTECAVE M.
Laboratoire de Chimie et Biochimie
Pharmacologiques et Toxicologiques
Université René Descartes
45, rue des Sts Pères
75006 PARIS - FRANCE

FORESTIER S.
L'Oréal
1 av. Eugène Schueller
93600 AULNAY S/BOIS - FRANCE

FREI B.
Department of Biochemistry
University of California
BERKELEY CA 94 720 - USA

FULLER B. G.
Academic Dept. of Surrey
Royal free Hospital
Pond Street
LONDON NW3 2QG - ENGLAND

FRAISSE L.
G.R.L.
BP. 34 Lacq
64170 ARTIX - FRANCE

FRANCESCHI C.
Institute of General Pathology
53100 SIENA - ITALY

FUCHS J.
Zentrum der Dermatologie und Venerologie
Abteilung II
Iohan Wolfgang Goethe Universitat
FRANKFURT/M. - FRG.

FULLER B. J.
Academic Department Surgery
Royal Free Hospital
Pond Street
LONDON - NW3 2QG - U.K.

GALEY J.B.
L'Oréal
188, rue Paul Hochart
94150 CHEVILLY LARUE - FRANCE

GALIANO A.
I.Q.B.
17, Avda de la Industria
Alcobendas 28100
MADRID - SPAIN

GARDES A.
Laboratoire de Chimie Physique
45, rue des Sts Pères
75270 PARIS CEDEX 06 - FRANCE

GAUDEZ H.
Laboratoire de Botanique et Biologie
Cellulaire
4 UE Médicament
35043 RENNES - FRANCE

GERBER M.
Centre Paul Lamarque
INSERM
34094 MONTPELLIER CEDEX - FRANCE

GEY K.F.
Vitamin Research Department
Hoffmann la Roche Co.
CH 4002 BASLE - SWITZERLAND

GIANNICO M.
U. Mantova 44
TOMA - ITALY

GIORGI - RENAULT S.
Laboratoire Chimie Organique
Faculté de Pharmacie
4, av. de l'Observatoire
75270 PARIS CEDEX 06 - FRANCE

JESSUP W.
Brunel University
UXBRIDGE, U. - K.

JONGKIND
Erasmos Université
ROTTERDAM - NETHERLANDS

JORE D.
Laboratoire de Chimie Physique
Université René Descartes
45, rue des Sts Pères
75270 PARIS CEDEX 06 - FRANCE

JUPIN M.C.
Laboratoire d'Immunopharmacologie expérimentale
Insitut Biomédical des Cordeliers
15, rue de l'Ecole de Médecine
75270 PARIS CEDEX 06 - FRANCE

KAHL R.
Department of Clinical Pharmacology
University of Göttingen
Robert Koch Str. 40,
D - 3400 GOTTINGEN - FRG

KAISER
Groupe SIES
DUSSELDORF - F.R.G.

KAPPUS H.
Free University of Berlin
FB3, WE 15
Augustenbunger Platz 1
D 1000 BERLIN 65

KARAM L.
Nist 245/C214
Gaithersburg, MD 20899 - USA

KAY P.M.
Cottage Lab.
Manchester Royal Infirmary
Oxford Road
MANCHESTER M13 9 WL.

KELLY F.
Departments of Human Nutrition
and Child Health
University of Southampton
SOUTHAMPTON S09 3 TU - U.K.

KETTERER B.
Department of Biochemistry, University
College and Middlesex School of Medecine
LONDON WIP 6DB - U.K.

KIHLSTROM M.
Department of Cell Biology
University of Jyvaskyla
Vapaudenkatu 4
SF 40100 JYVASKYLA - FINLAND

KLEINVELD H.
Erasmus University
ROTTERDAM - NETHERLANDS

KOLODZIEJZYK P.
University of Edmonton
E 3-43 Chemistry Building East
EDMONTON - CANADA

KOLVENBACH
Department of vascular Surgery and Kidney
transplantation
University of Düsseldorf - Medical School
DUSSELDORF - RFA

KOUKAY Nizameddin
Lab. de Biochimie C - Service Pr. Favier -
Hôpital Nord - B.P. 217 X -
38043 GRENOBLE - FRANCE

KOZAK Y.
Immunopahtologie de l'Oeil
Institut Biomédical des Cordeliers
15, rue de l'Ecole de Médecine
75006 PARIS - FRANCE

LAMBOEUF Y.
INSERM U. 87 - Institut de Physiologie -
2, rue François Magendie
31400 TOULOUSE - FRANCE

LAMBOTTE L.
Laboratoire de Chirurgie Expérimentale
UCL 5570
1200 BRUXELLES - BELGIQUE

LANDVIK S.
5101 Bernard Place
VERIS
EDINA, MN 55436 - U.S.A.

LAPLUYE G.
Université Paris VII
Lab. de Chimie Physique
2, Place Jussieu
75251 PARIS CEDEX 05 - FRANCE

LARRAS - REGARD E.
Lab. de Biologie des Vertébrés
Université Paris Sud
91400 ORSAY - FRANCE

LE BLAY J.
Guerlain S.A.
Rue Ch. Teillier
Z.I. de Beaulieu
28000 CHARTRES - FRANCE

LE DOAN T.
Lab. de Biophysique
CNRS. U. A481
43, rue Cuvier
75005 PARIS - FRANCE

LEFORT D.
L.E.S.C.O. - CNRS -
2, rue H. Dunant
94320 THIAIS - FRANCE

LEGEAI
Lab. Jacques Logeais
71, av. du Général de Gaulle
92130 ISSY-LES-MOULINEAUX - FRANCE

de LEIRIS J.
Université de Grenoble I
BP 53X
38041 GRENOBLE - FRANCE

LEMONNIER F.
Hôpital Bicètre
Unité de Recherche d'Hépatologie Pédiatrique
U. 56
94270 LE KREMLIN BICETRE - FRANCE

LENAERS A.
FONDAX (Lab. Servier)
7, rue Ampère
92800 PUTEAUX - FRANCE

LENOBLE G.
40, Place Jules Ferry
92120 MONTROUGE - FRANCE

LIBON C.
Laboratoire de Virologie Immunologie
Expérimentale
Centre d'études Pharmaceutiques
5, rue J.B. Clément
92296 CHATENAY - MALABRY

LIENERS C.
Ludwig Boltzmann Institute for
Experimental and Clinical Traumatology
Windmuhlg 1
A - 7060 VIENNA - AUSTRIA

LIDEMAN
Hooigracht 81 lq
2312 KP LEIDEN - NETHERLANDS

LINDENBACH
Laboratoire de Biochimie
Hôpital A. Béclère
157, rue de la Porte Trivaux
92141 CLAMART - FRANCE

LINDENBAUM
Lab. de Biochimie Appliquée
Centre d'études pharmaceutiques
Rue J. B. Clément
92290 CHATENAY MALABRY - FRANCE

LIPPA S.
Dept. of Human Physiology
Clinical Chemistry
Catholic University
ROME - ITALY

LIU XU HUI
I C I G.
Hôpital Paul Brousse
14-16, av. Paul Vaillant Couturier
94800 VILLEJUIF - FRANCE

LOWN J.W.
University of Alberta Edmonton
Dept. of Chemistry
E3-43 Chemistry Building East
EDMONTON/ALBERTA - CANADA

LUNEC
Rheumatology Unit
Dept. of Biochemistry, Selly oak Hospital
Birmingham, B29 6 JD
Raddlebarn Road - U.K.

MAC ELROY M.
Depts. of Human Nutrition and Child Health
University of Southampton
SOUTHAMPTON S09 3 TU - U.K.

MAEDA H.
Dept. of Microbiology
Kumamoto University medical school
KUMALOTO - JAPAN

MANNING A.S.
Dept. of Biology
Roche Products Ltd.
Welwyn Garden City
HERTS AL7 3 AH - U.K.

MARAK G.E.
2059 HUNTINGTON Avenue
ALEXANDRIA - VIRGINIA 22303 - USA

MARIE B.
Lab. Biochimie
Hôpital Laennec
Rue de Sèvres
75015 PARIS - FRANCE

MARIKOVIC D.
AMC. Univ. Hospital
AMSTERDAM - NETHERLANDS

MARKLUND S.L.
Dept. Clinical Chemistry
UMEA University Hospital
3 901 85 UMEA - SWEDEN

MAROTTI T.
Department of Experimental Biology
and Medecine
Ruder Boskovic Institute
P.O. BOX 1016
BIJENICKA 54 YU-41000
ZAGREB - YUGOSLAVIA

MARRIOTT A.
Brunel University
UXBRIDGE MIDDX - U.K.

MARTEL P.
INRA
Laboratoire des Sciences de la Consommation
78350 JOUY EN JOSAS - FRANCE

MARX J. J. M.
Dept. of Haematology
University Hospital Utrecht
Catharᗡnesingel 101
UTRECHT - NETHERLANDS

Di MASCIO P. D.
Institut für Physiologische Chemie I
Universität Desseldorf
Moorenstrabe 5,
D 4000 DUSSELDORF 1 - FRG

MASSOL M.
INSERM U. 305
University Paul Sabatier
55, rue Bayard
31400 TOULOUSE - FRANCE

MAUPOIL - DAVID V.
Laboratoire de Pharmacodynamie
Faculté de Pharmacie
7, Bd. Jeanne d'Arc
21000 DIJON - FRANCE

MAZIERE J. C.
Laboratoire de Biochimie
Faculté de Médecine St Antoine
27, rue Chaligny
75012 PARIS - FRANCE

MEIER B.
Chemisches Institut Tierärztliche
Hochshule Hannover
3000 HANNOVER 1 - FRG

MESCHTER C.
American Health Foundation
I DANA KCP.
VALHALLA NY. 1050 - USA

MEYBECK
LVMH Recherche
48-50, rue de Seine
92704 COLOMBES - FRANCE

MICHAEL B.
Cancer Research Campaign Gray Lab.
Mount Vernon Hospital
Northwood
MIDDLESEX HA6 2 RN - U.K.

MIYAZAWA T.
Department of Food Chemistry
Tohoku University
Amamiyamachi 1-1
SENDAI 980 - JAPAN

MOISON R.
Kruisweg 22 bis
UTRECHT - NETHERLANDS

MONBOISSE J. C.
Laboratoire de Biochimie - CNRS. UA610
UFR. Medecine
51, rue Cocgnac Jay
51095 REIMS - FRANCE

MONTEIL A.
RL - CERM -
63203 RIOM CEDEX - FRANCE

MOREAU P.
10, rue Oswaldo Cruz
75016 PARIS - FRANCE

MORLIERE P.
Laboratoire de Recherche Clinique
en dermatologie - U312 INSERM -
Hôpital Henri Mondor
94010 CRETEIL - FRANCE

MSIKA P.
L. R. T. P. H.
Fondation Rotschild
25, rue Manin
75019 - PARIS - FRANCE

MUIRHEAD R.
Dept. of Biochemistry
University College London
National Westminster P. L. C.
Tottenham Court RD. Branch
LONDON - ENGLAND

MULLER D.
Institute of Child Health
30 Guilford St.
LONDON WCIN IEH - U. K.

MURPHY M. E.
Institut für Physiologische Chemie I
Universität Düsseldorf
Moorenstrabe 5
D-4000 DUSSELDORF 1 - FRG

MULLERTZ A.
Dept. of Biochemistry and Nutrition
Building 224
The Technical University of Denmark
2800 LYNGBY - DENMARK

NADEUF
RL - CERM
63203 RIOM CEDEX - FRANCE

NASCIMBENI R.
Dept. of Surgery
Istituto di Patologia Chirur.
University of Brescia
Via Valsabbina 19
BRESCIA 24124 - ITALY

NAITO Y.
First Department of Medicine
Kyoto Prefectural University of Medicine
Kamigyo - Ku
KYOTO 602 - JAPAN

NITSCHMANN W.
Zentrum der Dermatologie und Venerologie
Abteilung II
Iohan Wolfgang Goethe Universitat
FRANKFURT /M. - FRG

NOEL H.
17, rue de Pontoise
95520 OSNY - FRANCE

NOHL H.
Institute of Pharmacology and Toxicology
Veterinary University Vienna
Linke Bahngasse 11
A 1030 VIENNA

NORDMANN R.
Dept. de Recherches Biomédicales
sur l'alcoolisme
Faculté de Médecine Paris-Ouest
45, rue des Sts Pères
75006 PARIS - FRANCE

OLIVIER L.
Vice - Président de l'Université Paris VI
4, Pl. Jussieu
75005 - PARIS - FRANCE

OLIVIERO
Génétique Moléculaire & Hématologie
Hôpital Henri Mondor - 51, av. Ml de Lattre de
Tassigny - 94010 CRETEIL - FRANCE

PACKER L.
Dept. of Physiology Anatomy
University of California
BERKELEY - CALIFORNIA - 94720

PAHLMAN R.
Plilopolku 3A3
SF 02130 ESPOO - FINLAND

PAOLETTI C.
Institut Gustave Roussy
Rue Camille Desmoulins
94800 VILLEJUIF - FRANCE

PAPACONSTANTIN
Parfuns Rochas
75, rue d'Aigremont
78300 POISSY - FRANCE

PARANT Monique
Lab. Immunopharmacology
CNRS. UPR. 35
Institut Biomedical des Cordeliers
15, rue de l'Ecole de Medecine
75270 PARIS Cédex 06 - FRANCE

PARCK M. K.
Département de Recherches Biomédicales
sur l'alcoolisme
Université René Descartes
45, rue des Sts Pères
75270 PARIS Cédex 06 - FRANCE

PARNHAM J. M.
Rhône-Poulenc Nattermann
Cologne Research Centre
PO BOX 350120
D5000 COLOGNE 30 - FRG

PARSONS B. J.
Research Division
North East Wales Institute
Connale's way, Deeside
CLWYD CH5 4BR - U. K.

PASQUIER C.
CHU. Bichat - U. 294
16, rue H. Huchard
75018 PARIS - FRANCE

PATERSON C.
University of Louisville
LOUISVILLE - USA

PELEN F.
Laboratoires Labcatal
7, rue Roger Salengro
92120 MONTROUGE - FRANCE

PERDRIX L.
Sté IRIS
27, rue du Pont
92200 NEUILLY S/SEINE - FRANCE

PETERHANS E.
Institute of veterinary virology
University of Berne
Länggass strasse 122
3012 BERNE - SWITZERLAND

PHARMA 2000 S. A. R. L.
584, rue Fourny
78530 BUC - FRANCE

PICCININI F.
Instituto di Farmacologia
Applicata
Universita di Milano
20133 MILANO - ITALY

PIERCE B.
Brunel Univ.
Middx UB8 3PH
UXBRIDGE - ENGLAND

PINCEMAIL J.
Laboratory of Radiobiology and Biochemistry
University of Liège
Institut of chemistry
B6, Sart Tilman
B 4000 LIEGE - BELGIUM

PIRIOU A.
UFR. Medecine Pharmacie
86240 POITIERS - FRANCE

POLI G.
Departmento di Medicina oncologia
sperimentale
Sezione di Patologia Generale
Corso Rafaello 30
TORINO 10125 - ITALY

POMMIER J.
INSERM. U. 96
Hôpital Bicètre
78, av. Général Leclerc
94270 BICETRE - FRANCE

POOT M.
Department of Human Genetics
University of Wuerzburg
Koellikerstrasse 2
D 8700 WURZBURG - FRG

POTOKAR M.
Henkel K Ga A
Postfach 1100
D 4000 - DUSSELDORF - FRG

PRADIER F.
Société L'Oréal
41, rue du Martre
92117 CLICHY - FRANCE

PRAYER S.
Universitate HNO Klinik
Joseph Stelzmann Str.
D 5000 KOLN 41

PRIGENT D.
Lab. Logeais
15, rue Denis Papin
78190 TRAPPES ELANCOURT - FRANCE

PUIG PARELLADA
Falcultad de Medicina
Catedro de Farmacologia
Casanova 143
08036 BARCELONA - SPAIN

RADEKE H.
Institut für Molekularpharmacologie
Medezinische Hochschule
D 3000 HANNOVER 61 - FRG

RAGUENEAU N.
SHD Vichy
6-10, avenue Ste Anne
92600 ASNIERES - FRANCE

RAGUENEZ VIOTTE G.
INSERM U. 295
U. E. R. Medecine Pharmacie
76800 ROUEN - FRANCE

RANSON M.
7, Place du Général Catroux
75017 PARIS - FRANCE

RAPF B.
Doerenkampf
A1107 VIENNA

RENAUD A.
Laboratoires Logeais
15, rue Denis Papin
78190 TRAPPES - FRANCE

REDL H.
Ludwig Boltzmann Institute for experimental
and clinical Traumatology
Donan Eschingerstr. 13
A1200 VIENNA - AUSTRIA

RENAUD de la FAVERIE
Service - Fondax
7, rue Ampère
92800 PUTEAUX - FRANCE

REYFTMAN
Musuem d'Histoire Naturelle
43, rue Cuvier
75231 PARIS CEDEX 05 - FRANCE

RHODES I. M.
Dept. of Medicine
University of Liverpool
P. O. BO X 14 - LIVERPOOL - U.K.

RIBIERE C.
Departement de Recherches Biomedicales sur
l'alcoolisme
Université René Descartes
45, rue des Sts Pères
75270 PARIS CEDEX 06 - FRANCE

RICE - EVANS C.
Dept. of Biochemistry
Royal Free Hospital School of Medicine
Rowland Hill Street
LONDON NW 3 2 PF - ENGLAND

RICHARD C.
Laboratoire Chimie Radicalaire
Université Nancy I
Faculté des Sciences - BP. 239
54506 VANDOEUVRE - FRANCE

RICHARD P.
Réanimations Brulés
Hôpital Trousseau
26, av. du Dr. A. Netter
75012 PARIS - FRANCE

RICHARD M. J.
Laboratoire de Biochimie C.
C. H. R. U. G.
B. P. 217 X
38043 GRENOBLE CEDEX - FRANCE

RICKETT G.
School of Biochemical and Physiological
Sciences - University of Southampton
S09 3TU - SOUTHAMPTON - U.K.

RIZZO A.
Dept. of pharmacology and Toxicology
National Veterinary Institute
PO BOX 6
HELSINKI 55 - FINLAND

ROC Laboratoires
48-50, rue de Seine
92700 COLOMBES - FRANCE

ROCHE-ARVEILLER
Pharmacologie - CNRS. UA595
Hôpital Cochin
27, rue du Fg. St Jacques
75014 PARIS - FRANCE

RODGER S.
INSERM U. 201
Laboratoire de Biophysique
N. N. H. N. - 43, rue Cuvier
75231 PARIS CEDEX 05 - FRANCE

RODOLA F.
Dpt. of Anesthesiology
Catholic University
ROME - ITALY

ROMAN V.
Centre de Recherches du Service
de Santé des Armées
24, av. du Maquis de Grésivaudan
38700 - LA TRONCHE - FRANCE

ROMMAIN M.
Laboratoire Cassenne
Immunological Research Center
17, rue de Pontoise
95520 - OSNY - FRANCE

ROSE S.
Atte Meichsstrasse 3
665 HAMBURG - W. GERMANY

ROSENBAUM G.
Société L'Oréal
41, rue du Martre
92117 CLICHY - FRANCE

ROUACHE H.
Dept. de Recherches Biomédicales sur l'alcoolisme
45, rue des Sts Pères
75006 PARIS - FRANCE

ROUGEE M.
Laboratoire de Biophysique
M. N. H. N. - 43, rue Cuvier
75231 PARIS CEDEX 05 - FRANCE

ROUSSEAU -RICHARD
Laboratoire de Chimie Radicalaire
Université de Nancy I - BP. 239 -
54506 VANDOEUVRE CEDEX - FRANCE

RUF J.C.
INSERM. U. 63
22, rue du Doyen Lépine
69576 BRON - FRANCE

RUSSEL V.
University of Southampton
SOUTHAMPTON S09 3TU - ENGLAND

SAFFAR C.
Departement de Recherches Biomédicales sur
l'alcoolisme
Faculté de Médecine Paris-Ouest
45, rue des Sts Pères
75006 PARIS - FRANCE

SAINT BLANQUAT (de) G.
INSERM U. 87
Institut de Physiologie
2, rue F. Magendie
31400 TOULOUSE - FRANCE

SAMUNI A.
Radiation Oncology Branch. NCI. NIH.
Bldg 10/B3B69
BETHESDA - MD. 20892 - U.S.A.

SANTAMARIA J.
Laboratoire de Recherches Organiques de
L'E S P C I.
10, rue Vauquelin
75231 PARIS CEDEX 05 - FRANCE

SANTUS R.
U. 312 - Service de Dermatologie
Hôpital Henri Mondor
94010 CRETEIL - FRANCE

SCHEIDL H.
Institute of Pharmacology and Toxicologie
Veterinary - University Vienna
Linke Bahngasse 11
1 1030 VIENNA

SHER D.
C/o Zambon Group SpA
PO BOX 10180
20110 MILAN - ITALY

SCHESCHONKA A.
Institut für Physiologische Chemie I
Universitat Dusseldorf
Moorenstr. 5
DUSSELDORF 4000 - WEST GERMANY

SCHMEDES A.
Department of Biochem. & Nutrition
Building 224 DTH
2800 LUNGBY - DENMARK

SCHMIDT K.H.
Eberhard - Karls - Universität
TUBINGEN - FRG

SCHOLICH H.
INSTITÜT FÜR PHYSIOLOGISCHE CHEMIE I
UNIVERSITÄT DÜSSELDORF
MOORENSTRABE 5
D 4000 DUSSELDORF 1 - RFG

SCHONEICH C.
CHIRURGISCHE KLINIK
ULM - RFA

SCHREUDER J.C.P.
PO BOX 430
3740 AK BAARN - HOLLANDE

SCHRIJVER J.
DEPT. OF CLINICAL BIOCHEMISTRY
INSTITUTE CIVO-TOXICOLOGY AND NUTRITION TNO
UTRECHTSWEG 48
PO BOX 360
3700 AJ ZEIST - NETHERLANDS

SELLE S.
CHEM. INST.
TIERÄRZTLICHE HOCHSCHULE
D 3000 HANNOVER

SEUFFER R.H.
D 7410 REUTLINGEN
LASSALLESTR. 40

SHIGEKAZU K.
FUJIMICHO 1-11-2 CHIYODA KU
TOKYO - JAPAN 102

SIES H.
INSTITUT FÜR PHYSIOLOGISCHE CHEMIE I
UNIVERSITAT DÜSSELDORF
MOORENSTR. 5
D 4000 DUSSELDORF - FRG

SIESS M.H.
LAB. DES ALIMENTS DE L'HOMME - INRA
17, RUE SULLY
21034 DIJON CEDEX - FRANCE

SIMIC M. G.
RADIATION INTERACTIONS AND DOSIMETRY GROUP
NATIONAL INSTITUTE OF STANDARDS AND TECHNOLOGY
C214/BLDG 245
GAITHERSBURG - MD 20899 - USA

SIMONOFF M.
CENTRE ÉTUDES NUCLÉAIRES
33170 BORDEAUX - FRANCE

SIMON - SCHNASS I.
GEORG KALB STR. 5-8
D 8023 GROSSHESSE LOHE

SINET P.
CNRS. U.R.A. 1335
LAB. DE BIOCHIMIE GÉNÉTIQUE
HÔPITAL NECKER
149, RUE DE SÈVRES
75743 PARIS - FRANCE

SINISALO M.
3-26, SQUARE PAUL ELUART
94000 CRETEIL - FRANCE

SLUITER W.
ERASMUS UNIVERSITY
ROTTERDAM - NETHERLANDS

SMITS E.
BOUHAAVRLAAN 28
2334 EP LEIDEN - NETHERLANDS

SORENSON J. R. J.
COLLEGE OF PHARMACY
UNIVERSITY OF ARKANSAS
PINE BLUFF
ARKANSAS 71601 - U.S.A.

STERIER M.
HNO KLINIK
UNIVERSITÄT KÖLN
JOSEF STELZMANUSTR. 9
5 KOLN 41 - FRG

STOCKER R.
INST. VET. VIROLOGY
UNIVERSITY OF BERNE
LANGGASS - STASSE 122
3012 BERNE - SWITZERLAND

SUCK
LABORATOIRE LABCATAL
7, RUE ROGER SALENGRO
92120 MONTROUGE - FRANCE

TACHON P.
1, AV. EUGÈNE SCHUELLER
L'ORÉAL
93600 AULNAY SOUS BOIS -FRANCE

TAKEMURA T.
FIRST DEPT. OF MEDICINE
KYOTO PREFECTURAL UNIVERSITY OF MEDICINE
KAMIGYO-KU
KYOTO 602 - JAPAN

TAMBA M.
INSTITUTO F. R. A. E. (CNRS)
VIA CASTAGNOLI 1
40126 BOLOGNA - ITALY

TANIGAWA T.
FIRST DEPT. OF MEDICINE
KYOTO PREFECTURAL UNIVERSITY OF MEDICINE
KAMIGYO-KU
KYOTO 602 - JAPAN

TEISSEIRE B.
FONDAX ACTAM -
GROUPE DE RECHERCHE SERVIER
7, RUE AMPÈRE
92800 PUTEAUX - FRANCE

THEROND P.
INSERM. U. 56
Hôpital de Bicêtre
78, rue du Général Leclerc
94270 KREMLIN BICETRE - FRANCE

THOMAS N.
INSERM. U. 295
UER. Médecine et Pharmacie
76800 St ETIENNE DE REVRAY - FRANCE

TILL G. O.
Dept. of Pathology
University of Michigan Medical Center
1301 Catherine Street
Ann Arbor - MICHIGAN 48109-0602

TISSIE G.
Centre de Recherche - Laboratoire Chauvin
104, Rue de la Galera
34009 MONTPELLIER - FRANCE

TISSOT
Pharmacologie - Hôpital Cochin
27, rue du Fg. St Jacques
75014 PARIS - FRANCE

TOLONEN M.
Ylisriune 16B4
University of Helsinki
SF 02210 ESPOO - FINLAND

TORDJMAN C.
Institut de Recherches Servier
11, rue des Moulineaux
92150 SURESNES - FRANCE

TOUMI M.
Chez UBRAC
22, avenue du Lautaret
05100 BRIANCON - FRANCE

TURKKI A.
University of Helsinki
Dept. of Nutrition
Vukki B-talo
SF 00710 HELSINKI - FINLAND
-
TURNHAM D.
Dunn Nutritional Laboratoires
Downhams Lane
Milton Road
CAMBRIDGE CB4 1XJ - U.K.

UDEN S.
Dept. of Gastroenterology
Royal Infirmary
Oxford Road
MANCHESTER M 13 9WL - U.K.

UEDA S.
First Dept. of Medicine
Kyoto Prefectural University of Medicine
Kamigyo - Ku
KYOTO 602 - JAPAN

ULRICH
Asta Farma Ag
6000 FRANKFURT - ALLEMAGNE OUEST

ULVICK R. J.
Laboratory of Clinical Biochemistry
University of Bergen
Haukeland Sykehus
5016 BERGEN - NORWAY

VAN ASBECK B. S.
Dept. Int. Medicine
Univ. Hospital Ultrecht
Catharynesingel 101
3511 GV ULTRECHT - NETHERLANDS

VAN DER WAL
Med. Microbiology
Univ. of Utrecht
UTRECHT - NETHERLANDS

VAN JAARSFELD H.
Dept. of Chemical Pathology
University of the Orange Free State
BLOEMFONTEIN - REP. of SOUTH AFRICA

VAN LIER J. E.
MRC Group in the Radiation Sciences
Univ. of Sherbrooke
Fac. of Med.
SHERBROOKE - QUEBEC JIH 5 NA - CANADA

VAN ZOEREN
Ding Van Zoeren
University Hospital
LEIDEN - NETHERLANDS

VARMA S. D.
Dept. Ophtalmology - Univer. of Maryland -
School of Medecine
BALTIMORE - MD 21201 - USA

VERDETTI J.
Lab. de Physiologie Cellulaire
URA CNRS 56
University Joseph Fourier
BP 53X
38041 GRENOBLE CEDEX - FRANCE

VINCENT F.
EPHE Paris
15, rue de l'Ecole de Médecine
75006 PARIS - FRANCE

VIRION A.
U. 96 - INSERM -
Unité Thyroide
78, rue du Général Leclerc
94275 LE KREMLIN BICETRE - FRANCE

WALLAT
Marie Curie Str. 9
D 4019 MONHEIM

WATANABE T.
3 rd. Medical Clinic of Kyoto
University Hospital
KYOTO - JAPAN

WEGLECKI W.
Dept. of Medicine
The George Washington University Medical Center
Room 439, Ross Hall - 2300 Eye Street N. W.
WASHINGTON D. C. 20037

WESTERMARCK T.
Helsinki Central Institute for the mentally
retarded
KIRKKONUMI - FINLAND

WICKENS D. G.
Dept. of Chem Paht. Whittington Hosp.
LONDON N195NF - U. K.

WILLSON R. L.
Dept. Biology and Biochemistry
Brunel University
UXBRIDGE MIDDX. UB8 3PH - U.K.

YOSHIKAWA T.
Kyoto prefectural University
of Medicine
Kamigyo - Ku
KYOTO 602 - JAPAN

ZOPHEL A.
Fa. Bender + Co GmbH.
Ernst Boehringer - Institut
Dr. Boehringer - Gasse 5-11
A1121 VIENNA - AUSTRIA

INDEX

Iron chelators, 229, 444
Ischemia-reperfusion, 3, 8, 13,
    187, 247, 353, 361, 367,
    373, 383, 389, 397, 401,
    411, 419, 465

Lipid peroxidation, 79, 111, 133,
    203, 227, 292, 302, 424,
    494, 529, 575
Lipoic acid, 111, 212
Lipoxygenase inhibitors, 257, 261,
    441

M-AMSA, 317
Metal chelators, 294
Methionine, 210, 563
Mitoxantrone, 31
Muscular dystrophy, 485

N-acetyl-cysteine, 210, 349, 444
N-acetyl-dehydroalanines, 339
N-centered free radicals, 323
Neurologic disorders, 475, 493,
    499
Neuronal ceroid lipofuxinosis, 480
Nitroxides, 91
NSAID, 265, 468

Octylgallate, 283
Opioid peptides, 271
Organ failure, 17
Orgotein, 468
Oxypurinol, 349, 367

Pancreatitis, 555
Parkinson's disease, 475, 479, 493
Penicillamine, 294, 468, 488
Periventricular haemorrhage, 477
Peroxidases, 233
Peroxides, 235, 329
Phospholipase inhibitors, 441
Photic retinopathy, 515
Platelet activating factor, 275
Premature infants, 477
Propylgallate, 283, 345
Prostaglandins, 203, 328, 399, 491
Pyrogallol, 345

Quercetin, 171, 519
Quinone toxicity, 37-43

Radiation Protection, 210, 291
Radiation therapy, 64, 311, 430
Renal failure, 383, 389, 393, 397,
    428
Retinal degeneration, 513
Retrolental fibroplasia, 478, 513

Selenium, 175, 179, 183, 294, 353,
    456, 487, 563
Semiquinone radicals, 40, 340

Shock, 17
Sodium benzoate, 517, 548
SOD mimicks, 45, 51, 59, 64, 85
Spinal cord damage, 13
Superoxidase dismutase, 1, 13, 17,
    29, 37, 45, 52, 92, 294,
    547

Thalassemia, 426
Thiol derivatives, 210, 257
Thymine glycol, 301, 315
Thymine hydroperoxides, 301
Tocopherols, 117, 133, 159
Topoisomerase inhibitors, 317
Transplantation, 393, 397
Trimetazidine, 247, 373, 377, 388
Tumor necrosis factor, 275

Ultraviolet irradiation, 179, 534
Urate, 159
Uveitis, 259, 513

Vitamine A, 295
Vitamine C, 155, 456, 495, 507,
    515, 563
Vitamine E, 93-103, 105, 125, 129,
    143, 148, 151, 292, 295,
    358, 423, 428, 456, 468,
    475, 487, 495, 517, 535,
    548, 553, 563
Vitamine E free radical reductase,
    111
VP-16, 31

WR-2721, 211, 293

Zinc, 217, 291, 294, 411, 487
Zinc-Carnosine, 223